STRUCTURE OF MATTER

STRUCTURE OF MATTER

BY

WOLFGANG FINKELNBURG

DR. PHIL., PROFESSOR OF PHYSICS
UNIVERSITY OF ERLANGEN-NUREMBERG

TRANSLATED FROM THE 9th/10th EDITION
OF "EINFÜHRUNG IN DIE ATOMPHYSIK" BY THE
AUTHOR IN COOPERATION WITH
DR. OTTILIE MATOSSI-RIECHEMEIER

WITH 279 FIGURES

SPRINGER-VERLAG BERLIN HEIDELBERG GMBH 1964

Springer-Verlag Berlin Heidelberg GmbH

ISBN 978-3-662-23038-1   ISBN 978-3-662-25001-3 (eBook)

DOI 10.1007/978-3-662-25001-3

Softcover reprint of the hardcover 1st edition 1964

Library of Congress Catalog Card Number 64-23401

Title No. 1205

Preface

This book deals comprehensively with our present-day knowledge of the structure of matter from elementary particles to solids. It has been written for students of physics, chemistry, biology, and engineering, in fact for all who wish to keep abreast with the rapid progress being made in this important field. The author's aim has been to provide the reader with the grounding essential to a true understanding of this vast sphere of science which during the past few decades has led to so great a change of our fundamental concepts of nature and which has found such amazingly manifold application. Accordingly, first place is given to the interpretation of experiments and theories, and to the interrelationship of such seemingly divergent fields as atomic, molecular, nuclear, and solid-state physics. Mathematical or experimental details take second place and are treated only to the extent believed necessary for proper understanding of the subject matter. The author has nevertheless endeavored to take the reader right into the front line of atomic research and to impart to him the knowledge essential to an appreciation of the far-reaching problems involved and the implications at stake.

The book is a translation of the 9th/10th German edition of "Einführung in die Atomphysik", which was first published in 1948. The need for a work of this kind is indicated by the fact that so far a total of almost 50,000 copies has been printed in three languages.

The author is indebted to the publishers for their generous co-operation and special thanks are due to Dr. O. Matossi-Riechemeier for her invaluable help in translating and proofreading. The problems, which, in accordance with German practice, were not included in the original edition, were kindly supplied by the author's former co-worker, Professor S. K. Sen of the University of Manitoba, Winnipeg, Canada.

Erlangen, January 1964   Wolfgang Finkelnburg

Table of Contents

I. Introduction

II. Atoms, Ions, Electrons, Atomic Nuclei, Photons

III. Atomic Spectra and Atomic Structure

V. Physics of Atomic Nuclei and Elementary Particles

VI. Molecular Physics

VII. Solid-State Physics from the Atomistic Point of View

I. Introduction

In this book the structure of matter is presented in a very broad sense. It deals with our knowledge of the structure of matter, its constituent elementary particles, their properties and reactions. It attempts to understand the immense number of observed phenomena of our physical world in a unified way, and this with as small a number as possible of elementary particles, of general basic laws, and of fundamental constants. This science of the "atomistic" structure of nuclei, atoms, molecules, and crystals, though it has its roots in the last century, is the result of our twentieth-century physics. The beginning of this century is highlighted by PLANCK's discovery of the elementary quantum of action h and by the realization of its universal importance. This led to the development of quantum theory which provided the key to the understanding of all phenomena of atomic physics. A large percentage of all basic physics research of our century is in some way concerned with problems of the structure of matter.

1. The Significance of Atomic Physics in Science and its Practical Applications

The new concept of matter, and hence of the foundations of physics and chemistry, has greatly altered our concepts of many older and well-known fields of physics. For this reason we speak of *modern physics* in contrast to *classical physics*. "Classical physics" in this sense means our knowledge of physics as scientists knew it very completely at the dawn of our century without having taken into account any atomic or quantum phenomena. "Modern physics", however, aims at explaining all phenomena of physics and its related fields of science from the atomistic point of view.

Let us see how fundamentally our outlook on entire fields of physics has been changed and clarified by looking at its phenomena from an atomistic point of view. In the last century, for instance, a formally complete thermodynamic theory of heat had been developed. Nevertheless, the basic thermal processes were not understood until the thermodynamic theory was supplemented by an essentially atomistic theory, the kinetic theory of heat. In a similar way modern atomic physics explained one of the most fundamental problems of electricity, viz. metallic conductivity, and opened the way for a deeper understanding of elasticity, plasticity, strength, and similar properties of solid matter, moreover of all the complex phenomena near the absolute zero of temperature. Furthermore, atomic physics created an entirely new concept of all aspects of radiation. These are only a few examples of problems from different fields of classical macrophysics. In addition to contributing to a more comprehensive understanding of many of the long-known fields of physics, atomic physics opened up entirely new fields of science. These new fields which will be treated in detail in this book are concerned with the structure of elementary particles, nuclei, atoms, molecules, and crystals. We will show how atomic physics explains and coordinates wide and important groups of seemingly unrelated physical phenomena, and how it unifies the picture of our physical world.

However, the importance of atomic physics is not confined to physics proper. The development of many other fields of science and industry, in fact our whole concept of the universe, has been deeply changed and influenced by it. The closely allied science of chemistry has received fundamental contributions in the form of the explanation of the Periodic System of the elements and the theory of the chemical bond. Moreover, it uses to an ever increasing extent new methods of research developed in molecular physics. Major parts of astrophysics today are actually applied atomic physics, in particular applied spectroscopy. Spectroscopic methods enable the astronomer to study stellar atmospheres, temperatures, and distances. The problem of stellar energy production was solved by the discovery of exothermal thermonuclear reactions, and this explanation opened the road to a better understanding of stellar evolution. Crystallographers and mineralogists utilize experimental methods of atomic physics as well as theoretical ones. By means of optical spectra they study the constitution of crystals and minerals, by means of X-ray and neutron diffraction they investigate their structure. They use theoretical results on atomic properties and on physics of the chemical bond in order to understand the binding forces, especially in mixed crystals. Biologists use the electron microscope as a tool of undreamed-of efficiency. They irradiate organisms with ultraviolet light, X-rays, or nuclear radiation and, on a large scale, investigate the influence of the radiation on organisms and their mutations. JORDAN's "biophysical amplification theory" added interesting, though partially controversial, stimuli from the field of quantum theory to the field of biology; it still seems premature to judge the true significance of all the aspects of quantum biology. There seems to be little doubt, however, that in a future physicochemical explanation of the autocatalytic processes of cell proliferation (chromosome duplication) the results of molecular physics will play a decisive role. Even mathematicians have been stimulated by the latest developments of quantum mechanics to discuss more thoroughly the possibilities of a mathematical formalism using finite smallest quantities, which would be better adapted to the discontinuous quantum phenomena than the differential equations which are based on the concept of infinitely small changes of all quantities.

The influence of atomic physics on most fields of industrial technology is growing beyond any expectation. We can cite only a few examples here. The military as well as the industrial utilization of the huge energy bound in atomic nuclei and released in nuclear fission or fusion is so much under discussion that many an engineer or layman seems to consider atomic physics to be of interest only for nuclear engineering. But, actually, the influence of atomic physics reaches into all branches of modern technical science. Modern technology of gas discharges and light sources is almost completely applied atomic physics. The same is true for modern communication techniques and electronics since the development of the numerous electronic instruments from oscilloscopes, image tubes, television equipment, image scanners, thyratrons, and transistors up to the electron microscope is based on pure atomic research. Many fields of electrical engineering (such as amplification techniques, the technology of switchgear and contacts as well as modern semiconductor technology) are an outgrowth of atomic physics. For other fields its results give hints for improvements, as for instance for ferromagnetic or ferrimagnetic materials so abundantly used in electrical engineering. The knowledge of band spectra as interpreted by the molecular physicist is used in calculating the molecular radiation of flame gases in industrial boiler design or for measuring the temperature of rocket jets. The physical properties of all solid materials depend on the interatomic forces, so that all progress made in atomic physics of the solid state has its bearing on this important field of technology. Modern materials research, fabrication control and testing depend increasingly on the different methods of non-destructive testing by spectroscopy or X-rays and on an ever increasing number of electronic devices of all types. There is no doubt whatsoever that the extent of the application of atomic physics will increase greatly, so that a sound understanding of atomic physics will be a stringent requirement for the future engineer.

We conclude this survey with a few remarks about the bearing of atomic physics on our general concepts of the natural sciences and of philosophy itself. Its most important effect is that it has fundamentally changed our all too rigidly mechanical concepts of matter, elementary particles, and energy; of the meaning of forces, and of the meaning and importance of the philosophical categories: matter and causality. Already now the deep influence of this change on our whole physical picture of the world, on natural philosophy based on it, and on the theory of knowledge is unmistakable. The interested reader may be referred to the magnificent book by BAVINK and also to numerous other books and papers about this subject by PLANCK, HEISENBERG, JORDAN, MARGENAU, REICHENBACH, and VON WEIZSÄCKER. The basic difficulties of these philosophic consequences of quantum physics, which will be discussed later, are, no doubt, considerable, especially for the experimental scientist. Nevertheless, it is necessary to study these problems seriously for anyone who really strives to be a natural *scientist*, just as, on the other hand, anyone who wants to discuss modern philosophy intelligently must necessarily study atomic physics seriously.

2. Methods of Atomic Physics Research

Atomic physics, since it deals with the structure of matter, really is the foundation of all macrophysics, which takes matter and its properties simply for granted. We know, however, that atomic physics actually was conceived after classical macrophysics had been nearly completely developed. Because of this fact, the methods of research, reasoning, and verification in atomic physics are in various respects quite different from those formerly employed in individual fields of macrophysics. Experimental and theoretical methods and results from practically all domains of macrophysics are used extensively in exploring atomic processes and in forming a "classical" picture of them. This "model" then is further developed by constant cooperation between experiment and theory in order to achieve a maximal approximation to "reality", an improved picture, which in most cases is very different from the initial classical picture. In contrast to most objects in macrophysics, our objects of study (electron, nucleus, atom, ion, or molecule) cannot be observed *directly*. Only what may be deduced indirectly from the physical effects can be observed.

The general procedure in atomic research is the following: From the first results of accidental observations or deliberate experiments, we conceive a pictorial model of the particle, *e.g.*, an atom, in order to understand the observed effects. From this first very crude model we try to draw conclusions about the effects of the particle in other experiments. Such experiments are then performed, if possible, and their results lead to a confirmation, change, or refinement of the first model. As soon as a certain agreement on the model of the particle and its properties has been reached, it is studied theoretically. It is important to realize that this first theory does not claim to describe the behavior of the *particle itself*, but is only concerned with the properties of its *model*, which we want of course to agree with the real particle in some essential points. In most cases, a quantitative theory of this model allows to draw a large number of qualitative and quantitative conclusions.

An experimental test of these shows the points of agreement with reality as well
as the points of disagreement where another change of the concept is required.
Thus, step by step, the model is improved and a better approximation to physical
reality is gained. Each new or improved theory stimulates new experiments, the
conception of which, by the way, often requires considerable ingenuity. Each exact
experiment, on the other hand, which does not result in equally exact agreement
with the theoretical prediction instigates a further improvement of the theory.

Moreover, we shall see that in several critical phases during the development
of atomic physics new experiments forced the theoreticians to realize that a very
radical and fundamental change of existing theories and basic concepts was
unavoidable. Physicists thus were forced to realize that many laws of classical
macrophysics whose applicability to atomic phenomena had been considered
self-evident lost their validity and applicability in the realm of atomic dimensions.
The interaction of theory and experiment thus resulted in the development of
fundamentally new theories that considerably changed our physical reasoning.
However, only the exact experimental confirmation can verify a theory, just as the
experiment also determines the limits of validity of the theory. We shall discuss
later a number of impressive examples of this close interaction between experiment
and theory which is so characteristic of atomic physics. It implies also that the
atomic physicist in this field cannot be *exclusively* an experimenter or a theoretician. An experimenter has to understand at least enough theory to be able to
draw conclusions from it and to explain his experimental results correctly, whereas
a theoretician has to know enough of experiments and experimental possibilities
to participate in the discussion of experimental results or to judge for himself the
possibilities of testing his theories experimentally.

These particular methods of atomic research imply methods of verification
which are somewhat different from those in macrophysics. The correctness of an
assertion in the field of atomic physics cannot be *proved* as directly and unambiguously as in macrophysics because of the unobservability of atomic objects. For
example, the correctness of an assertion such as that of the existence and universal
importance of PLANCK's quantum of action h is based on the fact that a smallest
quantity of action, h, appears in *all* decisive formulas of the entire field of atomic
physics again and again. Moreover, mechanical, optical, photoelectrical, or
X-ray-spectroscopic experiments, though differing widely from each other, yield
the *same* value of this smallest quantum of action, h. In a similar way the
"correctness" of an atomic theory like quantum mechanics cannot be proved
directly. Its validity can rather be concluded from the fact that in the range of
its asserted applicability (i.e., in all atomic physics with the possible exception of
certain domains of nuclear and elementary-particle physics) all experimental
results can be accounted for quantitatively, while all verifiable conclusions from
this theory have been confirmed by experiments. This complete consistency and
agreement is exactly what we mean when we speak of the "correctness" of a
theory. This kind of reasoning in atomic physics as described above seems to be
well justified by the consistency of its results throughout its field; it does not
seem less well founded than the logical methods of reasoning that are possible
and often used in macrophysics.

3. Difficulties and Methods of Presentation of Atomic Physics

The particular relation of atomic physics to the fields of macrophysics, which
we indicated briefly, implies certain difficulties which the beginner encounters
in his attempt to study this field and which will be apparent in this book. The

reader who is familiar with classical physics, especially the engineer, may be consciously or subconsciously inclined to consider as satisfactory only those explanations of observed phenomena that are evident, pictorial, or even concrete. When studying the structure of matter, however, it is necessary to change this attitude very basically. We mentioned already that nearly all objects of atomic physics are invisible. Atomic phenomena can be described, i.e., explained, only by means of the abstract methods of quantum mechanics (for details, see Chapter IV); it will therefore be most difficult, if not impossible, to give pictorial explanations which are nevertheless correct. It is our concern to present in this book our subject as understandably as possible; but nevertheless there are phenomena which ask for abstract methods by necessity. We already mentioned that the explanation of such phenomena by modern atomic physics is not less correct and precise than that of macrophysics. A thorough study will convince the reader of the correctness of any given explanation. Through repeated reasoning, he will get used to abstract thinking and to conclusions more indirect than those he got used to in classical science, and with this he will gain a deeper understanding of the structure of matter. A philosophical analysis of this problem will be given later; but we want to point out here already that it is worth-while to re-adjust our thinking to this abstract way. For including this non-pictorial thinking into the concept of the understanding of nature means an immense widening of our intellectual horizon.

In addition to this very basic difficulty let us list a few less fundamental ones. We have already mentioned that atomic physics makes continual use of results of nearly all fields of classical physics. A good knowledge of classical physics, therefore, should be the basis for studying atomic physics. The kinetic theory, the phenomena of electrolysis, and the basic facts of chemistry form the foundation on which atomic physics has been developed. The concept of electron orbits in the atom was derived from the planetary orbits. A knowledge of gyromechanics was necessary in order to recognize the atom as a small gyroscope or, more correctly, as a system of coupled microscopic gyroscopes. The orbital electrons form electric convection currents and produce magnetic fields. An understanding of the effects of these fields is impossible without some basic knowledge of electromagnetism, whereas an understanding of the emission of radiation by atoms requires knowledge of electromagnetic waves and optics. These few examples may show how experiments and theoretical concepts of atomic physics originate in all fields of macrophysics, and that theory and experiment are more closely interwoven than in most other domains of physics. Undoubtedly this implies difficulties for the beginner. This forced versatility, on the other hand, makes atomic physics so attractive and interesting that it justly has been called the outstanding field of all physics.

It is evident from our introductory discussion that atomic physics is not a well-defined field. The definition of its borderline is highly arbitrary. We can either confine our interest to the basic laws of quantum physics and the structure of atoms in a narrow sense, or we may go to the other limit and try to derive nearly all phenomena and laws of present-day physics from atomic and nuclear physics. Because of the ambiguity in the definition of atomic physics, courses and textbooks vary more widely in their subject matter and manner of presentation than in most other fields of physics, a fact that may confuse the beginner. Some authors prefer, for this very reason, to treat single domains of atomic physics rather than the entire field. Such presentations, excellent and complete as they may be in themselves, fail to reveal the internal relations and cross connections between the single domains, so that the student is not easily able to realize the wonderful harmony which, in spite of so many unsolved problems, today unifies the entire physics of elementary particles, nuclei, atoms, molecules, and crystals.

It is the aim of this book to point out as clearly as possible this unity and harmony of all domains of atomic physics. It is assumed that the student has a knowledge of undergraduate physics including the essential facts of the kinetic theory of gases and of statistics. A further prerequisite is some basic knowledge of theoretical physics, unless the reader is willing to *believe* the correctness of the mathematical derivations and theoretical explanations.

The order of presentation of atomic physics in this book follows essentially from didactic considerations. However, it actually follows largely the historic development. Chapter II presents briefly, besides the evidence for the atomistic nature of matter and electricity and for the existence of electrons, atoms, ions, nuclei, protons, and photons, those facts about these particles (including isotopy) which are a prerequisite for their further treatment. In Chapter III, the basic relations between atomic structure and atomic spectra (which also hold for molecules as discussed in Chapter VI) are dealt with on the basis of the Bohr theory. It is in agreement with the historical development that quantum mechanics (treated in Chapter IV) is not yet used in this chapter. Moreover, this arrangement makes it easy to *prove* to the reader the necessity of introducing wave and quantum mechanics and makes it evident that every new theory does not "overthrow" the old one but expands, refines, and finally includes it as a special case. In agreement with the historical development and our didactic intention, elementary-particle physics and nuclear physics, which in a systematic treatment should form the basis for atomic physics proper, are discussed in Chapter V, after the treatment of quantum mechanics. A real understanding of nuclear processes presumes some knowledge of quantum mechanics, whereas quantum mechanics itself is presented as a logically unavoidable outgrowth from research on atomic electron shells, in agreement with the historical development. In this order of presentation, the energy level diagrams, the quantum numbers, the selection rules of nuclei and their bearing on γ radiation and nuclear processes are easily understandable from their analogy with the phenomena in atomic shells, which were treated before. The same is true for the bremsstrahlung of fast electrons and for many other processes. In Chapter VI, molecular physics is discussed. It seems unduly far separated from the corresponding treatment of the atoms in Chapter III; but again historical as well as didactic reasons favor the discussion of molecular structure after that of quantum mechanics to which frequent reference has to be made. Chapter VII, finally, presents the multitude of phenomena and theoretical explanations of the huge field of solid state physics from the atomistic point of view. According to the present state of development, this chapter cannot be as theoretically concise as the other chapters. For this very reason, however, it may give an impression of the variety of problems dealt with and of the success of atomistic explanations.

It was pointed out in the preface that we shall avoid discussing too many experimental or mathematical details. Rather it has been our intention to stress the fundamental concepts and interrelations between the different subjects, but not to present completely all details. The *applications* of atomic physics will be mentioned at the appropriate places, though lack of space forbids a detailed discussion. For their treatment, the reader may be referred to the books mentioned below.

But it is our intention to lead the reader up to the front line of research and to point out to him numerous still unsolved problems, since it is this insight into active research which will stimulate the student more than anything else.

Literature

Because of the rapid progress of atomic physics, books about this subject become obsolete very soon. In addition to the basic theoretical works, we therefore list here only rather recent books about the entire field of atomic physics which supplement our presentation.

Theory:

SCHAEFER, CL.: Quantentheorie. Vol. III/2 of ,,Einführung in die theoretische Physik". 3rd. Ed. Leipzig: W. de Gruyter 1951.
SOMMERFELD, A.: Atombau und Spektrallinien. 2 vols. 7th and 8th Eds. Braunschweig: Vieweg 1949.
WEIZEL, W.: Struktur der Materie. Vol. II of ,,Lehrbuch der Theoretischen Physik". 2nd Ed. Berlin-Göttingen-Heidelberg: Springer 1959.

General Presentations:

LEIGHTON, R. B.: Principles of Modern Physics. New York: McGraw-Hill 1959.
RICE, F. O., and E. TELLER: The Structure of Matter. New York: Wiley and Sons 1949.
RICHTMYER, K. F., E. H. KENNARD and T. LAURITSEN: Introduction to Modern Physics. 5th Ed. New York: McGraw-Hill 1955.
SLATER, J. C.: Quantum Theory of Matter. New York: McGraw-Hill 1951.

Applications to Astrophysics:

CHANDRASEKHAR, S.: Principles of Stellar Dynamics. Chicago: University of Chicago Press 1943.
JORDAN, P.: Die Herkunft der Sterne. Stuttgart: Wissenschaftliche Verlagsgesellschaft 1947.
ROSSELAND, S.: Astrophysik auf atomtheoretischer Grundlage. Berlin: Springer 1930. Improved English edition: Oxford 1936.
UNSÖLD, A.: Physik der Sternatmosphären. 2nd Ed. Berlin-Göttingen-Heidelberg: Springer 1956.

Applications to Biology:

DESSAUER, F.: Quantenbiologie. Berlin-Göttingen-Heidelberg: Springer 1954.
JORDAN, P.: Physik und das Geheimnis des organischen Lebens. 6th Ed. Braunschweig: Vieweg 1949.

Philosophical Consequences:

BAVINK, B.: Ergebnisse und Probleme der Naturwissenschaften. 8th Ed. Stuttgart: Hirzel 1948.
EDDINGTON, A. S.: Philosophie der Naturwissenschaften. Bern: Franke 1949.
HEISENBERG, W.: Wandlungen in den Grundlagen der Naturwissenschaft. 8th Ed. Stuttgart: Hirzel 1948.
JEANS, J. H.: Physik und Philosophie. Zürich: Rascher 1944.
JORDAN, P.: Die Physik des 20. Jahrhunderts. 8th Ed. Braunschweig: Vieweg 1949.
MARCH, A.: Natur und Naturerkenntnis. Wien: Springer 1948.
MARGENAU, H.: The Nature of Physical Reality. New York: McGraw-Hill 1950.
PLANCK, M.: Wege zur Physikalischen Erkenntnis. 5th Ed. Stuttgart: Hirzel 1948.
WEIZSÄCKER, C. F. v.: Zum Weltbild der Physik. 6th Ed. Stuttgart: Hirzel 1954.
WEYL, H.: Philosophy of Mathematics and Natural Philosophy. Princeton: University Press 1949.

II. Atoms, Ions, Electrons, Atomic Nuclei, Photons

1. Evidence for the Atomistic Structure of Matter and Electricity

We begin our discussion of the structure of matter by demonstrating the evidence for the existence of individual atoms and their constituent parts: electrons, ions, and nuclei. For it is the properties and behavior of these particles with which atomic physics deals and from which atomic physicists hope to derive all other properties of matter. When we speak of the atom as being the smallest particle of matter — when we say, for instance, that the iron atom is the smallest particle into which a piece of iron can be divided — we mean *that the parts of an iron atom do not show any more the chemical characteristics typical for that element.* We make this distinction because it has been known for a long time now that the atom can be broken down into smaller particles. It was in the present century that the existence of atoms as the smallest units in the meaning mentioned above was conclusively proved by experiments. These will be presented in this book. As recently as in the latter part of the last century a vehement scientific debate was waged over the question whether the atom existed as an actual physical entity, or whether the atomic hypothesis was merely a convenient working hypothesis for explaining many observations which had been made on the behavior of matter. In the following pages we shall discuss the evidence for the atomic structure of matter and, at the same time, the proof for the atomicity of the electric charge and thus of electricity itself since it is now known that matter and electricity are inseparably related.

The question of the structure of matter, whether it is homogeneous or atomistic, was considered and philosophically discussed many centuries ago. However, it was not until about 1800, when DALTON investigated the structure of chemical compounds, that the atomistic structure of our material world was becoming clear. DALTON found that in a chemical compound the relative masses of the constituent materials, the elements, are always constant (law of definite composition); and that when two elements can unite with varying masses, these masses are always integral multiples of a smallest equivalent mass (law of multiple proportions). For example, the masses of oxygen in N_2O, NO, N_2O_3, NO_2, and N_2O_5 are — always referred to the same mass of nitrogen — as $1:2:3:4:5$. DALTON's laws are very surprising and difficult to understand if matter is assumed to be homogeneous and arbitrarily divisible. However, from the viewpoint of atomic physics these laws are self-evident, for they are a logical consequence of the fact that always the same number of atoms of the various constituent elements unite to form a particular compound. That is, the ratio of oxygen atoms to nitrogen atoms is always one, two, three, four, or five to two. There are no fractions of atoms.

The concept of the kinetic theory of heat by KRÖNIG and CLAUSIUS and its further development, especially by MAXWELL and BOLTZMANN, in the second half of the nineteenth century, brought to light completely independent but equally conclusive evidence for the existence of individual atoms and molecules. The conclusion that the gas pressure and its increase with temperature is the result of the collisions of gas atoms or molecules and the increase of their velocity with temperature, and that heat conduction as well as internal friction of a gas are the result of transfer of energy and momentum by collisions of the gas atoms or molecules, was such a convincing proof of the kinetic theory of gases and thereby evidence for the existence of individual atoms and molecules that the theory of atomic structure of matter received new support. The consequences of the kinetic theory of gases were especially convincing because, some decades earlier, the discovery of Brownian molecular motion had already produced clear experimental confirmation of the most important assertions of the kinetic theory, i.e., of the temperature-dependent random thermal motion of atoms and molecules.

However, only the present century provided the direct and conclusive proof for the atomic structure of matter, and this proof follows from research in atomic physics proper. Experiments with canal rays definitely proved that these positive rays consist of individual, charged atoms. The investigation of tracks of individual atoms or ions in the Wilson cloud chamber (see V, 2), the discovery of the diffraction of X-rays by the regularly spaced atoms in a crystal which serve as a diffraction grating, and the large number of spectroscopic experiments, which will be presented in the following chapter, removed the last possible doubt of the existence of the atom. From all this evidence we are convinced that the atom is the smallest particle into which matter can be divided by chemical methods; whereas by the methods of physics the atom can be divided into its constituent parts which, however, no longer have the characteristics of the atom itself.

If we assume the existence of the atom, we can regard FARADAY's law of electrolysis, discovered in 1833, as evidence for the existence of an elementary quantum of electricity, or, in other words, of an atom of electricity. If the quantity of a substance transported in an electrolyte depends only on the quantity of charge transported, then each of these atoms migrating as ions carries the same charge. And if, in the case of a monovalent element, the quantity deposited by a unit quantity of electricity is proportional to the atomic weight (see next section), we must conclude that monovalent atoms carry, regardless of their mass, always the same charge. If a bivalent electrolyte transports the same quantity of electricity and only half as much of the substance is deposited, we conclude that a bivalent atom always carries two units of electric charge. The existence of an elementary electric charge, designated as e and quantitatively discussed on page 22, though appearing extremely probable from FARADAY's laws, was not yet definitely proved. The direct proof was provided by atomic physics proper, i.e., by the discovery and investigation of the free elementary electric charge called the electron. We shall discuss it in detail in section II, 4.

2. Mass, Size, and Number of Atoms. The Periodic System of Elements

a) Atomic Weights and the Periodic Table

Now that we have discussed the empirical evidence for the existence of individual atoms, we discuss their properties, especially their masses and diameters. In order to determine the mass of the atom in absolute units, we must know the number of atoms in a mole, called AVOGADRO's number.

When talking about the masses of atoms in physics and chemistry we usually do not mean their absolute masses but relative ones, referred to the mass of carbon which is arbitrarily taken to be 12.0000. This relative number is called the *atomic weight*. This unfortunate misnomer is due to a confusion in the conception of mass and weight when science was in its infancy. In this scale, hydrogen has an atomic weight of 1.008 and the uranium atom, which until the recent discovery of "transuranium" elements was the heaviest atom, has an average atomic weight of 238. The reason for speaking of the average atomic weight will soon become clear.

AVOGADRO's law is used to determine the atomic weights of gases. This law states that under the same conditions of temperature and pressure equal volumes of all gases contain the same number of gas particles (atoms or molecules). Since chemists have shown that the inert gases are monatomic but that hydrogen, nitrogen, and oxygen are diatomic, atomic masses can be determined by weighing equal volumes of various gases under the same conditions and comparing their weights with that of oxygen. In order to determine the atomic weight of the non-gaseous elements, it is necessary to form compounds with gases of known atomic weight. If the chemical formula of the compound is known, e.g., CO_2, the relative atomic weight of the solid element can be computed by measuring the exact mass of the compound and of all of the oxygen liberated after completely decomposing the compound. Relative atomic weights have been measured in this way with the highest precision. They are published annually in the International Table of Atomic Weights (see Table 3) according to the latest state of the art.

The determination of the relative atomic masses stimulated the idea of listing all elements according to their chemical behavior in the so-called Periodic System of Elements by LOTHAR MEYER and MENDELEJEFF (1869). In this system (see Table 1) the elements are arranged in a sequence of increasing order of atomic mass to form periods, and the periods are arranged to form columns or groups so that chemically similar elements such as the alkali metals, the halogens, or the inert gases, each form a *group*. Thus the monovalent elements form the first group, the bivalent elements the second group, and the chemically inactive, inert gases the eighth and last group of the system. On the left side of the Periodic Table we have the electropositive elements, so called because they are charged positively in an electrolyte, while in the next to the last group on the right we find the halogens which are the most electronegative ones. The Periodic Table broke off for no apparent reason in the middle of a period with uranium, the heaviest element found in nature. It has been one of the important tasks of atomic physics to explain the chemical and physical properties of the elements of the Periodic Table, only briefly indicated here, as well as its whole arrangement on the basis of the internal structure of the atoms. We shall deal extensively with the theory of the Periodic System in III, 19.

We mentioned already that the sequence of elements in the Periodic Table corresponds in general to increasing atomic mass. In several cases, however, the chemical properties of an element made it necessary to deviate from this rule. For example, tellurium with an atomic mass of 127.60 has properties which definitely place it under selenium and in front of iodine of atomic weight 126.90, which in turn quite clearly belongs with the other halogens and thus below bromine. Therefore, some departures from the sequence of atomic masses are necessary. Furthermore, it was evident from the Periodic Table that some elements had not yet been discovered. Between radium emanation (radon) of atomic weight 222, which definitely belongs to the inert gases, and radium of the atomic weight 226, which just as definitely belongs to the alkaline earth metals, an element belonging to the alkali metal group (first group) was missing. The Periodic Table thus gave information not only about the known but also about the unknown elements, showing, in addition to their number, their types and properties from the location of the gaps within the table. All missing elements were indeed later identified by looking for these predetermined properties. Since the atomic weight evidently does not always determine the position of an element in the Periodic Table, the elements were arranged according to their chemical behavior and then were numbered successively beginning with hydrogen and temporarily ending with uranium. The numbers assigned to the elements of the Periodic Table are called their *"atomic numbers"*. We shall soon see that these atomic numbers have a very important physical significance.

An inspection of the relative atomic masses of the elements reveals that in a surprising number of cases they are very close to integral numbers when referred to carbon equal to 12.0000. However, there are a number of outstanding exceptions to this rule. It was PROUT who had, already shortly after 1800, proposed the hypothesis (which for a long time lay in oblivion) that all elements might be built up of hydrogen atoms. This idea of the construction of all elements from one basic substance received new support, when at the beginning of this century the study of radioactive elements (see V, 6) led to the discovery of isotopes. These

Table 1. *Periodic Table of the Elements*

	I	II	III	IV	V	VI	VII	VIII		
1	1 H 1.008									2 He 4.003
2	3 Li 6.939	4 Be 9.012	5 B 10.81	6 C 12.011	7 N 14.007	8 O 15.999	9 F 18.998			10 Ne 20.183
3	11 Na 22.990	12 Mg 24.31	13 Al 26.98	14 Si 28.09	15 P 30.97	16 S 32.06	17 Cl 35.45			18 Ar 39.95
4	19 K 39.10	20 Ca 40.08	21 Sc 44.96	22 Ti 47.90	23 V 50.94	24 Cr 52.00	25 Mn 54.94	26 Fe 55.85	27 Co 58.93	28 Ni 58.71
4	29 Cu 63.54	30 Zn 65.37	31 Ga 69.72	32 Ge 72.59	33 As 74.92	34 Se 78.96	35 Br 79.91			36 Kr 83.80
5	37 Rb 85.47	38 Sr 87.62	39 Y 88.91	40 Zr 91.22	41 Nb 92.91	42 Mo 95.94	43 Tc 99	44 Ru 101.1	45 Rh 102.91	46 Pd 106.4
5	47 Ag 107.87	48 Cd 112.40	49 In 114.82	50 Sn 118.69	51 Sb 121.75	52 Te 127.60	53 I 126.90			54 Xe 131.30
6	55 Cs 132.91	56 Ba 137.34	57 La 138.91 [58-71]	72 Hf 178.49	73 Ta 180.95	74 W 183.85	75 Re 186.2	76 Os 190.2	77 Ir 192.2	78 Pt 195.09
6	79 Au 197.0	80 Hg 200.59	81 Tl 204.37	82 Pb 207.19	83 Bi 208.98	84 Po 210	85 At 210			86 Rn 222
7	87 Fr 223	88 Ra 226.03	89 Ac 227 [90-103]							

6	58 Ce 140.12	59 Pr 140.91	60 Nd 144.24	61 Pm 149	62 Sm 150.35	63 Eu 152.0	64 Gd 157.25	65 Tb 158.92	66 Dy 162.50	67 Ho 164.93	68 Er 167.26	69 Tm 168.94	70 Yb 173.04	71 Lu 174.97
7	90 Th 232.04	91 Pa 231	92 U 238.03	93 Np 237	94 Pu 239	95 Am 243	96 Cm 245	97 Bk 245	98 Cf 248	99 Es 255	100 Fm 252	101 Mv 256	102 No 253	103 Lw 257

are atoms which have different masses in spite of the fact that they belong to the same element and thus have the same atomic number and chemical properties. The atomic masses of these isotopes, which were found in the meantime for most of the elements, are actually close to being integral numbers. The outstanding nonintegral atomic weight, chlorine with its atomic weight of 35.453, arises from the mixture of several isotopes of different masses, in the case of chlorine 35 and 37. We shall go into the important phenomenon of isotopes more fully in II, 6.

If we consider molecules instead of atoms, we have the *molecular mass* instead of the atomic mass. If we know the chemical formula, we can compute the molecular mass from the masses of the constituent atoms; for example, the molecular mass of methane, CH_4, is, in round numbers, $12 + 4 \times 1 = 16$. The number of grams of a substance equal to the atomic or molecular mass of the substance is called a *mole*. One mole of molecular hydrogen, H_2, is 2 gm; 1 mole H_2O, 18 gm of water; 1 mole Hg, 200.6 gm of mercury. By applying the mole concept to AVOGADRO's law, which states that equal volumes of gases under the same conditions contain the same number of atoms or molecules, we see that *a mole of any gas has the same number of atoms or molecules. This statement, first found valid for gases only, can be proved to be correct for any other substance as well. Thus, we can compute the absolute mass of an atom or molecule from the relative atomic mass if we know the number of atoms in a mole, i.e.,* AVOGADRO's *number N_0*. It was determined in 1865 by LOSCHMIDT to be about 10^{23}.

b) The Determination of Avogadro's Number and the Absolute Masses of the Atoms

At present there are many completely different independent methods by which AVOGADRO's number may be determined. The fact that the results agree excellently with each other is the best proof of the physical reality of this important constant.

LOSCHMIDT's own method made use of the kinetic theory of gases and was based upon measurements of gas viscosity. We do not go into any details because the method is rather indirect and not very exact. The same applies to a number of other methods of determining this important fundamental constant of atomic physics. Three of the methods depend on measuring BOLTZMANN's constant k which is equal to the gas constant R divided by AVOGADRO's number,

$$k = \frac{R}{N_0}. \tag{2.1}$$

To measure k, PERRIN (in 1909) made use of the fact that minute suspended particles tend to settle out of a liquid due to gravity, but that by diffusion they tend to maintain a uniform density. The decrease in density with the height of the liquid corresponds to the decrease of the density of an isothermal atmosphere with height. According to the kinetic theory, the decrease of the particle density n between the heights 0 and h is given by the formula

$$n_h = n_0 e^{-mgh/kT} \tag{2.2}$$

where n_h, n_0, the mass of the particle, m, and the absolute temperature T can be measured. By substituting these values in Eq. (2.2) the value of k can be calculated.

Closely related to this method of PERRIN's is the determination of k (and N_0 from Eq. (2.1)) by measuring the average lateral displacement of suspended

particles due to their Brownian motion. The theory, developed first by EIN-
STEIN, shows that the mean square displacement $\overline{x^2}$ of spherical particles of
radius r_0 moving in a gas of viscosity η at an absolute temperature T after an
observation time τ is given by

$$\overline{x^2} = \frac{kT\tau}{3\pi\eta r_0}. \tag{2.3}$$

$\overline{x^2}$ can be measured and, since all other constants are known, k can be computed.

The Boltzmann constant k and, with Eq. (2.1), N_0 also can be determined
from radiation measurements. According to the Stefan-Boltzmann law, the
total radiation from a square centimeter of a blackbody at absolute temperature
T is given by

$$S = \frac{2\pi^5 k^4}{15c^2 h^3} T^4, \tag{2.4}$$

whereas the energy radiated in the wavelength interval λ to $\lambda + d\lambda$ is given by
PLANCK's law

$$J(\lambda) = \frac{2hc^2}{\lambda^5} \frac{1}{e^{hc/k\lambda T} - 1}. \tag{2.5}$$

After measuring S and J, the two unknown constants h and k can be computed
by using Eqs. (2.4) and (2.5), and then N_0 again follows from Eq. (2.1).

Another optical method makes use of the scattering of light by very small
particles (Rayleigh scattering). The intensity J_0 of light with the wavelength λ
decreases by passing through the scattering medium according to the formula

$$J = J_0 e^{\frac{CN\alpha^2}{\lambda^4}}. \tag{2.6}$$

J depends upon the number N of scattering particles per cm^3, e.g., air molecules
in the atmosphere. C is a numerical factor $(128\,\pi^5/3)$, and α represents the polarizability (see VI, 2) of the scattering molecules. So, the measurement of the
attenuation due to Rayleigh scattering provides a method for determining the
number N of scattering molecules per cm^3. With the measurable density δ
(gm/cm^3) and the known molecular weight M $(gm/mole)$ of the gas we compute
AVOGADRO's number N_0 from

$$N_0 - \frac{NM}{\delta}. \tag{2.7}$$

One of the two most exact methods of determining N_0 is based on measuring
the molecular weight M and the density δ of crystals with as perfect a structure
as possible. After it has become possible to determine X-ray wavelengths most
accurately by means of ruled optical gratings, the method of X-ray diffraction
(see VII, 4) allows to measure distances of atoms in a crystal lattice with high
precision. From these distances the volume V available to a single structure element
of a crystal may be computed. We then obtain AVOGADRO's number from these
data to be

$$N_0 = \frac{M}{\delta V}. \tag{2.8}$$

The second precision method follows from FARADAY's laws of electrolysis
according to which a quantity of electricity

$$F = 96{,}487 \pm 1 \text{ amp sec (coulomb)} \tag{2.9}$$

is required to deposit one mole of a monovalent substance. Then, if we know the elementary electric charge e, which we shall consider later, we use the relation

$$N_0 = \frac{F}{e} \tag{2.10}$$

for determining N_0. The last two methods give the best value for N_0 that we have at the present time. It is

$$N_0 = (6.0225 \pm 0.0003) \times 10^{23} \text{ (molecules/mole)} . \tag{2.11}$$

From the relative atomic mass of an atom A based on the value of carbon $= 12$ we can obtain the absolute mass of the atom:

$$M_A = \frac{A}{N_0} . \tag{2.12}$$

From Eq. (2.12) we can calculate the absolute mass of the hydrogen atom to be

$$M_H = 1.67329 \pm 0.00004 \times 10^{-24} \text{ gm} . \tag{2.13}$$

c) The Size of the Atom

In asking for the size of an atom we encounter the fundamental difficulty how to define the radius of an atom. The same problem applies to the radius of the electron and of a nucleus. We visualize this difficulty by starting from two extreme cases. The exact definition of the radius of a billiard ball can be given as follows: Two equal balls pass each other undeflected if the distance of their centers while passing remains larger than twice their radius. They are deflected, however, as soon as this distance would become smaller. So the impact radius r and the impact cross section πr^2 can be clearly defined in the case of two billiard balls: The force exerted by one ball on the other falls off very suddenly at a definite radius, i.e., with a high power of $1/r$. If we consider the other extreme case, that of electrically charged particles, such as electrons and ions, then the particles passing each other at any arbitrary distance exert Coulomb forces of attraction or repulsion so that it seems impossible to speak of any definite impact radius. The atom, as we shall see in the next section, lies between the two extreme cases: It cannot be regarded as a rigid ball with an exactly defined radius; nevertheless, its radius is fairly well defined since the forces between two colliding atoms decrease rather fast with increasing distance. Thus, if the atomic radius or atomic volume $4\pi r^3/3$ is determined by different methods, the results will differ somewhat, depending on the value of the forces acting between the atoms during the measurement.

The probably best method of determining atomic radii makes use of the density of crystals or fluids (e.g., liquified rare gases). With the mass of the atoms known, it is possible to compute from the density the atomic volumes and thence the atomic radii. We obtain reliable values only if we consider for each case the specific structure which determines the percentage of atom-filled space. In the close-packed structure 74% of the space is filled. But we shall learn in VII, 1 that very few fluids have a close-packed structure. Even more complicated are crystals and solids because of the numerous possibilities of arranging their atoms spatially. Fig. 1 shows the radii of atoms as well as those of positive and negative ions obtained from such data. The ionic radii are dealt with in detail in VI, 2.

It can be proved by many independent methods that the obtained values which, according to Fig. 1, range from 0.5 to 2.5×10^{-8} cm, are of the right order of magnitude. For instance, from the well-known van der Waals equation for real gases

$$\left(p + \frac{a}{v^2}\right)(v - b) = RT \quad (2.14)$$

we may determine radii of atoms or molecules since the constant b is four times the molecular volume. However, Eq. (2.14) is only an approximative equation; the atomic radii obtained from it may differ up to $\pm 30\%$.

Atomic radii may furthermore be determined by measuring the internal friction of gases because this depends upon the mean free path λ of the gas molecules. λ in turn is a function of the density N of the gas and the atomic radius r_0:

$$\lambda = \frac{1}{4\sqrt{2} N \pi r_0^2}. \quad (2.15)$$

As mentioned before, it is impossible to define the radius of an atom accurately. In other words, the surface of an atom may easily be deformed, i.e., shows a certain "softness". Therefore, the r_0 values from Eq. (2.15) depend considerably on the relative velocity of the colliding particles.

Among the more indirect methods for determining atomic radii we mention only the so-called atom form factor method which, in a complicated way, evaluates X-ray or electron scattering by atoms (see VI, 1 and VII, 4). It permits to determine the distance from the center of the atom at which the interaction between the atom and X-rays or electrons

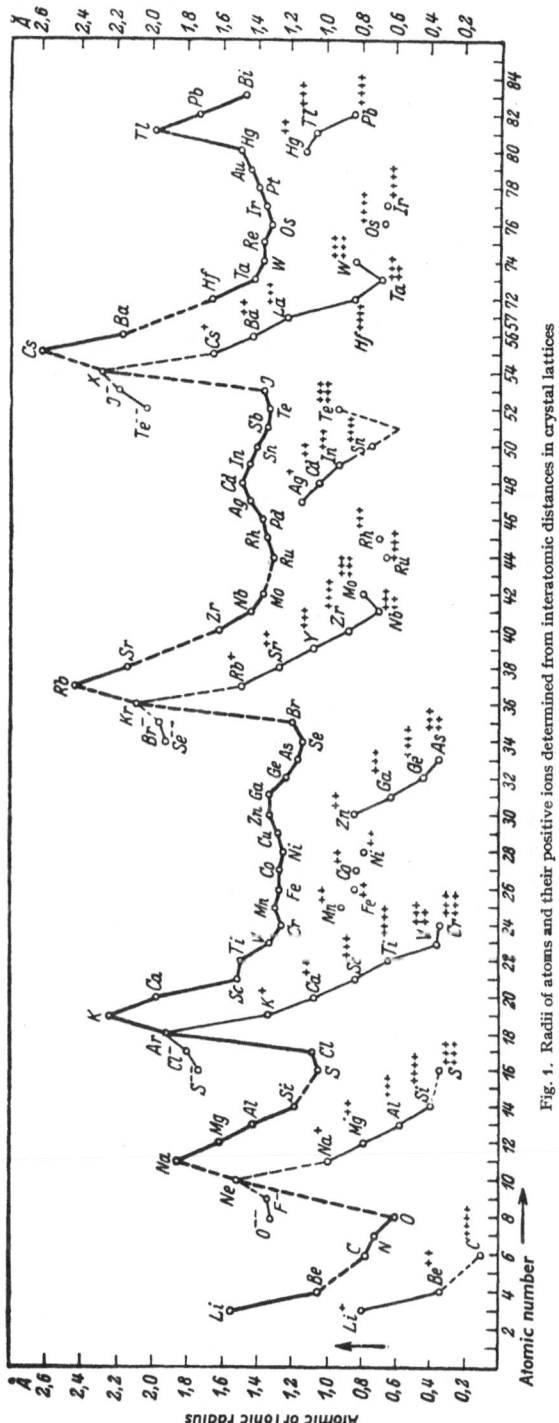

Fig. 1. Radii of atoms and their positive ions determined from interatomic distances in crystal lattices

falls below a certain limit. This distance may be called the radius of the atom.

The important result of all these independent measurements is that all values without exception have the same order of magnitude of some 10^{-8} cm. Also the dependence of the atomic radii on the atomic number (to be seen in Fig. 1), the smaller values of the radii of positive ions, and the relatively higher values of the negative ions result from all methods consistently and can therefore be considered reliable.

3. Nuclei and Electron Shells as Parts of the Atoms. Atomic Models

Originally the atom, as its name implies, was assumed to be the smallest particle into which matter could be divided, and it was generally assumed that our entire material world was built from about 90 different kinds of atoms which number today 103. We know now that the atom can be divided, not by chemical methods, but by the methods of physics. Atomic physics actually starts with the realization that the atom is divisible. The first evidence of the structure of the atom came from gas-discharge physics from which it was evident that neutral atoms and molecules were split into electrically charged particles, called electrons and ions.

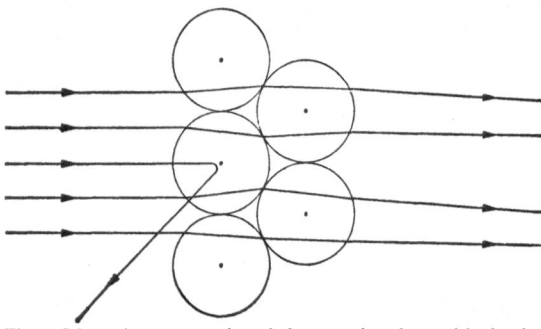

Fig. 2. Schematic representation of the scattering of α-particles by the atoms of a thin metal foil, illustrating small-angle scattering by the electrons of the shells, large deflections only by collisions with atomic nuclei

The study of atomic structure proper began with LENARD's experiments on the passage of cathode rays, i.e., fast electrons, through thin metal foils. LENARD discovered first that fast electrons can penetrate a large number of atoms without noticeable deviation from their original direction. From this he drew the correct conclusion *that atoms could not be thought of as massive balls, but that they consist mostly of empty space.* He investigated how the radius of the scattering atom depends on the velocity of the impinging electrons and found for slow electrons the atomic radius to be equal to that determined from kinetic theory, about 10^{-8} cm. But with electrons of increasing velocity the effective atomic radius ultimately decreases to about 10^{-12} cm, one ten-thousandth of the original value! As a result of these experiments LENARD came to the conclusion that the atom consists of a small massive nucleus surrounded by a force field of 10^{-8} cm radius, and that this force field deflects (scatters) slow electrons but does not affect fast electrons very much. LENARD thought this force field to come from an equal number of positive and negative charges regularly arranged within the atom, thus making the whole atom electrically neutral.

These fairly correct conclusions of LENARD's (in 1903) initiated an entirely new concept of the atom. They were confirmed and quantitatively extended by RUTHERFORD (from 1906 until 1913) who also experimented with thin layers of scattering matter. Instead of electrons, RUTHERFORD used α-particles which are about 7,000 times as heavy as electrons and carry a double positive charge. He found that α-particles could penetrate thousands of atoms without any noticeable deviation and that only very rarely an α-particle was deflected (scattered) under a large angle (Fig. 2). A few years later, Wilson developed his cloud chamber (see V, 2) with which he was able to make the scattering of single α-particles directly visible (Fig. 3). Since a large deflection occurred only rarely, Rutherford concluded that the radius of the scattering center of the atom actually was of the order of only 10^{-12} to 10^{-13} cm. Furthermore, almost the entire mass of the atom must be concentrated in the small "nucleus" of the atom, for otherwise the heavy α-particle could not be scattered through such a large angle. RUTHERFORD thus conceived the atom as consisting of a positively charged nucleus containing almost all of the mass of the atom, surrounded by almost massless electric charges (electrons). These must revolve around the nucleus with such a velocity that the centrifugal forces compensate the electrostatic attractive forces. According to this conception, the α-particles can be scattered only by the nucleus because the very light electrons cannot deflect the heavy α-particles.

To test the correctness of this atomic model RUTHERFROD treated the scattering process theoretically and then compared the theoretical results with the measurements. Since the α-particle of charge $+2e$ is scattered by the nucleus of charge $+Ze$, there acts between the two particles a Coulomb force (Fig. 4)

$$F = \frac{2Ze^2}{r^2} \qquad (2.16)$$

which depends on r in the same manner as does the gravitational force, derived from planetary motion, mM/r^2. The orbit of the α-particle thus is a conic section with the nucleus at a focal point, and because of the repulsive force between the two equal charges, the orbit is a hyperbola as in Fig. 4, degenerating into a straight line for a central collision with the particle following this line twice, in opposite directions. Let p be the perpendicular distance between the nucleus and the original direction of the α-particle (impact parameter), d_0 be the closest distance to which the α-particle with the initial velocity v_0 in a central collision actually approaches the nucleus, and ϑ the angle of deflection, which becomes 180° for a central collision. Then the value of d_0 follows from

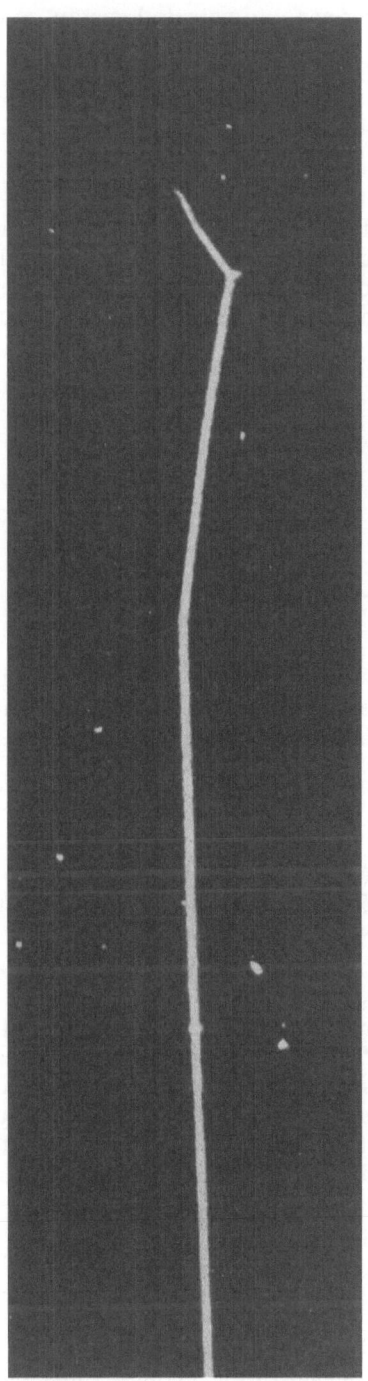

Fig. 3. Cloud-chamber photograph by C. T. R. WILSON. Double deflection of an α-particle by collisions with nuclei of air molecules. The short track toward the right from the point of the second deflection marks the recoil nucleus

the condition that at this distance all the initial kinetic energy has been transformed into potential energy, so that we have

$$\frac{m}{2} v_0^2 = \frac{2Ze^2}{d_0}$$

(2.17)

or

$$d_0 = \frac{4Ze^2}{m v_0^2}.$$

(2.18)

It follows from the laws of conservation of energy and angular momentum that this minimum distance d_0 is equal to the distance between the vertices of the two branches of the hyperbola. Since, for the degenerated hyperbola, the focal point and the vertex coincide, this distance d_0 is reached only in a central collision. From Fig. 4 we get the deflection angle ϑ as a function of the impact parameter p,

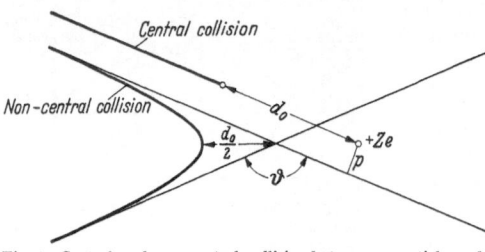

Fig. 4. Central and non-central collision between α-particle and nucleus of charge $+Ze$

$$\cotan \frac{\vartheta}{2} = \frac{2p}{d_0} = \frac{p m v_0^2}{2Ze^2}.$$

(2.19)

Since the number of α-particles scattered into a definite angle can be measured, Eq. (2.19) is used to calculate the fraction dn/n of α-particles scattered through the angle ϑ into the solid angle $d\Omega$ by a foil of thickness d with N atoms per cm³, each with a positive charge $+Ze$. By assuming that in a thin foil an α-particle is scattered only once, RUTHERFORD's formula

$$\frac{dn}{n} = \frac{2dNZ^2e^4}{m^2 v_0^4} \frac{1}{\sin^4(\vartheta/2)} d\Omega$$

(2.20)

follows from statistical methods. The experimental data have proved this famous formula essentially correct, and thereby also RUTHERFORD's atomic model from which the formula was derived.

The question of the absolute magnitude of the positive charge Ze of the scattering nuclei was left open. At first it was only assumed to be approximately half of the atomic weight A of the scattering atom. In 1920 CHADWICK, by simultaneously measuring dn and n, determined the absolute charge Ze of the scattering nuclei and thereby established the important law that the charge of the nucleus Ze of an atom is exactly equal to its atomic number N in the Periodic Table. Because the atom is electrically neutral, the number of electrons circling around the nucleus is equal to the positive charge of the nucleus. CHADWICK's law thus may be written

Atomic number N = number of positive charges Z of the nucleus
= number of electrons of the atom.

Thus, the atomic number, which first was a purely formal result of counting the elements in the Periodic Table, had gained a very decisive physical significance for the structure of the atom.

MOSELEY, about 1913, in an entirely different way obtained essentially the same results. He investigated the X-ray spectra of all available elements — their origin and significance will be discussed in detail in III, 10 — and found that the wavelength of the shortwave principal line, K_α, became shorter in a regular manner in going from the light to the heavy elements (Fig. 5). He determined the frequency $\nu = c/\lambda$ of the K_α lines of all elements (atoms) quantitatively to be

$$\nu = \frac{3R}{4}(N-1)^2 \quad (2.21)$$

where N increases by one in going from one element to the next one in the Periodic Table, and was identified with the atomic number N of the element. R represents the so-called Rydberg constant (see III, 2). Thus MOSELEY (by improving earlier measurements of BARKLA) had, for the first time, found a purely physical method for determining the atomic numbers of the elements and could justify the arrangement of tellurium and iodine in the Periodic Table which we discussed on page 10 as contradictory to the sequence of the atomic weights (Fig. 6). A little later VAN DEN BROECK showed that MOSELEY's atomic number N was identical with the charge of the nucleus Z, a conclusion later confirmed by CHADWICK's scattering experiment.

From the experiments described in this section it became clear that an atom, in contrast to the naive concept of a homogeneous particle of impenetrable matter, actually consists mostly of empty space with a very small, but extremely heavy,

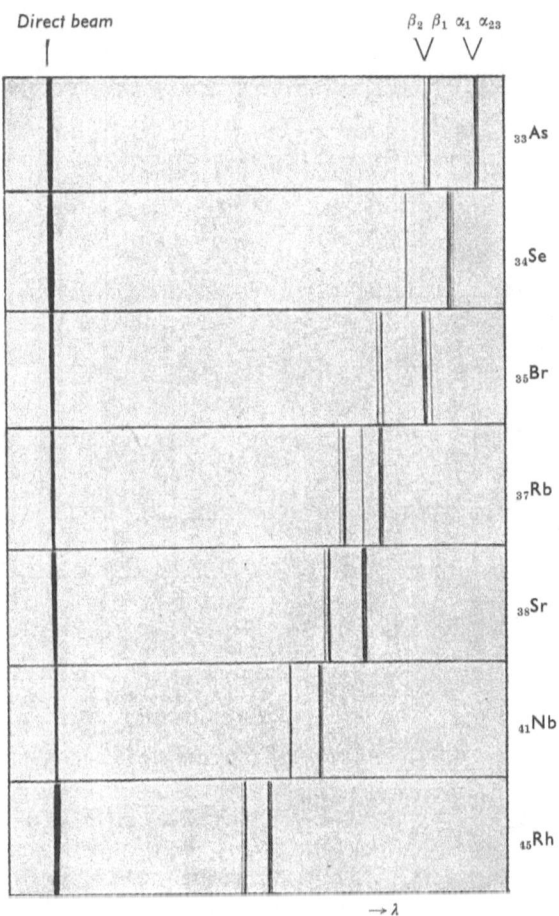

Fig. 5. Shortwave X-ray spectra of the elements $_{33}$As, $_{34}$Se, $_{35}$Br, $_{37}$Rb, $_{38}$Sr, $_{44}$Ne, and $_{45}$Rb illustrate the increase of the frequency with increasing atomic number (MOSELEY's law). (After SIEGBAHN)

positively charged nucleus surrounded by a shell of almost massless very small negative electrons, held together as a system by electrical forces. "Impenetrable matter" in the meaning of the naive concept does not exist. Also it became clear *that the Periodic Table is more than a practical table of by now 103 elements for chemists, but that it is very closely related to the structure of the atom. In particular, the atomic number of the atom in the Periodic Table, because it is the number of positive charges of the nucleus and identical with the number of electrons in the atomic shell, turns out to be one of the most important figures for the structure of the atom.*

The pictorial concepts of the atom, derived from experimental results in the way discussed above and to be continued later, are called *atomic models*. The various models are being improved and developed constantly by new research results, and it is our belief that this process of continuous improvement of the atomic model leads to an increasingly better agreement of the properties of the theoretical model with those of the actual atom. LENARD started from the old model of the atom as a small homogeneous ball, based on chemical knowledge and the kinetic theory. He improved it by using the results of his scattering experiments, thus arriving at a model consisting of positive and negative charges held together by electrical forces with a nucleus one ten-thousandth the diameter of the atom. RUTHERFORD extended the model of LENARD by showing that practically all the mass and positive charge of the atom is concentrated in the nucleus and that the revolving electrons are held in equilibrium by centrifugal and electrostatic forces, thus arriving at a dynamically but not statically stable atom. We shall deal with the further extensions of this model by BOHR, SOMMERFELD, HEISENBERG, and SCHRÖDINGER in the following chapters.

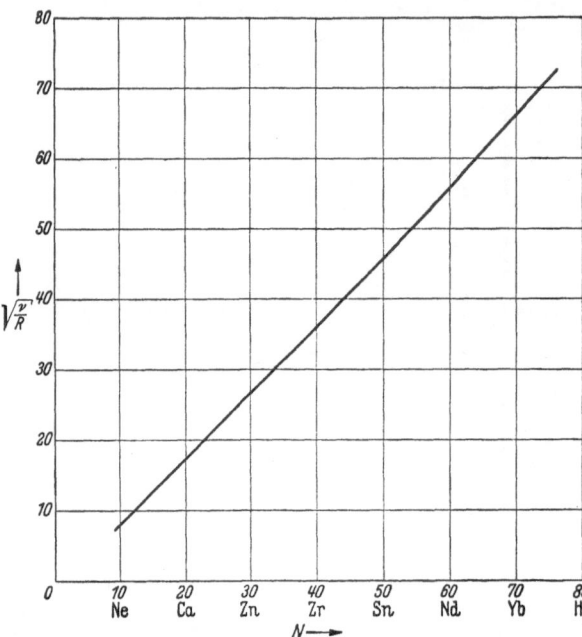

Fig. 6. The linear increase of the frequency of the X-ray line K_α with the square of the atomic number of the elements. (Moseley curve)

4. Free Electrons and Ions

In the last section we considered the evidence for the discontinuous structure of the atom and learned that it consists of positive and negative charges. We also learned that the electron [the name given in 1891 by STONEY to the elementary negative charge in the atom] has very little mass, and that most of the mass of the atom is concentrated in the nucleus. In this section we shall be concerned with the production and characteristics of free electrons and ions (as atoms or molecules with an excess negative or positive charge are called.).

Electrically charged atoms, i.e., ions, were first observed in electrolysis. When an electric field is applied to a sodium chloride solution, the sodium atoms migrate to the negative pole and the chlorine atoms to the positive pole, demonstrating that they are positively and negatively charged. We speak of Na^+ and Cl^- ions; and we understand the process of their formation. The electropositive sodium atom, whose place is in the first group of the Periodic Table, gives up an electron to the chlorine atom which is in the seventh group. In this process, an electron changes place from one atom to another, but it does not turn up as a free electron. Quite generally the liquid state is not well suited for the study of free ions or electrons because of the mutual interaction of the closely packed atoms or ions. In the following discussions we therefore deal with electrons and ions in gases.

a) The Production of Free Electrons

We begin by examining the possibilities of producing free electrons. The most important methods are: the release of electrons from gas atoms by impact ionization; the "evaporation of electrons" from an incandescent metal surface; and the release of electrons from atoms or solid surfaces by irradiation with short-wave light, i.e., the photoelectric effect.

Ionization is the process of separation of an electron from a neutral atom or molecule, which thus becomes a positive ion, or, under certain conditions, from an ion, which then becomes a doubly charged ion. Ionization by impact is the separation of an electron by collision with a fast particle, may it be a radioactive particle, an electron or ion accelerated by an electric field, or an electron, ion, or neutral atom with random high velocity due to sufficiently high temperature. Impact ionization by radioactive particles serves as a method of detection of such particles in a cloud chamber, ionization chamber, or counter tube, as we shall discuss in detail when considering the experimental methods of nuclear research (see V, 2). Ionization by impact in a gas at high temperature (together with the simultaneous photoionization to be discussed in III, 6c) is called thermal ionization. The processes of thermal ionization together with the counteracting formation of neutral atoms by recombination of ions and electrons (see III, 6) yield an exponentially increasing degree of ionization α of all atoms, dependent on the temperature T. α can be calculated from the Eggert-Saha equation

$$\frac{\alpha^2}{1-\alpha^2} p = \frac{(2\pi m)^{3/2}}{h^3} (kT)^{5/2} e^{-\frac{E_i}{kT}} \qquad (2.22)$$

with p = pressure of the gas, E_i = ionization potential of the atoms. This thermal ionization plays an essential role only at temperatures above several thousand degrees Kelvin, i.e., in arc discharges, in the hottest flames, and in the atmosphere as well as in the inaccessible interior of the fixed stars. The impact ionization by electrons accelerated by an electric field is important for the mechanism of glow discharge and of spark breakdown. Electrons accelerated by a high voltage in a vacuum tube leave the negative cathode with very high velocity and are called cathode rays (GOLDSTEIN, 1876). We have already discussed LENARD's scattering experiments with cathode rays on page 16. Related to impact ionization of gas atoms is the release of electrons from a metal by positive ions impinging on the cathode of a glow-discharge tube, or by fast electrons, the so-called secondary electron emission. We shall return to these phenomena in Chap. VII.

Table 2. *Ionization Energies (in Electron Volts) of Some Important Atoms and Molecules*

Gas or vapor	Ionization potential (ev)
H	13.6
H_2	15.4
C	11.2
N	14.5
N_2	15.8
O	13.6
O_2	11.2
He	24.5
Ne	21.5
Ar	15.7
Hg	10.4
Na	5.1
Cs	3.9

For ionizing an atom, the ionization energy must be expended. It varies from taom to atom and depends on the binding force between the atom and the electron to be released. Table 2 shows the ionization energies of some of the more important atoms and molecules. The ionization energy for impact ionization is supplied by the kinetic energy of the colliding electrons. The energy can be measured in ergs, but usually it is given in terms of the accelerating potential in volts. The kinetic energy of an electron with the charge e, when accelerated by the potential V, is given by the equation

$$eV = \frac{m}{2} v^2. \qquad (2.23)$$

In this sense it is customary to speak of the ionization potential in volts or, since charge times potential has the dimension of energy, of the ionization energy in eV=electron volts. For computing the energy in ergs and other energy units see Eqs. (3.10) and (3.11), page 58.

In the second method for producing free electrons, the photoelectric effect, the energy is supplied by the incident radiation. LENARD's discovery that the kinetic energy of the released electrons is independent of the intensity of the incident radiation, and EINSTEIN's explanation of this phenomenon played an essential role in the foundation of quantum theory. The release of electrons requires incident light of short wavelength. The higher the energy of the bond between the electrons to be released and their ions or the surface of the metal, the shorter is the effective wavelength. However, the number of electrons released per second per square centimeter is dependent upon the intensity of the radiation. Photoelectric electrons can be released from isolated atoms and molecules as well as from large atom complexes, i.e., solid surfaces. In the first case, to which we shall return in III, 6, we speak of *photoionization* of the atom or molecule. Only in the second case do we use the term photoeffect in its restricted sense. In addition to this "external" photoeffect we know the closely related release of electrons by absorption of radiation in the interior of a crystal, the "internal" or crystal photoeffect, which we shall consider further in connection with solid state physics (VII, 10). A practical application of the photoeffect is the photoelectric cell which consists of a light-sensitive layer on a portion of the inside of a glass bulb; electrons are released from this layer by the incident radiation. Such photocells have gained wide scientific and technical significance as receivers of radiation. The electrons freed from the layer are drawn to the anode by a suction voltage. The resulting photocurrent is measured by an amplifier or multiplier.

The method of producing free electrons that is most important for technical purposes is the evaporation of electrons from an incandescent metal or semiconductor. We shall discuss the atomistic mechanism of this phenomenon in Chap. VII. The saturation current density j, obtainable by electrically sucking the electrons off the emitting surface, depends on its absolute temperature T and on the so-called "work function" W according to the Richardson equation

$$j = A\,T^2 e^{-\frac{W}{kT}} \text{ amp/cm}^2. \tag{2.24}$$

The work function W is the energy necessary for the release of an electron from the metal against the binding forces. It is a characteristic figure of every solid and is related to the ionization energy of the atoms which form the solid, whereas A is another material constant, which for most metals has values between 60 and 100. This thermoionic emission is used in a large number of electronic devices.

b) The Measurement of the Charge and Mass of the Electron

Now that we have learned how free electrons are produced, we discuss the properties of this important constituent of the atom. We begin with the mass of the electron and its charge, which we have already defined to be equal to the elementary quantum of electricity e.

By reversing the last mentioned method (page 14) of determining AVOGADRO's number N_0 and using the most accurate value of FARADAY's constant F, which is the electrical charge transported by a mole of a monovalent electrolyte, we find the elementary quantum of electricity e to be

$$\left.\begin{aligned} e = \frac{F}{N_0} &= (1.60210 \pm 0.00007) \times 10^{-19} \text{ amp sec (coulomb)} \\ &= (4.8030 \pm 0.0002) \times 10^{-10} \text{ esu.} \end{aligned}\right\} \tag{2.25}$$

Historically, e was first determined in a very similar manner by measuring the total charge transported by a known number of charge carriers; for example, by counting a large number of α-particles and measuring with an electrometer the charge transported by them. The most ingenious and direct method, however, for measuring the charge of a single electron was invented by EHRENHAFT (in 1909) and improved by MILLIKAN (in 1911). In this famous *oil-drop method*, the motion of microscopic oil drops is being observed between the horizontal plates of a condenser after electrons, which had been produced photoelectrically in a gas, attached themselves to the droplets. If the upper plate is charged positively and the field between the plates is $E = V/d$, where V is the potential difference and d the plate separation, then the forces acting on the droplet are: the downward force of gravity, mg, and the oppositely directed electrostatic force between the charged plates, eE. If the field in the condenser could be adjusted so that the droplet is held stationary between the plates, the charge e could be directly computed from the equilibrium condition

$$eE = mg . \tag{2.26}$$

The weight of the droplet could then be calculated from its radius r, measured micrometrically, and the densities of the oil δ and air δ_a (to account for the buoyancy of the oil drop in air) by the expression

$$mg = \frac{4\pi r^3 (\delta - \delta_a) g}{3} . \tag{2.27}$$

Actually, however, the droplet cannot be held in equilibrium nor can the radius be measured accurately enough. Consequently, the vertical downward or upward motion of the droplet with and without an electric field is measured. With no field the droplet falls with a constant velocity resulting from the equilibrium between the force of gravity and the force of friction of air (viscosity η), which can be computed from STOKES' law

$$6\pi\eta r v_0 = mg . \tag{2.28}$$

The constant velocity v_0 of the falling particle in a field-free condenser results from Eqs. (2.27) and (2.28) as

$$v_0 = \frac{mg}{6\pi\eta r} = \frac{2r^2 g (\delta - \delta_a)}{9\eta} . \tag{2.29}$$

Since all quantities in Eq. (2.29) except r are known, r can be computed from measuring the velocity of fall of the droplet without an electrical field in the condenser. If now the condenser is charged so that a field E exists between the plates, an additional force eE acts on the droplet; its direction depends on the polarity of the plates. Under these conditions the droplet falls (or rises) with a velocity given by

$$6\pi\eta r v_1 = \frac{4\pi}{3} r^3 (\delta - \delta_a) g \pm eE . \tag{2.30}$$

Since now the radius r is known from having measured the velocity v_0 of the same droplet falling in a field-free space, only the charge e is unknown in Eq. (2.30). After making measurements on many droplets, Millikan found that the preponderant value of e was, within the limits of the experimental error, that given in Eq. (2.25). In some cases the charge appeared twice or three times as large because two or three electrons were attached to one droplet. Thus the existence of an elementary quantum of electricity was demonstrated by direct measurement of the charge of individual particles.

We next discuss the mass m_e of the electron. It cannot be measured directly. In the various methods, of which we discuss only one, the specific charge e/m_e of an electron, i.e., its charge per unit mass, is measured. By using the value of e from Eq. (2.25) we then obtain m_e. As early as at the end of the last century, e/m_e was determined by observing the electrostatic and magnetic deflection of fast cathode-ray electrons in crossed electrostatic and magnetic fields. Fig. 7 is a schematic illustration of a cathode-ray tube built by F. BRAUN, in which electrons originating at K pass through the electrostatic field between the condenser plates and simultaneously through the magnetic field between the external coils. The electrostatic and magnetic fields are perpendicular to each other. In the electric field, the electrons are accelerated according to the equation

$$b = \frac{eE}{m_e}, \qquad (2.31)$$

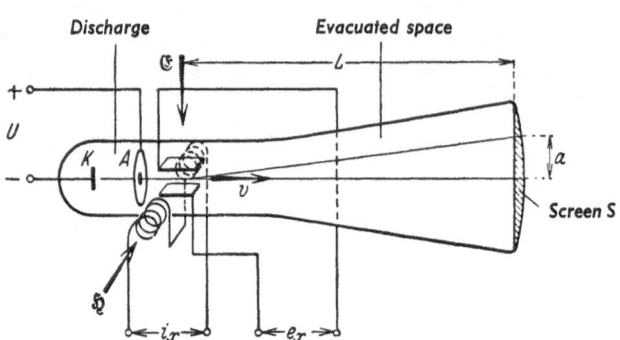

Fig. 7. Cathode-ray oscilloscope (Braun tube) with electric and magnetic deflection of the beam for measuring the specific charge e/m_e of electrons

where b is the acceleration. A time $t = l/v$ is required for electrons of velocity v to pass through the whole length l of the condenser. The electrons are deflected through a distance

$$y = \frac{1}{2} bt^2 = \frac{eEl^2}{2m_e v^2} \quad (2.32)$$

perpendicular to the initial direction. The deflection a (Fig. 7) on the luminescent screen S in the distance L from the condenser is L/l times larger than y. For determining the specific charge e/m_e, we need a second equation since the velocity v of the electrons is not known either. The magnetic field H exercises an additional force K_m on the electrons according to

$$K_m = eHv. \qquad (2.33)$$

With appropriate polarity, K_m may counteract the electric force and compensate it. H_k may be the magnetic field strength where magnetic and electric forces are in balance. Then the deflection of the electrons, visible on the screen S, is zero and we obtain

$$v = \frac{E}{H_k}. \qquad (2.34)$$

By substituting this value into formula (2.32), an e/m_e value results which according to the most recent measurements is

$$e/m_e = (1.75880 \pm 0.00002) \times 10^8 \text{ amp sec/g}. \qquad (2.35)$$

From this and Eq. (2.25), the mass of the electron is obtained as

$$m_e = (9.1091 \pm 0.0004) \times 10^{-28} \text{ g}. \qquad (2.36)$$

From Eqs. (2.13) and (2.36) we see that the mass of the electron is only 1/1,836 of the mass of the lightest atom, that of hydrogen. The small inertia of the electron resulting from its small mass is an important factor in most electronic instruments.

The mass of the electron given in Eq. (2.36) actually is its *rest mass* because the velocities in the experiment are small compared to the velocity of light. The mass of the electron depends, as every mass does, on its velocity v relative to the measuring device (laboratory). According to the theory of relativity, the mass of the moving electron is given by

$$m_e = \frac{m_{e0}}{\sqrt{1 - \dfrac{v^2}{c^2}}} . \tag{2.37}$$

This relativistic mass increase as a function of the velocity was confirmed at the beginning of this century when KAUFMANN and BUCHERER determined e/m of β-ray electrons. These electrons which are emitted in the radioactive decay (see V, 6) may reach 90% of the velocity of light and thus their "dynamic mass" exceeds their rest mass by more than 100%. We will learn in V, 3 that modern synchrotrons accelerate electrons to so high a velocity that their dynamic mass is many thousand times as large as their rest mass.

In addition to its charge and mass the electron has, as GOUDSMIT and UHLEN-BECK found in 1925, a constant mechanical angular momentum (spin)

$$|\vec{s}| = \frac{1}{2} \frac{h}{2\pi} = (5.2721 \pm 0.0002) \times 10^{-28} \text{ g cm}^2 \text{ sec}^{-1} \tag{2.38}$$

and an associated magnetic moment

$$\mu_B = \frac{e h}{4\pi m_e c} = (9.2732 \pm 0.0006) \times 10^{-21} \text{ gauss cm}^3. \tag{2.39}$$

This quantity is called the Bohr magneton; it is a consequence of the rotation of the electric charge about its own axis. But the experimentally determined magnetic moment is larger than the Bohr magneton (2.39) by 0.1160%. Therefore, an additional "intrinsic magnetic moment" is attributed to the electron, for which SCHWINGER gave a quantum-electrodynamic explanation. Its computed value, in a first approximation, is $e^2 \mu_B / hc$.

In addition to the properties discussed, we might ask about the size of the electron just as we did on page 14 for the atom. However, here we run into the same, or even greater, difficulty how to define and determine the radius of the electron. In collisions, the electron acts through its Coulomb field which varies as $1/r^2$, so that no reasonable definition of the impact radius is possible. And we have no other experimental evidence of the structure or radius of the electron. Only in a rather artificial manner can a so-called "classical radius" of the electron be defined. If we compute from the mass-energy equivalence equation

$$E = m_e c^2 \tag{2.40}$$

the rest energy corresponding to the rest mass of the electron, and equate this to the electrostatic potential energy of a sphere having a charge e,

$$E_{\text{pot}} = \frac{e^2}{r} , \tag{2.41}$$

we obtain the "classical radius" of the electron

$$r_e = \frac{e^2}{m_e c^2} = (2.8178 \pm 0.0001) \times 10^{-13} \text{ cm}. \tag{2.42}$$

However, in the denominator of this we would get a factor 2, (so that r_e would equal 1.41×10^{-13} cm) if we assume in a purely classical calculation that a sphere of the radius r be electrically charged by infinitesimal steps up to the charge

e of the electron. On the other hand, THOMSON found from the energy of the electric and magnetic field of a rotating electron that the value r_e should be $\frac{2}{3}$ of that of Eq. (2.42), i.e., $r_e = 1.88 \times 10^{-13}$ cm. We see, there is a certain arbitrariness. Furthermore, two doubtful assumptions were made in this derivation. The first is the assumption that COULOMB'S law is valid down to a distance of 10^{-13} cm, and the second, that the entire rest energy (2.40) of the electron consists of electrostatic energy. So, the classical radius of the electron is of rather dubious significance. We shall come back to this problem in V, 24 when dealing with the theory of elementary particles.

c) Applications of Free Electrons. Electronic Instruments

Now that we know the properties of normal electrons except for their behavior at extremely high impact energies which we shall discuss in V, 20, we turn to a brief survey of the most important applications of free electrons. First we mention the use of electron beams as almost inertialess switches. Generally known is the high-vacuum electron tube which, in addition to a hot cathode and an anode, has one or more grids as control electrodes. A high negative potential on the grid forces the electrons back to the cathode, preventing a current flow to the anode. It thus interrupts the current. By applying a less high and periodically varying negative potential to the grid, a continuous flow of electrons is generated, increasing and decreasing with the grid potential. The lower the potential the higher is the electronic current and vice versa. So, the vacuum tube can be used as an amplifier, receiver, or transmitter, depending on the specific circuit. A vacuum tube with only a hot cathode and an anode is an excellent rectifier. Since the cold anode cannot emit electrons, the current flows only when the hot cathode has a negative potential.

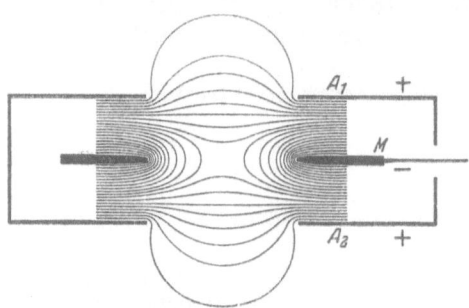

Fig. 8. Schematic sketch of the potential field of an electrostatic electron lens. A_1 and A_2 are external electrodes, M is a central electrode

Another instrument which makes use of the exceedingly small inertia of electrons has gained utmost importance for all kinds of electrical measurements, the cathode-ray oscilloscope. In this instrument, which is a modern development of BRAUN'S tube (page 24), the cathode ray passes successively through two mutually perpendicular electrostatic fields. The deviation of the electron beam is made visible on a fluorescent screen; the light spot, produced at the intersection of beam and screen, describes a curve which is observed directly or may be photographed. Because electrons have almost zero inertia, extremely fast processes can be investigated in detail, for instance a spark discharge which occurs in less than 10^{-8} sec. The television tube is another variation of the Braun tube. In this tube the electron beam covers the entire screen line by line with varying intensity so that the electrically transmitted picture becomes visible.

Since electrons are influenced by appropriate electric or magnetic fields (e.g., Fig. 8) in the same way as light rays are influenced by lenses, *electron optics* represents a wide area of applied electronics. The best known representative is the *electron microscope*. In the most common form of this instrument, instead of the light beam in a light microscope, a beam of electrons from a hot cathode (Fig. 9) penetrates the object and is attenuated to varying degrees depending upon the absorption of the material which is used as the object. The divergent electron beam which leaves the object is then made convergent by electrostatic or magnetic fields (called electrostatic or magnetic lenses, Fig. 8) to form an image of the object. Later, a second lens combination magnifies this electron image which is then projected on a fluorescent screen or photographed directly. Since the useful magnification of a light microscope is limited by its resolving power, which is approximately equal to half a wavelength of the used radiation, it cannot

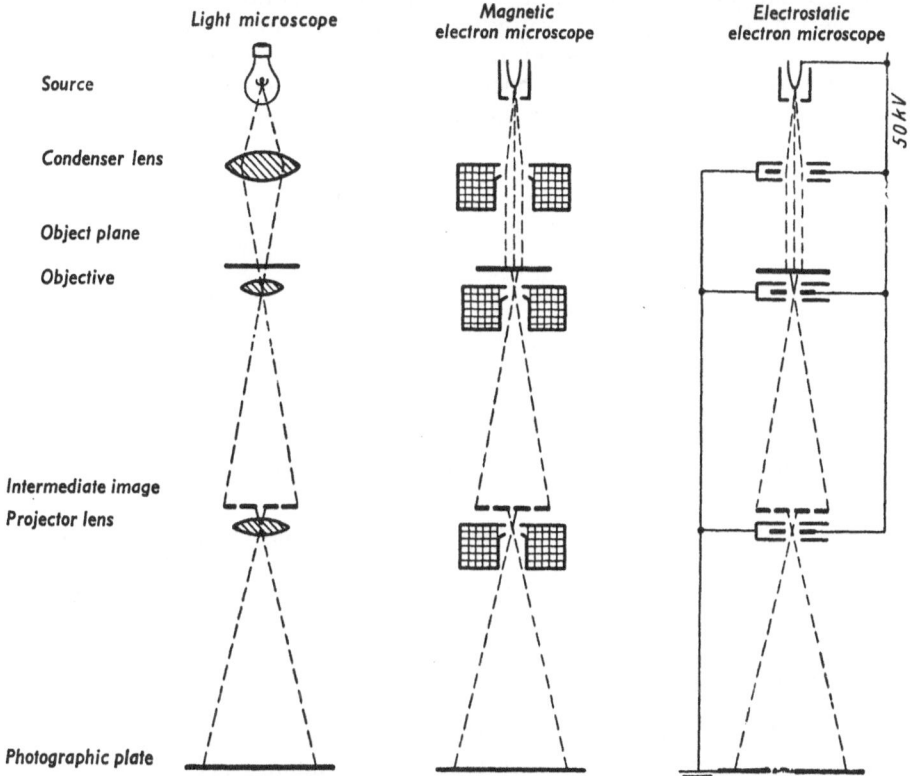

Fig. 9. Schematic sketches of light microscope, magnetic and electrostatic electron microscope

resolve objects smaller than $0.2\,\mu$. As we shall learn in IV, 2, moving electrons also exhibit wave properties with a wavelength depending on the velocity of the electrons. With the potentials used in electron microscopes (i.e., with the corresponding velocities of electrons), this wavelength is smaller by nearly the factor 10^{-4} than the wavelength of visible light. So, the electron microscope permits in principle to resolve much finer details, e.g., even large molecules, and to use magnifications by far higher than the light microscope.

E. W. MÜLLER developed an instrument similar to this electron microscope but nevertheless of an entirely new kind, the so-called *field electron microscope*. It is based on the phenomenon of field emission which we shall discuss in VII, 14. Its most essential part is an extremely fine, strongly negative metal point which creates a high electric field around itself. This field extracts, in high vacuum, electrons from the point which cause a fluorescent screen, surrounding the point in a certain relatively large distance, to emit light. Since the electron emission as we shall learn in VII, 14, depends on the atomistic structure of the surface, we may study the surface by observing its electron emission on the surrounding fluorescent screen. Furthermore, if single molecules of certain organic substances are brought upon the point by evaporation, the electron emission in their environment is changed in such a way that the fluorescent screen shows actual images of these molecules. There certainly exist possibilities of further development of this field electron microscope which may supplement the conventional electron microscope in a most interesting way.

Among the particularly interesting applications of electron optics is the *image converter*. The image of an object which emits only infrared or X-ray radiation is projected on an infrared-sensitive layer (photo-cathode). According to the varying intensity of the incident radiation, a varying number of electrons is emitted from every point of the photo-cathode. These electrons are electrically accelerated and concentrated on a fluorescent screen by means of electron optics.

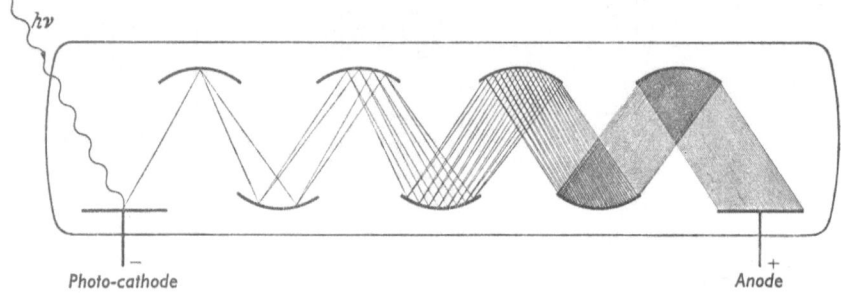

Photo-cathode   Anode

Fig. 10. Schematic sketch of a photomultiplier. (The circuit for the electric potentials of the reflecting electrodes is not indicated)

There, the electronic image excites fluorescence and thus creates a visible image of the infrared or X-ray object.

Photoelectric cells are being used increasingly in combination with a new type of electronic amplifier, the *electron multiplier*. This amplifier (Fig. 10) consists of a high-vacuum tube with a large number of electrodes (often called dynodes) with specifically treated surfaces so that every single electron hitting the surface with a sufficiently high energy releases several secondary electrons from it. The theory will be treated in VII, 21. If the number of secondary electrons released by a single primary electron is for instance six, the number of primary electrons released photoelectrically at the photo-cathode is multipled by six at every multiplying step. Amplifications up to 10^{10} have been obtained by such multiplier tubes. Their resolution is excellent (up to 10^{-10} sec); the background noise is small.

The modern field of *solid-state electronics* opens other entirely new possibilities of application, topped by the *transistor* in all its varieties. We postpone its discussion to VII, 22 because we need for its understanding a knowledge of solid-state atomic physics.

In concluding we mention briefly that when fast electrons strike solid matter, such as an anticathode put into their path, X-rays are produced, so that X-ray tubes are electronic instruments also. This short treatment will be sufficient to demonstrate how very important free electrons are for the entire field of physics and industry, and that we can say with certainty that they will play an even more important role in the future.

d) Free Ions

We have already mentioned that positive ions are atoms or molecules which have lost one, and, occasionally, two or more electrons from their shells. Negative ions are formed when electrons attach themselves to neutral atoms. Unfortunately, the binding energy between electrons and neutral atoms (the so-called *electron affinity*) is well known for a few atoms only. For the majority of atoms its value is small or zero; it reaches 2 to 4 ev only for the elements of the sixth and seventh group of the Periodic Table. Because the electron affinity is so much smaller than the ionization energy of the atoms, negative ions play, in general, a less important role in gaseous processes than positive ions. We designate ions by adding the appropriate number of $+$ or $-$ signs as superscripts to the symbol of the atom, thus speaking of a Ca^{++} ion.

The same methods as used for producing electrons are available for producing positive ions. Electrons, and with them positive ions, are formed by impact ionization in the cloud chamber, ionization chamber, and the counter tube as well as in all gas discharges. They are also produced by thermal ionization in highly heated gases and vapors (plasma) from which the ions can be withdrawn by a suitable arrangement of electric fields. The production of fast, directed ion beams in low-pressure gas discharges has gained special importance. At sufficiently low pressures the ions formed by impact in the gas volume travel toward the cathode perpendicularly to its surface without being deviated by too frequent

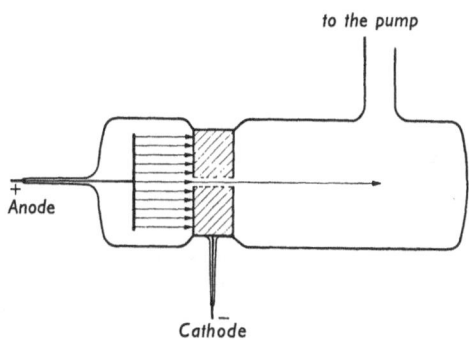

Fig. 11. Schematic representation of a canal-ray tube

collisions with gas atoms. If the cathode is perforated according to Fig. 11, the accelerated ions pass through the cathode into an evacuated field-free chamber as so-called *canal rays* or *positive rays*; they can be studied there without being disturbed by electrons or other plasma particles. In a condenser discharge through an evacuated capillary tube, highly charged ions can be produced, i.e., atoms with 3, 4, 5, and even up to 23 electrons removed. These have been called "stripped atoms" (page 79) because they have lost a considerable fraction of their original electron shell.

The production of positive ions at a metal surface by impact of fast electrons, as the counterpart of the production of electrons at a metal surface by impact of fast ions, occurs to a small extent at the anode in many discharges. In general, this process has little significance. On the other hand there are two important methods of producing positive ions thermally. According to Kunsman, certain mixtures of alkaline earth compounds with similar oxides emit positive ions when heated. These "Kunsman ion sources" are very practical for producing ions in a high vacuum. The second method of thermal production of positive ions applies only to the alkali and heavier alkaline-earth atoms. If these atoms impinge on an incandescent platinum sheet, they are reflected as positive ions. We shall discuss this phenomenon more thoroughly in VII, 24.

In concluding, a special case of positive ions may be mentioned, the α-particles which are emitted by radioactive substances. These particles have been identified as doubly charged helium atoms, He^{++}. That means, they are bare atomic nuclei. In nuclear physics, in addition to the α-particle, the hydrogen atomic ion is used because, like the α-particle, it is a bare nucleus too and, when accelerated, has a greater effect on other nuclei than ions which have an electron shell and therefore a size comparable to that of normal atoms, which makes their "penetrating power" much smaller than that of atomic nuclei.

We based our discussion of ions on our knowledge of electrons and of the formation of electrons and positive ions by ionization of atoms, so that there was no doubt about the nature of the positive ion. Historically, the direction of approach was reversed: positive charges of unknown nature were found in discharges, isolated as canal rays, and their nature determined by measuring their mass and charge. As the charge can be only one or a small integral multiple of the elementary quantum of electricity, the determination of the specific charge e/M was sufficient to determine the mass M of the ion. The measurement of e/M was made by measuring simultaneously the deflection of canal rays in crossed electrostatic and magnetic fields. By means of such measurements, W. WIEN proved that free ions from gas discharges or other ion sources were positively charged atoms or molecules. This discovery served also as proof for the existence of individual atoms. In II, 6, we shall discuss the measurement of the masses of isotopes, the so-called mass spectroscopy.

5. Brief Survey of the Structure of the Atomic Nucleus

In this section we present a brief survey of the structure of the nucleus, anticipating results from a more thorough discussion in Chap. V. We need this information in order to understand the phenomena of isotopes, which will be discussed in the following section.

We have already learned from the scattering experiments (II, 3) that the diameter of the nucleus is approximately 10^{-12} cm, i.e., only one tenthousandth of that of an atom, and that almost all of the mass of the atom is concentrated in the nucleus. From the nuclear volume and the absolute mass of the atom (II, 2b) one can thus easily compute that the nucleus has an inconceivably large density, of the order of 10^{14} g cm^{-3}. One cubic centimeter of nuclear material weighs approximately 100 million tons! It is not improbable that in the interior of certain giant stars there exist large masses of matter of comparable density.

Each atomic nucleus is characterized by its atomic weight (in general non-integral), its mass number (the integral number of mass units, i.e., of $\frac{1}{12}$ of the carbon atom), and the atomic number which indicates the number of positive charges that the nucleus carries. We mentioned already PROUT's hypothesis that all elements were built up of hydrogen atoms. Up to 1932 it was the general belief that a nucleus of mass number A was made up of A nuclei of hydrogen, called protons. In order to reduce the positive charge of A protons in a nucleus to the correct nuclear charge Z, which is about half as large as the mass number A, the nucleus had to be regarded as having $A-Z$ built-in electrons. But there are arguments against the existence of electrons in the nucleus which we will discuss in V, 4a. The discovery of the neutron, the uncharged nuclear particle of mass and size approximately equal to the proton [for details see V, 13], made the hypothesis of nuclear electrons avoidable. Today we are certain that the nucleus consists, according to HEISENBERG, of Z protons and $A-Z$ neutrons which are bound together by exchange forces of a specific type that was first explained by quantum mechanics (see V, 25). We shall discuss all the details about the structure of the nucleus in Chapter V.

6. Isotopes

a) The Discovery of Isotopes and their Significance
for the Atomic-Mass Problem

In discussing atomic masses (II, 2a) we showed that neither PROUT's hypothesis of H atoms as the constituents of all atoms nor the new concept of the nucleus as consisting of protons and neutrons is compatible with atomic masses which are non-integral multiples of $\frac{1}{12}$ of the carbon atom mass. However, there are a number of such elements in the Periodic Table (page 11) whose atomic masses are not integral numbers, for example chlorine with an atomic mass 35.457. This difficulty was cleared up by discovering the phenomenon of isotopy which means that atoms exist with the same nuclear charge and, therefore, of the same element, but with different atomic masses.

Isotopy was discovered after 1907 when experiments with radioactive elements and their disintegration products revealed many elements which had different decay periods and disintegration products, but which could not be separated chemically. Thorium, radiothorium, and ionium, all having the atomic number 90, were the first three atoms discovered which differed in their physical properties but obviously were chemically identical. SODDY called them isotopes. That non-radioactive elements may have isotopes as well, was first proved in 1912 by THOMSON in a mass-spectroscopic analysis of neon, later by HÖNIGSCHMID who found that lead refined from uranium ores has a considerably smaller atomic mass (206.05) than that from thorium ores (207.90). We shall learn in V, 6 that the lead isotope derived from the radioactive uranium family actually has an atomic mass of 206, whereas the isotope derived from the thorium family has an atomic mass of 208. In the following years ASTON, using his famous mass spectrograph, proved that isotopes are not exceptions but that most elements are actually mixtures of isotopes.

With this discovery, the question of atomic masses appeared in an entirely new light. The atomic mass as determined chemically is no longer a constant of the atom. It is an average value computed from the atomic masses and relative abundances of the isotopes of each element. Since this also applies for the reference element carbon, all atomic masses now are measured, according to an international agreement, each in units of $\frac{1}{12}$ of the mass of the carbon isotope $C^{12}=12.000\,000$.

In general, the relative abundance of the isotopes of the elements is constant on our earth so that the chemical atomic weight is the average atomic mass of the natural mixture of isotopes and as such it is a characteristic property of each particular element. The only exceptions to the above rule are the end products of the radioactive elements which result from nuclear transformations. The outstanding example is lead, which has an average atomic weight that varies with the locality where the metal is found because lead may originate from three different radioactive series.

b) The Explanation of Isotopes and their Properties

From what we learned in II, 5 about the structure of the nucleus there can be no doubt about the explanation of isotopes. *Since the number of protons in a nucleus determines the atomic number and with it the element, the isotopes of one element have the same number of protons in a nucleus, but differ in the number of neutrons.* For instance, the nucleus of the heavy stable isotope of hydrogen consists of one proton and one neutron; it has therefore the *"mass number"* 2, whereas the common hydrogen nucleus with the charge 1 and the mass 1 is the proton itself. In V, 11, when considering the systematics of atomic nuclei, we shall find out which elements have many stable isotopes and which ones have only a few or none at all. When dealing with artificial radioactivity (see V, 7), we shall see that nuclei with too large an excess or deficit of neutrons may spontaneously decay into stable nuclei by ejecting an electron or a positron (see later), respectively. Such unstable isotopes may also be produced in so-called artificial nuclear transformations which will be discussed later. Thus there exists, besides the stable isotopes of an element (e.g., oxygen with O^{16}, O^{17}, and O^{18}), an additional number of unstable isotopes (O^{14}, O^{15}, and O^{19} of oxygen). We shall treat this subject in detail in V, 7.

In order to understand completely the behavior of any chemical element, it is no longer sufficient to know just the chemical atomic weight. It is also necessary to know the exact masses and relative abundances of all its stable isotopes. Later on we shall take up in detail the methods for determining these figures. At this time we mention only one very important fact: that *all isotopes, relative to* $C^{12} =$ 12.000000, *have atomic masses which are integral numbers within* 1 *per cent*. This important result is, as will be pointed out in V, 5, in complete agreement with the theory that the nucleus is built up of protons and neutrons.

Now that we know that the isotopes of an element differ from one another only in the number of neutrons in their nuclei, we can draw several important conclusions about the general properties of isotopes. First it is evident that a variation in the number of neutrons in the nucleus, because of the neutron's electric neutrality, has no essential influence on the binding forces between the positive nucleus and the negative electron shell, so that the electron shell, the chemical properties, and those physical properties that depend on the shell *alone* are practically the same for all isotopes of one element. However, we expect a different physical behavior of isotopes in those properties for which the mass plays an important role. They offer a possibility for distinguishing between the various isotopes of the same element. This difference in properties is the more pronounced the greater the relative difference in the masses of the isotopes of a particular element is; isotopes of the light nuclei thus will be easiest to distinguish. Hence the specific significance of the heavy hydrogen isotopes H^2 and H^3 becomes clear; they have been given seperate names, viz. deuterium D and tritium T, respectively; their nuclei are called deuteron d and triton t. Their masses are twice and three times that of the normal H^1atom, whereas the two oxygen isotopes O^{18} and O^{16} have a relative mass difference of only 12% and the two uranium isotopes U^{238} and U^{235} of hardly more than 1%. The specific charges e/M of the positive ions of the isotopes vary by corresponding amounts. The measurement of e/M therefore provides a method for determining the masses of isotopes.

Among the properties of the atom that depend upon the mass are the evaporation velocity and the coefficient of diffusion, which can be used to distinguish between isotopes and to separate them. We shall investigate the small but significant differences that different masses of isotopes cause in atomic and molecular spectra in III, 20 and VI, 12. The large difference in the masses of H and D is the reason for the substantial difference in the moments of inertia as well as in the zero-point energy of the molecules H_2, D_2, and HD, and also of the hydrogen compounds as NH_3 and HD_3, CH_4 and CD_4. It also explains the difference in the heats of vaporization, molar heats, molar volumes, chemical constants, dissociation energies, and vibration frequencies of these molecules. As an example, the heats of vaporization of H_2 and D_2 are 220 cal and 362 cal, respectively; and even the corresponding values for light and heavy water, H_2O and D_2O, still differ by 3.5%.

So far we dealt mainly with stable isotopes whose numbers and masses are listed in Table 3. In the nuclear physics Chapter V, we will deal extensively with the unstable, β-active isotopes which originate from nuclear transformations.

c) Mass Spectroscopy

For determining the mass and relative abundance of isotopes, we may use the methods of optical spectroscopy. The difference in mass can be determined from the difference in the wavelengths of the spectral lines belonging to the various isotopes, and the relative abundance by measuring the relative intensities of these lines. We shall discuss these optical methods of isotope investigation in detail in III, 20 and VI, 12. The classical method for investigating isotopes, introduced by THOMSON and ASTON and perfected by MATTAUCH, however, is the method of measuring e/M, the so-called mass spectroscopy.

The parabola method used by THOMSON in 1913 is the oldest method used which gave reliable results. Instead of the usual crossed fields, THOMSON used superimposed electrostatic and magnetic fields (Fig. 12) with the condenser plates between the pole faces. Let the direction of the canal rays be the z axis and the direction of the field be the y axis, then the deviation caused by the electrostatic field acting over a distance l follows from Eq. (2.32)

$$Y = \frac{e\,E\,l^2}{2\,M\,v^2}.\qquad(2.43)$$

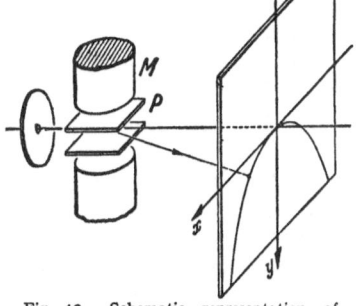

Since the magnetic force on the ions acts perpendicular to the velocity vector, the ions are deflected into circles in the xz-plane. The radius of these circles may be computed by equating the magnetic force according to Eq. (2.33) and the centrifugal force $\dfrac{M\,v^2}{R}$, so that

$$R = \frac{M\,v}{e\,H}.\qquad(2.44)$$

Fig. 12. Schematic representation of THOMSON's parabola method for determining ion masses. $P=$ condenser plates, $M=$ magnetic pole pieces

After the ions have passed the distance l in the magnetic field, they are deviated in the x direction by

$$X = \frac{1}{2}\,b\,t^2 = \frac{1}{2}\,\frac{v^2}{R}\left(\frac{l}{v}\right)^2 = \frac{l^2}{2R} = \frac{e\,H\,l^2}{2\,M\,v}.\qquad(2.45)$$

By eliminating the unknown velocity v, which is different for different ions, we obtain from Eqs. (2.43) and (2.45)

$$Y = \frac{2E}{l^2 H^2}\,\frac{M}{e}\,X^2.\qquad(2.46)$$

This means that ions of equal mass and charge but varying velocity form parabolas on the fluorescent screen (Fig. 13, page 40). The slope of an individual curve is used for determining the value e/M and from it the mass of a particular ion. Ions of known masses are used for calibrating the system. The intensities of the individual parabolas correspond to the relative abundance of the various ions in the mixture.

The second mass-spectroscopic method for determining e/M, and from it the masses of isotopes, uses crossed electrostatic and magnetic fields as presented on page 24. It has been developed into a precision method. Fig. 14 (page 40) shows the first mass spectrograph built by ASTON in 1919. It corresponds to an optical spectrograph in that it has an entrance slit, splits the ion beam up in an electrostatic and magnetic field, and finally collects the beams of different mass on different points of a photographic plate or receiver. The canal rays pass through a fine hole S_1 in the cathode of a discharge tube, and through a second collimating hole, S_2. The beam then passes through the electric field E and is deflected through an angle α. After passing a comparatively wider slit, K_2, it traverses the magnetic field H, perpendicular to the plane of Fig. 14 (page 40), wherein it is deflected in

Table 3. *Nuclides*

(Included are all stable isotopes and the most stable ones of the naturally radioactive
isotopes. Unstable nuclei are characterized by ⟨ ⟩)

Z		Element	Mass number, A	Neutron number, A−Z	Relative abundance, per cent	Isotope mass, $C^{12} = 12.000000$	Atomic weight (1962)
0	n	Neutron	1	1	—	1.0086654	—
1	H	Hydrogen	1	0	99.986	1.00782522	1.00797
	(D)	Deuterium	2	1	0.014	2.0141022	
2	He	Helium	3	1	$1.3 \cdot 10^{-4}$	3.0160299	4.0026
			4	2	100	4.0026036	
3	Li	Lithium	6	3	7.30	6.015126	6.939
			7	4	92.70	7.016005	
4	Be	Beryllium	9	5	100	9.0121858	9.0122
5	B	Boron	10	5	18.83	10.0129389	10.811
			11	6	81.17	11.0093051	
6	C	Carbon	12	6	98.892	12.0000000	12.01115
			13	7	1.108	13.0033543	
7	N	Nitrogen	14	7	99.635	14.0030744	14.0067
			15	8	0.365	15.0001081	
8	O	Oxygen	16	8	99.759	15.9949149	15.9994
			17	9	0.0374	16.9991334	
			18	10	0.2036	17.9991598	
9	F	Fluorine	19	10	100	18.9984046	18.9984
10	Ne	Neon	20	10	90.92	19.9924404	20.183
			21	11	0.257	20.993849	
			22	12	9.823	21.9913845	
11	Na	Sodium	23	12	100	22.989773	22.9898
12	Mg	Magnesium	24	12	78.98	23.985045	24.312
			25	13	10.05	24.985840	
			26	14	10.97	25.982591	
13	Al	Aluminium	27	14	100	26.981535	26.9815
14	Si	Silicon	28	14	92.18	27.976927	28.086
			29	15	4.71	28.976491	
			30	16	3.12	29.973761	
15	P	Phosphorus	31	16	100	30.973763	30.9738
16	S	Sulfur	32	16	95.018	31.972074	32.064
			33	17	0.750	32.971460	
			34	18	4.215	33.967864	
			36	20	0.017	35.967091	
17	Cl	Chlorine	35	18	75.40	34.968854	35.453
			37	20	24.60	36.965896	
18	Ar	Argon	36	18	0.337	35.967548	39.948
			38	20	0.063	37.962724	
			40	22	99.600	39.9623838	
19	K	Potassium	39	20	93.0800	38.963714	39.102
			40	21	0.0119	39.964008	
			41	22	6.9081	40.961835	
20	Ca	Calcium	40	20	96.92	39.962589	40.08
			42	22	0.64	41.958627	
			43	23	0.129	42.958780	
			44	24	2.13	43.955490	
			46	26	0.003	45.95369	
			48	28	0.178	47.95236	
21	Sc	Scandium	45	24	100	44.955919	44.956
22	Ti	Titanium	46	24	7.95	45.952633	47.90
			47	25	7.75	46.951758	
			48	26	73.45	47.947948	
			49	27	5.51	48.947867	
			50	28	5.34	49.944789	
23	V	Vanadium	50	27	0.23	49.947165	50.942
			51	28	99.77	50.943978	

Table 3. *Nuclides* (continued)

Z	Element		Mass number, A	Neutron number, A−Z	Relative abundance, per cent	Isotope mass, C¹² = 12.000000	Atomic weight (1962)
24	Cr	Chromium	50	26	4.31	49.946051	
			52	28	83.76	51.940514	51.996
			53	29	9.55	52.940651	
			54	30	2.38	53.938879	
25	Mn	Manganese	55	30	100	54.938054	54.9381
26	Fe	Iron	54	28	5.81	53.939621	
			56	30	91.64	55.934932	55.847
			57	31	2.21	56.935394	
			58	32	0.34	57.933272	
27	Co	Cobalt	59	32	100	58.933189	58.9332
28	Ni	Nickel	58	30	67.77	57.935342	
			60	32	26.16	59.930783	
			61	33	1.25	60.931049	58.71
			62	34	3.66	61.928345	
			64	36	1.16	63.927959	
29	Cu	Copper	63	34	68.94	62.929594	
			65	36	31.06	64.927786	63.54
30	Zn	Zinc	64	34	48.89	63.929145	
			66	36	27.81	65.926048	
			67	37	4.07	66.92715	65.37
			68	38	18.61	67.924865	
			70	40	0.62	69.92535	
31	Ga	Gallium	69	38	60.16	68.92568	
			71	40	39.84	70.92484	69.72
32	Ge	Germanium	70	38	20.52	69.92428	
			72	40	27.43	71.92174	
			73	41	7.76	72.92336	72.59
			74	42	36.54	73.92115	
			76	44	7.76	75.92136	
33	As	Arsenic	75	42	100	74.92158	74.9216
34	Se	Selenium	74	40	0.87	73.92245	
			76	42	9.02	75.91923	
			77	43	7.58	76.91993	
			78	44	23.52	77.91735	78.96
			80	46	49.82	79.91651	
			82	48	9.19	81.91666	
35	Br	Bromine	79	44	50.53	78.91835	
			81	46	49.47	80.91634	79.909
36	Kr	Krypton	78	42	0.354	77.920368	
			80	44	2.266	79.91639	
			82	46	11.56	81.913483	
			83	47	11.55	82.914131	83.80
			84	48	56.90	83.911504	
			86	50	17.37	85.910617	
37	Rb	Rubidium	85	48	72.20	84.91171	
			87	50	27.80	86.90918	85.47
38	Sr	Strontium	84	46	0.55	83.91337	
			86	48	9.75	85.90926	
			87	49	6.96	86.90889	87.62
			88	50	82.74	87.90561	
39	Y	Yttrium	89	50	100	88.90543	88.905
40	Zr	Zirconium	90	50	51.46	89.90432	
			91	51	11.23	90.9052	
			92	52	17.11	91.9046	91.22
			94	54	17.40	93.9061	
			96	56	2.80	95.9082	
41	Nb	Niobium	93	52	100	92.9060	92.906

Table 3. *Nuclides* (continued)

Z	Element		Mass number, A	Neutron number, A−Z	Relative abundance, per cent	Isotope mass, $C^{12} = 12.000000$	Atomic weight (1962)
42	Mo	Molybdenum	92	50	15.84	91.9063	
			94	52	9.04	93.9047	
			95	53	15.72	94.9057	
			96	54	16.53	95.9045	95.94
			97	55	9.46	96.9057	
			98	56	23.78	97.9055	
			100	58	9.63	99.9076	
43	⟨Tc⟩	Technetium	99	56		98.9064	
44	Ru	Ruthenium	96	52	5.68	95.9076	
			98	54	2.22	97.9055	
			99	55	12.81	98.9061	
			100	56	12.70	99.9030	101.07
			101	57	16.98	100.9041	
			102	58	31.34	101.9037	
			104	60	18.27	103.9055	
45	Rh	Rhodium	103	58	100	102.9048	102.905
46	Pd	Palladium	102	56	0.80	101.9049	
			104	58	9.30	103.9036	
			105	59	22.60	104.9046	106.4
			106	60	27.10	105.9032	
			108	62	26.70	107.9039	
			110	64	13.50	109.9045	
47	Ag	Silver	107	60	51.92	106.9050	107.870
			109	62	48.08	108.9047	
48	Cd	Cadmium	106	58	1.215	105.9059	
			108	60	0.875	107.9040	
			110	62	12.39	109.9030	
			111	63	12.75	110.9041	
			112	64	24.07	111.9028	112.40
			113	65	12.26	112.9046	
			114	66	28.86	113.9036	
			116	68	7.78	115.9050	
49	In	Indium	113	64	4.23	112.9043	114.82
			115	66	95.77	114.9041	
50	Sn	Tin	112	62	0.94	111.9049	
			114	64	0.65	113.9030	
			115	65	0.33	114.9035	
			116	66	14.36	115.9021	
			117	67	7.51	116.9031	
			118	68	24.21	117.9018	118.69
			119	69	8.45	118.9034	
			120	70	33.11	119.9021	
			122	72	4.61	121.9034	
			124	74	5.83	123.9052	
51	Sb	Antimony	121	70	57.25	120.9037	121.75
			123	72	42.75	122.9041	
52	Te	Tellurium	120	68	0.09	119.9045	
			122	70	2.43	121.9030	
			123	71	0.85	122.9042	
			124	72	4.59	123.9028	
			125	73	6.98	124.9044	127.60
			126	74	18.70	125.9032	
			128	76	31.85	127.9047	
			130	78	34.51	129.9067	
53	I	Iodine	127	74	100	126.90435	126.9044

Table 3. *Nuclides* (continued)

Z	Element	Mass number, A	Neutron number, A−Z	Relative abundance, per cent	Isotope mass, $C^{12} = 12.000000$	Atomic weight (1962)
54	**Xe** Xenon	124	70	0.096	123.90612	
		126	72	0.020 (90)	125.90417	
		128	74	1.919	127.90354	
		129	75	26.44	128.90478	
		130	76	4.075	129.903510	131.30
		131	77	21.18	130.90508	
		132	78	26.89	131.904162	
		134	80	10.44	133.905398	
		136	82	8.87	135.90722	
55	**Cs** Cesium	133	78	100	132.9051	132.905
56	**Ba** Barium	130	74	0.102	129.90625	
		132	76	0.098	131.9051	
		134	78	2.42	133.9043	
		135	79	6.59	134.9056	137.34
		136	80	7.81	135.9044	
		137	81	11.32	136.9056	
		138	82	71.66	137.90501	
57	**La** Lanthanum	138	81	0.89	137.90681	138.91
		139	82	99.911	138.90606	
58	**Ce** Cerium	136	78	0.19	135.9071	
		138	80	0.25	137.90572	140.12
		140	82	88.49	139.90528	
		142	84	11.07	141.90904	
59	**Pr** Praseodymium	141	82	100	140.90739	140.907
60	**Nd** Neodymium	142	82	26.80	141.90748	
		143	83	12.12	142.90962	
		144	84	23.91	143.90990	
		145	85	8.35	144.9122	144.24
		146	86	17.35	145.9127	
		148	88	5.78	147.9165	
		150	90	5.69	149.9207	
61	⟨**Pm**⟩ Promethium	149	88		148.9181	
62	**Sm** Samarium	144	82	2.95	143.9116	
		147	85	14.62	146.9146	
		148	86	10.97	147.9146	
		149	87	13.56	148.9169	150.35
		150	88	7.27	149.9170	
		152	90	27.34	151.9195	
		154	92	23.29	153.9220	
63	**Eu** Europium	151	88	47.77	150.9196	151.96
		153	90	52.23	152.2909	
64	**Gd** Gadolinium	152	88	0.2	151.9195	
		154	90	2.16	153.9207	
		155	91	14.68	154.9226	
		156	92	20.36	155.9221	157.25
		157	93	15.64	156.9239	
		158	94	24.95	157.9241	
		160	96	22.01	159.9271	
65	**Tb** Terbium	159	94	100	158.9250	158.924
66	**Dy** Dysprosium	156	90	0.0525	155.9238	
		158	92	0.0905	157.9240	
		160	94	2.297	159.9248	
		161	95	18.88	160.9266	162.50
		162	96	25.53	161.9265	
		163	97	24.97	162.9284	
		164	98	28.18	163.9288	
67	**Ho** Holmium	165	98	100	164.9303	164.930

Table 3. *Nuclides* (continued)

Z		Element	Mass number, A	Neutron number, A−Z	Relative abundance, per cent	Isotope mass, $C^{12} = 12.000000$	Atomic weight (1962)
68	Er	Erbium	162	94	0.154	161.9288	
			164	96	1.606	163.9283	
			166	98	33.36	165.9304	167.26
			167	99	22.82	166.9321	
			168	100	27.02	167.9324	
			170	102	15.04	169.9355	
69	Tm	Thulium	169	100	100	168.9343	168.934
70	Yb	Ytterbium	168	98	0.13	167.9339	
			170	100	3.03	169.9349	
			171	101	14.27	170.9365	
			172	102	21.77	171.9366	173.04
			173	103	16.08	172.9383	
			174	104	31.92	173.9390	
			176	106	12.80	175.9427	
71	Lu	Lutecium	175	104	97.40	174.9409	174.97
			176	105	2.60	175.94274	
72	Hf	Hafnium	174	102	0.199	173.9403	
			176	104	5.23	175.94165	
			177	105	18.55	176.94348	178.49
			178	106	27.23	177.94387	
			179	107	13.73	178.9460	
			180	108	35.07	179.9468	
73	Ta	Tantalum	180	107	0.0123	179.94752	180.948
			181	108	100	180.94798	
74	W	Wolfram (Tungsten)	180	106	0.16	179.94698	
			182	108	26.35	181.94827	
			183	109	14.32	182.95029	183.85
			184	110	30.68	183.95099	
			186	112	28.49	185.95434	
75	Re	Rhenium	185	110	37.07	184.95302	186.2
			187	112	62.93	186.95596	
76	Os	Osmium	184	108	0.018	183.9526	
			186	110	1.582	185.95394	
			187	111	1.64	186.95596	
			188	112	13.27	187.95597	190.2
			189	113	16.14	188.9582	
			190	114	26.38	189.95860	
			192	116	40.97	191.96141	
77	Ir	Iridium	191	114	38.5	190.96085	192.2
			193	116	61.5	192.96328	
78	Pt	Platinum	190	112	0.012	189.95995	
			192	114	0.8	191.96143	
			194	116	30.2	193.96281	
			195	117	35.2	194.96482	195.09
			196	118	26.6	195.96498	
			198	120	7.2	197.9675	
79	Au	Gold	197	118	100	196.96655	196.967
80	Hg	Mercury	196	116	0.15	195.96582	
			198	118	10.12	197.96677	
			199	119	17.04	198.96826	
			200	120	23.25	199.96834	200.59
			201	121	13.18	200.97031	
			202	122	29.54	201.97063	
			204	124	6.72	203.97348	
81	Tl	Thallium	203	122	29.46	202.97233	204.37
			205	124	70.54	204.97446	

Table 3. *Nuclides* (continued)

Z	Element		Mass number, A	Neutron number, A−Z	Relative abundance, per cent	Isotope mass, $C^{12} = 12.000000$	Atomic weight (1962)
82	**Pb**	Lead	204	122	1.54	203.97307	
			206	124	22.62	205.97446	
			207	125	22.62	206.97590	207.19
			208	126	53.22	207.97664	
83	**Bi**	Bismuth	209	126	100	208.98042	208.980
84	⟨**Po**⟩	Polonium	210	126		209.98287	
85	⟨**At**⟩	Astatine	210	125		209.9870	
86	⟨**Rn**⟩	Radon	222	136		222.01753	
87	⟨**Fr**⟩	Francium	223	136		223.01980	
88	⟨**Ra**⟩	Radium	226	138		226.02536	
89	⟨**Ac**⟩	Actinium	227	138		227.02781	
90	**Th**	Thorium	232	142	100	232.03821	232.038
91	**Pa**	Protactinium	231	140	100	231.03594	
92	**U**	Uranium	234	142	0.006	234.04090	
			235	143	0.720	235.04393	238.03
			238	146	99.274	238.05076	
93	⟨**Np**⟩	Neptunium	237	144		237.04803	
94	⟨**Pu**⟩	Plutonium	242	145		239.05216	
95	⟨**Am**⟩	Americium	243	148		243.06138	
96	⟨**Cm**⟩	Curium	245	152		245.06534	
97	⟨**Bk**⟩	Berkelium	245	150		247.07018	
98	⟨**Cf**⟩	Californium	248	151		249.07470	
99	⟨**E**⟩	Einsteinium	255	153		254.0881	
100	⟨**Fm**⟩	Fermium	252	153		252.08265	
101	⟨**Mv**⟩	Mendelevium	256	155			
102	⟨**No**⟩	Nobelium	253	151			
103	⟨**Lw**⟩	Lawrencium	257	154			

the opposite direction through an angle β, after which it impinges on a photographic plate W. ASTON realized that sufficient intensity could be obtained only if the slit K_2 is not too narrow, and that sharp lines and thus a good resolution of neighboring masses was possible only if by the focussing action of the magnet field all ions of the same mass, regardless of their velocity, arrived at the same point of the photographic plate. He accomplished this "velocity focussing" by a certain relation between the deflection angles α and β of the ion beam in the electric and magnetic fields, respectively, and the distances l and r between the centers of the two fields and between the center of the magnetic field and the photographic plate, respectively.

Fig. 15 shows photographs by ASTON with mass numbers which permit to judge the resolution of his instrument. These photographs have led to the discovery of a large part of the isotopes known so far.

In the meantime, ASTON's mass spectrograph has been modified in various ways. DEMPSTER showed that the magnetic field can also be used to focus ions of different initial directions on one point of the plate (direction focussing). A decisive improvement was accomplished by MATTAUCH in 1934, when by using carefully computed distances and angles of deflection he succeeded in simultaneously focussing ions of different initial velocities as well as directions on a plane photographic plate (Fig. 16). The ion beam here is first spread out with respect to velocities in a radial electrostatic field of $31° 50'$ and then, after passing through a reasonably wide diaphragm B, enters the magnetic field where it is deflected through an angle of $90°$ and spread out with respect to the different masses. At the same time, the magnetic field focusses the individual beams again with respect to both velocity and direction. Fig. 17 shows an especially fine example for the high resolving power of the instrument. This mass spectrograph allows today to determine ion masses and, therefore, atomic weights with an accuracy of better than 10^{-7} mass units. We shall understand the significance of these measurements in V, 5, where we discuss the phenomenon of mass defect. From the differences in intensity of the mass lines, the relative abundance of the isotopes can be determined (see column 5 in Table 3).

The absolute resolution of the described mass spectrograph decreases with increasing ion mass. This disadvantage is avoided by using the *"time-of-flight spectrograph"* of GOUDSMIT. This instrument is therefore particularly qualified for determining the mass of heavy ions as found at the end of the Periodic Table. According to Eq. (2.44), ions in a homogeneous magnetic field follow circles or more or less narrow helical curves, if they have a velocity component parallel to the magnetic field. Ions with a single positive charge require, for completing a full circle with the radius R, a time of flight τ, which according to Eq. (2.44) is

12   13   14   15   16
C+   CH+ CH$_2$+ CH$_3$+ CH$_4$+

Fig. 13. Mass-spectroscopic analysis of a mixture of hydrocarbon ions by the Thomson parabola method. (After CONRAD)

$$\tau = \frac{2\pi R}{v} = \frac{2\pi M}{eH}. \qquad (2.47)$$

In GOUDSMIT's mass spectrograph, individual ion groups, sharply limited with respect to time, enter the instrument which is adjusted for velocity focussing at 360° angle. The time which the ions need for 7 full circles (7τ) is then measured by a method borrowed from radar technique. The accuracy is 10^{-8} sec.

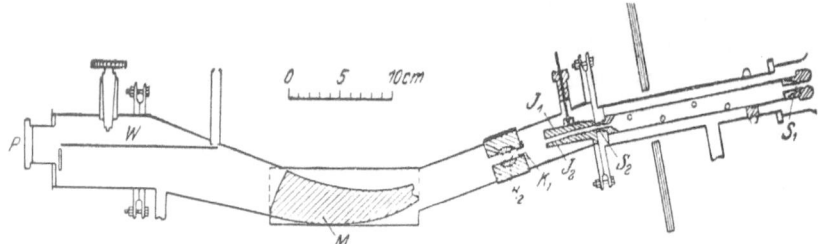

Fig. 14. Mass spectrograph of ASTON (schematic). S_1, S_2, K_1, and K_2 = diaphragms, J_1 and J_2 = condenser plates, M = magnetic pole pieces, W = photographic plate

Since τ is proportional to the ion mass [see Eq. (2.47)], the accuracy of measurement is independent of the mass and equal to about 0.001 atomic mass units. This time-of-flight spectrograph was successfully used for heavy ions. Fig. 18 shows an oscillogram of 7 xenon isotopes arriving one after the other after 8 revolutions.

An entirely different type of instrument which stands out for its simplicity is the *high-frequency mass spectrograph*, developed by BENNET. In principle, it consists simply of a high vacuum tube that has four or more grids between cathode and collector electrode. The high-frequency grid potentials are chosen in such

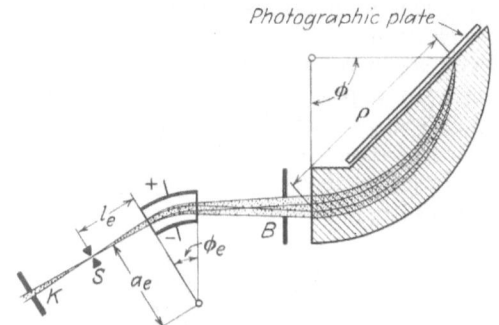

Fig. 16. Improved mass spectrograph of MATTAUCH with velocity and direction focussing. Φ_e = deviation in the electric field, Φ = deviation in the magnetic field

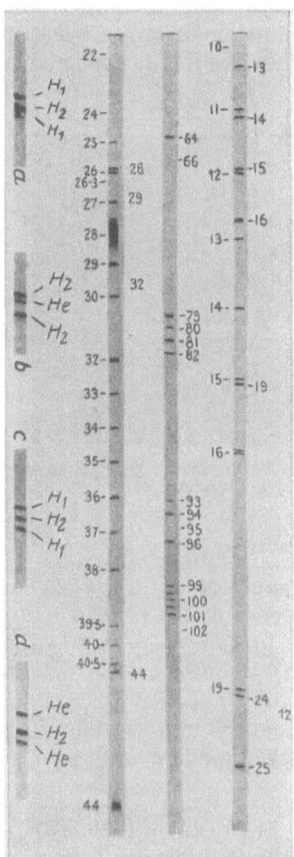

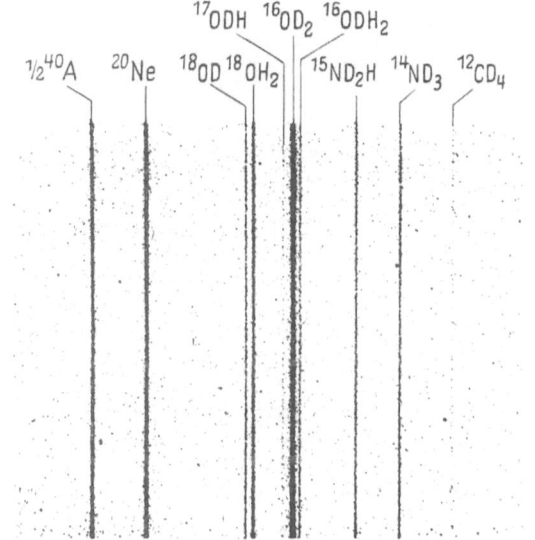

Fig. 15. Mass spectra by ASTON which were used for determining masses and relative abundances of isotopes

Fig. 17. Mass spectrum with fine structure by BIERI, EVERLING, and MATTAUCH shows the excellent resolution of very small mass differences. (Ten different ions of the mass number 20 are separated whose atomic or molecular masses are between 19.9878 and 20.0628)

a way that only ions of a certain specific charge e/M and, therefore, mass M are moving in the right rhythm to pass all the grids and to reach the collector electrode. Thus, a purely electric method of measurement is used here.

Based on a resonance effect is also the mass spectrometer that is known as "omegatron". It is frequently used for gas analysis. Basically it is a tiny cyclotron [see V, 3] with a radius of beam curvature close to 1 cm. Under the combined action of an axial magnetic field of several thousand gauss and of a transverse high-frequency field (e.g., 0.1 volt/cm at some 10^6 cps), only ions of the "correct" specific charge e/M and mass M are accelerated on a spiral track until they reach a collector and are measured there. Ions of "incorrect" mass, however, do not reach the collector. The resolving power of the omegatron reaches 40,000 for a small ion mass and changes inversely proportional to the ion mass.

The *mass filter* developed by PAUL is based on a similar resonance effect. Here the ion beam moves in the center between and parallel to four long rod-shaped electrodes. In a cross section view, these electrodes seem to be placed at the corners of a square. High-frequency voltage is applied to them so that always two diametrically opposite electrodes have the same polarity. It can be shown that an appropriate alternating voltage and frequency allows ions of only a very limited mass range to pass this mass filter. Ions of smaller or larger mass are excited to oscillations of growing amplitude so that they are finally captured by the rod electrodes.

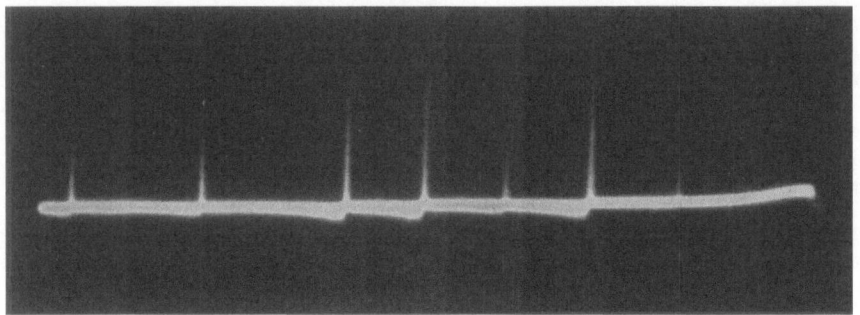

Fig. 18. Recording of the Xenon isotopes 128, 129, 130, 131, 132, 134, and 136 with the "time-of-flight spectrograph". The distance of neighboring isotopes is 8 μsec. (Recorded by S. GOUDSMIT)

d) Methods of Isotope Separation

Originally, isotopes were studied for the sake of increasing our knowledge of the structure of matter. Today, isotopes are an indispensable aid to industry as well as to molecular physicists, chemists, and biologists, for isotopes are being used for identifying and tracing individual atoms among a large number of chemically similar ones. Therefore, science and technology are greatly interested in obtaining pure isotopes, i.e., in separating as completely as possible the various isotopes of an element. This interest was strongly stimulated by the modern utilization of atomic energy [see V, 16].

The best method of separation, though suited only for small quantities, is that of mass spectroscopy. For this purpose, mass spectrographs are used with large high-intensity ion sources and with collectors for the various isotopes. Some tenths of 1 gram of pure isotopes can be produced per hour in such "electro-magnetic separators".

For technical purposes however, relatively pure isotopes are needed in ton quantities. To satisfy this need, methods of *enrichment* by numerous repeated steps are applied. Each of these steps produces a relatively small change in the relative abundance of isotopes; but sufficiently numerous repetitions lead to high enrichment of the desired isotope. In this way it was possible to produce the technically very important heavy water, D_2O, in a concentration of better than 99.8%.

Essentially, four different methods are applied:

1) In a high-rpm centrifuge the heavy isotope is moving outward, the lighter one towards the axis.

2) Through filters with very small pores light isotopes diffuse better than heavy ones [pore diffusion method 2a]. In an isotope mixture with a temperature gradient,

the lighter isotopes diffuse to the side of high temperature, while the heavier ones move towards the low-temperature side [thermo-diffusion method 2b].

3) When a mixture of isotopes is evaporated, the lighter component gets enriched in the vapor; the heavier one in the liquid.

4) Nearly all chemical exchange reactions, particularly between the liquid and vapor phase, cause a change in the relative abundance of isotopes. Thus, the exchange between NO and aqueous nitric acid (HNO_3) produces gaseous NO with an enrichment of the lighter N isotope. H_2S gas absorbs deuterium from water of high temperature and releases it in an exchange reaction to water of low temperature (hydrogen sulfide method).

For the separation of the easily fissionable uranium 235 from uranium 238 which is not fissionable by thermal neutrons [see V, 14], the pore diffusion method (2a) of G. HERTZ is being used in huge plants. Tests are also being made with the centrifuge method (1) for separating gaseous uranium hexafluoride UF_6.

The production of pure heavy water (D_2O) was originally carried out only by means of repeated electrolysis of water. Through a multiplicity of interacting mass-dependent processes (differences in ion mobility, in the discharge velocity at the electrodes, etc.) preferably the H_2O molecules are decomposed, while HDO and D_2O molecules remain in the water. At present, exchange reactions (4) are being used, such as the hydrogen sulfide method or catalytically initiated exchange reactions between liquid water, steam, and hydrogen gas. In addition, the fractional distillation of liquid hydrogen, by means of method (3), is used.

Thermo-diffusion (2b) was developed by CLUSIUS and DICKEL to serve as an effective separation method. Their separation tube consists of a long vertical glass tube (of a length up to 20 meters) with a diameter of several centimeters. A wire, heated electrically to several hundred degrees centigrade, extends along the axis of the tube. The lighter component of the gaseous isotope mixture rises upward near the hot wire, whereas the heavier isotopes sink down near the cooler wall of the tube. The lighter isotopes can be drawn off from the top of the tube, the heavier isotopes from the bottom.

7. Photons

So far we have spoken only of the atomic constituents of matter and of their properties. However, one of the most important problems with which atomic physics deals is that of the interaction between matter and electromagnetic radiation, e.g., the emission or absorption of radiation by atoms, molecules, etc. By radiation we mean its manifestations in the entire wavelength region from the shortwave γ radiation found in cosmic rays and nuclear physics through X rays, ultraviolet, visible, infrared, and heat rays, up to and including what, in a restricted sense, are called electromagnetic waves. These various wavelength regions, which phenomenologically differ to a great extent, are plotted in Fig. 19 with the corresponding scales of wavelengths, frequencies, and energies of the radiation quanta or photons which will be discussed below.

The transition from classical to atomic physics is intimately linked with the transition from continuous to discrete atomic phenomena. This is as true for radiation as it is for matter. Classical physics is concerned with light waves which are continuous in space and time, e.g., a spherical wave emitted by a light source. As we shall see in the next chapter and throughout this book, the phenomena of atomic physics, such as the interaction between matter and radiation, can be explained only by assuming that radiation is emitted and absorbed as single *light quanta* which we call *photons*. According to PLANCK's fundamental relation,

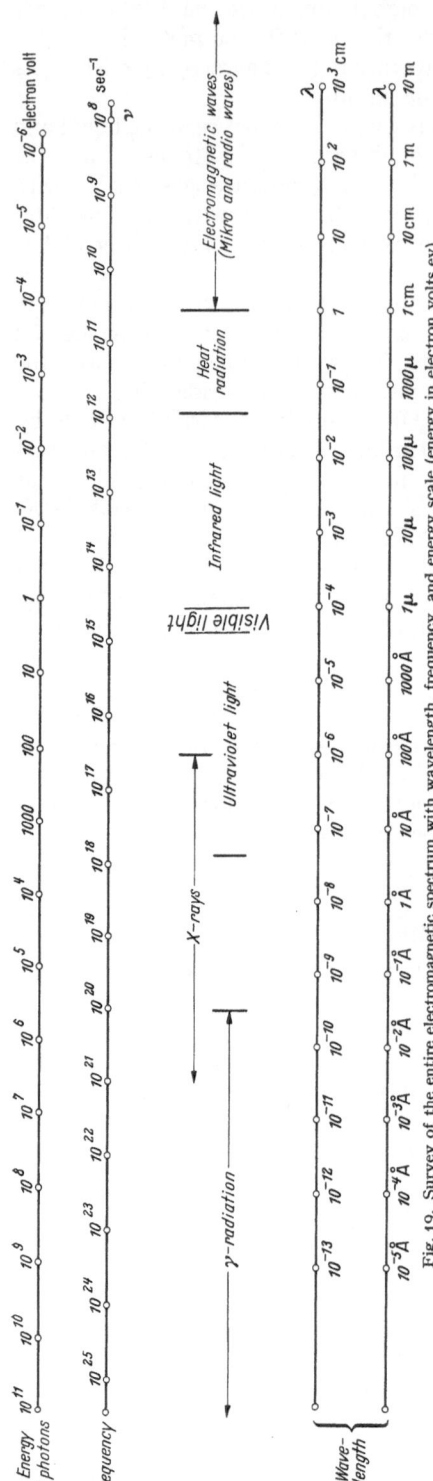

Fig. 19. Survey of the entire electromagnetic spectrum with wavelength, frequency, and energy scale (energy in electron volts ev)

the energy of these photons is proportional to the frequency ν (and thereby to c/λ) of the corresponding wave,

$$E = h\nu. \tag{2.48}$$

h is called PLANCK'S constant and is equal to

$$h = (6.6256 \pm 0.0005) \times \left. \begin{array}{c} \\ \\ \end{array} \right\} \tag{2.49}$$
$$\times 10^{-27} \text{ erg-sec.}$$

From Eq. (2.48) it follows that the energy of the quanta of electromagnetic radiation is inversely proportional to its wavelength. This explains many phenomena, among others why photographic plates react more to violet and blue light than to red light, and why ultraviolet light, X-rays, and γ rays are more penetrating than longer-wave radiation which has less energy. We shall see that Eq. (2.48) proved to be *the* fundamental relation of all modern physics.

Next to the energy, we are most interested in the inertial mass and in the momentum of photons, because these properties are important for their interaction (emission, absorption, collision processes) with atomic particles. Photons can have no rest mass, because they do not "exist" unless traveling with the velocity of light. However, according to the mass-equivalence equation (2.40), every energy E has a mass equivalent E/c^2. Therefore, a photon of energy $h\nu$ has an inertia corresponding to a mass

$$m_{\text{Ph}} = \frac{h\nu}{c^2} \tag{2.50}$$

and a momentum (mass times velocity c)

$$p_{\text{Ph}} = \frac{h\nu}{c}. \tag{2.51}$$

This momentum is causing the radiation pressure, a phenomenon well-known to classical physics.

The fundamental relation (2.48) was discovered by PLANCK in 1900, when he tried to derive theoretically his formula (2.5) for the spectral energy distribution of blackbody radiation. A blackbody radiator is one which completely absorbs incident radiation of every wavelength. Such a radiator emits continuous radiation with a spectral energy distribution which depends exclusively on the absolute temperature T, in a manner illustrated by Eq. (2.5) and by Fig. 20. PLANCK had originally obtained his formula, which agrees very well with experimental results, by empirically modifying the less exact formula of WIEN. He did not succeed, however, in rigorously deriving his formula (2.5) on the basis of the assumption, self-evident in classical physics, that the radiating dipoles ("resonators") may vibrate with any amplitude, the square of which is proportional to the radiation energy in classical physics. The correct relation (2.5) he could derive only on the basis of the revolutionary assumption that the vibrational energy of a resonator is proportional to its frequency, Eq. (2.48), and is always an integral multiple of $h\nu$. This assumption of *quantized energy states* of the radiating oscillators is the beginning of quantum physics. Some years later, EINSTEIN applied PLANCK's formula to radiation itself and conceived the idea of the light quanta or photons. Quite a number of phenomena furnish evidence for the corpuscular nature of the photons. One of these, the photoelectric effect, was mentioned in II, 4a. Another pertinent experiment, the Compton effect, will be treated in IV, 2 together with the complicated question of the relation between the wave and quantum properties of radiation.

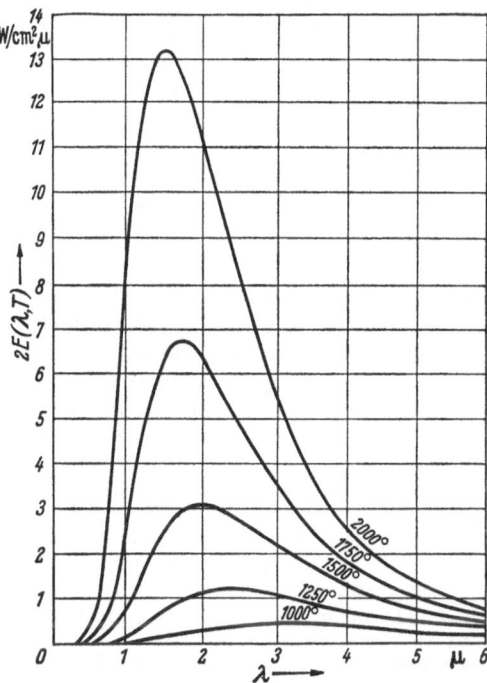

Fig. 20. The spectral energy distribution of a black-body radiator at temperatures between 1000 and 2000 °K, plotted against wavelengths. The ordinate measures the unpolarized radiative power per unit solid angle and a wavelength interval of 1 μ

In the realm of quantum physics, instead of a spherical wave being radiated from a source, we have an irregular statistical emission of photons into all directions of space. Only if we consider an average over a sufficiently long time, do we arrive at the classical result. Every surface element of a sphere surrounding the radiation source then receives an equal number of photons and thus an equal amount of energy. Impressive evidence of this emission of single photons is given by an experiment of JOFFE in 1925. As a radiation source he used a very small anticathode of an X-ray tube, and as a receiver a small metal ball from whose surface photoelectrons were released by the incident photons. The small collecting sphere was placed at such a distance from the source that its cross section was only about one millionth of the surface of an imaginary sphere with its center at the source. If the X-ray tube was excited to emit very short pulses of X-rays, only one out of about 10^6 pulses caused the emission of a photoelectron from the small receiver sphere. *The overwhelming majority of the emitted photons did not reach the receiver.*

Even more impressive is the experiment of BRUMBERG and VAVILOW. Their arrangement consists of a screen with a very small hole that is intermittently illuminated. If the illumination of this small hole is kept so low that a "flash" of only 50 photons reaches the eye within the observation period of 0.1 sec, the opening is not seen at every revolution of the rotating sector. It is seen or not seen at completely irregular intervals because of the statistical fluctuations of the number of photons that reach the eye. If the illumination is increased and with it the number of photons reaching the eye, the percentage of "visible flashes" increases; if the illumination decreases, the "visible" percentage decreases too. This experiment does not yet completely exclude the possibility of interpreting the observed phenomenon as a physiological effect of the eye's threshold of stimulation. But the following variation of the experiment does not permit such an explanation. If a double prism is inserted between sector and eye, the light beam coming from the small opening is separated into two equally intense beams so that two equally bright images of the hole should be seen according to classical physics. But actually, in a statistically irregular sequence, now one image of the illuminated hole is seen, then the other, occasionally both or none at a time; for the number of photons necessary for the excitation of our eye passes statistically either through one part of the double prism or through the other, occasionally through both parts or through none of them. This result is in striking contrast to the wave theory of light, according to which the double prism produces two coherent beams of the same intensity. *Thus, the wave theory gives the correct result only in the statistical average.* It is interesting that, as an average, only *one* out of 50 photons reaching the eye is absorbed by a light-sensitive rod of the retina. This means that the absorption of a single photon seems to be able to activate the complicated mechanism in our brain that results in the perception of "light". In IV, 2 we shall discuss further evidence for the quantum character of light.

Problems

1. By what fraction would the density of nitrogen gas (molecular mass = 28.016) at 27 °C decrease in the earth's gravitational field as the altitude increases by 0.5 km ? Assume that the acceleration due to gravity is constant.

2. The mean free path of the molecule of a gas at a pressure of 100 atmospheres and temperature 20 °C is 1.02×10^{-7} cm. Find the diameter of the gas molecule, assuming that the molecules have a Maxwellian velocity distribution.

3. An electron beam is led through crossed electric and magnetic fields for a distance of 5 cm. When the magnetic field of 3.75 gauss alone is applied, the beam is deflected through an angle of 3° 40'. But when the electric field of 200 volts per cm is simultaneously applied, the beam is undeflected. Find e/m of the electrons.

4. A beam of 5 Mev α-particles is incident normally on a gold foil of thickness 10^{-4} cm. (a) What is the closest distance to which an α-particle approaches a gold nucleus ? (b) What fraction of the incident α-particles that undergo single Rutherford type scattering through an angle of 60° with respect to the incident direction would be received by a detector of area 2 cm² at 20 cm from the point of traversal of the foil ?

5. In an experiment similar to that of Millikan, a charged oil drop rises 5 mm in 25 sec at constant air velocity, when 2000 volts is applied between the plates 1 cm apart. Without the electric field, the same oil drop falls 5 mm in 20 sec at constant speed. If the drop carries one unit of charge and the viscosity of air is 1.8 dyn sec/cm², what is the radius of the drop ? Neglect the buoyancy of the oil drop in air.

6. Find the mass of an electron which is bent into a circle of radius 10 cm by a magnetic field of 1000 gauss.

Literature

Section 4:

GLASER, W.: Grundlagen der Elektronenoptik. Wien: Springer 1952.
KLEMPERER, O.: Einführung in die Elektronik. Berlin: Springer 1933. Improved English Ed.: Cambridge 1953.
MASSEY, H. S. W.: Negative Ions. 2nd Ed. Cambridge: University Press 1950.

SIMON, H., R. SUHRMANN, and others: Der lichtelektrische Effekt und seine Anwendungen. 2nd Ed. Berlin: Springer 1959.
ZWORYKIN, V. K., and E. G. RAMBERG: Photoelectricity and its Application. New York: Wiley 1949.

Section 6:

ASTON, F. W.: Isotopes. Leipzig: Hirzel 1923. Modern 2nd English Ed.: New York: Longmans 1942.
BARNARD, G. B.: Modern Mass Spectrometry. London: Institute of Physics 1953.
EWALD, H., u. H. HINTENBERGER: Methoden und Anwendungen der Massenspektroskopie. Weinheim: Verlag Chemie 1953.
MATTAUCH, J., u. A. FLAMMERSFELD: Isotopenbericht. Tübingen: Verlag der Zeitschrift für Naturforschung 1949.
RIECK, G. R.: Einführung in die Massenspektroskopie. Berlin: Deutsch. Verlag d. Wiss. 1956.

III. Atomic Spectra and Atomic Structure

"Ever since the discovery of spectral analysis, no doubt was possible that the language of the atom could be understood if we learned how to interpret atomic spectra. The tremendous amount of spectroscopic data accumulated during the past 60 years seemed at first to be too diverse and too complex to be disentangled. Seven years of X-ray spectroscopy have contributed more to a clarification, because here the problem of the atom is attacked at its root and the interior of the atom is revealed. If we listen today to the language of the spectra, we hear a true 'music of the spheres' of the atom, chords of integral proportions, an increasing order and harmony in spite of all diversity. For all times, the theory of spectral lines will bear the name of BOHR. Yet another name will be permanently linked with his, the name of PLANCK. All integral laws of spectral lines and atomistics are basically consequences of the quantum theory. This is the mysterious organ on which nature plays her music of the spectra and according to whose rhythms the structure of the atom and the nucleus are arranged." SOMMERFELD expressed these thoughts in 1919 in the preface of the first edition of his famous work "Atombau und Spectrallinien". Since in 40 years they have not lost anything of their significance and beauty, they may be placed at the beginning of the chapter on atomic spectra and atomic structure. They characterize, in a manner unequaled, the fundamental importance of spectroscopy for atomic physics. Therefore, we begin our discussion with a survey of the methods of spectroscopy.

1. Recording, Evaluation, and Classification of Spectra

a) Experimental Methods of Spectroscopy in the Various Spectral Regions

The spectroscopist is interested in the entire wavelength region in which spectra of atoms and molecules are observed, i.e., the wavelength region from 10^{-10} cm up to many meters. This whole region is subdivided into the region of γ-rays and X-rays, the ultraviolet, the visible, the infrared, the microwave, and the radiowave regions. Each of these ranges requires a specific experimental technique of recording and investigation. Wavelengths are usually measured in units of 10^{-8} cm, known as the Ångstrom unit and abbreviated Å. However, in the infrared region the unit generally used is the micron $1\,\mu = 10^{-4}$ cm, and in the X-ray region the unit is the X unit, $XU = 10^{-3}$ Å $= 10^{-11}$ cm[1]. Table 4

---

[1] The historical X unit, still often used, is 1.00203×10^{-3} Å, because it is related to an obsolete, too low value of the electronic charge e.

Table 4. *The Different Spectral Regions and the Spectroscopic Equipment Used in Them*

Wave length	Spectral region	Spectroscopic equipment	Radiation detector	
$\lambda < 100$ Å	X-ray	Crystal lattice spectograph	Ionization chamber, photographic plate, Geiger counter	
100—1800 Å	Vacuum ultraviolet	Vacuum concave grating ($\lambda >$ 1050 Å fluorite prism spectrograph)	Schumann photographic plate	In addition, photomultiplier or Geiger counter
1800—4000 Å	Quartz ultraviolet	Quartz spectograph or grating	Photographic plate	
4000—7000 Å	Visible region	Glass spectrograph or grating	Photographic plate	
7000--10000 Å	Photographic infrared	Glass spectrograph or grating	Infrared-sensitive photographic plate	
1—5 μ	Near infrared	Grating	Photoresistive cell	
5—40 μ	Medium infrared	Prism or grating spectrograph	Thermocouple, bolometer, Golay cell	
40—400 μ	Far infrared	Echelette grating		
$\lambda > 400$ μ	Micro- and radiowaves	Special techniques for high and highest frequencies		

presents a brief survey of the various spectral regions and the spectroscopic equipment and detectors used in each region.

The best accessible spectral regions are the visible and ultraviolet between 7000 and 2000 Å where simple spectrographs and standard photographic plates can be used. Glass optics are used in the visible region, from 7000 to 4000 Å. Because glass absorbs in the ultraviolet, quartz prisms and lenses have to be used in that region. Small single-prism spectrographs (Fig. 21) are used for surveying the entire spectral region on one photographic plate. For detailed studies of a limited spectral region at high dispersion and high resolution, instruments with three or more prisms and a long focal length are used. For precision investigations, especially in the long wavelength region where the dispersion of optical media is small, large grating spectrographs are used.

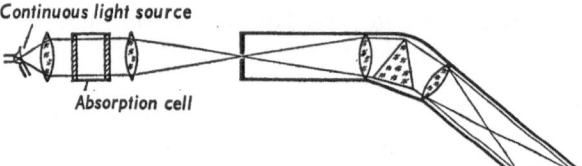

Continuous light source

Absorption cell

Fig. 21. One-prism spectrograph with an arrangement for photographing absorption spectra

In the spectral region below 1800 Å the atmosphere begins to absorb, and quartz ceases to transmit. Furthermore, the gelatin used in the usual photographic plate begins to absorb. For wavelengths below 1800 Å, therefore, the entire apparatus including the light source and photographic plate must be placed in a highly evacuated chamber. This region therefore is called vacuum ultraviolet. Besides, Schumann photographic plates, which do not contain gelatin in the photographic layer, are used. Photographic recording of the spectra may be replaced by recording with a photo-multiplier (page 28) that scans the whole spectrum point after point. Since fluorite transmits down to 1050 Å, small vacuum prism spectrographs using fluorite optics may be used for surveying the spectrum in this region. To obtain high dispersion in the short wavelength region down to X-rays, a concave grating ruled on highly reflecting metal, which makes additional focussing optics unnecessary, is used.

The grating as well as the source, slit, and plate are placed in a large vacuum chamber (Fig. 22). The grating constant of the optical gratings, mostly about 10^{-4} cm $= 1\,\mu$, is too large for the shortwave vacuum ultraviolet and, furthermore, the reflectivity of metals decreases considerably. In this case, the radiation is directed at the grating in grazing incidence so that the effective grating constant

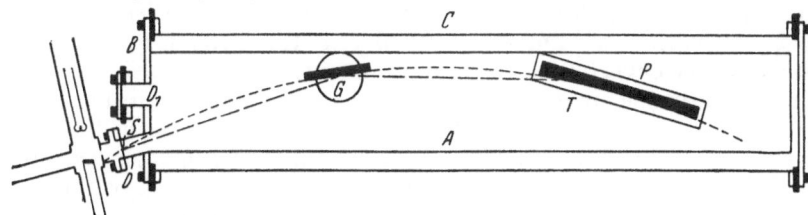

Fig. 22. Vacuum-grating spectrograph for spectroscopy in the extreme ultraviolet or in the very-long-wave X-ray region. $S =$ slit, $G =$ grating, $P =$ photographic plate. (After SIEGBAHN)

is the projection of the actual grating constant on the direction perpendicular to the incident light (Fig. 23). Interferometers such as the FABRY-PEROT, LUMMER-GEHRCKE etc., are required for high-dispersion and high-resolution work on the fine structure of spectral lines (see III, 20).

The experimental technique in the photographic infrared does not differ fundamentally from that in the visible, except that infrared-sensitized plates have to be used. The present limit for this technique is about 13,600 Å; but extremely high exposure times are required to reach much beyond 11,000 Å.

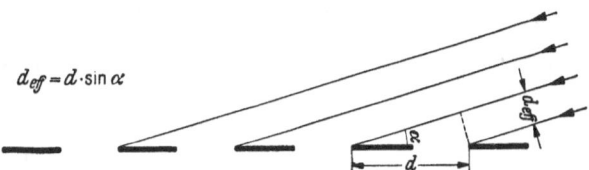

$d_{ef} = d \cdot \sin \alpha$

Fig. 23. The effective grating constant at grazing incidence

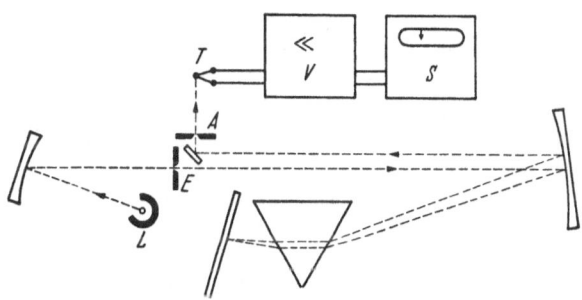

Fig. 24. Schematic arrangement of an infrared spectrograph with prism and concave mirrors. $L =$ light source, $E =$ entrance slit, $A =$ exit slit, $T =$ thermocouple, $V =$ amplifier, $S =$ recorder

The so-called near infrared between 1 and 6 μ was opened to precision measurements with optical gratings by the development of sensitive infrared receivers (lead sulfide and indium antimonide photoresistive cells). Also for the longer-wavelength infrared, up to 40 μ, optical materials for prisms, lenses, and windows are available nowadays. But concave mirrors are preferred for focussing the radiation in order to avoid absorption losses and chromatic errors (Fig. 24).

In the far infrared, gratings are used with grating constants correspondingly larger than those used in the visible region. With suitably shaped grating lines almost all the diffracted radiation can directed into *one* order. In this manner the superposition of various orders of the spectrum is avoided and, at the same time, more energy is made available at the so-called blazing wavelength.

Until recently it was not possible to photograph the far infrared spectrum. CZERNY has developed a method in which an infrared spectrum is projected upon a thin oil film which is evaporated according to the energy in the different parts of the spectrum. The relief structure of the oil film is then photographed with suitable illumination (evaporography). In the customary methods, however, the intensity of the infrared spectrum is automatically recorded. The above-mentioned semiconductor cells (see VII, 22c) are used as receivers in the near infrared, whereas in the far infrared region thermocouples, bolometers of different types, and Golay cells are applied. In the Golay cell, the radiation indirectly heats a small gas volume in a membrane-covered cell; the increase of the volume is measured by the deflection of the thin membrane.

Only recently the spectral range from millimeter waves to meter waves has been opened to investigation by radar technique. Absorption methods are exclusively used in connection with a monochromator. A generator for very high frequencies emits monochromatic "radiation" of known and, within certain limits, mechanically or electrically adjustable frequency. In the most important wavelength region between 1 millimeter and several centimeters, reflexklystrons are used. To cover a larger range of frequencies or wavelengths a whole set of calibrated klystrons is necessary. Fig. 25 shows the principle of the simplest microwave spectrometer. The shortwave radiation of the klystron is partially absorbed in an absorption tube of several meters of length that serves as wave guide. The radiation that is not

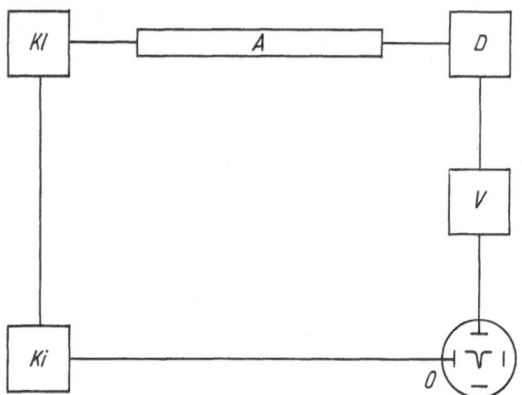

Fig. 25. Arrangement for recording microwave spectra (schematic). Kl = klystron, A = absorption tube, D = detector, V = amplifier, Ki = cut-off frequency generator, O = oscillograph

absorbed is measured by means of a crystal detector, amplified by a low-frequency wide-band amplifier and applied to the vertical plates of an oscilloscope. If a low-frequency voltage applied to its horizontal plates simultaneously modulates the radiation frequency of the klystron, the oscilloscope's abscissa becomes a frequency scale and the electron beam on the screen writes the klystron radiation that leaves the absorption tube as a function of frequency or wavelength. In this way, an absorption line of the absorbing gas within the modulation range of the klystron appears directly on the oscilloscope's screen. Since the klystron frequency may be determined by comparison with a standard frequency with an accuracy of 10^{-7}, and since the high constancy of the klystron frequency permits an equally high resolving power, microwave spectroscopy in this respect is by far superior to infrared spectroscopy. But this resolving power can be exploited for sharp absorption lines only. Therefore the gas pressure in the absorption tube must be kept below 0.1 torr in order to avoid broadening of the lines. At such a low gas pressure and with handy absorption lengths, the absorption strength is very small and its measurement becomes even more difficult by the unavoidable low-frequency noise of detectors and amplifier tubes. Therefore, the so-called Stark effect method is used more and more instead of the simple arrangement of Fig. 25. An electric field in the absorption tube changes, according to III, 16, the frequency of the absorption line. If, therefore, the tube is made into a condenser with a high-frequency field (e.g. 1000 v/cm, 100 kcps) at its plates, the absorption as received by the detector is modulated while the klystron frequency remains constant. The modulated absorption frequency is amplified by a resonance amplifier, rectified and applied to the vertical plates of the oscilloscope. The electron beam then writes the absorption line with its Stark components as described in III, 16d. This method is sensitive enough for measuring an absorption of $10^{-9}\,\mathrm{cm}^{-1}$; that means a tube of 10,000 km of length would be necessary to reduce the intensity to $1/e$ of its original value.

It is obvious that microwave and radiowave spectroscopy does not use dispersing elements such as prisms and gratings, because the high-frequency generator acting as a radiation source provides monochromatic radiation.

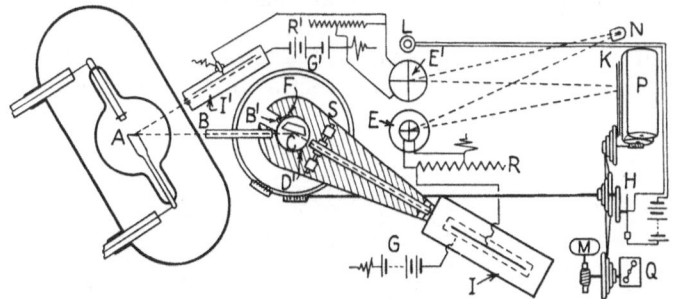

Fig. 26. X-ray spectrograph with rotating crystal C, ionization chamber I, and recorder for the dispersed radiation of the anticathode A. In order to check the constancy of the radiation, the undispersed radiation is recorded simultaneously. (After COMPTON)

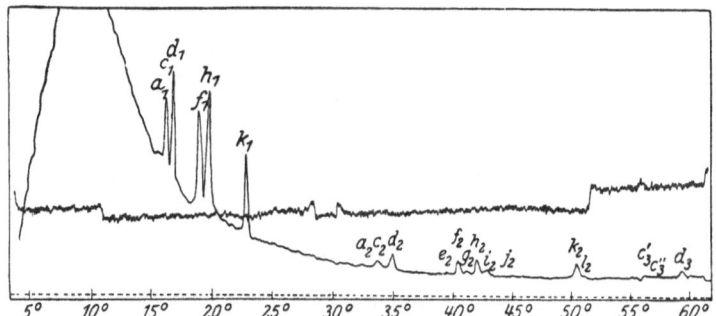

Fig. 27. X-ray spectrum of a tungsten anticathode (continuous bremsspectrum with superposed X-ray lines of tungsten) recorded with the instrument shown in Fig. 26

A special case of high-frequency spectroscopy is the so-called *paramagnetic resonance* of electrons or nuclei. With a method that in principle is the same as that used in high-frequency spectroscopy, those frequencies are measured for which resonance occurs between the incident high-frequency radiation and flip-flop processes of the angular momenta of electrons or nuclei. This phenomenon, also called *spin resonance*, will be treated in more detail in V, 4.

For the extreme shortwave region of the spectrum, that of X-rays and γ-rays ($\lambda = 0.01$ to $100\,\text{Å}$), refracting substances for prisms and lenses do not exist. Optical concave gratings with grazing incidence of the radiation (Fig. 22 and Fig. 23) are used for the longer X-rays as well as for the shortest wavelengths of ultraviolet radiation. Since VON LAUE's discovery in 1912, the diffraction of X-rays by regularly spaced atoms of a crystal lattice is utilized in designing crystal grating spectrometers for the whole X-ray region. A crystal, however, is not a *line* grating but a threedimensional *point* lattice. This has the consequence that incident radiation of a specific wavelength can leave the crystal only under a definite angle between lattice planes and incident radiation. The diffracting crystal therefore is slowly rotated (Fig. 26) in order to get photographs or recordings of the entire X-ray spectrum (lines and continuous spectrum) of the anticathode material (see Fig. 27). Sharp lines and high intensity are obtained by using the focussing effect of bent crystals, similar to that of concave gratings. In this way, DuMond opened even the region of γ-rays to precision measurements.

b) Emission and Absorption Spectra

We distinguish between emission and absorption spectra according to whether the atom under consideration itself emits the wavelengths in question or absorbs them out of an incident radiation containing all wavelengths of that region (continuous radiation). In the first case, under direct observation, the spectrum appears bright on a dark background, in the second case it is dark on a bright background. Examples of emission line spectra are the spectra of all incandescent

Hα   ↑ series limit

Fig. 28. The Balmer line spectrum emitted by the hydrogen atom (Courtesy Schlüter and Vidal)

Fig. 29. Spectral line series of the sodium atom, absorbed by sodium vapor. (After Kuhn)

gases (Fig. 28), while the best known absorption line spectrum is that of the Fraunhofer lines in the sun's spectrum. These lines, which are a result of the absorption of gas atoms in the sun's atmosphere, appear dark against the bright continuous emission spectrum of the lower layers of the sun, the photosphere. Fig. 29 is an example of an atomic absorption spectrum. A special case of an absorption spectrum is the so-called self-absorption of spectral lines. This phenomenon is based on the fact that an atom, as we shall see in III, 5, can absorb many of the lines which it emits. Since, according to III, 21, the width of a spectral line depends on the disturbance of its emitting or absorbing atoms by its surroundings, the Cu lines emitted by a copper spark are broader than the Cu lines absorbed by the cooler copper vapor of the outer envelope of the spark. The Cu atoms of the outer envelope thus absorb the center of the broader Cu emission line, so that a sharp absorption line is seen against the background of the broader emission line. Fig. 30 is an example of a density curve of a pure emission line of copper and the same line "self-reversed" because of self-absorption.

In order to observe an emission spectrum, the atoms under investigation must be "excited". We shall return to the physical meaning of this excitation in III, 4. X-rays are mostly excited by fast electrons (cathode rays) impinging on the anticathode. *Optical* spectra are emitted by incandescent gases or vapors that are excited by impacts between electrons and gas atoms. Glow discharges, high-frequency discharges, arc discharges, and spark discharges as well as flames may be used to excite the emission of optical spectra. Arcs and sparks have the additional advantage of very high temperatures at which even traces of solids contained in the electrodes may evaporate so that these atoms can be excited.

The atmosphere of the sun and of fixed stars are examples for *thermal* excitation. Finally, atoms can be excited by irridiating them with light of sufficiently short wavelength; these spectra are called *fluorescence spectra*.

For recording an *absorption spectrum* we need a continuous emission spectrum, i.e., one which contains all wavelengths in the spectral region under consideration. The absorbing substance (e.g., gas or vapor atoms), which absorbs its characteristic spectral lines, is placed between the continuous source and the spectrograph. In the visible and infrared regions, an incandescent solid such as a tungsten filament lamp or the positive crater of a carbon arc are used as continuous emission sources.

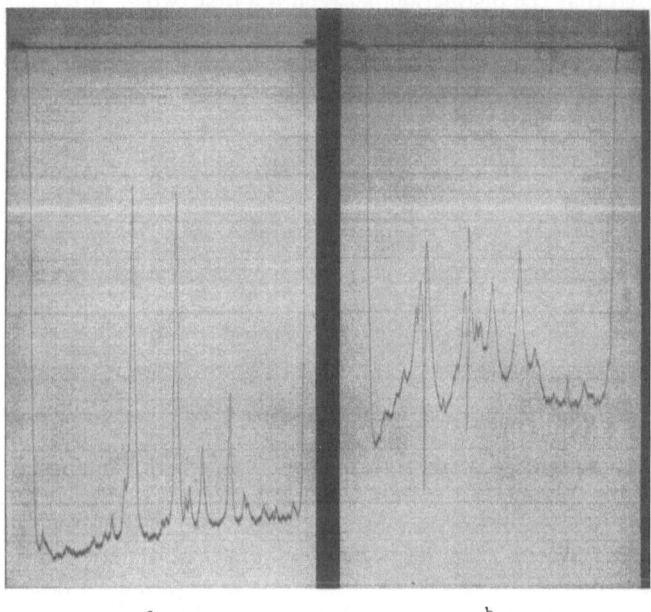

a   b

Fig. 30. Densitometer record as an example for the evaluation of a spectrum by means of a recording microphotometer. Of the four most intensive Cu lines, the two at the left are resonance lines which appear in (a) without self-absorption and in (b) with self-absorption. Photograph by the author, 1929

In the ultraviolet, the continuous spectra of various discharges, especially that of a xenon high-pressure arc or the hydrogen molecular continuum, may be used, in the shortwave ultraviolet the high-vacuum capillary spark or the helium molecular continuum (VI, 8). We shall deal with the X-ray absorption spectra later (III, 10d) because here the situation is more complicated.

c) Wavelength and Intensity Measurements

The theoretical evaluation of atomic spectra is based on the measurement of wavelengths and intensities of spectral lines. For the continuous spectra we have to know the spectral intensity distribution, i.e., the intensity as a function of the wavelength.

As mentioned before, wavelengths are usually given in Ångstrom units, $1 \text{ Å} = 10^{-8}$ cm, and in the infrared in microns, μ. Because of the relation $\lambda = c/\nu$, where c is the velocity of light in the medium under consideration, the wavelengths depend on the refractive index of the medium in which the light is propagated. In order to have comparable values, the measured wavelengths are reduced to those they would have in vacuum. As we shall see, the rules of line spectra are simplest not in terms of wavelengths but in terms of frequencies, $\nu = c/\lambda$, or wave numbers, $\bar{\nu} = 1/\lambda$ (the number of waves per centimeter). This computation of $1/\lambda$ is usually combined with the reduction to the vacuum. In the microwave and radiowave spectroscopy, not the wavelength but the frequency of the spectral lines is measured. The unit for this frequency is 1 Hertz (Hz) = 1 cycle per second (cps).

d) Lines, Bands, and Continuous Spectra

Very early, spectra were classified according to their appearance as line, band, or continuous spectra. Fig. 31 shows examples of the three types. Much later it was realized that the distinction between the first two groups is directly related to their different origin; *line spectra are always emitted or absorbed by atoms or atomic ions, while the emitters of band spectra are always molecules.* We shall deal in some detail with band spectra in Chap. VI. Line spectra were further classified

Fig. 31. Examples of (a) a continuous spectrum, (b) a molecular band spectrum, and (c) an atomic line spectrum

as arc or spark spectra according to the source in which they occur with highest intensity. A more precise terminology was introduced when it became known that arc spectra belong to neutral atoms, whereas spark spectra are emitted by ionized atoms, i.e., ions. The term arc spectrum is now understood to mean the spectrum of a neutral atom and is labeled by I, so that Fe I designates the arc spectrum of iron. In the case of spark spectra the terms first, second, etc., spark spectrum of iron are used. The spectra are classified according to the ions, Fe^+, Fe^{++}, etc., which emit them and are labeled Fe II, Fe III, etc. Thus the Fe IV spectrum is the third spark spectrum of iron and is emitted by the Fe^{+++} ion.

With respect to the third group, the continuous spectra, an unambiguous classification as to their origin is not possible without a detailed investigation. As we shall see, atoms as well as molecules can emit continuous spectra, but they differ clearly in their excitation conditions. Into this class of the atomic spectra belongs also the important continuum which is emitted when free electrons collide with positive ions (electron retardation continuum or, using the German term, bremscontinuum). We shall return to it in III, 6e. Above all, continuous spectra are emitted by all incandescent solids. The prototype of these continuous thermoemitters is the black body which absorbs all incident light; its emitted intensity and its intensity distribution are unique functions of the temperature only (see II, 7).

2. The Analysis of Line Spectra.
Series Formulae and Term Diagrams

An inspection of a simple line spectrum such as that of the hydrogen atom (Fig. 28) reveals that a relationship between the wavelengths of its spectral lines must exist. A regular sequence of lines in a spectrum is called a *series*, and the analytical expression of the wavelengths of the lines of a series is called a series

formula. The first of them was found in 1885 by BALMER for the first four lines H_α, H_β, H_γ, and H_δ of the hydrogen series. This series lies in the visible and near ultraviolet and is named after him. In the following years, KAYSER and RUNGE analyzed the accumulated mass of wavelength data of all available atoms which had been measured with great precision by them and their coworkers. They deduced a large number of series rules and showed that *all* atomic spectra consist of series of lines in the visible and ultraviolet regions, the only spectral regions which were accessible at that time. Furthermore, KAYSER and RUNGE realized that the series formulae became much simpler, if instead of the wavelength they used the *wave number*, $\bar{\nu} = 1/\lambda$, that is the number of waves per centimeter of light path in a vacuum, measured in reciprocal centimeters, cm⁻¹.
RYDBERG finally found the correct form for the series rule concealed in BALMER'S formula. In this form the wave numbers of the spectral lines are written as the difference of two quantities, which he called "terms". The physical significance of these terms, whose dimension is the same as that of wave numbers, i.e., reciprocal centimeters, will become clear later. With RYDBERG'S modifications, the Balmer formula for the visible hydrogen series becomes:

Table 5. *Comparison of Computed and Observed Wavelengths of Several Lines of the Balmer Spectrum of the H-Atom*

Name of line	n	λ observed, Å	λ computed, Å
H_α	3	6562.793	6562.78
H_β	4	4861.327	4861.32
H_γ	5	4340.466	4340.45
H_δ	6	4101.738	4101.735
H_ε	7	3970.075	3970.074
H_ζ	8	3888.052	3889.057
H_η	9	3835.387	3835.397
H_ϑ	10	3797.900	3797.910
H_ι	11	3770.633	3770.634
$H_\varkappa$	12	3750.154	3750.152
H_λ	13	3734.371	3734.372
H_μ	14	3721.948	3721.948
H_ν	15	3711.973	3711.980

$$\bar{\nu} = \frac{R}{2^2} - \frac{R}{n^2} \qquad n = 3, 4, 5, \ldots . \quad (3.1)$$

Here R is a constant with the dimension cm⁻¹, the so-called Rydberg constant, and n runs through the series of integers. We shall see in the next section that the representation of a spectral line as a difference of two terms has considerable significance, and we shall understand the physical meaning of a "term". The accuracy with which the simple Balmer formula (3.1) describes the wave numbers and thus the wavelengths of the hydrogen spectrum is shown in Table 5, where observed and computed values of the first 13 lines of the Balmer series are compared.

In general, a series of spectral lines is represented, according to RYDBERG, by the formula

$$\bar{\nu} = \frac{R}{(m+a)^2} - \frac{R}{(n+b)^2} \quad \text{with} \quad n > m . \quad (3.2)$$

a and b are characteristic constants of a particular series, m is a small integral number, characteristic for the specific series of the spectrum, and n again runs in each series through all the integers.

An important step forward was accomplished when RITZ in 1908 formulated his *combination principle*. It states that by additive or subtractive combination of frequencies of spectral lines (or their corresponding terms) new spectral lines or terms of the atom under consideration are obtained. RITZ' combination principle led to the discovery of a large number of new spectral lines and thus contributed much to our knowledge of spectroscopy, although actually not every combination of two observed lines can be observed. All these relations which at first appeared rather mysterious will be explained in the next section on the basis of BOHR's theory. Only then shall we understand their full significance.

We know already that the H atom as a source of the Balmer spectrum is characterized by a series of term values

$$T_n = \frac{R}{n^2} \ [\text{cm}^{-1}] \quad \text{with} \quad n = 1, 2, 3, \ldots \tag{3.3}$$

from which by subtractive combination, according to the combination principle, the wave numbers of the H atom lines (3.1) are obtained.

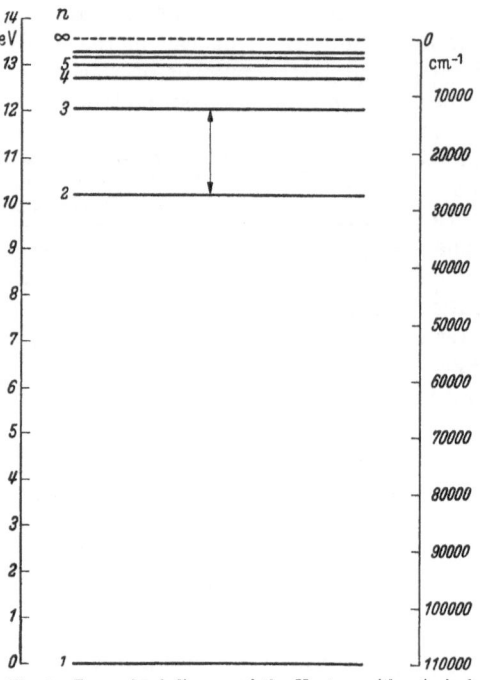

Fig. 32. Energy level diagram of the H atom with principal quantum numbers n, with term-value scale in cm^{-1} (right-hand side) and excitation-energy scale in ev (at the left)

For the following theoretical treatment of the spectra, a diagram representing the terms of an atom has proved valuable. We shall call it the *term diagram* of the atom. Each term is represented in the diagram by a horizontal line, and these lines are arranged in the order of increasing term number n, as shown in Fig. 32. Since the term value T_n is inversely proportional to the square of the term number n, the term values increase in going from the top to the bottom of the diagram, different from the normal ordinate direction. For this reason they are called *negative* term values. The term value $T_\infty = 0$ corresponds to $n = \infty$, and the term value $T_1 = -R$ corresponds to $n = 1$. Fig. 32 shows the terms (3.3) of the H atom. The series of term values converges with increasing term number n toward a limit. The physical significance of this limit will be explained in the next section. We mentioned already that the Rydberg formula expresses the wave number of each spectral line as the difference of two terms of the atom. This is illustrated in Fig. 32 by the arrow connecting the two terms under consideration, e.g. in our case:

$$\bar{\nu}_{23} = T_3 - T_2 = -\left(\frac{R}{9} - \frac{R}{4}\right) = \frac{R}{4} - \frac{R}{9}. \tag{3.4}$$

3. The Fundamental Concepts of Bohr's Theory

Since the work of KAYSER, RUNGE, and RYDBERG, it was clear that, in a certain sense, the spectrum is a picture of the atom. There was no doubt that the spectrum is intimately associated with some kind of motion of the electrons revolving round a positive nucleus, and that the dynamic atomic model of RUTHERFORD (see II, 3) was essentially correct. However, it was not until 1913 that NIELS BOHR found the key which opened the way for understanding the relations between atoms and atomic spectra.

BOHR started from the Rutherford model in which the Coulomb attraction between the nucleus and the electrons is compensated by the centrifugal force acting on the revolving electrons. Former theoretical attempts at explaining the spectra had always been based on the classical concept that the nucleus and its revolving electrons form varying electric dipoles and consequently must radiate. A relation between the frequency of revolution of an electron and the frequency of a spectral line, however, was sought for in vain. There were two main difficulties which no classical theory was able to solve: First, the existence of a stable atom could not be explained because the atom must continually radiate and thus lose energy as long as its electrons revolve. Second, there was no way of understanding the emission of a spectrum of discrete frequencies with no apparent relation to the frequency of revolution of the electron.

BOHR eliminated these difficulties in a manner similar to severing the Gordian knot by formulating his famous postulates which limited the validity of classical physics so far as the realm of atoms is concerned. These postulates, conceived with as much boldness as with admirable physical intuition, were theoretically justified 12 years later when HEISENBERG and SCHRÖDINGER developed their quantum-mechanical theory which will be discussed in Chap. IV. Meanwhile BOHR had already proved in his first paper that his postulates were extremely useful for understanding the atomic spectra. In the following years, BOHR's theory achieved success after success and won fame by explaining the Periodic Table.

What are the fundamental concepts of this new theory? BOHR argued that, because of the existence of stable atoms, there *must* exist at least certain electronic orbits where the electrons revolve without radiating, contrary to the requirements of classical electrodynamics. Each of these unique "quantum orbits" corresponds to a definite energy state E. The innermost quantum orbit with the smallest possible radius belongs to the normal atom. In order to bring an electron into an outer orbit, a certain amount of energy, the "excitation energy" for the particular orbit, must be expended. After an average time of 10^{-8} sec in an orbit of higher energy, the electron "jumps" spontaneously to an orbit of lower energy and eventually to the ground orbit. The energy difference between the higher energy of the initial orbit E_i and the lower energy of the final orbit E_f is emitted as a spectral line of frequency ν given by BOHR's frequency condition

$$E_i - E_f = h\nu \tag{3.5}$$

where h again is the famous Planck constant (see II, 7).

According to BOHR, the "allowed" orbits are distinguished by a quantum condition: the product of the circumference of the orbit $2\pi r$ and the momentum of the electron mv (a product which has the dimensions of action, ergsec) is equal to an integral multiple of PLANCK's constant h,

$$2\pi r m v = n h \qquad n = 1, 2, 3, \ldots . \tag{3.6}$$

Here, for the first time the concept of *quantization* appears, a concept that is instrumental all through the field of atomic physics. It sounds rather mysterious at first; we shall deal in detail with its physical significance in quantum mechanics, Chap. IV. But already here we may illustrate its meaning. According to PLANCK's discovery, the *action, with the dimension ergsec = g cm² sec⁻¹, occurs in nature not in arbitrarily small amounts but atomistically, i.e., in integral multiples of the quantum of action h* (cf. II, 7). This suggests the hypothesis that *all* physical quantities with the dimension of action may occur only in discrete amounts, as for instance the left-hand side of Eq. (3.6) or, in general, every angular momentum. PLANCK's constant enters into both of the fundamental equations of the Bohr theory (3.5) and (3.6), and therefore is characteristic of all atomic processes. Details will be presented in the theoretical treatment of the H atom in III, 5.

By application of Bohr's postulates, it actually proved possible to interpret the atomic spectra without too much difficulty. By substituting in the Bohr frequency condition (3.5) the wave number $\bar{\nu} = \nu/c$ for the frequency and solving the equation for $\bar{\nu}$, we obtain

$$\bar{\nu} = \frac{E_i}{hc} - \frac{E_f}{hc}. \tag{3.7}$$

By comparing this with the Rydberg series formula (3.2)

$$\bar{\nu} = T_2 - T_1, \tag{3.8}$$

the significance of the "terms" T in the empirical series formula becomes evident. *A term is equal to the energy of a stationary energy state that corresponds to a particular electron orbit divided by hc.* The term diagram shown in Fig. 32 thus may be regarded as an energy level diagram. Each term value may be converted to an energy value by multiplying it by hc. Such an energy level diagram represents the possible energy states of the quantum orbits of an atom. We shall show when treating the H atom that the quantum number n associated with the quantum condition (3.6) is identical with the term number n of the terms in Eq. (3.3).

We may characterize a certain atomic state with a quantum number n either by specifying its term value in reciprocal centimeters $\left(\text{the unit of } \dfrac{E}{hc}\right)$, with the numbers progressing from the highest term T_∞ to the lowest (Fig. 32), or by specifying its energy value $E_n = hcT_n$ in ergs. Instead of measuring the energy in ergs, it is customary in atomic physics to measure it in electron volts, the potential through which a charge $-e$ must fall in order to acquire a kinetic energy E. This new energy unit is called an electron volt or ev and is defined by the equation

$$E = \text{ev}. \tag{3.9}$$

These three energy units are related as follows:

$$1\,\text{ev} = 1.602 \times 10^{-12}\,\text{erg} \triangleq 8067.5\,\text{cm}^{-1}. \tag{3.10}$$

We shall continually make use of these relations. There is still another energy unit which is frequently used. Instead of dealing with the energy of a single atom, it is possible to refer to the energy of 1 mole $= 6.022 \times 10^{23}$ atoms and measure it in thermal energy units, i.e., kilocalories (kcal). Measuring energy in kilocalories per mole is generally preferred by chemists. The relation between this new energy unit and the electron volt (3.10) is

$$1\,\text{ev} \triangleq 23.04\,\text{kcal/mole}. \tag{3.11}$$

After this digression into the energy units which are used in atomic and molecular physics we return to the energy level diagram, Fig. 32, and inquire into its physical significance.

For the normal atom, the energy of the ground state is $-E_0$; the energy (as a binding energy of the electron) is negative. If the electron (we speak here of an H atom with only one electron) is brought into an outer orbit, work must be done in order to overcome the electrostatic attractive force, the Coulomb force, between electron and nucleus. By this transfer of excitation energy, the binding energy $-E_0$ of the atom is reduced. However, an atom cannot absorb any arbitrary amount of energy, but, according to Bohr's quantum condition (3.6), only those energies which correspond to differences between its own and any higher energy state in the energy level diagram. If we continue to transfer excitation energy to the electron and thus remove it to orbits farther and farther away from the nucleus, the consecutive excitation energies decrease more and more. *The energy levels of the atom converge toward a limit* where, in our notation, the energy of the electron is zero. This limit will be reached when the electron is no longer bound to its nucleus, or in other words, when it has been removed to infinity. The term or energy value zero thus corresponds to the state of ionization where the electron is completely removed from the nucleus. From this consideration we see that the negative term or energy values correspond to the binding energy between the electron and its nucleus. In the ground state of the normal atom, where the electron occupies the innermost orbit, the electron is bound most strongly to the atom, i.e., the atom has its largest negative energy value. This relationship should be understood very clearly in order to avoid any confusion about negative term and energy values. The relationship may be reversed. If we start with the normal atom as it is most frequently found in nature and denote the energy of its ground state by zero, the values of its energy levels go according to the left scale in Fig. 32. In this case we speak of excitation energies. In other words, the negative energies are binding energies, and the positive energies are *excitation energies*. State E_2 in Fig. 32 may be regarded as having an excitation energy of 10.15 ev, or a binding energy of -3.39 ev, corresponding to a term value $-27,420$ cm^{-1}. Excitation energies have a clear physical significance: *Only after we have transferred a certain excitation energy to an atom, a spectral line of frequency*

$$\nu = \frac{E_i - E_f}{h} \qquad (3.12)$$

can be emitted, caused by a spontaneous energy transition from the excited level to the ground level.

Since, according to Ritz' combination principle, each difference between two energy levels in an energy level diagram corresponds to a spectral line (we consider exceptions later), it can be seen from Fig. 32 that an atom in its first excited level can emit only one line and that by increasing the excitation energy gradually the entire spectrum can be emitted. The Bohr theory thus indicates that *there is a direct relationship between the excitation energy and the spectrum emitted by each atom.* We shall consider the evidence for this in the next section.

We know that by exciting the electron of an atom to higher and higher energy levels the electron is eventually separated from the atom. In other words, *the excitation energy of the term limit E_∞ is the ionization energy of the atom.* An experimental proof of this consequence of Bohr's theory will be presented in the next section. Since excitation and ionization energies are usually given in electron volts, the energy states of an atom, as computed from the quantum condition (3.6), can also be given in electron volts and then are called "critical potentials" of the atom.

In concluding this section, we return to a brief survey of Bohr's concepts and assumptions. Bohr assumed classical electrodynamics invalid in so far as he postulated that an electron should not radiate when revolving on a "stationary" orbit defined by his quantum condition (3.6). He correlated the energy change of the atom, which occurs when an electron "jumps" from one orbit to another, with the emission or absorption of radiation by his frequency condition (3.5). By this postulate he gave up the old idea that the frequency of the emitted or absorbed radiation must be equal to the orbital frequency of the electron. The Bohr theory does not demand any direct relationship between the orbital frequency of the electron around the nucleus and the emitted or absorbed frequency. However, we shall show in III, 22 that for the limiting case of large quantum numbers and small energy changes classical and quantum-theoretical predictions become identical. The Bohr theory makes no assertions about the *mechanism* of the emission or absorption of radiation as a consequence of electron transitions. It only relates the energy change of an atom and the wavelength of the emitted or absorbed spectral line.

Before we treat the Bohr theory of the hydrogen atom, we present the direct experimental confirmation of the theory by the experiments by J. FRANCK and G. HERTZ.

4. Quantum Transitions due to Collisions

BOHR'S frequency condition

$$E_i - E_f = h\nu = hc\bar{\nu} = \frac{hc}{\lambda} \tag{3.5}$$

postulates the existence of a relation, to be tested experimentally, between the excitation energy E_i of an atom and the emission of a spectral line of wavelength λ.

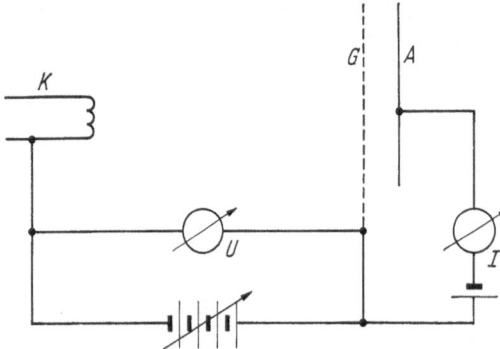

Fig. 33. Schematic arrangement of FRANCK and HERTZ for detecting the discrete energy levels of atoms. $K=$ thermionic cathode, $G=$ accelerating grid, $A=$ receiving electrode, $I=$ amperemeter, $U=$ voltmeter

After preliminary experiments by RAU, J. FRANCK and G. HERTZ in 1914 obtained conclusive experimental proof of the basic assumption (3.5) of the Bohr theory. A schematic diagram of their arrangement is shown in Fig. 33. Electrons emitted by the thermionic cathode K are accelerated by a variable potential difference U between K and the grid G of a gas-filled vacuum tube. In the space between K and G, these electrons collide with gas atoms, e.g., Hg atoms. Electrons arriving at G pass through the grid and approach the collector electrode A which is charged negatively to about 0.5 volt relative to the grid potential. If the energy of the electrons exceeds 0.5 volt, they arrive at A and are being measured. Those electrons which have lost all or a large part of their energy in a collision with an Hg atom cannot overcome the opposing potential difference of 0.5 volt. Consequently, they do not arrive at A and are not measured.

Neglecting contact potentials and the initial velocity of the electrons since they are of no basic importance for the understanding of the experiment, we get the following results with Hg vapor. As the potential difference between K and G is gradually increased up to about 4.5 volts, the current flowing through the galvanometer increases rapidly. This means that the collisions between electrons and Hg atoms are purely elastic. Because of their small mass, the electrons lose only a negligible part of their kinetic energy to the atoms, according to the principles of conservation of energy and momentum. Evidently it is not possible in this energy range, to transfer sufficient energy to an atomic electron to raise it to a higher Bohr orbit (so-called *inelastic collision*). However, at a potential difference of 4.9 volts between K and G, the current suddenly drops (Fig. 34), indicating that much fewer electrons arrive at A. This can only mean that the majority of the electrons has lost their kinetic energy by inelastic collisions with the Hg atoms, transferring their kinetic energy as excitation energy to the atoms.

They then do not have sufficient kinetic energy to overcome the 0.5 volt potential difference between G and A. If the potential is increased further, the current again increases until a potential difference of 9.8 volts is reached. There the current again drops. At this potential difference, the electrons excite twice in traveling from K to G. The first inelastic collision occurs somewhere in the middle between K and G, the second one close to G. Fig. 34 shows that a third drop of the current occurs at three times the excitation energy. The first "critical potential" (see page 59) of the Hg atom thus is 4.9 volts.

Bohr postulated that this energy of the excited Hg atom should be emitted some 10^{-8} sec later in the form of a light quantum (photon) whose frequency or wavelength is determined by BOHR's frequency condition (3.5), while the atom returns to the ground state. According to the energy relation (3.10), an excitation energy of 4.9 volts corresponds to a wave number $\bar{\nu}=4{,}9\times 8067.5=39530$ cm^{-1}, or a wavelength $\lambda=1/\bar{\nu}\approx 2530$ Å, which is a line in the ultraviolet region. FRANCK and HERTZ pointed a spectrograph at the space between K and G and actually found a single spectral line in the ultraviolet. The line was measured to have a wavelength $\lambda=2537$ Å, in agreement with the computation within the limits of error. The existence of the critical excitation potentials, predicted by the Bohr theory, was thus proved for the case of the Hg atom. At the same time the validity of the frequency relation (3.5), as a basis of BOHR's theory of the atom, was

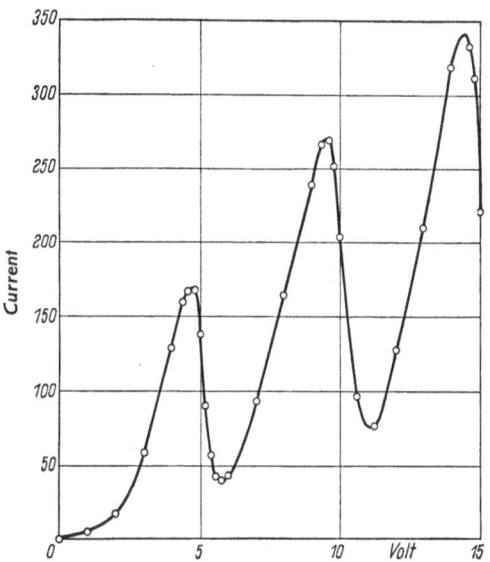

Fig. 34. Curve measured with mercury vapor by FRANCK and HERTZ by means of an arrangement as in Fig. 33, establishing evidence for the first excitation potential of the Hg atom at 4.9 ev

established in a very impressive way. *By electron impact, an excitation energy of 4.9 electron volts was transferred to the Hg atom, and when the electron jumped back to the ground state, this excitation energy was emitted as a photon of the theoretically expected wavelength $\lambda=2537$ Å.* Shortly after this first experiment, many similar ones were performed with a large number of atoms.

By a slight modification of their experimental arrangement, FRANCK, KNIPPING, and EINSPORN succeeded in demonstrating the higher excitation levels of atoms also. By their arrangement, the electrons are accelerated in the short distance between K and G_1, Fig. 35. The pressure in the chamber is greatly reduced to avoid exciting collisions in this region. In the simple arrangement of Fig. 33 the electrons are accelerated within the collision region. The arrangement of Fig. 35, however, allows to work in the impact region between G_1 and G_2 with electrons of a well defined energy. In this way, the curve for Hg vapor (Fig. 36) was obtained and the indicated new critical potentials of the mercury atom were found. Among these newly found energy levels of the mercury atom are those at 4.68 and 5.29 ev. Since transitions from these levels to the ground state are not observed as spectral lines, these transitions are called "forbidden". These energy states are also called metastable (see III, 14) because they have a longer life and do not immediately revert to a lower state by emitting their energy. The metastable states were first discovered by this electrical method.

As a consequence of the excitation of higher energy levels, the emission of an increasing number of spectral lines is expected. From the relation (3.10)

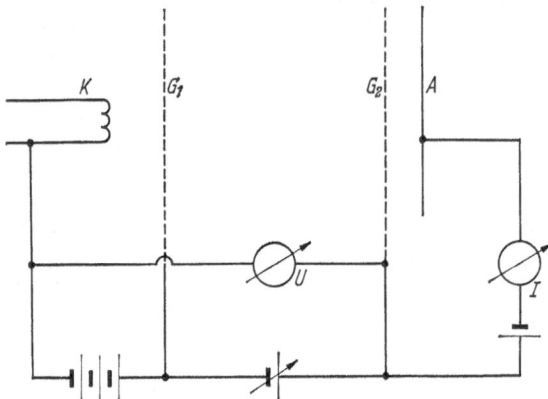

we compute that the transition from the 6.73 ev level to the ground state is accompanied by the emission of the shortwave Hg line 1849 Å. However, as will be understood later, the transition which is expected from the combination principle, $\Delta E = 6.73 - 4.9$ ev, is forbidden.

With a qualitative optical method, GEHRCKE and SEELIGER had already observed the step-by-step excitation of the various lines of an atomic spectrum before the publication of the Bohr theory. They directed a cathode-ray beam

Fig. 35. Schematic representation of the improved experiment by FRANCK and KNIPPING for detecting higher excitation potentials. K = thermionic cathode, G_1 and G_2 = accelerating grids, A = receiving electrode, I = amperemeter, U = voltmeter

obliquely into a low-pressure gas and stopped it by an opposing field in the manner illustrated in Fig. 37. Near the vertex of the resulting parabola, the electron energy was lower than the lowest excitation energy of the atom so that no excitation could occur there, and that part of the beam was invisible. How-

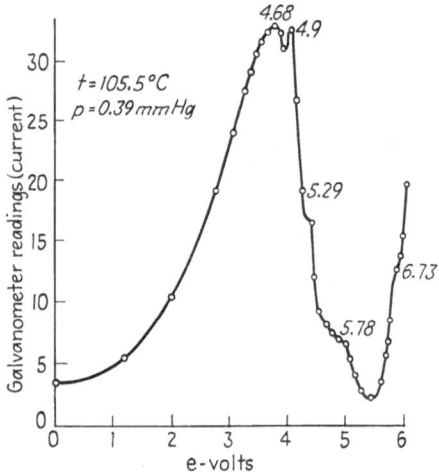

ever, near this invisible part, the cathode ray displayed regions of different color, caused by the gradual appearance of the lines corresponding to increasing excitation potentials. Later a similar experiment utilizing properties of a glow discharge was performed by LAU and REICHENHEIM and used successfully in an investigation of the H_2 spectrum, carried out in cooperation with the author. Finally, G. HERTZ measured the step-by-step excitation of a spectrum very exactly. As an example, Fig. 38 shows the increase of the number of spectral lines in the neon spectrum by increasing the (non-corrected) electron energy from 23.6 to 24.4 ev.

Fig. 36. Curve measured by FRANCK and EINSPORN with an arrangement as in Fig. 35 showing evidence for higher (partly metastable) energy levels of the Hg atom. Abscissa values give uncorrected ev

The emission mechanism discussed so far is called excitation emission. Fundamentally different from it is the recombination emission, which occurs when an electron is "captured" by an ion and then emits radiation by making a transition to a lower state. We shall consider details of this process in III, 6d.

So far we have discussed only the optical and electrical evidence of *excited* states and the measurement of excitation energies, but we have not discussed the measurement of *ionization energies*. Fundamentally, in a plot like Fig. 36, we expect a sudden decrease in the current at the ionizing potential also. However, no distinction between excitation and ionization seems possible by this method. Therefore, almost all collision experiments carried out with the purpose of determining ionization potentials use the space-charge effect of the resulting positive ions. The positive ions reduce the negative space charge of the electrons and thus cause a sharp increase of the current when the ionization potential is reached. An illustration of this method by G. HERTZ is shown in Fig. 39. The large auxiliary thermionic cathode G emits electrons which build up a high negative space charge. Therefore, only a small electron current, independent of the suction voltage, reaches the wall and is measured. However, as soon as the electrons from the thermionic cathode D are accelerated by a voltage exceeding the ionization potential, they are able to ionize the gas atoms and thus produce positive ions in the impact space. Because of their small mobility, these ions compensate the electronic space charge that inhibits the electron current between G and the wall so that the current now increases according to Fig. 40. For calibration purposes, the first excitation potential of the gas is measured in addition to the ionization potential. The result is given as a sharp maximum in Fig. 40.

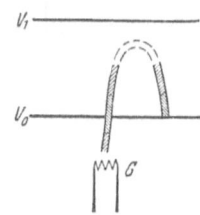

Fig. 37. Arrangement of GEHRCKE and SEELIGER for stepwise excitation of spectral lines by retarding a cathode ray in an opposing electric field

We recapitulate: From electron collision experiments which were carried out in various ways, BOHR'S relationship between the excited states, the term values of the

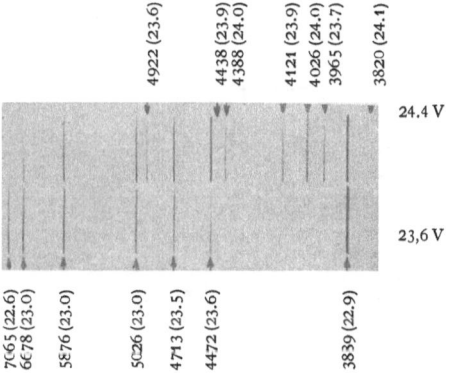

Fig. 38. Stepwise excitation of the line spectrum of the Ne atom. (After HERTZ)

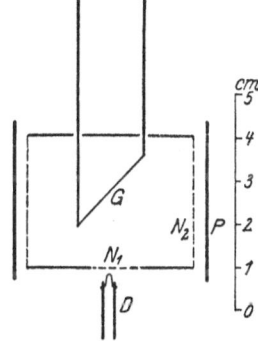

Fig. 39. Arrangement by HERTZ for measuring the excitation and ionization potential of atoms

spectral lines, and the ionization energies was completely verified. These measurements thus are a striking confirmation of the postulates of BOHR'S theory.

Because excitation and ionization by electron collisions plays an essential role in the whole field of atomic physics and in many applications such as gas discharges and light production, we discuss briefly the yield of these collision processes. So far we have only asked which electron energies are necessary for producing excitation or ionization. Now we ask, with what probability an electron collision of given energy actually results in an excitation or ionization. The dependence of excitation probability or ionization probability on the energy of the colliding electrons is called the electrical excitation or ionization function, respectively. Fig. 41 and 42 present examples. Excitation probabilities may in principle be measured electrically by the above methods, if the number of collisions and the number of slow electrons resulting from the inelastic collisions

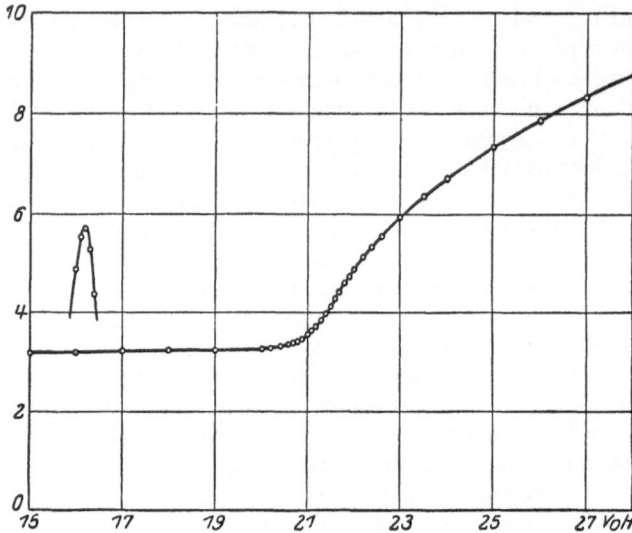

Fig. 40. Determination of the ionization potential and the first excitation potential of the Ne atom with the arrangement shown in Fig. 39

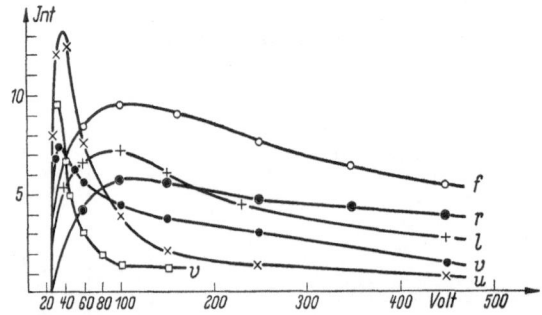

Fig. 41. Optical excitation functions: Dependence of the excitation probability of some He lines on the energy of the colliding electrons; f, r, l = singlet lines; u, v = triplet lines. (After HANLE)

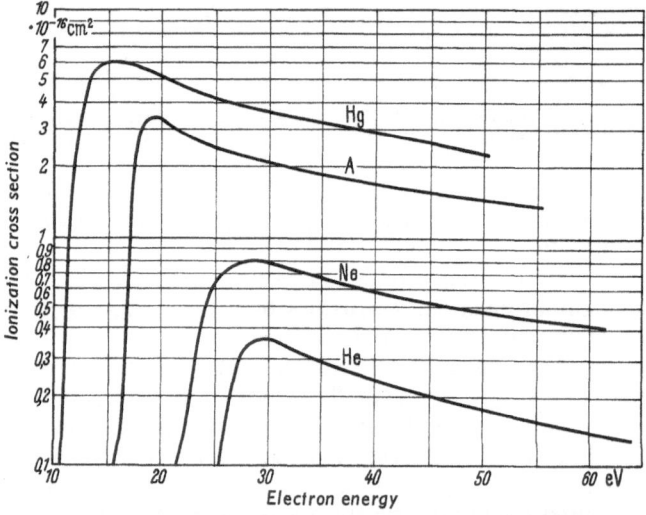

Fig. 42. Ionization functions of several noble gases and of mercury vapor. Logarithm of the ionization cross section versus electron energy

are measured. However, the values obtained from this method are not very accurate because of several sources of error which are difficult to eliminate. By this method the excitation probabilities for the most easily excitable states of various atoms have been found to be between 0.1 % (Ne) and almost 100 % (K). More accurately known is the *relative shape* of excitation functions since the intensity of the spectral line arising from a certain excited state can be measured and plotted as a function of the energy of the exciting electrons. An example of several such *optical excitation functions* is shown in Fig. 41. Generally, the excitation probability rises steeply beyond the critical potential and, after passing through a maximum, decreases more or less rapidly. The position of the maximum and the slope of the curve beyond the maximum vary considerably for different types of lines. This is illustrated in Fig. 41 for the singlet and triplet lines (for details see III, 11) of a two-electron system such as the helium or mercury atom. These differences of the excitation functions will be explained in III, 11.

There are several methods of determining *ionization functions* also. Basically, in all of them the number of ions produced per centimeter of electron path is measured as a function of the energy of the ionizing electron. Some results of such measurements are plotted in Fig. 42.

5. The Hydrogen Atom and its Spectra According to Bohr's Theory

In the last section, we discussed the experimental evidence for BOHR's postulates of the existence of discrete, stationary atomic energy states which converge toward the ionization energy. We also discussed the relationship between these excited states and the emission of spectral lines as a consequence of transitions between these states, i.e., the experimental evidence for the correctness of BOHR's frequency condition (3.5). We now apply BOHR's theory to the simplest atom, that of hydrogen, and to its sepctra.

The line spectrum of the hydrogen atom in the visible and near ultraviolet region is shown in Fig. 28. Its wave numbers are described with high accuracy, according to III, 2, by BALMER's formula

$$\bar{\nu} = R\left(\frac{1}{2^2} - \frac{1}{n^2}\right) \qquad n = 3, 4, 5, \ldots . \tag{3.1}$$

How does the Bohr theory explain these facts?

As the first atom in the Periodic Table, the hydrogen atom has the atomic number 1. Consequently, it consists of only one proton as nucleus and, according to BOHR's concept, one electron moving around it in a circular orbit. If we denote the radius of this circular orbit of the electron by r, the mass of the electron by m, and its angular velocity by ω, the system will only be in equilibrium if the attractive Coulomb force e^2/r^2 is equal to the centrifugal force on the electron:

$$\frac{e^2}{r^2} = mr\omega^2 . \tag{3.13}$$

In order to determine the two unknown quantities r and ω, we need a second equation. This is provided by BOHR's quantum condition, according to which the action, integrated over one complete revolution of the electron, is equal to an integral multiple of the quantum of action h,

$$\oint p\, dq = nh \qquad n = 1, 2, 3, \ldots . \tag{3.14}$$

We saw in III, 3 that in its simplest form this integral is the product of the electron momentum and the circumference of its orbit, thus leading to Eq. (3.6). If we substitute in Eq. (3.14) for the momentum p and the displacement dq of the electron in its circular orbit the expressions

$$p = mv = mr\omega ,\tag{3.15}$$

$$dq = r\, d\varphi ,\tag{3.16}$$

we obtain the quantum condition

$$\oint p\, dq = mr^2\omega \int_0^{2\pi} d\varphi = 2\pi mr^2\omega = nh \qquad n = 1, 2, 3, \dots ,\tag{3.17}$$

in agreement with our earlier Eq. (3.6). We point out again that we will arrive at a physical understanding only by wave mechanics (Chap. IV) but that we may consider the quantum condition (3.17) as a description of the following empirical fact: *In nature, angular momenta $mr^2\omega$ have an "atomistic" nature and occur only as integral multiples of $h/2\pi$ or, occasionally, of $h/4\pi$. In any case, they can change only by integral multiples of $h/2\pi$.* From the equilibrium condition (3.13) and the quantum condition (3.17), we compute the two unknown quantities r and ω and obtain

$$r_n = \frac{h^2\, n^2}{4\pi^2\, me^2} ,\tag{3.18}$$

$$\omega_n = \frac{8\pi^3\, me^4}{h^3\, n^3} .\tag{3.19}$$

Here r_n is the radius of the nth quantum orbit corresponding to the quantum number n. If we substitute in Eq. (3.18) the values for m, e, and h, we obtain for $n=1$ the radius of the first Bohr orbit of the normal hydrogen atom,

$$r_1 = 0.529167 \times 10^{-8}\ \text{cm} .\tag{3.20}$$

This value, computed theoretically from BOHR's postulates, is in so satisfactory an agreement with our empirical knowledge of atomic radii that we may regard this result as a first triumph of BOHR's theory.

We now compute the energy states of the H atom which belong to the other quantum orbits. We write for the total energy, which is the sum of the kinetic and the potential energy[1],

$$E_n = \frac{1}{2} I_n \omega_n^2 - \frac{e^2}{r_n} ,\tag{3.21}$$

where I_n is the moment of inertia of the atom in the state n. By substituting Eqs. (3.18) and (3.19) and remembering that $I = mr^2$, we get the quantized energy values of the H atom

$$E_n = -\frac{2\pi^2 me^4}{h^2 n^2} \qquad n = 1, 2, 3, \dots .\tag{3.22}$$

We consider now the spectral lines which are emitted as a consequence of transitions between these stationary energy states. Let the quantum numbers of the initial and final state be n_i and n_f, respectively. Then the expression for the wave numbers of the spectral lines of the H atom is, from Eq. (3.22),

$$\bar{\nu} = \frac{1}{hc}\, (E_i - E_f) = \frac{2\pi^2 me^4}{h^3 c}\left(\frac{1}{n_f^2} - \frac{1}{n_i^2}\right)\tag{3.23}$$

with the condition

$$n_i > n_f .\tag{3.24}$$

---

[1] The potential energy is negative because it represents a binding energy (see III, 3). The potential is taken to be zero for the electron removed to infinity.

By comparing Eq. (3.23) with the empirical Balmer formula (3.1) we see that the term number n of BALMER's formula is identical with the quantum number n in BOHR's theory. We also see that BALMER's formula (3.1) follows from the general theoretical formula (3.23) by setting

$$n_f = 2 \quad \text{and} \quad n_i = 3, 4, 5, \ldots . \tag{3.25}$$

Thus, from BOHR's theory, the value of the Rydberg constant R (see Eq. (3.2)) follows as

$$R_\infty = \frac{2\pi^2 m e^4}{h^3 c} = 109{,}737.309 \pm 0.004 \text{ cm}^{-1}. \tag{3.26}$$

Our computations so far are not exact, however, since they were made for the limiting case of an infinitely large nuclear mass at rest, with the electron moving around it. This is indicated by the index ∞ in Eq. (3.26). Actually, both the electron and the nucleus revolve around their common center of mass to which the moment of inertia in Eqs. (3.17) and (3.21) is referred. The deviation is not large because of the large mass ratio of $1:1{,}836$ of the electron to the proton. Nevertheless, to be exact we have to use, instead of the Rydberg constant R_∞ (3.26), which is valid for infinite nuclear mass, the value

$$R = \frac{2\pi^2 m e^4}{c h^3 (1 + m/M)} = \frac{R_\infty}{1 + m/M} \tag{3.27}$$

where m is the mass of the electron and M the mass of its nucleus. The value of the Rydberg constant R_H for the H atom computed from Eq. (3.27) is in best agreement with the empirical spectroscopic value,

$$R_H = 109{,}677.576 \pm 0.012 \text{ cm}^{-1}. \tag{3.28}$$

This reduction of the Rydberg constant, computed from the hydrogen spectrum, to the universal constants m, e, h, and c ist another outstanding success of BOHR's theory. The extraordinary accuracy with which the Rydberg constant of hydrogen (3.28) was determined is further evidence for the high accuracy of spectroscopic data which is so valuable for any quantitative check of a theory.

If we now plot the energy level diagram of the hydrogen atom, using the energy values obtained from (3.22), we see that because of the identity of Eq. (3.22) with the empirical formula (3.3) we arrive at a diagram which, except for a factor hc, is identical with Fig. 32. From the empirical Balmer formula and Eq. (3.23), we conclude that the lines of the Balmer series (Fig. 28) have as their common final state the state with the quantum number 2. The Balmer series thus is emitted by transitions of the atom (i.e., its electron) from the higher quantum states $n = 3, 4, 5, \ldots$ to the $n = 2$ state. These transitions are represented in the energy level diagram Fig. 43 by arrows, the direction of which represents the direction of the transitions. The longer the arrow the larger is the energy and thus the wave number of the emitted spectral line, and correspondingly, the smaller is the wavelength, since $\bar\nu = 1/\lambda$. The wavelengths of the Balmer lines of hydrogen converge, as can be seen from Fig. 43, toward a limit which corresponds to the binding energy of the electron in the $n = 2$ state. Since the energy values (Eq. (3.22)) of the hydrogen atom are binding energies, they have been assigned negative values as discussed on page 59. They are numbered downward from the ionization limit which is shown as a dotted line in Fig. 43. The energy corresponding to the Balmer series limit, according to Eq. (3.22), is equal to $hc\, R_H/4$. From this and the value (3.28) for R_H, we compute the short wavelength limit of the Balmer series:

$$\lambda_\infty = \frac{1}{R_H/4} = 3646 \text{ Å},\qquad(3.29)$$

which is in agreement with the observed value. *Thus, $hc\,R_H/4$ is the binding energy of the $n=2$ electron; or, conversely, it is the energy required to release an electron from this $n=2$ state of the H atom, i.e., the ionization energy of the H atom excited to the energy state $n=2$. In the same way we can explain the physical significance of R_H.*

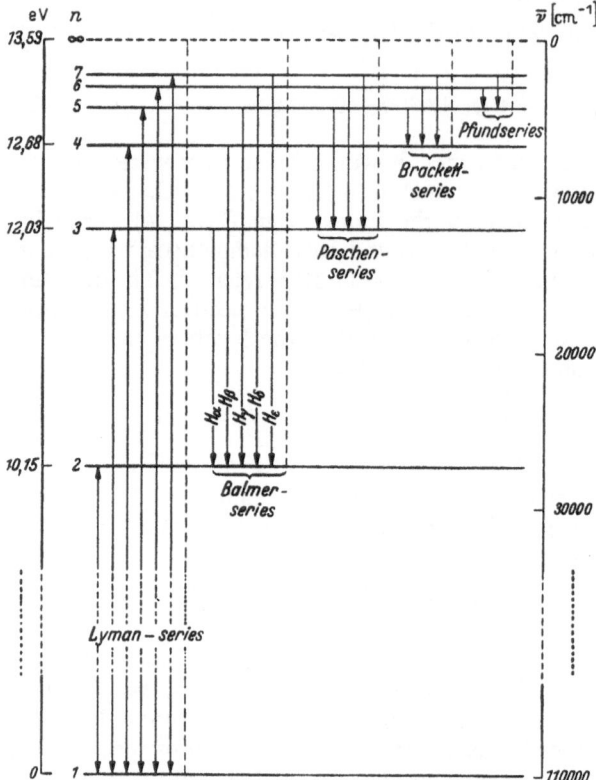

It is logical to inquire about the spectral lines of the H atom which correspond to the theoretical formula (3.23) when the values

$$\left.\begin{array}{l} n_f = 1 \\ n_i = 2,\,3,\,4,\,\dots \end{array}\right\}\quad(3.30)$$

are used. This spectral series

$$\left.\begin{array}{l} \bar\nu = R_H\left(1 - \dfrac{1}{n^2}\right) \\ n = 2,\,3,\,4,\,\dots \end{array}\right\}\quad(3.31)$$

corresponds to the transitions from the various excited states, enumerated on the left in Fig. 43, to the *ground* state, $n=1$, of the hydrogen atom. We

Fig. 43. Energy-level diagram of the H atom with its different spectral series

may compute the wavelength of the longest line of this series, of the transition $2\to1$. From Eq. (3.31) follows for $n=2$

$$\bar\nu = \frac{3}{4}\,R_H \qquad \lambda = \frac{1}{\bar\nu} = 1215 \text{ Å}.\qquad(3.32)$$

The hydrogen series represented by Eq. (3.31) thus must lie in the vacuum ultraviolet and, actually, was discovered there by LYMAN. The wave number $\bar\nu_\infty$ of the series limit of the Lyman series (see Fig. 43) is equal to the Rydberg constant R_H. Consequently, its wavelength is

$$\lambda_\infty = \frac{1}{R_H} = 911 \text{ Å}.\qquad(3.33)$$

It lies in the far ultraviolet, where it was found.

Thus the physical significance of the Rydberg constant is clear. *R_H is, as may be seen from Fig. 43, the distance of the ground state $n=1$ from the ionization limit, i.e., the binding energy of the hydrogen electron in the normal state $n=1$, or, conversely, the ionization energy of the hydrogen atom, i.e., the energy necessary* *to completely separate the electron from the ground-state atom.* This important atomic constant which can be measured experimentally by the Franck-Hertz experiments as discussed in the last section, thus can be determined spectroscopically with much higher accuracy. *The ionization energy is equal to the energy of the series limit of that spectral series of the atom which has the shortest wavelengths.* By means of the energy relation (3.10), the ionization energy of the hydrogen atom follows from the R_H value (3.28) as 13.595 ev, in agreement with the less accurate results of electron impact experiments.

We know also some higher spectral series of the hydrogen atom, which have as end states the energy levels $n=3$, $n=4$, and $n=5$ and are plotted at the right in Fig. 43. These series, computed from formula (3.23), were found and named for their discoverers as the PASCHEN, BRACKETT, and PFUND series. Although the higher lines of the PASCHEN series lie in the photographic infrared, an estimate shows that the other series lie entirely in the far infrared.

Thus we know the following spectral series of the hydrogen atom:

(1) $\bar{\nu} = R_H \left(1 - \dfrac{1}{n^2} \right)$ $n = 2, 3, 4, \ldots$ (Lyman series)

(2) $\bar{\nu} = R_H \left(\dfrac{1}{2^2} - \dfrac{1}{n^2} \right)$ $n = 3, 4, 5, \ldots$ (Balmer series)

(3) $\bar{\nu} = R_H \left(\dfrac{1}{3^2} - \dfrac{1}{n^2} \right)$ $n = 4, 5, 6, \ldots$ (Paschen series)   (3.34)

(4) $\bar{\nu} = R_H \left(\dfrac{1}{4^2} - \dfrac{1}{n^2} \right)$ $n = 5, 6, 7, \ldots$ (Brackett series)

(5) $\bar{\nu} = R_H \left(\dfrac{1}{5^2} - \dfrac{1}{n^2} \right)$ $n = 6, 7, 8, \ldots$ (Pfund series) .

Recently, also the first line of the sixth series was found. Its wavelength is 12.37 μ. All these series are emitted by hydrogen atoms, for example in a glow discharge in which the H_2 molecules are dissociated into H atoms, and these are then excited to the various states by electron collisions.

According to BOHR's theory, the H atom should also be able to *absorb* those spectral lines which it emits. The energy corresponding to a line, when absorbed by the atom, excites it to the corresponding higher level. In Fig. 43, the transitions corresponding to absorption are indicated by arrows pointing upward. Is it actually possible to observe in absorption all lines of the hydrogen series mentioned above? We neglect for the moment the fact that hydrogen is normally a molecular gas and we think of it as atomic hydrogen. Then, at normal temperature, we have only normal atoms in their ground state $n=1$. From Fig. 43 we see that the normal atom can absorb only *those* spectral lines the lower state of which is the ground state. For the hydrogen atom, this is the LYMAN series. Such spectral lines which correspond to transitions between upper states and the ground state of an atom are called resonance lines for historical reasons. *Only resonance lines thus appear as absorption lines at normal temperature.* Their arrows, therefore, may be shown double-headed. *All higher series normally appear only in emission.*

This result seems to be in disagreement with the fact that in the spectra of many fixed stars the Balmer lines, though not ending on the ground state, are observed in absorption (see Fig. 44). The answer to this apparent contradiction lies in the fact that the absorbing hydrogen in the star's atmosphere is by no means at a normal low temperature, but has a temperature of several thousand degrees, as shown clearly by the dissociation of the H_2 molecules into H atoms. For these high temperatures, it follows from MAXWELL's velocity distribution that a computable percentage of the hydrogen atoms is always excited, because of thermal collisions, to the first excited state $(n=2)$. Such hydrogen atoms of course may absorb the Balmer series. *Whereas at normal temperature only resonance lines are absorbed, at a sufficiently high temperature higher series or lines can also be absorbed.*

To recapitulate: Not only are the wavelengths of the long-known Balmer lines and the value of the ionization energy of the hydrogen atom, as given by BOHR'S theory, in excellent quantitative agreement with experimental results, but the theoretically predicted existence of a large number of additional spectral series was verified also. Finally, the behavior of the various series with respect to their appearance in emission and absorption was satisfactorily explained by the theory.

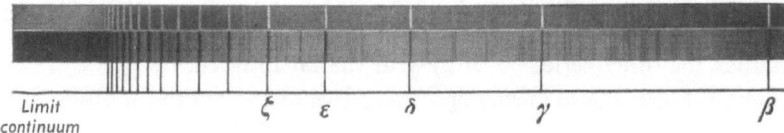

Fig. 44. Photograph of spectral lines of the Balmer series (without H$_\alpha$), emitted and absorbed by the fixed stars γ Cassiopeiae and α Cygni. Below, drawing of the Balmer series of the hydrogen atom. (After POHL)

6. Atomic Processes and their Reversal. Ionization and Recombination. Continuous Atomic Spectra and their Significance

In the last section, we became acquainted with a number of important atomic processes such as excitation by electron impact and the subsequent emission of radiation, and ionization by electron impact. We now discuss these atomic processes in a somewhat different form, i.e., as reaction equations, and at the same time broaden and deepen our knowledge of these processes and the corresponding spectroscopic phenomena.

a) Collisions of the First and Second Kind. Emission and Absorption

We begin by treating the process in which an atom A is excited by electron impact and where, after excitation, the electron jumps back to the ground state under emission of a photon with an energy equal to the excitation energy. We consider only the simplest case of *one* excited state. We may write the process of exciting an atom A by electron impact in the form of a chemical equation:

$$A + el_{\text{fast}} \rightarrow A^* + el_{\text{slow}} \tag{3.35}$$

where A^* is the excited atom. The fast electron gives up kinetic energy to the normal atom in an inelastic collision. After the collision, we have an excited atom A^* and an electron with correspondingly smaller kinetic energy. As a result of the excitation, the excited atom A^* spontaneously emits a photon of the energy $hc\bar{\nu}$ while returning to the ground state. Consequently, this process may be written

$$A^* \rightarrow A + hc\bar{\nu}. \tag{3.36}$$

Now there is a well-established rule in physics that every process may be reversed, and so we ask whether our equations (3.35) and (3.36) may be reversed and what this means. The most obvious reversal is that of (3.36). A normal atom A in its ground state "collides" with a photon of the proper energy $hc\bar{\nu}$, absorbs it, and is converted into an excited atom A^*: *By reading Eq. (3.36) from left to right we have emission, and from right to left, absorption of radiation.*

The reversal of the excitation by electron collision (3.35) means that an excited atom $A*$ collides with an electron and gives up its excitation energy in the collision so that we get an atom in its ground state and an electron of correspondingly higher kinetic energy. This process is called a *"collision of the second kind"*, as compared with excitation by electron collision which is called a collision of the first kind. Collisions of the second kind, in which excited atoms (or molecules) give up their excitation energy without radiation, do not occur in collisions with electrons only, but they may happen with other particles as well. For instance, an excited atom $A*$ may collide with another normal atom or molecule B which has a possible excited state of the same energy as the excitation energy of $A*$. In a collision of the second kind, $A*$ may then give up its excitation to B. We write this process

$$A*+B \rightarrow A+B*. \tag{3.37}$$

In contrast to the reverse process of (3.35), viz.

$$A*+el_{slow} \rightarrow A+el_{fast}, \tag{3.38}$$

the process (3.37) can occur only if A and B have approximately the same excitation energy, i.e., if there is resonance between the energy $A*$ and one of the excitation states of B.

The radiation from atoms or molecules B as a result of collisions of the second kind with excited atoms $A*$ according to (3.37) is called, after FRANCK, *sensitized fluorescence*. It may be observed in a mixture of different metal vapors with mercury vapor. Collisions of the second kind with electrons, according to (3.38), were studied first by KLEIN and ROSSELAND. They occur with high probability (yield) whenever collisions between excited atoms (or molecules) and electrons occur at all. It is, of course, necessary for both collisions of the second kind (3.37) and (3.38) that the collision occurs before the excited atom $A*$ releases its energy in the form of radiation. It can be shown from the kinetic theory of gases how the probabilities of the concurrent processes of radiation and collision of the second kind depend on the gas pressure since, in general, the emission of radiation occurs after a "mean lifetime" of the excited state of about 10^{-8} sec. We mentioned in III, 4, however, and shall learn in detail in III, 14 that there exist certain states of the atoms and molecules from which emission of radiation is impossible. Such states are called *metastable*, and the corresponding excited particle is called a metastable atom or molecule. Collisions of the second kind therefore occur predominantly in the presence of atoms or molecules with metastable states, such as the noble gas atoms or the mercury atom.

b) Impact Ionization and Recombination in Triple Collisions

We now turn our attention from *exciting* collisions and their reverse processes to the *ionization* by electron collisions in which an electron with sufficiently high kinetic energy collides with an atom and ionizes it. After the collision, we have a positive ion A^+, an electron resulting from the ionization, and the colliding electron with a correspondingly reduced kinetic energy. The ionization by electron impact thus may be written:

$$A+el_{fast} \rightarrow A^++el+el_{slow}. \tag{3.39}$$

This process can also be reversed, i.e., we may read the above equation from right to left: A positive ion undergoes a so-called triple collision with two electrons and recombines with one of them by forming a neutral atom, while the second electron carries away the released binding energy of the atom as kinetic energy. This released binding energy is equal to the ionization energy of the atom if the recombined atom is in its ground state. However, if the recombined atom is still excited (A^*), the released binding energy equals the difference between the ionization energy and the excitation energy of the excited atom A^*. Later on, the excited atom will give up its energy as radiation by emission of spectral lines and will finally jump back to the ground state. This *triple-collision recombination* of an ion and electron into a neutral (or an excited) atom is one of the most important processes in all electric gas discharges, in which charge carriers are eliminated mainly by this process if the density of electrons is not too small.

Since in this reversed process (3.39), viz.

$$A^+ + el + el \rightarrow A + el_{fast},\tag{3.40}$$

one of the two electrons has no other task than to carry away the binding energy (and the corresponding momentum) when the ion recombines with the other electron, it may be replaced by any other particle B, i.e., an atom, molecule, or ion. Recombination in triple collisions, therefore, can also occur in a process

$$A^+ + el + B \rightarrow A + B_{fast}.\tag{3.41}$$

Instead of colliding with a third particle B, the recombining ion and electron may finally collide with any larger body such as the wall of a discharge tube or electrode and may release their binding energy there, thus heating up the body. Recombination of ions and electrons occurs at the walls of almost all discharge tubes and accounts, at least partly, for their high temperature.

In the processes (3.40) and (3.41), recombination occurs in triple collisions because the released binding energy has to be carried away by a third particle. Recombination in a two-particle collision is possible only if the binding energy can be carried away without the aid of a third particle, that is, if it could be emitted as radiation. This process is also possible, and its discussion leads to a more detailed understanding of the relation between spectra and atomic processes.

c) Photoionization and Series-Limit Continua in Absorption

We already discussed photoexcitation, i.e., the excitation of electrons by absorption of the corresponding energy according to

$$A + hc\bar{\nu} \rightarrow A^*.\tag{3.42}$$

We also know that the series of excitation states of an atom converges toward the ionization energy. Thus the idea suggests itself that ionization of an atom should also be possible by absorption of the ionization energy in the form of a photon. In this case, which indeed occurs, we speak of *photoionization*. In order to understand these correlations, we extend the energy level diagram of an atom, viz. Fig. 45. So far we discussed only the stationary energy states of the atom. The higher the energy state, the weaker is the bond between the nucleus and the electron. The dashed line in Fig. 45 represents the limit to which the energy states converge and thus corresponds to a free electron completely removed from its atom. If we impart more energy to the ground state than is necessary to ionize the atom, the electron will be released and will take with it the excess energy as kinetic energy, $mv^2/2$. By extending our usual concept of the atom, we may regard this state of the system, i.e., ion + electron with kinetic energy, as a state of the atom also. We then speak not of a *stationary state of the bound electron*, but of a *non-stationary state of the free electron*. Since the energy of the free electron is larger than the ionization energy, its state must lie above the ionization limit. Naturally, these free-electron states are *not* quantized since the free electron may have any arbitrary amount of kinetic energy. Our energy level diagram of the atom, Fig. 45, thus shows, in addition to the discrete energy levels of the atom, the continuous energy range of the non-stationary states of the free electron above the ionization limit.

What is the result of this discussion with respect to the absorption spectrum of the atom? Transitions from the ground state to the different excited states of the atom, plotted as arrows in Fig. 45, correspond to the absorption process (3.42). They result in a line series (for the H atom: the Lyman series according to page 68) which converges toward the ionization limit. But the spectrum does not end here. Beyond the series limit, extending towards shorter wavelengths, we expect and find a continuous spectrum (see the absorption spectrum of the Na atom, Fig. 29) the so-called *series-limit continuum*. It is the result of transitions from the stationary ground state to the free states of different kinetic energy lying above the ionization limit. That the absorption intensity of this continuum beyond the series limit falls off toward shorter waves, follows from intensity rules to be discussed later. It means that the probability of absorption for high-energy photons is smaller than for those energies which exceed the ionization energy by only small amounts.

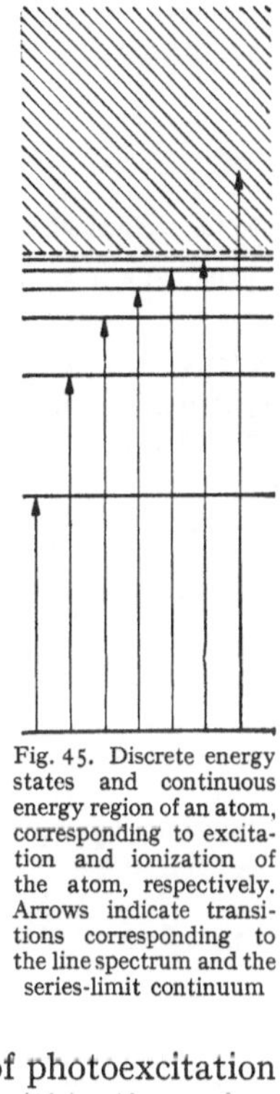

This process, the ionization of an atom by absorption of a photon with a wave number corresponding to the series limit continuum, is the *photoionization* process. It may be written

Fig. 45. Discrete energy states and continuous energy region of an atom, corresponding to excitation and ionization of the atom, respectively. Arrows indicate transitions corresponding to the line spectrum and the series-limit continuum

$$A + hc\bar{\nu}_c \rightarrow A^+ + el + \frac{mv^2}{2}. \qquad (3.43)$$

From observations of this series-limit continuum in absorption we infer the existence of an important atomic process, i.e., photo-ionization. Quantitatively, the intensity ratio of the atomic lines to the series-limit continuum is equal to the probability of photoexcitation compared to that of photoionization. The intensity distribution within the series-limit continuum, on the other hand, gives us the probability of photoionization as a function of the kinetic energy of the released electron.

d) Radiative Recombination and the Series-Limit Continua in Emission

The photoionization process (3.43) may also be reversed:

$$A^+ + el + \frac{m}{2} v^2 \rightarrow A + hc\bar{\nu}_c. \qquad (3.44)$$

This means that a positive ion A^+ collides with an electron and recombines to form an atom A. In most cases, A is first excited and then reverts to the ground state by line emission. The sum of the atomic binding energy ($=$ ionization energy of the capturing atom A) and the relative kinetic energy of ion and electron is emitted as a photon with a wave number in the series-limit continuum. In the language of the energy level diagram, this means that, in addition to transitions from the higher stationary states to the ground state, transitions from the continuous energy region above the ionization limit to the ground state are possible also. As a consequence of these transitions, the continuum is emitted.

From the ionization energy of 24.6 ev as measured for the He atom by electron impact, we compute, using Eq. (3.10), the wave number of the limit of the resonance series to be about 200,000 cm⁻¹ and from it the wavelength to be about 500 Å. The series-limit continuum emitted according to (3.44) during the radiative recombination of the He⁺ ion with an electron into the ground state of the normal

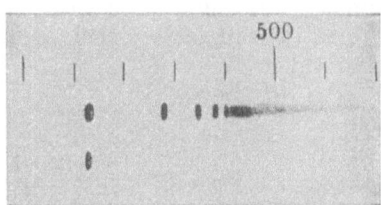

atom thus should extend from this wavelength to shorter waves. This continuum, as shown in Fig. 46, was actually found in a helium discharge by LYMAN with the vacuum spectrograph. Its long-wavelength limit, 509 Å, is in good agreement with our estimate. *We conclude from the observation of series-limit emission continua that recombination of atoms and electrons in two-particle collisions with emission of the binding energy according to (3.44) actually occurs. From the intensity distribution within these continua we can draw conclusions about the probability of the recombination of ions with electrons of various kinetic energies.* The decrease of intensity toward shorter waves indicates that radiative recom-

Fig. 46. Emission lines and series-limit continuum of the principal series of the He atom in the extreme ultraviolet near 500 Å. (Photograph by LYMAN)

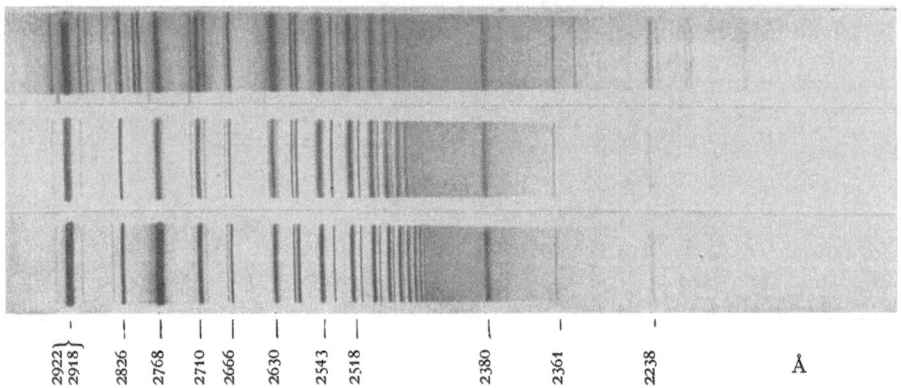

Fig. 47. Series-limit continuum of the subsidiary series of the Tl atom emitted by a thallium vapor discharge. With increasing current density the continuum penetrates more and more across the series limit into the series, caused by the perturbing fields of electrons and ions. Simultaneously, forbidden lines appear. (Photograph by KREFFT)

bination occurs predominantly with slow electrons but, with increasingly smaller probability, with fast electrons also.

Observations of the emission and absorption of series-limit continua and measurements of their intensity distribution are important for astrophysicists because from them they may draw conclusions about the atomic processes within stars or nebulae, and the state of the radiating or absorbing layers of them. Here, at very low pressures, the radiative recombination (3.44) is more frequent than the triple-collision recombination (3.40) or (3.41), which on the other hand predominates in gas discharges. The high intensity of the series-limit continuous spectra of a discharge in thallium vapor, Fig. 47, illustrates that the radiative recombination (3.44) is a very important process in many discharges also.

The radiative recombination need not necessarily lead directly to the ground state of the atom. Electron capture often occurs in an excited state and then is followed by the emission of spectral lines when the electron jumps to the ground state. We already mentioned this process of line emission under the name of *recombination radiation* in III, 4. Since transitions from the higher excited states of the hydrogen atom to the second quantum state according to Fig. 43 correspond to the Balmer series, transition of *free* electrons to the second quantum state result in the emission of the Balmer continuum beyond the short-wave limit of the Balmer series. Similarly, the continuum beyond the short-wave limit of the Paschen series is emitted upon capture of electrons of various kinetic energies into the third quantum state of the hydrogen atom. This spectrum may be observed in a condenser discharge in hydrogen. For such observations, discharges of high current density are best suited because a high ionization degree is necessary for a sufficiently large number of collisions between electrons and ions. Fig. 52 shows the spectrum of such a discharge in potassium vapor in which the emission continua of the various series limits may be distinguished.

To be complete, we want to mention that in some discharges, according to recent experiments, radiative recombination of electrons predominantly occurs with *molecular* ions instead of *atomic* ions. From this process (3.45) mostly excited molecules result; these either return to the ground state by emitting their characteristic band spectrum (see Chap. VI) or they dissociate, according to VI, 7, into a normal and an excited atom. These two possibilities thus may be written:

$$(AB)^+ + el \rightarrow (AB)^* \begin{cases} AB + hc\bar{\nu} \text{ (band)} \\ A + B^* \rightarrow A + B + hc\bar{\nu} \text{ (line).} \end{cases} \quad (3.45)$$

e) Electron Bremsstrahlung

Series-limit continua correspond to transitions from the continuous energy region to the different stationary states of atoms. For the sake of completeness, we expect that transitions between two states of a free electron within the continuous energy region, Fig. 48, should be possible also. They should correspond to either absorption or emission of radiation. Since the initial and final states now lie in a continuous energy region, the resulting spectrum has to be continuous and should have no direct relation with any line spectra. To what process do the transitions shown in Fig. 48 correspond?

If we consider transitions from a higher to a lower state, then the free electrons have a higher kinetic energy in the initial state than after the transition to the lower state. Each electron then emits the energy difference as a photon with a wave number of the continuous spectrum. We thus are dealing with a process, *by which a fast electron is stopped in the field of a positive ion.* The electron is not captured, however, in a stationary state but, after collision with the ion, continues on, though with reduced kinetic energy. This process may we written as

$$A^+ + el_{\text{fast}} \rightarrow A^+ + el_{\text{slow}} + hc\bar{\nu}_c. \quad (3.46)$$

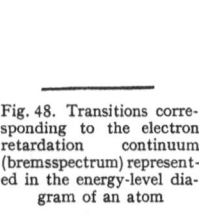

Fig. 48. Transitions corresponding to the electron retardation continuum (bremsspectrum) represented in the energy-level diagram of an atom

A continuous spectrum resulting from this process is called an *electron retardation continuum*, or, by the German expression, an *electron bremsspectrum*. Its short-wave limit depends upon the kinetic energy of the fastest colliding electrons. Evidently not more than the total kinetic energy, $(m/2) v^2$, can be emitted in this retardation process (3.46) so that the wave number $\bar{\nu}_\infty$ of the shortwave limit is given by the relation

$$h c \bar{\nu}_\infty = \frac{m}{2} v^2 . \qquad (3.47)$$

From this limit, the continuum extends infinitely toward longer wavelengths since of course arbitrarily small energy changes may occur in the retardation process.

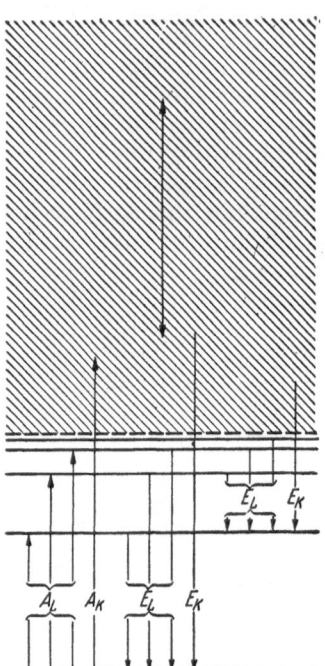

As in all atomic processes treated so far, the process of retardation radiation (bremsstrahlung) may be reversed. Reading (3.46) from right to left means that an electron colliding with an ion absorbs continuous radiation and gains kinetic energy. In that case the transition arrow in Fig. 48 is pointing upward. This last process plays an important role in astrophysics.

An electron retardation continuum (3.46) was first found *with cathode-ray electrons retarded by colliding with Z-fold charged nuclei of the metallic atoms in the anticathode of an X-ray tube.* For example, if an electron of 50,000 ev is retarded in the electric field of the nucleus of an anti-cathode atom, then from the energy relation (3.10) the wave number $\bar{\nu}$ of the limit is about 3×10^8 cm^{-1} and the corresponding wavelength limit $\lambda = 1/\bar{\nu}$ is 2.5×10^{-9} cm $= 0.25$ Å, i.e., it belongs to the X-rays (see Fig. 19). The well-known continuous X-ray spectrum is due to the process (3.46). This spectrum is called the *X-ray retardation continuum.* In our illustration and our scale in Fig. 48, the upper end of the arrow would lie 250 meters (!) above the lower end since the ionization energies of the atoms are of the order of 10 ev only. Electron retardation continua in the optical region were unknown for a long time, until it was realized that a large number of continuous spectra, especially of arc and spark discharges of high current density, are retardation continua. Retardation radiation in the visible region is also of importance in astrophysics.

Fig. 49. Schematic representation of all possible transitions in the energy-level diagram of an atom, corresponding to emission and absorption, respectively. (A_L= absorption of spectral lines; A_K= absorption of series-limit continuum; E_L= emission of spectral lines; E_K= emission of series-limit continua; uppermost double arrow= electron retardation continuum)

In Fig. 49 an energy level diagram is shown with all possible transitions corresponding to emission as well as absorption of lines, series-limit continua, and retardation continua (bremscontinua).

7. The Spectra of Hydrogen-like Ions and the Spectroscopic Displacement Law

Before the publication of BOHR'S theory, two further series were ascribed to the hydrogen atom, in addition to the spectral series (3.34) because they resemble the Balmer series in their structure. One of these series was found by FOWLER in a condensed hydrogen-helium discharge, the other by PICKERING in the spectra of certain stars. The Balmer series of hydrogen, as measured by PASCHEN, and the Pickering series are shown in Fig. 50. Each second line of this series falls close to a Balmer line, whereas the other lines fall in between two Balmer lines.

The two series in question may be described by the following Rydberg formulae:

$$\text{Fowler series:} \quad \bar{\nu} = 4R\left(\frac{1}{3^2} - \frac{1}{n^2}\right) \qquad n = 4, 5, 6, \dots, \tag{3.48}$$

$$\text{Pickering series:} \quad \bar{\nu} = 4R\left(\frac{1}{4^2} - \frac{1}{n^2}\right) \qquad n = 5, 6, 7, \dots. \tag{3.49}$$

How can these formulae be explained by BOHR's theory? The H atom definitely cannot be responsible for the spectra, since from our computations on page 66 there is no room in the theory for series of this type. Since the new series were observed by PASCHEN in pure helium also, the idea suggested itself to ascribe them to a positively charged helium ion, He⁺. This explanation was proved correct and led to a number of important conclusions.

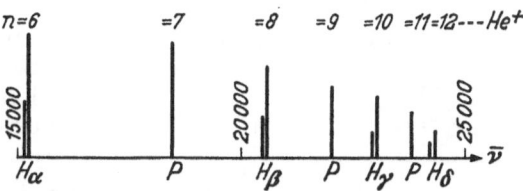

Fig. 50. Comparison of the wavelengths of the Balmer series of the H atom and the Pickering series (indicated by P) of the He⁺ ion. (After SOMMERFELD)

We designate the He⁺ ion as hydrogen-like in a rigorous sense because, like the H atom, it has a nucleus and *one* orbital electron. Because of the double charge on the helium nucleus we have to use in the Bohr theory of the He⁺ ion, instead of the equilibrium condition (3.13), the condition

$$\frac{2e^2}{r^2} = mr\omega^2. \tag{3.50}$$

The further theoretical treatment is analogous to that of III, 5. Instead of the energy values of the hydrogen atom

$$E_n(\text{H}) = -hc\frac{R}{n^2} \qquad n = 1, 2, 3, \dots, \tag{3.51}$$

because of the doubly charged nucleus of helium we now get the energy values

$$E_n(\text{He}^+) = -hc\frac{4R}{n^2} \qquad n = 1, 2, 3, \dots. \tag{3.52}$$

Quite generally, we find for the energy states of a nucleus with charge $+Ze$ and one orbital electron:

$$E_n(Z) = -hc\frac{Z^2R}{n^2} \qquad n = 1, 2, 3, \dots. \tag{3.53}$$

The energy level diagram of the He⁺ ion thus is exactly the same as that of the hydrogen atom except that all energy differences are increased by a factor 4 (or in general Z^2). The ionization energy necessary to remove the only remaining electron of the He⁺ ion, therefore, is $4R = 54.38$ ev. The wave numbers $\bar{\nu}$ of the spectral lines of He⁺ follow from Eq. (3.53) and are

$$\bar{\nu} = 4R\left(\frac{1}{n_f^2} - \frac{1}{n_i^2}\right) \qquad n_i > n_f. \tag{3.54}$$

We thus have again the Balmer formula (3.23) of page 66 except that the factor R is replaced by $4R$. All lines of He⁺ have only one fourth the wavelength of the corresponding lines of the hydrogen atom. This means that they are shifted into the ultraviolet. A comparison of Eq. (3.54) with Eqs. (3.48) and (3.49) shows that the Fowler series and the Pickering series are clearly series of He⁺. Why each second line of the Pickering series coincides with a Balmer line, may be understood if the formula for the Pickering series (3.49) is rewritten as

$$4R\left(\frac{1}{4^2}-\frac{1}{n^2}\right)=R\left[\frac{1}{2^2}-\frac{1}{(n/2)^2}\right] \qquad n=5,6,7,\dots \qquad (3.55)$$

For even n, this gives those lines of the Pickering series which coincide with Balmer lines, for odd n those between the H lines.

We pointed out in connection with Fig. 50 that in contrast to our computations the lines with even n do not coincide exactly with Balmer lines. Table 6 shows the approximate wavelengths of the first lines of both series from which the wavelength differences may be seen. This non-coincidence is caused by the difference between the Rydberg constants R for hydrogen and helium, which in turn is due to the effect of the different masses of H and He on the revolution of nucleus and electron around their common center of mass. The Rydberg constant R_∞ [see Eq. (3.26)] was deduced on page 67 for an infinitely large nuclear mass M. By taking into account the actual motion of the nucleus and electron around their common center of mass, this value was reduced in the ratio

Table 6. *Comparison of the Wavelengths of the Lines of the Pickering Series of He⁺ and the Balmer Series of the H-Atom*

Upper quantum number of Pickering lines	He⁺	H
6	6560.1	6563.1 (H$_\alpha$)
7	5411.6	—
8	4859.3	4861.3 (H$_\beta$)
9	4561.6	—
10	4338.7	4340.6 (H$_\gamma$)
11	4199.9	—
12	4100.0	4101.9 (H$_\delta$)

$$(1+m/M):1 . \qquad (3.56)$$

Consequently, since the mass of the helium nucleus is about four times that of the hydrogen nucleus, the Rydberg constant of He⁺ is

$$R_{He}=\frac{R_\infty}{1+(m_e/M_{He})}\approx\frac{R_\infty}{1+(m_e/4M_H)} . \qquad (3.57)$$

From the spectra of He⁺, i.e., from the Rydberg formulae (3.48) and (3.49), the value of R_{He} has been found to be in agreement with (3.57). Instead of the R_H value (3.28), we have

$$R_{He}=109,722.267\pm0.012\ \text{cm}^{-1}. \qquad (3.58)$$

From the agreement between computed and spectroscopic results, the important conclusion may be drawn that the center-of-mass law, which was used for computing the motion of the nucleus, remains valid in atomic systems.

Let us now return to the spectral series of He⁺. Of the several series predicted by (3.54) we have already discussed the Fowler series ($n_f=3$) and the Pickering series ($n_f=4$). The two short-wave series with $n_f=2$ and $n_f=1$ were found by LYMAN in the short-wave vacuum ultraviolet in quantitative agreement with (3.54). Consequently, analogous to the five series (3.34) of the H atom, we have the following series for the hydrogen-like He⁺ ion:

First Lyman series: $\bar\nu=4R_{He}\left(1-\frac{1}{n^2}\right)$ $\quad n=2,3,4,\dots$

Second Lyman series: $\bar\nu=4R_{He}\left(\frac{1}{2^2}-\frac{1}{n^2}\right)$ $\quad n=3,4,5,\dots$

Fowler series: $\bar\nu=4R_{He}\left(\frac{1}{3^2}-\frac{1}{n^2}\right)$ $\quad n=4,5,6,\dots$

Pickering series: $\bar\nu=4R_{He}\left(\frac{1}{4^2}-\frac{1}{n^2}\right)$ $\quad n=5,6,7,\dots$

$$(3.59)$$

The next strictly hydrogen-like ion is the Li^{++} ion which consists of a triply charged positive nucleus and one orbital electron. Its spectra are given by the Rydberg formula

$$\bar{\nu} = 9 R_{Li}\left(\frac{1}{n_f^2} - \frac{1}{n_i^2}\right) \quad n_i > n_f \tag{3.60}$$

in which the Rydberg constant is determined by substituting the mass of the lithium nucleus in Eq. (3.27). The first five lines of the resonance series of Li^{++} were found in the extreme vacuum ultraviolet; they are in excellent agreement with those computed from the theoretical formula (3.60). By now, also spectral lines of higher hydrogen-like ions were observed, up to some of the sevenfold ionized oxygen atom. The line of shortest wavelength that was observed by normal ultraviolet technique is the spectral line of the elevenfold ionized aluminum atom at 6.31 Å. This ion, however, is not hydrogen-like since it still has two electrons. The general formula (3.53) for the energy values of a hydrogen-like ion with the nuclear charge Z leads to an interesting conclusion. If all electrons but one could be stripped, for example, from the calcium atom with $Z = 20$, the ionization energy of this ion would be $400R = 4.4 \times 10^7$ cm$^{-1} \hat{=} 5400$ ev. The value for the wavelength of the series limit would be $\lambda_\infty = 1/400R = 2.3$ Å. This wavelength is beyond the optical region, in the X-ray region. We see that *X-rays are emitted by transitions of the innermost Ca electron in the strong Coulomb field of the twentyfold positive Ca nucleus, and we* see that *there is a continuous transition from the optical to the X-ray spectra.* We will discuss this in further detail in III, 10.

So far we discussed only those ions and their spectra which are hydrogen-like in the most precise sense. The spectra of all other atoms and ions with more than one electron have a rather complex structure. The electrons of these atoms and ions are arranged in shells, as we shall discuss in detail, and we speak of hydrogen-like ions in a wider sense in the case of the alkali atoms and of those ions that have only one electron in their outermost shell, which upon excitation is responsible for the spectra. Energy states of these atoms and ions that are hydrogen-like in a wider sense can no longer be computed according to Eq. (3.53) because the electrostatic binding between the outermost "valence electron" and the nucleus is disturbed by the presence of the other electrons. We shall learn details later while dealing with the spectra of the alkali atoms.

We saw that the spectra of the sequence H, He$^+$, Li^{++} were identical except for the shift towards the ultraviolet, caused by the increasing nuclear charge. We expect the same behavior for the spectra of the following ions, which are still hydrogen-like to some approximation:

Li, Be$^+$, B^{++}, C^{+++}, N^{++++}, O^{5+}, F^{6+}, Ne^{7+}

Na, Mg$^+$, Al^{++}, Si^{+++}, P^{++++}, S^{5+}, Cl^{6+}.

All these ions, like the alkali atoms lithium and sodium, have one electron in their outer shell. Their spectra and energy level diagrams therefore should be identical, except for an increasing shift toward the ultraviolet, which may be computed from Eqs. (3.53) and (3.27). These predictions have been confirmed by spectroscopic investigations. MILLIKAN and BOWEN, by discharging a condenser through highly rarefied gases, first produced such highly ionized atoms and called them "stripped atoms". They studied their spectra with a vacuum grating spectrograph, see Fig. 22. At the present time, as many as 23 electrons have been removed from some atoms (Sn^{23+}). These experiments confirmed the *spectroscopic displacement law* of SOMMERFELD and KOSSEL, which states that, except for some exceptions which will be cleared up in III, 19, *the spectrum of any atom is always similar to that of the singly charged positive ion which follows it in the Periodic Table. It is likewise similar to that of the doubly charged positive ion of the element which is two places to the right in the table, and so on.* We have shown for the special case of the hydrogen-like spectra that this law is theoretically understandable.

The confirmation of this spectroscopic displacement law is of special significance because it is evidence for the validity of the "construction principle" for the electron shells of the atoms. This principle states that each atom of the Periodic Table is built up from the atom which precedes it in the table by increasing the nuclear charge by unity and adding one additional electron. If this principle is correct, then, by stripping an outer electron from an atom, we get a positive ion which is similar in its electron arrangement and therefore in its level diagram and spectrum to the preceding element. This is exactly what the experimentally confirmed spectroscopic displacement law states.

8. The Spectra of the Alkali Atoms and their Significance. *S*-, *P*-, *D*-, *F*-term Series

So far, we discussed in detail only the spectra of single-electron systems. We now take up the next more complicated group of atoms, the alkali atoms Li, Na, K, Rb, Cs, and Fr. It is known from their chemical behavior that each of these atoms has one or more closed shells and *one* external electron, the first one that may be excited and also separated in an ionization process. In this respect, these atoms are similar to the H atom. The outermost electron, which is called the *emitting electron*, or, because of its role in chemical binding (cf. VI, 14), the *valence electron*, does not move in the simple Coulomb field of the nucleus. The Z-fold positive charge of the nucleus is screened off, i.e., partially compensated for by the closed inner shells of electrons, so that the effective charge is of the order of unity. The nucleus with its closed inner electron shells is called the *core* of the atom, so one speaks of the core and the emitting electron. From the electrostatical viewpoint, we may simply substitute a positive effective charge $Z_{eff}e$ for the core. *We must add, however, to this point charge a perturbation potential which depends on the orbit of the outer electron with respect to the core. This perturbation potential affects the energy states, and thus the spectra, of the alkali atoms in a very characteristic and decisive manner.*

We again begin by discussing the empirical data. KAYSER and RUNGE had already found that the spectra of each alkali atom could be resolved into three series. Later a fourth series was found in the long-wavelength region. These series are named principal series, sharp series, diffuse series, and Bergmann series. In absorption, at normal temperature, only the principal series is observed. The principal series thus terminates on the ground state of the atom. The wave numbers $\bar{\nu}$ of the four series may be represented by Rydberg formulae of the following form:

$$\text{Principal series:} \quad \bar{\nu} = \frac{R}{(1+s)^2} - \frac{R}{(n+p)^2} \qquad n = 2, 3, 4, \ldots$$

$$\text{Sharp series:} \quad \bar{\nu} = \frac{R}{(2+p)^2} - \frac{R}{(n+s)^2} \qquad n = 2, 3, 4, \ldots$$

$$\text{Diffuse series:} \quad \bar{\nu} = \frac{R}{(2+p)^2} - \frac{R}{(n+d)^2} \qquad n = 3, 4, 5, \ldots$$

$$\text{Bergmann series:} \quad \bar{\nu} = \frac{R}{(3+d)^2} - \frac{R}{(n+f)^2} \qquad n = 4, 5, 6, \ldots$$

$$(3.61)$$

Here, s, p, d, and f are constants which are characteristic for the different alkali atoms, and R again is the Rydberg constant. The s values increase from 0.4 for Na to 0.87 for Cs; the p values range from zero to 0.3, whereas the d and f values differ noticeably from zero only for the heaviest alkali atoms. If we compare formulae (3.61) with the Rydberg formula (3.34) for the hydrogen series,

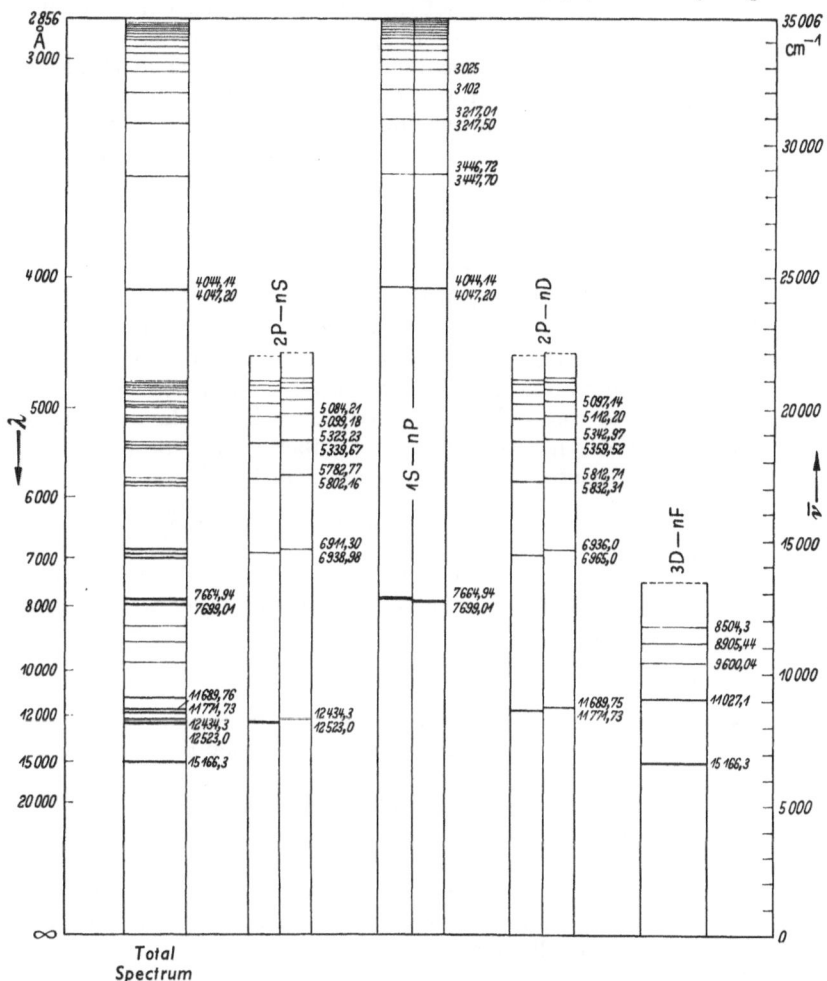

Fig. 51. Schematic representation of the spectrum of the K atom (at the left) consisting of the four series indicated at the right. (After GROTRIAN.) The significance of the fact that the three middle series of lines appear double with a small difference of wavelengths will be explained in III, 9. In Figs. 51, 65, 67, and 68, colons are to be read as points

page 69, we see that *for the hydrogen atom the values of s, p, d, and f are zero.* So we conclude that these characteristic constants are due to the perturbation of the emitting electron by the core. This perturbation, of course, is missing in the case of the H atom. The values of the constants indicate to what extent the different energy states of the alkali atoms deviate from those of the hydrogen atom, i.e., how much they are disturbed by the electrons of the core. We find here an explanation for the fact that this perturbation is smallest for lithium with its two core electrons only and is largest for caesium with its 54 electrons in the core. If the value zero is substituted for s, p, d, and f in Eq. (3.61), the principal series becomes identical with the Lyman series of hydrogen, the sharp and the

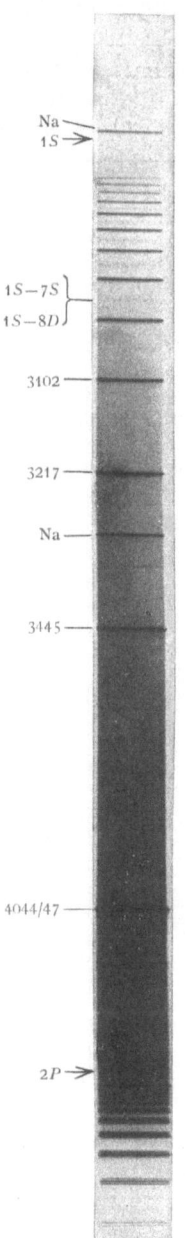

Na
1S

1S−7S
1S−8D

3102

3217

Na

3445

4044/47

2P

diffuse series then coincide with the Balmer series, and the Bergmann series with the Paschen series. The individual series of potassium are shown in Fig. 51 and, on the left, the total spectrum which results from the overlapping of the four series and therefore appears rather complicated. Fig. 52 is a photograph of a part of the emission spectrum of potassium. The last lines of the sharp and diffuse series, which converge to their common limit 2P, may be seen in the lower part of the photograph together with their common series-limit continuum. In the upper part (i.e., toward shorter waves), the principal series, which converges toward the limit 1S, may be seen.

From the series formulae (3.61), we see that the alkali atoms do not have one single term sequence as we know it from the H atom, but that there are four series resulting from the perturbation of the emitting electron by the core. The sequences corresponding to the constants s, p, d, and f are designated by the capital letters S, P, D, and F, respectively. They are represented graphically next to one another in Fig. 53. For the hydrogen spectrum, we know the relation between the principal quantum number n and the term values T_n, measured in wave numbers (cm⁻¹):

$$|T_n| = R/n^2; \qquad n = \sqrt{R/T_n}. \qquad (3.3)$$

For any non-hydrogen-like atom, an *effective principal quantum number* n_{eff} may be defined and computed from an analogous formula:

$$n_{\text{eff}} = \sqrt{R/T_n}. \qquad (3.62)$$

Values of n_{eff} for the various energy states of the Na atom are shown at the left in Fig. 53; they deviate from integers by the s, p, d, f values mentioned above. If we denote the ground state of each alkali atom by 1S and the higher terms according to the numbers of Fig. 53, the principal quantum numbers agree the better with the effective principal quantum numbers computed from Eq. (3.62) the more we approach the series limit. The Rydberg formulae (3.61) may be written in a symbolic form:

Principal series: $\bar{\nu} = 1S - nP,$ $n = 2, 3, 4, \ldots$

Sharp series: $\bar{\nu} = 2P - nS,$ $n = 2, 3, 4, \ldots$

Diffuse series: $\bar{\nu} = 2P - nD,$ $n = 3, 4, 5, \ldots$

Bergmann series: $\bar{\nu} = 3D - nF,$ $n = 4, 5, 6, \ldots$ (3.63)

The S state, being the lowest-energy state, has the principal quantum number $n = 1$. The lowest existing P state has $n = 2$; the lowest D state, $n = 3$; and the lowest F state, $n = 4$. From the level analysis of the spectra, an energy-level diagram, as shown in Fig. 53 for sodium, is obtained. The symbols S, P, D, and F have merely historical significance. P comes from *p*rincipal series, S and D designate the *s*harp and *d*iffuse series because of the different sharpness of their lines, which can be explained nowadays also, and F comes from *fundamental*. Originally, the Bergmann series was called the "fundamental series" because erroneously a special significance had been ascribed to it.

Unfortunately, the way in which the principal quantum number n is used in Eqs. (3.61) and (3.63) and in Fig. 53 contradicts the introduction of n in III, 5 in so far as here the ground state of the outermost electron of *all* alkali atoms is called $1S$ while actually, as we shall learn in III, 19, the valence electron of lithium belongs to the second shell, that of sodium to the third shell, and so on. Consequently, the ground state of the lithium atom should be called $2S$, that of the cesium atom $6S$. It is in this sense that the ground state of the mercury atom in Fig. 202 is called $6S$. Spectroscopists, unfortunately, use both these meanings of the principal quantum number side by side. The *effective principal quantum number*, however, is unambiguously defined by Eq. (3.62) and may be determined from the spectra. If the ground state of all alkali atoms is called $1S$, according to Fig. 53, n_{eff} is given by $(n+s)$ or $(n+p)$ etc., with the values of the constants s, p, d of page 81. However, if the shell number of the valence electron is designated by n ($n = 1, 2, 3, \ldots$ for a valence electron in the K, L, M, $\ldots$ shell, see III, 10), n_{eff} is $(n-\mu)$ with μ called the *quantum defect*.

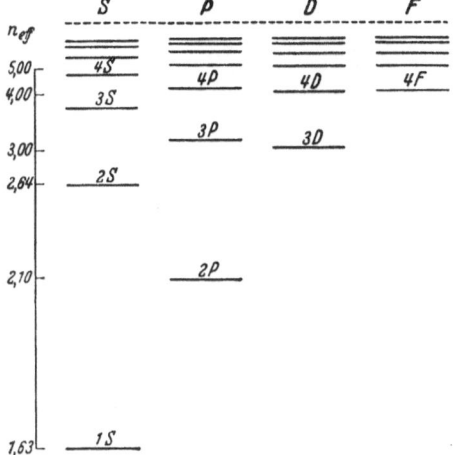

Fig. 53. Energy-level diagram of the sodium atom with the different term sequences and the effective principal quantum numbers of the terms

In Fig. 53 we see that the higher P, D, and F terms of the alkali atoms are very similar to the hydrogen terms, while the S terms and the lowest P term are quite dissimilar. That is what has to be expected: The energy states with the same principal quantum number n coincide in the case of the hydrogen atom. In the case of the alkali atoms, the various states, S, P, D, F, split up as a result of the electrostatic influence of the core. Therefore, it is easy to see that for a highly excited emitting electron the perturbation by the atomic core is mostly negligible. The S states are an exception because, in BOHR's theory, they correspond to elliptic orbits of largest eccentricity, so that they penetrate into the core and are thus considerably disturbed by its field.

By computing the perturbing potential of the core due to which the S, P, and D terms split up, we find that the decisive term which must be added to the Coulomb term e^2/r varies as $1/r^4$. This is easy to explain. The core is polarized by the emitting electron, i.e., the electron attracts the positive nucleus and repels the negative electron shells. The result is an electric dipole with a dipole moment

$$M = \frac{\alpha e}{r^2} \qquad (3.64)$$

where α is the polarizability of the atomic core. According to the theory of electricity, the potential of a dipole with a moment M is

$$U(r) = \frac{M}{r^2} = \frac{\alpha e}{r^4}. \qquad (3.65)$$

6*

The perturbing potential of the core, which varies as $1/r^4$ and which we must assume for explaining the alkali spectra, thus is a result of the polarization of the core by the field of the valence electron. The polarizability of the alkali ions as computed from their spectra is in quantitative agreement with the values of the polarizability found by other methods (see VI, 2).

By plotting an energy-level diagram with the transitions corresponding to the actually observed spectral lines of an alkali atom, we arrive at Fig. 54. We notice that the result is in contrast to that of the hydrogen atom and the original Ritz combination principle (see III, 2). In the alkali spectra, transitions do no longer occur between *all* energy states. Rather, there is a "selection rule", according to which only those spectral lines are observed that are due to transitions between *neighboring* terms. Transitions between states of the same term series ($S \rightarrow S$, $P \rightarrow P$, $D \rightarrow D$) are "forbidden" just as are transitions between non-neighboring series ($S \rightarrow D$, $F \rightarrow P$). However, we want to point out that these *selection rules are not strictly valid.* Forbidden transitions may occur if the emitting or absorbing atom is disturbed, e.g., by a strong electric field. For instance, in discharges of high current density in which line emission of atoms occurs in a disturbing field of neighboring electrons and ions, forbidden lines corresponding to $S \rightarrow S$ and $S \rightarrow D$ are observed, although with considerably reduced intensities. Fig. 52 shows an example.

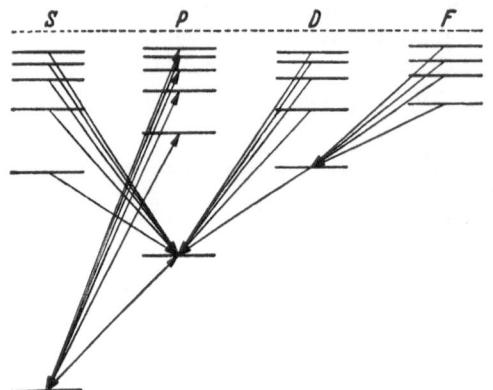

Fig. 54. Term diagram of the potassium atom with transitions corresponding to the different spectral series

In the case of the H atom, an electronic state is unambiguously determined by *one* number, the principal quantum number n. This does not hold for the alkali atoms. In addition to the principal quantum number n, it is necessary here to indicate the particular term series to which the energy state belongs. Sommerfeld introduced for this purpose the secondary quantum number, k. We shall use, instead of k, the orbital quantum number l because, according to quantum mechanics (Chap. IV), it has a clearer physical significance. l is so defined that for an atom with a single external electron, an S term is characterized by $l=0$, a P term by $l=1$, a D term by $l=2$, etc. Each electron state (in Bohr's model, each electron orbit) then is unambiguously determined by the two quantum numbers n and l, where

$$. \; l \leq n-1. \tag{3.66}$$

Table 7. *Electron symbols, electron quantum numbers, and the corresponding term sequences for one-electron atoms*

		S	P	D	F
n	l	0	1	2	3
1		$1s$			
2		$2s$	$2p$		
3		$3s$	$3p$	$3d$	
4		$4s$	$4p$	$4d$	$4f$

For labeling the electron, we use the small letters s, p, d, f following the principal quantum number, so that one speaks of a $1s$ electron, $3p$ electron, etc. Table 7 shows the relation between these electron symbols, the quantum numbers n and l, and the symbols of the corresponding term sequences. The possible n, l pairs indicated in the table result from the restricting condition (3.66).

The selection rule for optical transitions mentioned above now states that only those transitions are permitted for which the orbital quantum number l changes by ± 1:

$$\Delta l = \pm 1. \tag{3.67}$$

In the original Bohr-Sommerfeld theory of the atom, the principal quantum n determined the largest diameter of the particular electron orbit and, according to Eq. (3.22), the total energy of that electron. The secondary quantum number $k = l + 1$ determined the eccentricity of the orbit. The name "orbital quantum number" for l still reminds us of these orbits and is still appropriate for the emitting electron of the alkali atom has, according to quantum mechanics, a mechanical orbital momentum equal to $\sqrt{l(l+1)}\,(h/2\pi)$. In the case of the H atom, SOMMERFELD showed that, because of the purely central Coulomb field, the energy values of the orbits belonging to the same principal quantum number n are almost exactly equal so that we need only *one* quantum number, n. The energy state of the H atom, labeled as n, consists in this approximation of n coinciding energy states of different orbital quantum numbers. This is called n-fold degeneracy. As a result of the disturbance of the emitting electron by the atomic core, the energy values of orbits having the same n no longer coincide in the case of alkali and other atoms having more than one electron. The n-th quantum state therefore splits up into n different energy states ns, np, nd, nf, etc. Their energy differences are the larger the more the emitting electron is perturbed by the core. This splitting thus increases with the number of electrons in the core. The similarity to hydrogen is most marked in the case of lithium, whereas the departure from the hydrogen-like behavior and the splitting of the states having the same n is largest for cesium, which is the stable alkali atom with the largest number of electrons.

We mentioned before that the orbits of different eccentricity described by the different quantum numbers l for hydrogen have the same energy; in a first approximation, their levels therefore coincide. An exact coincidence of the levels with the same principal quantum number is theoretically expected only if no relativistic effects are taken into account. This was shown by SOMMERFELD in his famous theory of the fine structure of hydrogen lines. We may visualize its basis by considering that the electron in an orbit of high eccentricity approaches the nucleus in its perihelion so closely that its velocity is not small any more compared to the velocity of light. Similar to the case when the planet Mercury comes close enough to the sun, the relativistic effects cause a slow rotation of the longer elliptical axis; i.e., the elliptical orbit degenerates into a rosette-type orbit. Obviously, the amount of this relativistic energy depends on the eccentricity of the orbit. Thus, it differs for electrons of different orbital quantum numbers. Because of the relativistic correction, a small splitting of those energy states occurs which correspond to the same principal quantum number but different orbital quantum numbers.

This result is quantitatively described by SOMMERFELD's fine-structure formula

$$T_{n,l} = \frac{2R}{\alpha^2 \sqrt{1 + \dfrac{\alpha^2 Z^2}{K^2}}} + \frac{2R}{\alpha^2} \tag{3.68}$$

which replaces the simple Rydberg formula Eq. (3.3). R is again the Rydberg constant, which is connected with SOMMERFELD's fine-structure constant

$$\alpha = \frac{2\pi e^2}{hc} \tag{3.69}$$

by the relation

$$m_0 c^2 = \frac{2hcR}{\alpha^2}.$$ (3.70)

We shall encounter the fine-structure constant α again and again in the following sections. K is a function of the nuclear charge Z as well as of the quantum numbers n and l:

$$K = n - (l+1) + \sqrt{(l+1)^2 - \alpha^2 Z^2}.$$ (3.71)

By expanding the fine-structure formula (3.68) in powers of α^2 and substituting Eq. (3.71), we obtain

$$T_{n,l} = -\left[\frac{Z^2 R}{n^2} + \frac{Z^4 R \alpha^2}{n^4}\left(\frac{n}{l+1} - \frac{3}{4}\right) + \text{ terms with } \alpha^4, \alpha^6 \ldots\right]$$ (3.72)

From Eq. (3.72) we obtain the non-relativistic Rydberg formula (3.3) by neglecting $\alpha^2 = 5.3 \times 10^{-5}$, which is small compared to 1. This explains why for hydrogen the s, p, d, f terms with the same principal quantum number indeed coincide in a first approximation and why the measurement of the fine-structure of the Balmer lines with their small term splitting is rather difficult. The fine-structure formula has still to be changed in another decisive point before we shall get complete agreement with the observations. We come back to this point in III, 9c. After this modification, the energy states of electrons even in strong nuclear fields are well described.

We do not carry through the quantitative theory, i.e., the computation of the electron orbits by which all details of the alkali terms mentioned above, including the particular deviation of the S terms from those of the hydrogen atom, are satisfactorily described. For in Chap. IV, in which the further development of quantum physics will be discussed, we shall see that the concept of electrons revolving in well defined orbits does not have the significance which was originally attributed to it. For instance, all attempts at describing the orbits of the two helium electrons led to basic difficulties since none of the mechanically reasonable models of the helium atom was able to explain the diamagnetism of helium (see III, 15).

The existence of $S, P, D, \ldots$ term sequences and their corresponding line series, which we explained by introducing the orbital quantum number l is of course not restricted to the spectra of alkali atoms.

The spectra of other atoms and ions with *one* emitting electron, such as Cu, Ag, and Au, resemble those of the alkali spectra. The same applies, according to the spectroscopic displacement law, to the singly positive ions of the earth alkalies Mg, Ca, and Sr, etc., which have lost one of their two outermost electrons by ionization, to the doubly positive ions of the elements in the third column of the Periodic Table, etc. We showed already in III, 7 that the spectra of these ions are shifted toward shorter wavelengths, because the square of their effective nuclear charge (4 or 9 or, in general, Z^2) appears as a factor in their term formulae (3.53).

At first sight it seems surprising that the spectra of atoms with two external electrons such as helium, the earth alkali atoms of the second group of the Periodic Table, as well as the metals Hg, Cd, Zn, display a great similarity to the alkali spectra in so far as they too have principal, sharp, diffuse, and Bergmann series. On the other hand, they exhibit phenomena (two separate term systems) by which the two-electron spectra differ from the alkali spectra in a very characteristic way. We shall discuss the explanation of this fact in III, 11. The existing similarity can be explained only by the assumption that *in atoms with two equivalent external electrons in general only one is excited and acts as an emitting electron whereas the other electron may be regarded as part of the core.*

9. The Doublet Character of the Spectra of One-Electron Atoms and the Influence of the Electron Spin

Our previous discussion of the alkali spectra and their explanation did not account for one important phenomenon. The lines which in Fig. 54 are presented as single lines actually are closely spaced double lines. This fact is well known, for example, for the yellow D line of sodium, corresponding to the transition $2P \rightarrow 1S$, which consists of two lines D_1 and D_2 that are 6 Å apart from each other. A spectroscopic analysis shows that all energy states of the alkali atoms (Fig. 51) as well as those of all other atoms with *one* valence electron are doublet states with the only exception of the S states. An important extension of our atomic model is necessary in order to understand this doublet character of the one-electron terms.

a) Orbital Momentum, Spin, and Total Momentum of One-Electron Atoms

From the fact that an atom which is characterized by the quantum numbers n and l can exist in two states of somewhat different energy we conclude that these two quantum numbers are not sufficient for a complete and unambiguous description of an atomic state. We need at least one additional quantum number. It was introduced by SOMMERFELD in 1920 and was called "inner quantum number" j. All spectroscopic results are in accordance with the assumption that j is equal to $l + \frac{1}{2}$ or $l - \frac{1}{2}$ for all one-electron atoms. The same may be expressed in a different way by introducing a new quantum number s with only two possible values $(+\frac{1}{2})$ and $(-\frac{1}{2})$ for one-electron atoms. The difference of these two s values is unity in agreement with the general rule that quantum numbers always change by integers. In trying to understand the physical significance of these results and of the quantum numbers in general, we go back to BOHR's original quantum condition for the principal quantum number n, according to Eq. (3.17),

$$\oint p\,dq = 2\pi r p = 2\pi m r^2 \omega = nh \qquad n = 1, 2, 3, \dots . \qquad (3.73)$$

If we write this equation

$$pr = mr^2\omega = J\omega = n\frac{h}{2\pi}, \qquad (3.74)$$

we see that actually it is not the momentum $\vec{p}$ of the electron but the angular momentum $\vec{J\omega}$ of the atom that is quantized; it has the dimension of an action. Quite generally atomic angular momenta therefore are measured in units of $h/2\pi$ according to Eq. (3.74). In quantum theory this quantity $h/2\pi$ is often designated by the symbol $\hbar$ and is called "h bar". So far we have only spoken of circular orbits. If, however, the orbits are elliptical, then both r and φ vary simultaneously. For the orbital momentum $\vec{l}$ of the electron in which only φ varies, we can then write an expression which corresponds to the quantum condition (3.17) or (3.73):

$$\oint |\vec{l}|\,d\varphi = 2\pi|\vec{l}| = lh \qquad l = 0, 1, 2, \dots . \qquad (3.75)$$

Eq. (3.75) is identical with the quantum condition (3.17) for the principal quantum number of the hydrogen electron because (3.17) dealt with circular orbits only.

For circular orbits, the formula for the variation dq of the position coordinate q of the electron, given generally by

$$dq^2 = dr^2 + r^2 d\varphi^2, \tag{3.76}$$

is reduced to

$$dq = \text{const } d\varphi. \tag{3.77}$$

Thus, with an orbital momentum $\vec{l}$ that does not contain a variation of r, we have the relation

$$|\vec{l}| = l \frac{h}{2\pi} \tag{3.78}$$

where l is the orbital quantum number which was introduced on page 84 in order to explain the alkali spectra. The orbital momentum of an electron revolving around the nucleus may be represented by a vector drawn perpendicular to the plane of the orbit at its center (Fig. 55). Its length is equal to the orbital momentum and its direction is that of a right-handed screw rotating in the same sense as the electron.

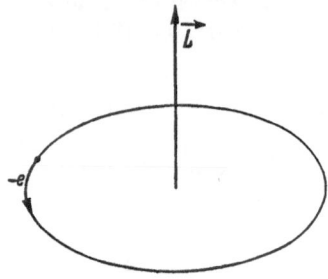

However, we shall see in quantum mechanics (IV, 8) that Eq. (3.78), which follows from Eq. (3.75), is not quite correct, and that Eq. (3.78) has to be replaced by

$$|\vec{l}| = \sqrt{l(l+1)}\, \frac{h}{2\pi}. \tag{3.79}$$

Fig. 55.   Orbital momentum and orbital motion of an electron

This formula holds quite generally for *all* angular momenta and the corresponding quantum numbers. Although the new fomula deviates considerably from Eq. (3.78), this classical formula remains valid for the maximum component of any angular momentum in the direction of an external orienting field. This is always an integral multiple of $h/2\pi$, exactly what we would expect from the original formula (3.78).

The necessity of introducing the inner quantum number j or the quantum number s is evidence for the fact that an additional angular momentum must exist beside that of the revolving electron. It was discovered in 1925 by GOUD-SMIT and UHLENBECK who, for reasons to be discussed later, ascribed to the electron itself a mechanical angular momentum or spin $\vec{s}$ of the amount of $\frac{1}{2} \times \frac{h}{2\pi}$. In the classical picture, it corresponds to a rotation of the electron around its own axis.

The orbital momentum $\vec{l}$ and the spin $\vec{s}$ of the electron are coupled by their corresponding magnetic fields (see III, 15) and therefore combine vectorially to form the total angular momentum $\vec{j} = \vec{l} + \vec{s}$. This total angular momentum is related to the inner quantum number j, in analogy to Eqs. (3.78) and (3.79), by

$$|\vec{j}| = j \frac{h}{2\pi}, \qquad j = \frac{1}{2}, \frac{3}{2}, \frac{5}{2}, \dots \tag{3.80}$$

or, quantum-mechanically correct,

$$|\vec{j}| = \sqrt{j(j+1)}\, \frac{h}{2\pi}. \tag{3.81}$$

Like $\vec{l}$, also $\vec{j}$ and $\vec{s}$ can change only by integral multiples of $\frac{h}{2\pi}$.

b) The Doublet Structure of the Alkali Atom Levels

The doublet structure of the terms of the alkali atoms and their spectra is a direct consequence of the facts presented above. We begin by discussing the S terms. They correspond, according to page 84, to the orbital momentum $\vec{l}=0$ of the emitting electron. According to III, 15, the electron thus does not have a magnetic moment with respect to which the spin $\vec{s}$ might orient itself. All orientations of $\vec{s}$ correspond to the same energy. *For that reason S terms are always single.* In the case of a P term, the orbital momentum $\vec{l}=1$ is coupled with the spin $\vec{s}=\frac{1}{2}$ of the emitting electron in such a way that the values of the total angular momentum differ by integral numbers. It is evident that only two possibilities satisfy this condition, namely

$$\left.\begin{aligned} j_1 &= l+s = 1+\tfrac{1}{2}=\tfrac{3}{2} \\ j_2 &= l-s = 1-\tfrac{1}{2}=\tfrac{1}{2}. \end{aligned}\right\} \tag{3.82}$$

In the case of the D terms with $l=2$, we have the two possibilities:

$$\left.\begin{aligned} j_1 &= l+s = 2+\tfrac{1}{2}=\tfrac{5}{2} \\ j_2 &= 2-s = 2-\tfrac{1}{2}=\tfrac{3}{2}. \end{aligned}\right\} \tag{3.83}$$

Except for the S terms, all terms of the alkali atoms, because of the two possible orientations of the spin, are doublet terms. The S terms also may be regarded as doublets, but because of $l=0$, no spin orientation is possible so that the S terms appear single. In spite of this, they are called "doublet S terms" because they belong to a doublet system.

The existence of the electron spin of the constant value $\frac{1}{2}(h/2\pi)$ is the reason for the doublet character of the alkali terms. In the spectroscopic symbols, this fact is expressed by adding a 2 as a superscript on the left side of the term symbol which characterises the orbital quantum number $l(S, P, D, F, \ldots$ for $l=0, 1, 2, 3, \ldots)$. The symbol ^2P is then read "doublet P". The inner quantum number j of the total angular momentum of the electron, which distinguishes the two doublet components, is added as a subscript to the right of the term symbol. The two doublet term components of an alkali atom, which belong to $n=3$ and $l=1$, are thus written, according to Eq. (3.82), $3\,^2P_{\frac{3}{2}}$ and $3\,^2P_{\frac{1}{2}}$. Correspondingly the ground state of an alkali atom is $1\,^2S_{\frac{1}{2}}$.

The spectroscopic analysis reveals by which energy the two doublet terms split up. Within each group of the Periodic Table, this energy difference increases strongly with increasing atomic number. The two doublet components $^2P_{\frac{3}{2}}$ and $^2P_{\frac{1}{2}}$ of the ground state of lithium, for instance, differ by only 0.34 cm^{-1}, whereas this same energy difference is 554 cm^{-1} for cesium, i.e., more than thousand times the value of lithium. The doublet spacing decreases strongly with increasing principal and orbital quantum numbers. We will treat the theory of this phenomenon in III, 17 after presenting the magnetic behavior of electrons and atoms (see III, 15).

From the spectra, we infer the selection rule for the total angular quantum number j:

$$\Delta j=0 \quad \text{or} \quad \pm 1. \tag{3.84}$$

We will show in IV, 9 how this selection rule for optical transitions follows from quantum mechanics. If we consider the possible transitions according to this selection rule by making use of a section of the alkali level diagram, Fig. 56, we notice that all $P \to S$ transitions are double, resulting in doublet lines such as the well-known sodium D lines at 5890/5896 Å. In the case of transitions between P, D, and F terms, Fig. 56b, however, we expect theoretically three lines since, for example, for $D \rightarrow P$ the transitions $\frac{5}{2} \rightarrow \frac{3}{2}$, $\frac{3}{2} \rightarrow \frac{3}{2}$, and $\frac{3}{2} \rightarrow \frac{1}{2}$ are allowed according to Eq. (3.84). However, an intensity rule, derived from the spectra and theoretically understandable, favors transitions where l and j change in the same direction (in Fig. 56b: $P_{\frac{1}{2}} \rightarrow D_{\frac{3}{2}}$ and $P_{\frac{3}{2}} \rightarrow D_{\frac{5}{2}}$) compared to those with constant j where

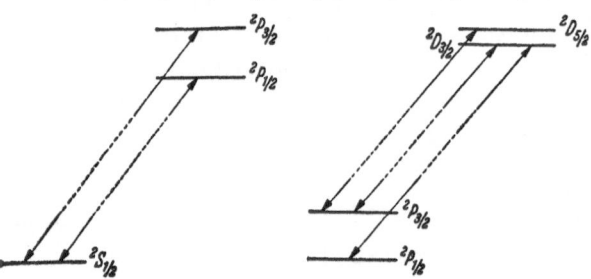

l and s change in opposite directions. One of the spectral lines corresponding to transitions between doublet terms with $l \neq 0$ (in Fig. 56b: $P_{\frac{3}{2}} \rightarrow D_{\frac{3}{2}}$) therefore appears with very low intensity and is called a *satellite*. In simplifying the actual situation, one thus speaks of the doublet spectra of one-electron atoms.

Fig. 56a. Schematic energy-level diagram and transitions resulting in a $^2P \rightarrow {}^2S$ doublet of an alkali atom

Fig. 56b. Schematic energy-level diagram and transitions resulting in a $^2D \rightarrow {}^2P$ doublet of an alkali atom

c) Doublet Character and Fine Structure of the Balmer Terms of the Hydrogen Atom

It follows from our presentation that all one-electron atoms should have a doublet-term system because of the spin $\frac{1}{2}$. What about the hydrogen atom which is the most important one-electron atom? Actually, a fine structure of the

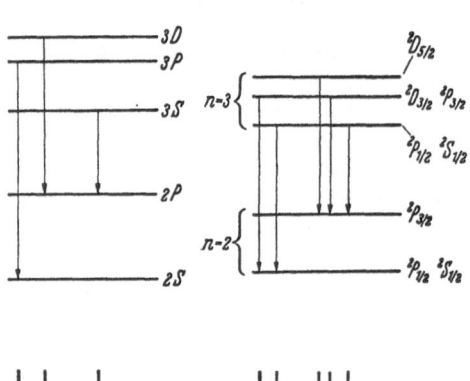

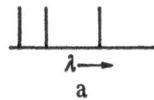
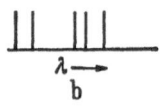

a   b

Fig. 57. Half-schematic diagram of the splitting of terms and the corresponding components of the hydrogen line $H\alpha$. a) According to SOMMERFELD's original theory. b) According to the theory taking into consideration the electron spin

Balmer lines of hydrogen was found when the spectrum was studied with high resolution. At first, SOMMERFELD thought that (see page 85) this was due to the fact that the S, P, and D terms of the same principal quantum number split up a little because of relativistic effects. On the other hand, the spin theory requires that all these energy states of the hydrogen atom, like those of all one-electron systems, split into doublets with the inner quantum numbers $j = l \pm \frac{1}{2}$. The amount of this splitting depends, for all doublets, on the amount of interaction between orbital momentum and spin of the electron. Detailed computations by HEISENBERG and JORDAN gave a very surprising result: *The hydrogen levels of the same quantum number j always coincide in spite of their different orbital quantum numbers l. In other words, the energies of the two terms $^2P_{\frac{1}{2}}$ and $^2S_{\frac{1}{2}}$ are the same, just as the two terms $^2D_{\frac{3}{2}}$ and $^2P_{\frac{3}{2}}$ do not differ in their energy value. Thus, in spite of the doublet structure, the number of fine-structure components of each hydrogen term still is equal to its principal quantum number n, exactly as it was in* SOMMERFELD's *original theory.* The correct fine-structure formula is obtained by replacing $(l+1)$ by $(j+\frac{1}{2})$ in Eqs. (3.68) or (3.71). But there is *one* qualitative and essential difference between the original and the new explanation of the hydrogen fine-structure. We explain it by making use of Fig. 57. In a half-schematic way, the fine structure of the hydrogen terms $n=2$ and $n=3$ which give rise to the Balmer line H_α is presented, at the left (Fig. 57a) according to the original theory by Sommerfeld, at the right (Fig. 57b) according to the doublet theory. From Fig. 57a and the selection rule for l, Eq. (3.67), the number of components of H_α results as three, whereas the doublet theory by making use of the selection rule for j, Eq. (3.84), according to Fig. 57b results in *five* components of H_α so that a decision between these two cases is possible. Very careful measurements of the Balmer line H_α, carried out by Hansen, and of the corresponding He^+ line 4686 Å by Paschen confirmed the fine structure of the lines to be expected from Fig. 57b. The influence of the electron spin on the fine structure of H and He^+ was thus proved.

Investigations by Lamb and Retherford, using the method of high-frequency spectroscopy mentioned in III, 1, yielded a very important new result: The terms $2\,^2S_\frac{1}{2}$ and $2\,^2P_\frac{1}{2}$, in Fig. 57b plotted as one line, actually do not coincide. The first one is about 0.03 cm^{-1} (i.e., $\frac{1}{11}$ of the distance between the components of the $(n=2)$-terms) higher. This discovery activated a thorough re-investigation of the behavior of the electron itself and led to the discovery of the small deviation of the electron's magnetic momentum from the amount of Bohr's magneton Eq. (2.39), which was mentioned in II, 4b. New developments in quantum mechanics yielded an explanation of this effect, which is caused by an interaction between electron and radiation field. This effect is a good example for teaching us how inconspicuous quantitative differences between experimental and theoretical results may lead to the discovery of fundamentally important effects.

10. X-ray Spectra, their Theoretical Interpretation, and their Relation to Optical Spectra

We showed in III, 7 that the spectra of strictly hydrogen-like ions, i.e., those with a nucleus and only one electron, for elements of atomic number above 20 have wavelengths of the order of or below one angstrom unit, i.e., they lie in the X-ray region. Highly ionized atoms of this type cannot yet be produced in the laboratory. They are found only in the interior of stars where the atoms, because of the extremely high temperatures, are ionized thermally. The X-rays which we use in experiments are emitted by atoms which still have all or nearly all their electrons. In order to understand the emission of these X-ray spectra which are characteristic of the particular atom, we have to postulate a number of facts about the structure of the electron shells of atoms of higher atomic number. These facts actually have been deduced by Kossel from the study of X-ray spectra.

a) Electron Shell Structure and X-ray Spectra

As we shall show in detail in III, 19, the electrons of atoms of higher atomic number are arranged around the nucleus in individual shells. These shells are called the K, L, M, N, O, and P shell. Only 2 electrons can occupy the K shell; 8 the L shell; 18 the M shell; and in the N shell there is room for 32 electrons, as may be seen from Table 10, page 131. Since the outermost electron shells correspond to the outermost Bohr orbits and therefore to the highest energy states, we may plot the complete energy-level diagram, as in Fig. 58, for the Cu atom with atomic number 29. The K, L, and M shells with principal quantum numbers $n=1, 2$, and 3, respectively, are filled with electrons, while in the 4-quantum N shell there is only one electron, the emitting electron of the copper atom. The optical spectra which may be excited in an electric arc between copper electrodes arise from the excitation of this electron from its $n=4$ state into higher, normally unoccupied energy states and the subsequent transitions of the electron from the higher states to the $n=4$ state. These higher energy states, which are partially or completely unoccupied, are therefore called *optical levels*. We want to prove that transitions of electrons between these outer levels of the copper atom result in the emission or absorption of optical spectra. Let us consider one of the outermost electrons of the atom. It is bound to the nucleus by electrostatic forces, but most of the high nuclear charge of 29 positive charges is shielded by the inner, occupied electron shells. The outermost electron, available for optical transitions similar to those of the hydrogen atom, moves in a field of a nucleus with an effective charge of the order of unity. Thus its ionization and excitation energies must be of the same order of magnitude as those of the hydrogen atom.

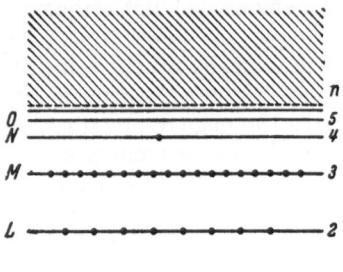

Fig. 58. Simplified energy-level diagram of the Cu atom, indicating also the inner electron states which are normally filled with electrons and therefore mostly omitted in diagrams referring to optical spectra

On the other hand, we expect the emission of X-rays as a consequence of transitions between the L and K shells, because here the electron moves in the strong field of either the full nuclear charge or that charge diminished by the screening of only one unit. Electron transitions between these levels next to the nucleus which give rise to the emission of X-rays are not possible under normal conditions, because all these levels are normally occupied by electrons. Emission of X-rays is therefore possible only if there is an unoccupied position in one of the inner electron shells, in other words, if one of the two K electrons is removed from the K shell by ionization.

If we neglect the disturbance of the surrounding shells and the screening of the nuclear charge by the other K electron, the energy necessary for ionizing one of the K electrons is equal to $Z^2 Rhc$, according to Eq. (3.53). For the innermost copper electron this energy is about 11,400 ev, since Z equals 29. It is thus possible to ionize a copper K electron by an impinging cathode ray electron which has been accelerated in the X-ray tube by more than 12 kv.

b) The Mechanism of X-ray Line Emission

The mechanism of the emission of X-ray lines, for instance of copper, was first explained by Kossel as follows: Electrons are emitted by a cathode and accelerated by the voltage between the cathode and anticathode. The electrons impinging on the copper atoms of the anticathode penetrate the outer electron shells without collision and ionize the atom by kicking out one of the two (innermost) K-shell electrons. To be exact, it is not necessary to kick out the K electron in question completely. It is sufficient to lift it to one of the unfilled optical levels. Since the separation of the unfilled levels from the ionization level is negligible compared with the distance between the L and K levels (greatly distorted in Fig. 58), the entire process is usually called impact ionization.

The hole that is left in the K shell is refilled (as indicated in Fig. 59) by a transition of an electron from the L, M, or N shell, accompanied by the emission of the energy difference in the form of an X-ray line. All X-ray lines the common end state of which is the first quantum state are called the K series. Its lines are designated, in decreasing order of their wavelength, by K_α, K_β, and

K_γ. When the K_α line is emitted, an electron hole is created in the L shell which in turn is filled by an electron from the M or N shell. This transition is coupled with the emission of "softer" X-ray lines, i.e., of longer wavelength, called L_α, L_γ. When the hole in the M shell is filled by transition of an electron from the N shell which is the highest occupied shell of copper, the very soft M_α line is emitted.

Actually, the situation is much more complicated, since each X-ray line of Fig. 59 consists of a number of closely spaced lines of different intensity. We shall discuss this fine structure of X-ray lines soon. Fig. 60 presents a photograph of the L spectrum of tungsten showing *all* lines. The wavelength unit used here is the $XU = 10^{-3}\,\text{Å} = 10^{-11}$ cm. It is evident that the L spectrum may also be emitted without the K series. This happens whenever the cathode-ray electrons do not have sufficient kinetic energy for ionizing the K electron, and thus only L electrons are ionized by impact.

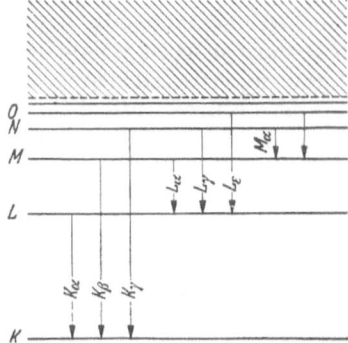

Fig. 59. Energy-level diagram of an atom with transitions of inner electrons corresponding to the emission of X-ray lines

The X-ray line spectra thus correspond in every respect to the optical spectra. However, their emission is possible only after the ionization of an electron from an inner shell makes "room" in a normally filled inner shell for an electron transition. The excitation energy of an X-ray spectrum thus is practically equal to the energy required to throw out an electron from its final state, i.e., the ionization energy of the atom from the energy state in question. The wavelengths of the X-ray lines, just as those in the optical spectrum, depend on the difference of the energy levels (in this case of the innermost occupied ones) and thus are

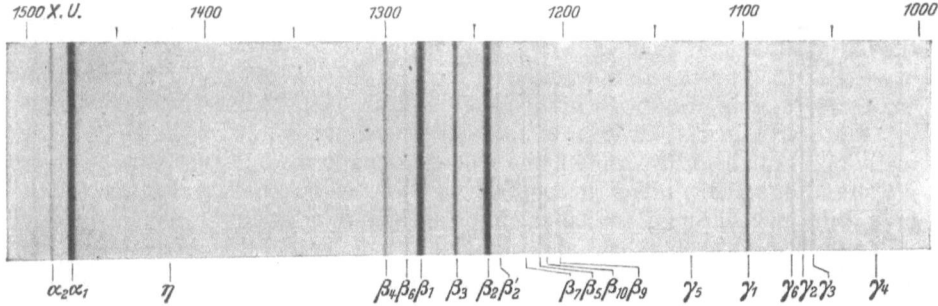

Fig. 60. Example of an X-ray line spectrum: the L spectrum of the tungsten atom. (After Siegbahn)

characteristic for any particular atom. The X-ray lines for this reason are called the *characteristic X-ray radiation*. The larger the atomic number, i.e., the nuclear charge, of a radiating atom, the shorter are the wavelengths of its characteristic radiation, as may be seen from Figs. 5 and 6. We mentioned this law, which was discovered by BARKLA and MOSELEY, as a possibility of determining the atomic numbers of the elements in II, 2. The relation between X-ray lines and optical lines may become clearer by mentioning that the K_α line of the hydrogen atom is identical with the longest-wave line of the Lyman series of hydrogen (see III, 5). There are also spectra which are interesting transition cases between optical and X-ray spectra. In the far ultraviolet namely, there are spectra, of Zn for instance, which correspond to the excitation of an electron of the inner closed $3d$ shell, whereas the normal optical spectra are due to transitions of the outer $4s$ electrons.

The continuous X-ray bremsstrahlung, in contrast to the characteristic X-ray radiation, is independent of the anticathode atoms. It is emitted when cathode-ray electrons are stopped in the nuclear field of the anticathode atoms. The continuum extends, as was explained in III, 6e, from very long waves down to a sharp shortwave limit with an energy $hc\bar{\nu}_\infty$ equal to the maximum energy of the stopped cathode-ray electrons. The absolute intensity of the continuum depends on the nature of the retarding atom, in particular on its atomic number which determines the strength of the retarding nuclear field. Because of the existence of the X-ray continuum, every X-ray spectrum is superimposed on a continuous background, as may be seen in Fig. 27.

c) The Fine Structure of X-ray Spectra

In our discussion of the emissive transitions, we have simplified matters to the extent that by the transition of an electron from the L to the K shell only one line, K_α, was emitted. Actually, each such transition involves a number of lines, as shown in Fig. 60 for the L spectrum of tungsten. The simplification of Fig. 59 corresponds, as we see by comparing with Fig. 43, to the assumption that every energy state of the hydrogen atom is determined by one quantum number only. We learned in the last section that this simplification is not quite correct since it cannot explain the fine structure of the Balmer lines. For the hydrogen atom, the relativistic influence and the electron spin had to be taken into consideration. Like the hydrogen spectra, all other one-electron spectra cannot be understood without introducing the orbital momentum quantum number and the spin or the total momentum quantum number, which are necessary for describing the energy levels of one-electron systems.

Now X-ray spectra have to be looked at as being *one*-electron spectra also so that we expect their energy states to resemble those of the alkali atoms since the initial state of an atom emitting an X-ray line is characterized by the absence of *one* electron in a closed inner electron shell. HEISENBERG showed that such an energy state of an atom with a filled shell and one electron missing is analogous to a state with a single electron in an otherwise empty shell. We shall encounter many applications of this rule in the following chapters, especially in solid-state physics. For understanding this equivalence, we mention the fact that the resulting orbital momentum as well as the resulting spin of all electrons in a completely occupied electron shell is zero, and the same is true for a completely empty shell. Therefore, a state with one electron missing differs from the value zero of the resulting angular momentum in the same way as a state with a single electron in an otherwise empty shell.

Just as for the alkali atoms, every energy state of the principal quantum number n quite generally consists of a group of $2n-1$ more or less closely spaced states. It is evident that the energy state which results from the ionization of the K shell with $n=1$ is a $1S_{\frac{1}{2}}$ level (as is the ground state of the alkali atoms). From an electron missing in the L shell with $n=2$, three atomic states result: $2S_{\frac{1}{2}}$, $2P_{\frac{1}{2}}$, and $2P_{\frac{3}{2}}$; for an electron ionized from the M shell, 5 energy states: $3S_{\frac{1}{2}}$, $3P_{\frac{1}{2}}$, $3P_{\frac{3}{2}}$, $3D_{\frac{3}{2}}$ and $3D_{\frac{5}{2}}$, and so on. Fig. 61 shows a schematic diagram of all levels and the corresponding X-ray transitions without taking into account the actual energy differences of the levels which belong to the same shell. The X-ray transitions follow, in analogy to the optical one-electron spectra, from the selection rule for j

$$\Delta j = 0, \pm 1 . \tag{3.84}$$

The spacing of the levels which belong to different l and j values but to the same principal quantum number n follows from SOMMERFELD's fine-structure

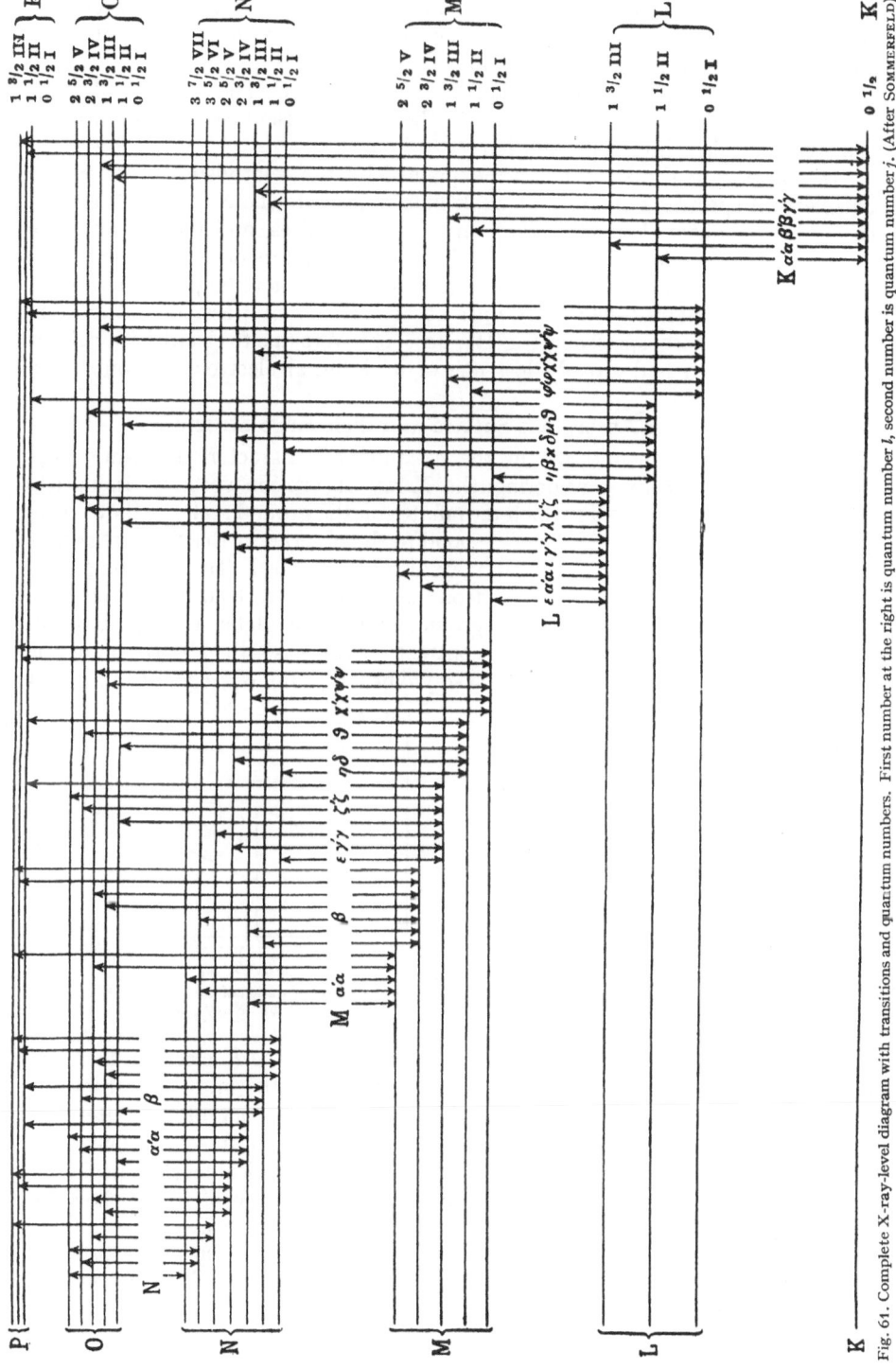

Fig. 61. Complete X-ray-level diagram with transitions and quantum numbers. First number at the right is quantum number l, second number is quantum number j. (After SOMMERFELD)

formula (3.72) (see III, 8 and III, 9c). The relativistic effects on which the formula is based lead to a considerable energy splitting for the X-ray spectra in contrast to the small effect for the Balmer lines. This is due to the fact that the inner electrons of multi-electron atoms move in the scarcely shielded strong electric field of the nucleus; they have velocities comparable to the velocity of light. Their orbits around the nucleus therefore are not closed elliptic orbits but rosette-like orbits with a notable precession of the perihelion. The energies of these orbits of different l differ very much if the various orientation possibilities of the spin are taken into consideration.

In addition to these X-ray lines explained by the relativistic theory of the one-electron spectra, a number of weak lines appears in the X-ray spectra, especially of the lighter elements. These are called satellites, and their significance is not yet entirely clear. They may be due to the simultaneous excitation of two electrons and the existence of electron shells which are not completely occupied. They are occasionally called *spark lines* because it was first assumed that, as in the case of the optical spark spectra (III, 2d), they were due to electron transitions in an atom which had already lost one of its internal electrons by ionization.

d) X-ray Absorption Spectra and their Edge Structure

The processes which result in the emission of characteristic X-ray lines become still clearer by considering the X-ray absorption spectra. In optical spectra, all lines of the series which terminates on the ground state — they correspond to the K series in the X-ray region — may also be absorbed. The absorbing electron then makes a transition from the ground state to a higher energy state. In contrast to this case, the absorption of the X-ray lines K_α and K_β, e.g., of Cu, is not

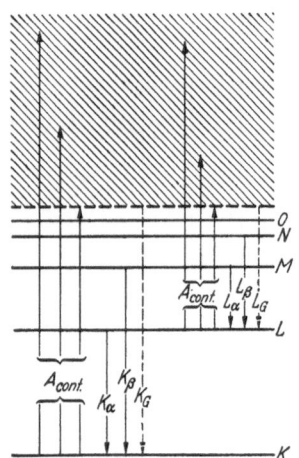

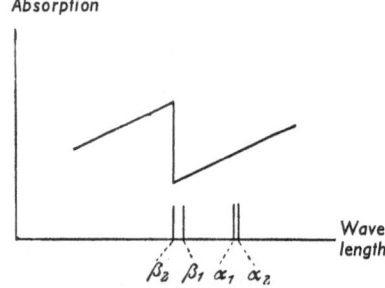

Fig. 62. Energy-level diagram indicating the relation between X-ray emission and absorption spectra

Fig. 63. Schematic representation of the relation between X-ray absorption edge and X-ray emission lines

possible because the L and M shells are filled up and the K electron therefore cannot jump into either of them by absorbing one of the lines. By absorbing radiation, the K electron can only make a transition to one of the outermost unfilled optical levels or into the continuous energy region above the ionization limit. If we consider the fact that the separation of the optical levels is negligible compared to the separation of the X-ray levels, an electron of the K shell can practically absorb only the K-series limit continuum, and an electron of the L series correspondingly only the L-series limit continuum. *Thus we expect and observe X-ray emission lines, but no X-ray absorption lines. The X-ray absorption spectra consist exclusively of the continua beyond the X-ray series, with some structure near their longwave limits, which we shall discuss later.* In the X-ray region, these continua are usually called absorption *edges* and they extend from the ionization limit of the particular levels with decreasing intensity toward shorter waves.

The absorption of a photon of the K edge produces an electron hole in the K shell which makes the emission of K lines possible. The relation between emission and absorption of characteristic X-rays is illustrated by the transition diagram, Fig. 62. *The X-ray lines which are emitted as a result of the absorption of the X-ray edge continuum lie on the long-wave side of the associated absorption edge.* In the case of Ag, for instance, the wavelength of the K absorption edge is 0.482 Å, whereas the shortest-wavelength emission line was measured as 0.485 Å. In Fig. 63 the absorption coefficient is plotted versus the wavelength in the vicinity of the K edge; also plotted are the positions of the K_α and K_β emission lines. The absorption coefficient in-

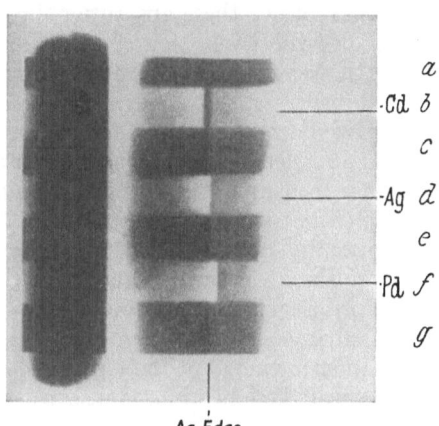

Fig. 64. Photograph of X-ray absorption edges. Detailed discussion in text. (After WAGNER)

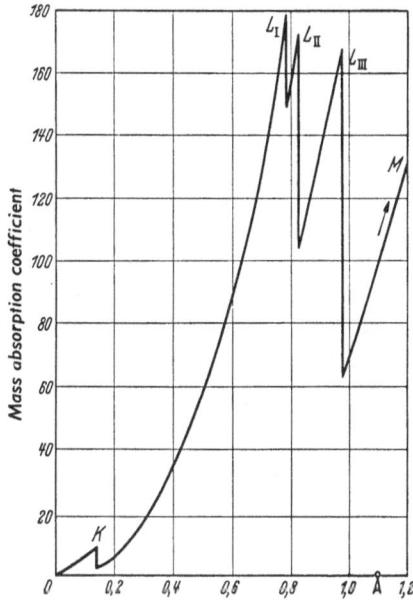

Fig. 65. Absorption coefficient near the K and L X-ray absorption edges of lead

creases rapidly near the wavelength of the ionization limit and gradually decreases toward longer waves. On the long-wave side of the absorption edge, the absorption coefficient is not zero because of the short-wave extension of the L absorption edge.

Fig. 64 is an example of X-ray absorption edges photographed by WAGNER who, simultaneously with M. DE BROGLIE, observed and explained these X-ray absorption edges for the first time. If a continuous X-ray bremsspectrum is photographed on a silver bromide plate, there is a discontinuity in the photographic density at the location of the absorption edges of Ag and Br. Increasing absorption is indicated by increased photographic density. The Ag edge is easily recognized on the records *a*, *c*, *e*, and *g* in Fig. 64. The photographic density decreases with the absorption from the edge toward shorter waves (at the left). At *b*, *d*, and *f*, an absorbing foil of Cd, Ag, and Pd was inserted in front of the photographic plate. These foils absorbed the continuum on the short-wave side of their metal absorption edges. The plate appears less dense at these places. The Ag edge coincides, as is to be expected, with one edge of the AgBr emulsion, whereas the Cd edge appears at the long-wave side and the Pd edge at the short-wave side of the Ag edge.

So far, we discussed the K, L, M, ... edges as if they were single sharp absorption edges. Actually though, they are often found as multiple edges (cf. Fig. 65) and, on photographs taken with sufficient dispersion, they may reveal a very complicated fine structure. The multiplicity of the edges is a consequence of the splitting of the $L, M, N, \ldots$ levels into $3, 5, 7, \ldots$ sublevels, which differ in orbital and total momentum quantum numbers (see Fig. 61). The three L absorption edges of Fig. 65 show clearly that the distances of the three L levels from the ionization limit of the atom are different from each other. The differences of wavelengths, if converted into energy units, give the energy differences between the three L levels. The K edge, of course, is single because there exists only *one* K level.

The fine structure of the absorption edges may be shown more or less easily, depending on the state of aggregation and the chemical structure of the absorbing substance. According to KOSSEL, the fine structure may in principle be understood by considering that in addition to the true photoionization of an inner electron, there is also an excitaticn possible, i.e., lifting the electron to an unoccupied optical level by absorption of a discrete line. Because the optical term differences are so small (only a few ev, compared to the ionization energy of the inner electrons of many thousands of ev) these absorption lines appear as a fine structure of the long-wave side of the absorption edge and are difficult to resolve. The existence of these lines, however has been proved unambiguously by using argon and nitrogen gas as absorbers of soft X-rays.

If the X-rays are absorbed by *solid* matter, we encounter new complications since the optical levels here are disturbed and broadened because of the close packing of the atoms (cf. III, 21). Moreover, since transitions may occur between optical levels of neighboring lattice atoms, the picture becomes very complicated. In addition to this fine structure, X-ray absorption by *solids* reveals a secondary structure which in some cases extends over more than a hundred electron volts. According to KRONIG, this phenomenon is due to the fact that in a solid we have no completely free electrons, as will be discussed in solid-state physics. There exist certain allowed and forbidden regions of kinetic energy of the electrons so that the ionization continuum (absorption edge) near the ionization limit consists of individual energy bands.

We mention in concluding that the exact position of the long-wave edges of the X-ray absorption continua, just as those of the long-wave X-ray lines, depends upon the binding state of the absorbing atom. For example, the Cl edges and Cl lines in Cl_2, HCl, NaCl, $NaClO_4$ have somewhat different wavelengths because of the different bond of the chlorine atom in these compounds. Since the X-ray spectrum originates in the innermost electron shells, its characteristics are determined *essentially* by the atomic number, which is equal to the number of nuclear charges of the corresponding atom. However, there is evidence in the X-ray spectra of the influence also of the outer electron shells of the atom, which determine the chemical behavior of the atom.

11. The Spectra of Many-Electron Atoms.
Multiple Term Systems and Multiple Excitation

We already discussed briefly the spectra of atoms with two outermost electrons. We shall now discuss, in a general way, atoms with several external electrons. Empirically (cf. the Fe spectrum, Fig. 66) two phenomena are evident: an increasing complexity of the spectra concomitant with the gradual disappearance of the Rydberg series, as soon as more than two outer electrons exist, and the appearance of several term systems which do not combine with each other.

The empirical term analysis of the helium spectrum, which originates from the simplest two-electron atom, yielded two term systems (Fig. 67) with the most essential difference that in the system on the right-hand side the ground term $1S$ is missing. The two term systems are completely independent of each other. Lines which correspond to a transition from a level in one system to a level in the other, so-called *intercombination lines*, do not appear. At first, it was believed that the two term systems belong to two different kinds of helium atoms which were called orthohelium and parhelium. We shall see that this is not the correct explanation when we discuss in III, 13 the reason for the two term systems and the rule which forbids intercombinations. Investigation of these spectra with high resolution showed, furthermore, that the energy states of parhelium are single, whereas those of orthohelium consist, with exception of the S states, of three states lying very close together. This different multiplicity of the terms is denoted (as was discussed in III, 9c) by a superscript to the left of the corresponding symbol (^1S, ^3P). Parhelium and orthohelium today are called the singlet and triplet system of the He atom, respectively.

This phenomenon of singlet and triplet series is also found in the spectra of the earth alkali atoms, the metals mercury, cadmium, and zinc, as well as in the spectra of all ions with two outer electrons. It seems to be *typical for all two-electron systems*. It also occurs in the spectra of the two-electron *molecules* such as the H_2 molecule, as we shall see in Chap. VI.

One-electron atoms have a doublet-term system, as we showed in III, 9, and the two-electron atoms have singlet and triplet systems which do not intercombine. For atoms with three outermost electrons, which are found in the third column of the Periodic Table, we again find two term systems which do not intercombine. They are called a doublet system and a quartet system because their levels consist of two and four components, respectively. A four-electron atom has three term systems which do not intercombine, a singlet, a triplet, and a quintet system, whereas a five-electron atom has a doublet, a quartet, and a sextet system. Each atom has only *one* ground state ($n=1$) (for helium 1^1S), which in general belongs to the system of lowest multiplicity.

The multiplicity of terms and the occurence of several term systems which do not intercombine in the spectra of all many-electron

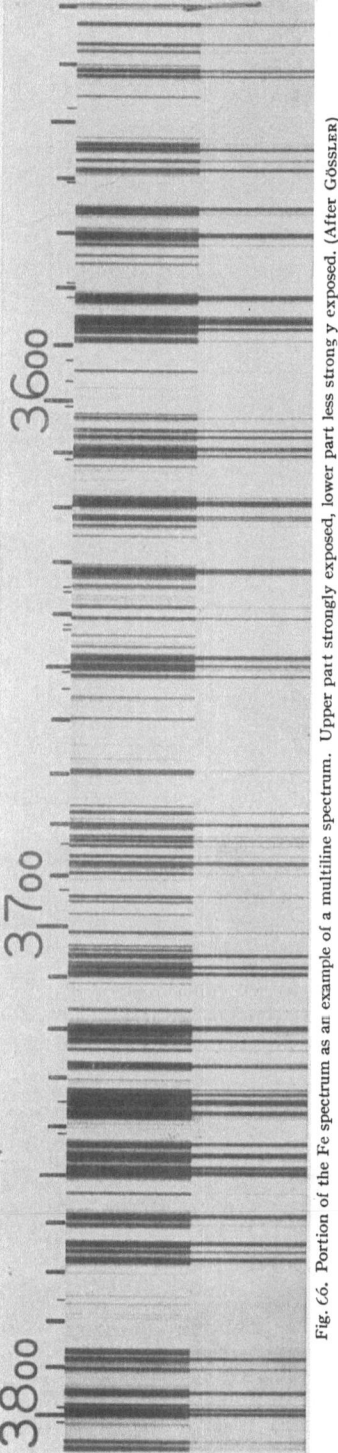

Fig. 66. Portion of the Fe spectrum as an example of a multiline spectrum. Upper part strongly exposed, lower part less strong y exposed. (After Gössler)

atoms thus seems to depend on the number of their outermost electrons. A so-called multiplicity rule has been established from experience: *In going from element to element in the Periodic Table, even and odd multiplicities alternate with*

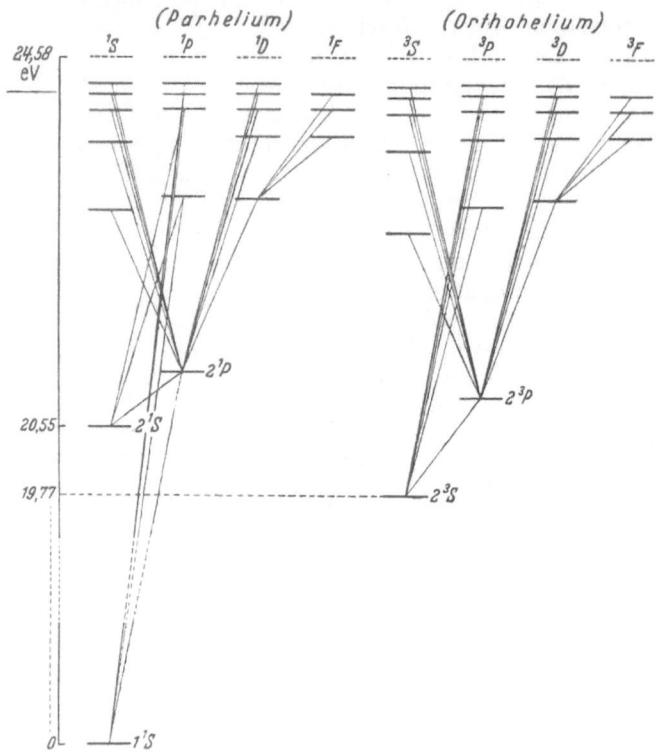

Fig. 67. Energy-level diagram of the helium atom with its two separate systems (singlet and triplet term system)

each other in such a manner that an odd multiplicity is associated with an atom or molecule having an even number of electrons and conversely.

It has also been empirically established that for atoms with the same number of external electrons the complexity of the spectra increases with increasing atomic mass. For example, the spectrum of the light two-electron atom helium is easily resolved into Rydberg series, in contrast to the spectrum of the heavy two-electron atom mercury. This difference is caused by the fact that the spacing of the multiplet levels in the light atoms is so small that it appears only as a fine structure of the spectral lines

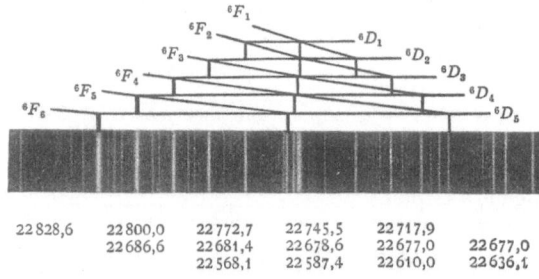

Fig. 68. Sextet in the spectrum of the vanadium atom as an example of a multiplet. (After FRERICHS). Wavelength figures in Å

which is difficult to resolve. On the other hand, the term differences in the heavy atoms with a many-electron core, such as Hg, are so large that transitions between these levels lead to widely separated spectral lines, which make it extremely difficult to recognize the series character of the spectrum which appears very complex. The complexity of the iron spectrum, Fig. 66, is due to the superposition of numerous multiplets. Fig. 68 shows another example, a sextet of vanadium, which gives an impression of the structure of these many-electron spectra.

However, the complexity of the spectra of atoms with several outermost electrons does not depend on the multiplicity alone. Even more important is the fact that in these cases we have no longer one single electron which may be excited and which accounts for the energy states of the atom. Rather, several electrons may be excited at once. The earth-alkali atom Ca with its two outermost electrons provided the first experimental proof of double excitation. In analyzing its spectrum, a particular term series, the P' series, was found which does not fit into the normal term system and which behaves peculiarly in many respects. These P' terms converge, as shown in Fig. 69, toward a term limit which lies above the ionization limit. In order to separate the electron which is responsible for the P' terms from its Ca atom, an amount of energy $\Delta E_i = 1.7$ ev in excess of the ionization energy of the normal atom is required. However, since the ionization of an atom must be independent of the way in which it is produced, this amount of energy can mean nothing else than that after having been ionized by absorption of the P' term series, the Ca$^+$ ion is not in its ground state but in an excited state. Actually, the energy difference ΔE_i is equal to the first excitation potential of the Ca$^+$ ion. Conse-

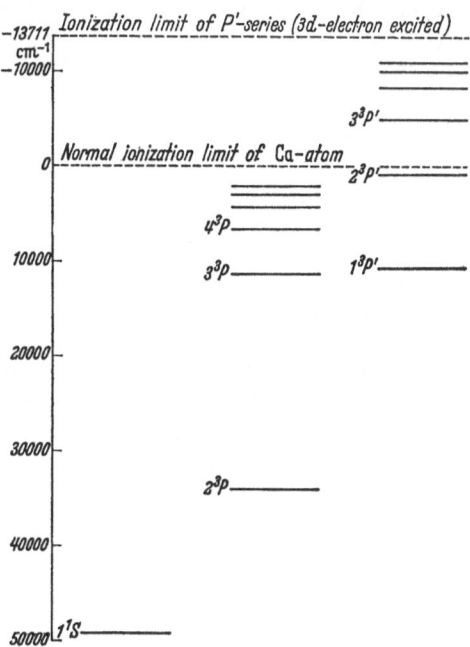

Fig. 69. Energy-level diagram of the Ca atom with P' term sequence of the doubly excited Ca atom converging toward a limit which lies above the normal ionization limit by an amount equal to the excitation energy of the second Ca electron

quently, the P' terms of the earth-alkali atoms correspond to energy states in which one electron is in the first excited state while the second may be excited to any energy state. In the transition from a P' term to an unprimed term *both* electrons change their energy states (in the old conception, they change their orbits). That such a double transition actually occurs, is evidence for the strong coupling of the two electrons.

In the case of atoms with several outermost electrons, this coupling has the effect that the Rydberg series, so prominent in the spectra of hydrogen and the alkali atoms, tend to disappear. When a certain amount of excitation energy is supplied, it seems more probable that several electrons are excited than that one is excited to a very high energy state. Multiple excitation and term multiplicity thus are responsible for the confusing abundance of lines and the complexity of the spectra of many-electron atoms.

12. Systematics of Terms and Term Symbols of Many-Electron Atoms

After this survey of the spectroscopic phenomena in the many-electron spectra, we consider in more detail the *coupling of the orbital momenta of different electrons* of an atom and the character of the terms resulting from this coupling. Here we restrict ourselves to the orbital momenta and shall deal in the next section with the influence of the intrinsic angular momenta of the electrons (spins) and the resulting multiplet structure.

We have characterized each electron by assigning to it a principal quantum number n, and one of the symbols s, p, d, or f corresponding to its orbital quantum number $l=0$, 1, 2, or 3. An energy state of an atom with *one* electron is described sufficiently by the symbols of this electron, which determines the *entire* energy state of the atom. The quantum numbers of the energy states of many-electron atoms may be determined from the spectra, and in some cases from the influence of a magnetic field to be discussed in III, 16. We describe the different possible term characters of an atomic energy state by the capital letters S, P, D, or F. These term characters are determined *by the arrangement and behavior of all the outermost electrons of the many-electron atom.* Theoretically, the term character is determined by the resulting orbital momentum $\vec{L}$ of all electrons. The S, P, D, F terms correspond to the resulting orbital momentum $L=0$, 1, 2 or 3, measured in units of $h/2\pi$. Because the alkali atoms are one-electron atoms, the term quantum number L in this case is identical with the orbital momentum l of the only electron. Therefore, it makes no difference in the case of the one-electron atom whether the atomic state is characterized by the capital letters S, P, D, F, or by the symbol of its single electron s, p, d or f. What is the situation with many-electron atoms, especially those with several electrons in the outermost shell? Evidently the total orbital momentum of the electron shell, $\vec{L}$, is the vector sum of the orbital momenta $\vec{l_i}$ of the individual electrons. By adding vectorially the $\vec{l_i}$, we may disregard the closed electron shells since their resulting orbital momentum is always zero. The theoretical reason will be given in III, 18; it follows empirically from the spectroscopic results that the ground states of all noble-gas atoms (which are characterized by closed electron shells) are ^1S states and thus have a resultant orbital momentum $\vec{L}=0$. We have, consequently, to consider only the orbital momenta $\vec{l_i}$ of the valence electrons in the outer shell. Their vectorial sum, i.e., their resultant $\vec{L}$, depends on the mutual orientation of their orbital planes. Quantum theory requires that not only the individual orbital momenta $\vec{l_i}$ but also the resultant orbital momentum $\vec{L}$ of the atom or electron shell be quantized, i.e., that they be integral multiples of $h/2\pi$. Consequently, not every orientation of the individual electron orbits with respect to each other is possible. They must be so arranged that the resulting momentum is an integer. Two p electrons, each of which has the orbital momentum $\vec{l}=1 \times h/2\pi$, therefore can produce an S, P, or a D term of the atom corresponding to $L=0$, 1, or 2. Fig. 70 shows how the individual two vectors add up to give these three possible atomic states. In the same manner, by adding vectorially the individual orbital momenta of all electrons, the possible atomic terms can be determined for any number of valence electrons.

In order to characterize an atomic state, the symbols of all electrons, or at least of all outer electrons, are written in front of the symbol of the whole atomic state, which in turn has the multiplicity of the state as a superscript on the left. Several equivalent electrons are denoted by writing the corresponding number as a superscript on the right. For example, the symbol for three $2p$ electrons is $2p^3$. The ground state of the helium atom is written $1s^2\,^1S$, a certain excited helium state of the triple system $1s3p\,^3P$. In order to abbreviate, the symbol for the individual electrons is frequently omitted, and only the principal quantum number of the electron with the highest energy is written in front of the term symbol. Thus the last-named helium state is written $3\,^3P$.

The individual electrons of an atom interact electrostatically as well as through the magnetic fields which correspond to their orbital momentum and their spin. As a result of this coupling of the electrons, the entire atom behaves like a system of coupled gyroscopes, since each electron rotating around its own axis and moving in its orbit acts like a small gyroscope. Such a system is subject to the law of conservation of the angular momentum, which requires that the resultant angular momentum remains constant in space with respect to magnitude and direction, if no external forces are acting. The individual orbital momenta of the electrons $\vec{l}_i$, therefore, precess about $\vec{L}$ (Fig. 71). The angular velocity of this precession is the larger the larger the interaction of the individual electrons is. If the interaction of the individual electrons is sufficiently large, the velocity of precession may be of the same order of magnitude as the velocities of the individual electrons in their orbits. In that case, the individual momenta $\vec{l}_i$ lose their physical significance, and only the total orbital momentum $\vec{L}$, which determines the term character, always retains its meaning, that is, it remains exactly defined.

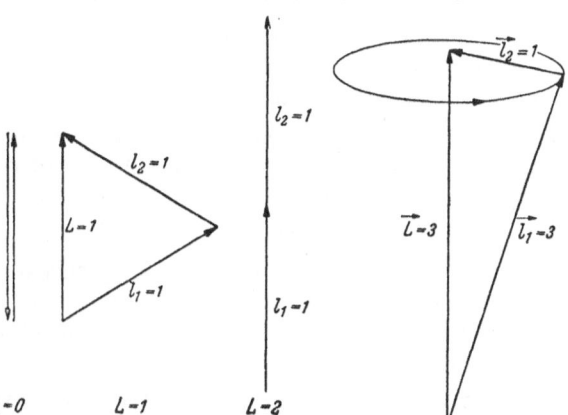

Fig. 70. The three possibilities of vectorial addition of the orbital momenta of two p electrons ($l=1$) forming the resulting orbital momentum L of the whole atom, corresponding to the S, P, and D state of the atom, respectively

Fig. 71. Precession of the orbital momenta $\vec{l}_i$ of the individual electrons about the resulting orbital momentum $\vec{L}$ of the atom

We shall refine this picture in the next section by taking into account the angular momenta of the electrons around their own axes.

We had found the selection rule for transitions of the electron of one-electron atoms to be

$$\Delta l = \pm 1 . \tag{3.67}$$

Under consideration of what we learned about the orbital momenta of the electrons of an atom, this selection rule means that the orbital momentum of the emitting or absorbing electron is changed by one unit $(h/2\pi)$ in a transition. Consequently, it follows from the general law of conservation of the angular momentum which is valid for all isolated systems (i.e., without external momenta acting) that *the photon hv itself which is emitted or absorbed during the transition of an electron from one energy state to another one has an angular momentum of the amount of h/2π.* This selection rule also remains valid for transitions of the excited electron in a many-electron atom. There is, however, an additional selection rule which applies to the resulting orbital quantum number L of the entire atom, which requires

$$\Delta L = 0 \quad \text{or} \quad \pm 1 . \tag{3.85}$$

This means that transitions without a change in L are also possible. Because of Eq. (3.67), these may occur only in the case where transitions of two electrons occur simultaneously so that their orbital momenta change in such a manner that L remains constant. The selection rule $\Delta L = 0$ thus holds only for heavy atoms with strong electron interaction, where several electrons may simultaneously change their energy states.

13. The Influence of the Electron Spin and the Theory of the Multiplets of Many-Electron Atoms

In dealing with one-electron atoms in III, 9, we learned already that the existence of a constant electron spin $\vec{s}$ of magnitude $h/4\pi$ has to be postulated for explaining the multiplicity of term systems, and that the total angular momentum $\vec{j}$ is the vectorial sum of the electron spin $\vec{s}$ and the orbital angular momentum $\vec{l}$ of the emitting electron.

We now investigate this coupling of orbital momentum and spin for many-electron atoms. Depending on the size of interaction of the electrons among themselves on the one hand, and of the momenta $\vec{l}_i$ and $\vec{s}_i$ of each individual electron on the other hand, two limiting cases of coupling may be distinguished. These have to be distinguished in computing the resulting total angular momentum $\vec{J}$ of the electron shell.

We know from the study of atomic spectra that in the majority of all atoms and, without exception, in not too heavy atoms, the orbital and spin momenta of the outer electrons follow the so-called Russell-Saunders LS coupling. In this case, the interaction of all orbital momenta $\vec{l}_i$ of the electrons and that of all spins $\vec{s}_i$ is large compared with the interaction of the orbital momentum and the spin momentum of the individual electrons. Thus, the orbital momenta $\vec{l}_i$ of all outer electrons combine to form the resulting orbital momentum $\vec{L}$ of the atom, which determines the character of the energy state (S, P, D, or F) according to III, 12. On the other hand, all spin momenta $\vec{s}_i = \pm \frac{1}{2}$ of the individual electrons combine vectorially to form a resultant spin momentum $\vec{S}$ of the electron shell, which for an even number of electrons is integral and for an odd number of electrons is half-integral. Finally, $\vec{L}$ and $\vec{S}$ combine vectorially to form a quantized total momentum $\vec{J}$ of the atom, about which $\vec{L}$ and $\vec{S}$ precess because of their gyroscopic properties.

The other coupling case, the so-called jj coupling, predominantly applies to the excited states of the heaviest atoms. Here the interaction of the orbital momentum $\vec{l}$ and the spin $\vec{s}$ of each individual electron is large compared to the interactions of the $\vec{l}_i$ among each other and of the $\vec{s}_i$ among each other. The orbital momentum and the spin of each electron combine therefore to form a total momentum $\vec{j}$ of the electron, and the various $\vec{j}_i$ of the outer electrons in turn combine vectorially to form the resultant quantized total momentum $\vec{J}$, about which they precess. The essential point is that in jj coupling a resultant orbital momentum $\vec{L}$ cannot be defined any more, and there is no possibility of designating the terms by symbols, S, P, D, or F. In the case of jj coupling, a term is defined by its J quantum number only. In the following, we shall discuss details only for the more important case of LS coupling. For details of the jj coupling and the transition from LS coupling to jj coupling the reader is referred to the special spectroscopic texts.

The selection rule for the total quantum number J, which limits the possible optical transitions in many-electron atoms, is the same as that which applies to the total momentum quantum number j of the single electron,

$$\Delta J = 0 \text{ or } \pm 1, \tag{3.86}$$

again with the restriction that the transition $0 \rightarrow 0$ is forbidden. Thus, optical transitions *without* a change of the total angular momentum of the atom are possible only for momentum values which differ from zero.

The spectroscopically observed multiplicity of the various many-electron atoms may be derived theoretically from LS coupling as discussed before. In the case of a two-electron atom, such as helium, we have two possibilities for the orientation of the electron spins; parallel orientation which gives $S=1$, and antiparallel orientation with $S=0$. We shall show that the first case corresponds to the triplet-term system, the second to the singlet-term system of the helium atom. *That intercombinations between two term systems are forbidden, i.e., never occur in the case of light atoms and in the case of heavier atoms only with small intensities (see III, 4), evidently means that a reversal of the electron spin from $+\frac{1}{2}$ to $-\frac{1}{2}$, corresponding to a change of the rotational direction of the electron, is very improbable even if it is combined with a change of other electron properties (quantum numbers).* This understandable difficulty of a change of the term multiplicity accounts for the different shapes of the electron excitation functions which we mentioned on page 65 for excitation of singlet and triplet states from the singlet ground state of the atom. Since a "reversal" of the spin is very unlikely, excitation of the triplet state (spins parallel) from the singlet ground state (spins antiparallel) can occur only if the colliding electron is exchanged for another electron of the atom, one of an opposite spin. This exchange requires a certain time (in other words, a stronger interaction than for a simple energy transfer!) and is therefore possible only if the colliding electron does not have too much kinetic energy: The excitation function of triplet lines, excited from the singlet ground state, decreases very fast beyond the maximum (cf. Fig. 41).

Now we consider in detail the terms corresponding to the values $S=0$ and $S=1$ of the total spin quantum number. We have to distinguish here two different meanings of the letter S. As used by spectroscopists, S denotes as well the total spin quantum number as also the S state of the resulting orbital momentum $\vec{L}=0$ (see III, 12). Since in the case of S terms the resulting orbital momentum of the atom is zero, there is no resulting orbital mangetism either (see III, 15) and therefore no magnetic field with respect to which a resulting magnetic spin (for $S=1$) may orient itself. *Thus, S terms are always single, no matter whether the atom has a resulting spin or not.* However, if one electron of a two-electron atom has an angular momentum $\vec{l}=1$, i.e., if it is a p electron, while the other electron is an s electron, we have $L=1$ and the atom is in a P state. If the spins of the two electrons have opposite directions, i.e., if the resulting spin is $\vec{S}=0$, a spin-magnetic field does not exist so that no orientation of the orbital magnetic momentum is possible. *The total spin $\vec{S}=0$ therefore always leads to singlet terms.* If the spins of the electrons are parallel, so that the resulting spin is $\vec{S}=1$, we have three orientation possibilities for $\vec{L}$ and $\vec{S}$ which result in J values differing by $h/2\pi$. Fig. 72 shows how these three orientations lead to the values 0, 1, and 2 of the resultant total momentum $\vec{J}$, for example of the helium atom. All P states, which would be single states if no electron spin would exist, split up, because of $S=1$, into the three states ${}^3P_0$, ${}^3P_1$, and ${}^3P_2$. The resultant spin $|\vec{S}|=1$ thus leads to a triplet term system. In the same way D terms with $|\vec{L}|=2$, if combined vectorially with $|\vec{S}|=1$, result in the three terms $^3D_1$, $^3D_2$, and $^3D_3$.

The resultant spin $\vec{S}=0$ thus produces a singlet term system with multiplicity 1; the spin $|\vec{S}|=\frac{1}{2}$ (alkali atoms), a doublet term system with multiplicity 2; and the spin $|\vec{S}|=1$, a triplet term system with multiplicity 3. From these examples we conclude that the expression relating the multiplicity of the terms with the total spin quantum number S of the electron shell of an atom may be written

$$\text{Multiplicity} = 2S+1. \qquad (3.87)$$

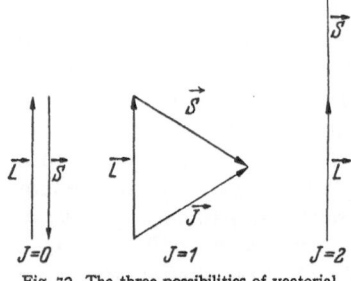

Fig. 72. The three possibilities of vectorial addition of the resulting orbital momentum $|\vec{L}|=1$ and the resulting spin $|\vec{S}|=1$ forming the resulting total momentum $\vec{J}$ of the atom

For three outer electrons we have as possible values of the total spin of the atom $S=\frac{1}{2}$ and $S=\frac{3}{2}$, and consequently doublet and quartet term systems. Atoms with four outer electrons may have the S values 0, 1, and 2 and consequently singlet, triplet, and quintet term systems. For the case of the vanadium atom with five outer electrons, a sextet of which is shown in Fig. 68, the total spin S may have the values $\frac{1}{2}$, $\frac{3}{2}$, and $\frac{5}{2}$, and the spectrum reveals three term systems with doublet, quartet, and sextet terms. From Fig. 73 we see that this highest term multiplicity, six, can be obtained by combining $\vec{L}$ and $\vec{S}$ only if $L \geq 3$, i.e., for F terms. From Fig. 74 it follows that, in agreement with the J

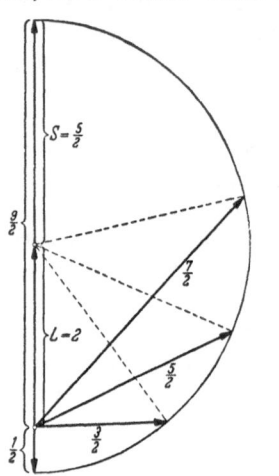

Fig. 73. Vectorial addition of the orbital momentum $|\vec{L}|=2$ and the resulting spin $|\vec{S}|=\frac{5}{2}$. The five possibilities yield the total momentum quantum numbers $J=\frac{1}{2},\frac{3}{2},\frac{5}{2},\frac{7}{2}$, and $\frac{9}{2}$ of the atom

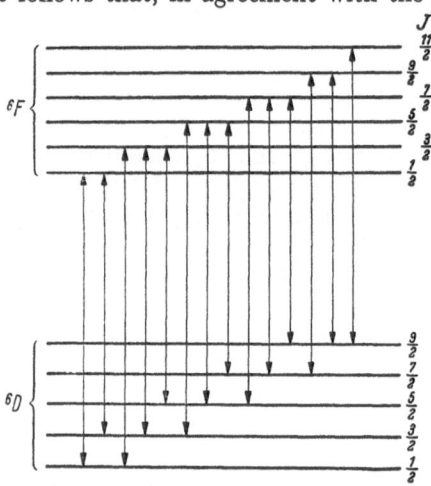

Fig. 74. Energy-level diagram with transitions (schematic) for the vanadium sextet $^6F \rightarrow {}^6D$, shown in Fig. 68

selection rule (3.86), the 14 lines marked in Fig. 68 correspond to the combination $^6F \rightarrow {}^6D$. In Fig. 68 the upper terms are represented as oblique lines and the lower terms by horizontal straight lines, according to MEGGERS; each point of intersection corresponds to a spectral line, as shown in the photograph below.

For the sake of better understanding we repeat the symbols of atomic terms which were introduced in the last sections step by step: The principal quantum number of the term takes the first place in the symbol, e.g., $n=3$ for the arbitrarily chosen term $3\,^2P_{\frac{1}{2}}$. The symbol for the resulting orbital momentum L of the atom (see III, 12) comes next. In our example it is P for the orbital momentum $1 \times h/2\pi$. The superscript at the left indicates the multiplicity which is, according to Eq. (3.87), $(2S+1)$ with S being the value of the resultant spin of the electron shell. In our example, the multiplicity has the value 2 for $S=\frac{1}{2}$ of the alkali atom. Finally, there is the quantum number of the resultant total angular momentum of the atom added as a subscript. It is $\frac{3}{2}$ for our example with the total angular momentum $|\vec{J}|=\frac{3}{2}\,h/2\pi$.

In summarizing, we may state that even the most complicated phenomena in the spectra of many-electron atoms, such as the existence of several non-intercombining term systems and the occurrence of multiplets, which at first could not be understood, have been explained satisfactorily by the introduction of the half-integral electron spin and the vectorial combination of electron spin and electron orbital momenta in forming the quantized total momentum $\vec{J}$ of the atom.

14. Metastable Atoms and their Effects

After the survey of the spectra of many-electron atoms and their explanation which we presented in the last sections, we now briefly discuss an important consequence of the existence of the different term sequences and of term systems which do not intercombine. It is the occurrence of metastable states and their important role in atomic physics.

From the term diagram of the helium atom, Fig. 67, we see that the lowest state of the triplet system, the $2\,{}^3S$ state, lies 19.77 ev above the $1\,{}^1S$ ground state of the atom. A spectral line corresponding to the transition $2\,{}^3S \rightarrow 1\,{}^1S$ has not been observed: *Intercombinations are forbidden.* A He atom in the $2\,{}^3S$ state thus cannot give up its high excitation energy of almost 20 ev by radiation. This energy may be given up in a collision of the second kind, as discussed in III, 6a. We mentioned already that this type of excited states, which do not combine with the ground state, is called *metastable,* in contrast to the normal excited states, which, after about 10^{-8} sec, spontaneously return to the ground state, either directly or in steps. Metastable states cannot be excited from the ground state by absorption of radiation either since this, just as the reverse process of emission, is forbidden also. However, metastable states may be excited by collisions (electron collisions or collisions of the second kind) since there are no sharp selection rules for collisions.

There are other types of metastable states also. In the case of helium, the $2\,{}^1S$ state with its excitation energy of 20.55 ev (see Fig. 67) is metastable, since the selection rule $\Delta l=\pm 1$ forbids the transition $2\,{}^1S \rightarrow 1\,{}^1S$, and a lower P state to which a transition might occur does not exist. In the case of the alkali atoms, according to Fig. 54, from the $2\,{}^2S$ state a transition is possible to the lower $2\,{}^2P$ state, and from this state to the $1\,{}^2S$ state so that alkali atoms have no metastable states. *Quite generally, metastable states are excited states of atoms or molecules which cannot combine, directly or indirectly, with the ground state.*

The rule according to which transitions from a metastable state are forbidden does not hold strictly. All selection rules are valid only with restrictions: In the treatment of electrodynamics, in general only the radiation by electric dipoles is discussed in detail, and electric dipole radiation is not possible from metastable states. However, there are other possibilities of radiation, for example by electric quadrupoles or by magnetic dipoles, and this type of radiation often is possible from metastable states. The intensity of this radiation, however, is several orders of magnitude smaller than that of electric dipoles. Moreover, we mentioned already that most selection rules are modified by an electric field so that forbidden transitions become less improbable if the electric field of electrons and ions around a metastable atom is sufficiently strong. A perturbation of the atom which may be called an internal one in contrast to an external perturbation occurs in heavy many-electron atoms such as mercury. Here the perturbation of the two external electrons by the core with its many electrons is so strong that singlet-triplet intercombinations, which are strongly forbidden for light and undisturbed He atoms, occur with considerable intensity. The well-known Hg line 2537 Å is such an intercombination line $1\,{}^1S \leftrightarrow 2\,{}^3P$.

Such modifications of the selection rules by inner perturbations and the existence of quadrupole and magnetic dipole radiation make the emission of excitation energy possible and lead to emission of forbidden lines if collisions of the second kind are not too frequent. For this reason, the mean life of metastable states is by no means infinite, but it is large compared to the mean life of normal excited states of about 10^{-8} sec. The mean life of metastable atoms in gas discharges is mostly limited by collisions of the second kind (see III, 6a), if investigations are not carried out in a very high vacuum. Only under the most extreme conditions of gaseous and planetary nebulae with their very low gas density, and in highly rarefied atmospheres, such as the upper layers of the earth's atmosphere, it is possible to observe undisturbed radiation from metastable atoms. According to BOWEN, the mysterious "nebulium lines" are due to such forbidden transitions in the ions O^+, O^{++}, and N^+. Also, some long disputed lines of the aurora borealis were explained as being due to similar forbidden transitions, this time within the neutral O atom.

We have been using the term *"metastable atom"* for an atom in the metastable state. This expression has come into use because atoms in the metastable state have large amounts of energy which they can give up in collisions of the second kind, and because, in addition to normal atoms and ions, they are important constituents of each gas discharge (plasma). The mean life of the usual excited states of atoms is so short, 10^{-8} sec, that they occur only in comparatively small concentrations. In most cases, their effects are thus negligible compared with those of metastable atoms. Many mysterious phenomena in noble-gas discharges are now understood as being caused by metastable atoms. Carrying energy, these may diffuse from one region of the discharge tube to another without being influenced by an electric field. Among their effects are the following:

Metastable atoms may ionize other gases mixed with them if their excitation energy is larger than the ionization energy of the other atoms. Metastable He atoms with their high energy of about 20 ev as well as those of the other noble gases are particularly effective when mixed with metal vapors, which all have lower ionization energies. For example, this ionizing effect explains that the breakdown potential of a neon discharge is lowered by adding a small amount of mercury vapor to the neon. The Hg atoms are ionized in collisions of the second kind with metastable neon atoms and act as new charge carriers. Furthermore, if *two* metastable atoms collide, one of them may be ionized while the other returns into the ground state and the remaining energy is transferred into kinetic energy.

If the energy of the metastable atoms is not sufficient to ionize the other atoms, it may still be high enough for exciting them so that we may observe the excitation of gases and vapors as a second effect of metastable atoms. Especially interesting is the possibility of exciting atoms or molecules to a definite energy level which is equal to the energy of the exciting metastable atom.

As a third effect, metastable atoms are able to excite molecules or to dissociate them into their atoms or atom groups (see VI, 7) with a high probability. This is due to the fact that molecules have a large density of energy states (including vibrational and rotational states) so that there is a higher probability for energy resonance between metastable atoms and colliding molecules than between them and other colliding atoms.

A fourth effect of metastable atoms is the release of secondary electrons from metal surfaces (cf. VII, 21 c). Such secondary electrons, which may also be released from metal parts of a noble-gas discharge tube other than the actual electrodes, are responsible for many a mysterious effect in gas-discharge measurements.

We mentioned already that metastability is not limited to atoms but is also found with molecules. Metastable nitrogen molecules are well known because of their role in the mechanism of the surprisingly long afterglow of certain nitrogen discharges, the so-called active nitrogen.

15. The Atomistic Interpretation of the Magnetic Properties of Electrons and Atoms

Magnetism is one of the properties of matter which can be understood only by taking into account atomic properties. Since it is caused by the orbital revolution and the spin of electrons, we discuss it here. Of the three kinds of magnetism — *paramagnetism*, *diamagnetism*, and *ferromagnetism* — the first two are decidedly properties of single atoms, whereas ferromagnetism is a property of crystals and therefore will be treated in VII, 15 c. The fact that iron *atoms* as well as ions of an iron compound in solution show paramagnetism only is evidence that ferromagnetism is not an atomic property. On the other hand, certain mixed crystals of the non-magnetic metals copper and manganese are ferromagnetic.

In experimental physics, often two types of magnetic fields are distinguished: one which is the result of an electric current, and another which has its origin in the magnetic material itself. The results of atomic physics described in the last section leave no doubt that the latter kind of magnetic field, in terms of our pictorial model, arises from electric currents set up by electrons moving in their Bohr orbits or by the rotation of electrons around their own axes, i.e., by the electron spin. The molecular currents, introduced by AMPÈRE for explaining the atomic magnetism, thus found their atomistic interpretation. Whereas ferromagnetism is due to a parallel alignment of the spin momenta of most of the outer electrons of the atoms within a relatively large crystal domain, we must explain diamagnetism and paramagnetism from the structure of the atom itself.

The fundamental facts and definitions are well-known: In the material under investigation, an external field generates a *magnetization* $\mathfrak{P}$ that, in general, is proportional to the field strength $\mathfrak{H}$. The magnetization $\mathfrak{P}$ is defined as the magnetic moment that is induced per unit volume. The constant of proportionality χ of Eq. (3.88), i.e., the magnetic moment per unit volume induced by *unit* field strength is called the *magnetic susceptibility*:

$$\mathfrak{P} = \chi \mathfrak{H}. \tag{3.88}$$

If χ is positive, the induced magnetic moment and the inducing field are parallel; the behavior of the material is then called *paramagnetic* (or *ferromagnetic* if $\chi > 1$). For negative χ, the induced magnetic moment and the inducing field are antiparallel; we call the material *diamagnetic*.

Whether an atom is diamagnetic or paramagnetic, depends, according to the theory of atomic structure, on the arrangement of its electrons and may be seen immediately from the term symbols of the ground state (Table 10 in III, 19). From the singlet ground state of the helium atom, for instance, we see that the two spin momenta of the electrons, and consequently their magnetic moments, are oppositely directed and thus cancel each other. Because the ground state is an S state, the resultant orbital momentum $\vec{L}$, and the corresponding magnetic moment, are zero. Consequently, *the helium atom and all other atoms with a* $^1S_0$ *ground state have no magnetic moment.* The same holds for certain diatomic molecules, such as H_2, which are formed from atoms that have an S ground state. Although the resultant spin moments of the individual atoms differ from zero (for the H atom, $S = \frac{1}{2}$), they compensate each other in the molecule so that the molecule has a singlet ground state.

Thus, in all these cases paramagnetism is not possible. We might at first expect that these atoms and molecules exhibit no magnetic behavior at all, i.e., we might expect a susceptibility $\chi = 0$. That their susceptibility is actually negative and that they are *diamagnetic*, is due to a secondary effect of the external magnetic field which is used to study the magnetic properties of the atoms. If we have two electrons with oppositely directed orbital momenta which would normally cancel, one of them is accelerated in the magnetic field while the other is slowed down. Since the quantum orbits are not changed, a magnetic moment is thus induced in the atom. The direction of this moment is such that it reduces the magnetic field which produces it. This effect of reducing an external field, however, is the characteristic property of diamagnetic materials. This accelerating or decelerating action of the external field on the electrons, which causes the diamagnetic behaviour of materials without magnetic moments, is also present in the case of atoms and molecules *with* magnetic moments, but here the corresponding small reduction of the magnetic moments is only a small secondary effect. The diamagnetic behavior of atoms with several normally compensating orbital momenta is thus explained. That a diamagnetic moment can be induced by an external magnetic field in atoms such as the He atom with its two $1s$ electrons at first sight seems strange since we know that s electrons have an orbital momentum zero (cf. page 84). This difficulty will be explained in the chapter on quantum mechanics. Diamagnetism is thus caused by the induction of a magnetic moment in an atom that itself is non-magnetic. The induction is caused by the external magnetic field applied in the investigation. In contrast to this, paramagnetic atoms and molecules have a magnetic moment even without any external field. In an external magnetic field, however, these magnetic moments are oriented so that a macroscopic magnetic moment of the whole sample results. *This magnetic moment of paramagnetic atoms is due to the revolution of the electrons in their orbits and to their spin.* The so-called magneto-mechanical parallelism enables us to distinguish between these two contributions to atomic magnetism.

We compute the magnetic moment $\mathfrak{M}$ which is produced, according to classical theory, by an electron circling on an orbit of radius r with an angular velocity ω or a linear velocity $v = \omega r$. This revolution is equivalent to an electric current of intensity

$$i = \frac{e\omega}{2\pi} = \frac{ev}{2\pi r}. \tag{3.89}$$

The negative circular current has a magnetic moment of value

$$|\mathfrak{M}| = -\frac{i\pi r^2}{c}. \tag{3.90}$$

Since the mechanical orbital momentum of the revolving electron is given by

$$|\vec{L}| = m_a r^2 \omega, \tag{3.91}$$

we obtain with Eqs. (3.89) and (3.90) the important relation between the me-chanical orbital momentum $\vec{L}$ and the magnetic moment $\mathfrak{M}_L$ of the electron revolving on its orbit:

$$\mathfrak{M}(L) = -\frac{e}{2m_e c}\vec{L}, \tag{3.92}$$

which is the so-called magneto-mechanical parallelism. Thus, the magnetic moment which corresponds to the unit quantum-theoretical angular momentum $h/2\pi$ is

$$\left|\mathfrak{M}\left(\frac{h}{2\pi}\right)\right| = \mu_0 = \frac{eh}{4\pi m_e c}, \tag{3.93}$$

which was introduced in II, 4b as BOHR's magneton.

The corresponding computation for a sphere with radius r_e and electric charge e, rotating around its own axis (classical model of the electron), leads to the same result.

But the phenomena of the Zeeman effect and the result of the Stern-Gerlach experiment, both to be treated in the following section, do not leave any doubt *that the magnetic spin moment $\mathfrak{M}_S$ of the electron which corresponds to the unit mechanical spin $\vec{S}$ is twice the magnetic orbital moment $\mathfrak{M}_L$ which corresponds to the unit orbital momentum $\vec{L}$.* In other words: The magnetic eigenmoment of the electron belonging to the spin quantum number $s=\frac{1}{2}$, in a first approximation, is equal to the magnetic moment which belongs to the quantum number $l=1$ of the orbital angular momentum. Both magnetic moments are equal to one Bohr magneton μ_0 (3.93).

Instead of Eq. (3.92) we thus have Eq. (3.94):

$$\mathfrak{M}(S) = -\frac{e}{m_e c}\vec{S}. \tag{3.94}$$

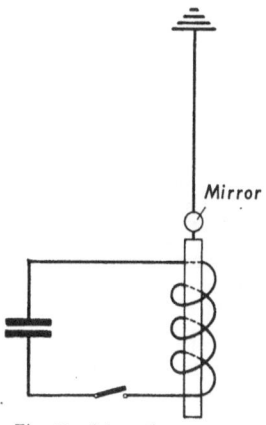

Fig. 75. Schematic arrangement for measuring the Richardson-Einstein-de Haas effect

This fact is called the magneto-mechanical anomaly of the rotating electron. The same result follows from DIRAC's relativistic wave-mechanical theory of the electron.

The relation (3.94) between the mechanical and the magnetic moment can be tested experimentally. We shall show in IV, 15b, when dealing with the magnetism of *solids*, that by the conglomeration of atoms in forming a solid the orbital motion of the electrons is disturbed to such an extent that the orbital magnetism practically disappears and the observed magnetism is almost entirely due to the spin of the electrons. We now apply the conservation of the total angular momentum to an iron rod which consists of magnetic atoms. According to this con-servation law, a change of the mechanical angular momenta of the electrons, caused by a change of the magnetization, produces a change of the resulting angular momentum of the iron rod in the opposite direction. This phenomenon is called, after its discoverers, Richardson-Einstein-de Haas effect. The reverse effect, i.e., a magnetization produced by rotating the iron rod, is called Barnett effect. The first effect can be studied by an arrangement such as in Fig. 75: An iron rod magnetized to saturation is suspended on a thin thread inside a magnet coil. If a current pulse is sent through the coil so that the magnetism of the rod is reversed, the electrons responsible for the magnetism of the iron reverse the direction of their rotation, i.e., their spin changes by $2|\vec{S}|\,h/2\pi$. This change of the electron spin is compensated, according to the law of conservation of the angular momentum, by a recoil rotation of the whole rod, which can be measured by means of the rotating mirror. In this way, the macroscopic magnetic moment of the iron rod is measured simultaneously with the total spin of the electrons which generate this magnetic moment. Since we know from the results of solid-state physics (see VII, 3) how many electrons contributing to the magnetism are contained in the unit volume of iron, such measurements can be used for checking Eq. (3.94).

In the general case of atoms with an orbital angular momentum $\vec{L}$ and a resulting spin $\vec{S}$ of the electron shell, which are vectorially added to give the total angular momentum $\vec{J}$ of the electron shell (see Fig. 76), the magnetic moment consists of contributions due to orbital *and* spin magnetism of the valence electrons. When computing the resulting magnetic moment $\mathfrak{M}$ of such a paramagnetic atom, we must take into account that the directions of the magnetic components $\mathfrak{M}_L$ and $\mathfrak{M}_S$, caused by the revolution or the spin of the electrons, respectively, coincide with $\vec{L}$ or $\vec{S}$, respectively. If the units in Fig. 76 are chosen so that the amount of $\mathfrak{M}_L$ is equal to $|\vec{L}|$, then $\mathfrak{M}_S$ is twice as large as $|\vec{S}|$ because of the magneto-mechanical anomaly of the spin. *Thus, the total magnetic moment $\mathfrak{M}$ of an atom which is vectorially composed from $\mathfrak{M}_L$ and $\mathfrak{M}_S$ does not coincide with the direction of $\vec{J}$.* According to the law of conservation of the angular momentum, the resulting total angular momentum $\vec{J}$ remains constant in time and space. The magnetic moment $\mathfrak{M}$ of the atom therefore precesses around the direction of $\vec{J}$ in such a way that *only its component $\mathfrak{M}_J$ parallel to $\vec{J}$ manifests itself as the magnetic moment of the atom.* Because of Eqs. (3.81) and (3.92) to (3.94), we have the following relations for the magnetic moments due to orbital and spin motion, respectively:

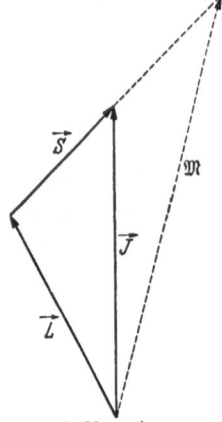

Fig. 76. Magnetic moment and mechanical angular momenta of an atom with orbital and spin magnetism. Since the magnetic moment per unit mechanical spin $\vec{S}$ is twice as large as that per unit $\vec{L}$, the resulting magnetic moment $\mathfrak{M}$ does not coincide with the resulting mechanical angular momentum $\vec{J}$

$$|\mathfrak{M}(L)| = \sqrt{L(L+1)}\,\frac{eh}{4\pi mc},\qquad (3.95)$$

$$|\mathfrak{M}(S)| = 2\sqrt{S(S+1)}\,\frac{eh}{4\pi mc}.\qquad (3.96)$$

We obtain the components of these partial moments in the direction $\vec{J}$ by multiplying Eqs. (3.95) and (3.96) by $\cos(\vec{L}\vec{J})$ and $\cos(\vec{S}\vec{J})$, respectively, and thus can write the effective total moment:

$$|\mathfrak{M}(L, S, J)| = \left[\sqrt{L(L+1)}\,\cos(\vec{L},\vec{J}) + 2\sqrt{S(S+1)}\,\cos(\vec{S},\vec{J})\right]\frac{eh}{4\pi mc}.\quad (3.97)$$

Applying the cosine theorem to the triangle formed by the vectors $\vec{S}, \vec{L}, \vec{J}$ (see Fig. 76), we transform Eq. (3.97) to

$$|\mathfrak{M}(L, S, J)| = \frac{3\,J(J+1)+S(S+1)-L(L+1)}{2\sqrt{J(J+1)}}\,\frac{eh}{4\pi mc}\qquad (3.98)$$

and write this relation as

$$|\mathfrak{M}(L, S, J)| = \sqrt{J(J+1)}\, g(L, S, J)\mu_0, \tag{3.99}$$

with μ_0 representing BOHR's magneton (3.93) and $g(L, S, J)$ being the so-called Landé factor

$$g(L, S, J) = \frac{3\,J(J+1) + S(S+1) - L(L+1)}{2\,J(J+1)}. \tag{3.100}$$

Here we did not take into account, however, that the magnetic moment of the electron is by 0.116% larger than BOHR's magneton (cf. II, 4b).

The ratio

$$\frac{|\mathfrak{M}(L, S, J)|}{\sqrt{J(J+1)}} = g(L, S, J)\,\mu_0 \tag{3.101}$$

is called the *gyromagnetic ratio* of the atom in the state characterized by L, S, and J. For the case of pure orbital magnetism, i.e., $S=0$, J equals L, and from Eq. (3.100) follows $g=1$, which corresponds to the normal gyromagnetic ratio (3.92). On the other hand, for pure spin magnetism ($L=0$) we obtain $J=S$ and the Landé factor $g=2$, corresponding to the magneto-mechanical anomaly of the spin. All other atomic states with both orbital and spin magnetism have g-values different from 1 and 2. Their g-values may be computed, together with the corresponding magnetic moments, from Eq. (3.98) by using the spectroscopically determined quantum numbers L, S, and J.

Although the *magnitude* of the magnetic moment of a paramagnetic atom is given by Eq. (3.98), it is not the entire moment that becomes effective if such atomic magnets are oriented in an external magnetic field. Observations of the directional quantization (see III, 16) agree with quantum mechanics in that the component of every mechanical angular momentum in the direction of an external field be an integral or half-integral multiple of $h/2\pi$ depending on whether J is integer or half-integer, respectively. This means that *the component of any magnetic moment in the field direction is always an integral multiple of* BOHR's *magneton* μ_0. The value $\sqrt{J(J+1)}$, which in general is not integral, has thus always a component M in the field direction that is integral or half-integral just as J is, and the component $\mathfrak{M}_H$ of the magnetic moment (3.99) of an atom in the direction of an external magnetic field is given by

$$|\mathfrak{M}_H(L, S, J)| = M g(L, S, J)\mu_0, \tag{3.102}$$

with M being called the *magnetic quantum number* and representing the integral component of J in the direction of the magnetic field, measured in units of $h/2\pi$. All this will become clearer in the next section.

Let us now consider a paramagnetic gas in an external magnetic field $\mathfrak{H}$. The resulting magnetic moment per unit volume, i.e., the magnetization $\mathfrak{P}$, depends on the competition between the magnetic field favoring parallel orientation and the counteracting thermal random motion. Since the energy differences of the different orientations of the atomic magnets toward the field direction are, in general, small compared to the thermal energy kT, only a small fraction of the paramagnetic atoms is oriented in the normally attainable magnetic fields. Such a gas is thus far from a saturated state of paramagnetism. The magnetization $\mathfrak{P}$ under these conditions is given by CURIE's law

$$|\mathfrak{P}| = \frac{J(J+1)\,g^2\mu_0^2 N}{3kT}\,|\mathfrak{H}|, \tag{3.103}$$

with N being the number of atoms per cm³. Consequently, $|\mathfrak{P}|$ is proportional to the field strength $|\mathfrak{H}|$, and inversely proportional to the absolute temperature T. Atomic physics thus is able to explain the para- and diamagnetism of isolated atoms. The magnetism of solids will be discussed in VII, 17.

16. Atoms in Electric and Magnetic Fields.
Space Quantization and Orientation Quantum Number

So far we have used three quantum numbers for describing the properties of atomic electrons: the principal quantum number n which approximately indicates the energy of the electron state; the orbital momentum quantum number l which, in the Bohr theory, specifies the excentricity of the orbit; and the spin quantum number which indicates the orientation of the spin momentum. Spectroscopic evidence requires the introduction of a fourth quantum number for completely describing the state of an atomic electron. This quantum number, which specifies the orientation of one of the angular momenta ($\vec{l}$ or $\vec{j}$) with respect to an external, or in some cases internal, electric or magnetic field, may be called the *orientation quantum number m* (or M, if it refers to the entire electron shell). Changes in atomic spectra caused by a magnetic field were discovered by ZEEMAN in 1896 and for atoms in an electric field by STARK in 1913. These spectroscopic phenomena, i.e., the splitting and displacement of spectral lines caused by magnetic or electric fields, are called Zeeman effect and Stark effect, respectively.

In both cases we have a precession about the field direction of the small atomic gyroscopes, characterized by the resultant orbital momentum $\vec{J}$. In the Stark effect this is due to a permanent electric moment or due to one which is induced by polarization. In the Zeeman effect it is due to the magnetic orbital and spin moments of the atom. The important point here is that, according to quantum theory, not every angle between $\vec{J}$ and the field direction is possible, but only those angles (orientations) for which the components of $\vec{J}$ in the field direction, called $\vec{M}$, are integral or half-integral multiples of $h/2\pi$, depending on whether J itself is integral or half-integral. This fact, discovered by SOMMERFELD, is called *space quantization or directional quantization.* The difference between Stark and Zeeman effect consists in the type and magnitude of the influence of the electric or magnetic fields on the energy states. *The importance of the Zeeman effect is due to the possibility of empirically determining the quantum numbers L, S, and J of the atomic terms from the splitting of a spectral line in a magnetic field.* The Stark effect was one of the first test cases for the ability of quantum theory to account for a fairly complicated atomic process. The Stark effect is also of considerable importance for the theory of the electrons in molecules.

a) Space Quantization and the Stern-Gerlach Experiment

Let us first consider an electron circling in its orbit around the atomic nucleus (cf. Fig. 77) in such a way that its orbital momentum $\vec{L}$ is inclined by an angle α toward the magnetic field $\mathfrak{H}$. We compute the potential energy $U(\alpha)$, taking the energy for perpendicular orientation of $\vec{L}$ and $\mathfrak{H}$ as zero, i.e., $U(\pi/2)=0$. If $\mathfrak{M}$ designates the magnetic moment of the atom and $\mathfrak{M}_H$ its component in the field direction, we obtain

$$U(\alpha)=-\mathfrak{M}H\cos\alpha=-\mathfrak{M}_H H. \tag{3.104}$$

The magnetic field attempts to turn the axis of the angular momentum into the direction of $\mathfrak{H}$. Since the atom is a gyroscope, its axis is deviated perpendicularly to the acting force so that the atom precesses about the field direction (see Fig. 77) with the so-called Larmor frequency

$$\nu_L = \frac{e}{4\,\pi m\,c}\,|\mathfrak{H}|\,. \qquad (3.105)$$

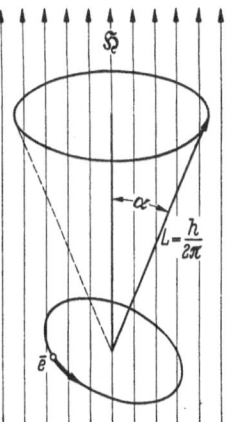

Corresponding considerations apply for the total angular momentum $\vec{J}$ of every electron system that is exposed to central forces only. In this general case (J instead of L), the precession frequency equals the Larmor frequency (3.105) multiplied by the Landé factor (3.100).

The essential point is that according to quantum theory not every angle α between the angular momentum $\vec{J}$ and the direction of the field is possible. Only those angles are "allowed" for which the components $\vec{M}$ of $\vec{J}$ in the field direction differ by an integral multiple of $h/2\pi$. M is half-integral or integral, depending on whether J is so, respectively. Consequently, we define a magnetic or orientation quantum number M by

Fig. 77. Precession of the orbital momentum $\vec{L}$ of an orbiting electron about the direction of a magnetic field

$$|\vec{M}| = M \times h/2\pi. \qquad (3.106)$$

The possible values of M are

$$M = J, J-1, J-2, \ldots, -J \qquad (3.107)$$

so that there are $2J+1$ different values of M.

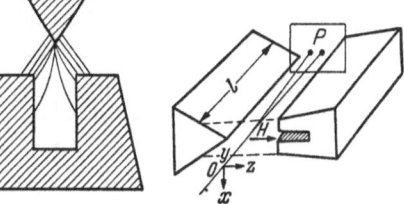

STERN and GERLACH demonstrated this directional quantization by an impressive experiment. They shot a beam of Ag atoms through an *inhomogenous*

Fig. 78. The Stern-Gerlach experiment. At the left, the pole pieces and field lines of the strongly inhomogeneous magnetic field. At the right, the splitting of the atom beam, injected from the front, by the inhomogeneous magnetic field. P = photographic plate

magnetic field (Fig. 78), and observed the deviation of the Ag atoms on the photographic plate P. Since the Ag atom with its ground state $^2S_{\frac{1}{2}}$ has a total angular momentum $J = \frac{1}{2}$, the only possible orientations in the magnetic field are $M = \pm\frac{1}{2}$, with no intermediate positions, according to space quantization. In the Stern-Gerlach experiment, the magnetic field produces not only a quantized orientation of the atomic magnets. Because of the inhomogeneity of the field, the field strength has different values at the two poles of the atomic magnets, and therefore atoms of different spin orientation are moving in different directions.

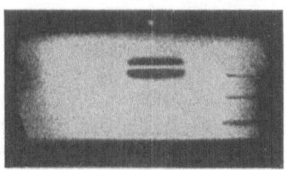

Fig. 79. Photograph of a Stern-Gerlach experiment with lithium atoms, taken by means of an arrangement according to Fig. 78. Separation of an atomic beam into two distinct beams corresponding to the two possible spin orientations. (After TAYLOR)

Without space quantization *all* orientations of $\vec{J}$ with respect to the field would be possible, and the trace of the Ag atoms which impinge on the plate P would be a broad band. In contrast to this, only two separate traces are actually found (see Fig. 79). Space quantization thus causes a splitting of the atomic beam into two beams which correspond to the two possible orientations $M = +\frac{1}{2}$ and $M = -\frac{1}{2}$. Fig. 79 is a photograph of a Stern-Gerlach experiment which demonstrates this space quantization.

b) The Normal Zeeman Effect of Singlet Atoms

According to Eq. (3.104), atoms in a magnetic field have different energies depending on their orientation to the direction of the field. An atomic state which is characterized by the total angular momentum quantum number J therefore splits up into $2J+1$ different energy levels in a magnetic field. The corresponding splitting of the spectral lines is called the Zeeman effect. We know that there are two types of atomic magnetism, the orbital magnetism and the spin magnetism. The behavior of singlet states is comparatively simple. In this case we deal exclusively with the *orbital magnetic moment* of the atom, causing the so-called *normal* Zeeman effect. The complexity of the anomalous Zeeman effect of the non-singlet states, on the other hand, is caused by the magneto-mechanical anomaly of the spin magnetism (treated on page 111), i.e., by the fact that the magnetic spin moment, compared to the value of the mechanical spin, is twice as large as expected classically and that its direction does not coincide (see Fig. 76) with the direction of the mechanical spin.

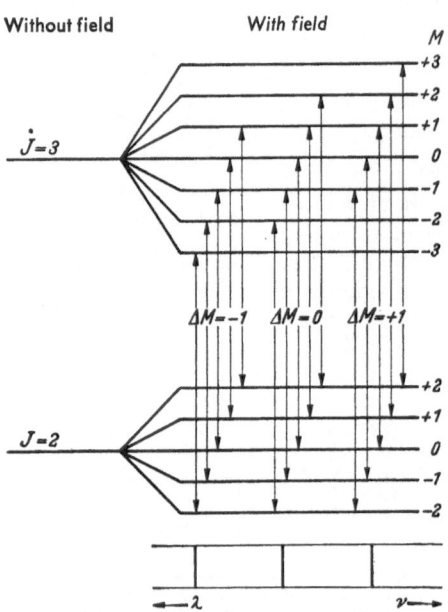

Fig. 80. Term splitting, transitions, and resulting normal Zeeman triplet in the case of the normal Zeeman effect. The transitions of each of the indicated groups coincide and thus form one component of the observed triplet

First we deal with the normal Zeeman effect, i.e., with singlet states for which $S=0$ and $J=L$. Although the total magnetic moment of the atom

$$|\mathfrak{M}(L)|=\sqrt{L(L+1)}\,\mu_0 \quad (3.108)$$

according to Eq. (3.99), is not an integral multiple of the Bohr magneton μ_0, this is true for its maximal component in the direction of an external field [see Eq. (3.102)].

For each of the $2L+1$ possible orientations of $\mathfrak{M}(L)$ in the field $\mathfrak{H}$ we have a different energy value of the space-quantized atom since the potential energy (3.104) in the magnetic field is

$$U=\mathfrak{M}_H H=M\mu_0 H. \quad (3.109)$$

The components $\mathfrak{M}_H$ of neighboring orientations of $\mathfrak{M}$ differ by one unit of the magnetic quantum number M so that the energy difference of neighboring term components in a magnetic field,

$$\Delta E=\mu_0 H, \quad (3.110)$$

is proportional to the magnetic field strength H. The normal Zeeman splitting of neighboring term components ($\Delta M=1$), computed in wave numbers $\bar{\nu}$, is

$$\Delta\bar{\nu}_{\text{norm}} = \frac{\mu_0}{h\,c}\,H = 4.67\times 10^{-5}\,H\,[\text{cm}^{-1}]. \quad (3.111)$$

It can be measured directly by means of high-frequency spectroscopy (see III, 1 a).

Fig. 80 shows the Zeeman splitting of two atomic states combining with each other. Since the selection rule which holds for J also applies to the magnetic quantum number M:

$$\Delta M = 0 \quad \text{or} \quad \pm 1, \tag{3.112}$$

we always find three lines, the so-called *normal Zeeman triplet*, regardless of the number of term components. Evidently this is due to the fact that the spacing of the components in the upper and lower state has the same value so that transitions with the same ΔM coincide in the spectrum. In Fig. 80, these co-inciding transitions are grouped together.

c) The Anomalous Zeeman Effect of Non-Singlet Atoms and the Paschen-Back Effect

The simplicity of the term diagram for the normal Zeeman effect is due to the fact that for singlet states the term separation in the magnetic field is independent of the quantum numbers and, therefore, is the same for each of the two combining states. This, in turn, may be seen from the fact that the Landé factor, which contains the quantum numbers, is missing in the expression (3.110) for the magnetic energy. The magnetic moments of all non-singlet atoms with magnetic orbital *and* spin moments, however, contain the Landé factor $g(L, S, J)$, according to Eq. (3.102), and the same is true for the term separation in a magnetic field. For the Zeeman effect of these non-singlet atoms, which for historical reasons only is called "anomalous", the spectral-line splitting is very complicated since the two states have different quantum numbers L, S, and J. The Zeeman effect splitting of non-singlet states thus is empirically characterized by its large number of components and by variable distances between the components. The latter ones are, however, always rational multiples of the normal Zeeman splitting (RUNGE's rule).

Also in the case of the anomalous Zeeman effect, there is a precession of $\vec{J}$ with its quantized component $\vec{M}$ about the direction of the field. $\vec{J}$ is the vectorial sum of $\vec{L}$ and $\vec{S}$, and this Russell-Saunders coupling is not disturbed if the magnetic field is not too strong. What was mentioned in III, 15 holds for the vectorial addition of the magnetic moments $\mathfrak{M}_L$ and $\mathfrak{M}_S$, which correspond to $\vec{L}$ and $\vec{S}$ respectively, so that the total angular momentum $\vec{J}$ which precesses about the direction of the external field, causes a magnetic moment $\mathfrak{M}_H$ in the field direction:

$$|\mathfrak{M}_H| = M g(L, S, J) \mu_0 . \tag{3.113}$$

A non-singlet atomic state with total momentum quantum number J thus also splits into $2J+1$ term components which are distinguished by their M-values. The energy distance of the components from the undisplaced term and the energy difference of neighboring terms, however,

$$\Delta E = \mu_0 \, g(L, S, J) H \tag{3.114}$$

now depends on the quantum numbers L, S, and J, in contrast to the normal Zeeman effect. These quantum numbers are, in general, not the same for the upper and lower states so that very complicated line patterns with many com-ponents may occur if we take into account the selection rule (3.112). Since LANDÉ's g-factor is always a rational number, the term separations in the case of the anomalous Zeeman effect are also rational multiples [see (3.114)] of the normal splitting (3.111). Thus, RUNGE's rule is theoretically explained.

LANDÉ first derived his g-factor for the anomalous Zeeman effect on the basis of BOHR-SOMMERFELD's quantum theory and found the expression

$$g(L, S, J) = \frac{3J^2 + S^2 - L^2}{2J^2},\qquad (3.115)$$

whereas we learn from quantum mechanics (see IV, 8) that we have to substitute J by $\sqrt{J(J+1)}$. *Only the new expression* (3.100) *for the Landé factor agrees with the*

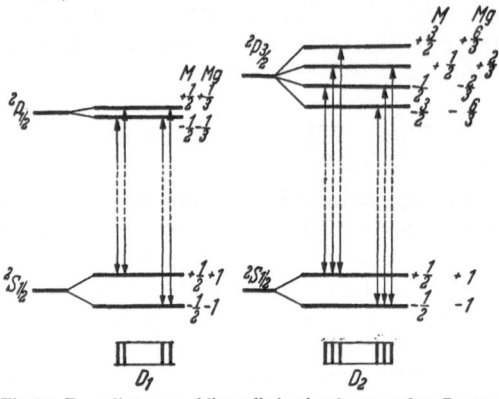

Fig. 81. Term diagram and line splitting for the anomalous Zeeman effect of the two sodium D lines. (After HERZBERG)

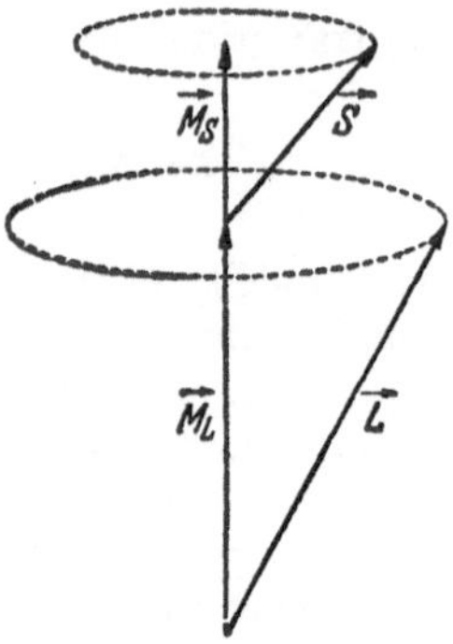

Fig. 82. Vector combination in the case of the Paschen-Back effect

experimental results of the anomalous Zeeman effect, a point in favor of the new quantum theory and against the old one which was presented so far.

Since the term spacing (3.114), *which we find from the line pattern in the anomalous Zeeman effect, depends on the quantum numbers L, S, and J* (3.100), *the anomalous Zeeman effect is a most important aid in determining the quantum numbers of atomic states empirically and, for this reason, is essential for the theoretical term analysis.* Fig. 81 shows the term pattern and the resultant line splitting for the two sodium D lines.

If the magnetic field is so strong that the resulting Zeeman splitting (3.114) is as large as the normal multiplet splitting due to the $L\,S$ interaction (see III, 17), a new phenomenon, found by PASCHEN and BACK, occurs. The Russell-Saunders coupling between $\vec{L}$ and $\vec{S}$ is then uncoupled by the magnetic field, and $\vec{L}$ and $\vec{S}$ no longer precess together about $\vec{J}$, nor $\vec{J}$ about the direction of the field. $\vec{L}$ and $\vec{S}$, having been uncoupled, then precess independently about the direction of the field with the quantized components M_L and M_S, respectively (Fig. 82). When the Paschen-Back effect has developed completely, the term splitting becomes again an integral multiple of the normal splitting (3.111), because M_L, like L, is always an integral number, whereas M_S, like S, may be a half-integer. How-

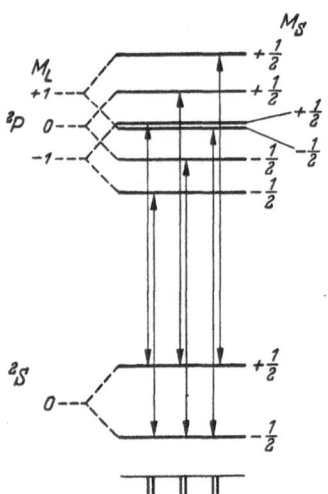

Fig. 83. Term splitting, transitions, and line structure for the Paschen-Back effect of a $^2S \rightarrow {}^2P$ transition. (After HERZBERG)

ever, because of the magnetic anomaly of the spin, we get an integral contribution of $\mathfrak{M}_S$ to the magnetic moment and, consequently, to the formula for $\varDelta E$:

$$\varDelta E = \mu_0 H M_L + 2\mu_0 H M_S.\qquad (3.116)$$

Because of the selection rule (3.112), the Paschen-Back effect thus results in a normal Zeeman triplet, though with an additional fine structure of each component due to the still existing $L\,S$ interaction. Fig. 83 shows an example of the term splitting and the resulting spectrum for the Paschen-Back effect. For transitions between the anomalous Zeeman effect and the Paschen-Back effect, the line patterns become extremely complicated and rather difficult to treat theoretically.

d) The Stark Effect

In the case of the Stark effect, the relations are less simple than in the Zeeman effect because, in general, an atom does not have an electric moment. The Stark effect of the hydrogen atom, and of hydrogen-like atomic states in general, is different from that of all other atomic states, because in the hydrogen case an atomic state of principal quantum number n is n^2-fold degenerate, in so far as terms of the same quantum number n but different orbital quantum numbers l normally coincide (see III, 8). This degeneracy is removed by the electric field, which causes a quantized orientation of the different orbital momenta $\vec{l}$ with respect to the direction of the field. As a consequence of this directional quantization, each hydrogen term splits up symmetrically into $2n-1$ term components. The energy differences between these term components, computed from a theory by SCHWARZSCHILD and EPSTEIN, were found to be in excellent agreement with the results of STARK, and are integral multiples of

$$\Delta E = \frac{3h^2 n}{8\pi^2 m e}\, F \qquad\qquad (3.117)$$

where, in this case, we use F for the electric field strength in order to distinguish it from the energy E. This result can be interpreted as follows: The external electric field disturbs the electrons revolving on orbits of different eccentricity in a different manner, and thus the energy state of quantum number n splits up into terms of different orbital quantum numbers l, analogous to the effect of the core of the alkali atoms. Because the term spacing is proportional to the electric field strength, the Stark effect of the hydrogen lines is called *linear*, in contrast to the more general Stark effect which we shall consider below. The term splitting of the hydrogen states is, according to Eq. (3.117), proportional to the principal quantum number of the atomic state. The explanation of the Stark effect of the Balmer lines with all its details was quite a triumph for the old quantum theory. However, since essential conclusions for atomic physics in general are not derived from it, we do not go into further details.

In the case of all other atoms, such a degeneracy of energy states which might be removed by the electric field does not exist, because the energy states associated with the different orbital quantum numbers are already split up by the field of the atomic core (see III, 8).

The observed Stark effect of the non-hydrogen-like atoms is due to a polarization of the atoms by the electric field, by which an electric dipole moment proportional to the field strength is produced. Just as in the case of the Zeeman effect, the total angular momentum $\vec{J}$ of the atom then precesses about the direction of the electric field, and it is so oriented with respect to the field (space quantization) that its components $\vec{M}$ in the direction of the field differ by integral multiples of $h/2\pi$. The energy displacement of the terms, compared with the field-free case, as well as the energy splitting of the term components which are characterized by different M's depend, as in the Zeeman effect, on the product of the dipole moment and the field strength. As the moment is proportional to the field strength, in this general case *the displacement and splitting of terms and lines are proportional to the square of the field strength. For this reason, we speak of the quadratic Stark effect.* The details of the Stark effect are rather complicated because, in addition to the quantum numbers of the terms, their distances from the neighboring terms play an important role. In very strong fields, in which the Stark effect splitting is large compared to the normal multiplet splitting (or the velocity of precession of $\vec{J}$ about the field direction is large compared to that of $\vec{L}$ and $\vec{S}$ about $\vec{J}$), $\vec{L}$ and $\vec{S}$ are uncoupled in a similar way as in the magnetic case, and we get an electric analogy of the Paschen-Back effect.

All these relationships have comparatively little significance for atomic structure because, in contrast to the Zeeman effect, it is not possible to draw direct conclusions about the quantum numbers of the atoms from their electrical line splitting. However, the Stark effect is important for the theory of the electronic states of the molecules (cf. VI, 5). In diatomic molecules, for instance, the axis joining the two positive nuclei is an electrically distinguished direction about which the orbital momenta of the electrons precess. The Stark effect is also essential for understanding the perturbation of atoms by neighboring electrons and ions (line broadening, cf. III, 21). This perturbation can be regarded as an interatomic Stark effect due to the electric microfields of surrounding electrons and ions which rapidly change in space and time. This plays a considerable role in astrophysics as well as in physics of gas discharges and of high-temperature plasmas. For example, the density of electrons in a plasma can be determined by measuring the Stark-effect broadening of certain atomic lines.

17. Multiplet Term Splitting as a Magnetic Interaction Effect

Now that we know the magnetic properties of electrons and atoms, we can explain the energy splitting of the term multiplets treated in III, 13, i.e., the energy differences between states belonging to different quantum numbers J but to the same quantum numbers L and S, such as the three components 2^3P_0, 2^3P_1, and 2^3P_2 of the He or Hg atoms. This energy splitting is due to the magnetic interaction of orbital and spin momenta of the valence electrons. For a single valence electron, we consider the doublet splitting of the alkali terms (cf. III, 9b). We therefore ask how the energy of an atomic state computed without taking the spin into account is changed by the fact that the electron with its spin is moving in a magnetic field that is produced by its own orbital motion (analogy to the Zeeman effect). We designate the magnetic field produced by the revolving valence electron as $\mathfrak{H}_l$; its direction coincides with that of the orbital angular momentum $\vec{l}$. The magnetic spin moment of the electron has the same direction as the spin $\vec{s}$ and is designated as $\mathfrak{M}_s$. The magnetic interaction energy of orbital momentum and spin of this electron then depends on the angle between these two momenta so that [cf. Fig. 77 and Eq. (3.104)]:

$$E_m(l, s) = |\mathfrak{H}_l| |\mathfrak{M}_s| \cos(\vec{l}, \vec{s}) . \tag{3.118}$$

According to page 111, $\mathfrak{H}_l$ is proportional to the orbital angular momentum $\vec{l}$, and $\mathfrak{M}_s$ proportional to the spin $\vec{s}$. Therefore, instead of Eq. (3.118), we may write

$$E_m(l, s) = a|\vec{l}| |\vec{s}| \cos(\vec{l}, \vec{s}) \tag{3.119}$$

where a is a proportionality factor, which was computed by LANDÉ to be

$$a = \frac{8\pi^4 m e^8 Z_{\text{eff}}^4}{c^2 h^4 n^3 l (l + \frac{1}{2})(l+1)} . \tag{3.120}$$

Here n is the principal quantum number, l is the orbital quantum number, and Z_{eff} the effective nuclear charge acting on the valence electron. The cosine term in Eq. (3.119) can be expressed by the quantum numbers j, l, and s in the same way as in computing the Landé factor with the aid of the cosine theorem. Thus we obtain

$$E_m(l, j, s) = \frac{a}{2} \left[j(j+1) - l(l+1) - s(s+1) \right]. \tag{3.121}$$

The spectroscopically observed energy splitting $\varDelta E$ of the doublet components corresponding to the two possible spin orientations therefore is

$$\varDelta E = \frac{a}{2} \left[j_1(j_1+1) - j_2(j_2+1) \right], \tag{3.122}$$

where j_1 and j_2 are the quantum numbers for the two orientations of the total angular momentum, while l and s are the same for both orientations. From Eq. (3.122) the doublet splittings can be computed; the result is in agreement with spectroscopic observations. Some empirically known rules follow directly from Eq. (3.120). Since the quantum numbers n and l occur in the denominator in the third power, the doublet splitting decreases sharply with increasing principal quantum number and is substantially larger for P terms than for D terms or even F terms. Furthermore, since Z_{eff} increases with increasing atomic number within each group of the Periodic Table (cf. Table 9, in III, 19), the coupling factor a, being proportional to Z_{eff}^4, increases with Z very strongly. We thus have explained the previously mentioned empirical fact (see page 100) that the multiplet splitting quite generally increases strongly from lighter to heavier elements. It is therefore difficult to observe multiplet splitting of lithium or helium. For cesium or mercury, however, the splitting is so large that components of the same multiplet can often hardly be recognized as such since they overlap with the components of other multiplets.

In treating the general case of atoms with more than one valence electron, we remark first that the electrons of the closed shells do not contribute to the interaction energy since their angular momenta as well as their magnetic moments compensate each other. Furthermore, in general we can neglect the interaction between an orbital momentum of one electron and the spin momenta of the others. Therefore, we only have to add the contributions (3.122) for all the valence electrons of the outermost shell. For the normal case of Russell-Saunders coupling, we introduce the resulting orbital, spin, and total momenta $\vec{L}$, $\vec{S}$, and $\vec{J}$ and obtain for the total interaction energy the expression

$$E(J) = \frac{J(J+1) - L(L+1) - S(S+1)}{2} \sum_i a_i \frac{|l_i|}{|L|} \frac{|s_i|}{|S|} \overline{\cos(\vec{l}_i, \vec{L})} \; \overline{\cos(\vec{s}_i, \vec{S})}. \tag{3.123}$$

This is the energy difference between a term characterized by the total momentum $\vec{J}$ and the "center of gravity" of the entire multiplet. Eq. (3.123) permits, for example, to compute directly the splitting within one multiplet by inserting the different possible J values, since L and S remain constant. We thus conclude the treatment of the theory of multiplets, begun in III, 9c and III, 13, which is the most complicated part of the theory of atomic spectra.

18. Pauli's Principle and Closed Electron Shells

From the spectroscopical data as interpreted in the previous sections, significant conclusions may be drawn about the structure of the atom. The classification of the observed atomic spectra in series or multiplets together with the splitting of spectral lines in a magnetic field (Zeeman effect) allows us to determine the quantum numbers of all energy states of an atom. The most remarkable result of this analysis is that the helium atom, as well as every two-electron atom, possesses only *one* ground state, 1^1S (Fig. 67); the 1^3S ground state of the triplet system is missing. In contrast to this, all higher states occur in both the singlet and the triplet system.

The interpretation of this observation was given by PAULI in 1925; it led to the principle named after him. This principle proved to be fundamental for the whole of atomic physics. If the ground state of the helium atom is a $1\,S$ state, the two atomic electrons must have the quantum numbers $n=1$ and $l=m=0$. Such electrons that agree in the three quantum numbers n, l, and m are called *equivalent* electrons. Since the actually observed ground state 1^1S is a singlet state, the spin vectors of the two electrons must have opposite direction with spin quantum numbers $s=+\frac{1}{2}$ and $-\frac{1}{2}$. The electrons in this observed 1^1S ground state of the He atom differ, therefore, in their spin quantum numbers. The *missing* 1^3S state, as a triplet state, would correspond to electrons with parallel orientation of their spins. These two electrons would agree in *all* four quantum numbers n, l, m, and s. The non-occurrence of this energy state led PAULI to assume that quite generally *only such arrangements of electrons in atoms or molecules occur in nature in which the two electrons (or more generally all atomic electrons) differ in at least one of their four quantum numbers.* That the higher triplet states do occur together with the corresponding singlet states (cf. Fig. 67) is obviously due to the fact that one of the two He electrons has a higher principal quantum number n than the other. Therefore the remaining three quantum numbers (including the spin) may well be identical for the two electrons without resulting in identical electrons. A survey of all atomic energy states which result, according to III, 12, from the vectorial addition of the angular momenta of all valence electrons shows that quite generally *only such energy states of atoms occur which result from the vectorial addition of the momenta of non-identical valence electrons.* The whole set of empirical data of spectroscopy is, therefore, in agreement with the Pauli principle, which states that *electrons that agree in all four quantum numbers n, l, m, and s do not occur in any atomic system (atom, molecule, or some larger internally bound complex).* We may roughly illustrate this by saying that, according to our experience, there is no "room" in the atom for electrons identical in all properties (quantum numbers). In IV, 10 we shall come back to the Pauli principle in its much more general quantum-mechanical form. Only then will we fully realize the importance of this fundamental principle for the build-up of matter from elementary particles.

Before dealing, in the next section, with the theoretical interpretation of the Periodic Table of the elements which again reflects the Pauli principle, we consider first the role of the "inner" electrons of the atoms, which constitute the atomic core (see page 80). They were not taken into account in treating the spectra, and this was justified since a survey of the spectra proved that only the outermost electrons (valence electrons) contribute their angular momenta to that of the total electron shell. Thus, according to their spectra, the resulting angular momenta of the alkali atoms are equal to the angular momentum of their *one* valence electron. From this we necessarily infer that the two inner electrons of lithium, the ten of sodium, the 18 of potassium, the 36 of rubidium, etc., are arranged in so-called *closed shells*, and that all angular momenta of electrons arranged in closed shells compensate each other to zero. This conclusion from the empirical spectroscopical data is in best agreement with PAULI's principle. To understand this fact, we first demonstrate that the numbers of non-identical electrons with principal quantum numbers $n = 1, 2, 3$, and 4 are 2, 8, 18, 32, respectively.

The largest orbital momentum of an electron with principal quantum number n is $l_{max} = n-1$ (cf. page 84). This leads, for the different n values, to the possible electrons given in column 3 of Table 8. For every l value, however, there are, according to page 115, $2l+1$ different values of m (column 4) and each of the electrons thus characterized by the values of n, l, and m can exist with the two spin orientations $s = \pm\frac{1}{2}$ (column 5). For the total number of non-identical atomic electrons, we thus have the numbers as shown in column 6 for a given l, and in column 7 for a given n (determining a shell).

Table 8. *Number of non-identical electrons in the individual electron shells*

n	Shell	Possible electrons	Number of different m values	Number of spin orientations	Number of non-identical electrons	
					for given l	for given n
1	K	$1s$	1	2	2	2
2	L	$2s$	1	2	2	$\left.\begin{array}{c} \\ \end{array}\right\}$ 8
		$2p$	3	2	6	
3	M	$3s$	1	2	2	
		$3p$	3	2	6	$\left.\begin{array}{c} \\ \\ \end{array}\right\}$ 18
		$3d$	5	2	10	
4	N	$4s$	1	2	2	
		$4p$	3	2	6	$\left.\begin{array}{c} \\ \\ \\ \end{array}\right\}$ 32
		$4d$	5	2	10	
		$4f$	7	2	14	

According to Table 8, the electron core of the lithium atom consists of the two possible electrons with principal quantum number $n = 1$; that of sodium, of the $2 + 8$ electrons of the two closed shells with $n = 1$, and $n = 2$. The next heavier alkali atoms with the core electron number $2+8+8$ and $2+8+8+18$ do not follow the scheme of Table 8. This points to the existence of closed *sub*-shells in these atoms instead of closed full shells. We return to this question on page 128.

As an example for the compensation of all angular momenta to a resulting zero angular momentum in each closed sub-shell and, therefore, the more so in a closed full shell, we consider the shell of eight electrons with $n = 2$. In this shell, there is first the sub-shell with two $2s$ electrons. Being s electrons, both of them have zero orbital angular momentum and thus also zero components of the orbital momentum in the direction of an external field ($l = m = 0$). Their spin orientations must, however, be opposed to each other so that the spins compensate to zero. All of the remaining six $2p$ electrons forming the second sub-shell of the ($n = 2$)-shell possess the orbital angular momentum $l = 1$, but they are oriented in such a way that we obtain $L = 0$. Viz., two each of the $2p$ electrons (with opposite, compensating spins) have, in the direction of an external field, the components $h/2\pi$, 0 and $-h/2\pi$, corresponding to the directional quantum numbers $m = 1$, 0, -1. This automatically results in a compensation of all angular momenta. It can be shown similarly for all other sub-shells that their resulting total momentum, orbital momentum, and spin are always zero.

19. The Atomistic Interpretation
of the Periodic Table of the Elements

The crowning achievement of the Bohr theory was, undoubtedly, the satis-factory explanation of the Periodic Table. Through it, the chemical properties of the elements which form the basis of the table found their atomistic explanation and foundation.

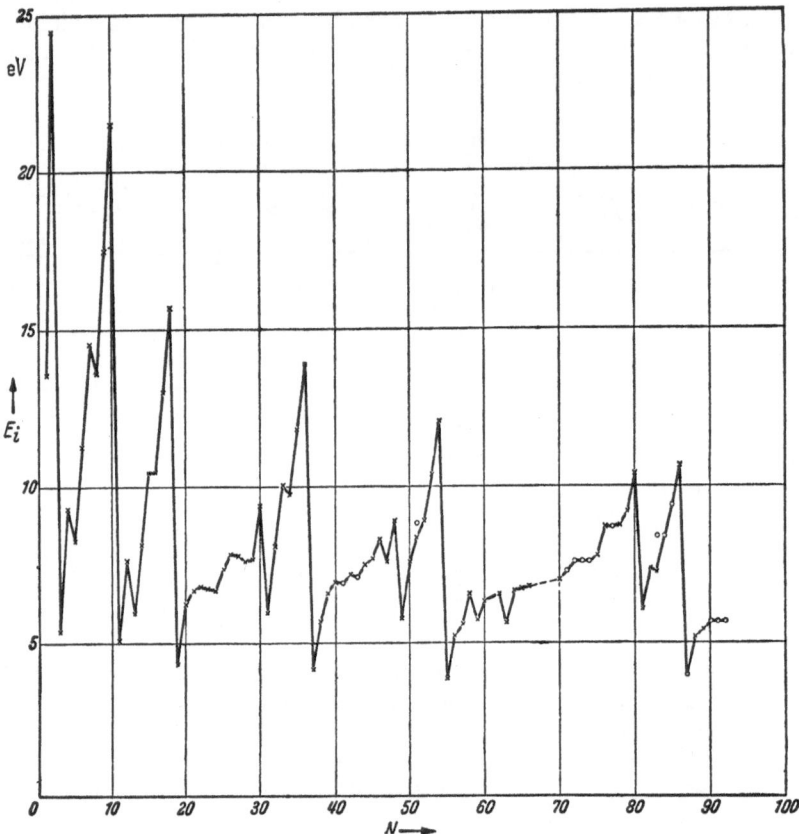

Fig. 84. Ionization potentials of the atoms of the Periodic Table plotted against their atomic numbers

KOSSEL realized already in 1916 that the non-occurrence of X-ray absorption lines (cf. page 96) could be due only to closed electrons shells with a maximal number of electrons not to be exceeded by any means. Their configuration follows from the spectra, especially from X-ray spectra. He succeeded in the following years in proving that the configuration of these atomic electron shells also ac-counts for the periods of the Periodic Table according to the so-called *construction principle, which states that the electron configuration of each atom is derived from that of the preceding atoms by the addition of one more electron.* We have already shown on page 79 that the spectroscopic displacement law of SOMMERFELD and KOSSEL may be regarded as a direct empirical confirmation of this build-up principle. By increasing the nuclear charge in unit steps, and by adding one electron with each increase of the positive charge, the atoms of the entire Periodic Table could be built up from the H atom. Also the change of the ionization energies of the atoms with increasing atomic numbers (see Fig. 84) offers clear evidence in favor of the shell structure of the electrons, which we utilized already when dealing with X-ray spectra. The second electron, which is built into the second atom, helium, has an ionization energy of 24.58 ev and therefore is much more closely bound than that of the H atom. Also the second electron must consequently be situated in the innermost $(n=1)$ K shell, and both electrons are more closely bound to the nucleus because of its double positive charge. However, the third electron which is built into the lithium atom must be at a considerably greater distance from the nucleus and, consequently, must be more loosely bound to it because of its small ionization energy of only 5.39 ev. With lithium the construction of the second shell (L shell) begins. This L shell is completely filled for neon with its second-largest ionization potential of 21.56 ev. We conclude from the smaller ionization potential (5.14 ev) of the next element, sodium, that here the construction of the third shell (M shell) begins. We see that *each period of the Periodic Table corresponds to one electron shell of the atom.* The distinct chemical behavior of the extremely active alkali atoms on the one hand and of the chemically inactive noble gases on the other hand thus is understandable from atomic physics: All alkali atoms have *one* electron outside their closed electron shells, and it is the effect of this electron, the valence electron, which determines the chemical behavior of the atom. The noble gases, however, have closed shells which exert no external forces and thus are not susceptible to chemical binding. We shall return to a more detailed discussion of these problems in VI, 15.

This electron-shell structure, which was first deduced from X-ray spectra and the empirical ionization potentials, will be more quantitatively discussed below. It is in full agreement with the Pauli principle. We note, however, that this principle cannot yet be derived theoretically but was deduced from the totality of the spectroscopic data. The number of non-identical atomic electrons which correspond to the series of principal quantum numbers $n=1, 2, 3, 4$ is exactly equal to the number of elements in the closed periods of the Periodic Table as is shown in Table 8 which was deduced from the Pauli principle. The construction of the entire Periodic Table now follows automatically from the construction principle and Table 8 even though, as we shall see, the order of the consecutive electrons is not in complete agreement with that of Table 8.

First of all, both electrons of the normal He atom are $1s$ electrons. Consequently, each has $n=1$ and $m=l=0$. Therefore, in order to agree with the Pauli principle, they must have different spin directions. Their spin quantum numbers are $+\frac{1}{2}$ and $-\frac{1}{2}$, as is known from the fact that the ground state of the He atom is a singlet state. There is no room in the innermost shell, the K shell, for a third electron since the K shell contains only $1s$ electrons with $l=m=0$. Moreover, because there are only two possible spin orientations (quantum numbers $+\frac{1}{2}$ and $-\frac{1}{2}$), there can be no third electron which is not identical with one of the others with respect to all four quantum numbers. Consequently, it necessarily follows from the Pauli principle that with lithium a new shell must be formed. In a similar manner, the whole Periodic Table can be derived from the Pauli principle.

According to WEIZEL, the electron-shell structure, and with it the Periodic Table, can be more clearly and quantitatively understood by making use of the empirical ionization energies. This also leads to a deeper understanding of the behavior of the atomic electrons.

According to Eq. (3.53) (see page 77) the energy necessary for ionizing a single-electron atom or ion, if the electron is in the shell with number n, is given by

$$E_i = \frac{Z^2 R}{n^2} \tag{3.124}$$

where R is measured in the same units as E_i.

If we want to apply Eq. (3.124) to atoms with more than one electron, we must remember that not the whole nuclear charge $Z e$ acts on the valence electron, but that some part of the nuclear charge is shielded by the inner electrons. Let Z_{eff} be the effective nuclear charge acting on the outermost electron after taking into account the shielding effect. The empirical ionization energy E_i can then be used to compute Z_{eff} from Eq. (3.124),

$$Z_{eff} = n \sqrt{\frac{E_i}{R}}, \qquad (3.125)$$

if we know the quantum number n, i.e., the shell number of the external electron. The effective nuclear charge Z_{eff}, computed from Eq. (3.125), the shielding constants

$$s = Z - Z_{eff}, \qquad (3.126)$$

and the ionization energies E_i in ev are presented in Table 9 for the first 36 elements of the Periodic Table.

Table 9. *Principal Quantum Number of the Emitting Electron, Ionization Potential, Effective Nuclear Charge, and Shielding Constants of the First 36 Elements of the Periodic Table*

Element	Atomic number, $Z = N$	Principal quantum number, n	Ionization energy, E_i [ev]	Effective nuclear charge, $n\sqrt{\frac{E_i}{R}} = Z_{eff}$	Shielding constant, $s = Z - Z_{eff}$	Shielding increment, Δs
H	1	1	13.595	1.00		
He	2	1	24.580	1.35	0.65	0.65
Li	3	2	5.390	1.25	1.75	1.10
Be	4	2	9.320	1.66	2.34	0.59
B	5	2	8.296	1.56	3.44	1.10
C	6	2	11.264	1.82	4.18	0.74
N	7	2	14.54	2.07	4.93	0.75
O	8	2	13.614	2.00	6.00	1.07
F	9	2	17.42	2.26	6.74	0.74
Ne	10	2	21.559	2.52	7.48	0.74
Na	11	3	5.138	1.84	9.16	1.68
Mg	12	3	7.644	2.25	9.75	0.59
Al	13	3	5.984	1.99	11.01	1.26
Si	14	3	8.149	2.32	11.68	0.67
P	15	3	10.55	2.64	12.36	0.68
S	16	3	10.357	2.62	13.38	1.02
Cl	17	3	13.01	2.93	14.07	0.69
A	18	3	15.755	3.23	14.77	0.70
K	19	4	4.339	2.26	16.74	1.97
Ca	20	4	6.111	2.68	17.32	0.58
Sc	21	4	6.56	2.78	18.22	0.90
Ti	22	4	6.83	2.84	19.16	0.94
V	23	4	6.738	2.82	20.18	1.02
Cr	24	4	6.76	2.82	21.18	1.00
Mn	25	4	7.432	2.96	22.04	0.86
Fe	26	4	7.896	3.05	22.95	0.91
Co	27	4	7.86	3.04	23.96	1.01
Ni	28	4	7.633	3.00	25.00	1.04
Cu	29	4	7.723	3.01	25.99	0.99
Zn	30	4	9.391	3.32	26.65	0.69
Ga	31	4	5.97	2.66	28.34	1.66
Ge	32	4	8.13	2.09	28.91	0.57
As	33	4	9.81	3.40	29.60	0.69
Se	34	4	9.750	3.38	30.62	1.02
Br	35	4	11.84	3.73	31.27	0.65
Kr	36	4	13.996	4.06	31.94	0.67

By assuming that the second helium electron fits into the K shell, there follows from its ionization energy of 24.58 ev and from Eq. (3.125) an effective nuclear charge of 1.35 and a shielding of 0.65 nuclear charges. Since the two electrons are in the same shell, i.e., at the same distance from the nucleus with its charge of $+2e$, it might be expected that the order of magnitude of the shielding would be $\frac{1}{2}$. Consequently, the theoretical result of 0.65 appears to be reasonable. We may mention in this connection that the assumption of a "spread-out electron" (to be discussed in the following chapter, cf. IV, 7d and Fig. 103) instead of the point-like electron revolving in Bohr circles is more appropriate for the understanding of the shielding effect. We must pay attention to this fact in what follows.

If we carry out the computations for the third electron of lithium under the assumption that it is included in the innermost $(n=1)$ shell, we obtain, because of the small ionization potential of lithium, an effective nuclear charge of 0.63. Under these circumstances, the two electrons already in the K shell must shield 2.37 of the 3 positive nuclear charges, a completely unreasonable conclusion. We therefore conclude again that it is not possible for a third electron to be in the K shell. On the other hand, if we compute Z_{eff} for the third electron of lithium under the assumption that it is in the next higher shell, the L shell with $n=2$, we get from Eq. (3.125) the value $Z_{eff}=1.25$. In this case each of the two inner electrons of the K shell shields the third electron in the outer shell by 0.87 nuclear charges. This is a very reasonable result. In the case of beryllium we expect a shielding of 0.87 from each of the two inner K-electrons; and a shielding of 0.65, as for helium, for the third $(2s)$ electron. Together the three electrons should provide a shielding of 2.39 as compared with 2.34 obtained by computing the shielding from Eqs. (3.125) and (3.126). The agreement is excellent and indicates that the fourth electron is also a $2s$ electron. In the case of the fifth atom, boron, the shielding is increased by 1.10 according to Eqs. (3.125) and (3.126)[1]. Thus the fifth electron cannot be a $2s$ electron, since then it would shield only 0.65 nuclear charges. It must be situated farther away from the nucleus. Nor can it be an $n=3$ electron, since for this electron the shielding would be about 3.7 instead of 3.44. So we conclude that the fifth electron of boron is built in in such a way that in the time average it is farther away from the nucleus than a $2s$ electron but is closer than a $3s$ electron. According to the quantum-mechanical results (see IV, 7d) this is the expected behavior of a $2p$ electron. The conclusion that the first $2p$ electron appears in boron is in agreement with the spectroscopic result (see Table 10) that the ground state of boron is a $^2P_{\frac{1}{2}}$ state. The next two electrons, found in the atoms C and N, are also $2p$ electrons. In agreement with spectroscopic results (increasing multiplicity of the ground state) they are p electrons with their spins in the same direction, but they have different orientation quantum numbers m. Three additional $2p$ electrons are built into the atoms O, F, and Ne, whose spin is directed oppositely to that of the electrons added to the atoms B, C, and N. This follows from the decreasing multiplicity of the ground state and the jump in the shielding constant in the transition from N to O (due to the addition of the first electron with oppositely directed spin). With neon the second electron shell, the L shell, is completed; there is no possibility of adding another electron of principal quantum number 2 which is not identical with another electron with respect to all four quantum numbers. All together the orbital and spin momenta

---

[1] The author has shown that by introducing the shielding increment Δs in the Periodic Table very informative regularities are obtained which make it possible to extrapolate unknown Δs values and to use Eqs. (3.125), (3.126), and (3.124) for determining unknown ionization potentials.

of the electrons compensate each other; neon has the $^1S_0$ ground state of the chemically inactive noble gases.

The large change in the shielding constant by going from neon to the first element of the third period of the Periodic Table, sodium, indicates that here the construction of a new shell begins. The ground state $^2S_{\frac{1}{2}}$ tells us that the first electron included in this shell is a $3s$ electron. The further building up of the shell to the next noble gas, argon, corresponds exactly to the second period. The third shell contains a second $3s$ electron (of opposite spin) and six $3p$ electrons. In a preliminary way, with argon the third electron shell (M shell) is filled. We say *preliminary* because we have not yet made use of the fact that there is still room for $3d$ electrons in the ($n=3$) shell. In fact, because of the five different possibilities of orientation (quantum numbers m) of the orbital momentum, each with two spin directions (Table 9), there may be altogether ten $3d$ electrons. However, we conclude from the shielding numbers and the $^2S_{\frac{1}{2}}$ ground state that, before the ten $3d$ electrons appear, the construction of the fourth quantum shell ($n=4$) has started with the alkali atom potassium, having one outer $4s$ electron, and calcium with a second $4s$ electron. For the next element, scandium, Sc, the construction does not go on, as it did in the case of the other periods or shells, by including a p electron. This can be seen from the small increase in the shielding. Moreover, the shielding evidence tells us that here an electron is added farther inside. For this, only one of the $3d$ electrons is available, in agreement with the spectroscopic result that scandium has a $^2D_{\frac{3}{2}}$ ground state. Up to nickel, seven more $3d$ electrons are added. Then there follows an especially interesting process in the case of copper. By adding the ninth $3d$ electron we would get a bivalent Cu atom with two outermost $4s$ electrons, which, except for one missing $3d$ electron, has its three inner shells completed. This electron configuration, however, is not stable. One of the two external $4s$ electrons changes over to the free $3d$ position, whereby the number of external valence electrons, and thus the chemical valency of the atom, decreases by unity. This state, the $^2S_{\frac{1}{2}}$ state, is the ground state of copper, whose doublet character shows that there is only one external electron, while the S term shows that the number of d electrons must be completed since there is no resultant orbital momentum. However, the Cu state mentioned first, with the two external $4s$ electrons and the one missing $3d$ electron, is energetically so close to the ground state that it is reached from the ground state under the influence of very small external forces. Consequently, from this theoretical evidence we expect that copper may possibly have two valencies. In fact, chemists have long known that copper can be mono- or bivalent. The theory of the Periodic Table thus easily explains even detail features of the chemical behavior of individual atoms. Referring back to page 79, we want to point out that for such an irregular arrangement of the electron shells, the spectroscopic displacement law cannot remain valid. The Cu$^+$ ion, resulting from the ionization of the copper atom, because of the altered electron arrangement does not have a term diagram and spectrum similar to those of the preceding atom, nickel. However, this exception to the spectroscopic displacement law is not only understandable from atomic theory, but it is a necessary consequence of it.

In the case of zinc the second $4s$ electron, which in copper had been drawn into the d shell, is added. For the further six elements up to krypton the six $4p$ electrons are added, by which the fourth shell is closed in a preliminary form. There are lacking, however, not only the ten $4d$ electrons but also the fourteen f electrons which should appear for the first time in the ($n=4$) shell according to Table 8. With the elements rubidium and strontium, the construction of the ($n=5$) shell, O shell, begins with two $5s$ electrons, whereupon, as in the preceding period, the inclusion of the ten $4d$ electrons begins with the third element, yttrium. The construction now proceeds somewhat differently than in the fourth period, since after the inclusion of the third $4d$ electron in niobium, Nb, one of the external $5s$ electrons is drawn into the d shell and thus Nb, like the following elements, is monovalent. After the inclusion of the ninth $4d$ electron in palladium, the last $5s$ electron is used to fill the d shell, as shown by the ground term $^1S_0$ of Pd. Thus the particularly noble, i.e., chemically inactive, character of palladium is explained by the atomic theory. For the following eight elements from silver to xenon, the two $5s$ electrons are again included, and with the inclusion of six $5p$ electrons a preliminary completion of the fifth shell is obtained, as indicated by the noble-gas character of xenon.

In the large sixth period, which begins with the alkali metal cesium and ends with the noble gas radon=emanation, Rn, the behavior is especially complicated. However, even here the interpretation is in excellent agreement with the chemical behavior. In the case of the first elements, Cs and Ba, the sixth shell is started with the inclusion of the $6s$ electrons. With the next element, lanthanum, as at the corresponding place in the fifth period, the supplementary addition of the $5d$ electrons begins. However, with the next element, cerium, there begins a series of elements which are very much alike chemically and which therefore are difficult to isolate, the rare earths. They are built up by the supplementary inclusion of the $4f$ electrons which we mentioned above. Since these are built into the $(n=4)$ shell, beyond which the $(n=5)$ and $(n=6)$ shells are already partly completed, these latter shells determine the chemical behavior of the atoms, which is not noticeably changed any more by the successive inclusion of the inner $4f$ electrons. This chemical similarity and the difficulty of separation of the rare-earth atoms could not be explained from a purely chemical point of view. However, this explanation follows as an evident consequence from the electron shell structure. The $4f$ shell is completed in ytterbium, and in the following elements up to platinum the $5d$ electrons are included along with some of the external $6s$ electrons. From gold to radon the two $6s$ electrons and six $6p$ electrons are built in as usual. For the last elements of the Periodic Table, the construction follows that of the preceding period. The construction of the $(n=7)$ shell begins with the two $7s$ electrons (Fr and Ra); then in actinium (as in La of the preceding shell) a $6d$ electron is included. In the same way as in the preceding period the rare earths, also called lanthanides, follow lanthanum, incorporating inner $4f$ electrons, so is Ac followed by the last 14 elements of the Periodic Table, correspondingly called actinides (including the recently discovered "transuranium elements", neptunium, plutonium, americium, curium, berkelium, californium, einsteinium, fermium, mendelevium, nobelium, and lawrencium), in which inner $5f$ electrons are built in. Chemical and magnetic evidence, however, favors the assumption that, at least for Th, Pa, and U, these elements in the bound state do *not* have $5f$ electrons but additional $6d$ electrons. Only the spectroscopical investigation of the free atoms will finally settle this problem.

The step-wise completion of the atomic electron shells described above and, correspondingly, of the Periodic Table is indicated in Fig. 85. We must note, though, that it is impossible here to take into account the irregularities that do not obey the construction principle. The most important data on atoms, the atomic number, the term symbol of the ground state, and the arrangement of the electrons in the different shells, are shown in Table 10.

Let us consider now what has been achieved and what remains to be done. The fundamental idea that the periodic structure of the Periodic Table corresponds to the electron-shell structure of the atoms was derived from the construction principle which itself was suggested by the spectroscopic displacement law. On this basis, we were able to explain the Periodic Table and the chemical properties

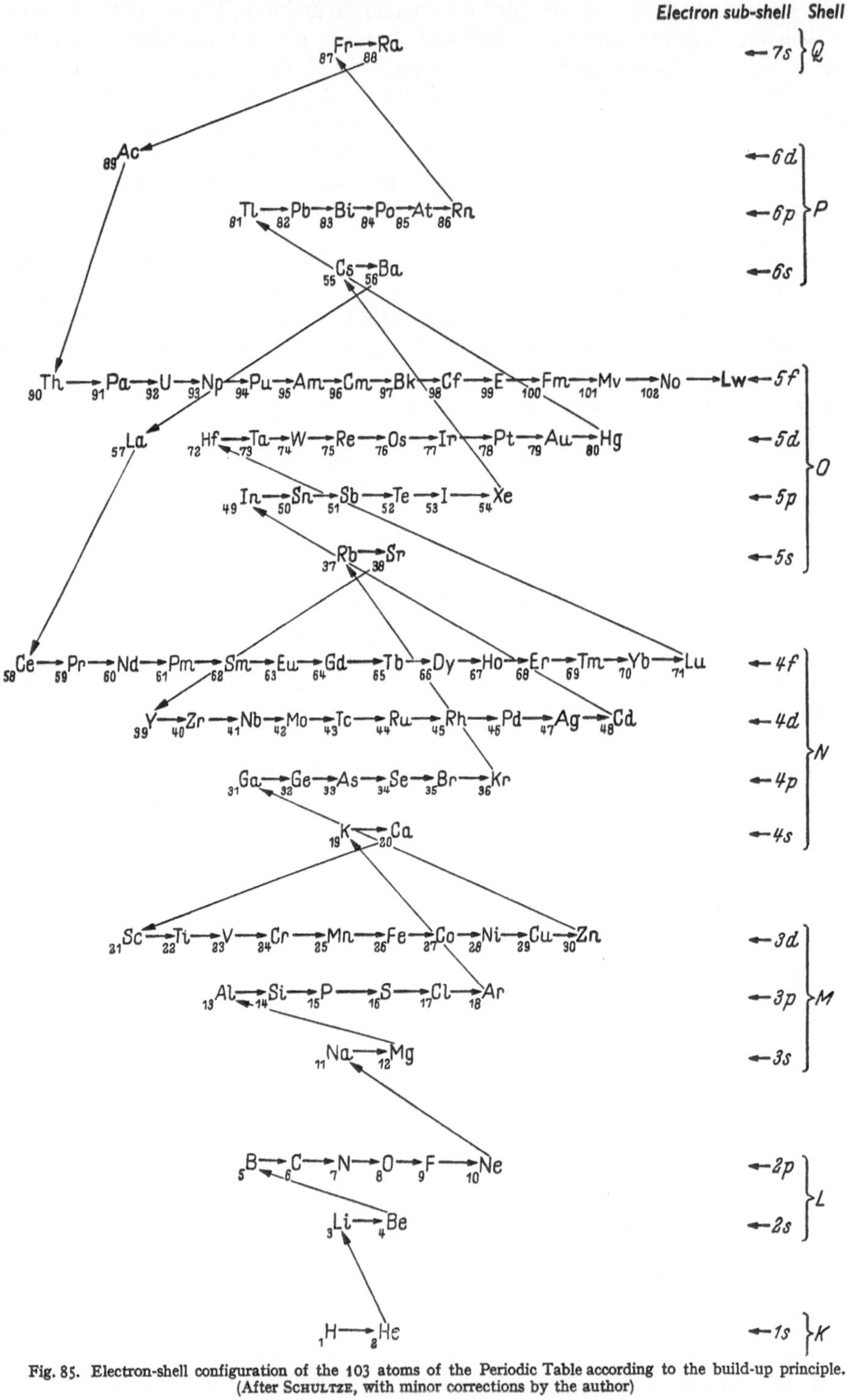

Fig. 85. Electron-shell configuration of the 103 atoms of the Periodic Table according to the build-up principle. (After SCHULTZE, with minor corrections by the author)

Table 10. *Ground States and Electron Configurations of the Elements*

Z			K	L	M	N	O	P	Q
			1s	2s 2p	3s 3p 3d	4s 4p 4d 4f	5s 5p 5d 5f	6s 6p 6d	7s
1	H	$^2S_{1/2}$	1						
2	He	$^1S_0$	2						
3	Li	$^2S_{1/2}$	2	1					
4	Be	$^1S_0$	2	2					
5	B	$^2P_{1/2}$	2	2 1					
6	C	$^3P_0$	2	2 2					
7	N	$^4S_{3/2}$	2	2 3					
8	O	$^3P_2$	2	2 4					
9	F	$^2P_{3/2}$	2	2 5					
10	Ne	$^1S_0$	2	2 6					
11	Na	$^2S_{1/2}$	2	2 6	1				
12	Mg	$^1S_0$	2	2 6	2				
13	Al	$^2P_{1/2}$	2	2 6	2 1				
14	Si	$^3P_0$	2	2 6	2 2				
15	P	$^4S_{3/2}$	2	2 6	2 3				
16	S	$^3P_2$	2	2 6	2 4				
17	Cl	$^2P_{3/2}$	2	2 6	2 5				
18	Ar	$^1S_0$	2	2 6	2 6				
19	K	$^2S_{1/2}$	2	2 6	2 6	1			
20	Ca	$^1S_0$	2	2 6	2 6	2			
21	Sc	$^2D_{3/2}$	2	2 6	2 6 1	2			
22	Ti	$^3F_2$	2	2 6	2 6 2	2			
23	V	$^4F_{3/2}$	2	2 6	2 6 3	2			
24	Cr	$^7S_3$	2	2 6	2 6 5	1			
25	Mn	$^6S_{5/2}$	2	2 6	2 6 5	2			
26	Fe	$^5D_4$	2	2 6	2 6 6	2			
27	Co	$^4F_{9/2}$	2	2 6	2 6 7	2			
28	Ni	$^3F_4$	2	2 6	2 6 8	2			
29	Cu	$^2S_{1/2}$	2	2 6	2 6 10	1			
30	Zn	$^1S_0$	2	2 6	2 6 10	2			
31	Ga	$^2P_{1/2}$	2	2 6	2 6 10	2 1			
32	Ge	$^3P_0$	2	2 6	2 6 10	2 2			
33	As	$^4S_{3/2}$	2	2 6	2 6 10	2 3			
34	Se	$^3P_2$	2	2 6	2 6 10	2 4			
35	Br	$^2P_{3/2}$	2	2 6	2 6 10	2 5			
36	Kr	$^1S_0$	2	2 6	2 6 10	2 6			
37	Rb	$^2S_{1/2}$	2	2 6	2 6 10	2 6	1		
38	Sr	$^1S_0$	2	2 6	2 6 10	2 6	2		
39	Y	$^2D_{3/2}$	2	2 6	2 6 10	2 6 1	2		
40	Zr	$^3F_2$	2	2 6	2 6 10	2 6 2	2		
41	Nb	$^6D_{1/2}$	2	2 6	2 6 10	2 6 4	1		
42	Mo	$^7S_3$	2	2 6	2 6 10	2 6 5	1		
43	Tc	$^6S_{5/2}$	2	2 6	2 6 10	2 6 5	2		
44	Ru	$^5F_5$	2	2 6	2 6 10	2 6 7	1		
45	Rh	$^4F_{9/2}$	2	2 6	2 6 10	2 6 8	1		
46	Pd	$^1S_0$	2	2 6	2 6 10	2 6 10			
47	Ag	$^2S_{1/2}$	2	2 6	2 6 10	2 6 10	1		
48	Cd	$^1S_0$	2	2 6	2 6 10	2 6 10	2		
49	In	$^2P_{1/2}$	2	2 6	2 6 10	2 6 10	2 1		
50	Sn	$^3P_0$	2	2 6	2 6 10	2 6 10	2 2		
51	Sb	$^4S_{3/2}$	2	2 6	2 6 10	2 6 10	2 3		
52	Te	$^3P_2$	2	2 6	2 6 10	2 6 10	2 4		
53	I	$^2P_{3/2}$	2	2 6	2 6 10	2 6 10	2 5		
54	Xe	$^1S_0$	2	2 6	2 6 10	2 6 10	2 6		

Table 10 (continued)

Z			K	L	M	N	O	P	Q
			1s	2s 2p	3s 3p 3d	4s 4p 4d 4f	5s 5p 5d 5f	6s 6p 6d	7s
55	Cs	$^2S_{1/2}$	2	2 6	2 6 10	2 6 10	2 6	1	
56	Ba	$^1S_0$	2	2 6	2 6 10	2 6 10	2 6	2	
57	La	$^2D_{3/2}$	2	2 6	2 6 10	2 6 10	2 6 1	2	
58	Ce	$(^3H_4)$	2	2 6	2 6 10	2 6 10 1	2 6 1	2	
59	Pr	$^4I_{9/2}$	2	2 6	2 6 10	2 6 10 3	2 6	2	
60	Nd	$^5I_4$	2	2 6	2 6 10	2 6 10 4	2 6	2	
61	Pm	—	2	2 6	2 6 10	2 6 10 5	2 6	2 ?	
62	Sm	$^7F_0$	2	2 6	2 6 10	2 6 10 6	2 6	2	
63	Eu	$^8S_{7/2}$	2	2 6	2 6 10	2 6 10 7	2 6	2	
64	Gd	^9D	2	2 6	2 6 10	2 6 10 7	2 6 1	2	
65	Tb	—	2	2 6	2 6 10	2 6 10 9	2 6	2	
66	Dy	$^5I_8$	2	2 6	2 6 10	2 6 10 10	2 6	2	
67	Ho	$^4I_{15/2}$	2	2 6	2 6 10	2 6 10 11	2 6	2	
68	Er	$^3H_6$	2	2 6	2 6 10	2 6 10 12	2 6	2	
69	Tm	$^2F_{7/2}$	2	2 6	2 6 10	2 6 10 13	2 6	2	
70	Yb	$^1S_0$	2	2 6	2 6 10	2 6 10 14	2 6	2	
71	Lu	$^2D_{3/2}$	2	2 6	2 6 10	2 6 10 14	2 6 1	2	
72	Hf	$^3F_2$	2	2 6	2 6 10	2 6 10 14	2 6 2	2	
73	Ta	$^4F_{3/2}$	2	2 6	2 6 10	2 6 10 14	2 6 3	2	
74	W	$^5D_0$	2	2 6	2 6 10	2 6 10 14	2 6 4	2	
75	Re	$^6S_{5/2}$	2	2 6	2 6 10	2 6 10 14	2 6 5	2	
76	Os	$^5D_4$	2	2 6	2 6 10	2 6 10 14	2 6 6	2	
77	Ir	^4F	2	2 6	2 6 10	2 6 10 14	2 6 7	2	
78	Pt	(^3D)	2	2 6	2 6 10	2 6 10 14	2 6 9	1 ?	
79	Au	$^2S_{1/2}$	2	2 6	2 6 10	2 6 10 14	2 6 10	1	
80	Hg	$^1S_0$	2	2 6	2 6 10	2 6 10 14	2 6 10	2	
81	Tl	$^2P_{1/2}$	2	2 6	2 6 10	2 6 10 14	2 6 10	2 1	
82	Pb	$^3P_0$	2	2 6	2 6 10	2 6 10 14	2 6 10	2 2	
83	Bi	$^4S_{3/2}$	2	2 6	2 6 10	2 6 10 14	2 6 10	2 3	
84	Po	$^3P_2$	2	2 6	2 6 10	2 6 10 14	2 6 10	2 4	
85	At	$^2P_{3/2}$	2	2 6	2 6 10	2 6 10 14	2 6 10	2 5	
86	Rn	$^1S_0$	2	2 6	2 6 10	2 6 10 14	2 6 10	2 6	
87	Fr	$^2S_{1/2}$	2	2 6	2 6 10	2 6 10 14	2 6 10	2 6	1
88	Ra	$^1S_0$	2	2 6	2 6 10	2 6 10 14	2 6 10	2 6	2
89	Ac	$(^2D_{3/2})$	2	2 6	2 6 10	2 6 10 14	2 6 10	2 6 1	2 ?
90	Th	$(^3F_2)$	2	2 6	2 6 10	2 6 10 14	2 6 10	2 6 2	2 ?
91	Pa	—	2	2 6	2 6 10	2 6 10 14	2 6 10 2	2 6 1	2 ?
92	U	—	2	2 6	2 6 10	2 6 10 14	2 6 10 3	2 6 1	2
93	Np	?	2	2 6	2 6 10	2 6 10 14	2 6 10 4	2 6 1	2 ?
94	Pu	?	2	2 6	2 6 10	2 6 10 14	2 6 10 6	2 6	2 ?
95	Am	?	2	2 6	2 6 10	2 6 10 14	2 6 10 7	2 6	2
96	Cm	?	2	2 6	2 6 10	2 6 10 14	2 6 10 7	2 6 1	2 ?
97	Bk	?	2	2 6	2 6 10	2 6 10 14	2 6 10 9	2 6	2 ?
98	Cf	?	2	2 6	2 6 10	2 6 10 14	2 6 10 10	2 6	2 ?
99	Es	?	2	2 6	2 6 10	2 6 10 14	2 6 10 11	2 6	2 ?
100	Fm	?	2	2 6	2 6 10	2 6 10 14	2 6 10 12	2 6	2 ?
101	Md	?	2	2 6	2 6 10	2 6 10 14	2 6 10 13	2 6	2 ?
102	No	?	2	2 6	2 6 10	2 6 10 14	2 6 10 14	2 6	2 ?
103	Lw	?	2	2 6	2 6 10	2 6 10 14	2 6 10 14	2 6 1	2 ?

of the atoms. From this explanation there follow necessarily, without any special assumptions, not only the general rules of the chemical behavior of the elements, but also the finer features such as the different valencies of copper and the special behavior of the noble metals palladium and platinum as well as the particular properties of the rare earths. This whole explanation of the Periodic Table by means of the electron shell theory is based on the empirically measured ionization potentials and the spectroscopically determined properties of the ground states of the atoms (multiplicity and L values). This explanation can be deduced from the Pauli principle without the aid of empirical data, but not in all details. This principle itself was derived from experience, and up to the present time it has not been possible to derive it as a logical consequence of the theory.

The success of the Bohr theory in explaining the Periodic Table is so startling that it may be ragarded as its crowning achievement. Nevertheless, our discussion shows very clearly what has yet to be done. BOHR's theory provides an *explanation* of the Periodic Table; but it is not a consistent and rigorous *theory*. Such a general theory would have to explain why each consecutive electron takes *just that* and no other place in the atom. This complete theory of the Periodic Table, which by necessity would contain a derivation of the Pauli principle, is still lacking. However, by applying quantum statistics (see IV, 13) to the electrons of the atomic shells, FERMI has been able to show that the arrangement of Table 10 is the theoretically most stable one of all possible arrangements. On the other hand, the question which we left open in our presentation, why the Periodic Table breaks off with the ninety-second element, uranium, or now with the transuranium elements which have been recently discovered, has been answered in the meantime. The reason for this breaking-off is not, as was believed for a long time, due to an instability of the electron shells beginning with uranium. It is due to an instability of the heaviest *nuclei*. We shall consider this when we discuss the subject of nuclear physics in V, 14.

20. The Hyperfine Structure of Atomic Spectral Lines. Isotope Effects and Influence of the Nuclear Spin

The splitting of atomic energy states and spectral lines into multiplets was discussed on page 104 as being due to the quantized coupling of orbital momentum and spin momentum of the electrons. It is usually called the *fine structure of the spectral lines*, although for heavy atoms this splitting of the terms is so large that the fine-structure components become widely separated lines. In contrast to this fine structure of the spectral lines, which depends entirely on processes within the electron shells, the so-called *hyperfine structure of the spectral* lines is due to the interaction of the electron shells with the *nucleus*. The hyperfine structure can be measured only with highly resolving interference spectroscopes or with the methods of high-frequency spectroscopy.

This hyperfine structure may be caused by two fundamentally different effects. In the first place, the atom under consideration may have several isotopes. The effect of the different masses and different structure of the isotopic nuclei on the electron shell then causes small wavelength differences of the spectral lines which belong to the different isotopes, so that under high resolution what appear to be single lines actually are revealed as very closely spaced lines belonging to the different isotopes. Secondly, many nuclei have a mechanical angular momentum $\vec{I}$ and a small magnetic moment (see V, 4e). It is the interaction of this nuclear magnetic moment with the resulting magnetic moment of the electron shell that causes term and line splittings, the so-called hyperfine-structure multiplets. The overlapping of both effects naturally complicates the explanation of the observed hyperfine structure.

We consider first the isotope effect. When we discussed the hydrogen atom, we learned that RYDBERG's constant R and thus the wavelengths of the spectra lines depend on the mass of the nucleus as the result of the motion of the nucleus and the electron about their common center of mass. If we replace the nucleus of the H atom, the proton, by that of the heavy hydrogen isotope, the deuteron, which consists of one proton and one neutron, we get from Eq. (3.27) (page 67) a change in the wavelength of the first Balmer line H_α by 1.79 Å. UREY demonstrated the existence of heavy hydrogen in 1932 by measuring this wavelength difference. In the case of heavy atoms with small mass differences between their isotopes, the λ difference rapidly becomes so small that it falls into the region of hyperfine structure. Small differences of this isotopic shift, which is caused by the motion of the nucleus, occur also as a result of the rotation of the external electrons of the many-electron atoms in the same or opposite directions. We shall not consider the details of the effect though. For heavy atoms there is a second isotope effect which is worthy of notice. For higher masses it exceeds the *nuclear-motion effect* and is called the *volume effect*. By addition of one or more neutrons to the nucleus, the spatial arrangement of the protons in the nucleus is changed. Consequently, since in the case of large nuclei the positive charge can no longer be regarded as a point charge, the energy of the electrostatic field between the nucleus and the electron shells, which determines the energy state of the atom, depends upon the arrangement of protons and neutrons in the nucleus. It varies from isotope to isotope, and this results in a small λ difference. The number, masses, and relative abundances of the isotopes of an element can thus be determined by measuring the number, separation, and intensities of the hyperfine-structure components resulting from the isotope effect. Such spectroscopic investigations of isotopes therefore supplement the mass-spectroscopic research on isotopes, treated in II, 6c. We shall discuss the spectroscopic effects of isotopes on molecular band spectra in VI, 12.

The actual hyperfine structure of terms and lines of individual isotopes, i.e., after separation from the isotope effect, is treated theoretically in the same way as was the multiplet structure in III, 13. Like the electron shells, each nucleus may have a mechanical angular momentum $\vec{I}$:

$$|\vec{I}| = \sqrt{I(I+1)}\ \frac{h}{2\pi} \tag{3.127}$$

with I values between 0 and 9/2. This mechanical angular momentum causes a magnetic moment of the nucleus which is often measured in units of the so-called *nuclear magneton*

$$\mu_n = \frac{eh}{4\pi Mc}. \tag{3.128}$$

It is deduced from the Bohr magneton (3.93) by substituting the proton mass M for the mass of the electron m. For all details see V, 4e. The value of the nuclear magneton is thus 1/1836 of the Bohr magneton, i.e., of the magnetic moment of the electron. Because the value of the nuclear magnetic moment is so small, the interaction between nucleus and electron shell is also small, and this is the reason for the very small separation of the hyperfine-structure components.

The angular momentum $\vec{I}$ of the nucleus is space-quantized in the magnetic field of the electron shell and vectorially combines with the total momentum $\vec{J}$ of the electron shell to form the quantized total angular momentum $\vec{F}$ of the entire atom. Therefore, the number of hyperfine-structure terms is $2J+1$ or $2I+1$, depending on whether $I > J$ or $J > I$, respectively. The *number* of the hyperfine-structure terms thus depends, as in the case of the normal fine structure, on the

values of the angular-momentum quantum number. The magnitude of the *spacing*,
on the other hand, depends on the value of the nuclear magnetic moment $\mathfrak{M}_n$ and
the field strength H_n of the magnetic field set
up by the electron shell at the location of the
nucleus. The energy differences between the
hyperfine structure components may be computed, in principle at least, from the formulae
of the multiplet splitting of III, 17.

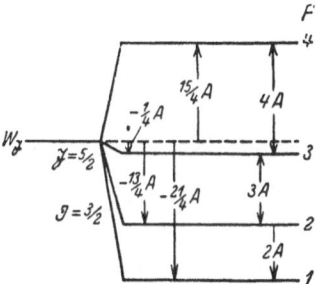

As an example, we show in Fig. 86 a hyper-
fine-structure term multiplet for $J = \frac{5}{2}$ and $I = \frac{3}{2}$
with the values 1, 2, 3, and 4 of the total angular
momentum F of the atom. The same selection rule

$$\Delta F = 0 \quad \text{or} \quad \pm 1 \qquad (3.129)$$

Fig. 86. Hyperfine-structure term multiplets with quantum numbers and term
differences. (After Kopfermann)

which applies to the quantum number J is valid
for transitions between hyperfine-structure term
components. Fig. 87 shows as an actual example the densitometer recording of
the hyperfine structure of the bismuth line $\lambda = 4122$ Å from which the term
diagram and transitions shown in Fig. 88 result. The term
differences of the two upper states, δ, and the two lower
ones, Δ, occur twice in Fig. 87 and may be recognized at
a glance.

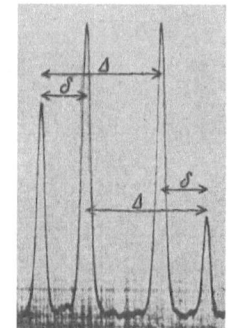

The influence of an external magnetic field on the
hyperfine-structure terms may be used to prove the accuracy of a hyperfine-structure term analysis. All hyperfine-
structure terms are split up in a magnetic field into $2F+1$
components as a result of space quantization. As in the
case of ordinary terms (page 118), we distinguish in the
case of hyperfine structure between the Zeeman effect and
the Paschen-Back effect, according to whether the spacing
of the term components due to the external magnetic field
is small or large compared with the natural hyperfine-
structure spacing. By measuring the number, orientation,
and separation of the hyperfine-structure components we

Fig. 87. Densitometer record
of the Bi line $\lambda = 4122$ Å.
(After Zeeman, Back, and
Goudsmit)

can determine the nuclear spin $\vec{I}$ and the sign of the
magnetic moment $\mathfrak{M}_n$. In favorable cases, $\mathfrak{M}_n$ itself
may be determined. By using the deviation of the
measured hyperfine-structure components from those
computed under the assumption that the nuclear charge
distribution is spherically symmetric, Schüler succeeded in computing even electric quadrupole moments
(see V, 4b). Thus, hyperfine-structure experiments
have become an important aid to nuclear physics.

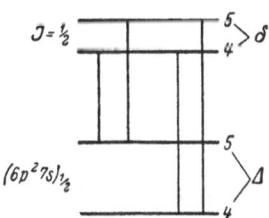

Fig. 88. Hyperfine-structure term
diagram of the Bi line $\lambda = 4122$ Å

21. The Natural Width of Spectral Lines
and the Influence of Internal and External Effects

Our discussion of the relation of line spectra to atomic structure and to
atomic processes would not be complete if we would not discuss briefly the interesting question of the width of spectral lines. All spectral lines have a "natural"
width which is determined by atomic processes and is independent of the

resolution of the spectroscopic equipment. If, furthermore, the spectral line is not emitted or absorbed by an atom at rest, but by atoms in thermal motion, or by atoms exchanging energy with their environment by kinetic collisions or by the action of an electric field due to electrons or ions, then the effect of these factors can be observed by the change in the width and intensity distribution of the spectral lines. Vice versa, experiments on these variations permit conclusions to be drawn on the state of motion of the atoms and their perturbation by their surroundings. These relations are of great importance for detailed studies of electric gas discharges as well as in astrophysics. For example, research on the physical conditions in the various layers of the sun's atmosphere (pressure, temperature, particle density) depends to a great extent on the theory of line broadening.

It can be shown classically that spectral lines have a certain natural width, because the radiation consists of individual wave trains of finite length. This is identical with the concept that in the classical picture we have an emission of damped waves by Hertzian oscillators. Then a Fourier analysis of the oscillation results not in *one* frequency, but in a more or less broad band of frequencies or wave lengths, as indicated in Fig. 89. In speaking of the width of a spectral line, we use the expression "half width" when we mean the width at half the maximum intensity. For electric dipol radiation, classical computations show this natural width of a spectral line to be

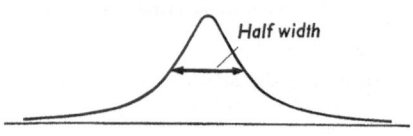

Fig. 89. Intensity distribution of a spectral line (schematic) with half width

$$b_0 = 1.19 \times 10^{-8} \text{ Å}, \tag{3.130}$$

independent of the wavelength. Its value is so small that it is not easy to measure the natural width of a line, and thus, in general, the normal lines of undisturbed atoms appear very sharp. The width of high-frequency lines may be even smaller for reasons to be discussed below.

In quantum mechanics, the line width is determined by the widths of the two combining energy levels. If follows from the uncertainty principle, which we shall consider in the next chapter (IV, 3), that the energy level of an unstable atomic state has a finite width, since the uncertainty principle states that the energy of a state is the less accurately determined the shorter its mean life τ, or expressed in a formula,

$$\Delta E \approx \frac{h}{\tau}. \tag{3.131}$$

Here ΔE is the uncertainty of the energy, i.e., the width of the energy level under consideration. For resonance lines it can be shown that the formula for the line width which follows from Eq. (3.131) is identical with Eq. (3.130), which was arrived at by considering classical radiation, if we use in Eq. (3.131) the average life of 10^{-8} sec. For forbidden lines, which play an important role, e.g., in the microwave region (see page 50), the width according to Eq. (3.131) is much smaller than that from Eq. (3.130). However, it is a consequence of the quantum-mechanical formula (3.131) that, in contrast to the classical formula, *the width of the energy level and thus of the spectral line depends not only on the mean life with respect to radiation but on the actual mean life, which also may be limited by processes that are not accompanied by radiation.*

In addition to collisions with neighboring particles and similar disturbances, there are two more non-radiative processes which may limit the mean life of an atomic state. Because of their importance, we shall mention them briefly. They are *auto-ionization* and ionization of an atom in a strong electric field *(field ionization)*. Auto-ionization is possible if, as a result of double excitation (page 101), a stationary atomic state exists above the normal ionization limit. If certain selection rules are fulfilled, then, according to Fig. 90, a transition from the doubly excited state to a lower state accompanied by radiation *or* a non-radiative transition to a neighboring continuous energy region can occur. The latter means that the second excited electron goes into the ground state and the first one, using the released energy, leaves the atom entirely and the atom is ionized. Under certain conditions, ionization of an atom from a doubly excited state can thus occur instead of emission of a photon. Therefore, we call this process auto-ionization. Another term which is used often is *pre-ionization*, which means that ionization occurs

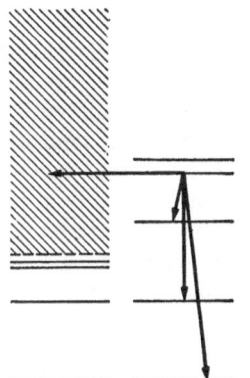

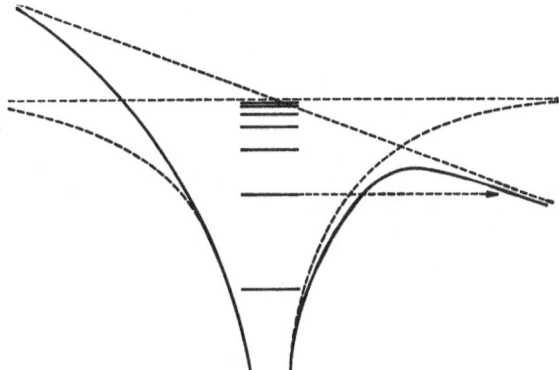

Fig. 90. Term diagram and transition possibilities for auto-ionization of an atom

Fig. 91. Potential curve $U(r)$ and energy levels of a hydrogen atom in a strong external electric field. Dotted curve = undisturbed potential of the proton-electron system. Solid curve = potential disturbed by the external electric field

before the ionization limit of the particular series is reached. The mean life of the excited stationary state in this case is limited not only by the emission of radiation, but under certain conditions with higher probability by a non-radiative transition into the continuum. According to Eq. (3.131), this shortened mean life corresponds to an increased width of the atomic state and thus to a broader line being emitted by the transition to lower states. The width of such a line, which is sometimes very large, may be measured and serve as evidence for the existence of pre-ionization and as a measure of its probability.

A similar case is the ionization in a strong external electric field. If a hydrogen atom is brought into a strong homogeneous electric field E in x-direction, its potential $-eEx$ overlays that of the Coulomb field of the nucleus $-e^2/r$ so that the hydrogen electron moves in the potential:

$$U(r) = -\frac{e^2}{r} - eEx. \tag{3.132}$$

In Fig. 91 the Coulomb field is plotted with the superimposed homogeneous field shown as a straight dotted line; by adding both fields, we get the "distorted" potential curve shown as a solid line. It can be seen from Fig. 91 that the stationary states of the hydrogen atom which lie above the maximum of the potential curve can no longer be stable. The electron which is normally bound to the nucleus is torn from the atom by the external field, i.e., the atom is ionized. E. W. MÜLLER succeeded in increasing the field strength in the vicinity of extremely thin metal points to the extent that even the ground state of the hydrogen atom was

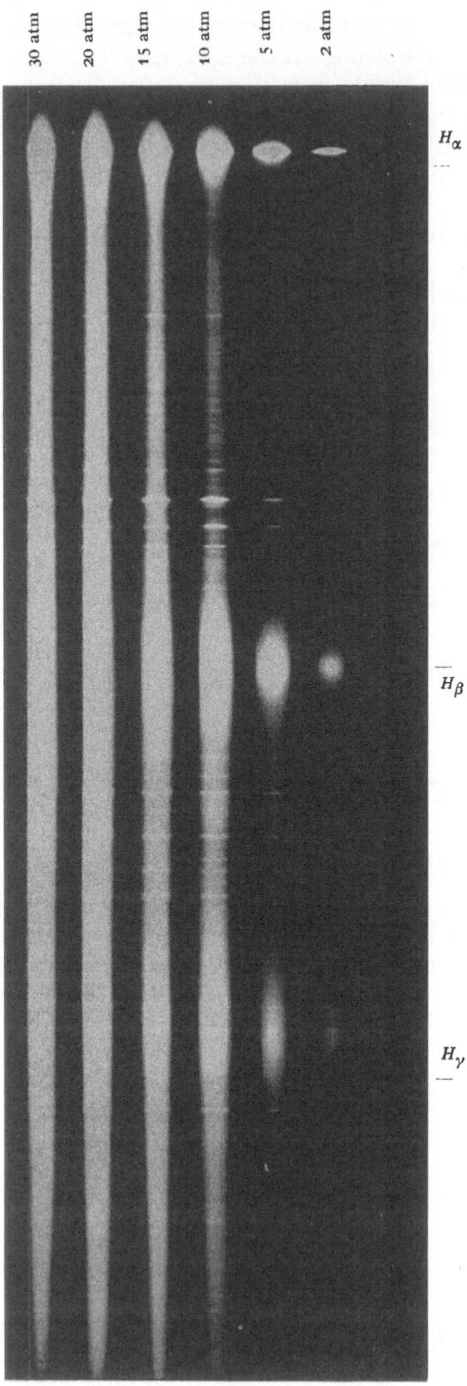

Fig. 92. Transition of the Balmer lines H_α, H_β, and H_γ into a uniform continuum as a result of the effect of strong interatomic electric fields. Spectrum of a spark in hydrogen of 2 to 30 atmospheres pressure. (Photograph by the author, 1930)

not stable anymore so that the normal hydrogen atom was field-ionized. According to an important effect, which will be considered in quantum mechanics (tunnel effect, see IV, 12), there is a certain probability that even electrons in stable energy states are ionized: If the potential wall is not too high or too broad, the electron can penetrate it by means of the tunnel effect. For a stationary state just below the potential maximum there is, thus, in addition to the possibility of a transition to [a lower state with emission of radiation, the possibility of the electron passing through the potential wall without radiating, leading to an ionization of the atom. This process tends to further reduce the mean life τ of a stationary state and thus to increase its width and that of the corresponding line. As a consequence of the processes which have been described, the last series lines disappear in the continuum, and the last lines which can still be recognized show a noticeably increased width because of the possibility of non-radiative ionization from their upper states. Fig. 47 shows this effect, which is called "overlapping of the continuum into the series lines", very clearly in the spectrum of a thallium discharge. The electric field, which is inhomogeneous with respect to time and space, originates in this case from electrons and ions of the discharge.

In conclusion, we discuss the third possibility of limiting the mean life of stationary states, that of collisions with gas particles of all kinds, by which the energy is carried away in collisions of the second kind or by which, quite generally, the emission process is disturbed. The result of the reduction of the mean life of the stationary states by collisions is the so-called *collision broadening of spectral lines*. The investigation of this phenomenon has developed into an independent branch of spectroscopy, which proved particularly important for astrophysics. Because of the difficulty in defining the radius of electrically charged colliding particles (as discussed on page 25) there is no sharp distinction between the direct collision broadening resulting in a reduction of the mean life of the excited states, and the perturbation or broadening of atomic states by the interatomic Stark effect due to inhomogeneous electric fields of neighboring electrons and ions. Such interatomic electric fields change the shift and separation of the energy states and thus cause a varying shift and splitting of the spectral lines, which appears as line broadening. The density of the perturbing electrons may in principle be directly computed by means of the above theory from measurements of the line widths. Since in thermal equilibrium, e.g., in an arc discharge, the electron density is a function of the temperature, the measurement of line widths furnishes a method of determining high temperatures.

The perturbability of different atoms is very different, and it is evident that the H atom and the alkalies are especially perturbable because their one external electron is at a great distance from the nucleus or core, respectively. Consequently, the spectra of these atoms exhibit the largest line broadening. The broadening can become so extensive with these atoms that the different lines flow together to form one continuous spectrum. Fig. 92 is an example of a spectrum of a spark discharge in compressed hydrogen in which the broadened Balmer lines merge into a single continuous spectrum.

The extreme case of such a perturbation of an atom by its neighbors exists in liquids and solids, which will be discussed from the atomic viewpoint in Chap. VII. Disregarding certain exceptions, which will be explained later, in these cases we no longer have sharp spectral lines but only more or less extended continuous spectra or spectral bands as a result of perturbation by the surroundings.

22. Bohr's Correspondence Principle and the Relation between Quantum Theory and Classical Physics

The theoretical considerations of this chapter so far dealt exclusively with the problem of spectral frequencies or wavelengths, whereas their intensities were not yet taken into account. The basic reason for this procedure lies in an interesting limitation of the statements of BOHR-SOMMERFELD's quantum theory. We shall understand this fact to its full extent by considering the relation between BOHR's quantum theory and classical physics. In doing so we shall also find a sensible approach to quantum mechanics which will be treated in the following chapter.

Our theoretical considerations started, as is evident from the brief presentation of the theory of the hydrogen atom on page 65, from purely classical concepts, e.g., from the equilibrium between electrostatic attraction and centrifugal force. These were adapted, in a more or less forced way, to the actual atomic events by introducing *two* quantum conditions, *both* of which contain PLANCK's constant h. The first assumption is the quantization of the atomic angular momentum or, for more complicated cases, of the angular momenta of the individual electrons. It leads automatically to the quantized energy states of the orbiting electrons. The second condition postulates a proportionality between energy and emitted or absorbed frequency ($E = h\nu$). It was suggested by photoelectric experiments and PLANCK's theory of heat radiation (see page 45). With these quantum postulates the theory of BOHR-SOMMERFELD predicts quantitatively the frequencies or wavelengths emitted or absorbed when the energy states of the atoms or electrons change; but nothing follows about their intensities.

BOHR himself repaired the shortcomings of his quantum theory with his remarkable intuitive feeling for physics. He connected quantum theory with classical physics by his famous correspondence principle which is based on pre-quantum-mechanical assumptions and yet allows the computation of intensity and polarization of the emitted radiation. He proved that there exists a far-reaching correspondence between classical and quantum physics. The latter becomes identical with classical physics in the limiting case of $h \to 0$ or, which is essentially the same, for very high quantum numbers.

We learned on page 73 that the periodic motion of bound atomic electrons around the nucleus corresponds to stationary energy states and thus to discrete line spectra, whereas the non-periodic motion of free electrons corresponds to continuous energy ranges and therefore to emission or absorption of continuous spectra. This feature of quantum theory is closely related to classical mechanics: We know that Fourier analysis reveals every periodic motion as composed of a discrete number of purely harmonic motions with definite frequencies, corresponding to a discrete frequency spectrum. The Fourier analysis of *non-periodic* motions, however, results always in a *continuous* frequency spectrum, in complete analogy to the continuous energy spectrum of free electrons. The continuous X-ray retardation spectrum (see page 76) is a typical example. It corresponds, except for its sharp limit towards short wavelengths, precisely to the classical "shock spectrum", whereas the short-wavelength limit is determined by the discrete quantum of action h according to Eq. (3.47). With $h \to 0$ the limit would move to infinitely high frequencies, in agreement with the classical shock spectrum.

If we now assume that, except for the quantization of energy because of the existence of the quantum of action, atomic systems follow the laws of classical physics, we may draw a conclusion of fundamental interest which was first pointed out by JORDAN: We know that two identidal coupled oscillators in classical physics are able to exchange their energies in arbitrary amounts. According to the correspondence principle, we expect such an energy exchange to be possible also for oscillators of atomic dimensions. This exchange is possible, however, only in integral multiples of a smallest energy E_0 because of the quantization of energy. From the fact that only *identical* oscillators exchange their energy, i.e., that frequency resonance is necessary, we conclude that the size of the energy quanta E_0 depends on the characteristic frequency of the resonator; we may therefore write it $E_0(\nu)$. In order to make the exchange of arbitrarily large amounts of energy between identical oscillators possible, the magnitude of these energy quanta $E_0(\nu)$ must be identical so that the energy of an oscillator may be written as

$$E = \text{const} + n E_0(\nu); \qquad n = 0, 1, 2, 3 \ldots . \qquad (3.133)$$

The constant energy steps of the oscillator which were postulated by PLANCK (cf. page 45) can thus be understood from the correspondence principle. We may even go a step further. The simplest assumption with correct dimensions for the dependence of the energy quantum E_0 on the characteristic frequency of the resonator is the proportionality $E_0 = h\nu_0$, and it can be shown that this form is the only one which is relativistically invariant, i.e., compatible with the theory of relativity. *The quantization of the oscillator (equality of its energy steps) and the $h\nu$ relation thus follow from the correspondence principle, if relativity theory is taken into account, without any special assumptions.*

There is also a clear and unambiguous correspondence between quantum theory and classical physics with regard to emitted or absorbed frequencies of spectral lines in so far as the quantum-theoretical frequencies for high quantum numbers and small quantum jumps converge towards the classical frequencies. Let us consider the spectral lines of the hydrogen atom as an example.

According to classical electrodynamics, the frequency of a light wave emitted by a hydrogen atom equals the revolution frequency of the electron; therefore [from Eq. (3.19) on page 66]

$$\nu_{cl} = \frac{\omega}{2\pi} = \frac{4\pi^2 m e^4}{h^3 n^3}.$$ (3.134)

The frequency emitted in a quantum transition $n_i \to n_e$ in Bohr's quantum theory [cf. Eq. (3.23)], however, is given by

$$\nu_{cl} = \frac{2\pi^2 m e^4}{h^3} \left(\frac{1}{n_e^2} - \frac{1}{n_i^2} \right) = \frac{2\pi^2 m e^4}{h^3} \frac{(n_i^2 - n_e^2)}{n_i^2 n_e^2}.$$ (3.135)

For high quantum numbers and small quantum jumps, i.e., for

$$n_i - n_e = \Delta n \ll n_i,$$ (3.136)

Eq. (3.135) becomes

$$\nu_{qu} = \frac{4\pi^2 m e^4}{h^3 n_i^3} \Delta n.$$ (3.137)

For the transition between two neighboring quantum states ($\Delta n = 1$) the classical form (3.134) and the quantum-theoretical Eq. (3.137) thus result in exactly the same frequency. The transition $\Delta n = 2$ corresponds, in classical physics, to the first harmonic $2\nu_{cl}$, etc.

Furthermore, Bohr concluded from this correspondence that also the intensity and polarization of spectral lines may be correctly computed from classical theory for sufficiently high quantum numbers and approximately even for intermediate quantum numbers. Experience confirmed this daring extrapolation. The excellent agreement between measured and computed intensities and polarization properties of the Stark effect components of the Balmer lines left no doubt about the value of the correspondence principle. This is the more remarkable since the intensity problem has entirely different aspects in classical and quantum physics. Classically, the line intensity is determined by the amplitude of an electromagnetic wave. Frequency and intensity are, therefore, characteristic properties of one and the same phenomenon, viz. the wave. According to quantum theory, on the other hand, the frequency of the photon, i.e., the wavelength of the radiation process, depends on the energy states involved, whereas the intensity of a spectral line is given by the number of emission processes per unit volume and unit time, and thus is apparently a statistical problem quite independent of the individual emission process. This viewpoint makes the correspondence principle appear as a kind of miracle, and only quantum mechanics has made the correspondence principle understandable by its fundamental clarification of the relation between classical and quantum physics.

The correspondence principle makes it also possible to understand the selection rules which, at first sight, look so arbitrary. While their explanation with the help of the correspondence principle is somewhat complicated for the atomic quantum numbers L, J, and M (the only ones treated so far), we can easily understand the selection rule for a special case, namely the total angular momentum $\vec{J}$ of a polar molecule rotating around its main axis of inertia (e.g., HCl). Comparison of Eqs. (3.134) and (3.137) has shown that the quantum transition $\Delta n = \pm 1$ corresponds to the classical fundamental frequency; the transitions $\Delta n = \pm 2, 3, 4, \ldots$ correspond to emission or absorption of the higher harmonics of the fundamental frequency. *In classical physics, these higher harmonics can be emitted only if they exist in the corresponding vibration of the electric dipole, i.e., if the emitting or absorbing vibrating system does not carry out a purely harmonic motion.* Because of the conservation of the total angular momentum, our molecule rotates with a constant angular velocity. Such a constant rotation can be described as a superposition of two purely harmonic vibrations with a phase difference of 180°. Therefore only the fundamental frequency can be emitted or absorbed classically by this rotating molecule. The corresponding quantum-theoretical selection rule therefore must state that only transitions between neighboring rotational energy states are possible. This is the explanation of the experimentally proved selection rule $\Delta J = \pm 1$ [cf. Eq. (6.44)] on the basis of the correspondence principle. Quite generally, the correspondence principle allows us to compute atomic processes first with the aid of classical physics and then to adapt the results to the discrete atomic events by a suitable quantization. In this sense the correspondence principle has a certain significance even in quantum mechanics.

23. Transition Probability and Line Intensity. Life Time and Oscillator Strength. Maser and Laser

We now consider in some more detail the transitions between an excited and a non-excited state, E_m and E_n, respectively. Closely related to this is the problem of the intensity of emission or absorption lines.

We know that excited atoms may spontaneously go over into the ground state by emitting a photon of frequency ν_{nm} according to

$$E_m - E_n = h\nu_{nm}. \tag{3.138}$$

The probability of this spontaneous emission may be called A_{nm}. If such atoms are irradiated by radiation of the same frequency ν_{nm}, then there is, according to EINSTEIN, a probability $B_{nm} u(\nu)$ for a transition from the normal state E_n to the excited state E_m under absorption of photons $h\nu_{nm}$, where $u(\nu)$ is the energy density of the incident radiation and

$$B_{nm} = \frac{c^3}{8\pi h\nu^3} A_{nm}. \tag{3.139}$$

EINSTEIN also proved that this same probability $B_{nm} u(\nu)$ governs a process by which excited atoms are "induced" by the incident radiation to emit photons $h\nu_{nm}$, thus returning to the ground state. This latter process is called *induced emission* or *stimulated emission*. The total probability for a transition from the upper to the lower state with simultaneous emission thus is composed of the probability A_{nm} for spontaneous emission and that for induced emission, B_{nm}, which is proportional to the radiation density as is the absorption probability.

The energy I_ν emitted spontaneously per sec by one cm³ at the frequency ν is equal to the number of atoms per cm³ in the initial quantum state n multiplied by the probability of the corresponding transition and by the energy of each emitted photon, viz.

$$I_\nu = N_n A_{nm} h\nu. \tag{3.140}$$

The transition probability A_{nm} for the transition from state n to state m can be computed only by quantum mechanics (see IV, 9). It is, however, possible to determine $A_{nm} h\nu$ classically by making use of the correspondence principle. The occupation number of the initial quantum state n, i.e., N_n, is dependent on the conditions of excitation. With thermal excitation, N_n may be determined from Boltzmann statistics, according to which the occupation of a state with excitation energy E_n at the absolute temperature T is given by

$$N_n = N_0 e^{-\frac{E_n}{kT}}. \tag{3.141}$$

This formula holds under the condition of strictly single states E_n (statistical weight equal to one), but not for degenerate states, which actually represent the superposition of several states of equal energy. An energy state with quantum number J, for instance, has a statistical weight $g_J = 2J+1$, i.e., it consists of $2J+1$ coinciding energy states which may be split by a magnetic field (Zeeman effect, see III, 16) into as many components. If the statistical weights of ground state and excited state are designated by g_0 and g_n, respectively, then we have instead of Eq. (3.141)

$$N_n = N_0 \frac{g_n}{g_0} e^{-\frac{E_n}{kT}}, \tag{3.142}$$

and Eq. (3.140) for the emitted intensity becomes

$$I_\nu = N_0 \frac{g_n}{g_0} e^{-\frac{E_n}{kT}} A_{nm} h\nu. \tag{3.143}$$

The different parameters which occur in Eq. (3.143) are not always of equal importance, and this simplifies the situation. We mention a few typical cases only. For absorption at moderate temperatures, usually only the ground state of the atoms is occupied since the energy distance of the first excited state from the ground state is normally very large compared to kT. Therefore the excited state is practically unoccupied. Furthermore, the statistical weights of the states participating in the transitions may be computed from the respective quantum numbers and are constant within a series. Therefore, the intensity distribution in an absorption series, i.e., the decrease of intensity with increasing principal quantum number, shows directly the dependence of the transition probability on the quantum number of the variable term and thus allows the determination of such transition probabilities from intensity measurements. At sufficiently *high* temperatures, as in stellar atmospheres, also absorption from excited states may occur (see the Balmer series absorption, Fig. 44). In this case, the absorption intensity within such a series depends on the temperature as the Boltzmann factor $e^{-\frac{E_n}{kT}}$. Consequently intensity measurements may be used for determining the distribution N_n of the atoms in the corresponding states (e.g., that of H atoms in the lower state $n=2$ of the Balmer series), and from this, the temperature of the absorbing layer.

We now turn to an application of induced emission that more and more becomes technically important. It is the *atomic or molecular amplifier* in the several forms of the so-called maser, invented by Townes. "Maser" is coined from the initials of "*M*icrowave *A*mplification through *S*timulated *E*mission of *R*adiation". Its principle is as follows: A system whose excited state may have a higher occupation than its ground state (in contrast to the normal case) is subjected to intensity-modulated radiation of the frequency corresponding to the transition between these two states. Then this radiation will stimulate such transitions to the lower state in the rhythm of the modulation, and thus the stimulating radiation is amplified.

It is, for instance, possible to separate, by methods not to be discussed here, excited atoms or molecules from a beam containing a mixture of excited and unexcited particles. These selected excited atoms enter a cavity oscillator that is tuned to the frequency to be amplified. If, as is usually the case with masers, the transition frequency belongs to the microwave region and if, furthermore, spontaneous transitions are forbidden because the transitions may correspond to spin-reversal processes, then spontaneous transitions may be neglected compared to induced emission processes. This is important, because spontaneous emission, being independent of the stimulating frequency-modulated radiation, gives rise to "noise" which is of low intensity only if the above-mentioned conditions are met. The modulated radiation to which the cavity is tuned may thus be efficiently amplified through the stimulated emission, if the induced transitions correspond to the proper frequency. It may be mentioned that systems in which more atoms or molecules occupy an excited state than the unexcited one may be described formally by a *negative temperature* as is clear from Eq. (3.142).

Of higher importance for applications than the atomic or molecular beam amplifier just dealt with is the *solid-state* maser. In certain crystals such as $K_3Cr(CN)_6$, a magnetic field splits the state of the Cr ion into three states 1, 2, 3, the energy differences of which may be varied by adjusting the magnetic field. The transitions between these states correspond to microwaves (e.g., 2.8—2.9 Gcps at 2000 oersteds).

At room temperature, the three states are about equally occupied because of their very small energy differences of the order of 10^{-5} ev. At a temperature of 1 °K, however, at which solid-state masers are normally used, the higher states 2 and 3 are markedly less occupied than the lowest state 1. Irradiation of the frequency ν_{13} then results in about equal occupation of states 1 and 3 when saturation is reached, while state 2 remains nearly unoccupied since it cannot be reached by transitions induced with ν_{13}. This "filling-up" of the excited state 3 by ν_{13} is called *optical pumping*. If the crystal is brought into a cavity tuned to both the pumping frequency *and* to the frequency ν_{2-3}, and if then modulated radiation of frequency ν_{23} is sent into the cavity, this radiation will stimulate induced emission of ν_{23}, i.e., transitions from the continuously filled-up state 3 to the less occupied state 2.

Such molecular amplifiers thus are amplifiers of low noise for the microwave region, in which normal electronic amplification so far was hampered by the shortness of the wavelengths. The solid-state maser, if compared with the atomic-beam maser, has another advantage. The energy states of the Cr ion in the crystal lattice are relatively broad, according to III, 21, so that, even without varying the magnetic field, there is a technically satisfactory width of the frequency band around ν_{23} in which amplification is possible.

More recently, the optical maser or laser (*l*ight instead of *m*icro-waves) has become increasingly important. Here, the stimulated radiation in most cases corresponds to transitions from a metastable to the ground state, e.g., in the ruby laser of chromium atoms in solid solution. The metastable state is populated by transitions from higher excited states which are reached by optical pumping. Semi-transparent coatings of the plane end faces of the crystal have the effect that some part of radiation is reflected back into the interior of the crystal and here stimulates further transitions. This produces an amplification so that a very intense monochromatic light beam with a very narrow opening angle leaves the crystal. For the *gas laser*, low-pressure gas discharge tubes with plane and semi-transparent windows are used with helium-neon mixtures. The metastable neon state which serves as initial state of the stimulated transitions is populated by collisions of the second kind with excited helium atoms, and this is possible because an excited helium state has the same energy as the metastable neon state. In the gas laser, consequently, the population is not the result of optical pumping, but of electric pumping.

The difference between maser and laser is that the maser serves as micro-wave amplifier, whereas the laser, due to its auto-amplification, serves as a most intense monochromatic light source which, however, may be modulated and thus used for transmission of information.

In VII, 22b, we shall discuss the most interesting and important semi-conductor laser in which also the pumping is done electrically.

We return to the problem of line intensities and deal briefly with their relation to the mean life of the corresponding excited states, which we so far considered to be about 10^{-8} sec. But in discussing metastable states, we learned that much longer life times are possible, in principle even infinitely long ones (for strictly forbidden transitions). How can we determine life times and how are they related to the transition probability?

In classical physics, the life time of an excited atom is simply the span of time necessary for the atom to emit its excitation energy as a radiating dipole. Therefore, we expect the mean life to be the shorter the larger the amplitude of the radiation is or the higher the light intensity (for equal numbers of atoms), according to Eq. (3.143). This relation of classical physics remains valid in quantum mechanics as far as the correspondence principle is applicable. Quantum-theoretically, there exists for an excited atom a certain probability that it may emit its energy within a given time as a photon. This emission occurs, if no other transitions are possible, in the average after the "mean life" τ of the excited state. The number of photons emitted per second by a certain number of excited atoms (or the intensity of the corresponding spectral line) is therefore inversely proportional to the mean life if the latter is determined by just this *one* transition possibility. If N is the number of excited atoms present at a certain time, then the number of atoms decaying per sec by emitting photons is proportional to the number of atoms not yet decayed; therefore

$$dN/dt = -\gamma N \tag{3.144}$$

or

$$N(t) = N_0 e^{-\gamma t}. \tag{3.145}$$

This relation is valid for every decay governed purely by probability, and we will encounter it again in discussing the radioactive decay of atomic nuclei (see V, 6b). The constant γ is called the decay constant; we shall explain its connection with our transition probability A_{nm} below. If there is only *one* decay possibility, the mean life τ is defined as the reciprocal of γ,

$$\tau = 1/\gamma. \tag{3.146}$$

If, for instance, excited atoms are moving away from their region of excitation with high velocity (about 10^8 cm/sec) in one direction, as may be observed for canal rays, then the intensity of the canal ray will decrease with the distance from the point of excitation according to the exponential law (3.145). By measuring this intensity decrease along the beam, we can determine γ and thus the mean life τ from Eq. (3.146).

It is furthermore possible to compute, according to Eq. (3.143), the transition probability A_{nm} from measuring the absolute intensity of a spectral line if the number of atoms is known. A_{nm} would be equal to the decay constant γ_n of the excited state n or equal to the reciprocal of its mean life, if only the *one* transition $n \rightarrow m$ would be possible.

In the general case, however, we have

$$\tau_n = \frac{f_{nm}}{A_{nm}}, \tag{3.147}$$

where f_{nm} is the so-called oscillator strength of the transition in question, which is equal to 1 if there is only one transition possible, but which in general is given by

$$f_{nm} = \frac{A_{nm}}{\sum_m A_{nm}}. \tag{3.148}$$

With the aid of the relations (3.139), (3.147), and (3.148), it is possible to determine oscillator strengths or mean life times also from the frequency dependence of the refractive index n, since dispersion theory relates the absorption probability coefficient B_{nm} defined by Eq. (3.139) to the refractive index through the formula

$$n^2 - 1 = \frac{3hN}{4\pi^3} \sum_m \frac{g_m \nu_{nm} B_{nm}}{\nu_{nm}^2 - \nu^2} . \tag{3.149}$$

Our discussion of the intensity problems will be taken up again in IV, 9 in connection with the quantum-mechanical theory of radiation.

Problems

1. A one megawatt transmitter sends radiowaves at a frequency of 10^6 cycles per second. How many photons does it emit per second?

2. The radius of the first Bohr orbit of the normal hydrogen atom is 0.529×10^{-8} cm. What is the radius of the Bohr orbit of the doubly ionized lithium atom in the state $n = 3$?

3. Determine the energy in electron volts necessary to remove the single remaining electron from a lithium ion in its ground state.

4. What is the highest energy state to which a hydrogen atom initially in its ground state can be excited by an electron of 11 ev energy? What spectral line would one observe from such an excited atom and what is its wave number?

5. Derive the expressions of the Rydberg constants for hydrogen and singly ionized helium, taking into account the effect of the finite mass of the nucleus in Bohr's theory. If the Rydberg constants for $_2$He^4 and $_1$H^1 are 109,722.264 and 109,677.576 cm^{-1} respectively, calculate the ratio of the mass of the proton to the mass of the electron. Assume $M_{He} = 3.9717 M_H$.

6. The transition of a hydrogen atom from the state with $n = 4$ to the state with $n = 2$ corresponds to the emission of a spectral line of wavelength equal to 4861 Å. What would be the wavelength of the line emitted by positronium (positron-electron system) for the same transition?

7. What modifications of the optical line spectra are caused by placing an atom in an external weak magnetic field? A silver atom in its ground state is placed in an external magnetic field of strength 10 kilogauss. Calculate (a) the energy of separation of the adjacent levels in ev; (b) the line frequencies emitted by the atom as it makes a transition from the first excited state to the ground state.

8. Ultraviolet light of wavelength 800 Å, when incident on hydrogen atoms in the ground state, liberates electrons. What is the maximum kinetic energy of the electrons?

9. What is the short wavelength limit of the continuous X-ray spectrum produced by an X-ray tube operating at a potential difference of 100 kilovolts?

10. Find the mass and momentum of a photon of wavelength 0.10 Å.

11. Using Moseley's law, determine the wave number of the K_α lines of helium and uranium.

Literature

General Literature

CONDON, E. U., and G. H. SHORTLEY: The Theory of Atomic Spectra. Cambridge: University Press 1935.

GROTRIAN, W.: Graphische Darstellung der Spektren von Atomen und Ionen. 2 Vols. Berlin: Springer 1928.

HERZBERG, G.: Atomic Spectra and Atomic structure. 2nd. Ed. New York: Dover 1944.

HUND, F.: Linienspektren und Periodisches System. Berlin: Springer 1927.

PAULING, L., and S. GOUDSMIT: The Structure of Line Spectra. New York: McGraw-Hill 1930.

SOMMERFELD, A.: Atombau und Spektrallinien. 7th Ed., Vol. I. Braunschweig: Vieweg 1949.

WHITE, H. E.: Introduction of Atomic Spectra. New York: McGraw-Hill 1934.

Section 1:

BAUMAN, R. P.: Absorption Spectroscopy. New York: Wiley 1962.

BOMKE, H.: Vakuumspektroskopie. Leipzig: Barth 1937.

GORDY, W., W. V. SMITH and R. F. TRAMBARULO: Microwave Spectroscopy. New York: Wiley 1953.
HARRISON, G. R., R. C. LORD and J. R. LOOFBOUROW: Practical Spectroscopy. New York: Presentice-Hall 1948.
JAFFÉ, H. H., and M. ORCHIN: Theory and Applications of Ultraviolet Spectroscopy. New York: Wiley 1962.
KAYSER, H., u. H. KONEN: Handbuch der Spektroskopie. 8 Vols. Leipzig: Hirzel 1900—1934.
SAWYER, R. A.: Experimental Spectroscopy. 2nd Ed. New York: Prentice-Hall 1951.
SIEGBAHN, M.: Spektroskopie der Röntgenstrahlen. 2nd Ed. Berlin: Springer 1931.
TOLANSKY, S.: High-Resolution Spectroscopy. New York: Pitman Publ. Co. 1949.
TOWNES, C. H., and A. L. SCHAWLOW: Microwave Spectroscopy. New York: McGraw-Hill 1955.

Section 2:

BACHER, R. F., and S. GOUDSMIT: Atomic Energy States. New York: McGraw-Hill 1932.
This rather obsolete but still indispensable work is being newly edited by C. E. MOORE.
MOORE, C. E.: Atomic Energy States. Vols. I, II. Washington D. C.: National Bureau of Standards Publications 1949/1952.
FOWLER, A.: Report on Series in Line Spectra. Fleetway Press 1922.
PASCHEN, F., u. R. GÖTZE: Seriengesetze der Linienspektren. Berlin: Springer 1922.

Section 4:

FRANCK, J., u. P. JORDAN: Anregung von Quantensprüngen durch Stöße. Berlin: Springer 1926.
MASSEY, H. S. W.: Excitation and Ionization of Atoms by Electron Impact. Encyclopedia of Physics. Vol. 36. Heidelberg: Springer 1956.
MASSEY, H. S. W., and E. H. S. BURSHOP: Electronic and Ionic Impact Phenomena. London: Oxford University Press 1952.

Section 6:

FINKELNBURG, W.: Kontinuierliche Spektren. Berlin: Springer 1938.
MITCHELL, A. C. G., and M. ZEMANSKY: Resonance Radiation and Excited Atoms. New York: McMillan 1934.
MOTT, N. F., and H. S. MASSEY: The Theory of Atomic Collisions. 2nd Ed. Oxford: Clarendon Press 1949.

Section 10:

SIEGBAHN, M.: Spektroskopie der Röntgenstrahlen. 2nd Ed. Berlin: Springer 1931.

Section 13:

MOORE, C. E.: Multiplet Table of Astrophysical Interest. Princeton: University Press 1945.

Section 15:

BATES, L. F.: Modern Magnetism. 2nd Ed. Cambridge: University Press 1948.
STONER, E. C.: Magnetism and Matter. London: Methuen 1934.
VLECK, J. H. VAN: The Theory of Electric and Magnetic Susceptibilities. Oxford: Clarendon Press 1932.

Section 17:

BACK, E., u. A. LANDE: Zeeman-Effekt und Multiplettstruktur. Berlin: Springer 1925.

Section 20:

KOPFERMANN, H.: Kernmomente. 2nd Ed. Frankfurt: Akademische Verlagsgesellschaft 1956.
TOLANSKY, S.: Hyperfine Structure in Line Spectra and Nuclear Spin. London: Methuen 1948.

Section 21:

BURHOP, E. H. S.: The Auger Effect and other Radiationless Transitions. Cambridge: University Press 1952.
FINKELNBURG, W.: Kontinuierliche Spektren. Berlin: Springer 1938.
TRAVING, G.: Druckverbreiterung von Spektrallinien. Karlsruhe: Braun 1959.

IV. Quantum Mechanics

1. The Transition from Bohr's Theory
to the Quantum-mechanical Theory of the Atom

The theoretical treatment of atomic physics as presented in the preceding pages was based on the introduction of simple atomim codels. The behavior of the models was assumed to be governed by the known laws of classical physics. These laws were then used to obtain the results described in the last chapter. These classical theoretical results were "forced" into agreement with experimental results by introducing seemingly arbitrary and basically incomprehensible quantum conditions.

The amazing success of this "naive" theory presented in the last chapter is indisputable and shows that, in spite of its simplicity, the Bohr theory contains a great amount of truth. On the other hand, not only is the arbitrary imposition of quantum conditions a very unsatisfactory feature of the Bohr theory, but the theory fails completely in a number of outstanding examples, such as the model of the helium atom; the derivation of the Landé factor for the anomalous Zeeman effect (page 118); in the quantitative theory of molecular spectra (see VI, 6) and of the chemical bond (cf. VI, 14). Since about 1924 physicists became more and more aware that all these difficulties and inadequacies could be removed only by a new atomic theory which, though leading to the correct results of the naive theory, could not be built on any classical basis. This new atomic theory was developed simultaneously and independently in 1926 by SCHRÖDINGER, who based his work on that of DE BROGLIE, and by HEISENBERG, who used a more fundamental approach. Instead of starting from concepts of classical physics as BOHR did, the new theories try to arrive at a correct description of atomic physics by making use of observed phenomena only, such as the existence of discrete energy states and transition probabilities. Both theories which are called wave mechanics and quantum mechanics were, as was soon realized, basically identical in spite of their different initial assumptions and mathematical treatment. They constitute a definite departure from the earlier, seemingly self-evident conception that microphysical phenomena must be described in the same manner and with the same laws which are used in classical physics for describing the behavior of systems which are large compared to atomic dimensions. We shall consider later, in IV, 15, how this new concept of microphysics modified all our scientific thinking and, in its consequence, natural philosophy.

At first, HEISENBERG's and SCHRÖDINGER's quantum mechanics suffered from three deficiencies. It did take into account neither the electron spin nor the theory of relativity, and it only dealt with matter without considering its interaction with light, i.e., with the electromagnetic field and the other wave fields which will be mentioned later. Soon, however, the first defect was corrected by HEISENBERG, JORDAN, and PAULI; later DIRAC developed the theory to encompass the more complete quantum mechanics and quantum electrodynamics. DIRAC's theory is beyond the scope of this book; so we shall restrict ourselves to the simpler form of quantum mechanics, i.e., we shall disregard the theory of relativity and discuss only briefly the fundamental ideas of quantum electrodynamics, taking the electron spin into account subsequently.

The rigorous development of quantum mechanics, as we shall call the new theory, led to a more complicated but conceptually consistent atomic theory which, to "normal" thinking, may appear rather strange. But the new theory not only completely overcame the objections to the old atomic theory, but also led to many new results which were confirmed by experiment. We shall deal with these results in all the following chapters. Thus we have good reason to believe that this theory provides an accurate theoretical description of atomic processes and so may be considered the "correct" atomic theory. Only with the most extreme phenomena of elementary particle physics, e.g., very-high-energy collision processes which we will consider in V, 20, do we find that even quantum mechanics is not general enough. So we may expect further research to lead to an even more general theory which includes quantum mechanics as a special case.

In the following discussion, the fundamental reasoning and basic equations of quantum mechanics will be presented in the wave-mechanical form especially suited for calculating atomic phenomena. In the mathematical development of the theory we restrict ourselves to considering only what is necessary to understand the conclusions. We shall emphasize, however, as clearly as we are able to, the basic new features of the theory and its implications with regard to the wider field of atomic physics including nuclear physics, molecular physics, and physics of the solid state.

Since the new atomic theory is based on the wave-particle dualism of light and matter, we shall discuss this next.

2. The Wave-Particle Dualism of Light and Matter

From the time of FRESNEL up to the establishment of the Bohr atomic theory, there was not the slightest doubt that light was a wave phenomenon and that, on the other hand, matter in the form of molecules, atoms, and electrons had to be regarded as particles which, in a first approximation, could be treated as point masses. The development of the quantum-mechanical atomic theory began when it was realized that the conception of waves and particles as being unrelated entities, once considered self-evident, was too restricted. Depending on what experiments we perform with radiation and matter, they display wave or particle characteristics. Waves and particles are but two manifestations of the same physical reality.

The wave nature of light seemed so definitely proved by numerous interference and diffraction experiments that the demonstration of the diffraction of X-rays by VON LAUE was regarded as an undeniable proof of the wave nature of X-rays and, consequently, of the identity of X-rays and light. The first observed phenomenon of light which could not be explained by the wave theory was the photoelectric effect, the release of electrons from a metal surface when irradiated by light.

The experimental result that the velocity of the electrons which escape from the metal depends only on the color (wavelength or frequency) of the incident light and not on its intensity, could not be understood from the standpoint of the wave theory of light, according to which the energy of any radiation is proportional to the square of the amplitude of the waves but has nothing to do with their frequency. The correct relation between the mass and velocity of the escaping electrons, on the one hand, and the frequency v of the incident light on the other, was derived by EINSTEIN from experimental results. He found

$$\frac{m}{2} v^2 = hv - W \tag{4.1}$$

where W, the work function, is a characteristic constant for each metal (cf. VII, 14), whereas the constant h proved to be identical with the elementary quantum h introduced in 1900 by PLANCK in his theory of heat radiation. As mentioned on page 45, EINSTEIN concluded from this fact that light actually consists of individual quanta of energy $h\nu$, now called photons. Accordingly, photons must be emitted from a light source into all directions of space, and they must also be capable of being absorbed as quanta by atoms. For the absorption of light also proved to be a phenomenon which was difficult to understand from the point of view of the wave theory. One has only to ask how it is possible that a spherical wave of total energy $h\nu$, expanding into space with the velocity of light, can be absorbed in its entirety by *one* atom. The photon hypothesis automatically removes this obstacle, since, according to this theory, the total energy is concentrated in *one* photon and is therefore absorbed in its entirety as a photon. The wave theory would lead us to expect that a spherical wave of sufficient energy might be absorbed *simultaneously* by two or more absorbants at equal distance from the point of origin, whereas the quantum hypothesis of light requires that the photons are emitted one after the other statistically into the different spatial directions so that *no* coincidences of two absorption processes are to be expected beyond the rate of incidental occurences of this type. BOTHE's experiment decided this question in 1926 in favor of the photon hypothesis. Two counters (cf. V, 2) located at equal distances from an X-ray source did *not* show any simultaneous photon absorptions exceeding the expected statistical coincidences.

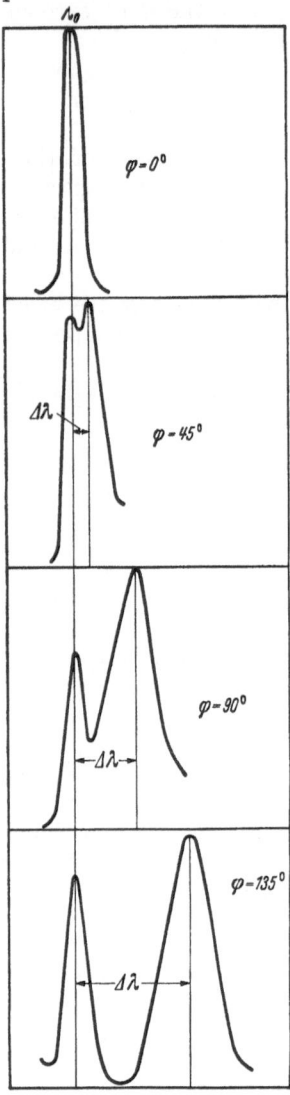

Fig. 93. Dependence of the wavelength shift $\Delta\lambda$ between scattered Compton line and primary line on the angle φ between direction of observation and direction of incident ray. (After measurements by A. H. COMPTON.) Abscissa scale: 0.28 Å/mm

The decisive experiment, however, with respect to the wave or particle nature of electromagnetic radiation was the discovery of a famous effect in 1922 by COMPTON. If short-wave X-rays or γ-rays of definite wavelength are scattered by a suitable solid such as graphite, spectral analysis of the laterally scattered radiation reveals that, in addition to a line corresponding to the scattered primary wavelength λ_0, another spectral line appears which is shifted in the direction of longer wavelengths. The wavelength difference $\Delta\lambda$ (see Fig. 93) between primary line and scattered Compton line increases with increasing angle of observation, as measured from the direction of the incident beam. The interpretation of this effect in the framework of the wave theory of electromagnetic radiation met unsurmountable difficulties. From the particle point of view, however, one may regard the process as a collision of a photon whith an electron of an atomic shell, whereby the photon transfers momentum and energy to the electron. By the impact of the photon, the electron is kicked out of its atom, whereas the photon is deflected into a direction which is defined by the laws of elastic impact (Fig. 94). From the mass-energy equivalence equation

$$E = m c^2 = h \nu \tag{4.2}$$

we had computed (cf. page 44) mass and momentum of a photon to be

$$m = h\nu/c^2 \quad \text{and} \quad p = mc = h\nu/c.$$

Let ν_0 and ν be the frequency of the photon before and after its collision with an electron of mass m_e, respectively, which may have a velocity v after the impact. Then we obtain the following equations for the energy and the components of the momentum parallel and perpendicular to the incident ray:

$$h\nu_0 = h\nu + \frac{m_e}{2} v^2, \qquad (4.3)$$

$$\frac{h\nu_0}{c} = \frac{h\nu}{c} \cos \varphi + m_e v \cos \vartheta, \qquad (4.4)$$

$$0 = \frac{h\nu}{c} \sin \varphi + m_e v \sin \vartheta. \qquad (4.5)$$

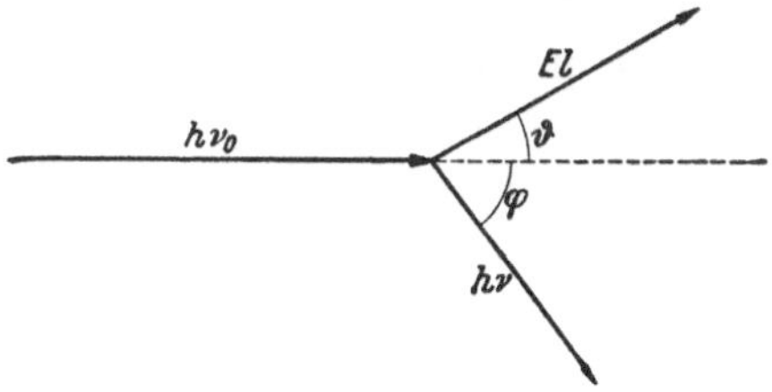

Fig. 94. Compton scattering of a photon $h\nu_0$ through the angle φ in a collision with an electron (Compton effect)

By eliminating the angle ϑ and the electron velocity v, which are of no interest, from these three equations, we get

$$h\nu_0 = h\nu + \frac{h^2}{2 m_e c^2} (\nu_0^2 + \nu^2 - 2\nu_0\nu \cos \varphi). \qquad (4.6)$$

By solving for $\nu_0 - \nu = \Delta\nu$ and substituting ν for ν_0 in the expression in parentheses, since $\Delta\nu$ is small compared to ν and ν_0, we arrive at the equation

$$\Delta\nu = \frac{h\nu_0^2}{m_e c^2} (1 - \cos \varphi) = \frac{2 h\nu_0^2}{m_e c^2} \sin^2 \frac{\varphi}{2}. \qquad (4.7)$$

It gives the frequency difference between the incident γ-ray and that scattered by the Compton effect as a function of the angle between the incident and the scattered photon. By converting $\Delta\nu$ into wavelength differences, we finally get

$$\Delta\lambda = \frac{2h}{m_e c} \sin^2 \frac{\varphi}{2}. \qquad (4.8)$$

This last equation is in complete agreement with the experimental results. The quantity $h/m_e c = \lambda_e$, which has the dimension of a length, is called the "Compton wavelength" of the electron. The wavelength difference between the scattered Compton line and the primary line is, from Eq. (4.8), independent of the wavelength of the primary line and depends only on the angle of deflection φ. Although this derivation did not take into consideration the relativistic mass dependence, it agrees exactly with the results of the more rigorous relativistic derivation. Bothe and Geiger proved in 1925 that, in agreement with this theory, the scattering of the photon and the recoil of the electron occur *at the same time*. They measured the coincidences between scattered photons and recoil electrons by means of suitable counters.

The foregoing treatment of the Compton effect did not take into account that the scattering electrons are not at rest before being hit by the photons, but are orbiting around the nucleus. The initial momenta of the scattering electrons, statistically distributed with respect to the incoming photons, cause a broadening of the Compton line as expected from computations so that even in these fine details there is complete agreement between experiment and photon theory. In the interest of later application we mention already here that, in principle, also an inverse Compton effect should exist in which a fast, high-energy electron transfers energy and momentum by impact to a low-energy photon, thus "displacing" the photon from the visible spectrum into the range of γ-radiation, whereas the electron loses the corresponding energy. According to V, 20a, it may well be that this effect plays a certain role in areas of the universe with high photon density so that it has to be considered in theoretical discussions of cosmic radiation.

It is essential for our discussion that the theory of Compton scattering proves to be in quantitative agreement with the experimental results *only if the photon is assumed to have particle properties and may transfer energy and momentum to the electron with which it collides, just like a billiard ball.* All consequences of this photon theory are experimentally confirmed, namely the simultaneous occurence of scattering and recoil, the correlation between the angles ϑ and φ which corresponds to the law of conservation of momentum, with $h\nu/c$ always being the momentum of the photon, and finally the wavelength shift $\varDelta\lambda$ of the scattered radiation corresponding to the energy loss of the photon as computed from the conservation laws of energy and momentum.

We thus know of light phenomena, such as interference and diffraction, which can be explained only by the wave theory, and of other phenomena, such as the photoelectric effect, the absorption of light, and the Compton effect, which can be explained easily only by the particle concept. *The earlier assumption that wave and particle concepts were mutually exclusive has yielded today to the conviction that light may be regarded as an extended field of waves or as point-like particles depending upon the experiments performed: An elementary process can always be explained by the particle concept, while wave phenomena (e.g., an interference pattern) follow from the statistical behavior of the particles.* Later we shall have to discuss the meaning and significance of the complementarity of wave and particle, which was stressed especially by BOHR. At present we accept it as a mere *fact* derived from experiments.

The converse of our statements about radiation holds for matter. The phenomena of the kinetic theory of gases seem to provide such clear evidence for the corpuscular nature of molecules and atoms, and numerous cathode-ray experiments have so plainly shown that the electron with its mass and velocity appears as a definite corpuscular particle that it seemed almost absurd, before 1924, to question the particle nature of matter.

However, following the theoretical discussions of DE BROGLIE, to be reviewed presently, DAVISSON and GERMER as well as G. P. THOMSON in 1927 obtained diffraction patterns from electron beams scattered by crystals. These patterns, except for small deviations due to a difference of the refractive index, are completely analogous to the X-ray diffraction patterns of VON LAUE or DEBYE-SCHERRER. Fig. 95 shows an example of electron diffraction by a thin metal foil, and, for comparison, an X-ray diffraction photograph of the same material. These experiments have to be regarded as evidence of the wave nature of the electron in the same way as LAUE patterns are regarded as proof for the wave nature of X-rays. The technique of electron diffraction, in the meantime, has become common knowledge to physicists and chemists, who use electron as well as X-ray diffraction as a means of determining molecular and crystal structures (cf. VI, 2a and VII, 4). From electron diffraction measurements with crystal lattices of known lattice constants, the wavelength of the "material waves" corresponding to a beam of electrons may be determined as from X-ray diffraction. These measurements showed that the wavelength of particle waves is inversely proportional to the momentum p of the particles, i.e., for a given mass m inversely proportional to their velocity v, and that the proportionality constant is PLANCK'S constant of action;

$$\lambda = h/p = h/mv. \tag{4.9}$$

This formula is one of the most important relations of modern physics. It is called the de Broglie relation for reasons to be presented in IV, 4. For electrons having a velocity v after having been accelerated through a potential difference V according to

$$eV = mv^2/2 \tag{4.10}$$

we get; by substituting the values for m_e and h,

$$\lambda_e = \frac{12.3}{\sqrt{V}} \, \text{Å}, \tag{4.11}$$

where V is measured in volts. For example, an electron which has fallen through a potential difference of 10,000 volts has a wavelength of 0.12 Å, corresponding to that of hard X-rays.

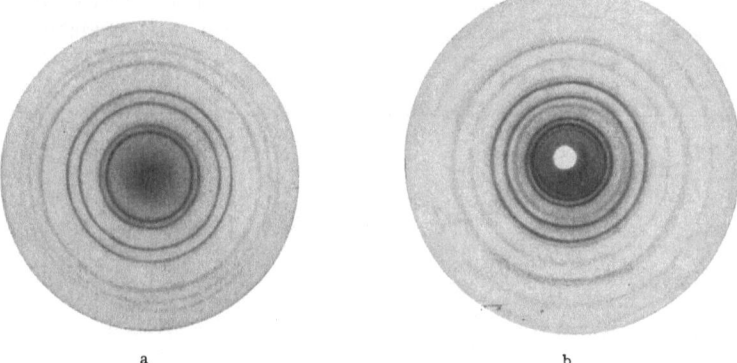

a   b

Fig. 95. Comparison of electron diffraction and X-ray diffraction by a silver foil. (After MARK and WIERL.) (a) Diffraction of 36,000-volt electrons ($\lambda = 0.06$ Å). (b) Diffraction of X-rays (copper K radiation, $\lambda = 1.54$ Å)

Diffraction experiments illustrating the wave nature of matter can be performed successfully not only with electrons but also with atomic and molecular beams. However, since the mass appears in the denominator of Eq. (4.9), the corresponding wavelength of these heavier particles for the same velocity is smaller by a factor of about 10^{-4}. Using the comparatively small velocity which atoms and molecules have due to their random thermal motion, STERN carried out diffraction experiments with these heavier particles and was able to verify Eq. (4.9). Recently, the diffraction of neutron beams, which was discovered in 1936, meets increasing interest for research on molecules and crystals.

All these diffraction experiments provide definite evidence that in suitable experiments particle beams of electrons, atoms, and molecules exhibit a wave-like character. That this wave character of matter remained undiscovered so long and that it plays no part in macrophysics, follows from Eq. (4.9) for the de Broglie wavelength. Macrophysics deals throughout with such large masses that moving bodies have, according to Eq. (4.9), associated wavelengths by far too short to be detectable. Macrophysics and its laws, therefore, are not changed by the new results or the theoretical conclusions which will be drawn from them in the next sections, in spite of the fundamental change in our conception of physics.

The apparently so fundamental difference between light waves on the one hand and material particles on the other has largely disappeared. *Light (electromagnetic radiation) as well as matter exhibit wave or particle properties complementary to each other, depending on the kind of experiments which are performed. For both of them there is no either/or, but characteristic for both is the complementarity of phenomena as described above. This is the basis of modern atomic physics.*

For clarifying the mutual relations we want to point out that duality *practically* ceases and that only wave *or* particle phenomena can be detected in the limiting cases of infinitely large or infinitely small wavelengths, respectively (corresponding to infinitely small or infinitely large quantum energy). Photons of highest energy as those found in cosmic rays manifest themselves exclusively as particles and do not show measurable wave qualities, whereas long radio waves appear only as waves, the quantum character of which, expected as complementary to their wave character, cannot be detected any more.

3. Heisenberg's Uncertainty Principle

Before familiarizing ourselves with the characteristic features of the wave-mechanical formalism based on the complementarity between waves and particles, we discuss a fundamentally important consequence of this dualism which was

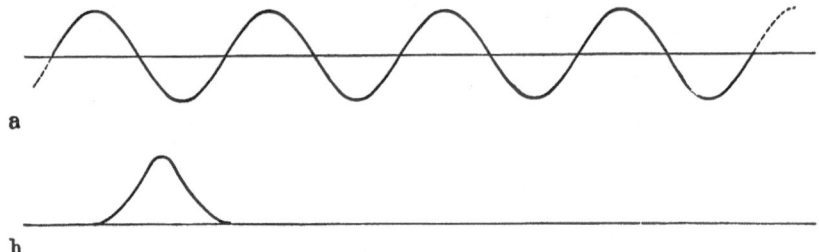

a

b

Fig. 96. (a) Infinitely extended wave train and (b) single wave pulse

described above. We ask the question: How can a physical system, which requires for its complete physical description wave and particle properties, be described unambiguously with respect to its behavior in space *and* time? In investigating this problem, we find a very peculiar and significant uncertainty, discovered by Heisenberg, which applies to *simultaneous* statements on position and momentum or energy and time of the system.

Let us begin with the wave picture and think of an infinitely long monochromatic wave train (cf. Fig. 96a). Its wavelength λ can be measured accurately by means of a spectrograph. From the wavelength we obtain, by means of the well-known wave relation

$$\nu = c/\lambda, \qquad (4.12)$$

with corresponding accuracy the frequency ν and the momentum

$$p = h\nu/c = h/\lambda \qquad (4.13)$$

of the photons which, in the particle picture, correspond to the wave. According to Eq. (4.13), the momentum is connected with the wavelength in the same manner as is the electron momentum with the corresponding de Broglie wavelength (4.9) according to the experiments described above. This is true quite generally, although Eq. (4.12) is valid only for light waves, as we shall learn on page 159. We see that for waves with an infinitely long monochromatic wave train, the wavelength and thus, by means of the fundamental relation (4.13), the momentum p of theas sociated particles may be determined accurately. The momentum, from the standpoint of our derivation, is to be regarded as a *wave property*. However, we note that no statement is possible in this case about the *position* (which according to Bohr is complementary to the momentum) of any particular photon, since the wave train (or by transition to three coordinates, the three-dimensional wave field) evidently does not permit any coordination between the corresponding photons and their particular positions in space.

Conversely, for determining the position of a photon we must, when using the wave presentation, go from an infinitely extended wave train (Fig. 96a) to an increasingly shorter wave train, which in the limiting case is reduced to a single maximum (Fig. 96b). Evidently then the corresponding photon must be looked for at the location of this maximum, so that its position is thus known. However, the attempt to determine *simultaneously* the momentum of the photon from Eq. (4.13) by measuring the wavelength now fails since the Fourier analysis, which is necessary for this measurement, by means of a spectrometer reveals a continuous spectrum with all wavelengths between 0 and ∞, though of course with varying intensity. The wavelength and the momentum of the photons, associated with the wave pulse according to Eq. (4.13), thus cannot be determined in this case in which the position of the photon is known accurately.

Fourier analysis teaches us in detail that the representation of a single pulse or "wave package", as shown in Fig. 96b, requires a band of wavelengths or frequencies that is the wider the more sharply the wave maximum is limited in space, since the relation holds

$$\Delta x \approx \frac{1}{\Delta(1/\lambda)}.\tag{4.14}$$

For dimensional reasons we here measure the band width in wave *numbers:* Since the uncertainty of position Δx has the dimension of length, the same must be true for the right-hand side. By substituting for the uncertainty in wave number the uncertainty in momentum with the aid of the de Broglie relation (4.13), we obtain

$$\Delta x \Delta p \approx h.\tag{4.15}$$

Consequently, in agreement with the result of our simple treatment above, position and momentum (velocity) of a particle cannot be simultaneously determined with accuracy. *The product of the error in measuring the position and that of measuring the momentum is at least[1] of the order of magnitude of* Planck's *constant h*. This is Heisenberg's famous *uncertainty principle* which generally applies to so-called canonically conjugated variables such as position and momentum. Canonically conjugated variables are such related quantities whose product has the dimension of an action (energy $\times$ time $=$ g cm^2 sec^{-1}). The uncertainty principle always relates two variables one of which is better described by the wave concept, the other one by the particle concept (cf. our previous example). It thus avoids contradictions which might arise from the existence of the wave-particle dualism in describing experiments. For in each of the two complementary descriptions, exact statements can be made only about such properties about which the other description does not allow any or only inaccurate statements.

Different methods of deducing the uncertainty principle lead to different figures at the right-hand side of Eq. (4.15), namely h, $h/2\pi$, or $h/4\pi$, depending

---

[1] In this connection we speak of "at least" because in addition to this fundamental uncertainty we always have an uncertainty caused by the inevitable errors of measurement.

on the definition of the uncertainty Δ. We chose for our example the maximal uncertainty and obtained h, whereas the smaller values correspond to the mean or the most probable uncertainty, respectively.

Energy and time are canonically conjugated variables also. Because these have a particular significance in atomic physics, the validity of the uncertainty principle shall be derived briefly for these variables also. In our example with wave trains of different lengths, the accuracy with which the time can be determined, for instance the transit time of a photon through a fixed plane, is the greater and, accordingly, the time error Δt the smaller the shorter the length of the wave train is. However, the shorter the length of the wave train the wider, according to the Fourier analysis, is the frequency range $\Delta\nu$ of the waves forming the wave train. Consequently

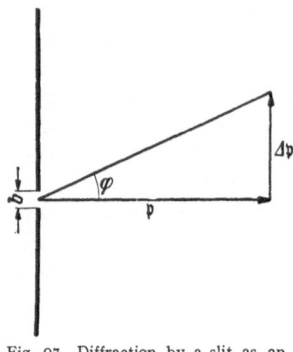

$$\Delta t \approx \frac{1}{\Delta\nu}. \qquad (4.16)$$

Fig. 97. Diffraction by a slit as an example of the uncertainty principle. b = slit width, φ = diffraction angle, p = initial momentum of the photon, Δp = change of momentum of the photon by the diffraction

We combine here the uncertainty of time with the corresponding uncertainty of frequency in such a way that Eq. (4.16) is dimensionally correct. From the basic Eq. (4.2) of quantum theory we get

$$\Delta E = h\Delta\nu, \qquad (4.17)$$

and thence

$$\Delta t \Delta E \approx h. \qquad (4.18)$$

The uncertainty principle thus holds in this case also: The more exactly the time of an atomic process is determined, or the shorter the duration of the process, the less exact is the knowledge we can obtain on the energy change of the atomic system under consideration.

An important example for this form of the uncertainty principle is the natural width of spectral lines, which we have considered on page 136. The experiment proves only that an excited atom emits its excitation energy within a time interval called the life time τ of the excited state. We thus have to accept τ as the uncertainty of time, Δt, of the emission process. For atoms, τ has in general the order of magnitude of 10^{-8} sec, but it may be much larger for forbidden transitions, and much smaller if the excited state is very unstable and if radiationless transitions (see page 137) may occur. To this uncertainty of time there corresponds, according to (4.18), an uncertainty of the emitted energy which manifests itself as a finite "natural" width of the emitted spectral line. For normal optical lines this width is close to 10^{-4} Å, corresponding to $\tau \approx 10^{-8}$ sec. If the states are unstable, however, or for nuclear γ spectra (cf. V, 6d) with an extremely short life time of the excited nuclei (usually $<10^{-14}$ sec) it is very large and may easily be measured. Conversely, the long life time of metastable states (see III, 14) causes very sharp lines; only recently has it become possible to determine these small widths with the aid of high-frequency spectroscopy. These investigations on the relation between the life time of excited atomic or molecular states and the width of the corresponding spectral lines may be considered a direct confirmation of the uncertainty principle (4.18).

Because of its basic importance, we want to look at the uncertainty principle a little further by discussing a simple experiment with waves of wavelength λ that are diffracted by a slit of width b (see Fig. 97). It does not matter here whether we deal with light waves or de Broglie waves of material particles. From elementary diffraction theory we know that the diffraction angle φ is related to the width of the slit b and the wavelength λ by

$$\lambda \approx b \sin \varphi. \qquad (4.19)$$

Thus the smaller the slit width b the larger is the diffraction angle φ. If we go from the wave picture of diffraction to the particle picture, the slit width b determines the position (coordinate q) of a photon, which passes the slit, up to an uncertainty Δq of the order of the slit width,

$$\Delta q = b. \qquad (4.20)$$

The diffraction of a wave by an angle φ corresponds in the particle picture to a change of the vectorial particle momentum $\vec{p}$ by the amount $\overrightarrow{\Delta p}$. If we assume that

$$|\overrightarrow{\Delta p}| \ll |\vec{p}|, \qquad (4.21)$$

then φ is small and

$$\Delta p = p \tan \varphi \approx p \sin \varphi. \qquad (4.22)$$

By substituting $\sin \varphi$ from (4.19) and using (4.13) we get

$$\Delta p = \frac{p\lambda}{b} = \frac{h}{b} \qquad (4.23)$$

and finally from Eq. (4.20)

$$\Delta p \, \Delta q = h. \qquad (4.24)$$

The previous derivations of the uncertainty relations by means of Fourier analysis and from the diffraction experiment show unambiguously that the uncertainty in question originates from the wave-particle dualism and is not caused by the disturbance of the quantity to be measured by the measurement of its complementary magnitude. This impression might be created by regarding exclusively an imaginary experiment which HEISENBERG first discussed: One might hold against the uncertainty principle that both variables, e.g., position and momentum of a projectile, can be determined by two successive snapshots. This is indeed true for *macro-physical* objects, because here the experimental uncertainty of the determination of position and momentum is by far larger than the quantum-mechanical uncertainty (4.24). If we apply, however, the same method to an electron, we meet *fundamental* difficulties. According to an elementary law of optics, the inaccuracy of the position, Δq, the so-called resolving power, e.g., of a microscope, depends on the aperture angle $2u$ (cf. Fig. 98) of its lens and on the applied wavelength λ by

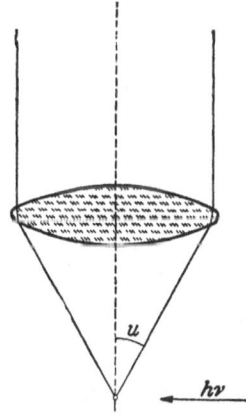

Fig. 98. Thought experiment concerning the uncertainty principle: microscopic observation of an electron. Arrow $h\nu$ indicates the direction of the incident light; u = half angle of the microscope lens

$$\Delta q = \frac{\lambda}{\sin u}. \qquad (4.25)$$

The position thus may be determined more accurately the shorter the wavelength used for "illuminating" the electron is. On the other hand, the inaccuracy of the momentum to be determined by a second observation then becomes the larger since the original momentum of the electron under consideration is changed uncontrollably by the impact of the "illuminating" photon. The position of the electron is determined by deflecting (scattering) at least one photon by the electron from its primary direction into the microscope. Because of the Compton effect, the electron by this deflection process receives a recoil of the order of magnitude $h\nu/c$. This change of the momentum $\overrightarrow{\Delta p}$ cannot be determined precisely, because all that is known about the scattered photon is that it is scattered into the solid angle $2u$ (cf. Fig. 98). So the uncertainty of the momentum is

$$\Delta p = \frac{h\nu}{c}\sin u, \qquad (4.26)$$

and from Eqs. (4.25) and (4.26) we get again Eq. (4.24). The decisive point here is that the Compton effect causes a change of the momentum which cannot be avoided or determined with sufficient accuracy. This is in sharp contrast to macrophysical measurement, e.g., of the temperature of a liquid which is changed by the thermometer used. This change, namely, can be kept small by using a small thermometer, and it can be computed from the heat capacity of the thermometer. This is *not* possible in this or any similar atomistic experiment.

Thus, HEISENBERG'S uncertainty principle is not in the least concerned with any inaccuracy of experimental results because of imperfect measuring devices, but it states a fundamental limit for the possibility of measuring complementary properties of an atomic system, as a consequence of the wave-particle dualism.

We finally mention another surprising and important consequence of the uncertainty principle. It is known from a number of independent observations that vibrational systems such as molecules or crystals do not exist in a state of complete rest but possess a certain amount of vibrational energy even at zero temperature. This so-called zero-point energy is a direct consequence of the uncertainty principle: Let us consider a mass point vibrating about an equilibrium position. According to classical physics, at zero temperature this mass would be expected at the point of lowest potential energy ($U=0$), i.e., in its equilibrium position, where its kinetic energy E_k would be zero also. The uncertainty principle, however, does not allow $U=0$ and $E_k=0$ simultaneously: $U=0$ would mean infinitely sharp positioning of the mass in the potential minimum, i.e. $\Delta q=0$. But then, according to Eq. (4.24), the momentum p and with it the kinetic energy $E_k=p^2/2m$ would be completely indefinite, i.e., they might have any value between 0 and ∞. On the other hand, $E_k=0$ means $p=0$ and therefore exact knowledge of the momentum ($\Delta p=0$). According to the uncertainty principle, then the *position* of the mass would be completely undetermined and the particle could not be found in the potential minimum. In IV, 7c we shall learn that the wave-mechanical treatment of the harmonic oscillator automatically yields a lowest energy state of the system with a certain finite vibrational energy (zero-point energy), and that the position of the particle is given only within a certain statistical uncertainty with maximum probability at the potential minimum. This wave-mechanical result, which is in complete agreement with the results of molecular spectroscopy (see VI, 6), may thus be considered as further direct proof of the uncertainty principle. A very impressive consequence of the zero-point energy is the phenomenon of the superfluid helium II which shall be treated in VII, 17b.

4. De Broglie Waves and their Significance for Bohr's Theory

By presenting the wave-particle dualism in IV, 2 we considerably overtook the historical development. The wave qualities of matter were completely unknown before 1924, and the decision between wave or particle nature of light was still looked for. In this year L. DE BROGLIE expressed the idea of a general wave-particle dualism and, for the first time, attributed wave qualities to material particles also. Three trains of thought led him to this startling assumption which was a little later so splendidly confirmed by the electron diffraction experiments (see page 152). First, it was his metaphysical belief in a general harmony of nature that made him ask whether the clearly established wave-particle dualism of light should not be valid for matter also. Secondly, he found the idea promising to regard the stationary states of the atoms, which so far had been unsatisfactorily "explained" by BOHR's postulates (see page 57), as energy states of stationary standing waves, in analogy to the stationary discrete vibrational states of a violin string or a drum. Thirdly, DE BROGLIE wanted to prove that the basic quantum equation $E=h\nu$, if applied to material systems, in combination with the theory of relativity logically leads to the assumption of waves that are connected with the particle motion, and that their wavelength as a function of the particle's mass and velocity is given by his celebrated formula

$$\lambda = \frac{h}{mv}. \tag{4.9}$$

For the case of *light*, the relation between energy E and momentum p of photons to the frequency ν and wavelength λ of the corresponding waves follows from a small modification of the previously used equations as

$$E = h\nu, \tag{4.27}$$

$$p = mc = \frac{h\nu}{c} = \frac{h}{\lambda}. \tag{4.28}$$

By applying the theory of relativity to Eq. (4.27) ,i.e., by considering an observer in motion with respect to the vibrating system, DE BROGLIE showed that Eq. (4.28), if applied to a particle of mass m and velocity v, remains valid in the form

$$p = mv = h/\lambda \tag{4.29}$$

if only the general proportionality between energy and frequency is taken into account as in Eq. (4.27). The statement $p = h\nu/c$, however, which was included in Eq. (4.28), applies to photons only. Eq. (4.29) evidently leads to the de Broglie wavelength (4.9), which was proved valid for moving particles (see page 153).

The propagation velocity u of a monochromatic wave to be ascribed to a beam of particles of definite velocity has to be well distinguished from the particle velocity v. This so-called phase velocity u may be computed from Eq. (4.29) with $\nu = E/h$:

$$u = \lambda\nu = \frac{\lambda E}{h} = \frac{E}{p} = \frac{E}{mv} = \frac{mc^2}{mv} = \frac{c^2}{v}. \tag{4.30}$$

Thus u is inversely proportional to the particle velocity v. According to Eq. (4.30), the real particle velocities, being smaller than the velocity of light, thus correspond to phase velocities u of the corresponding waves which are always larger than the velocity of light, c. This result does not contradict the theory of relativity as we might think at first sight since this theory requires only the non-existence of a velocity higher than c that can be used for transmitting signals. The phase velocity of waves, however, cannot be utilized for signal transmission because all wave peaks are identical. Another important result from Eq. (4.30) is the fact that the *phase velocity of the de Broglie waves depends on the wavelength, i.e., the waves show dispersion.* Since v is inversely proportional to λ, according to Eq. (4.29), u is directly proportional to λ.

By considering the wave-particle dualism from the viewpoint of DE BROGLIE'S theory, we arrive at an important consequence of the above result. A monochromatic wave train corresponds to a beam of particles of identical velocity v. The direction of propagation of the de Broglie waves corresponds to the beam direction in the particle picture. What was said in the section on the uncertainty principle holds for the relation of the wave characteristics to the particle properties. An extended wave field (or, in one dimension, a wave train) permits the exact determination of the wavelength of the particle wave and, consequently, the momentum (and the velocity) of the corresponding particles. A specification of the position of the particle in this case is impossible. A well localized particle, however, corresponds in the wave picture to a wave packet formed by superposition of many waves of somewhat different frequency. According to the general wave theory, such a wave packet is propagated with the so-called group velocity which is different from the phase velocity and can be proved to be identical with the particle velocity v. The more accurately it is possible to determine the position of the wave packet, the more numerous are the component frequencies which form the packet and are revealed by Fourier analysis, the less exactly, on the other hand, wavelength, momentum, and velocity of the associated particle can be determined. If such a wave package moves, the dispersion of its component waves [see Eq. (4.30)] causes a change of its form; it spreads the faster the more concentrated it originally was. This result corresponds exactly to the uncertainty relation, according to which the future position of a particle is the less determined, because of the uncertainty of momentum, the more exactly its position is given at the time $t=0$. *Since such a "packet" of de Broglie waves thus satisfies the uncertainty relation, we may use it as a good means for describing a "particle" in wave-mechanics.*

Let us try to understand the full significance of this correlation by considering it from a somewhat different point of view. The same physical phenomenon, e.g., a cathode ray consisting of electrons of given velocity v, can be described in two apparently irreconcilable ways: as a number of dimensionless point masses and as an infinitely extended field of de Broglie waves of wavelength $\lambda = h/p$. However, our experimental knowledgle tells us that in all practical cases this last conception is far too extreme. Evidently the location of an electron is at the very least confined to the discharge tube or the cathode ray. This confinement with respect to position, according to the uncertainty principle, implies an uncertainty Δp for the momentum which is inversely proportional to the positional confinement. This indefiniteness of p, according to the de Broglie relation (4.29), means a scattering of the wavelengths of the electron waves about a most probable mean value. By the superposition of these somewhat different wavelengths a wave group is formed, a wave packet, which is propagated with a group velocity v equal to the particle velocity. Such a wave packet has an amplitude differing from zero essentially only in a very limited region of space. Thus we obtain within the wave picture a much better representation of a beam of electrons than by the unsatisfactory infinitely extended de Broglie wave with exactly fixed wavelength at which we would have arrived by a superficial formal derivation. The wave-packet concept stands, as we see, to a certain degree between the two extremes of the dimensionless particle and the infinitely extended wave.

It is obvious that DE BROGLIE's theory, which started from light waves, must comprise also, as a limiting case, these electromagnetic waves. For these waves we get from Eq. (4.30) $u=c$; in other words, the propagation velocity of the photons is identical with the phase velocity of the corresponding electromagnetic waves. Furthermore, no dispersion exists for them, which makes it possible to use wave packets of electromagnetic waves in radar technique, where they do *not* spread, in contrast to de Broglie wave packets. We can also show that particles traveling with the velocity $v=c$ must have the rest mass zero and thus are, at least in this respect, identical with photons. To prove this we proceed from the momentum and energy formulae which, after taking into account the relativistic mass dependence, become

$$p = \frac{m_0 v}{\sqrt{1-\frac{v^2}{c^2}}}, \qquad (4.31)$$

$$E = \frac{m_0 c^2}{\sqrt{1-\frac{v^2}{c^2}}}. \qquad (4.32)$$

By eliminating v, solving for the rest mass, m_0, and substituting Eqs. (4.27) and (4.29), we find

$$m_0 = \frac{h}{c} \sqrt{\left(\frac{v}{c}\right)^2 - \frac{1}{\lambda^2}}. \qquad (4.33)$$

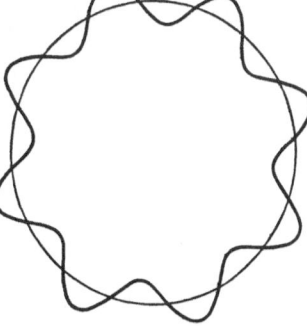

Fig. 99. Schematic representation of a two-dimensional standing electron wave in a Bohr orbit

For particles which are propagated with the velocity of light the wavelength is $\lambda=c/v$, and substitution in Eq. (4.33) gives the result

$$m_0 = 0. \qquad (4.34$$

Although the light waves are thus coordinated to the photons in the same way as de Broglie waves are to material particles, there still exists an essential difference between the two types of waves: Whereas the de Broglie waves determine only the behavior of the corresponding particles, the electromagnetic waves, being periodical changes in space and time of the electric and magnetic field strengths, have additional properties which may be observed, for example, by means of test antennae. Electromagnetic waves, therefore, are *more* than de Broglie waves which merely describe the spatial behavior of the corresponding photons. They show effects that are not described by DE BROGLIE's theory.

The wave properties of matter lead to an important new interpretation of the basic postulates of the Bohr theory with respect to the discrete quantum orbits of atomic electrons. For if the electron is regarded as a wave, a *stationary* orbit such as in Fig. 99 is distinguished from any arbitrary orbit by the fact that a *standing* wave is formed, i.e., that its circumference is an integral multiple of the electron's wavelength. Otherwise the electron wave would destroy itself in time by interference, and no stationary state of the atom would be possible. Indeed, DE BROGLIE was able to show that the arbitrarily introduced quantum condition for the principal quantum number n (page 57) is identical with the requirement that the circumference of the orbit be an integral multiple of the de Broglie wavelength of the revolving electron. For if we write BOHR's quantum condition for the special case of a circular orbit as

$$2\pi r m v = n h \qquad n=1, 2, 3, \ldots \qquad (3.6)$$

and substitute for the momentum mv of the electron the de Broglie wavelength λ according to Eq. (4.29), we obtain

$$2\pi r = n \lambda \qquad n=1, 2, 3, \ldots . \qquad (4.35)$$

Bohr's *quantum condition* (3.6), *at first seemingly arbitrary, thus is reduced to a physically significant condition of stationarity*, whereby the closed (standing) electron wave takes the place of the electron traveling around its orbit. It may be pointed out that this concept of wave mechanics also removes the much emphasized difficulty (page 57) that the electron travels in its orbit without radiating, contrary to the requirement of classical electrodynamics. If the electron actually does not, in the customary sense, revolve as a point charge and thus does not form an oscillating dipole, but as a standing wave forms a stationary time-independent system, the classical requirement of dipole radiation does not apply. The failure of the stationary orbits to radiate, and especially the stability of the ground orbits, thus is explained without contradicting the usual concepts of electrodynamics. We shall consider this in more detail in IV, 9.

In the above presentation we have not taken into account that an atom, according to observations, is a threedimensional, spherical system whereas we treated its electron as orbiting in a plane. The answer to this problem, together with all finer details of wave mechanics, follows from the quantitative development of the theory to be discussed in the following sections.

5. The Fundamental Equations of Wave Mechanics.
Eigenvalues and Eigenfunctions.
Matrix Mechanics and its Relation to Wave Mechanics

We shall now develop the formalism of wave mechanics, which is related to classical mechanics in a similar way as is wave optics to geometrical optics. In doing so, we cannot expect that the wave equation which we use for describing atomic systems and processes can be derived from the formalism of classical physics without special assumptions, since the laws of microphysics differ from the principles of classical macrophysics in essential points. Each of the numerous "derivations" of the Schrödinger equation therefore *necessarily* contains assumptions or empirical elements. As quite generally in theoretical physics, what matters is not the way leading to a formula, but *how a formalism, reached in whatever way, stands the test of describing all phenomena of atomic physics in agreement with observation.*

We thus associate with a particle or, more generally, with an atomic system a wave phenomenon Ψ. The exact physical meaning of this quantitiy Ψ which varies periodically in time and space and which travels with the phase velocity u will be discussed later. This Ψ wave, which Schrödinger introduced in further developing the ideas of de Broglie, must satisfy the general wave equation

$$\Delta \Psi = \frac{1}{u^2} \frac{\delta^2 \Psi}{\delta t^2} \qquad (4.36)$$

where

$$\Delta = \frac{\delta^2}{\delta x^2} + \frac{\delta^2}{\delta y^2} + \frac{\delta^2}{\delta z^2} \qquad (4.37)$$

is the so-called Laplace operator. For the velocity of propagation u of this Ψ wave we use, according to Eq. (4.30),

$$u = \lambda v = \frac{h v}{m v}. \qquad (4.38)$$

By this formulation we introduce, as the foundation of our derivation, the de Broglie postulate of the general validity of the basic equation $E = h v$ for the entire field of phyiscs. If we substitute for the particle momentum $m v$ an expression containing the total energy E and the potential energy U of the system according to the relation

$$\frac{m}{2}v^2 = E - U,\qquad(4.39)$$

we obtain for the phase velocity

$$u = \frac{h\nu}{\sqrt{2m(E-U)}}\qquad(4.40)$$

and by substituting Eq. (4.40) in the general wave equation (4.36), the equation for the Ψ wave

$$\Delta\Psi = \frac{2m(E-U)}{h^2\nu^2}\frac{\delta^2\Psi}{\delta t^2}.\qquad(4.41)$$

We now separate the quantity Ψ into two factors, one depending only on space coordinates and the other oscillating with a frequency ν, depending only on time,

$$\Psi(x, y, z, t) = \psi(x, y, z)e^{-2\pi i\nu t},\qquad(4.42)$$

in which ψ is assumed not to vary with time. By differentiating (4.42) twice with respect to time, we obtain

$$\frac{\partial^2\Psi}{\partial t^2} = -4\pi^2\nu^2\Psi\qquad(4.43)$$

and by substituting Eq. (4.43) in Eq. (4.41) under consideration of Eq. (4.42) we obtain *the famous Schrödinger equation*

$$\Delta\psi + \frac{8\pi^2 m}{h^2}(E-U)\psi = 0\qquad(4.44)$$

which describes the space dependence of Ψ. Because of the Laplacian operator Δ, this is a homogeneous partial differential equation of the second order. Since the equation does not contain the time, it cannot describe atomic *processes, but properties of atomic systems in stationary states*. Its solution yields, for a given potential energy $U(x, y, z)$ of the system, e.g., an atom, the stationary energy states E_n which determine the spectra according to Chap. III. Also the quantum numbers which characterise these states may be determined from the solutions $\psi(x, y, z)$. We will become acquainted with this procedure by treating some examples in IV, 7.

Differential equations of the type of Eq. (4.44) are thoroughly familiar to physicists and mathematicians. They occur, for example, in the treatment of the elastic standing waves of a violin string, of a stretched membrane (drumhead), or of an elastic sphere. It is characteristic for all of these cases that not all arbitrary vibrations corresponding to arbitrary solutions ψ of Eq. (4.44) can exist but only certain definite wave forms and frequencies are possible, the so-called eigen-vibrations and eigenfrequencies. This is caused by the "boundary conditions", i.e., the fact that both ends of a string are held fast, or that the drumhead is fastened at its circumference. This is easy to see in the one-dimensional case of the string. Since both ends of the string are held fast and thus are unable to move, they must be nodes of vibration, and a stationary vibration is possible only if the length L of the string is an integral multiple of a half wavelength:

$$L = n\frac{\lambda}{2}\qquad n = 1, 2, 3, \ldots.\qquad(4.45)$$

We recognize the formal similarity of this condition for the stationary vibration of a vibrating string with the quantum condition (4.35) for a standing electron wave on a Bohr orbit (see page 161). The case of the vibrating membrane is

more complicated but fundamentally the same. Here the circular fastening of
the membrane, as on a drum, acts as a boundary condition that permits only
discrete wave forms and frequencies. Mathematically it turns out that the differential equation with a given boundary condition (e.g., $\psi=0$ at the fixed circumference of the membrane) does not have solutions corresponding to stationary
wave forms for *all* arbitrary frequencies, but only for certain "eigenfrequencies"
of the vibrating membrane. The wave forms associated with these eigenfrequencies are called the eigenfunctions of the vibrating membrane.

From these macromechanical examples we understand why vibration problems which are characterized by certain boundary conditions are called *boundary*
value problems or *eigenvalue problems*. These mathematical characteristics, which
we shall find in all applications of the Schrödinger equation, are common to all
vibration problems and do not present, as the beginner may think, a mysterious
peculiarity of wave mechanics.

The analogy between the eigenvalues of the vibrating string and the quantum
conditions of the Bohr theory has played a great role in the historical development
of wave mechanics by SCHRÖDINGER, whose first paper bore the now understandable title " Quantization as an Eigenvalue Problem."

The Schrödinger equation (4.44) thus is, in general, an eigenvalue equation,
since for all stationary states of an atomic system we have the boundary condition
that the system and with it the corresponding wave phenomenon is limited in
space. This means that at infinity ψ must be zero. Thus in all stationary cases
the Schrödinger equation has solutions only for certain eigenfrequencies ν. Because of $E=h\nu$ these eigenfrequencies correspond to certain eigenvalues of the
energy which are the stationary energy states of the system. The solutions of the
Schrödinger equation, the eigenfunctions ψ which belong to these eigenvalues of
the energy, characterize completely the behavior of the system in these energy
states, e.g., that of an electron in the stationary states of an atom. We must
therefore distinguish between *wave functions* in general and the specific *eigen-*
functions which are associated with definite (quantized) eigenvalues of the energy.

Let us now consider a beam of electrons moving in free space with a velocity v.
The potential energy of this system then is zero ($U=0$), and the Schrödinger
equation (4.44) (by neglecting the interaction of the electrons) has the one-
dimensional form

$$\frac{\partial^2 \psi}{\partial x^2} + \frac{8\pi^2 m}{h^2} E\psi = 0 \tag{4.46}$$

where E is equal to the kinetic energy of the cathode-ray electrons. This differential equation is *not an eigenvalue problem since in this problem of free electrons*
no boundary condition limits the number of solutions. The theory of differential
equations tells us that this equation has a solution for *each* arbitrary value E
and that these solutions are periodic functions of the form

$$\psi(x) = e^{ikx}. \tag{4.47}$$

By introducing this expression into Eq. (4.46), we obtain for the constant k the
relation

$$k = \frac{2\pi}{h}\sqrt{2mE} = \frac{2\pi p}{h} = \frac{2\pi}{\lambda} \tag{4.48}$$

with p being the momentum and λ the de Broglie wavelength of the freely moving
electrons, so that the wave function describing these free electrons is given by

$$\psi(x) = e^{\frac{2\pi i x}{\lambda}}. \tag{4.49}$$

The treatment of the simple problem of freely moving electrons in the cathode ray by means of the Schrödinger equation consequently leads to the well-known result that the de Broglie wave corresponding to freely moving electrons is a sine wave with a wavelength given by Eq. (4.9). As there is no limitation for the electron velocity, there must be solutions for *all* E values.

For the case of stationary systems, which is of predominant interest in atomic physics, some further important statements can be made about the eigenfunctions ψ. First we notice that in general the eigenfunctions are complex, but that a real quantity, the "norm" of the eigenfunction, is obtained by multiplying ψ by its complex conjugate ψ^*. We shall discuss the physical significance of this norm $\psi\psi^*$ in the next section. If for each eigenvalue of the system we have only one eigenfunction, i.e., wave form, then the system is called non-degenerate. If, on the other hand, there are n eigenfunctions belonging to the same energy eigenvalue, we speak of an $(n-1)$ fold *degeneracy*.

We mentioned already that the Schrödinger equation (4.44) is a *homogeneous* differential equation. Therefore, its solution is given except for a constant factor. In order to compare various eigenfunctions, we determine this factor by the definition that the integral of the norm of the eigenfunction, taken over the whole space τ, be unity,

$$\int \psi\psi^*\, d\tau = \int |\psi|^2\, d\tau = 1. \qquad (4.50)$$

An eigenfunction determined in this manner is called *normalized*. We can discuss the physical significance of this normalization only in connection with the interpretation of the Ψ function to be given later.

Furthermore, our eigenfunctions are *orthogonal*. That means for two eigenfunctions ψ_m and ψ_n which belong to different eigenvalues E_m and E_n the following so-called orthogonality condition holds:

$$\int \psi_m\psi_n^*\, d\tau = 0. \qquad (4.51)$$

To prove this we start from the Schrödinger equations for ψ_m and ψ_n^*,

$$\Delta \psi_m + \frac{8\pi^2 m}{h^2}(E_m - U)\psi_m = 0, \qquad (4.52)$$

$$\Delta \psi_n^* + \frac{8\pi^2 m}{h^2}(E_n - U)\psi_n^* = 0. \qquad (4.53)$$

We multiply the first by ψ_n^* and the second by ψ_m and subtract the second from the first one. By integrating over the entire space we get

$$\int (\psi_n^*\Delta\psi_m - \psi_m\Delta\psi_n^*)\, d\tau = \frac{8\pi^2 m}{h^2}(E_m - E_n)\int \psi_m\psi_n^*\, d\tau. \qquad (4.54)$$

We transform the volume integral on the left by GREEN's theorem into a surface integral taken over an infinitely extended sphere

$$\int (\psi_n^*\Delta\psi_m - \psi_m\Delta\psi_n^*)\, d\tau = \oint (\psi_n^*\,\mathrm{grad}\,\psi_m - \psi_m\,\mathrm{grad}\,\psi_n^*)\, df \qquad (4.55)$$

which vanishes because the eigenfunctions ψ_m and ψ_n vanish at infinity. Therefore, we get from Eq. (4.54)

$$(E_m - E_n)\int \psi_m\psi_n^*\, d\tau = 0. \qquad (4.56)$$

This is the orthogonality condition since it was assumed that $E_m \neq E_n$. We shall not discuss here the degenerate case, $E_m = E_n$.

Another property of the wave functions or eigenfunctions that only recently became important is their parity. An eigenfunction is called *even*, or its parity *positive*, if its sign remains the same in case that the signs of all its coordinates are reversed. The eigenfunction is *odd*, however, or its parity *negative*, if the sign of the eigenfunction changes by this operation of *being reflected at the origin of the coordinates*. As an example, we mention the eigenfunctions (4.94) of the harmonic oscillator (cf. IV, 7c). From Eq. (4.94) we see that their parity is alternatingly positive and negative.

So far we have discussed only stationary states, which are of predominant interest in atomic physics; they do not change in time and are described by the Schrödinger equation (4.44). In our derivation, the time dependence of Ψ had been eliminated by assuming the validity of Eq. (4.42).

However, for all those problems of atomic physics in which the state of an atomic system changes in time, e.g., for the absorption or emission of radiation connected with changes of atomic states, a time-dependent Schrödinger equation is needed. We obtain it by eliminating from Eq. (4.44) the eigenvalue of the energy E which is characteristic for a stationary state. By substituting in Eq. (4.42) $\nu = E/h$ and differentiating with respect to time, we obtain

$$\frac{\partial \Psi}{\partial t} = -2\pi i \frac{E}{h} \Psi \quad \text{or} \quad E = -\frac{1}{2\pi i} \frac{1}{\Psi} \frac{\partial \Psi}{\partial t}. \tag{4.57}$$

By substituting this in the space-dependent Schrödinger equation (4.44), we obtain the *time-dependent* Schrödinger equation

$$\Delta \Psi - \frac{8\pi^2 m}{h^2} U\Psi + \frac{4\pi i m}{h} \frac{\partial \Psi}{\partial t} = 0, \tag{4.58}$$

which describes completely the behavior in space and time of the wave phenomenon that corresponds to the atomic system.

Eq. (4.58) is superior to the equally time-dependent equation (4.41) since Eq. (4.58) does not contain the energy E and frequency ν. It is therefore more general and better adaptable to special problems than Eq. (4.41).

Now we come back again to the relation of wave mechanics to classical mechanics by writing the complete time-dependent Schrödinger equation (4.58) in the form

$$-\frac{h}{2\pi i} \frac{\partial \Psi}{\partial t} = -\frac{h^2}{8\pi^2 m} \text{ div grad } \Psi + U\Psi \tag{4.59}$$

and comparing it with the fundamental equation of classical mechanics, according to which the total energy E of a system is composed of the kinetic energy $E_k = p^2/2m$ and the potential energy U:

$$E = \frac{p^2}{2m} + U. \tag{4.60}$$

We then see that a formal transition from classical mechanics to quantum mechanics is possible if we introduce, instead of the total energy E and the momentum p, certain differential operators that have to be applied to the wave functions Ψ.

They are

$$E \triangleq -\frac{h}{2\pi i} \frac{\partial}{\partial t}, \tag{4.61}$$

and

$$\vec{p} \triangleq \frac{h}{2\pi i} \text{ grad}. \tag{4.62}$$

By means of these relations which we have to regard as postulates, we can develop the entire formalism of quantum mechanics from classical mechanics.

The justification for this procedure lies in the experimental confirmation of the Schrödinger equation (4.58). We mention, however, that analogous equations play a similarly decisive role in the Hamilton-Jacobi theory of classical mechanics. The total energy (Hamilton function) of the system is here the negative partial differential with respect to time of the so-called action function, whereas the momentum appears as gradient of the action function with respect to the space coordinates, in full analogy with Eqs. (4.61) and (4.62).

Let us now look briefly at the differences between the classical and the wave-mechanical treatment of atomic problems. The difference begins already with the fundamental approach. Classical mechanics always proceeds directly or in-directly from the fundamental Newtonian equation

$$\vec{K} = m \frac{d^2 \vec{r}}{dt^2}. \tag{4.63}$$

It thus asks for the effects of a force applied to a mass and by integration of Eq. (4.63) arrives at the trajectory of the mass. This classical motion has no place in wave mechanics, because the simultaneous exact knowledge of position and velocity of an atomic particle, which is a fundamental concomitant of the classical theory, is in contradiction to the uncertainty relation. Quantum mechanics starts from a potential field $U(x, y, z)$, assumed to be known, in which a particle is moving. It asks first for the stationary energy states which are characteristic for the specific atomic system (e.g., for the electron in a Coulomb field), in other words, for the eigenvalues of the Schrödinger equation. These energy states are of primary interest because of their relation to the observable spectra (cf. Chap. III). Secondly, the state functions are determined, i.e., the solutions ψ_k of the Schrö-dinger equation, which describe the behavior of the system in its different energy states. In the next section, we shall come back to their significance and to their relation to the quantum numbers. The sequence of the energy states of an atomic system depends decisively on the shape of its potential $U(x, y, z)$ that, in the classical picture, determines the forces acting upon the system and within it. As an example, the parabolic potential of the harmonic oscillator (cf. IV, 7c), which depends on the square of the elongation, corresponds to a sequence of equidistant stationary energy states, whereas the Coulomb potential, being a function of $1/r$, leads to a sequence of energy states which converge against a limit (see Fig. 32). We may reverse this argument: From an observed sequence of energy states, such as those of a nucleon in a nucleus, we can derive the potential field in which the particle in question is moving. Here it is interesting to note that *deviations of wave-mechanical results from those found classically may be expected only if the potential U of the system changes considerably within a region comparable to the de Broglie wavelength* (4.9). This is the case, for instance, for slow electrons moving in an ionic crystal lattice; and here we find diffraction of electrons, in contrast to the expectation of classical mechanics. We shall deal with other examples of this important rule later.

Before taking up the meaning of the Ψ waves, we indicate in a few words the method used by HEISENBERG, BORN, and JORDAN to solve the difficulties of the Bohr theory. While SCHRÖDINGER arrived at wave mechanics by developing the ideas of DE BROGLIE's material waves, HEISENBERG realized that the difficulties of the Bohr theory were caused by an unscrupulous application of such classical physical concepts to atomic problems whose experimental determination, as we know, is fundamentally impossible. HEISENBERG therefore radically rejected, for his new quantum mechanics, the use of any physical quantities which can not be measured by an experiment. Among the rejected quantities are, for instance, the potential energy U (necessary in the simultaneous theory of SCHRÖDINGER) because it is derived by COULOMB's law from the assumption of point-like nuclei and electrons. What we know for certain about an atom are the frequencies of its spectrum and the intensities of its spectral lines. These quantities HEISENBERG considered as given and sought to deduce all other characteristics of atomic systems and processes from them. The theory built on this basis does not use continuously variable quantities such as coordinates, but matrices consisting of discrete numbers, such as the frequencies of line spectra. It is called *matrix mechanics*. With its help an atomic theory was developed which is equivalent to wave mechanics in its completeness, but undoubtedly clearer and less open to principal objections. Many problems such as transition probabilities and selection rules can be treated much easier and clearer by matrix mechanics than by wave mechanics. For the practical treatment of problems of atomic physics, on the other hand, it is mostly simpler to use wave mechanics because the treatment of differential equations is more familiar and simpler than that of matrices. For this reason it is of interest to note that the results of both theories not only agree quantitatively but that, in spite of their different starting points, wave and matrix mechanics are mathematically equivalent, as can be proved rigorously. In the following we shall use wave mechanics exclusively and thus need not go into matrix mathematics.

6. The Physical Significance of the Wave-mechanical Expressions. Eigenfunctions and Quantum Numbers

Now that we have laid the groundwork of the wave-mechanical formalism, we ask for the physical meaning of the expressions under consideration. We learned already *that the eigenvalues E of the Schrödinger equation are stationary energy states of the atomic system which is characterized by its potential energy U, in complete correspondence to the eigenvibrations which are determined by the mechanical forces acting on a mechanical system.* We have not yet discussed, however, the significance of the wavefunction Ψ. Whereas Ψ is a complex magnitude, as we already mentioned above, its norm $\Psi\Psi^* = |\Psi|^2$, obtained by multiplying Ψ by its complex conjugate Ψ^*, is always a real function and thus has a physical significance. *As was first shown by BORN, $\Psi\Psi^*$ represents a probability. For a one-electron system, e.g., a hydrogen atom, it is the probability of observing the particle described by $\Psi(x, y, z, t)$ at the time t at the point x, y, z or, more precisely, in the corresponding volume element $(x+dx, y+dy, z+dz)$.* The normalisation condition (4.50) then expresses the fact that the probability for finding the particle *somewhere* in space must be unity.

This probability $\Psi\Psi^*$ may be plotted in such a way that we obtain representations of the electron in space as shown in Fig. 103. One is therefore inclined to speak of *electron modes*, in analogy to the *vibration modes* of a string or a membrane. But we must realize that the concept of regarding $\Psi\Psi^*$ to be the electron density in space, as was first proposed by SCHRÖDINGER, is correct only for the time average and for one-electron systems. The first limitation is easily understood since the probability of finding an electron in a volume element is equal to the electron density in this volume element when averaged over not too small a time interval.

SCHRÖDINGER's interpretation, however, leads to difficulties if one considers many-electron systems and electrons in *non-stationary* states, whereas BORN's explanation of $\Psi\Psi^*$ as a probability always remains valid. For instance, the wave function of a two-electron system contains the coordinates of *both* electrons,

$\psi(x_1, y_1, z_1, x_2, y_2, z_2)$. This, obviously, cannot be represented any more in our three-dimensional physical space, but only in a figurative sense as density in a six-dimensional so-called *configuration space* (or, generally, in a $3N$-dimensional space for N electrons). Under certain circumstances, a particle in a non-stationary state may be described by a combination of two eigenfunctions, a fact to be discussed later. In BORN's interpretation, this simply means that the particle may experimentally be detected in the one or the other state with well defined probabilities. Finally, we shall show in IV, 12 that a wave representing, in SCHRÖDINGER's sense, *one* particle may be split into several partial waves by reflection or diffraction. This means that the probabilites for finding the particle on the one or the other side of the interface are given by the norms of the Ψ functions of the partial waves in question.

The statistical interpretation of the wavefunction also makes obvious its relation to the uncertainty principle (see IV, 3). Indeed, according to wave mechanics every state of an atomic system is unambiguously and completely determined by its Ψ function. Thus there is no room left in the formalism for additional statements beyond the wave-mechanical results, such as definite values of coordinates and momenta. The uncertainty principle, regarded from the standpoint of wave mechanics, is simply a consequence of the fact that a system is described by a Ψ function which permits only statements of probability with regard to position and momentum of the particle or its behavior in general. This clearly illustrates that the uncertainty principle is not connected in any way with a perturbation by the measuring process but is a logical consequence of describing an atomic system by a Ψ function.

BORN's interpretation of the wave-mechanical formalism also allows a concept of interpreting the wave-particle dualism, details of which we shall discuss in IV, 15. This concept is accepted by many physicists and we want at least to mention it here briefly: Interpreting the eigenfunction in the statistical way, we may consider the *particles* as the primary concept (some physicists would say: the realities). We then have to assume "guiding waves" Ψ described by SCHRÖDINGER's equations (4.44) or (4.58), which as probability waves describe the statistical behavior of the particles. These guiding waves, whose physical significance remains unknown, may interfere with each other, they may be reflected or diffracted; in other words, they behave like physically real waves. The square of their amplitudes averaged over a large enough number of particles represents the spatial and temporal density distribution of the particles. This interpretation, in which nearly point-like electrons are to be found around the nucleus distributed according to $\Psi\Psi^*$, is in accordance with the Compton effect in which the colliding photon hits a single atomic electron and ejects it from the atom.

But we must stress a rather important point. It makes a great difference whether a Ψ-wave field represents a single particle or a great number of them. In the first case, a significant difference between the quantum-mechanical interpretation and the classical concept manifests itself by the fact that the wave function depends on our subjective knowledge of the physical system so that it is impossible to distinguish clearly between object and subject of observation, in sharp contrast to classical physics. Let us look at a particle in a space which may be separated into two equal parts by a fluorescent screen with a hole in it. Since the particle may be anywhere in this space, it is described by a wave function $\psi(x, y, z)$ which everywhere differs from zero. But at the moment where the particle's existence in one of the parts is proved with the aid of the fluorescent screen, the probability of detecting it in the other part suddenly has dropped to zero and, thus, also the wave function in this second part is zero. Consequently, the process of observation suddenly changes our knowledge about the particle and, with it, the wave function Ψ that describes the particle. This so-called *reduction of the wave function* as the consequence of an observation illustrates that the wave function not simply describes objectively the probability behavior of the particle, but that it depends on our knowledge about the particle. The philosophical aspects of this fact will be treated in IV, 15. The relation of the quantum-mechanical description to the classical one becomes clear, as in III, 22, in the region of large quantum numbers, i.e., for a Ψ-wave field which describes a large number of particles. There, $\psi(x, y, z)$ really indicates the objective distribution of the particles in space and does not change appreciably if a single particle is detected on this side or the other of the screen. We thus see that *only in the limiting case of large quantum numbers, the quantum mechanical and the classical description agree with each other*.

What is the relation of the description of the behavior of an atomic electron by the norm of its eigenfunction to that using the four quantum numbers n, l, m, and s? We have described these quantum numbers in Chap. III as representing what might be called the size, shape, orientation, and spin of the atomic electron. We mentioned already that the electron spin is not yet included in pre-Dirac quantum mechanics but has to be added to the wave-mechanical results afterwards. However, the other three quantum numbers must be contained in the eigenfunction ψ, which represents the solution of the vibrational system corresponding to the electron, and we may designate its norm as the vibrational mode of the electron. This mode is characterized, just as in the mechanical case of the vibration of an elastic sphere, by the arrangement of stationary points, lines, or surfaces, the vibrational *nodes*. In analogy to the eigenvibrations of an elastic sphere, the three-dimensional ψ vibration has three series of nodal *surfaces*, namely nodal spheres (spherical shells), nodal planes going through the center, and nodal surfaces of double cones with the same axis and a common point identical with the center of the system. These three series of nodal surfaces of the ψ-function, which characterize the modes of vibration of the standing ψ wave in the atom, correspond to the three Bohr-Sommerfeld quantum numbers n, l, and m. In fact, the sum of all nodal surfaces is equal to $n-1$, the number of nodal spherical surfaces equals $n-l-1$, that of the nodal conical surface equals $l-|m|$, and that of the nodal planes equals m. All this will become clear when we treat the theory of the H atom. Knowing this, we can easily translate electron symbols into the language of wave mechanics. For example, a $4p$ electron with $m=0$ is described in wave mechanics by an eigenfunction which has two nodal spheres and one nodal cone but no nodal planes. The relationship of the old Bohr-Sommerfeld quantum symbols to the eigenfunctions of quantum mechanics is thus made clear. We will become acquainted with examples presently.

7. Examples of the Wave-mechanical Treatment of Atomic Systems

After having become familiar with the fundamental facts of the formalism and meaning of wave mechanics, we now discuss a few examples of the wave-mechanical treatment of atomic systems. Two of these examples, the rotator and the oscillator, are of special interest in molecular physics. The third example, which because of its relative complexity is treated last, is the hydrogen atom which serves as a test example for every atomic theory. In treating these examples, we shall again concern ourselves less with the mathematical difficulties, which occasionally may appear to the beginner somewhat confusing, than with the fundamentally interesting physical questions, especially the resaons for the occurrence of the quantum conditions.

a) The Rotator with Rigid Axis

We begin our discussion with the simplest model of a diatomic molecule, the rotator with a fixed axis. It consists of two equal masses M which are at a fixed distance $2r$ from each other and which rotate about an axis perpendicular to the line joining the two masses (Fig. 100). This is also called the dumbbell model. This system has no potential energy; its entire energy is kinetic. Thus the Schrödinger equation (4.44) reduces to

$$\Delta\psi + \frac{8\pi^2(2M)}{h^2}E\psi = 0. \tag{4.64}$$

Since we are dealing with rotation, it is more appropriate to use polar coordinates. If we assume the rotator to be fixed in space, then the only variable is the angular coordinate φ of the rotation. Thus Eq. (4.44) can be written

$$\frac{1}{r^2}\frac{d^2\psi}{d\varphi^2} + \frac{16\pi^2M}{h^2}E\psi = 0. \tag{4.65}$$

By using the substitution

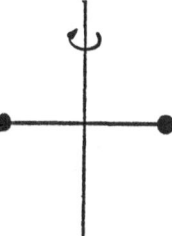

Fig. 100. Model of the rigid rotator, used as the simplest model for treating the rotation of diatomic molecules (cf. VI,9)

$$\frac{h}{16\pi^2cMr^2} = \frac{h}{8\pi^2cI} = B, \tag{4.66}$$

where $I = 2Mr^2$ is the moment of inertia of the molecule, we get the Schrödinger equation

$$\frac{d^2\psi}{d\varphi^2} + \frac{E}{hcB}\psi = 0. \tag{4.67}$$

The solution of this differential equation of the rotator with rigid axis has the form

$$\psi(\varphi) = e^{im\varphi}. \tag{4.68}$$

Substituting this solution into Eq. (4.67) we get

$$m^2 = \frac{E}{hcB} = \frac{16\pi^2Mr^2E}{h^2}. \tag{4.69}$$

As only those solutions (4.68) have a physical meaning which are single-valued, $\psi(\varphi)$ must have the same value after each complete revolution:

$$e^{im\varphi} = e^{im(\varphi + \pi)}. \tag{4.70}$$

Thus m must be an integral number. Because of condition (4.70), the quantized energy values of the rigid rotator follow from Eq. (4.69) to be

$$E_m = hcBm^2; \quad m = 0, \pm 1, \pm 2, \pm 3, \dots. \tag{4.71}$$

Because the solutions (4.68) of the Schrödinger equation (4.67) must be single-valued, the energy E of the rigid rotator cannot have any arbitrary value but the discrete eigenvalues given by Eq. (4.71), in contrast to the free non-periodic motion of electrons which is described by the same Schrödinger equation (4.46). The quantization of the rigid rotator thus is a consequence of the requirement that the eigenfunctions must be single-valued.

b) The Rotator with Free Axis

If we go from the rigid rotator to a rotator which is free in space, we need, in addition to φ, a second angular coordinate, ϑ, which describes the position of the rotation axis with respect to the coordinate system. The Schrödinger equation (4.65) of the rigid rotator therefore gets the more complicated form

$$\frac{1}{r^2}\left[\frac{1}{\sin\vartheta}\frac{\partial}{\partial\vartheta}\left(\sin\vartheta\frac{\partial\psi}{\partial\vartheta}\right)+\frac{1}{\sin^2\vartheta}\frac{\partial^2\psi}{\partial\varphi^2}\right]+\frac{8\pi^2(2M)}{h^2}E\psi=0. \tag{4.72}$$

As r is assumed to be constant (half of the rigid axis in Fig. 100), we may separate Eq. (4.72), after multiplying it by $\sin^2\vartheta$, into two parts depending respectively on φ or ϑ alone by writing

$$\psi(\varphi,\vartheta)=\Phi(\varphi)\Theta(\vartheta). \tag{4.73}$$

Eq. (4.72) thus becomes

$$\frac{\sin\vartheta}{\Theta}\frac{d}{d\vartheta}\left(\sin\vartheta\frac{d\Theta}{d\vartheta}\right)+A\sin^2\vartheta=-\frac{1}{\Phi}\frac{d^2\Phi}{d\varphi^2}. \tag{4.74}$$

From Eqs. (4.72) and (4.66) the value of the constant A follows as

$$A=\frac{16\pi^2Mr^2E}{h^2}=\frac{E}{hcB}. \tag{4.75}$$

Since the two sides of Eq. (4.74) are independent of each other, both of them may be set equal to the same constant C. Letting the right-hand side be equal to C leads to a differential equation analogous to Eq. (4.67):

$$\frac{d^2\Phi}{d\varphi^2}+C\Phi=0. \tag{4.76}$$

From its solution follows, with Eq. (4.68), the value of the constant C:

$$C=m^2 \qquad m=0,\ \pm1,\ \pm2,\ \pm3,\ \dots. \tag{4.77}$$

Here, however, m^2 is not related to E as it is in Eq. (4.69) because Eqs. (4.65) and (4.72) are different from each other. By setting the left-hand side of Eq. (4.74) equal to C and taking Eq. (4.77) into consideration, we obtain the differential equation for that part of $\psi(\varphi,\vartheta)$ which depends on ϑ:

$$\frac{1}{\sin\vartheta}\frac{d}{d\vartheta}\left(\sin\vartheta\frac{d\Theta}{d\vartheta}\right)+\left(A-\frac{m^2}{\sin^2\vartheta}\right)\Theta=0. \tag{4.78}$$

The solutions $\Theta(\vartheta)$ of this equation must be single-valued and continuous on the whole surface of the sphere. The somewhat wearisome polynomial method, to be found in every text-book of quantum mechanics, shows that Eq. (4.78) has single-valued and continuous solutions only if A has the discrete values

$$A=J(J+1)\quad\text{with}\quad J=0,\ 1,\ 2,\ 3,\ \dots. \tag{4.79}$$

Just as for the rigid rotator, *the quantization of the free rotator also is a consequence of the requirement that the solutions are single-valued and continuous.* The energy eigenvalues of the free rotator follow from Eqs. (4.79) and (4.75) to be:

$$E_J=hcBJ(J+1)=hcB(J+\tfrac{1}{2})^2-hcB/4\quad\text{with}\quad J=0,\ 1,\ 2,\ \dots. \tag{4.80}$$

In contrast to those of the rigid rotator (4.71), the energy states of the free rotator are thus characterized by *half-integral quantum numbers* $(J+\tfrac{1}{2})$. These half-integral rotation quantum numbers are in full accord with the empirical analysis of the rotational structure of the band spectra (see Chap. VI). Their explanation proved an insurmountable difficulty for the old quantum theory, it automatically follows from the wave-mechanical treatment of the rotator with an axis free in space.

The solutions $\Theta(\vartheta)$ of the differential equation (4.78), in combination with Eq. (4.79), are the so-called *associated Legendre functions* $P_J^{|m|}(\cos \vartheta)$ which may be taken from mathematical tables. Disregarding an inessential normalizing factor, we may thus write the complete eigenfunctions on the rotator with free axis in the form

$$\psi(\varphi, \vartheta) = \Phi(\varphi)\,\Theta(\vartheta) = e^{im\varphi}\, P_J^{|m|}(\cos \vartheta).\qquad(4.81)$$

Here, J is the quantum number of the total angular momentum, m is its integral component in the direction of a distinct axis, so that for J and m we have the conditions

$$J = 0, 1, 2, 3, \ldots \quad \text{and} \quad |m| \leq J.\qquad(4.82)$$

The eigenfunctions of the rotator have nodal surfaces through the center of mass. The number of the nodal surfaces equals the quantum number m or J of the corresponding energy state of the rotator. This is the wave-mechanical explanation of the rotational quantum numbers.

c) The Linear Harmonic Oscillator

The next example we discuss is the linear harmonic oscillator, which is the simplest model of the *vibrating* diatomic molecules (cf. VI, 6). A mass M which may move along the x axis may be bound to its rest position, taken as the origin of the coordinates, by an elastic force proportional to the displacement

$$K = -kx.\qquad(4.83)$$

It then vibrates through the origin with a frequency

$$\nu_0 = \frac{1}{2\pi}\sqrt{\frac{k}{M}}.\qquad(4.84)$$

The potential energy, which is taken to be zero at its minimum, i.e., at the equilibrium position, because of $K = -dU/dx$ and Eq. (4.84), is

$$U = \frac{1}{2}k x^2 = 2\pi^2 M \nu_0^2 x^2\qquad(4.85)$$

so that the Schrödinger equation of the linear harmonic oscillator is

$$\frac{d^2\psi}{dx^2} + \frac{8\pi^2 M}{h^2}(E - 2\pi^2 M \nu_0^2 x^2)\psi = 0.\qquad(4.86)$$

Following Schrödinger by substituting

$$\xi = 2\pi x \sqrt{\frac{M\nu_0}{h}}\qquad(4.87)$$

and

$$\frac{2E}{h\nu_0} = C,\qquad(4.88)$$

we get the differential equation

$$\frac{d^2\psi}{d\xi^2} + (C - \xi^2)\psi = 0.\qquad(4.89)$$

Assuming the solution to be

$$\psi(\xi) = e^{-\xi^2/2} H(\xi),\qquad(4.90)$$

we obtain for $H(\xi)$ the differential equation

$$\frac{d^2 H}{d\xi^2} - 2\xi \frac{dH}{d\xi} + (C-1)H = 0. \tag{4.91}$$

As a consequence of the physical nature of the problem, the distance of the vibrating mass from its rest position can never be infinite so that its wave function $\psi(x)$ or $\psi(\xi)$ must be zero at infinity. Because of the negative exponential function in Eq. (4.90) this is certainly the case if $H(\xi)$ stays finite through all space. That means, it must be possible to represent $H(\xi)$ by polynomials and not by infinite power series. This requirement for the solutions $H(\xi)$ of Eq. (4.91) is fulfilled only if

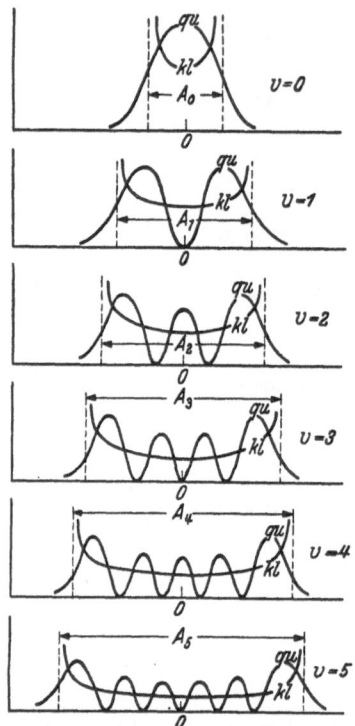

$$C = \frac{2E}{h\nu_0} = 2v+1 \qquad v = 0, 1, 2, 3, \ldots \tag{4.92}$$

is valid [cf. Eq. (4.88)], v being called the vibrational quantum number. From this we immediately obtain the possible energy values of the linear harmonic oscillator as

$$E = h\nu_0(v + \tfrac{1}{2}) \qquad v = 0, 1, 2, 3, \ldots . \tag{4.93}$$

In other words, *the physically obvious boundary condition that $\psi(x)$ must be zero at infinity leads to the quantum condition (4.93) for the energy eigenvalues.* This result is of threefold interest. First of all, we see that the energy difference of successive vibrational states, $h\nu_0$, is proportional to the eigenfrequency ν_0 of the system, while ν_0 itself is inversely proportional to the square root of the vibrating mass. *This is the reason why the quantization of the vibration energy is noticed only for the extremely small vibrating masses of atomic systems.* For macroscopic vibrators, the energy states are so closely spaced that they cannot be distinguished from the classical case where all vibrational energies would be possible. Secondly *we find, according to Eq.(4.93), again half-integral quantum numbers in agreement with the results of band spectroscopy*

Fig. 101. Classical and wave-mechanical treatment of a harmonic oscillator. The probability of finding the oscillator at a certain distance from the rest position is plotted for the first six quantum-mechanically allowed vibration states; kl = according to the classical theory, qu = according to quantum mechanics. With a small probability the classical vibration amplitudes A_1 to A_5 can be exceeded according to quantum mechanics as a consequence of the uncertainty principle. (After CL. SCHAEFER)

(cf. VI, 6). *Finally, Eq. (4.93) shows that the oscillator, and thus every system capable of vibrating, has a vibrational energy of the amount $E = h\nu_0/2$ even in the lowest possible energy state $v=0$, i.e., even at zero absolute temperature. This empirically known so-called zero-point energy could not be explained by the old quantum theory* (see, however, page 158).

The solutions $H(\xi)$ of Eq. (4.91) are HERMITE's polynomials to be taken from mathematical tables. They are for the first four v values:

$$\left. \begin{aligned} H_0(\xi) &= 1 \\ H_1(\xi) &= \xi \\ H_2(\xi) &= \xi^2 - 1 \\ H_3(\xi) &= \xi^3 - 3\xi. \end{aligned} \right\} \tag{4.94}$$

The vibrational eigenfunctions of the linear harmonic oscillator may thus be computed directly from Eqs. (4.90) and (4.87). Their norm $\psi\psi^*$ determines the probability density of finding the vibrating mass at different elongations; it plays an important role in molecular physics (see Chap. VI). The probability of finding the oscillator during a measurement at an elongation x is, according to classical theory, largest at the points where the oscillation reverses its direction because the mass remains there longest. This is shown for the first six vibrational states of the harmonic oscillator in the curves of Fig. 101 designated by "cl". In addition, Fig. 101 shows the values of the classical amplitudes of vibration, A_0 to A_5. The classical probability of finding the mass outside of the reversal points is zero since the potential energy there would exceed the total energy. On the other hand, the quantum-mechanical probability density is given by the norm of the eigenfunction $\psi\psi^*$ which, computed from Eq. (4.91) and designated by "qu", is also plotted in Fig. 101. We notice that the wave-mechanical character of the vibration phenomenon, neglected in the classical concept, manifests itself by the maxima and minima of $\psi\psi^*$, and that the number of maxima of $\psi\psi^*$ is equal to the quantum number v plus 1. We also see that quantum-mechanically a mass may, though with a small probability, be at a distance from the origin where it should never be according to the classical theory, i.e., outside of the reversal points. This is a consequence of the uncertainty principle (4.24) which relates the uncertainties of momentum and position in such a way that, for the oscillator, the latter one is determined by the width of the $\psi\psi^*$-maxima. The reason for the fact that the discrepancy between the quantum-mechanical result and the classical expectation is particularly large near the reversal points is that the de Broglie wavelength (4.9) of the vibrating mass is particularly large near these points because of the small velocity of the mass (cf. page 167). We finally see from Fig. 101 that in the vibration state $v=0$ the probability that the mass will be found at the center is largest, where classically the probability is smallest. With increasing v, on the other hand, the central maxima of $\psi\psi^*$ become less significant whereas the high outer maxima approach more and more the classical oscillation reversal points. Thus the higher the excited vibration state the smaller are the differences between the classical and quantum-mechanical behavior of the vibrating system. This is in agreement with our general law (treated in III, 22) that for large quantum numbers quantum physics merges into classical physics.

d) The Hydrogen Atom and its Eigenfunctions

As a last example, we discuss the wave-mechanical treatment of the hydrogen atom. It is closely related to that of the free rotator, but here the distance r of the rotating mass, i.e., that of the electron orbiting around the proton from its center, is variable also.

The potential energy of the proton-electron system is that of the Coulomb force of attraction

$$U = -\frac{e^2}{r} \tag{4.95}$$

so that the Schrödinger equation for the H atom is

$$\Delta\psi + \frac{8\pi^2 m_e}{h^2}\left(E + \frac{e^2}{r}\right)\psi = 0. \tag{4.96}$$

Written in polar coordinates r, ϑ, and φ, Eq. (4.96) becomes

$$\frac{\partial^2\psi}{\partial r^2} + \frac{2}{r}\frac{\partial\psi}{\partial r} + \frac{1}{r^2}\left[\frac{1}{\sin\vartheta}\frac{\partial}{\partial\vartheta}\left(\sin\vartheta\frac{\partial\psi}{\partial\vartheta}\right) + \frac{1}{\sin^2\vartheta}\frac{\partial^2\psi}{\partial\varphi^2}\right] + \frac{8\pi^2 m_e}{h^2}\left(E + \frac{e^2}{r}\right)\psi = 0. \tag{4.97}$$

For the solution we assume that ψ is the product of three functions, each of which depends on only one of the variables

$$\psi(r, \vartheta, \varphi) = R(r)\Theta(\vartheta)\Phi(\varphi).\tag{4.98}$$

Then Eq. (4.97) can be separated into three differential equations each of which is a function of only one coordinate.

$$\frac{d^2R}{dr^2} + \frac{2}{r}\frac{dR}{dr} + \left[\frac{8\pi^2 m_e}{h^2}\left(E + \frac{e^2}{r}\right) - \frac{A}{r^2}\right]R = 0,\tag{4.99}$$

$$\frac{1}{\sin\vartheta}\frac{d}{d\vartheta}\left(\sin\vartheta\frac{d\Theta}{d\vartheta}\right) + \left(A - \frac{m^2}{\sin^2\vartheta}\right)\Theta = 0,\tag{4.100}$$

$$\frac{d^2\Phi}{d\varphi^2} + m^2\Phi = 0.\tag{4.101}$$

The constants A and m^2, the so-called separation constants, are well known from page 172; but here we write A as

$$A = l(l+1)\qquad l = 0, 1, 2, \ldots\tag{4.102}$$

in order to indicate that l is the quantum number of the angular momentum of the electron moving around the proton, which was already introduced on page 84. The solutions $\Phi(\varphi)\Theta(\vartheta)$ of Eqs. (4.100) and (4.101), indicating the angular dependence of the eigenfunctions of the H-atom, are identical with Eq. (4.81) if we replace J by l. They are designated as *spherical harmonics*.

We still have to solve the most important of the three separated equations, the r-dependent Eq. (4.99). It is the most important part of the original Schrödinger equation because it alone contains the energy E. Making use of Eq. (4.102) we get

$$\frac{d^2R}{dr^2} + \frac{2}{r}\frac{dR}{dr} + \left[\frac{8\pi^2 m_e}{h^2}\left(E + \frac{e^2}{r}\right) - \frac{l(l+1)}{r^2}\right]R = 0.\tag{4.103}$$

This differential equation has finite, single-valued solutions R which are zero at infinity only for certain discrete eigenvalues of the energy E. They are the energy states of the H atom which we know already from the Bohr theory.

As the origin of our energy scale, we choose the state of ionization of the atom in which the proton and the separated electron are at rest relative to each other. With this stipulation the energy states of the bound electron have negative values (binding energy), while positive values of the energy E correspond to the kinetic energy of the separated free electron. In this energy scale, discussed on page 67, the bound electrons, in the elliptical orbits of the Bohr theory, have negative values and those in hyperbolic orbits (free electrons) have positive energy values.

We treat first the solutions of Eq. (4.103) for the stationary states $E < 0$. For large r-values the terms with $1/r$ and $1/r^2$ in Eq. (4.103) can be neglected and we have

$$\frac{d^2R}{dr^2} + \frac{8\pi^2 m_e}{h^2}ER = 0.\tag{4.104}$$

If we now substitute

$$\frac{h^2}{8\pi^2 m_e E} = -r_0^2\tag{4.105}$$

where r_0 turns out to be identical with the radius of the Bohr orbit of the ground state and if we use for r a new variable which is twice the radius r measured in units of r_0,

$$\varrho = \frac{2r}{r_0},\tag{4.106}$$

we obtain the "asymptotic solution" of Eq. (4.104) for large r,

$$R = \text{const } e^{-\varrho/2}. \tag{4.107}$$

For finite r or ϱ we can replace the constant in Eq. (4.107) by a function of ϱ which we shall consider later and write

$$R = e^{-\varrho/2} w(\varrho). \tag{4.108}$$

By putting this expression in Eq. (4.103), we get for $w(\varrho)$ the differential equation

$$\frac{d^2 w}{d\varrho^2} + \left(\frac{2}{\varrho} - 1\right) \frac{dw}{d\varrho} + \left[\left(\frac{\pi e^2 \sqrt{-2m_e}}{h\sqrt{E}} - 1\right) \frac{1}{\varrho} - \frac{l(l+1)}{\varrho^2}\right] w = 0. \tag{4.109}$$

We try to represent $w(\varrho)$ by a polynomial in order to be sure that R vanishes for large ϱ (each polynomial for $\varrho \to \infty$ approaches ∞ less rapidly than $e^{\varrho/2}$). For this purpose, we set

$$w(\varrho) = \varrho^l u(\varrho) = \varrho^l \sum a_p \varrho^p \tag{4.110}$$

and use this expression and its derivative in Eq. (4.109) to obtain the recursion formula which, in order to remain finite, must break off at some particular term. This will occur for the p-th term if

$$-(p+l+1) = -n = \frac{\sqrt{-2m_e} \pi e^2}{h\sqrt{E}}. \tag{4.111}$$

From this the energy eigenvalues of the hydrogen atom are

$$E = -\frac{2\pi^2 m_e e^4}{h^2 n^2} \qquad \begin{array}{l} n = 1, 2, 3, \ldots \\ n \geq l+1. \end{array} \tag{4.112}$$

By comparing this with Eq. (3.22) of page 66 we see that this is exactly the equation for the energy values of the Bohr quantum orbits of the H atom where n is the principal quantum number. The quantum condition thus is a logical consequence of the boundary condition that the solutions must vanish at infinity. If we express the degree p of the polynomial $u(\varrho)$ of Eq. (4.110) by n and l, we can write the solutions $R(\varrho)$

$$R(\varrho) = e^{-\varrho/2} \varrho^l u_{n-l-1}(\varrho). \tag{4.113}$$

Thus we have the complete eigenfunctions of the H atom, except for the normalization factors whose computation is complicated,

$$\psi(\varrho, \vartheta, \varphi) = e^{-\varrho/2} \varrho^l u_{n-l-1}(\varrho) P_l^{|m|}(\cos \vartheta) e^{im\varphi}. \tag{4.114}$$

These eigenfunctions describe completely the behavior of the electron in the stationary energy state of the H atom which is characterized by the three quantum numbers n, l, and m. Thus we have the following conditions for the quantum numbers

$$\left.\begin{array}{l} n \geq l+1, \\ l \geq |m| \geq 0, \\ m = 0, \pm 1, \pm 2, \pm 3, \ldots. \end{array}\right\} \tag{4.115}$$

The fourth quantum number, describing the spin s, according to page 148 does not follow from the simple form of quantum mechanics used here; it must be added as a supplement.

We notice that in Eq. (4.112) the quantum numbers l and m do not occur, but *only* the principal quantum number n. Accordingly, n^2 different eigenfunctions are associated with every energy eigenvalue E_n since l ranges from zero to $n-1$, and so to each l there exist $2l+1$ orientations of l, according to Eq. (4.115), i.e., $2l+1$ different m values and thus $2l+1$ eigenfunctions:

$$\sum_{l=0}^{l=n-1} (2l+1) = \frac{n}{2}(1 + 2n - 1) = n^2. \tag{4.116}$$

The energy eigenvalues of the H atom thus are (n^2-1)-fold degenerate (page 85). The degeneracy will be more or less removed by perturbations, e.g., as a result

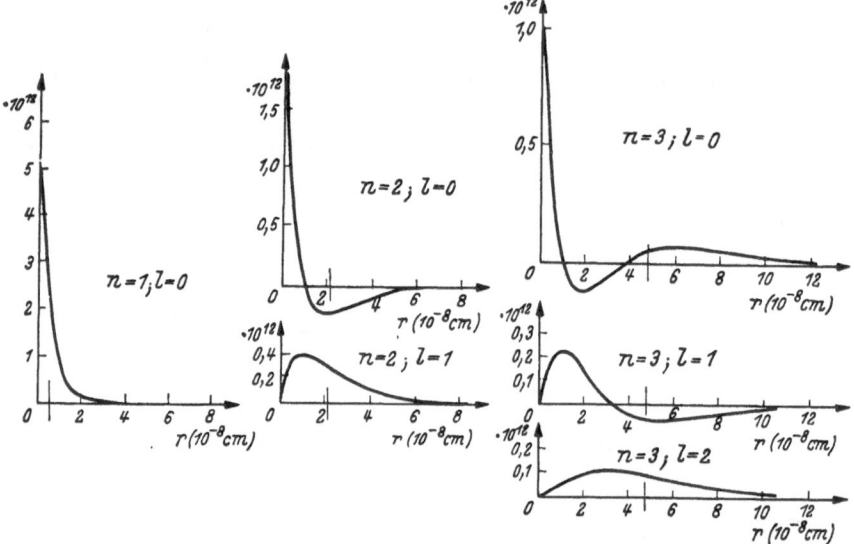

Fig. 102. The r-dependent parts of the electron eigenfunctions of the H atom for different values of the principal quantum number n and the orbital quantum number l. Note the different scales of the ordinates! The vertical marks on the abscissas indicate the radius of the corresponding Bohr orbit according to the old theory. (After HERZBERG)

of the motion of the nucleus (page 67) and in the case of the Stark and Zeeman effects (cf. III, 16).

The r-dependent portions of the eigenfunctions for the first three states of the H atom are shown in Fig. 102; the small vertical line in each case indicates the radius of the corresponding Bohr orbit. Fig. 103 is a pictorial representation of the norm $\psi\psi^*$ of the eigenfunctions of the hydrogen atom. The pictures show the value of the probability that an electron is at a certain point in space or, in other words, the time average of the electron density at this point. *They correspond completely to the vibrational modes of an elastic sphere.* These representations are very important for the understanding of chemical binding, especially for stereochemistry (cf. VI, 14d).

With this we conclude the wave-mechanical treatment of the *stationary* states of the H atom ($E<0$) and their eigenfunctions. We consider now briefly the states of positive energy ($E>0$), which correspond to hyperbolic Bohr orbits and which are responsible for the continuous spectra (page 74) of the H atom. Since these states correspond to an electron wave moving away from or toward the nucleus in the de Broglie sense, it is clear that there can be no quantum conditions. The wave function of an electron with kinetic energy ($E>0$) must, therefore, be a sine wave.

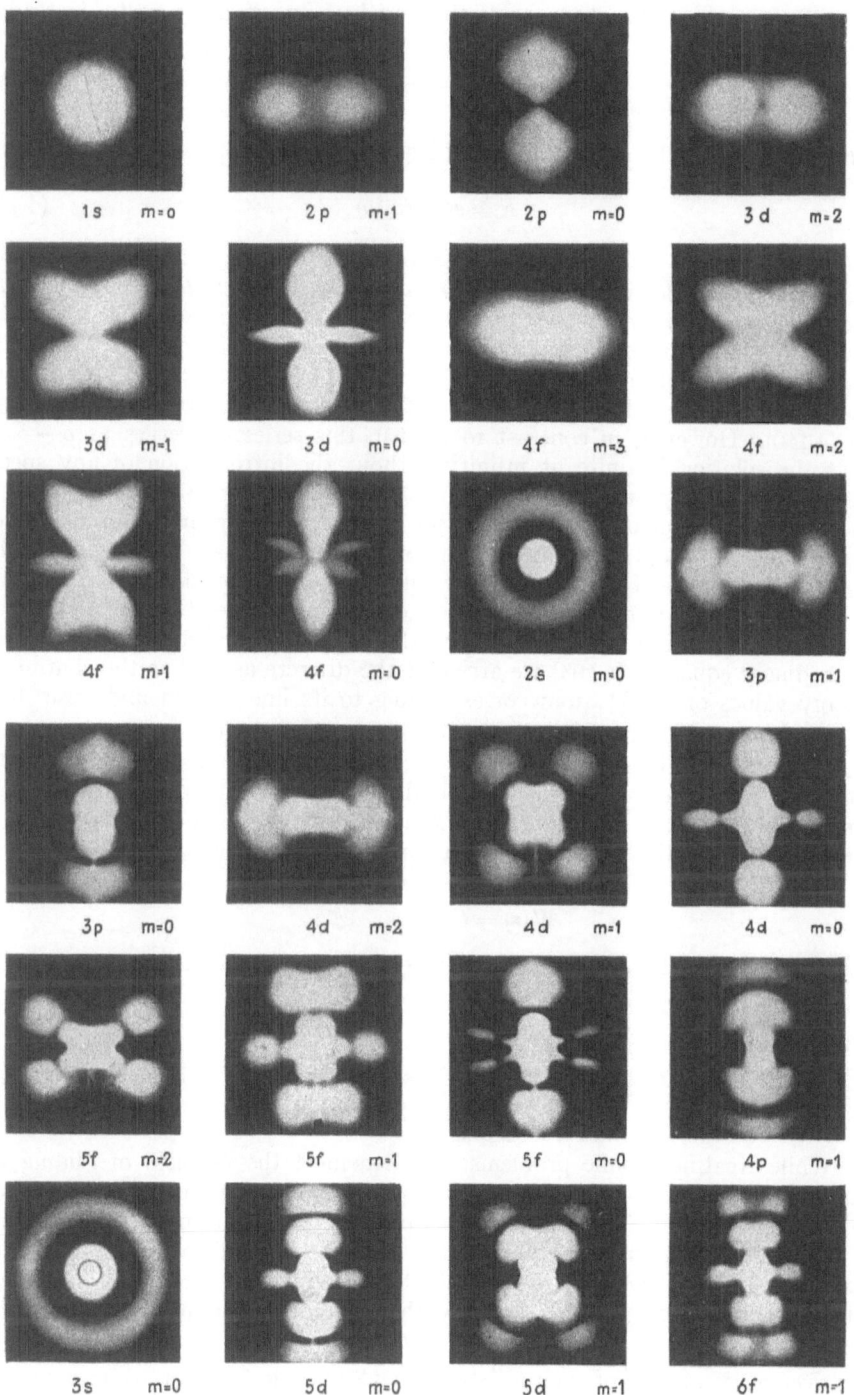

Fig. 103. Wave-mechanical "pictures" of the hydrogen electron in its excited states (vibrational modes of the "electron cloud") after WHITE. The brightness of light areas indicates the probability of finding the electron at that spot

The wave-mechanical computations are carried out in the same way as for the case $E<0$, except that because of the positive sign of E we write, instead of Eq. (4.105),

$$\frac{h^2}{8\pi^2 m_e E} = + r_0^2 .$$
(4.117)

Thus we have for the radial solution of Eq. (4.103) the expression

$$R = e^{\pm i\varrho/2} w(\varrho) .$$
(4.118)

The double sign in the exponent means than an electron wave can move away from as well as toward the nucleus[1]. The first case corresponds to the ionization of the atom, the second to the recombination of the electron and nucleus forming a neutral atom. By substituting Eq. (4.118) into the Schrödinger equation (4.103), we arrive at a differential equation for $w(\varrho)$, similar to Eq. (4.109), but with complex coefficients, whose solution can again be represented by a power series similar to (4.110). However, in contrast to (4.110), this series converges as $\varrho \to \infty$ so that the solution is finite at infinity without the introduction of any special finiteness condition such as Eq. (4.111). Since the finiteness condition resulted in the quantization of the E values for $E<0$, we see from the mathematical formalism that there are *no* quantum conditions for $E>0$. There exist, on the contrary, solutions of the Schrödinger equation (4.103) for all positive E values in agreement with our non-mathematical discussions above. It is one of the most satisfactory results of wave mechanics that from the mathematics of the one Schrödinger equation (4.103) we arrive at the discrete as well as the continuous energy values of the H atom corresponding to its line spectra and series limit continua, without making use of any arbitrary assumptions.

Since the solutions for the free electron $E>0$ are not stationary states but time-dependent waves traveling in either direction, we must consider the complete time-dependent wave function Ψ. The asymptotic solution (for large r) of an electron with kinetic energy $E>0$ therefore must be written as

$$\Psi(r) = C e^{i\left(\pm\frac{r}{r_0} - \frac{2\pi E t}{h}\right)} ,$$
(4.119)

corresponding to the expected progressing sine wave of frequency $\nu = E/h$ which is finite at infinity.

8. Quantum-mechanical Expressions of Observable Properties of Atomic Systems

While treating atomic problems, we often meet the problem of finding the quantum-mechanical expression of a measurable value or property of the system under consideration. The answer to this problem follows from the interpretation of the wave function ψ in combination with the normalizing condition (4.50), according to which $\psi\psi^* \, d\tau$ is the probability of finding the particle described by ψ within the volume element $d\tau$. We may, therefore, designate this expression as the *particle density averaged over a sufficiently long period of time.* From this follows the average distance $\vec{r}$ of such a "spread-out" particle from an arbitrary origin, or, in other words, the radius vector of its mass center averaged over time. If $\vec{r}$

---

[1] The photoelectric effect is treated quantum-mechanically in a corresponding manner.

is the radius vector of the volume element $d\tau$, then $\psi \vec{r} \psi^* \, d\tau$ indicates the probability of the occurrence of the radius vector $\vec{r} \, (d\tau)$, and thus

$$\vec{\bar{r}} = \int \psi \vec{r} \psi^* \, d\tau, \tag{4.120}$$

since the integral $\int \psi\psi^* \, d\tau$ by which Eq. (4.120) should be divided is unity because of (4.50). If we designate analogously by $\vec{\bar{p}}$ the time-averaged contribution of $d\tau$ to the momentum, we obtain an expression for the momentum of the particle which corresponds to Eq. (4.120),

$$\vec{\bar{p}} = \int \psi \vec{p} \psi^* \, d\tau. \tag{4.121}$$

In order to actually compute $\vec{\bar{p}}$, we must replace $\vec{p}$ by its corresponding differential operator (4.62):

$$\vec{\bar{p}} \,\hat{=}\, \int \psi \frac{h}{2\pi i} \, \text{grad} \, \psi^* \, d\tau. \tag{4.122}$$

Eq. (4.122) is the quantum-mechanical expression which permits us to compute $\vec{\bar{p}}$ from the known eigenfunctions ψ.

By using the well-known classical relation between momentum and kinetic energy, we obtain from Eqs. (4.122) and (4.62):

$$E_k = \frac{p^2}{2m} \,\hat{=}\, \frac{h^2}{8\pi^2 m} \, \text{grad}^2 = \frac{h^2}{8\pi^2 m} \, \Delta, \tag{4.123}$$

and thus a quantum-mechanical expression for the kinetic energy

$$\bar{E}_k = \int \psi \frac{h^2}{8\pi^2 m} \Delta \psi^* \, d\tau. \tag{4.124}$$

Eq. (4.124) allows the computation of E_k from the eigenfunction ψ. It may be derived from the Schrödinger equation (4.44) if the latter in its conjugated form is multiplied by $\psi d\tau$ and integrated over the entire space taking into account the normalizing condition (4.50).

We use this result for proving the correctness of the absolute value of an atomic angular momentum $|\vec{J}|$ introduced already on page 118, which as a function of the corresponding angular momentum quantum number J should be

$$|\vec{J}| = \sqrt{J(J+1)} \, h/2\pi, \tag{4.125}$$

according to wave mechanics; whereas classical theory requires the value $J h/2\pi$, in contradiction to spectroscopic results. For a rotating system, the kinetic energy is the rotational energy

$$E_k = \frac{1}{2} I \omega^2 = \frac{1}{2} m r^2 \omega^2. \tag{4.126}$$

Furthermore, from Eqs. (4.72), (4.78), and (4.79) we find for the rotator

$$\Delta \psi = - \frac{J(J+1)}{r^2} \psi. \tag{4.127}$$

From Eqs. (4.123), (4.126), and (4.127) we then get

$$\frac{1}{2} m r^2 \omega^2 \,\hat{=}\, \frac{h^2}{8\pi^2 m} \Delta = \frac{h^2}{8\pi^2 m} \frac{J(J+1)}{r^2}, \tag{4.128}$$

or

$$\omega = \frac{h}{2\pi m r^2} \sqrt{J(J+1)} = \frac{h}{2\pi I} \sqrt{J(J+1)}. \tag{4.129}$$

Since the angular momentum is the product of the moment of inertia I and the angular velocity ω, the result (4.125) follows directly from Eq. (4.129).

It is of interest for the understanding of atomic angular momenta to point out that angular momentum and angle of orientation are canonically conjugated variables (see page 115), so that they are subject to the uncertainty principle. Therefore the sharp quantization of the angular momentum (4.125) causes an uncertainty of its orientation with regard to a given direction (e.g., the direction of the magnetic field in the Zeeman effect). Because of this uncertainty, the component of $\sqrt{J(J+1)}$ in the field direction must be smaller than (4.125); the computation results in a maximal value of $Jh/2\pi$.

9. The Wave-mechanical Theory of Radiation. Transition Probability, Selection Rules, and Polarization

After we have studied the intensity problem for emission and absorption spectra by applying the correspondence principle in III, 23, we now develop a more systematic quantum theory of atomic radiation, based on wave-mechanical concepts.

From the considerations of the last section we know that the volume element $d\tau$ of a one-electron atom contains a fraction of the electron charge,

$$de = e\Psi\Psi^* d\tau. \tag{4.130}$$

Its integral over all space yields, of course, the entire electron charge e because of the normalizing condition (4.50). We used in Eq. (4.130) the complete time-dependent wave functions $\Psi(x, y, z, t)$ because emission processes are time-dependent and non-stationary.

Since an electric dipole moment is generally defined as the product of the charge by the distance of the two equal charges of opposite sign, the dipole moment of a positively charged atomic nucleus and an electron that, in the Bohr concept, orbits around this nucleus under consideration of Eq. (4.120) is given by

$$\mathfrak{M} = e\vec{r} = e\int\Psi\vec{r}\,\Psi^* d\tau, \tag{4.131}$$

with the Ψ values determined by Eq. (4.42). If we form the electric moment of an atom in a stationary state according to Eq. (4.131), the time-dependent factor $e^{-2\pi i\nu t}$ ist compensated by $e^{+2\pi i\nu t}$ due to multiplying Ψ by its conjugated complex state function Ψ^*. The charge distribution and with it the "stationary" dipole moment are thus constant with regard to time:

$$\mathfrak{M}_{stat} = e\int\psi\vec{r}\,\psi^* d\tau = \text{const}. \tag{4.132}$$

Electrodynamics requires, however, that energy is emitted only if an electric moment varies in time. Thus we have found the wave-mechanical explanation mentioned already on page 162 for the lack of emission by stationary atomic states.

In the same way as an elastic string is able to vibrate simultaneously with two different frequencies whose amplitudes superpose, also the wave function Ψ of an atom may be composed of several eigenvibrations, i.e., eigenfunctions. If we form the corresponding expression (4.131) of the dipole moment for the assumed case that Ψ and Ψ^* do not belong to the same eigenfunction but that two eigenfunctions Ψ_n and Ψ_m exist,

$$\Psi_n = \psi_n e^{-2\pi i\nu_n t} \quad \text{and} \quad \Psi_m^* = \psi_m^* e^{+2\pi i\nu_m t}, \tag{4.133}$$

whose superposition is supposed to represent the actual electronic vibrational state, we find the electric moment corresponding to the transition $E_n \rightarrow E_m$:

$$\mathfrak{M}_{nm} = e \int \Psi_n \vec{r}\, \Psi_m^* \, d\tau = e^{2\pi i (\nu_m - \nu_n)t} e \int \psi_n \vec{r}\, \psi_m^* \, d\tau. \tag{4.134}$$

In other words: If an electron vibrates with the two eigenvibrations Ψ_n and Ψ_m, the electric moment is not constant in time any more, but we have *an oscillation of the charge density with the beat frequency* $\nu_n - \nu_m$ which, according to classical electrodynamics, is connected with the emission of radiation of the same frequency. If we now apply Bohr's frequency condition and replace the frequencies in Eq. (4.134) by the energy difference corresponding to the associated quantum transition, we obtain

$$\nu_n - \nu_m = \frac{E_n - E_m}{h}. \tag{4.135}$$

Thus we find from Eqs. (4.134) and (4.135) that wave mechanics automatically yields what was postulated by Bohr, *namely the emission of a photon of frequency* $\nu_n - \nu_m = \nu_{nm}$ *for the energy change* $E_n - E_m$ *of the atomic state if the eigenfunctions* Ψ_n and Ψ_m *corresponding to the energy states* E_n and E_m *are excited*. The frequency of the time rate of change of the charge density equals, in wave-mechanics, the radiation frequency; no modification of classical electrodynamics is necessary.

According to electrodynamics, the energy emitted per unit time by a large number N of dipoles with the electric moment $\mathfrak{M}$ is

$$S = \frac{2N}{3c^3} \left(\frac{d^2 \mathfrak{M}}{dt^2} \right)^2. \tag{4.136}$$

Because of the amply confirmed correspondence between classical and quantum processes, we may apply Eq. (4.136) to our problem. By substituting for $\mathfrak{M}$ the expression (4.134) and differentiating twice, we obtain

$$S = \frac{2N}{3c^3} (2\pi \nu_{nm})^4 \, \overline{4 \cos^2 (2\pi \nu_{nm} - \delta)t} \, e^2 \left(\int \psi_n \vec{r}\, \psi_m^* \, d\tau \right)^2 \tag{4.137}$$

where the bar indicates the time average and δ is a phase constant which is of no special interest here because of the averaging over time. By substituting $\overline{\cos^2} = \frac{1}{2}$ and (4.132) into (4.137), we arrive at the radiation energy emitted per sec by the N excited atoms (in units of erg/sec):

$$S = \frac{64 \pi^4 N}{3c^3} \nu_{nm}^4 \, \mathfrak{M}_{nm}^2 \,(\text{stat.}) = \frac{64 \pi^4 N e^2}{3c^3 h^4} (E_n - E_m)^4 \left(\int \psi_n \vec{r}\, \psi_m^* d\tau \right)^2. \tag{4.138}$$

If we define the transition probability A_{nm} (cf. page 142) as the probability per unit time of a spontaneous transition of an atom from the state E_n to the state E_m under emission of radiation, then A_{nm} multiplied by the energy $h\nu_{nm}$ of the emitted photon and by the number N of excited atoms equals the average emission of radiation per second by the N excited atoms. Thus we have found the wave-mechanical expression for the transition probability A_{nm}:

$$A_{nm} = \frac{S}{Nh\nu_{nm}} = \frac{64 \pi^4 \nu_{nm}^3}{3hc^3} \mathfrak{M}_{nm}^2 \,(\text{stat.}) = \frac{64 \pi^4 e^2}{3h^4 c^3} (E_n - E_m)^3 \left(\int \psi_n \vec{r}\, \psi_m^* d\tau \right)^2, \tag{4.139}$$

from which the radiation of a given quantity of partially excited atoms may be computed, as well as the absorption probability and thus the intensity of emission or absorption spectra.

According to Eq. (4.139), the transition probability depends on the time-independent integral over the spatial part of the eigenfunctions. The components of this integral are

$$\left.\begin{aligned}
X_{ik} &= \int \psi_i \, x \, \psi_k^* d\tau, \\
Y_{ik} &= \int \psi_i \, y \, \psi_k^* d\tau, \\
Z_{ik} &= \int \psi_i \, z \, \psi_k^* d\tau.
\end{aligned}\right\} \qquad (4.140)$$

They are called the *matrix elements* of the corresponding transition and are significant for the selection rules and the polarization of the emitted radiation. If, for instance, the spatial symmetry (cf. Fig. 103) of the eigenfunctions of the combining states is such that all matrix elements are zero, the corresponding transition is "forbidden". If only X_{ik} differs from zero, the emitted radiation is polarized in a plane through the x axis and the direction of emission.

From the wave-mechanical standpoint, we may even say that forbidden transitions are the rule and allowed transitions are an exception; viz., if we substitute actual eigenfunctions into Eq. (4.140), all matrix elements and, thus, the transition probability (4.139) become in general zero. Only if the eigenfunctions ψ_i and ψ_k fulfill certain conditions, i.e., if the numbers of their nodes (quantum numbers) agree with the empirical selection rules of Chap. III, the matrix elements, or at least one of them, will differ from zero.

As an example we discuss the eigenfunctions of the free rotator (cf. IV, 7b), which are identical with the r-independent parts of the eigenfunctions of the H-atom. They are the spherical harmonics (4.81):

$$\psi = P_l^{|m|} (\cos \vartheta) \, e^{i m \varphi} \qquad (4.141)$$

where l designates the quantum numbers of orbital momentum and m is the orientation or magnetic quantum number. By substituting Eq. (4.141) into Eq. (4.140) we find, by a somewhat complicated computation, that the matrix elements differ from zero only if the quantum numbers l and m of the two interacting quantum states i and k differ by not more than unity. The specific results are:

$$\left.\begin{aligned}
X_{ik} &\neq 0 \quad \text{only for} \quad m_i = m_k \pm 1 \quad \text{and} \quad l_i = l_k \pm 1, \\
Y_{ik} &\neq 0 \quad \text{only for} \quad m_i = m_k \pm 1 \quad \text{and} \quad l_i = l_k \pm 1, \\
Z_{ik} &\neq 0 \quad \text{only for} \quad m_i = m_k \quad\quad \text{and} \quad l_i = l_k \pm 1;
\end{aligned}\right\} \qquad (4.142)$$

Eq. (4.142) obviously includes the selection rules (3.67) and (3.112) as well as the polarization rules. If, for instance, we have a magnetic field parallel to the z axis, the matrix elements X_{ik} and Y_{ik} are zero for $m_i = m_k$. The radiation is then linearly polarized with the electric field vector parallel to the z axis. If, however, the magnetic quantum numbers m of the two interacting states i and k differ by ± 1, then the matrix element Z_{ik} is zero, while the phases of X_{ik} and Y_{ik} show a difference of 90° so that a circularly polarized radiation results. Investigations about the polarization in the normal Zeeman triplet (see III, 16b) confirm these conclusions.

As a first example of the wave-mechanical computation of the intensity and polarization of spectral lines, SCHRÖDINGER calculated the Stark effect components of the Balmer lines of hydrogen and found complete agreement with observation. Beyond this, it can be shown that *all selection rules, which in Chap. III appeared so arbitrary, follow from the symmetry of the corresponding wave-mechanical eigenfunctions.* We are also able to understand that forbidden transitions may become more or less allowed by the action of an electric field. The electric field disturbs the eigenfunction of the electron, it "deforms" them in such a way that the product $\psi_n \vec{r} \psi_m$ in Eq. (4.134) on which the radiation intensity depends no longer is zero if integrated over all space, but will have a positive, though usually small value.

We still have to add a restricting remark: Our discussion was obviously confined to the radiation caused by the periodic changes of the electric dipole moment (4.134); the same is true for the selection rules. But aside from this electric dipole radiation, there may exist, as in classical physics, radiation of electric multipoles and magnetic dipoles although of an intensity much (down to 10^{-8} times) smaller. Our selection and polarization rules are not valid here; but computations of higher order may also explain these effects.

We see that wave mechanics has no need of the correspondence principle, as treated in III, 22, for intensity computations. On the other hand, the excellent agreement of the results of the wave-mechanical radiation theory with the experiments shows again the close correspondence between clasiscal and quantum physics; this agreement justifies the use of the radiation formula (4.136) of classical electrodynamics in wave mechanics.

However, this apparently so clear and successful radiation theory, based on the simultaneous excitation of the two electronic oscillations ν_n and ν_m, cannot explain how the radiation stops after the excitation energy of the atom has been emitted, i.e., how the atom finally reaches the stationary, non-radiating state E_m with the eigenfunction ψ_m. DIRAC succeeded in solving this difficulty by taking into account the interaction between the electric field of the electron and the electromagnetic alternating field of the light wave (or photon) emitted or absorbed as a result of the electron vibration. He was able to show quantitatively how a change of the energy state leads to the emission of electromagnetic energy, i.e., of a photon, because of the change of the electric field that is responsible for the bond between nucleus and electrons, and vice versa. For this purpose, however, quantum theory has to be introduced into electrodynamics by quantization of the electromagnetic field; thus quantum electrodynamics was created, the basic ideas of which will be treated in IV, 14.

10. The Wave-mechanical Treatment of the Pauli Principle and its Consequences

We dealt in III, 18 with the Pauli exclusion principle which, in the previously introduced form, states that no two electrons identical in all four quantum numbers can occur in any one atomic system.

We noted in III, 19 that this empirical principle controls the electron shell arrangement of the atoms, and we shall learn in V, 12 that the same is true for the configuration of the nucleons in the atomic nuclei. Since also the behavior of electrons in molecules and solids is governed by this principle, its significance can hardly be overestimated. We now ask for the wave-mechanical interpretation of the Pauli principle and will show that it consists of a characteristic restriction of the state functions describing the behavior of different elementary particles and that this restriction has important physical consequences.

Let us consider two electrons in different states i and k. Either they belong to the same atom or to two different ones, e.g., the two electrons of the He atom or those of two interacting hydrogen atoms. We designate by 1 and 2 the two particles with their coordinates and neglect the interaction of the electrons, i.e., treat them as independent particles. Then the total energy of the system consisting of the two particles equals the sum of the energies of the two individual particles,

$$E(1, 2) = E_i(1) + E_k(2), \tag{4.143}$$

whereas the state function of the total system is the product of state functions of the two single particles:

$$\Psi(1, 2) = \Psi_i(1)\,\Psi_k(2). \tag{4.144}$$

The latter equation follows from the theory of probability according to which the probability of a non-coupled double occurrence (here the occurrence of electron 1 in state i and of electron 2 in state k) is given by the product of the individual probabilities.

If both particles 1 and 2 are of the same type, e.g., both are electrons (or neutrons in a nucleus), $E(1, 2)$ is evidently degenerate since the exchange of both particles results in an energy state $E(2, 1)$ identical with $E(1, 2)$. In this case, we speak of *exchange degeneracy*. This degeneracy does not hold, however, for the state functions. If, for instance, Ψ_i is a sine function and Ψ_k a cosine function, while electron 1 may have the coordinate 0 and electron 2 the coordinate $\pi/2$, then we obviously have $\Psi(1, 2) = 0$, but $\Psi(2, 1) = 1$.

In other words, the Schrödinger equation describing the behavior of the system has two different solutions for the same energy eigenvalue $E(1, 2) = E(2, 1)$, which is a consequence of the exchange degeneracy. These solutions are (4.144) and

$$\Psi(2, 1) = \Psi_i(2)\Psi_k(1). \tag{4.145}$$

According to the theory of differential equations, not only (4.144) and (4.145) are solutions of the Schrödinger equation but also all linear combinations of them

$$\Psi = \alpha\Psi(1, 2) + \beta\Psi(2, 1) \tag{4.146}$$

with α and β being arbitrary constants. It is clear from physics that no different state of the system consisting of two independent particles can result from the exchange of two identical, non-distinguishable particles. Since the probability behavior of the system is given by the square of the absolute value of the corresponding state function, the relation

$$[\alpha\Psi(1, 2) + \beta\Psi(2, 1)]^2 = [\alpha\Psi(2, 1) + \beta\Psi(1, 2)]^2 \tag{4.147}$$

must be valid. This is obviously correct only for

$$\alpha = \pm\beta. \tag{4.148}$$

Therefore, only the two following solutions,

$$\Psi_s = \Psi(1, 2) + \Psi(2, 1) \tag{4.149}$$

and

$$\Psi_a = \Psi(1, 2) - \Psi(2, 1), \tag{4.150}$$

out of the total number of solutions (4.146) are qualified as physically meaningful state functions. The function (4.149) is called *symmetric* because it is not changed by the exchange of particle 1 and 2; function (4.150) is called *antisymmetric* because Ψ_a reverses the sign by particle exchange.

As always in wave mechanics, the eigenfunctions (4.149) and (4.150) refer only to the *orbital* motion of the electrons; they have to be supplemented by statements about the spin of the two electrons in order to completely describe the quantum states in question. For that purpose, the spin directions of the electrons 1 and 2 are indicated in parentheses behind the orbital eigenfunctions Ψ_s and Ψ_a. These directions may both be parallel or antiparallel to the resulting orbital momentum or to another distinguished direction, $[\Psi(\uparrow\uparrow)$ or $\Psi(\downarrow\downarrow)]$, or they may be opposite with respect to each other. Since the two electrons are non-distinguishable, the latter case represents a degeneracy of the two possibilities of orientation $(\uparrow\downarrow)$ or $(\downarrow\uparrow)$. In analogy to Eqs. (4.149) and (4.150) for the orbital functions, this state may be described by the sum or difference of the eigenfunctions for the two orientation possibilities, which may be indicated by the symbols $(\uparrow\downarrow\pm\downarrow\uparrow)$. In principle, the states of the discussed two-electron system thus can be written:

$$
\left.
\begin{aligned}
&\Psi_s(\uparrow\uparrow) \\
&\Psi_s(\uparrow\downarrow+\downarrow\uparrow) \\
&\Psi_s(\downarrow\downarrow) \\
&\Psi_a(\uparrow\downarrow-\downarrow\uparrow),
\end{aligned}
\right\} \tag{4.151}
$$

$$
\left.
\begin{aligned}
&\Psi_a(\uparrow\uparrow) \\
&\Psi_a(\uparrow\downarrow+\downarrow\uparrow) \\
&\Psi_a(\downarrow\downarrow) \\
&\Psi_s(\uparrow\downarrow-\downarrow\uparrow).
\end{aligned}
\right\} \tag{4.152}
$$

These eight eigenfunctions are subdivided into two groups (4.151) and (4.152) in order to demonstrate that the functions of group (4.151) remain unchanged if the two electrons are exchanged, whereas the functions of group (4.152) reverse their sign; they are therefore called symmetric and antisymmetric, respectively. We can learn only from experience which one of these two groups describes the behavior of a two-electron system correctly. Spectroscopy tells us that the two electrons of the He atom as well as those of the H_2 molecule (to be treated below) have the *same* quantum numbers n, l, and m in their ground state so that they are described by the orbital function (4.149) which is symmetric with regard to an electron exchange, whereas the spins must be oppositely directed since we have a singlet state. In other words, only the last function of (4.152) describes the ground state of the two-electron system correctly. According to page 99, every two-electron system possesses also a triplet system, the lowest state of which is characterized by the excitation of one of the two electrons and therefore is described by the antisymmetric orbital function (4.150). The spin directions of the two electrons, however, are parallel in the triplet states (cf. page 105) and, because of their three orientation possibilities, they result in the three triplet components of the 2^3S state, which are described correctly by the first three functions of group (4.152).

We thus learn from spectroscopic experience that *only the eigenfunctions (4.152) describe the behavior of a two-electron system correctly, but not the functions (4.151) which are symmetric with regard to exchange.* This result reaches far beyond the original Pauli principle, which only states that no two electrons can exist in one system that agree in all four quantum numbers. According to PAULI's formulation, an orbital function like (4.145) would be a valid description also, whereas we have seen that only the linear combinations (4.149) and (4.150) describe the orbital functions correctly. As a next approximation, we might assume that the correct eigenfunction simply would have to vanish if the two electrons agree in all four quantum numbers. Even this is not sufficient since the last of the four functions (4.151) fulfills this condition. *Only the above-stated condition of antisymmetry with regard to exchange selects those eigenfunctions among all possible ones that describe the two-electron system correctly. The principle of antisymmetry is thus the exact wave-mechanical formulation of the Pauli principle.*

We mentioned before that the Pauli principle cannot be derived from a more general basic principle, but that it has the character of an empirical rule despite of its far-reaching validity. The same is true for the wave-mechanical formulation of this principle. *We do not yet know why nature seems to prefer the antisymmetric state functions* (4.152) *rather than the symmetric functions* (4.151). But it is not mere coincidence that all particles obeying the Pauli principle have a half-integral spin measured in units of $h/2\pi$, while the spin of the photon and of the π mesons, which do not obey the Pauli principle, is integral (cf. V, 23).

In V, 13 we shall show that the statistical distribution of the energy among a great number of atomic systems decisively depends on whether these systems follow the Pauli principle or not. Since Fermi developed the statistics of particles following the Pauli principle, these are called *fermions*, while *bosons* are particles not obeying the Pauli principle but following a different statistics developed by BOSE and EINSTEIN. A system consisting of an even number of elementary particles is described by a *symmetric* state function and, thus, is a boson because an exchange of double particles corresponds to a twofold exchange of single particles and, therefore, to a twofold sign reversal of the state function describing the total system so that this function remains unchanged. *Particles composed of an even number of elementary particles are, therefore, bosons; their state functions are symmetric and their spin values are always integral.* The nucleus of the common Helium isotope $_2\text{He}^4$, which consists of two protons and two neutrons, thus is a boson, while the nucleus of the rare isotope $_2\text{He}^3$ is a fermion because it consists of two protons and one neutron. This leads to a most characteristic difference in the behavior of the two helium isotopes, which will be treated in VII, 17b.

We illustrate the effect of the antisymmetry principle on two neutrons with parallel spin orientation, moving in the x direction with the velocities v_1 and v_2, free and independent of each other. Their coordinates may be x_1 and x_2. Disregarding a constant we may write their wave functions as

$$\left.\begin{aligned}
\Psi_1(x_1) &= e^{2\pi i \frac{m}{h} v_1 x_1}, \\[4pt]
\Psi_2(x_2) &= e^{2\pi i \frac{m}{h} v_2 x_2}.
\end{aligned}\right\} \tag{4.153}$$

The probability of being found in the unit volume determined by the coordinate x for each individual neutron then is

$$\Psi\Psi^* = e^{2\pi i \frac{m}{h} v x}\, e^{-2\pi i \frac{m}{h} v x} = \text{const}. \tag{4.154}$$

This probability, which is constant in space, shows that the uncertainty principle does not permit any localization of the particle if its velocity is sharply defined. However, the wave function of the total system consisting of the two particles must be antisymmetric because of the validity of the Pauli principle; it reads therefore

$$\Psi = \Psi_1(x_1)\,\Psi_2(x_2) - \Psi_1(x_2)\,\Psi_2(x_1), \tag{4.155}$$

since the state described by (4.153) is degenerate with the other one in which particle 1 has the velocity v_2 and particle 2 the velocity v_1. The spatial distribution of the probability of finding the two particles at a spot in the configuration space determined by the two coordinates x_1 and x_2 according to IV, 6 is given by the expression $\Psi\Psi^*$:

$$\Psi\Psi^* = 2 - \Psi_1(x_1)\,\Psi_2(x_2)\Psi_1^*(x_2)\,\Psi_2^*(x_1) - \Psi_1(x_2)\,\Psi_2(x_1)\,\Psi_1^*(x_1)\,\Psi_2^*(x_2). \quad (4.156)$$

In consideration of (4.153) and the Euler formula, this leads to

$$\Psi\Psi^* = 2\left[1 - \cos 2\pi\,\frac{m}{h}\,(v_1 - v_2)\,(x_1 - x_2)\right]. \quad (4.157)$$

This result, which is a logical consequence of the antisymmetrization of the wave function, is most surprising. It means that for each particle of a given relative velocity $v_1 - v_2$, the probability of finding it is zero not only at the position of the other particle but also at a distance λ, 2λ, 3λ, ... of this position, with λ indicating the relative de Broglie wavelength $h/m\,(v_1 - v_2)$ of the second particle. Although we did not assume any forces between the two particles in question, they seem to reject each other because of the Pauli principle, i.e., the antisymmetry principle, as long as their distance in smaller than $\lambda/2$. In this distance range, the probability is the smaller the closer the particles are and converges to zero with decreasing distance. This behavior cannot simply be described by assuming a new type of force between the particles, since according to Eq. (4.157) the probability for finding the two particles is zero also at the distances λ, 2λ, etc., but has maxima for $(x_1 - x_2) = \lambda/2$, $3\lambda/2$, ... The probability density of the two particles thus shows maxima and minima whose distance depends on the relative velocity of the two particles. If their relative velocity is exactly zero, then the probability is zero everywhere in space, according to Eq. (4.157): *Two fermions with exactly the same velocity cannot exist.* Eq. (4.157) shows that quite generally the probability density for the two particles is always zero if the product of their relative velocity and their distance is an integral multiple of h/m.

Let us now consider a large number of fermions of different velocity (i.e., a Fermi gas) and inquire about the density distribution around a certain particle. The zero density at the position of this particle is of course retained since it occurs for all relative velocities, according to Eq. (4.157). But the probability zero at $x_1 - x_2 = \lambda$, 2λ, ... does not appear any more because of the different relative velocities and de Broglie wavelengths of the particles. The particle density is therefore constant in the more distant neighborhood of every selected particle, while *its position itself is avoided by all other particles as a consequence of the anti symmetry principle* (so-called Fermi hole).

The quantum-mechanical form of the Pauli principle thus requires a spatial distribution of neighboring fermions, such as electrons of an atom or molecule, that deviates from the classical expectation. This non-classical distribution causes a change also of the electrostatic energy of interaction, designated as *exchange energy*, for reasons that will immediately become understandable; and the forces that are based on it and, therefore, are of typically wave-mechanical character are called *exchange forces*.

11. The Interaction of Coupled Identical Systems.
Exchange Resonance and Exchange Energy

We now discuss in more detail the effects of these so-called exchange forces between coupled identical atomic systems. They play a decisive role in the theory of the energy differences between equivalent terms of different multiplicity of many-electron atoms (see Fig. 67) because of their identical electrons, furthermore for the interaction leading to molecular binding of two identical atoms (cf. VI, 14b), for the interaction of a large number of identical atoms or ions in a crystal (cf. VII, 11) and, finally, in ferromagnetism for the interaction of all the uncompensated electrons in a ferromagnetic crystal (cf. VII, 15e). *The problem in all these cases is what happens if two identical atomic systems with equal total energy E are coupled with each other, i.e., if their interaction is taken into account.* Actually, a certain interaction exists always due to electric and magnetic fields since the systems consist of charged elementary particles in motion.

In order to get an idea of the effects of the coupling of atomic systems, we consider the well-known macromechanical case of two coupled identical pendulums.

Fig. 104. Model of two pendulums coupled by a distance-dependent coupling force for demonstrating the homopolar chemical bond (after KOSSEL). The pendulums are tied to a spring-held thread fixed to the two poles

Although we here deal with a *frequency* resonance, while we find *energy* identity (degeneracy) in atomic systems, we may nevertheless transfer the results of the coupling of pendulums to atomic systems according to the correspondence principle (see III, 22).

If we couple the two pendulums of Fig. 104 elastically by a spring-held thread from pole to pole, we get (after KOSSEL) a good model for the coupling of the atomic systems, which is always *distance-dependent*. To begin, we disregard the possibility, shown in Fig. 104, that the suspension points of both pendulums move towards or away from each other; we assume their distance kept constant. If we start with one pendulum vibrating, the vibration energy is gradually transferred to the second pendulum until the first comes to rest. Then the energy is transferred back from the second to the first, and so on. Each of the two pendulums, which in the uncoupled state oscillate with a purely sinusoidal motion of constant amplitude (we disregard here any damping), oscillates in the coupled state as shown in Fig. 105. As a consequence of the coupling of the two pendulums of identical frequency, beat oscillations appear. Such beat oscillations usually occur only when two slightly dissimilar frequencies are superimposed, for example, by simultaneous excitation of two slightly different tuning forks. We thus find: By coupling two mechanical systems which in the uncoupled state have the same frequency, this original frequency splits into two different frequencies, one higher and one lower than the original one, and their difference is the larger the stronger the coupling and, consequently, the larger the exchange frequency of the energy between the pendulums. The latter equals the difference of the two frequencies into which the eigenfrequency of the uncoupled pendulum splits because of the coupling. This splitting of the unperturbed eigenfrequency follows very clearly from the pendulum model of Fig. 104. If both pendulums vibrate toward the same side (so-called symmetric consonance), the retroactive force is obviously smaller than if each pendulum would vibrate alone. A smaller vibration frequency ν_0 therefore follows from the equation for the eigenfrequency of a harmonic oscillator with mass m and force constant k:

$$\nu_0 = \frac{1}{2\pi}\sqrt{\frac{k}{m}}. \tag{4.158}$$

If, however, the coupled pendulums vibrate to opposite sides (so-called anti-symmetric consonance) the restoring force is greater than that acting upon the

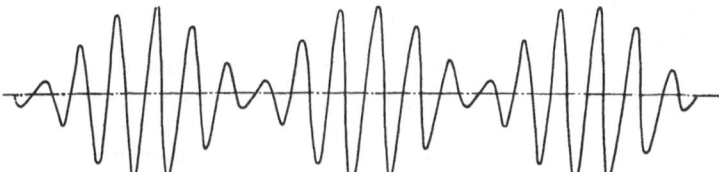

Fig. 105. Beat vibrations as a result of the coupling of two identical pendulums (schematic)

freely vibrating pendulums and, according to Eq. (4.158), the frequency of this antisymmetric coupled vibration is larger than the unperturbed eigenfrequency of each pendulum. Since the kinetic energy E changes proportional to the frequency ν (according to the adiabatic law which is the classical basis of the quantum relation $E = h\nu$), we see from these considerations that the two equal-energy states of two identical uncoupled atomic systems split into a higher and a lower energy state due to the coupling, and that the energy spacing increases with increasing coupling of the particles.

The exchange frequency of the vibrational amplitude (cf. Fig. 105) equals the difference $\Delta\nu$ of the two vibration frequencies which result from the coupling. Transferring this result to wave mechanics, we expect an exchange of the amplitude of the Ψ vibration among the coupled atoms, and this means an electron exchange since the square of Ψ corresponds to the statistical electron density according to IV, 6. Thus we have found the important result that *the spacing ΔE of the energy states caused by the coupling of the atoms is proportional to the frequency ν_e of the electron exchange between the two coupled atoms:*

$$\Delta E = h\nu_e. \tag{4.159}$$

In the subsequent chapters we shall present important applications of this result.

By the coupling of two atoms (as a consequence of the overlapping of their electron eigenfunctions) thus an energy state of the coupled system may occur that is *lower* than that of the uncoupled atoms, in analogy to the symmetrically consonant vibration of two coupled pendulums. This is the basis of the homopolar chemical bond between two equal atoms, details of which we shall discuss in VI, 14b. Since every physical system tends to the state of smallest potential energy, a system of two atoms will prefer the bound state of smaller energy to the higher energy state of isolated atoms. The two pendulums of KOSSEL's model experiment, whose coupling depends on their distance and which are arranged so that they can

move against each other, indeed approach each other in the symmetrically consonant vibration corresponding to the case of mutual attraction of the atoms in
the formation of a molecule, whereas they move away from each other in the
antisymmetrically consonant vibration corresponding to the mutual repulsion
of two atoms in the higher of two energy states.

We briefly discuss the wave-mechanical treatment of this problem for the
example of the H_2 molecule, first studied by HEITLER and LONDON. In order
to examine the interaction of two identical hydrogen atoms in the ground state,
we introduce into the Schrödinger equation of the system containing *two* electrons,

$$\Delta_1 \psi + \Delta_2 \psi + \frac{8\pi^2 m}{h^2} (E - U)\psi = 0, \tag{4.160}$$

the potential

$$U = -e^2 \left(\frac{1}{r_{a_1}} + \frac{1}{r_{b_2}} + \frac{1}{r_{b_1}} + \frac{1}{r_{a_2}} - \frac{1}{r_{12}} - \frac{1}{r_{ab}} \right). \tag{4.161}$$

The Laplace operators Δ_1 and Δ_2 [see Eq. (4.37)] refer, respectively, to the coordinates x_1, y_1, z_1 and x_2, y_2, z_2 of the two electrons. r_{a1}, r_{a2}, r_{b1}, and r_{b2} are the distances
of electrons 1 and 2 from the nuclei a and b

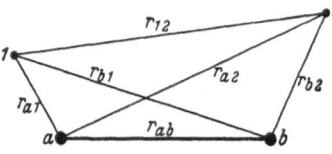

(Fig. 106), r_{12} is the distance between the two
electrons, and r_{ab} the distance between the two
nuclei. If we consider only the first two terms
in the parentheses, which describe the electrostatic potential of electrons 1 and 2 with respect
to "their own" nuclei a and b, the Schrödinger
equation separates into two equations for two
uncoupled H atoms. The remaining terms of
Eq. (4.161) are the interaction terms, and they represent the electrostatic
potentials between the electrons and the nuclei which originally were not
associated with them, between the two electrons, and between the two nuclei. In
order to take into account these interaction terms in the Schrödinger equation,
we have to go through a complicated "perturbation computation". Its result,
like that of our classically understandable model discussion, is that the identical
energy eigenvalue of the two uncoupled atoms, due to the interaction splits into
two eigenvalues one of which is higher and the other one lower than the original
energy state. Their distance is proportional to the strength of the interaction of
the atoms.

Fig. 106. Interaction of two hydrogen atoms
in the H_2 molecule. a and b = hydrogen
nuclei, *1* and *2* = electrons

If we begin by considering the two hydrogen atoms without interaction, then,
according to page 186, the energy E of the total system

$$E = E_0(1) + E_0(2) = 2E_0 \tag{4.162}$$

evidently is equal to the sum of the energies of the two atoms in the ground state,
whereas the eigenfunction of the system is equal to the *product* of the eigenfunctions of the individual electrons which belong to the nuclei a and b,

$$\psi_{12} = \psi_a(1)\psi_b(2). \tag{4.163}$$

Since the two electrons are indistinguishable, here we have again the exchange
degeneracy dealt with in IV, 10. Thus there exist two different orbital eigenfunctions for the system, corresponding to Eqs. (4.149) and (4.150):

$$\psi' = \psi_a(1)\psi_b(2) + \psi_a(2)\psi_b(1), \tag{4.164}$$

$$\psi'' = \psi_a(1)\psi_b(2) - \psi_a(2)\psi_b(1). \tag{4.165}$$

The Pauli principle allows *both* solutions, but antiparallel spin moments (singlet state) belong to Eq. (4.164) and parallel spin moments (triplet state) belong to Eq. (4.165). The electron distributions which correspond to the two eigenfunctions (4.164) and (4.165) are influenced by the distance-dependent electrostatic perturbation of the two atoms in *different* ways. This perturbation thus removes the exchange degeneracy. We therefore get, in analogy to the mechanical case of the coupled pendulums, two different energy states E_s and E_a of the system.

We have to consider this difficult point in more detail because of its important consequences by introducing into the Schrödinger equation not only the complete potential energy U from Eq. (4.161) including the interaction terms, but also by substituting in place of the eigenfunctions ψ' and ψ'' their actual perturbed values $\psi'+\varphi'$ and $\psi''+\varphi''$ (where φ is as yet undetermined). Similarly, we substitute the perturbed values of the energy $2E_0+\varepsilon'$ and $2E_0+\varepsilon''$ for the unperturbed eigenvalue $2E_0$. Then we have two *inhomogeneous* Schrödinger equations which, instead of zero, have perturbation terms on their right-hand sides. Then, from the orthogonality and normalizing conditions (cf. IV, 5) which apply to all eigenfunctions, we get the energy eigenvalues of the perturbed system

$$E_s = 2E_0 + e^2 C + e^2 A , \tag{4.166}$$

$$E_a = 2E_0 + e^2 C - e^2 A \tag{4.167}$$

where e, as usual, is the elementary charge. The significance of the quantities C and A will be explained immediately. First we note that in both equations a positive term is added to the unperturbed energy eigenvalue; it corresponds to the Coulomb interaction energy. Its constant is designated by C. There appears, furthermore, a term for the *exchange energy*, $e^2 A$. This term, however, may have either a positive or a negative sign. As a result of this, the single (but degenerate) eigenvalue E_0 now splits up into two values E_s and E_a, whose energy difference, the term spacing

$$\Delta E = 2e^2 A \tag{4.168}$$

depends only on the value of the exchange integral A. According to Eq. (4.159), it equals the frequency of the electron exchange between the two nuclei, multiplied by h.

We shall not go into the details of the derivation but only state the results of the computation of C and A, for which we find

$$C = \int \left(\frac{1}{r_{ab}} - \frac{1}{r_{a_1}} - \frac{1}{r_{b_1}} + \frac{1}{r_{12}} \right) \psi_a^2(1) \, \psi_b^2(2) \, d\tau \tag{4.169}$$

and

$$A = \int \left(\frac{1}{r_{ab}} - \frac{1}{r_{a_1}} - \frac{1}{r_{b_1}} + \frac{1}{r_{12}} \right) \psi_a(1) \, \psi_b(2) \, \psi_a(2) \, \psi_b(1) \, d\tau. \tag{4.170}$$

In Eq. (4.169) $\psi_a^2(1)$ and $\psi_b^2(2)$ are the probability densities which we designate by ϱ_1 and ϱ_2. Integrated over all space and multiplied by e they represent the total charges of the electrons 1 and 2, belonging to the nuclei a and b. We may therefore write for Eq. (4.169)

$$e^2 C = \frac{e^2}{r_{ab}} - \int \frac{e^2 \varrho_2}{r_{a_2}} \, d\tau_2 - \int \frac{e^2 \varrho_1}{r_{b_1}} \, d\tau_1 + \iint \frac{e^2 \varrho_1 \varrho_2}{r_{12}} \, d\tau_1 \, d\tau_2 . \tag{4.171}$$

From Eq. (4.171) we see that the term C actually is that part of the interaction energy which is due to Coulomb forces, since the first term is the repulsion between the nuclei a and b, the second and third the attraction between nucleus a and electron 2 and between nucleus b and electron 1, respectively, and the last term is the repulsion between electrons 1 and 2. From Eq. (4.170) it can be seen that the *exchange integral A* is very similar to the Coulomb integral (4.169), except that instead of the actual electron densities $e\psi_a^2(1)$ and $e\psi_b^2(2)$ we have the "mixed terms" representing the electron exchange, $e\psi_a(1)\psi_b(2)$ and $e\psi_b(1)\psi_a(2)$.

As a consequence of the interaction, the originally degenerate energy eigenvalues of coupled atomic systems split into a number of energy states equal to the number of particles which can be exchanged, whereby the magnitude of the energy spacing, the exchange energy, is dependent upon the value of the exchange integral (4.170). It is easy to see that this integral actually is a measure of the coupling of the two systems. The value of the integral is zero if in every volume element at least one of the four eigenfunctions is zero. On the other hand, the value of the integral is the larger the larger the values of all four eigenfunctions in every volume elementare, i.e., the more they overlap. If one eigenfunction vanishes at a point where the other ones are different from zero, it means that there is no coupling of the particles represented by these eigenfunctions. Large overlapping of the eigenfunctions, on the other hand, means that there is strong coupling.

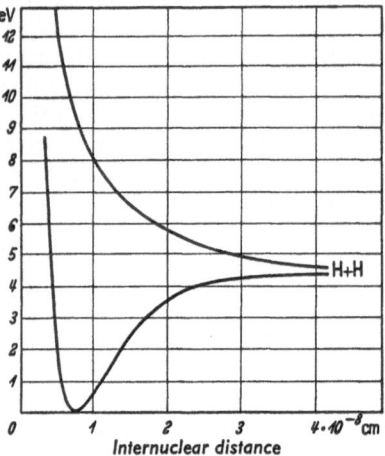

Fig. 107. Potential curves of the interaction of two H atoms after HEITLER and LONDON. The curve with a minimum corresponds to the stable H_2 molecule with opposite spin directions of the two electrons. The upper curve (identical spin directions of the two electrons) corresponds to repulsion for all internuclear distances, i.e., mutual elastic deflection of the H atoms

Consequently, with decreasing distance of the two H atoms their coupling must increase, and with it the energy spacing between the two energy states (4.166) and (4.167) of the originating H_2 molecule. This is in agreement with the result, shown in Fig. 107, of the wave-mechanical perturbation computation first done by HEITLER and LONDON. Details of the "potential curves" of Fig. 107 will be treated in molecular physics in Chap. VI, where we shall deal in detail with the fact that only the lower curve, representing the singlet ground state, has a potential minimum and thus gives rise, at a certain distance of the atoms, to a stable H_2 molecule.

12. The Refractive Index of Ψ Waves and the Quantum-mechanical Tunnel Effect (Penetration of a Potential Wall by a Particle)

In the last sections we met several times the general problem of the motion of a particle in a potential field. One example is the harmonic oscillator with its vibrating mass (cf. IV, 7c). This problem, occurring in numerous variations in all fields of atomic physics, can wave-mechanically be treated simplest and clearest by introducing the refractive index of the Ψ waves. We know from optics that the refractive index n equals the ratio of the phase velocities of the waves in free space, u_0, and in the specific medium, u: $n = u_0/u$. For Ψ waves, the free space corresponds to the case without potential, $U = 0$. Here the phase velocity follows from the general expression (4.40) as

$$u_0 = \frac{h\nu}{\sqrt{2mE}}, \qquad (4.172)$$

and the refractive index of the Ψ wave thus is, with (4.40):

$$n = u_0/u = \sqrt{\frac{E-U}{E}}. \tag{4.173}$$

This refractive index may be used quite generally and provides an easy way of surveying the behavior of particles in potential fields of all types. The well-known law of optics, for instance, that the wavelength is inversely proportional to the refractive index for a given frequency ν, applies to wave mechanics also. Since momentum p and kinetic energy $E_k = E - U$ satisfy the relation

$$E - U = p^2/2m \quad \text{and thus} \quad p = \sqrt{2m(E-U)}, \tag{4.174}$$

we have for the potential-free de Broglie wavelength

$$\lambda_0 = \frac{h}{p_0} = \frac{h}{\sqrt{2mE}} \tag{4.175}$$

and for the general case

$$\lambda = \frac{h}{\sqrt{2m(E-U)}} = \frac{\lambda_0}{n}, \tag{4.176}$$

as in optics, although we must note that, in general, the refractive index of wave mechanics is *smaller* than unity, according to Eq. (4.173).

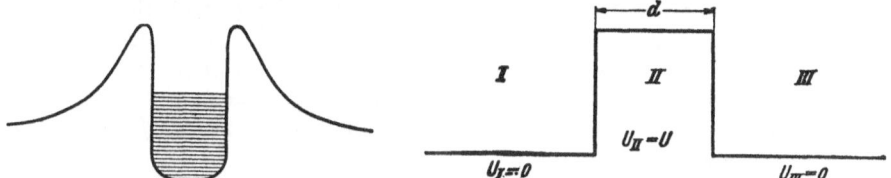

Fig. 108. Stationary states of the nucleons (protons and neutrons) in the potential well of a nucleus

Fig. 109. Rectangular potential wall used for explaining the tunnel effect

We are frequently concerned in atomic physics with the problem of whether or not a particle which is in a potential minimum (see Fig. 108) can surmount the surrounding potential wall and pass into outer space. In classical physics the answer is simple: Only if the kinetic energy of the particle is larger than the height U of the potential wall, can the particle get over the wall and "obtain its freedom". However, in the case of the radioactive α decay it is evident that α-particles whose kinetic energy is much smaller than the height of the enclosing potential wall do leave the nucleus (cf. V, 6c). GAMOW was able to show that a quantum-mechanical treatment of this problem leads to results entirely different from those obtained from the classical theory. Quantum-mechanically, there is a finite probability that a particle can escape through the potential wall. This is called the quantum-mechanical tunnel effect.

We discuss the simplest case in which a particle, e.g., an electron, approaches a rectangular potential wall of height U and width d from the left (Fig. 109). If the total energy E is larger than the height U of the potential wall, the case corresponds, according to Eq. (4.173), to the transition of a Ψ wave into a medium of smaller refractive index n. We expect and find wave-mechanically a partial reflexion of the wave at the boundary between the two media. The reflectance may be computed from the corresponding optical formula

$$R = \left(\frac{n-1}{n+1}\right)^2 \tag{4.177}$$

with the help of the refractive index (4.173). *While classically for $E > U$ we expect the particles to roll freely over the potential wall, wave-mechanically a partial reflexion takes place.* This phenomenon, however, plays a role only if $E - U$ is small compared to U. For $E = 1.5\ U$, for instance, this reflectance is only 7%.

If, however, the total energy E, which here is identical with the kinetic energy of the particles incident from region I, is *smaller* than the height U of the potential wall, the refractive index

$$n = \sqrt{\frac{E-U}{E}} = i\sqrt{\frac{U-E}{E}} \qquad (4.178)$$

becomes imaginary. There is obviously no direct optical analogy to this case. But the following computation will show that this imaginary refractive index leads to a total reflexion at the potential wall, even if the incident Ψ waves move perpendicularly onto the wall; at least if the wall is infinitely thick. We know from optics that also in the case of total reflexion a fraction of the light wave enters the medium with smaller refractive index; its amplitude, however, decreases so fast (strong damping!) that the penetration depth is only of the order of one wavelength. This behavior of optical waves is a consequence of the general continuity conditions at the boundary, which are valid in the same way for particle waves. Also the particle wave, incident from the left, is thus totally reflected at the boundary between region I and region II, although an exponentially decaying wave penetrates into the potential wall II. If the width d of the wall is not much larger than the particle wavelength λ computed from Eq. (4.176), then a wave of correspondingly small amplitude may also enter region III, in complete analogy to the optical case of the so-called total reflexion at a thin film. Translated into particle language, this result reads: *Contrary to classical mechanics, there exists a certain probability that particles hitting the potential wall from the left penetrate the wall and enter region III.*

The wave-mechanical computation confirms this result. The potential on each side of the potential wall may be zero (see Fig. 109), $U_{\mathrm{I}} = U_{\mathrm{III}} = 0$, while within the wall II the potential may be U, and we assume a stationary case with a constant particle flow traveling in the x direction with the constant kinetic energy E. Then we may eliminate the time dependence by using Eq. (4.42) and describe the problem in the different regions (cf. Fig. 109) by the three Schrödinger equations:

$$\frac{d^2\psi_{\mathrm{I}}}{dx^2} + \frac{8\pi^2 m}{h^2} E\psi_{\mathrm{I}} = 0, \qquad (4.179)$$

$$\frac{d^2\psi_{\mathrm{II}}}{dx^2} + \frac{8\pi^2 m}{h^2}(E - U)\psi_{\mathrm{II}} = 0, \qquad (4.180)$$

$$\frac{d^2\psi_{\mathrm{III}}}{dx^2} + \frac{8\pi^2 m}{h^2} E\psi_{\mathrm{III}} = 0. \qquad (4.181)$$

In region I as well as in region II, we expect as solutions an incident wave and a reflected wave, while in region III only a wave travelling towards the right is expected. Since the time dependence was eliminated, the solutions for I and III are

$$\psi_{\mathrm{I}} = a_{\mathrm{I}} e^{\frac{2\pi i x}{\lambda_0}} + b_{\mathrm{I}} e^{-\frac{2\pi i x}{\lambda_0}}, \qquad (4.182)$$

$$\psi_{\mathrm{III}} = a_{\mathrm{III}} e^{\frac{2\pi i x}{\lambda_0}}. \qquad (4.183)$$

Here the wavelength λ_0 of the incident particles is given as a function of their kinetic energy by Eq. (4.175) with coefficients yet to be determined. If the particles are to enter region III, they must pass region II, where we expect the same solution as for I, except that the wavelength is now λ_0/n, according to Eq. (4.176), with an imaginary refractive index n computed from Eq. (4.178). The exponentials of the expressions (4.182) and (4.183) now are free from i so that we get

$$\psi_{\mathrm{II}} = a_{\mathrm{II}}\, e^{-\frac{2\pi n' x}{\lambda_0}} + b_{\mathrm{II}}\, e^{\frac{2\pi n' x}{\lambda_0}} \qquad (4.184)$$

with

$$n' = n/i = \sqrt{\frac{U-E}{E}}. \qquad (4.185)$$

Thus in I and III we have processes *periodic* in space, because of the occurrence of i in the exponent, i.e., we have travelling waves. In region II, however, we do *not* find *any* wave, because of the imaginary refractive index n, but a *damping* in space which increases with the width d of the potential wall, in complete analogy to the corresponding optical case of "total reflexion at a thin film".

The still undetermined coefficients of the exponential functions in Eqs. (4.182) to (4.184) may be obtained from the continuity conditions. Since it is *one* process that takes place in all three regions, the following continuity conditions for the boundaries $x=0$ and $x=d$ must be satisfied:

$$\left.\begin{array}{ll} \psi_{\mathrm{I}}(0) = \psi_{\mathrm{II}}(0); & \psi_{\mathrm{II}}(d) = \psi_{\mathrm{III}}(d); \\[2mm] \left(\dfrac{d\psi_{\mathrm{I}}}{dx}\right)_{x=0} = \left(\dfrac{d\psi_{\mathrm{II}}}{dx}\right)_{x=0}; & \left(\dfrac{d\psi_{\mathrm{II}}}{dx}\right)_{x=d} = \left(\dfrac{d\psi_{\mathrm{III}}}{dx}\right)_{x=d}. \end{array}\right\} \qquad (4.186)$$

They allow to compute the coefficients of Eqs. (4.182) to (4.184). If we now define the "transmittance" D as the ratio of the probabilities for particles to be found in the outgoing wave in III relative to that for particles in the incoming wave entering II, then D evidently equals the ratio of the norms of ψ_{III} and of that part of ψ_{I} which describes the incoming wave. The exponential functions compensate each other because of the multiplication by the conjugated complex values so that we get

$$D = \frac{a_{\mathrm{III}}\, a_{\mathrm{III}}^{*}}{a_{\mathrm{I}}\, a_{\mathrm{I}}^{*}}. \qquad (4.187)$$

The computation of D from Eqs. (4.182) to (4.184) by means of Eq. (4.186) in general is rather complicated; but it is simple for the case of practical interest that $2\pi n' d$ is large compared to λ_0 [cf. Eq. (4.184)]. The transmittance of the potential wall is then represented by

$$D = \frac{16 E (U-E)}{U^2}\, e^{-\frac{4\pi d}{h}\sqrt{2m(U-E)}} \qquad (4.188)$$

or, still more simply, by introducing the refractive index (4.185) and the de Broglie wavelength λ_0 of the incident particles,

$$D = 16\,\frac{E^2}{U^2}\, n'^2\, e^{-\frac{4\pi n' d}{\lambda_0}}. \qquad (4.189)$$

According to Eq. (4.188), the probability for a particle to penetrate the potential wall is the greater the smaller the width d of the potential wall, measured in units of λ_0, and the lower its height $U-E$, measured from the level E.

This tunnel effect belongs to the most important results of quantum mechanics. Although it is in sharp contrast to the results of classical physics, it is quantitatively confirmed by experiments. We may qualitatively understand it also with the help of HEISENBERG's uncertainty principle: If the time interval Δt necessary for penetrating the potential wall is sufficiently small, then, according to Eq. (4.8) in (IV, 3), the corresponding uncertainty of energy ΔE is so large that there is a finite probability that E may exceed the height of the potential wall. From this presentation we again see the *fundamental* character and significance of the uncertainty principle which does not in any way depend on the accuracy of measurements. We also see that our pictorial expressions "penetrating" or "surmounting" the potential wall in the light of the uncertainty principle are only two different ways of expressing the same physical phenomenon. Our pictorial concepts thus prove not sufficient for an unambiguous description of the phenomena of atomic physics.

13. Fermi and Bose Quantum Statistics and their Physical Significance

In one of the two possible complementary methods of describing the phenomena of our physical world, atomic physics considers matter as consisting of individual particles such as electrons, atoms, molecules, photons, etc. It is therefore a frequent task in this field to compute the distribution of a state function, in particular the energy, among the particles of the system, e.g., as a function of temperature. This is the fundamental problem of *statistics*. In classical physics it was solved by MAXWELL and BOLTZMANN for the velocity distribution and energy distribution of molecules in a gas, and this is the basis of the kinetic theory of gases. Classical Boltzmann statistics is based on the assumption, in earlier times considered self-evident, that the probability for a certain molecule A to possess the energy E is independent of the energy of the other molecules $B, C, D, ...,$ which may have the same or different energies. This assumption is in complete analogy to the situation of throwing dice: The probability of each individual throw is independent of the result of the preceding throws. According to classical statistics, the probability that in a system just dN particles have energies between E and $E+dE$ is determined by the number of possibilities by which this state can actually be realized. In our imagination we think of all particles as numbered and determine how many different possibilities of distributing all N_0 particles in the different energy cells of size dE exist which fulfill the requirement that there are just dN particles in thermal equilibrium at absolute temperature T with energies between E and $E+dE$. The number of these possible distributions gives the probability of the total distribution. By computing the number of particles with energies between E and $E+dE$ as a function of T, we get the famous Maxwell-Boltzmann energy distribution

$$dN \sim E^{\frac{1}{2}} e^{-\frac{E}{kT}} dE . \tag{4.190}$$

According to this formula of classical statistics, at absolute zero of temperature all electrons in a metal would have zero kinetic energy, whereas at higher temperatures the energy would be distributed according to Eq. (4.190) so that the number of electrons with higher energy would not be zero but would decrease exponentially with increasing energy.

What is the answer to this problem from the standpoint of quantum theory? In the first place, there is no continuous energy distribution of the electrons, but the system has discrete, although very closely spaced energy states. The energy cells dE have a finite size. Above all, however, the two basic assumptions of classical statistics are incorrect: In the first place, the particles of a system can not be distinguished from each other. Secondly, according to quantum mechanics, they are not independent of each other but coordinated by "guiding waves". This applies, as we know, for photons as well as for electrons of cathode rays, or for molecules of a molecular beam. *The existence of a wave equation, the Schrödinger equation, is the clearest expression for the interdependence of the particles of an atomic system. This equation describes by its solutions, the eigenfunctions, the behavior of all particles which follow certain laws of physics and are not governed by mere chance. It thus follows from the existence of the interference phenomena of light, electrons, atoms, and molecules that classical Boltzmann statistics cannot be valid for any one of these particles and is, therefore, quite generally not valid.* If it is nevertheless successfully applied, as in the kinetic theory of gases, the reason lies in the fact that its results for gaseous molecules deviate noticeably from those of presently to be discussed quantum statistics only in exceptional cases and at extremely low temperatures.

In IV, 10 we have shown that all particles with half-integral spin are described by wave functions which are antisymmetric with respect to exchange of the signs of the coordinates and thus obey the Pauli principle, whereas particles with integral spin are represented by symmetric wave functions and are not subject to the Pauli principle.

The far-reaching significance of the distinction between symmetric and antisymmetric particles is illustrated by the energy distribution of a great number of particles in an atomic system at absolute zero. For instance, at absolute zero of temperature all the electrons in a metal crystal cannot be in the lowest energy state of the system since, according to the Pauli principle, each energy state can be occupied by not more than two electrons of opposite spin. For particles with symmetric eigenfunctions (e.g., α-particles), however, this restriction does not exist. The first case was studied by FERMI and DIRAC, the latter one by BOSE and EINSTEIN. This is the reason why we called particles obeying the Pauli principle and, therefore, Fermi statistics *fermions*, and vice versa the particles that follow Bose statistics and do not obey the Pauli principle, *bosons*.

Let us first look at the behavior of fermions, i.e., of all particles with half-integral spin, by discussing the most important example, the metal electrons. We shall learn in VII, 11 that every energy state of an individual atom in a metal crystal which consists of N equal atoms splits into N energy states because the degeneracy is eliminated by the interaction of the atoms. According to the Pauli principle, each of these N states can be occupied by two electrons at the most, which must differ in their spin directions, because otherwise there would be completely identical electrons of equal energy in the system, and exactly this is forbidden by the Pauli principle. Because of the existence of the quantized energy states and the validity of the Pauli principle, it is consequently not possible that all electrons have zero energy at absolute zero of temperature. Moreover, quantum theory requires that they be distributed in such a manner among the lowest possible energy states of the crystal that each state is occupied by only two electrons (of opposite spin). In contrast to classical theory, consequently, at absolute zero those electrons which occupy the highest states have considerable kinetic energy, the so-called *zero-point energy (Fermi energy)*. Its value follows with sufficient accuracy from a computation, not to be presented here, as

$$E_F = \frac{h^2}{8m}\, n^{\frac{2}{3}}. \tag{4.191}$$

E_F thus depends only on the electron mass m and the density n of the practically free conduction electrons per cubic centimeter of the metal, which we may obtain from measurements of the Hall effect according to VII, 13. The zero-point energies of most metals range from 3 to 7 ev. *As a result of the validity of the quantum laws, metal electrons thus have a kinetic energy at absolute zero which in the classical picture would require a temperature of several 10^4 °K.* Electrons in these states are called degenerate, and this phenomenon found by Fermi is called the degeneracy of the electron gas. It is exceedingly important for the electron theory of metals and especially for the theory of the electric conductivity of metals. It explains, for example, why contrary to classical expectation electrons contribute comparatively little to the specific heat of a metal, although they have to be regarded as practically free particles. Their zero-point energy is so large that small temperature changes produce only minor changes in the energy distribution of the electrons, though with increasing temperature gradually more and more electrons occupy energy levels which are vacant at $T=0$.

It is also clear that the degeneracy of the electron gas in a metal vanishes as the average thermal energy of the electrons becomes of the same order of magnitude as the zero-point energy of the degenerate electrons. At very high temperatures, the distribution of the electrons among the energy states of the metal no longer is determined by the quantum laws or the Pauli principle but by temperature. This limiting temperature, above which classical statistics again becomes valid, is called the degeneracy temperature T_0. It has been computed from Eq. (4.191) with $E = kT$ as

$$T_0 = \frac{h^2}{8mk} n^{\frac{2}{3}}. \tag{4.192}$$

For the common metals, T_0 ranges from 30,000 to 80,000 °K so that the conduction electrons are always degenerate. On the other hand, it follows from Eq. (4.192) that for arc discharges of very high electron density, with $n \sim 10^{17}$ per cm³, the degeneracy temperature is of the order of only 10 °K. The electrons of an arc plasma of about 10^4 °K thus are not degenerate and do obey classical statistics.

From our presentation it is obvious that because of the validity of the quantum laws an entirely different statistical theory applies to the degenerate gas. We need it if we want to represent the energy distribution of degenerate electrons by a formula similar to Eq. (4.190). We now no longer are dealing with the possibilities of distributing electrons which can be distinguished from another. The problem is rather how the occupation of the quantized energy levels by basically indistinguishable electrons depends on the temperature. The derivation of this *Fermi statistics* leads to an energy distribution formula which applies quite generally to fermions, i.e., to particles with half integral spin:

$$\frac{dN}{dE} = \frac{4\pi(2m)^{\frac{3}{2}}}{h^3} \frac{\sqrt{E}}{e^{\frac{E-E_F}{kT}} + 1}. \tag{4.193}$$

E_F is again the Fermi energy as determined by Eq. (4.191). In order to illustrate this formula, dN/dE is plotted in Fig. 110 against the energy E for $T=0$ and for a temperature considerably above zero, $T \gg 0$. For $T=0$, the number of electrons increases with the energy up to a certain limit, the energy of the highest occupied state. All higher energy states are completely vacant. At higher temperature, a fraction of the electrons, which at $T=0$ occupy the energy levels near E_F, moves into higher levels which were vacant at $T=0$. We thus obtain the energy distribution shown in the dotted curve of Fig. 110. The half width of the energy range with changed electron occupation (the range of the dotted line) has the value $4kT$. For $T > 0$ we thus have a "tail" of high-energy electrons which is important for a number of phenomena, e.g., the thermionic emission of electrons from heated solids, which will be treated in VII, 14.

The energy distribution looks entirely different for particles with integral spin (bosons) because these are not restricted by the Pauli principle. *Consequently, all bosons of a system may exist in the lowest energy state, in contrast to the behavior of fermions. The only reasons for the deviations of Bose statistics from classical Boltzmann statistics are that particles obeying quantum mechanics are indistinguishable and not independent from each other, and that we have discrete energy states.* This statistics, developed by BOSE, leads to an energy distribution formula which differs from Eq. (4.193) for fermions only by $E_F = 0$ and the substitution of -1 for $+1$ in the denominator. A comparison of this formula with Eq. (2.11) shows that it has the same structure as PLANCK's energy distribution formula for the black-body radiation. The latter was actually derived by EINSTEIN by applying Bose statistics to photons.

No further applications of Bose statistics were known until recently, since for gases deviations from classical statistics are to be expected for extremely low temperatures and for

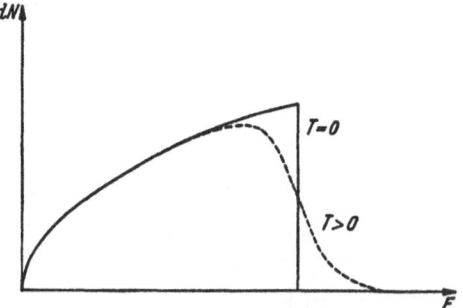

Fig. 110. Fermi distribution of the electrons in a metal. Ordinate = number of electrons, abscissa = electron energy. Solid curve for temperature $T = 0$, dotted curve for temperature $T > 0$ (schematic)

high pressures only. Here, they are hidden by the van der Waals deviations of the behavior of real gases from ideal gases. During the last years, however, it became clear that the particular behavior of bosons is responsible for the strange phenomenon of "superfluidity" of helium at temperatures below 2.186 °K. We shall come back to this in VII, 17b.

14. The Basic Ideas of Quantum Electrodynamics. Quantization of Wave Fields

Quantum mechanics as presented so far is incomplete with respect to one decisive point. We started from the complementarity of the wave-particle phenomena and arrived at quantum mechanics which clarified the relation between observable particles (e.g., electrons) and their corresponding Ψ wave fields to such an extent that experiments could be predicted and explained in an unambiguous way. Thus, the wave-particle dualism was actually solved, though not in an easily understandable way. Quantum mechanics proved not suitable, however, for similarly clarifying the relation between an electromagnetic field and the photons which correspond to it in the particle concept.

Let us compare two closed stationary systems, the hydrogen atom and a cavity with ideally reflecting walls, which may be filled with radiation. By quantization of the Ψ waves (cf. IV, 5), SCHRÖDINGER showed that for the hydrogen atom only certain discrete vibration modes, the eigenfunctions of the atom, are compatible with the boundary conditions, and he derived the discrete energy values of the electron which correspond to these vibration modes.

Quantization of the Ψ waves thus led to the properties of the electrons which correspond to the Ψ waves in the particle concept. For the second example, the radiation-filled cavity, the stationarity condition requires that only such waves occur for which the cavity dimensions are integral multiples of one half wavelength of its radiation.

With the help of the relation $E = h\nu$, quantum mechanics then yields the *energy* of the associated photons from the frequencies of the stationary waves. But *no* quantum-mechanical result states anything about the *number* of the photons which belong to a cavity eigenvibration of given frequency. A comparison with the first example shows that obviously something is still missing here which there led to the particle qualities, viz., a quantum condition. This is the quantization of the electromagnetic field. Since we have associated the wavelength (or frequency) of the quantized electromagnetic waves with the energy of the corresponding photons, we expect the second property of the cavity waves, i.e., their amplitude, to correspond to the *number* of photons. We thus come to the conclusion that *also the amplitudes of the individual discrete stationary waves of the cavity have to be quantized (so-called second quantization) so that only certain discrete amplitude values are possible.* For this purpose we need a Schrödinger equation whose solutions indicate the probability that the amplitude of a certain cavity vibration has precisely this or that value. This implies that only *probability* statements can be made about values of the field strength (viz., about amplitudes of the electromagnetic radiation). Indeed, field strength and number of photons prove to be complementary to one another in the sense of the uncertainty principle, since $\mathfrak{E}$ and $\mathfrak{H}$ belong to the wave concept, whereas the number n of the photons belongs to the particle picture. It is most interesting that the theory leads to the same result that had been *postulated* by PLANCK in 1900 (cf. II, 7) when he tried to describe theoretically the black-body radiation and thus initiated the entire development of quantum theory. The stationary electromagnetic waves of the cavity (black body) behave exactly like linear oscillators. Consequently, their energy can only be a discrete integral multiple of $h\nu_0$ if ν_0 is the eigenfrequency of the corresponding stationary wave, i.e., of its so-called virtual oscillator. In the particle picture, a cavity eigenvibration of the energy $nh\nu_0$ thus corresponds to n photons of energy $h\nu_0$.

In III, 22 we found that quite generally the differences between quantum-theoretical computation and classical computation disappear more and more with increasing quantum number n. If thus by field quantization the amplitude of a certain cavity vibration is associated to a *large* quantum number n, or in the particle language, if there are *many* photons with the corresponding energy $h\nu$ in the cavity, then no substantial deviations from the classical theory are expected, and none are found. This holds true for instance for the long-wave electric waves down to the infrared waves, where the energy $h\nu$ of the individual photons is small compared to the radiation energies used in radiation measurements, and where all phenomena can be described satisfactorily without making use of quantum theory. In the ultraviolet and the region of yet shorter wavelengths, however, the energy of the individual photon may be so large that considerable deviations from the classical theory are to be expected since even individual photons may be measured here. Indeed, it is known that the earlier formula for the spectral energy distribution of the black-body radiation by RAYLEIGH-JEANS agrees satisfactorily with the measurements in the region of long waves, but results in discrepancies which become the larger and more fundamental the shorter the wavelength of the particular radiation is.

Our interest, however, reaches considerably beyond the special case of a cavity with its stationary radiation field. In particular, we are interested in the interaction between the atoms and the quantized electromagnetic field. In other words, we are interested in the quantum theory of the electromagnetic-field changes which are related to those changes of the energy states of atoms which in the particle language are described as emission and absorption of photons. Both problems, the above-mentioned quantization of the electromagnetic field and its interaction with the atoms were treated by DIRAC in 1930; his theory is called *quantum electrodynamics*.

We can here present only a very brief sketch of DIRAC's theory. In treating the emission and absorption of radiation by atoms, DIRAC considers a system consisting of atoms that are imbedded in a radiation field. Whereas the atoms are characterized by the usual quantum numbers indicating their energy states, every frequency ν_i occurring in the radiation field is characterized by its corresponding amplitude quantum number n_i. In the particle concept we have a mixture of atoms and photons with just n_i photons of the energy $h\nu_i$ each. The interaction between atoms and radiation field is that of classical physics: The electric field strength of the electromagnetic radiation field induces in the particular atom an electric moment varying with the corresponding frequency; this moment couples field and atom. A particular frequency ν_k whose quantum energy $h\nu_k$ equals the energy difference ΔE_k of two stationary energy states of the atom in question causes a transition of the atom from one energy state into an other. During this transition the probability of which may be computed from the theory, the amplitude quantum number n_k of the electromagnetic radiation simultaneously increases or decreases by one. This is the quantum-electrodynamic description of the energy change of an atom under emission or absorption of a photon with the "correct" energy $h\nu_k$.

Thus, quantum electrodynamics describes the quantization of the electrodynamic field and, by coupling it with the atomic electrons, the emission and absorption of photons by atoms. With considerable success, it has recently been tried to treat the fields of nuclear forces in an analogous way. Here the mesons to be discussed in V, 23 are regarded as quanta of special meson fields, and one tries to understand the emission and absorption of mesons by nuclei from the interaction between the meson fields and the nuclei (cf. V, 25).

Presently DIRAC and others are studying the problem whether analogously to the development of quantum electrodynamics a theory of gravity may be developed which is called *gravity dynamics*. According to this concept, masses in motion produce gravity wave fields in a similar way as charges in motion give rise to electromagnetic wave fields. Just as in quantum electrodynamics quantization of the electromagnetic fields leads to photons, so quantization of the gravity waves should lead to quanta of the gravity field, called *gravitons*. The major unsolved difficulty of this theory lies in the fact that the energy density of the gravity field proves to be dependent on the specific coordinate system. This problem evidently is connected with the general theory of relativity. The experimental detection of gravitons, which according to the theory are expected to propagate with the velocity of light, seems to be difficult since their interaction with matter should be even smaller than that of the neutrinos, which we shall treat in V, 6f. Nevertheless, methods for their production as well as for their measurement are already under examination.

15. Achievements, Limitations, and Philosophical Significance of Quantum Mechanics

If we review quantum mechanics as discussed in this chapter and inquire about its achievements and limitations, we find that it describes practically the entire field of atomic, molecular, and solid-state physics, as well as many essential results of nuclear physics, and that it is in complete quantitative agreement with experience. The progress which it has accomplished in many domains in which the old Bohr-Sommerfeld quantum theory failed is tremendous.

The difficulties concerning the arbitrary introduction of BOHR's postulates, which were necessary in order to explain the atomic spectra, are removed. The fact that atoms do not radiate in their stationary states was explained without violating the laws of electrodynamics. Quantum mechanics gave us, as a necessary logical consequence of the mathematical theory, not only the quantum conditions for the energy levels, but it also explained the emission and absorption of radiation according to BOHR's frequency condition upon transition from one energy state to another, including the selection rules, the polarization of emitted spectral lines, and the line intensities. The anomalous Zeeman effect is explained just as quantitatively as are the finer features of the rotation and vibration of molecules including their interaction with the electron shell. This could not be computed from the old theory (Chap. VI). A large number of problems which were not understood at all by the old theory have been solved by means of the quantum-mechanical exchange theory. The problem of the helium atom and other atoms with several equivalent electrons belongs to this group, just as do homopolar binding and the energy bands of crystals with very many constituent atoms of the same kind (Chap. VII) or the phenomena of ferromagnetism (cf. VII, 15 c), which were so mysterious for many decades. Also a fundamental approach to the problem of nuclear forces was found on the basis of the exchange theory. The typically quantum-mechanical effect of the penetration of a potential wall by a particle (the tunnel effect) has made it possible to understand radioactive α-decay, the emission of electrons from a metal by a strong electric field (Chap. VII), as well as the finer features of predissociation, which will be discussed in VI, 7b. The Smekal-Raman effect (VI, 2d), which was predicted on the basis of quantum physics and later verified experimentally, is of special importance in molecular physics. Finally, it is a fundamental achievement of quantum mechanics that it explains the physical significance of the wave-particle dualism of light as well as of matter and, by the uncertainty principle, provides a means of determining the kind of statements and their accuracy which may be made in the field of microphysics.

Quantum mechanics and its extension by DIRAC (quantum electrodynamics) thus can be regarded as the correct theory for the quantitative description of phenomena and processes in the field of the atomic shells, including the interaction with radiation. Any statements and predictions made in these fields of physics on the basis of quantum-mechanical computations can be made with almost unlimited confidence. On the other hand, the limitations of the theory become evident. There are, in the first place, the difficulties which are associated with the unexplained "structure" of the electron and other elementary particles, which can be regarded as point charges only in a first, rough approximation. As will be shown in the next chapter, there are furthermore phenomena of nuclear physics which cannot be satisfactorily described by quantum mechanics as we have treated it here. Among these is the problem of the nuclear forces which hold the neutrons and protons together in the nucleus. It is possible, as we shall show in the next chapter, to draw an analogy between the nuclear forces and the quantum-mechanical exchange forces. However, it is evident that quantum mechanics in its present form does not give a satisfactory quantitative description of the nuclear forces. This failure of present-day quantum mechanics becomes especially clear in the case of such extreme nuclear processes as the generation of a large number of electrons and other elementary particles and their anti-particles in a collision of *one* very energetic proton or neutron with another nucleon or an entire nucleus (see V, 20/21). These materialization showers cannot be explained by our present quantum mechanics. This requires, according to HEISENBERG, an extension of the theory, which leads to the assumption of a smallest length l_0 of the order of 10^{-13} cm as a new universal constant, and a corresponding "smallest time", the time required to traverse the smallest length with the velocity of light, l_0/c. *It seems that present-day quantum mechanics is suitable to describe only those atomic processes which involve no essential change in the state of the system within this smallest time.* It is apparent that this condition is not fulfilled for such extreme processes as we have mentioned above so that an extension of quantum mechanics seems to be necessary. Work of theoretical physicists from all over the world is progressing towards this goal. In V, 24 we shall come back to the latest development of elementary-particle physics.

Before concluding this chapter, we must finally analyse the fundamental difficulties and problems which have been raised by quantum mechanics and which are closely related to the philosophy of science. These conceptual difficulties of quantum physics undoubtedly are so large that even today some scientists (and so many philosophers!) regard quantum mechanics and its description of the world with undisguised skepticism — a too complacent attitude since the results of the theory which we have described and their logical consequences should inspire them to a very serious analysis. Three objections are raised most frequently against the quantum-mechanical theory: its incomprehensibility, its indeterminacy, which is evidenced by the fact that its results can be described only in terms of statistics, and, related to this, its asserted acausality.

Now there is, in fact, no doubt that our new physical "picture" of the world is much less pictorial than that of classical physics of the year 1900, which we may regard as pictorial without entering into a discussion about the difficulty of its definition. The theory of relativity already had stripped our concepts of space and time of their seemingly so obvious absoluteness. Now quantum mechanics has changed greatly the meaning of mass, force, and energy, i.e., of concepts that so far had been relatively close to our classical experience: Mass and energy can be converted into each other, and the fundamental concept of classical physics, force, has actually been eliminated from the theory. The decisive role which PLANCK's elementary quantum of action h plays in all physics seems, moreover, to point very clearly to a dominant significance of the very unpictorial concept of action (energy times time) for all physics. Evidence for this central role of the concept action can also be derived from the relativity theory. Whereas the values of force, mass, and energy depend on the state of motion of the reference system, action is the only physical quantity which is independent of the reference system, time and position. It is "relativistically invariant". Both quantum theory and relativity theory thus agree on the central significance of action in physics, so that it seems appropriate to regard the quanta of action h as the last physical realities upon which our whole world of phenomena depends.

We have seen in IV, 3 that the uncertainty principle and the wave-particle dualism are also based on this dominant role which the basically unpictorial concept of the quantized action plays in modern physics. And it is the uncertainty principle and the wave-particle dualism which cause the fundamental conceptual difficulties in our new "picture" of the physical world.

The fact that all atomic particles, depending on the experiments we perform with them, display such contradictory properties as those of a particle well localized in space and time or those of an extended wave field, leads to only one logical conclusion: *The "atom as such" cannot be described within the usual concepts of space and time* since in these concepts a point in space and a wave field extended in space are simultaneously incompatible. Most theoretical physicists are well aware of the far-reaching significance of this statement. Although none of them was willing to accept it without serious misgivings, nevertheless they do not see any other way out. The interpretation of wave mechanics, as presented on page 169, according to which only the particles are "real" and the associated wave fields have "only" the task of guiding the particles statistically in a mysterious way to the correct positions, is no explanation either: Although only the particles may manifest themselves, for instance by ionizing the gas of a cloud chamber along their path, they are by necessity wave-mechanically associated with the probability waves that guide them so that we meet the wave-particle dualism and its logical difficulties in this explanation also. Thus we are forced to a conclusion that never before proved necessary in the entire field of science: *Although all experiments or observations with atomic systems have results that are pictorial in the classical sense, we cannot draw inferences from them with respect to the "real" existence of non-observed properties and, therefore, on "atoms as such".* This is a logical consequence of the experiments leading to the wave-particle dualism. *No statements can be made about atoms as such.* If the atoms are to appear in space and time, they must interact with other particles; then we always observe only one group of properties, whereas the properties which belong to the complementary side remain unobservable. Which properties may be observed, depends on our decision or our observational device. *The possibility of objectifying all phenomena and processes of the physical world, which is self-evident in classical physics, thus is lost in microphysics. This is a necessary consequence of the wave-particle dualism.*

We shall now discuss the problem of indeterminacy of atomistic events and, finally, the relation between indeterminacy and causality.

The significance of what is called indeterminacy or the existence of only statistical rules is illustrated by the example of the radioactive decay (see V, 6). There we shall learn that the half life of 1600 years is an essential and characteristic property of the nucleus of the radium atom. This means, that of 100 radium nuclei present at this moment, we know that 50 will decay within the next 1600 years. This decay is purely statistical, the atoms do not "age", and it is impossible to predict at what time a specific radium nucleus will actually decay. The decisive point here is the fact that our ignorance about the exact time of decay is not caused by an insufficient knowledge, but is a logical consequence of the well-confirmed formalism of wave mechanics, where only *probability statements* can be made about atomic events. This follows from our interpretation of $\Psi\Psi^*$ in IV, 6. Our entire present knowledge of physics indicates that this indeterminacy of individual atomic processes is a fundamental property of our world. Our present conception of microphysics (and we do not see, how this could be changed in the future!) allows in principle only statistical statements, which condense in macrophysics to seemingly complete determinacy, because statistical statements in the range of sufficiently large numbers become equivalent to statements of practically absolute certainty. Although we know exactly that one gram of radium emits 3.7×10^{10} α-particles in every second, we cannot say more about the decay of an *individual* nucleus than that it will occur with a probability of 50% within the next 1600 years.

This indeterminacy, the mathematical formulation of which proves to be the uncertainty principle, is closely related to the problem of the validity of the causality principle within the range of atomic physics. This is frequently misunderstood in public discussions. *If by interaction with some experimental arrangement we force an atom to reveal some of its properties, the exact relation between cause and effect applies to these properties and their dependence on space and time, just as we know it from classical physics, to which the experimental devices belong.* However, according to the above-mentioned "nature" of the atoms, no simultaneous statements can be made in this case about those properties which are complementary to the observed ones, or, if the exact determination of the properties of one side is refrained from, those statements are possible only within the limits of the uncertainty principle. *Consequently, with respect to an interconnected sequence of observations, the law of causality remains valid in microphysics also, but it cannot be applied to those properties "as such" of the atom which are not under observation.*

There is no doubt that we have found here a restriction of the causality principle, but mainly a restriction of its thoughtless application. In classical physics, causality is defined by the assertion: "If the state of a closed system is known completely and precisely at one moment, then it is possible, in principle, to compute the state of the system at any earlier or later time." From the standpoint of quantum mechanics, not even the presupposition can be fulfilled since it is not possible, according to the uncertainty principle, to determine the state of any system at any definite time precisely and completely. C. F. v. Weizsäcker, whom we have followed with regard to the presentation of these problems, has pointed out that also in classical physics always only certain properties are known and conclusions concerning these properties are drawn, while the unknown properties are without influence. This limitation of the causality principle, which actually applies also to classical physics, has been recognized by quantum mechanics as a fundamental restriction. Weizsäcker, therefore, formulates the principle in the following way: "If certain properties of a system are precisely known at a certain time, then all those properties of earlier or later states of the system can be computed which, according to classical physics, are causally connected with them".

While in macrophysics the uncertainty principle thus causes a fundamental restriction of the applicability of the causality principle only, we meet the problem of the so-called acausality in full when we try to causally understand *individual atomic events*. Two typical examples may illustrate these difficulties. The photon experiment of Joffé (cf. II, 7) can be understood without difficulty by means of the particle concept: Photons are emitted from the anticathode into all directions. If one of them hits the small detector (a question of probability), it may be absorbed and the absorbed energy causes the emission of an electron with a kinetic energy according to Eq. (4.1). This experiment, consequently, is well understood by making use of the photon concept. Its wave-theoretical explanation, however, leads to causal difficulties: In wave theory the energy is proportional to the square of the amplitude and is distributed all over the surface of the propagating spherical wave. The absorption of the *entire* wave at *one* point is causally not understandable since it would mean that the energy distributed over the whole sphere would have to concentrate momentarily at the absorbing point, which would be a process without any physical cause.

We find the reverse case for the diffraction of electrons at one or two slits (Fig. 111). Here, the wave-theoretical explanation offers no difficulties: We find on the screen a diffraction pattern which is caused by the interference of waves originating at different points of the one slit. This interference pattern changes of course if we add to the waves from the first slit other ones by opening the second slit. But there arise causal difficulties as soon as we try to apply the particle concept, which theoretically should be possible as well. Then the diffraction of the electrons by the slit is caused by interaction with the atoms of the slit itself, a process that we assume possible without detailed discussion. The diffraction

pattern of two slits should then be a superposition of the patterns of the two
individual slits, which is in contrast to the actual observation. The fact that the
diffraction pattern of two slits is actually quite different from that obtained from
two individual slits opened one after the other would mean that there should
exist a cause by which the electrons passing slit I learn (without delay) whether
slit II is open or closed. Evidently this assumption is physically impossible.
We see that here causal difficulties occur in the interpretation by means of the
particle concept.

REICHENBACH critically analysed these difficulties in his book listed below.
In agreement with v. WEIZSÄCKER, he states that we de not meet causal difficulties if we confine ourselves to considering connected sequences of observations
that actually were carried out or could have taken place. Causal difficulties arise
only if "intermediate phenome-
na" are discussed, i.e., if we want
to make statements about events
that are supposed to occur between the observations but never
can be actually observed, e.g.,
about the events between the
emission of a spherical wave and
the absorption of the entire energy at a particular point (cf. our
first example above). REICHEN-
BACH shows, furthermore, that
there exist satisfactory explanations for these intermediate
phenomena also, *if* we choose *that*
description which is suited to the particular experiment, either the wave picture
or the particle picture. Causal difficulties are met, however, as soon as we try to
describe the intermediate phenomena in the complementary way. According to
REICHENBACH, our world is so constructed that there does not exist any description which satisfies the causality principle and is suitable for *all* atomic processes;
but that for every particular atomic process a suitable presentation can be found
which avoids causal difficulties. Depending on the particular case, it makes use
of the wave concept *or* the particle concept.

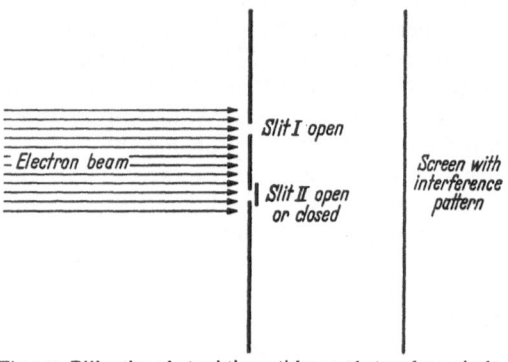

Fig. 111. Diffraction of atomistic particles or photons by a single
slit or a double slit

No doubt, this new situation is a fundamental restriction of the classical, rigid
causality according to which the whole course of the "world as such" was supposed to follow automatically and mechanically from a given initial state, like
a gigantic though extremely complicated clockwork. We will not discuss here
the problem, which may be of great interest to biologists and philosophers,
whether and to what extent room is left for a certain "freedom" in the course
of the world, though this point plays quite a role in the consequences of JORDAN'S
quantum biology. We want to stress our conviction, however, that the state of
quantum mechanics as presented here has to be regarded as final in so far as no
future extension of the atomic theory will be able to eliminate the field-particle
dualism.

It seems to be of general significance that we run into conceptual difficulties
("non-pictorial" concepts or concepts irreconcilable with fundamental philosophical ideas that so far had been considered as being self-evident) as soon as
the progress of physics leads us beyond the range of nature which is more or
less *directly* accessible to man. This holds true for velocities which are directly
(i.e., without complicated observational aids) inaccessible and are not small any

more compared to the velocity of light. These led to the theory of relativity with its conceptual difficulties. It is true also for the distances of the universe reaching beyond "our range" to billions of light years, where the apparently self-evident Euclidian geometry seems to fail ("curvature of the universe"). The same applies also to the atomic dimensions of microphysics, which are in the other direction beyond our direct reach. The phenomena in this region are correctly described only by quantum mechanics, and this description leads to the difficulties outlined above. Is it really incidental that we run into such difficulties whenever we go beyond the range of the world "given" to man? Could it not be that these conceptual and philosophical difficulties stem from the fact that all our elements ("tools") of thinking have been developed and derived from man's interaction (i.e., experience) with "his" region of the world and that we first have to adapt our methods of thinking and perception to the new expanded range of the world after we crossed the formerly given boundaries? *Quantum mechanics has taught us that it is possible and necessary to adapt our ability of conception to the new situation (e.g., v. WEIZSÄCKER's complementarity logic) and thus has opened new realms to our thinking. This seems to be one essential part of the philosophical significance of quantum mechanics.* It seems possible that from this angle new light might fall also on KANT's a-priori concepts, considered by him as a prerequisite of any science. Now MARGENAU's assumption appears to be justified that with the opening of new realms of our thinking also changes in the fundamental prerequisites of perception are to be expected so that these are stripped from their absolute character as assumed by KANT. No matter how all this may develop, further analytical and synthetical work on all the relations discussed in this chapter and on their consequences certainly should lead us to a more satisfactory scientific and philosophical picture of our physical world than was the old mechanistic picture of classical physics.

Problems

1. X-rays of 0.20 Å impinging on a carbon block undergo Compton scattering. Calculate the wavelength of the modified X-ray lines that are observed at a scattering angle of 60°.

2. A certain excited state of a nucleus is known to have a life time of 5×10^{-14} sec. What is the minimum error within which its energy can be measured?

3. Determine the de Broglie wavelength of the neutrons emerging from a nuclear reactor with an energy corresponding to the value for thermal equilibrium at 127 °C.

4. Representing a nucleus by a rectangular potential well of width 10^{-12} cm, calculate the number of attempts an α-particle emitted by U^{238} and having a kinetic energy of 4 Mev would make before it escapes through the potential barrier. The half-life of U^{238} is 4.5×10^9 years.

5. Calculate the height (U) of the potential barrier in Mev for a $_{90}Th^{234}$ nucleus as seen by an incoming α-particle. Assume that the radius of a nucleus is given by $R = 1.4 \times 10^{-13} A^{\frac{1}{3}}$ cm where A is the mass number.

6. If an α-particle of kinetic energy 5 Mev exists within the potential well of problem 5, what is the probability that it will penetrate the barrier? (hint: half-width of the well, $\frac{d}{2} = R_{Th} + R_\alpha$).

Literature

BLOCHINZEW, D. I.: Grundlagen der Quantenmechanik. Berlin: Deutscher Verlag d. Wiss. 1953.

BOGOLIUBOV, N. N., and D. V. SHIRKOV: Introduction to the Theory of Quantized Fields. New York: Interscience 1959.

BOHM, D.: Quantum Theory. New York: Prentice-Hall 1951.

BORN, M., u. P. JORDAN: Elementare Quantenmechanik. Berlin: Springer 1930.

BRILLOUIN, L.: Quantenstatistik. Berlin: Springer 1936.
BROGLIE, L. DE: Theories de la Quantification. Paris: Herman & Cie. 1932.
CORSON, E. M.: Perturbation Methods in the Quantum Mechanics of n-Electron Systems. New York: Hafner Publ. Co. 1951.
DIRAC, P. A. M.: The Fundamental Principles of Quantum Mechanics. 4th Ed. Oxford: University Press 1957.
FALKENHAGEN, H.: Statistik und Quantentheorie. Stuttgart: Hirzel 1950.
FLÜGGE, S., u. H. MARSCHALL: Rechenmethoden der Quantentheorie. 2nd Ed. Berlin-Göttingen-Heidelberg: Springer 1952.
FRIEDRICHS, K. O.: Mathematical Aspects of the Quantum Theory of Fields. New York: Interscience Publ. 1953.
GOMBAS, P.: Die statistische Theorie des Atoms und ihre Anwendung. Wien: Springer 1949.
GOMBAS, P.: Theorie und Lösungsmethoden des Mehrteilchenproblems der Wellenmechanik. Basel: Birkhäuser 1950.
HEISENBERG, W.: Die Physikalischen Prinzipien der Quantentheorie. 4th Ed. Leipzig: Hirzel 1944.
HEITLER, W.: Quantum Theory of Radiation. 3rd Ed. Oxford: Clarendon Press 1954.
HUND, F.: Materie als Feld. Heidelberg: Springer 1954.
JORDAN, P.: Anschauliche Quantentheorie. Berlin: Springer 1936.
KURZUNOGLU, B.: Modern Quantum Theory. San Francisco: Freeman 1962.
LANDAU, L. D., and E. M. LIFSHITZ: Quantum Mechanics. Reading: Addison Wesley Press 1958.
LANDÉ, A.: Quantum Mechanics. New York: Pitman Publ. Co. 1950.
LUDWIG, G.: Grundlagen der Quantenmechanik. Berlin-Göttingen-Heidelberg: Springer 1954.
MACKE, W.: Quanten. Leipzig: Akad. Verl. Ges. 1959.
MARGENAU, H.: The Nature of Physical Reality. New York: McGraw-Hill 1950.
MARCH, A.: Natur und Naturerkenntnis. Wien: Springer 1948.
MARCH, A.: Quantum Mechanics of Particles and Wave Fields. New York: Wiley 1951.
McCONNEL, J.: Quantum Particle Dynamics. 2nd Ed. Amsterdam: North Holland Publ. Co. 1959.
MERZBACHER, E.: Quantum Mechanics. New York: Wiley 1961.
MOTT, N. F., u. I. N. SNEDDON: Wave Mechanics and its Applications. Oxford: Clarendon Press 1948.
NEUMANN, J. v.: Mathematische Grundlagen der Quantenmechanik. Berlin: Springer 1932.
PERSICO, E.: Fundamentals of Quantum Mechanics. New York: Prentice-Hall 1950.
POWELL, J. L., and B. CRASEMANN: Quantum Mechanics. Reading: Addison-Wesley 1961.
REICHENBACH, H., u. M.: Die philosophische Begründung der Quantenmechanik. Basel: Birkhäuser 1949.
RICE, F. O., and E. TELLER: The Structure of Matter. New York: Wiley 1949.
ROSE, M. E.: Elementary Theorie of the Angular Momentum. New York: Wiley 1957.
RUBINOWICZ: Quantentheorie des Atoms. Leipzig: J. A. Barth 1959.
SCHAEFER, CL.: Quantentheorie, Vol. III/2 of "Einführung in die theoretische Physik". 2nd Ed. Leipzig: W. de Gruyter 1951.
SCHIFF, L. I.: Quantum Mechanics. New York: McGraw-Hill. 2nd Ed. 1955.
SCHRÖDINGER, E.: Abhandlungen zur Wellenmechanik. Leipzig: Hirzel 1928.
SLATER, J. C.: Quantum Theory of Matter. New York: McGraw-Hill 1951.
SOMMERFELD, A.: Atombau und Spektrallinien, Vol. II, 4th Ed. Braunschweig: Vieweg 1949.
THIRRING, W.: Einführung in die Quantenelektrodynamik. Wien: Deuticke 1955.
THOULESS, D. J.: The Quantum Mechanics of Many-body Systems. New York: Academic Press 1961.
WEIZEL, W.: Lehrbuch der Theoretischen Physik, Vol. II: Struktur der Materie. 2nd Ed. Berlin-Göttingen-Heidelberg: Springer 1959.
WEIZSÄCKER, C. F. v.: Zum Weltbild der Physik. 7th Ed. Stuttgart: Hirzel 1959.
WENTZEL, G.: Einführung in die Quantentheorie der Wellenfelder. Wien: Deuticke 1943.
WEYL, H.: Gruppentheorie und Quantenmechanik. Leipzig: Hirzel 1928.
WIGNER, E.: Gruppentheorie und ihre Anwendung auf die Quantenmechanik der Atomspektren. Braunschweig: Vieweg 1931.

V. Physics of Atomic Nuclei and Elementary Particles

1. Nuclear Physics in the Framework of Atomic Physics

From the standpoint of a rigid systematical approach, a discussion of the atomistic structure of matter should begin with a discussion of the elementary particles and atomic nuclei. Then atomic physics in its more restricted sense, molecular physics, and physics of the solid state would logically be built up on the theory of the structure of atomic nuclei. In presenting Bohr's pictorial atomic physics first, we have followed the historical development and, at the same time, had the advantage of proceeding from the simpler to the more difficult. Historically, up to about 1927, the first field to be studied and explained was physics of the atomic shells. Then further development led, on the one hand, to a theory of molecules which are composed of a number of atoms as well as to a theory of larger atomic complexes (liquids and solids) and, on the other hand, to nuclear physics. For this development, it was necessary to have the knowledge of Bohr's atomic theory and its quantum-mechanical refinements which we discussed in the last chapter. The same applies to our presentation in this book. Energy states and the transitions between them, accompanied by radiation, occur in the nucleus (though with correspondingly larger amounts of energy) just as in the electron shells of the atom. It is impossible to understand essential processes such as the decay of nuclei or the exchange forces which cause the bond between the nuclear constituents without a knowledge of quantum mechanics. It thus seems reasonable, both from the standpoint of an intelligible introduction as well as from that of the actual development of our science, to discuss nuclear physics at this point.

In our presentation, we shall discuss only briefly the experimental and technical questions and the abundance of the well-known nuclear reactions. For further details of these subjects the reader is referred to the monographs listed at the end of this chapter. However, we want to stress those problems which are of fundamental physical interest for nuclear physics and the physics of elementary particles, and which provide an entirely new and deeper insight into the meaning of matter and energy and their interrelationship.

Although the essential features of the structure of the nucleus, just as those of the atomic shell, may be represented pictorially, precise and quantitative statements about the behavior of the nucleus and about nuclear processes can be expected only from a quantitative theory. The problem of nuclear forces, the solution of which is being approached only gradually and slowly, has already shown us the limits of present-day quantum mechanics, which describes accurately all processes occuring in and with the electron shells. Therefore, very-high-energy nuclear reactions, as they are found in cosmic radiation and now may be studied by means of the large particle accelerators (cf. V, 3), are of special interest to us. They reveal new elementary particles and new processes such as the conversion of matter into radiation and vice versa. Physics of elementary particles has thus developed into a new field with which we shall deal in V, 19 to V, 24.

2. Methods of Detection and Measurement of Nuclear Processes and Nuclear Radiation

In order to understand the results of nuclear physics, it is necessary to present a brief survey of the experimental methods used for studying the nucleus. We disregard here highly specialized methods which will be treated later, as well as some experiments already discussed before. We shall first deal with the methods of investigating nuclear processes, viz., methods for detecting and identifying nuclear fragments, photons, and other elementary particles, and for measuring their energy. This will be done in the present section. In the next section, we shall discuss the methods of accelerating charged particles to such energies that they are able to induce nuclear transformations by penetrating into the nucleus.

Methods used for detecting and investigating natural or induced nuclear processes are always based on the detection and measurement of the energy of nuclear particles (α-particles, protons, neutrons, and larger fragments) or nuclear radiation (γ-radiation, electrons and mesons). Because of the high velocities or energies which prevail in nuclear physics, all charged particles ionize the matter through which they move. This ionization is used for detecting and also for measuring the energy of the charged particles. Being uncharged particles, neutrons cannot produce ionization directly. However, they can knock out ionizing particles, which may serve indirectly as a means of detecting neutrons and measuring their energy. The detection of neutrons will be taken up in detail in V, 13. Also the γ-radiation of the nuclei has no direct ionizing effect but manifests itself by secondary ionizing particles (see V, 6c).

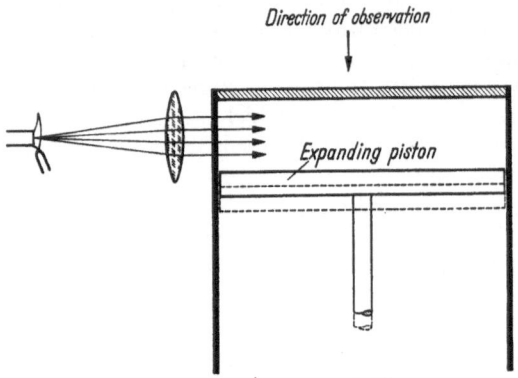

Fig. 112. The cloud Chamber of C. T. R. WILSON

We distinguish two groups of methods for the investigation of ionizing particles. First, there are those methods by which the track of a particle is directly made visible and is used for identifying the particle and for measuring its energy. Secondly, there is the group of methods by which we measure the total ionization that is produced in a certain volume by the particle.

The most important device for making particle tracks visible is the *cloud chamber* (Fig. 112), first built by C. T. R. WILSON. By rapid expansion, the gas in the chamber, which is saturated with water vapor and made dustfree in order to remove any condensation nuclei, becomes oversaturated. A charged particle of sufficient kinetic energy moving through this space produces ions which act as condensation nuclei for water droplets. The path of the particle thus becomes visible as a streak of mist and, if strongly illuminated, may be photographed. From the droplet density, which is obtained by counting the droplets, the specific ionization, to be discussed below, may be determined. For obtaining the energy of the ionized particles, the entire cloud chamber is placed in a magnetic field and the curved path of the particles is measured by means of stereoscopic photographs (cf. Fig. 173). Since the normal cloud chamber is in active working condition only for a short perid after each expansion (up to half a second), it is not suitable for recording nuclear processes. Therefore, continuously working cloud chambers have been developed in which a volume of supersaturated vapor is continuously maintained by chemical or thermodynamical methods.

In order to photograph the tracks of high-energy particles of long range, a much higher density of matter is necessary than is available in the cloud chamber. The bubble chamber or the photographic plate is used for this purpose. The *bubble chamber* (see Fig. 113) consists mainly of a volume filled with liquid hydrogen or another fluid with low boiling point. The liquid is heated close to the critical point but kept under such pressure that is does not begin to boil. By suddenly lowering the pressure, the fluid becomes superheated, and ionizing particles produce along their path in the fluid small vapor bubbles, which may be photographed (cf. Figs. 175 to 178). The presently largest cloud chamber contains 550 liters of liquid hydrogen and, together with its auxiliary devices, has a weight of more than 200 tons. For making particle tracks of very rare processes visible for which a space of many cubic meters has to be observed, the *spark chamber* has

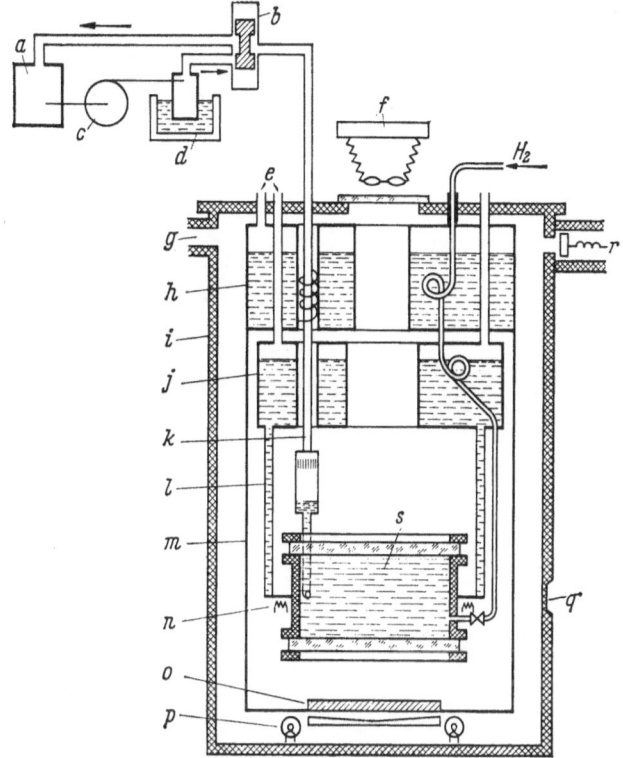

Fig. 113. Hydrogen bubble chamber; a = expansion tank, b = expansion valve, c = compressor, d = recompression tank in liquid N_2, e = outlet, f = photographic camera, g = to the vacuum pump, h = liquid N_2, i = vacuum tank, j = liquid hydrogen, k = expansion duct, l = heat-conducting connection, m = radiation shield at temperature of liquid nitrogen, n = heating system, o = diaphragm, p = illumination, q = particle window, r = outlet, s = liquid hydrogen

been developed. It consists of a large number of parallel condenser plates in series, which are charged up to nearly their breakdown voltage. An ionizing particle that passes through this condenser arrangement causes a series of sparks at the spots of its passage. Photographed from some distance in a direction along the large plates, the particle track becomes visible as a succession of sparks between the plates.Compared to these complicated and expensive devices, the *nuclear-emulsion method* is extremely simple: Stacks of photographic plates or films, sometimes interstratified with absorbing layers, are exposed to the radiation to be studied, and the developed films are examined under the microscope. Each charged high-energy particle that penetrates the photographic layer ionizes the silver bromide molecules in the same way as it would ionize gas molecules in a cloud chamber. Each ionized AgBr molecule then acts as a nucleus for a silver grain so that the particle's track becomes visible on the developed film. The tracks have to be examined by means of a microscope because the density of ionizable atoms in the photographic layer is about 1,000 times larger than in the gas of a normal cloud chamber. Thus each centimeter of a track in the cloud chamber corresponds to $\frac{1}{100}$ mm in a photographic layer. Furthermore, the time required for completely stopping a particle in the emulsion is about three orders of magnitude smaller than the time required to do the same in a gas. This proves to be very important for the investigation of unstable particles. Fig. 114 shows, as an example, an enlarged photograph of the tracks of electrons, mesons, and

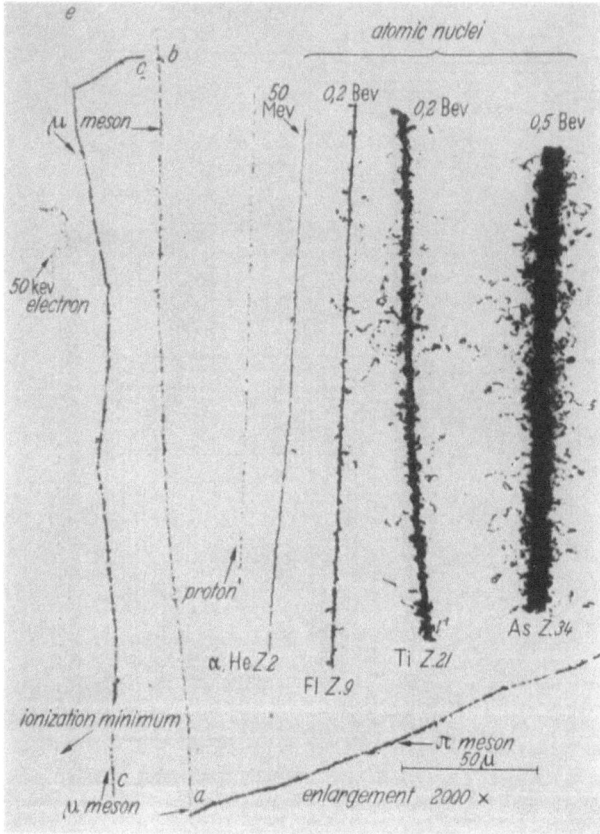

Fig. 114. Tracks of particles with different energies and masses between electron mass and 75 times the mass of a nucleon on an electron-sensitive plate (after LEPRINCE-RINGUET). Indicated are energies per nucleon

nuclei with atomic numbers between 1 and 34, taken by using a so-called electron-sensitive emulsion.

For identifying an ionizing particle and determining its energy in the cloud chamber, bubble chamber, or photographic layer, it is necessary to determine, in addition to the particle's charge, two unknown quantities, namely its rest mass m and, on the other hand, either its energy E or its velocity v. Of these two quantities only E or v is to be considered as an unknown since in the non-relativistic range they are connected with the mass by the relation $E = mv^2/2$, while for very high kinetic energy the velocity v is practically equal to the velocity of light c (relativistic range).

For a particle of charge Ze and velocity v which ionizes matter of nuclear charge Z', BETHE computed the energy loss per cm path length and obtained

$$\frac{dE}{dx} = -\frac{4\pi e^4 Z^2 Z' N}{m_e v^2}\left[\ln \frac{2m_e v^2}{E_i} - \ln(1-\beta^2) - \beta^2 - C\right]. \tag{5.1}$$

$\bar{E}_i$ is the average ionization potential of all electrons of the atoms or molecules which are ionized, N is their density, and β is the ratio of the particle's velocity to the velocity of light; the constant C finally is of no practical importance. This expression agrees well with measurements as long as the impact energy E is large compared to the ionization energy of even the innermost atomic electrons. For a given matter to be ionized, Eq. (5.1) becomes a function of the charge and the velocity alone; for air it has the form

$$\frac{dE}{dx}\,[\text{ev/cm}] = -\frac{182\,Z^2}{\beta^2}\left(\ln\frac{\beta^2}{1-\beta^2}-\beta^2+9.4\right). \tag{5.1a}$$

Since it is an experimental result that each ionization process in air, i.e., the formation of one ion pair, requires 32 ev, we obtain the number of ions formed by an ionizing particle per unit path length, the so-called *specific ionization*, by dividing expression (5.1a) by 32 ev. Obviously, the same formula, with different constants only, applies to the specific ionization in other substances, e.g., in crystals or in a photographic emulsion. Under favorable conditions we may even draw conclusions from the specific ionization of stripped nuclei or nuclear fragments on their *mass* via their charge Z as we shall show below.

For measuring the *energy* of an ionizing particle, three independent methods are available: First its total ionization, secondly the curvature of its path in a magnetic field, and thirdly deflections of the particle in nuclear collisions by so-called multiple scattering.

If the particle expends its entire energy in the cloud chamber, i.e., if it comes to rest there, the total number of ion pairs produced along the path may be counted. If we multiply this number by 32 ev (for air), we obtain the total energy of the particle. If, however, only part of the path can be observed, as it often happens for particles of high energy, then the curvature of the particle track in a magnetic field may be used for measuring the energy. According to the formula which we used in discussing the mass spectrograph, a particle of charge e, mass m, and velocity v is deflected by a magnetic field of the induction B perpendicular to the initial direction of the particle through a circular path with a radius of curvature given by

$$R = \frac{mv}{eB}. \tag{5.2}$$

If the mass and charge of the particle are known, its velocity v and thus its energy $mv^2/2$ $\left(\text{or, relativistically, } mc^2 = m_0c^2/\sqrt{1-(v^2/c^2)}\right)$ may be derived from its curved path in the magnetic field. The charge being known from the specific ionization, its sign may be determined from the direction to which the particle is deflected by the magnetic field. The third method (multiple scattering) is particularly suitable for measuring particle energies in photographic layers. It is based on the fact that the particle's path is a zigzag line caused by nuclear scattering. According to Eq. (2.19), the deflection angles are the smaller the larger the particle's velocity or its energy.

For ionizing *nuclei*, we thus obtain the two unknown quantities m and E from the specific ionization and from *one* of the above-mentioned methods for measuring E, whereas the distinction between particles of equal charge but different mass, such as electrons and protons, is based on the fact that, for the *same energy*, the specific ionization is proportional to the square root of the particle mass because of $v = \sqrt{2E/m}$. In the range of some 10^5 ev we find, in a rough approximation, that a proton produces per centimeter path in air about 10,000 ion pairs, compared to about only 200 per centimeter produced by an electron.

For the so-called relativistic range, i.e., for $v \approx c$ or $\beta \approx 1$, the specific ionization is independent of the particle mass. In air about 70 ion pairs are produced per cm path length. This is called *minimum ionization*. Fig. 115 shows a diagram in which for electrons, protons, and mesons (cf. V, 23) the specific ionization is plotted versus their energy. We see that minimum ionization begins as soon as the kinetic energy of the ionizing particle reaches the value of its rest energy $m_0 c^2$. In this range, the distinction of particles with the same charge but different mass is *impossible*. Only for stripped nuclei, the atomic number Z and from it the mass may be determined in some cases from Eq. (5.1). This proves important for the primary particles of cosmic radiation (see V, 20a). In the relativistic region, in general only inexact methods are available, viz., determining the mass from the sum of the masses and energies of the nuclear fragments origi-

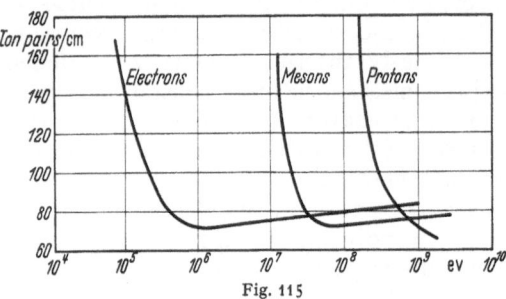

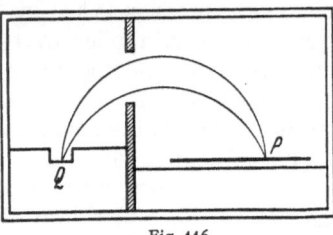

Fig. 115   Fig. 116

Fig. 115. Specific ionization (number of ion pairs produced in air per cm of path) of electrons, π mesons, and protons versus kinetic energy of the ionizing particles

Fig. 116. The most elementary β-spectrograph. Electrons from source Q are focussed on plate P by homogenous magnetic field perpendicular to drawing plane. Distance between trace in P and source Q measures velocity of electrons

nating from nuclear explosions, or from their angular distribution filling a solid angle that is the smaller the higher the kinetic energy of the primary particle is that initiates the process.

For particles of known mass and charge and of not too high energy, one often refrains from making their path visible but uses the *magnetic spectrograph* for measuring their energy. The principle of the simplest form of this instrument, which is related to the mass spectrograph (see II, 6c), is shown in Fig. 116. It is being used to an increasing extent for precision measurements of the energy of electrons and α-particles. Particles coming from the source are deviated into a circle by a magnetic field of known strength. Magnetic lenses make larger apertures possible. The orbital radius R may be determined from the distance $Q P$. Together with the known magnetic field, it allows to compute the particle energy. The photographic plate P may also be replaced by a slit with a detector behind it. In this arrangement, the orbital diameter $2R$ is kept constant and the magnetic induction B is varied until the detector shows a maximum.

For this and similar purposes, measuring devices are necessary that serve only for the detection of nuclear fragments or nuclear radiation, or for measuring their intensity or energy without indicating any detail of their path etc. If we only want to know how often and with what energy a particle is emitted from a substance or passes through a certain volume, we may use this volume as a gas condenser in which the radiation which we want to study produces ions. These ions the number of which is proportional to the energy of the ionizing particles, are attracted to the electrodes by an electric field. The resulting potential pulses, after appropriate amplification by a proportional amplifier, are fed into an oscillograph for recording, or in some cases simply registered by a counting device, after the energies have been sorted out by a discriminator. The chamber which contains the electrodes, the gas, and the electric field is called, after GREIN-ACHER, an *ionization chamber*. The interesting details of modern ionization chambers cannot be discussed here. They are able to register several hundreds of particles per second.

For more sensitive measuring devices, the *proportional counter* and the *Geiger-Müller counter* are used. These counters are essentially ionization chambers with a cylindrical metal tube as a negative and a thin axial wire as a positive electrode. The tube is charged several thousand volts positive against the wire electrode. A nuclear particle which enters the counter produces ions and electrons by ionization, just as in an ionization chamber. However, near the axial wire electrode the electrons are accelerated by the strong field to such an extent that they produce secondary ionization in the gas-filled tube. The total charge, greatly amplified by this secondary ionization, is measured as a voltage pulse by means of an amplifier (see Fig. 117). If the applied potential is not too high, the amplification of the primary ionization is proportional to the number of ions produced by the nuclear particle (proportional-counting range). Then particle *energies* can be measured. For detecting weakly ionizing particles such as electrons, the potential of the Geiger counter is increased to such an extent that the primary electrons start a discharge in

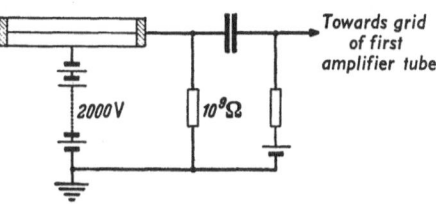

Fig. 117. Geiger-Müller counter with its circuit

the tube which extends along the entire wire and which, under proper conditions, is extinguished after about 10^{-4} sec. The resulting voltage pulses in this case do not permit any conclusions to be drawn about the type or energy of the primary ionizing particles but only about how often particles enter the counter. With this Geiger counter it is not difficult to measure the passage of every *single* particle into or through the counter.

The Geiger-Müller counter has become very popular for detecting and measuring cosmic rays and other ionizing particles. Since the utilization of nuclear fission has brought a steadily increasing number of scientists, technicians, and auxiliary personnel in contact with radioactive radiation, simple, reliable, and fast-recording pocket counters have found wide distribution (cf. V, 16/17).

Instead of the Geiger-Müller counter, *crystal counters* are presently being used in increasing quantities. These counters are simple and rugged, and they require very little space. Such a counter consists of a crystal to be penetrated by the radiation under investigation, with metal electrodes evaporated on two opposite surfaces. The electrons released in the crystal by ionization migrate in the applied field to the anode and are recorded or measured. These crystal counters are thus miniature ionization chambers in which the air is replaced by the crystalline matter. The purity of the crystal and its minuteness are critical since two conditions must be fulfilled. In the first place, in order to give a sufficiently large current pulse, the electrons released by an individual ionizing event must be able to migrate a sufficiently large distance before they are captured by lattice defects (cf. VII, 22c). In the second place, the migration of the electrons must occur so rapidly that the counter is ready for a new ionizing event after the shortest possible time interval. Recently, crystal counters from purest silicon with so-called $p\,n$ junctions have proved increasingly valuable. The explanation of the mechanism of these semiconductor detectors has to be postponed until we deal with solid-state physics in VII, 22. Good crystal counters have a resolving power of better than 10^{-8} sec. They can also be used as proportional counters for measuring energies since it is known that an energy loss of 3.5 ev in silicon corresponds to the release of one electron compared to an energy loss of 32 ev per ion pair in air.

Another detector for nuclear radiation which emerged from solid-state research is the *scintillation counter*, which recently has been developed into an extremely sensitive device and has found an ever wider field of application. It makes use of the fact that high-energy particles can locally produce visible light emission in certain solids or liquids. However, these minute light flashes are not observed any more by means of a microscope (as was done 50 years ago by RUTHERFORD'S scintillation method), but their intensity is amplified by means of a photomultiplier (cf. Fig. 10) and then recorded. The resolving power of modern scintillation counters is close to 10^{-10} sec. Since the radiation intensity of each light flash is proportional to the incident energy, the device may also be used for measuring energies. An excellent sensitivity, which often surpasses that of the Geiger counter, can be obtained if the luminescent substance used is specifically adapted to the radiation to be studied.

A new and very interesting detector for extremely rapid particles is based on an effect discovered in 1934 by CERENKOV. With the same device, also the velocity of particles may be measured. If a charged particle is moving in a transparent medium of the refractive index n, with a velocity v which exceeds the velocity of light v/n in this medium, the particle emits coherent light under an angle ϑ, measured from its direction of propagation, which is given by the relation

$$\cos \vartheta = c/nv. \qquad (5.3)$$

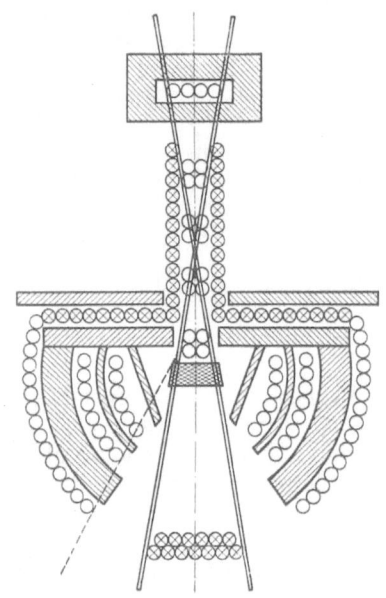

Fig. 118. Example of cosmic-ray telescope, consisting of Geiger-Müller counters and absorbers

Here, the radiation energy produced in a plexiglass cylinder of several cm of length by a single fast electron or meson is sufficient for detection with a photomultiplier. According to Eq. (5.3), the angle ϑ depends only on the particle velocity. If the refractive index of the medium is known, the Cerenkov detector may not only detect fast particles, but it may as well measure their velocity.

A combination of a cloud chamber with one or more Geiger counters according to BLACKETT is very useful for studying rare cosmic-ray events. An incident cosmic-ray particle releases a voltage pulse in the counter (or in a number of them connected for recording coincidences). The counter is so connected that this pulse simultaneously triggers, by means of relays, the expansion device of the cloud chamber, the illumination device, and the shutter of a photographic camera. If the counter is arranged in such a way that the incident particle must pass through the cloud chamber, the chamber is triggered only when a particle passes it so that unnecessary exposures are avoided. By applying a large number of suitably arranged and connected counters (so-called cosmic-ray telescopes), it is possible to record only those particles that come from a certain direction and have well-defined energies, or those which have penetrated known layers of absorbing substances. Suitable circuits allow to determine how often certain particles enter the counter *simultaneously* or to measure their emission under different angles. Fig. 118 shows an example of a rather complicated arrangement of counters and absorbers, here without a cloud chamber. It was used for investigating quantitatively the scattering of mesons by an iron plate in the center of the arrangement. Numerous variations of such counter telescopes are used in cosmic-ray research (cf. V, 20a).

Finally we mention that in connection with the industrial utilization of nuclear energy (cf. V, 16/17) a great need arose for devices called *dosimeters* which integrate the total radiation to which a person or instrument is exposed during a certain period of time. Photographic films, the density of which after development is proportional to the absorbed radiation, allow to read this directly by comparison with calibrated density scales. In addition, pocket-size ionization chambers are being used for this purpose. In them, the charge produced by ionization is utilized for discharging a condenser. The loss of charge of the condenser then is a measure of the incident radiation. In a third method, the ionizing radiation (particularly γ-radiation) produces absorbing centers in alkali halide crystals (cf. VII, 19). The density of these centers may be measured optically and is proportional to the incident radiation.

3. The Production of Energetic Nuclear Particles by Accelerators

Only a small part of nuclear physics of today deals with the investigation of nuclei occurring in nature and with the natural radioactive nuclear processes. Thus the possibility of artifically transforming atomic nuclei and todey even elementary particles has become one of the most important prerequisites of modern nuclear research. For their realization two methods exist that differ in principle from each other. We shall learn in V, 10 that numerous nuclear transformations occur in nuclear reactors. In addition, high-neutron-flux reactors may be utilized for transforming nuclei that are inserted into the reactor. In order to obtain nuclear transformations without reactors, however, energetic nuclear projectiles are needed, primarily protons, deuterons (the nuclei of the heavier hydrogen isotope $_1H^2$) and α-particles, but also heavier nuclei such as $_6C^{12}$ or $_6C^{13}$. Except for the α-particles from radioactive nuclei, they must be produced in discharges as "stripped atoms" (cf. page 79) and then be accelerated. Acceleration to a very high energy is required so that the positive projectiles are able to penetrate into the nuclei to be transformed against their electrostatic repulsion. Also high-energy electrons are of interest, primarily as a means for producing energetic photons (γ rays) which themselves are able to induce nuclear transformations.

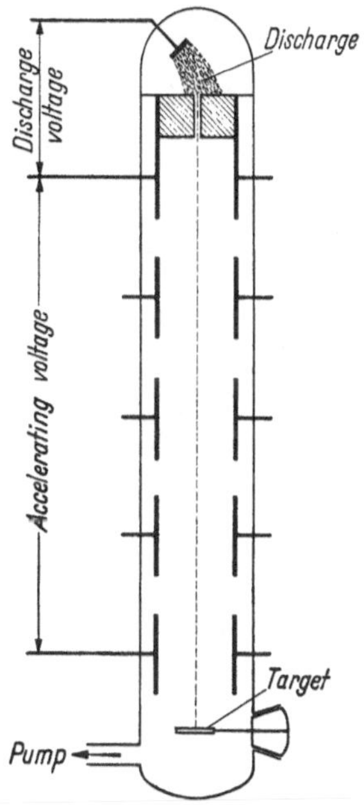

Fig. 119. Canal-ray tube with arrangement for further acceleration of canal-ray particles

Natural radioactive sources emit powerful α-rays of various energies (a maximum energy of 8.7 million ev is emitted by ThC'). The same sources emit energetic γ rays. Protons and deuterons are not emitted by radioactive sources, however. Neutrons may be produced, as we shall see in V, 13b, by bombarding beryllium $_4Be^9$ with α-rays from radioactive sources. A small tube filled with beryllium and an α-active radium sample thus is a convenient source of neutrons.

For producing artificial nuclear projectiles, the ions in question are pulled out of discharges by an electric field and are then accelerated to the required high energy by a strong electric field. Kinetic energies up to thirty billion ev have been obtained in this way. Acceleration up to several millions of volts is accomplished mostly in an acceleration tube (cf. Fig. 119) to which the entire potential difference is applied, whereas for acceleration to higher energies the particles are made to traverse the same relatively small potential difference many times (multiple accelerators).

For producing accelerating potentials of one to five million volts, an electrostatic van de Graaff generator or a cascade generator charged by a transformer is being used. In the electrostatic generator (Fig. 120), charge is sprayed at C on a rapidly revolving endless belt of insulating material. The charge is transported upwards by the belt and then removed from the belt at F in the fieldless interior of the large metal sphere A. A potential of only 20,000 volts is sufficient for

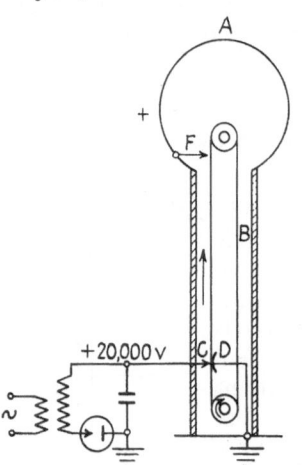

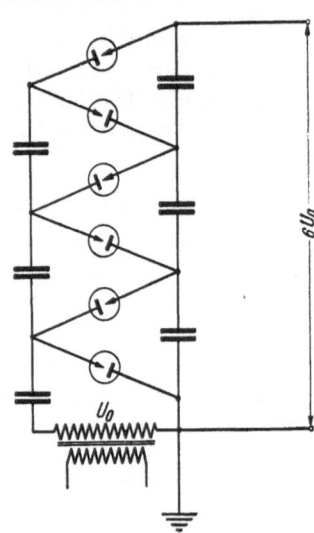

Fig. 120. Electrostatic van de Graaff high-voltage generator   Fig. 121. Electric circuit of a cascade generator (voltage multiplication)

charging the large metal sphere so highly that a corona discharge to the surrounding walls limits a further increase of the potential. The technological optimum of this accelerator type is the tandem VAN DE GRAAFF which is being used increasingly in nuclear research. In principle it consists of a combination of two horizontal van de Graaff accelerators. In the first one a beam of negative ions is accelerated by ten million volts. After two electrons have been stripped from the negative ions, which thus are converted to positive ions, these are accelerated again by the same voltage. This tandem accelerator thus yields 20 Mev ions which are practically monoenergetic and very well focussed.

In the cascade generator (see Fig. 121), a multiplication of the potential of about 100,000 volts furnished by the high-voltage transformer is effected by a suitable arrangement of condensers and rectifiers. Here again the available potential is limited by the size of the installation and the distance from the surrounding walls. In order to obtain a higher voltage, belt generators as well as cascade generators often are built into a pressure chamber which is filled with an electronegative gas (CCl_2F_2) of several atmospheres pressure in order to reduce the corona losses. The cascade generator is, in general, more expensive than the basically simple van de Graaff generator, which contains no expensive elements. However, the cascade generator produces much larger currents and, consequently, more intensive particle beams.

Among the machines for producing multiply accelerated particles we distinguish between the linear and circular accelerators, depending upon whether the particles are accelerated along a straight line or are deflected by a magnetic field so that they follow a circular or spiral path.

The linear accelerator consists of a large number of cylindrical electrodes in a long high-vacuum container (Fig. 122), which are connected to high-frequency

Fig. 122. Interior view of the 50 mev-proton linear accelerator of CERN near Geneva. (Courtesy CERN)

potential sources and are arranged in such a way that the electrons or ions always move in phase with the accelerating alternating field during their acceleration. Although the first linear accelerator was built *before* the first cyclotron at BERKELEY, interest in it soon decreased because it required too long a tube and because it was very difficult to obtain suitable very-high-frequency power sources. However, during the last war extensive experience in very efficient high-frequency generators such as those used in the development of radar was acquired and the interest in linear accelerators was revived. Fig. 122 shows a view into the 50 Mev-proton linear accelerator of CERN near Geneva. It is 50 meters long and its power is being applied to 111 drift tubes. An electron linear accelerator for more than 1000 Mev=1 Gev has been operating successfully for several years at Stanford University in California.

The oldest circular accelerator is the cyclotron, developed by LAWRENCE. A flat metal can split in the middle (Fig. 124), the halves called "dees" because of their shape, is placed in a highly evacuated chamber according to Fig. 123 (which shows a view in the direction of the field lines on the cross-hatched section of the dees). The chamber is placed in the magnetic field of a large electro-magnet, the field being homogeneous except for some carefully calculated devia-tions near the circumference (for a complete set-up see Fig. 125). The two dees are connected to the terminals of a high-frequency power source so that the electric field in the space between the dees varies periodically.

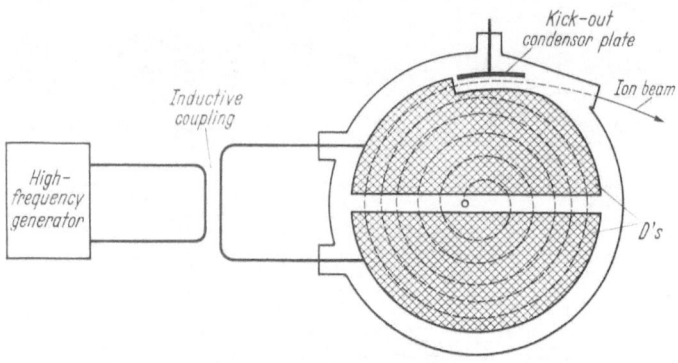

Fig. 123. Electric acceleration and ion orbit in normal cyclotron. More modern instruments occasionally use only one dee opposite to which a grounded electrode (rectangle of wire) is mounted

The ion source is placed in the center of the dees. The ion path is now curved by the vertical magnetic field into a circular path with a diameter which, according to Eq. (5.2), for a given magnetic field strength depends only on the velocity v of the ions. If an ion in its path comes to the slit between the dees, it is accelerated by the electric field.

If

$$\tau = \frac{2\pi r}{v} \tag{5.4}$$

represents the time needed by a particle of velocity v for passing once a circle of radius r, then, under consideration of Eq. (5.2), its angular velocity $\omega = 2\pi/\tau$ is given by

$$\omega = \frac{v}{r} = \frac{e}{M} B. \tag{5.5}$$

For constant ion mass M, τ and ω thus depend only on the magnetic induction B and not on the increasing orbital velocity v. Each following path then is traversed with a higher velocity, but the path also has a larger radius. The length of the orbits and the particle velocity in the orbits thus increase proportionally. The high-frequency source which produces the accelerating field and the magnetic field B is so adjusted that the electric field between the dees alternates precisely in phase with the revolution frequency of the ions. With v being the frequency of the accelerating field, the resonance condition then is

$$v = \frac{\omega}{2\pi} = \frac{eB}{2\pi M}. \tag{5.6}$$

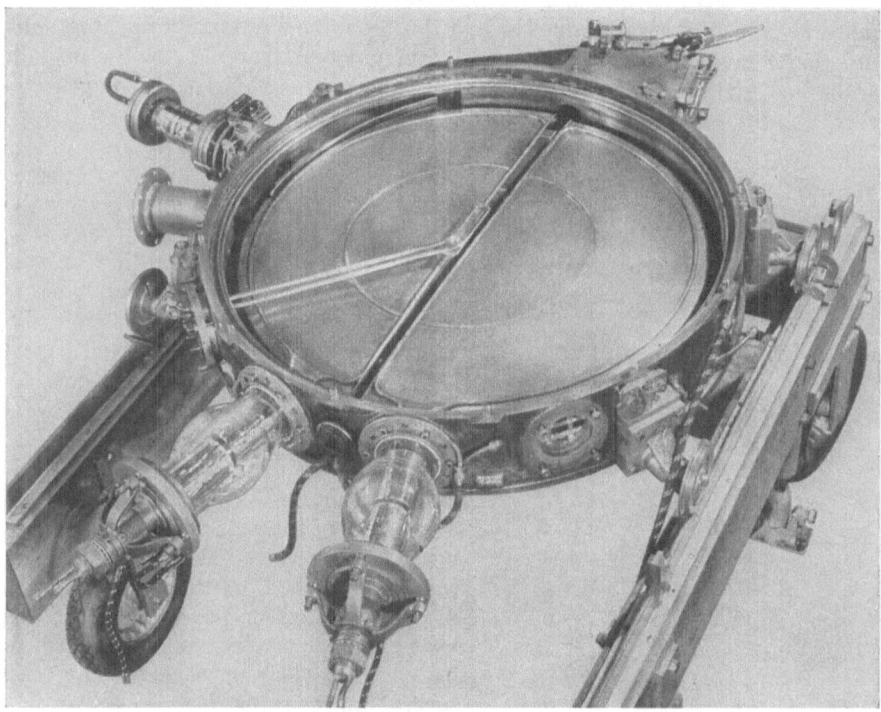

Fig. 124. Photograph of the dees of the cyclotron of Harvard University, 1939.
(Courtesy Harvard University)

Fig. 125. General view of the Berkeley synchrocyclotron of maximum energy 680 Mev. Weight of magnet 4000 tons, pole piece diameter 450 cm, diameter of oil diffusion pumps (only one visible at the left) 80 cm. Photograph was taken prior to the erection of an operating platform and installation of 300 cm thick concrete radiation shield. (Photo courtesy of Radiation Laboratory, Univresity of California, and the Atomic Energy Commisison)

If Eq. (5.6) is satisfied, the ions are accelerated by the voltage between the dees each time they pass through the slit and thus describe a gradually opening spiral until they arrive at the outer edge with an energy characteristic of the machine and limited by the diameter $2R$ of the dees and the magnetic field strength B,

$$E_{\lim} = \frac{M}{2} v_{\lim}^2 = \frac{e^2 B^2 R^2}{2M}. \tag{5.7}$$

The accelerated ions can either strike an inside target, or the ions may be deviated by a small electric condenser (cf. Figs. 123/124) so that they pass through a slit into the surrounding space where they may be used for bombarding external targets. Most cyclotrons operating according to this simple principle have dee diameters between 90 cm and 230 cm. The magnets of the larger machines of this type have a weight of several hundred tons and produce magnetic field strengths between 15,000 and 25,000 gauss. The voltage between the dees is up to 200,000 volts; its reversal frequency is of the order of 10^7 cps. The power of the high-frequency source is up to 100 kilowatts. The particle energies that are reached in this way go up to 15 Mev for protons, to 25 Mev for deuterons, and correspondingly higher for heavier ions. The ion currents may reach several milliamperes.

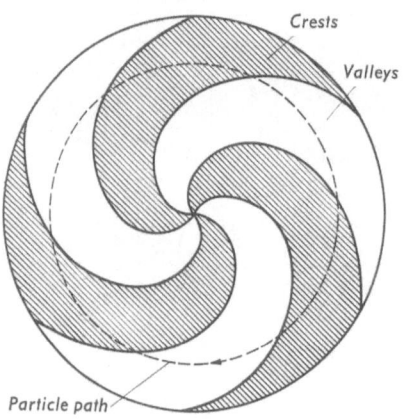

Fig. 126. Diagram of "crests" and "valleys" on pole pieces of the magnet of a spiral-crest isochron-cyclotron. (After Neu)

Attempts at accelerating ions by this simple method to even higher energies lead to difficulties. The method is based on the condition that the revolution time of the particles in the homogeneous magnetic field is independent of the particle energy, and that the reversal of the field direction in the slit therefore occurs for all particles always at the right time. This requires the particle mass in Eq. (5.6) to be constant, and this is true with sufficient accuracy for ions in the above-mentioned energy range. It is, however, not true any more for ions of higher energy whose mass increases relativistically. The largest cyclotrons built so far, with pole diameters up to 600 cm, which accelerate protons up to 720 Mev, therefore operate with frequency modulation as so-called *synchrocyclotrons* or FM-*cyclotrons*. In these machines only discrete groups of ions can be accelerated. With the magnetic field strength kept constant, the acceleration frequency here is varied during the acceleration of each particle group in such a way that its particles passing the slit between the dees are accelerated always at the right time, in spite of their relativistic mass increase. Thus the synchrocyclotron is a first step towards the even more complicated proton synchrotron, to be discussed below, which accelerates protons to even higher energies.

Since in the FM-cyclotron discrete ion groups are accelerated, the average particle flux is considerably smaller (viz., about one thousandth) than in cyclotrons with fixed acceleration frequency. This intensity decrease is avoided in the different types of *isochron-cyclotrons* in which all particles, in spite of their relativistic mass increase, take the same time for one revolution. They are therefore accelerated by a fixed high-frequency field. For $\omega = \text{const}$ (isochronous revolution) the orbital radius r increases, according to Eq. (5.5), proportional to the velocity v. The magnetic field must then increase towards the outside to the same extent as does the ion mass with increasing velocity. Such a field arrangement leads to an axial defocussing of the accelerated ions, details of which cannot be discussed here. THOMAS, however, was able to avoid this effect by varying the magnetic field periodically in the azimuthal direction. The focussing is still better in a so-called spiral-crest field produced by pole pieces which have spiral-shaped crests and valleys (see Fig. 126). This type of isochron-cyclotron, of which quite a number is under construction or is already operating, allows to accelerate protons isochronously up to more than 800 Mev.

The normal cyclotron cannot be used at all for the acceleration of *electrons* since their large relativistic mass increase (see II, 29) causes them immediately to run out of phase. The presently used machines for accelerating electrons are based on three different principles. They are the betatron, the electron synchrotron, and the electron cyclotron. This latter machine, proposed by VEKSLER, may be described first because of its relation to the cyclotron mentioned before. It is an ingeniously simple variation of the cyclotron principle. As in the normal cyclotron, magnetic field strength and acceleration frequency are kept constant. The acceleration voltage, however, is chosen to be exactly 511,000 volts since an acceleration by this voltage causes the electron's mass to increase by exactly its rest mass m_0, according to the equivalence equation (2.32). The mass of the accelerated electrons thus increases by m_0 each time they traverse the slit. It then follows from Eqs. (5.4) and (5.5) that the time needed for one revolution increases, from one acceleration to the next one, exactly by the time τ. As a consequence of this ingenious trick, the electrons reach the acceleration slit, which actually is a cavity resonator, always at the correct time, despite the constant high-frequency field. They thus remain in phase. The first model built in Ottawa reached an electron energy of more than 4 Mev with a useful current of 0.4 μamp, at only 8 revolutions and an orbital radius of 15 cm. In all electron accelerators, discrete electron groups are accelerated; in the model mentioned here, several hundred groups per sec.

The idea of the betatron, which has already found a wide field of application in industry and medicine, goes back to WIDEROE. STEENBECK first found the conditions for the guiding field which we shall discuss below and built a model of the machine, while the first efficient betatron was constructed in 1941 by KERST. The fundamental principle of the betatron is that of the transformer (cf. Fig. 127).

According to MAXWELL's second equation

$$\frac{\mu}{c}\frac{\partial H}{\partial t} = -\operatorname{curl} E, \qquad (5.8)$$

a varying magnetic flux induces a circular electric field, by which (just as in the secondary coil of a transformer) electrons, having been injected into a doughnut-shaped high-vacuum tube surrounding the magnet's core, are accelerated on circular tracks. The main difficulty in utilizing this simple idea based on the induction law is the problem of how to attain a stable circular orbit of the accelerated electrons, which must encircle the magnet's core about 10^6 times without being captured by the walls of the tube. The essential conditions for attaining a stable orbit were first recognized by STEENBECK and taken into account in his machine. The varying magnetic flux is produced by an alternating current in an electromagnet with a thinly laminated core. Between the ring-shaped pole pieces (see Fig. 127) a magnetic *guiding field* is set up the strength of which is always proportional to the actual electron velocity. The purpose of this guiding field is to keep the electrons during their acceleration on their circular path and to compensate for accidental deviations which may arise from incorrect initial

directions or from collisions with gas molecules. The guiding field then forces
the electrons back to their required orbit. When the electrons have reached their
maximum velocity (which depends on the size and field strength of the machine),
the guiding field is perturbed for a short moment so that the fastest electrons
are deviated outwards from their orbits. They are then used for experiments or
for exciting X rays. Since the direction of the accelerating flux changes 50 to
500 times per second along with the alternating current which excites the field,
the total acceleration of each electron group must be completed within one half

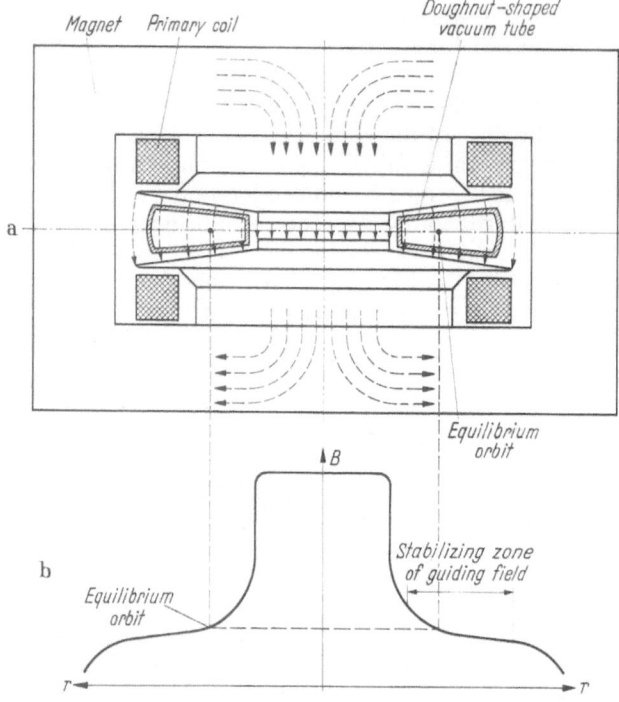

Fig. 127. a) Diagram of betatron with magnet, cross section of doughnut-shaped accelerator tube and theoretical electron
orbit. b) Magnetic flux density B along horizontal dot-dash line of (a) through magnetic center in order to show the
focussing radius region. (Courtesy Siemens-Reiniger-Werke AG)

of the period of the alternating current. Only if an electron group starts at the
right instant, can the maximum acceleration be attained. Fig. 128 shows a betatron
that has been developed and built by the Siemens-Reiniger-Werke according to
STEENBECK's patent; it produces electrons of 35 Mev. The largest instruments
built in the USA produce electrons up to 300 Mev; they serve for nuclear research,
while the 35 Mev betatrons are utilized mainly for testing materials and for
therapeutic purposes.

A particularly interesting variation is the ironless high-current *plasma betatron*,
in which the electrons of a pre-ionized plasma ring are accelerated within a few
microseconds to such a velocity that the probability of deviating collisions with
ions becomes very small. In a guiding field of 15,000 gauss and at a pressure of
10^{-4} torr it was possible to accelerate electrons corresponding to a current of
several hundred amperes (!).

An acceleration of electrons to considerably more than 100 Mev by a betatron
is prevented by the fact that, according to electrodynamics, the electrons, while
being accelerated on their circular orbits, radiate energy with the frequency of

their revolution and its higher harmonics. This energy emission increases with the fourth power of the particle energy, measured in units of its rest energy. The acceleration of electrons in a betatron therefore becomes less and less effective with increasing electron energy and finally becomes zero for a limit which according to the theory is about 500 Mev.

A further acceleration is possible, however, if the principle of the betatron is combined with that of the cyclotron. By this method the electrons, in addition to being inductively accelerated by the varying magnetic field, are accelerated by an alternating electric field as in the cyclotron. The electron synchrotron based on this idea thus combines in one instrument the principles of the betatron and the cyclotron: As in the betatron, individual electron groups are held in an orbit of constant radius R (in contrast to the cyclotron) by the guiding field and obtain their initial acceleration up to velocities close to that of light by the variation of the magnetic flux, while a further multiple acceleration to highest energies is produced by a cyclotron-like alternating electric field between two electrodes. In order to accelerate the electrons at exactly the right moment in spite of their increasing kinetic energy and the corresponding relativistic mass increase, the strength of the magnetic field must be modu

Fig. 128. 35 Mev-betatron of Siemens-Reiniger-Werke AG. Width about 60 cm. (Courtesy Siemens-Reiniger-Werke)

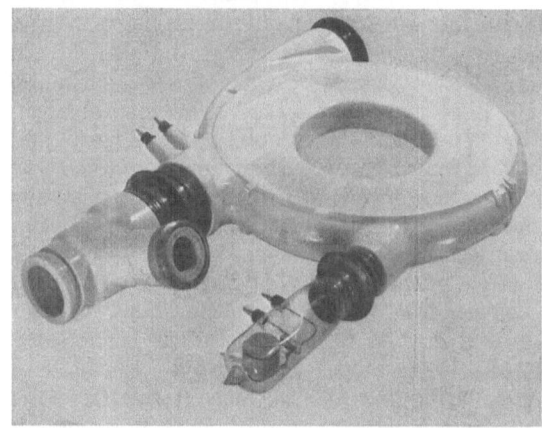

Fig. 129. Ceramic doughnut-shaped acceleration tube of a betatron. At right front: getter for absorbing freed gas remnants; at the left: electron injector; above: exit window for accelerated electrons. (Courtesy Siemens-Reiniger-Werke)

lated, according to Eq. (5.6), synchronically with the acceleration of each individual injected electron group, i.e., with the mass increase of the electrons. The acceleration *frequency*, however, is kept constant in this case.

So far, electron synchrotrons have been put into operation for energies up to 1000 Mev; machines for energies of 6000 Mev = 6 Gev are near completion. Their magnet is not different from that of a betatron, while their doughnut-shaped high-vacuum tube differs from that of a betatron (see Fig. 129) by the fact that the accelerating electrodes are missing in the betatron.

So far, we have discussed two applications of the synchrotron principle which was independently and simultaneously proposed by McMillan and Veksler. They are the synchrocyclotron and the electron synchrotron. In the first one, the magnetic field is kept constant and the frequency is being modulated; the ions are revolving in a widening spiral track. Vice versa, in the electron synchroton, the magnetic field strength is modulated in order to keep the electrons on a circular track, while a frequency modulation is unnecessary since the electrons at the beginning of the synchrotron acceleration are moving already with a nearly constant velocity, viz., that of light. By a combination of both principles, i.e., by machines which employ synchronous frequency modulation *and* modulation of the magnetic field strength, also heavy particles such as protons can be accelerated to highest energies in an orbit of constant radius, provided that they have been pre-accelerated in a different machine. For very large machines, this method has the advantage that because of the constant orbital radius only a ring-shaped magnet is necessary instead of a magnet with circular area. The amount of iron in such a proton synchrotron is thus decreased by orders of magnitude compared to a cyclotron of corresponding size. Nevertheless, these proton race tracks are gigantic and very expensive machines. The proton synchrotron, for instance, that in 1952 was put into operation under the name of *cosmotron* accelerates every 5 seconds a group of 4×10^9 protons to a maximum energy of 2.3 Gev. Since 1954, the *bevatron* produces protons of more than 6 Gev; an even larger device with an orbital radius of 20 meters in the USSR produces groups of 10^{10} protons of 10 Gev energy.

All these proton synchrotrons use normal focussing, similar to that of the betatron. However, by applying the so-called *alternating-gradient strong focussing*, in which the proton beam passes magnet segments with a radial inhomogeneity of the magnetic field which alternates between inside and outside directions, one has succeeded in compressing the proton beam to a cross section of few square inches so that the expense for the magnet stays within reasonable limits even for machines with huge orbital diameter. The first gigantic machine of this type was put into operation at the end of 1959 in Geneva. It is the proton synchrotron of the European Nuclear Research Institute CERN (see Fig. 130). Every 3 seconds, this machine accelerates 10^{11} protons to a maximum energy of 28.3 Gev. By the linear accelerator of Fig. 122, the protons obtain an energy of 50 Mev before they are injected into the magnet ring with a diameter of 200 meters which consists of 100 segments. This ring had to be adjusted to an accuracy of 0.1 mm. The acceleration of the protons lasts about one second during which the magnetic field is increased from 147 to about 12,000 gauss, and the high-frequency from 2.9 to 9.56 Mega cps. The protons pass the doughnut-shaped tube of more than 600 meters length 380,000 times during this second. A somewhat larger machine was put into operation meanwhile at Brookhaven, and machines for more than 1000 Gev are being designed.

The particle energies made available by these machines reach into the range of cosmic rays so that completely new realms are opened to the experimental physicist. The first results, such as the discovery of the anti-proton and all other heavy anti-particles, have already fulfilled the hopes fostered during the construction of these giant machines (cf. V, 21). On the other hand, the necessity of these installations for modern nuclear research led to a developement that may be regretted, though it is unavoidable; namely the fact that the decisive part of nuclear and elementary-particle research has shifted from the initiative and the laboratory of the individual scientist to the industrially organized research plants with sharply subdivided programs and team work.

We conclude this section with a short comparison of the advantages and disadvantages of linear and circular multiple accelerators. For particle energies up to several hundred million electron volts, the circular accelerators have the advantage of compact construction and a simpler high-frequency source, compared to the very long linear accelerator, which is difficult to adjust and requires a large number of high-frequency generators. The advantages of the latter, on the other hand, are that it does not need the expensive magnet and that it automatically produces well-collimated intense particle beams because of the focussing effect of

Fig. 130. View at some of the 100 magnets of the 28 Gev-proton synchroton of CERN near Geneva. Magnets are arranged on circle with a diameter of 200 meters. (Courtesy CERN)

the electrodes (see Fig. 122). The circular accelerator remains nevertheless superior for the acceleration of protons and heavier ions up to the Gev range. For the acceleration of electrons beyond 10 Gev, however, the radiation losses become so large (cf. our discussion of the betatron) that even the strongest high-frequency sources cannot compensate them anymore. For this purpose, only the linear accelerator is available. In Stanford, USA, a machine of this type is planned that will have a length of two miles and is to accelerate electrons to about 45 Gev. These extremely energetic electrons will be used for investigating the structure of the proton and the neutron.

4. General Properties of Nuclei

We begin our discussion of nuclear physics proper by summarizing the general properties of nuclei and the methods of their investigation.

a) Nuclear Charge, Nuclear Mass, and Composition of the Nucleus

The positive charge of the nucleus can be determined from X-ray spectra of the atoms (MOSELEY's law, page 19) or from the scattering of α particles by the nucleus (page 18). It is equal to the atomic number of the element in the Periodic Table and is measured in units of the elementary charge e. Thus, by knowing the element, we know the positive charge of the corresponding nucleus as well. This number is written as a subscript on the left of the symbol of the element, e.g., $_3$Li.

The mass of the nucleus is determined either by the methods of mass spectroscopy (dealt with in II, 6c) or, more frequently, from the energy of nuclear reactions (cf. V, 9a) in combination with mass-spectroscopic standards. It is usually measured in mass units mu which are 1/12 of the mass of the carbon atom C^{12}. The atomic weight, rounded-off to integers, is called the *mass number* of the nucleus; it is written as a superscript on the right of the symbol of the element, e.g., $_7N^{14}$ (frequently also on the left). We know from Table 3 on page 34 that most elements (which as such are unambiguously determined by their nuclear charge) exist with a number of different nuclei which are distinguished by their different masses. These are called *isotopic nuclei*. The *masses of all known isotopes* (referred to $C^{12}=12.000000$) *differ from integers by less than* 0.06 *mass units*. We shall explain this fact and the great significance of the small deviation from integers in V, 5.

The knowledge of charge and mass of the nucleus leads us to inquire what elementary "building blocks" form the nucleus. Before 1932, it was believed that all nuclei were built up from protons and electrons (A protons and $A-Z$ electrons). This idea, however, led to insurmountable theoretical difficulties. First, according to the uncertainty principle, an electron confined to the small space of the nucleus must have an exceptionally large momentum and thus a kinetic energy of about 10^9 ev, which is not consistent with our general knowledge of the nucleus. Secondly, the magnetic moment of the nuclei is about 1000 times smaller than that of the electron (see page 25). Thirdly, the mechanical spin of the electron which, like that of the proton, is $h/4\pi$, contradicts experimental evidence about the spin of some nuclei with odd charge number but even atomic mass number such as $_7N^{14}$. This nucleus, if composed of protons and electrons, would consist of 21 elementary particles of spin $h/4\pi$ each so that its own spin would be an odd multiple of $h/4\pi$. But its spin is actually integral, namely zero. HEISENBERG showed, immediately after the discovery of the neutron (V, 13), that all these difficulties can be avoided and complete agreement with the experiments is reached if we assume the nucleus to be made up of protons and neutrons instead of protons and electrons. This idea of the nucleus consisting of Z protons and $A-Z$ neutrons can be regarded as definitely proved. Since the protons because of their positive charge repel each other, a special nuclear force must act between neutrons and protons as well as between two protons or two neutrons. We shall consider the nature of this nuclear force in V, 25.

Since the specific nuclear forces are independent of the charge, i.e., practically identical for protons and neutrons, and since these two constituents of the nucleus may be transformed into each other, it is physically meaningful to consider them to be two different "states" of one heavy nuclear particle. In contrast to our common concept of different states of one system, these two states differ with respect to their charge. This particle whose two states are proton and neutron is called *nucleon* and thus is distinguished from other particles that occur only temporarily during nuclear transformations, such as especially mesons and hyperons. We shall treat these particles later.

It is convenient for the theory of the nucleus to distinguish the two states formally by the components $T_z=\frac{1}{2}$ (proton) and $T_z=-\frac{1}{2}$ (neutron) of a new quantum number called *isospin*. The β-decay (see V, 6f), where a neutron is transformed into a proton, then corresponds to the quantum transition $\Delta T_z = 1$ of the isospin. Its deeper physical significance becomes evident, for instance, when we regard the fact that the isospin of the total system remains unchanged when atomic nuclei undergo transformations. The isospin also determines the probability of nuclear reactions. We shall come back to it in V, 24.

b) Diameter, Density, and Shape of the Nuclei

The size or *diameter of the nuclei* is not well defined since a nucleus cannot be regarded simply as a rigid sphere. Rather, its outer boundary is determined by the decrease of the nuclear forces with the distance from the center of the nucleus. (Compare the analogous situation with regard to the atom, page 14.) We can determine the radius of the nucleus from RUTHERFORD's formula, page 18, by measuring the angular distribution of α-particles scattered by the nucleus, *if we define the radius of the nucleus as that distance from its center at which the deviation of the nuclear forces from that of the repulsive Coulomb force becomes noticeable.* Additional methods for measuring the radius of the nucleus depend on the scattering of fast neutrons by the nucleus and on the α-emission of radioactive nuclei, which will be considered in V, 6c. In the last method, the distance between the centers of gravity of the nucleus and the α-particle is computed at which the α-particle with its doubly positive charge must have been ejected by the nucleus in order to reach its actually measured final velocity. From these methods as well as from several other indirect ones we know that the radii of the nuclei are, in a good approximation, proportional to the cubic root of the mass number of the nuclei. *This means that the nuclear density is approximately constant.* In a next approximation, a density correction is introduced which takes into account that surface nucleons are unilaterally and, therefore, less strongly bound. By thus reducing the measured nuclear radii to a constant nuclear density, the different methods of determining the radii lead, with remarkable agreement, to the formula for the radius

$$r_n = 1.3 \times 10^{-13} A_n^{\frac{1}{3}} \text{ cm}, \tag{5.9}$$

A_n being the mass number of the nucleus. The author has called attention to the fact that the constant in Eq. (5.9) is, within the accuracy of the measurements, equal to the so-called Compton wavelength of the proton at rest, h/Mc. It can thus be expressed by the fundamental constants h and c and the mass of the proton M.

From Eq. (5.9), the average density of nuclear matter is computed to be 2×10^{14} g/cm³. Compared to the density of ordinary matter, this is inconceivably large. Because of their constant density, the nuclei may be compared to drops of a liquid whose density is also constant and independent of the droplet's radius. We shall frequently use this analogy between nuclei and liquid droplets in the following discussions. From several sources we also obtain information about the shape of the nuclei. In a first approximation, we may regard them as spherical. For light nuclei, this assumption of spherical force centers is in good agreement with experimental results. However, observations on the energy differences between hyperfine structure components (see III, 20) of elements of high atomic weight (heavy nuclei) indicate that the shape of the heavy nuclei shows deviations from a sphere. An ellipsoidal shape agrees best with the measurements, but the deviation from the spherical shape, in general, appears to be only of the order of one per cent. This deviation from spherical symmetry can be a prolongation or a shortening in the direction of the spin's axis. Numerically this asymmetry is expressed by associating an *electric quadrupole moment* with the nucleus and assigning a positive sign to the prolongation in the direction of the spin axis and a negative sign to the shortening of this axis. For the nuclei which have been investigated, the numerical values of the quadrupole moment lie between -0.5 and $+6.0 \times 10^{-24}$ cgs units (cm²). Most nuclei thus show a small prolongation in the direction of the spin axis. For the very heaviest nuclei we have further experimental evidence that the shape is ellipsoidal, namely the fact that the heaviest nuclei, when bombarded by neutrons, split into two parts of comparable mass (see V, 14); and such a decay by internal mechanical vibration can occur much easier if the nucleus has an ellipsoidal or pear-like shape. Neither is it surprising that among all the light nuclei the deuteron is characterized by a distinct quadrupole moment because the combination of a proton and a neutron must result in an elongated nucleus. The positive sign of its quadrupole moment indicates that the deuteron is rotating about the axis of its smallest momentum of inertia.

c) Angular Momenta and Isomerism of Nuclei

We learned in III, 20 that the hyperfine structure of atomic spectral lines can be explained in complete analogy to their multiplet structure. This is easily understood by assuming that each atomic nucleus has a constant mechanical angular momentum $\vec{I}$, measured in units of $h/2\pi$ like all atomic angular momenta, according to page 88:

$$|\vec{I}| = I h/2\pi. \tag{5.10}$$

For all known nuclei, the quantum number I of the nuclear angular momentum is an integral or half-integral number between 0 and $\frac{9}{2}$. The often used designation of I as nuclear *spin* is misleading. This designation is correct only for the elementary particles proton and neutron which, like the electron, have a spin of $\frac{1}{2} \times h/2\pi$. For nuclei consisting of protons and neutrons, $\vec{I}$ *represents the total angular momentum of the nucleus. Like the total angular momentum $\vec{J}$ of the electron shell, $\vec{I}$ is composed of the orbital momentum and the spin of all the protons and neutrons which form the nucleus.* The corresponding theoretical details will be treated in V, 12.

Two methods are available for determining the nuclear quantum number I. The first method is the analysis of the hyperfine structure of spectral lines (see III, 20), which is caused by the different possible orientations of the nuclear angular momentum $\vec{I}$ with respect to the resulting angular momentum of the electron shell. Secondly, we shall learn in VI, 9 that the intensity ratio of neighboring lines in the rotational bands of molecules with identical atoms (such as H_2, O_2, N_2, etc.) depends on the parallel or antiparallel orientation of the angular momenta of the two nuclei in the molecule. According to Eq. (6.49), the intensity ratio of a strong rotational line and the following weak one is equal to $(I+1)/I$. Intensity measurements in such molecular bands therefore allow us to determine nuclear angular momenta directly.

Such investigations of the hyperfine structure and the intensity of bands have produced an important result for nuclear theory (cf. V, 12): *Nuclei with an even number of nucleons, i.e., an even mass number, have an integral angular momentum, in most cases no angular momentum at all, $I=0$; whereas nuclei with an odd mass number always have a half-integral nuclear angular momentum.* These nuclear momenta, as deduced from spectroscopic investigations, always refer to nuclei in their ground states. We shall soon learn that atomic nuclei, like the electrons of the shell, may also exist in excited states of higher energy. Just as for the electron shell, the angular momenta of such excited nuclei may differ from those of the corresponding non-excited nuclei. Similar to the selection rules for atoms and molecules, the selection rules for transitions between different energy states of the same nucleus or from neighboring nuclei (cf. V, 10) depend on the nuclear angular momenta of the combining states.

The interesting phenomenon of nuclear *isomerism* is based on the fact that transitions between nuclear states with very different angular momenta are strongly forbidden, just as are the corresponding intercombination transitions of electrons (see page 105). Isomeric nuclei are nuclei which have the same nuclear charge and mass but different energies and degrees of stability, i.e., mean life times (we speak here not only of stable nuclei). Nuclear isomers consequently must differ in the arrangement of their constituent particles, just as in molecular physics molecules of the same composition but with different spatial arrangement of their atoms are usually called "isomers". A nuclear isomer differs from a normal excited nucleus in that the probability of its transition into a more stable state, particularly that into the ground state by emission of γ-radiation, is extremely small. Thus, referring to the terminology of atomic states, an isomeric nucleus may be called *metastable*. We shall come back to the phenomenon of isomerism in V, 7d.

d) Polarization of Atomic Nuclei in Particle Beams

Until recently, nuclear physics was concerned exclusively with nuclei or particle beams for which the orientation of the angular momenta varied statistically. But it has been known for some time that the spatial orientation of the angular momenta is of significance for the scattering process of nuclei or elementary particles by nuclei or, conversely, that scattering may create a specific orientation of the angular momenta. Every deviation of the spin directions from a purely statistical distribution is called *polarization* of these nuclei or elementary particles. It is therefore of equal interest for detailed studies of scattering processes to produce polarized particle beams, i.e., beams with preferred angular momentum or spin orientations, as well as to orient scattering nuclei. Furthermore, the connection between the orientation of the angular momentum and the direction of emission of electrons plays a decisive role with regard to parity investigations (see page 251). Thus for these cases we need oriented, in other words, polarized atomic nuclei. The nuclear angular momenta can be oriented, e.g., by a strong magnetic field at very low temperature since they are associated with nuclear magnetic moments which we shall discuss below. The production of polarized beams of all particles with magnetic moments can be accomplished by such methods as the Stern-Gerlach experiment or, for partial polarization, also by suitable scattering, e.g., by scattering of neutrons by magnetized iron.

e) Magnetic Moments of Protons, Neutrons and Complex Nuclei

As in the case of the electron shell, a magnetic moment $\mathfrak{M}_I$ is also associated with the mechanical angular momentum $\vec{I}$ of the electrically charged system of nucleons in a nucleus. In V, 12 we shall discuss in detail in which way the orbital motion and the spins of the nucleons contribute to this magnetic moment. The magnetic moments of protons and neutrons, however, show a theoretically significant anomaly. According to Eq. (3.94), there exists between the magnetic moment $\mathfrak{M}(S)$ of a rotating sphere of charge $-e$ and mass m and its mechanical spin $\vec{S}$ the relation

$$\mathfrak{M}(S) = -\frac{e}{mc}\vec{S}. \tag{5.11}$$

By substituting $h/4\pi$ for the spin $|\vec{S}|$ and for m the mass of the electron, we obtain BOHR's magneton μ_0 (2.30). If the mass of the electron is exchanged for the mass M of the proton, which is 1836 times larger, and if the positive charge of the proton is taken into consideration, we arrive at the so-called *nuclear magneton*

$$\mu_n = \frac{e h}{4 \pi M c}. \tag{5.12}$$

The relation (5.11) between mechanical spin and magnetic moment proved to be correct for the electron within less than 1%. Unexpectedly, for the proton it is correct only within an order of magnitude. Instead of the expected value of *one* nuclear magneton, the magnetic moment of the proton is equal to $+2.79$ nuclear magnetons, as taken from measurements to be discussed presently. It is even more surprising that the neutron has a magnetic moment of -1.91 nuclear magnetons although it does not carry any electric charge. The negative sign indicates that the magnetic moment corresponds to a rotating negative charge. The explanation of these two anomalies will be given in V, 25.

There are three different methods for measuring these magnetic moments. All three of them are based on an exact measurement of the frequency of a high-frequency field which causes reversing processes ("flip-flop"-processes), i.e., changes of the quantized (see III, 16a) orientation of the nuclear moments in an external magnetic field. The potential energy E of a magnetic dipole with the moment $\mathfrak{M}_I$ in the field of the strength H is, according to Eqs. (3.104) and (3.102),

$$E = (\mathfrak{M}_I)_H H = M_I g_I \mu_0 H. \tag{5.13}$$

Here $(\mathfrak{M}_I)_H$ is the quantized component of the magnetic moment $\mathfrak{M}_I$ in the field direction; M_I represents the quantized component of the mechanical angular momentum $\vec{I}$ in the same direction; while g_I is a factor corresponding to the Landé-factor (3.100). At present, it is not yet possible to compute this factor theoretically since (5.11) is not valid for nuclei, in contrast to the case of the electron shell. However, it is not the potential energy (5.13) itself which is of interest but the energy difference ΔE of neighboring quantized orientations of the moment $\mathfrak{M}_I$ in the magnetic field H. According to Eq. (3.112), these orientations are characterized by $\Delta M = 1$ so that the energy difference is given by

$$\Delta E = g_I \mu_0 H. \tag{5.14}$$

The magnetic moment in question, $\mathfrak{M}_I$, and the corresponding mechanical angular momentum I of the nucleus are connected [see Eq. (3.102)] by

$$\mathfrak{M}_I = \vec{I} g_I \mu_0 \tag{5.15}$$

so that Eq. (5.14) may be written as

$$\Delta E = \left| \frac{\mathfrak{M}_I}{\vec{I}} \right| H. \tag{5.16}$$

The ratio of magnetic moment and mechanical angular momentum, $\left| \dfrac{\mathfrak{M}_I}{\vec{I}} \right|$, is called the *gyromagnetic ratio*. It can be determined form the measurement of ΔE if the magnetic field is known.

The flip-flop energy (5.16) for the nuclear moment $\mathfrak{M}_I$ can be provided as energy quanta

$$\Delta E = h\nu \tag{5.17}$$

by a high-frequency field whose frequency can be determined from Eqs. (5.16) and (5.17). If we transform Eqs. (5.16) and (5.17) to

$$\frac{v}{H} = \frac{1}{h}\left|\frac{\mathfrak{M}_I}{\vec{I}}\right| = \frac{g_I \mu_0}{h}, \tag{5.18}$$

we notice that the measurement of $\left|\dfrac{\mathfrak{M}_I}{\vec{I}}\right|$ or g_I always amounts to measuring the ratio v/H.

For measuring v/H, RABI developed his molecular-beam method, which is an off-spring of the Stern-Gerlach experiment (page 115) but which demanded tremendous experimental skill. Because appropriate atoms are not available, such molecules were used for measuring the magnetic moment of nuclei whose electron shells do not have a resulting magnetic moment, e.g., Li^7Cl for studying the Li^7 nucleus. We will not go into the complication here that we have two more magnetic

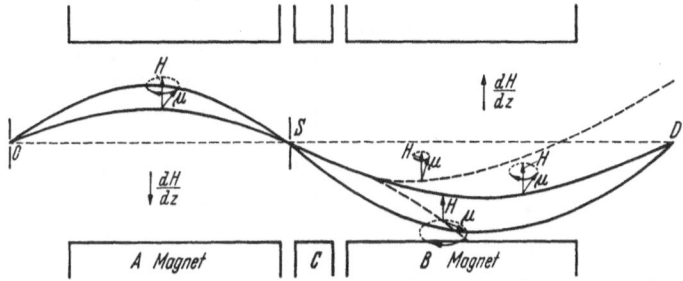

Fig. 131. Rabi method for measuring magnetic nuclear moments. Arrangement of the magnets, schematic path, and precession of nuclei originating at O and captured and measured at D. (After RABI and co-workers)

moments, namely one of chlorine and another one originating from the rotation of the heteropolar molecule (see VI, 2b). In the Rabi method, a narrow beam of LiCl molecules passes three different magnetic fields (Fig. 131). The strong fields of the magnets A and B have the same direction; but they are very inhomogeneous because of the specific shape of the pole shoes (similar to Fig. 78) and their inhomogeneities are in opposite directions. As in the Stern-Gerlach experiment, the nuclear magnetic moments are oriented by the field, the allowed directions following from space quantization. The inhomogeneity of the field deflects the molecules from their initial direction which, in general, is inclined toward the axis. Only LiCl molecules with the "right" initial velocity and orientation will pass through the slit S and then through the homogeneous field produced by the magnet C having the same orientation. Since the field direction of magnet B is the same also, the orientation of the nuclear magnetic moments is not changed in this field either; but the inhomogeneity of the field B deflects the beam back to the axis. It reaches a receiver D which measures the number of the arriving LiCl molecules in a manner that is not of interest here. The important conclusion is the following: If a high-frequency field is set up perpendicular to the constant homogeneous field H of the magnet C so that flip-flop processes occur, the molecules with the reversed nuclear spin are subject to a *different* force in B and they do *not* reach the receiver D. If we keep this high-frequency field constant (Fig. 132) and vary the field strength of C, the receiver current shows a minimum at a sharply defined v/H value. By introducing this v/H value into Eq. (5.18) we obtain the g_I value and, with known angular momentum $\vec{I}$, also the magnetic moment $\mathfrak{M}_I$ of the Li nucleus.

Since frequencies can be measured much more exactly than magnetic field strengths, the absolute method today is replaced by a relative one. With a constant but not necessarily known field H we measure the ratio of the flip-flop frequenceis of electron and proton. From Eq. (5.18) we obtain:

$$\frac{\nu_e}{\nu_p} = \frac{g_e}{g_p}. \qquad (5.19)$$

Now the g value of the electron is very well known to be 1.0011596. By measuring both spin-reversal frequencies, we can compute a very exact g value for the proton and, from Eq. (5.15), its magnetic moment. According to Eq. (5.18), the spin-reversal frequency is proportional to the field strength H; the known moment of the proton may thus be used for determining the strength H of an unknown magnetic field. This method proved to be the most exact one available at present for measuring magnetic field strengths. BLOCH has succeeded in determining also the magnetic moment of the neutron by applying a somewhat modified Rabi method.

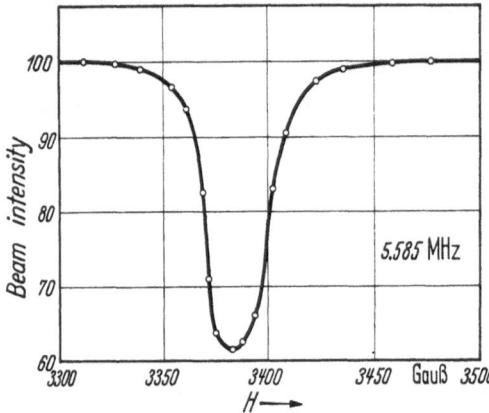

The two other methods for measuring the reversal frequency of nuclear moments, devised by BLOCH and PURCELL, do not work with individual atoms or molecules in a molecular beam but with matter in the fluid or solid state. Important conclusions may thus be drawn about the interaction between nuclear angular momentum or electron spin and the lattice structure of solids (cf. Chap. VII).

Fig. 132. Example of measurement by means of the apparatus Fig. 131. (After RABI and co-workers)

In PURCELL's method, a sample of the material (e.g., water for measuring the moment of the proton) is brought into a homogeneous constant magnetic field which causes an orientation of the nuclear moments. Perpendicular to it, a high-frequency field is set up that can produce the reversals of the spin orientations. If, by varying the frequency continuously, coincidence with the reversal frequency of Eq. (5.18) is reached, energy is taken out of the high-frequency field and a measurable reaction on the high-frequency circuit occurs from which g or $\mathfrak{M}_I/\vec{I}$ can be determined.

In the third method, developed by BLOCH and called the *nuclear induction method*, a detector coil is adjusted perpendicular to the two fields in PURCELL's method, and an electromotive force is induced in the coil by the reversing magnetic moments if the reversal frequency and the high frequency are in resonance. This induced emf serves as a means of determining the resonance frequency and thus the magnetic moment. These last two methods are used today, predominantly with a constant magnetic field, for measuring the ratio of the flip-flop frequencies of the electron or proton and the nucleus under investigation. Since the magnetic moments of the proton and neutron are non-integral multiples of the nuclear magneton (5.12), the same is true for all nuclear moments.

f) Parity

Parity belongs to the important properties of *all* atomic particles and, therefore, of nuclei too. According to IV, 5, in a system consisting of several elementary particles, we speak of even (positive) parity if the sign of the corresponding wave function remains unchanged if all coordinates of the system are reflected at the origin. Conversely, the parity is called negative if the sign changes. The best exemple for negative parity is a screw because reflection by a mirror changes it from a right-hand screw into a left-hand screw. Conversely, the parity of symmetric bodies such as a cylinder or a cube is positive because their mirror image is identical with the object itself. Since the parity of the total system is always conserved in nuclear reactions caused by strong interactions between nucleons and mesons (cf. V, 24), it is also possible to assign a parity to elementary particles by determining the parity of the initial state and of the end state of a nuclear reaction between different elementary particles. We cannot speak of a geometrical spatial parity *in this case* because the particles, in a first approximation, are point-like; hence we call it *intrinsic parity*. The latest development of this problem will be taken up again in V, 6g and V, 23.

5. Mass Defect and Nuclear Binding Energy. The Problem of Integral Isotope Masses

Since the atomic nuclei of the elements which form our material world prove to be very stable, the nucleons which are the constituents of the stable nuclei must be bound together by very strong forces. Even without a more thorough knowledge of the nuclear forces, we can plot the potential in the vicinity of the nucleus, since we know that outside the nucleus the potential due to the charge $+Ze$ falls off as $1/r$, while inside the nucleus, as a result of the binding forces, there must be a potential well. This well is separated from the Coulomb repulsion potential on the outside by a potential wall as shown in Fig. 140. Thus, the potential well corresponds to the stable nucleus, and each particle entering or leaving the nucleus must either overcome or penetrate the potential wall. We shall go into the details of this process when we discuss the radioactive decay in V, 6e.

What do we know about the binding energy of the Z protons and $A-Z$ neutrons in the nucleus? Information on this basic problem follows in a most interesting way from the so-called *mass defects of the nuclei*. If the mass of a nucleus is compared with the sum of the masses of its protons and neutrons, we find (in apparent contradiction to the chemical law of the conservation of masses) that *the mass of the nucleus is always smaller than the sum of the masses of its constituents*.

From the atomic weights A_a of the total atoms presented in the next to the last column of Table 3, page 34 and the atomic mass of the electron,

$$A_e = 0.000548597, \qquad (5.20)$$

the atomic mass A_n of the nuclei is computed as

$$A_n = A_a - 0.000549\,Z. \qquad (5.21)$$

The mass difference ΔM between the nuclear constituents and the nucleus, measured in units of $C^{12}/12$, is called the *mass defect*. Designating the atomic masses of the proton and the neutron A_P and A_N, respectively, we compute the value of the mass defect as

$$\Delta M = Z A_P + (A - Z) A_N - A_n, \qquad (5.22)$$

where

$$A_P = 1.0072765 \pm 0.0000002, \qquad (5.23)$$

and

$$A_N = 1.0086654 \pm 0.0000002. \qquad (5.24)$$

As an example, the mass defect of the α-particle ($_2\mathrm{He}^4$ nucleus) is obtained from Eqs. (5.21) to (5.24) as

$$\Delta M(\mathrm{He}^4) = (4.031884 - 4.001506)\ \mathrm{mu} = 0.030378\ \mathrm{mu}. \qquad (5.25)$$

Soon after the discovery of this result, physicists related the mass defect to the law of the equivalence of mass and energy,

$$E = mc^2, \qquad (2.32)$$

and interpreted the mass defect in the following way: When a nucleus is formed from protons and neutrons, the binding energy must be freed in order to get a stable nucleus. The mass equivalent of this energy is the mass defect. Vice versa, if the nucleus is to be split into its constituent nucleons, an energy that corresponds to the mass defect has to be *supplied*; it appears as mass increase (higher mass of the nucleons than that of the nucleus formed by them). The binding energy, corresponding to this mass defect, of the He nucleus is, from Eq. (5.25),

$$0.0304\ \mathrm{mu} \triangleq 28.3\ \mathrm{Mev}. \qquad (5.26)$$

Here the nuclear energy is measured in million electron volts (Mev). For converting mass units into energy we have the relation

$$1\ \mathrm{mass\ unit} \triangleq 931.441\ \mathrm{Mev} \quad \mathrm{and} \quad 1\,m_e \triangleq 0.511\ \mathrm{Mev} \qquad (5.27)$$

or, roughly: A mass defect of 1/1000 mass unit is equivalent to 1 Mev. The converted energies are, as the example of the He nucleus shows, of the order of magnitude of millions of electron volts and thus are about 10^6 times larger than the energies involved in processes of the electron shells of atoms and molecules. This is the reason why the binding energy of chemical molecules computed in mass units from Eq. (5.27) stays below the limits of observation so that, e.g., the mass of the CO_2 molecule within the accuracy of measurement is identical with that of the atoms $1\mathrm{C} + 2\mathrm{O}$ that form the molecule.

If the mass defects, computed according to Eq. (5.22) from the empirical nuclear masses A_n or the corresponding binding energies of all known nuclei, are plotted against their mass numbers, we get in a first approximation a straight line, Fig. 133. This means that the average binding energy of each nucleon (proton or neutron) is approximately the same and equal to about 8 Mev. This important result is in agreement with our droplet model of the nucleus, already mentioned in V, 4b, since in a liquid droplet the binding energy of each newly attached molecule is the same too.

The average binding energy of 8 Mev of each nucleon is the cause for the striking observation, mentioned above and read from Table 3, that the atomic masses of all known nuclides practically are integers in atomic units. According to Eq. (5.27), the average binding energy of 8 Mev corresponds to a mass defect of 0.0085 mass units. Since the average atomic mass of a free nucleon is 1.0080, according to Eqs. (5.23)/(5.24), the average atomic mass of every nucleon *bound* in a larger nucleus is equal to 1.000, with deviations of less than 0.1%.

The masses of the lightest nuclei, however, deviate up to 1% from integers. That is obviously due to the fact that the droplet model is only a poor approximation for systems consisting of only a few nucleons. In V, 11 we shall see in detail that theoretically we expect for each nucleon of such a system a smaller average binding energy, which varies from nucleus to nucleus. Actually, this is exactly the empirical result. For example, the above-mentioned α-particle with its binding energy of 28 Mev is the strongest-bound small unit of several nucleons in existence. In best agreement with this observation is the fact that $_2$He^4 nuclei, i.e., α-particles, are often ejected from atomic nuclei during nuclear reactions that we will discuss later. On the other hand, deuterons $_1$H^2 are comparatively less stable; their mass is 2.0136 with a mass defect of only 0.0024 mass units and a binding energy of only 10 per cent of that of the α-particle. Consequently, the α-particle is the only complex unit which may be preformed as such in the nucleus and therefore can be emitted as an entity in nuclear transformations. We shall go into all the details of this when we treat the systematics of atomic nuclei in V, 11. We shall take up next the internal processes in nuclei and their significance.

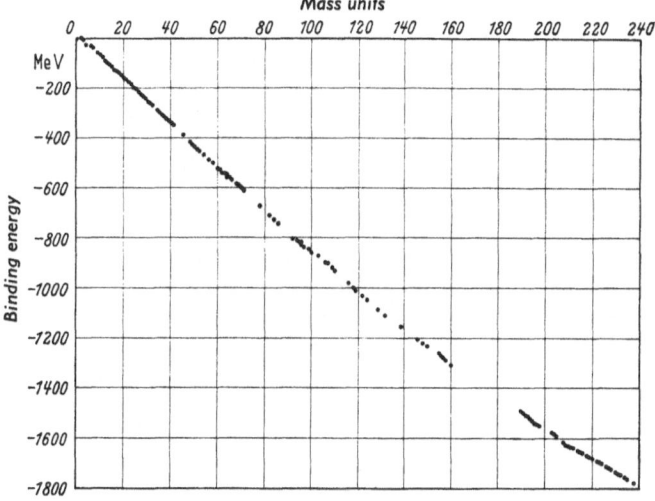

Fig. 133. Binding energies (in Mev) of all known stable nuclei versus their mass numbers

In concluding we may make the following point: *The fundamental significance of the relation between mass defect and binding energy is that we have found here for the first time a quantitative proof of the equivalence equation $E = mc^2$, which thus has been confirmed by nuclear physics exactly as required by relativity theory.* Further confirmations of the equivalence equation will be presented later.

6. Natural Radioactivity and the Nuclear Processes Disclosed by it

a) Natural Radioactive Decay Series

The first definite knowledge of the structure of the nucleus and its internal processes stems from the discovery of radioactivity in 1896 by BECQUEREL, and from the consequent experiments on its radioactivity by the CURIES. It was soon realized that this phenomenon of spontaneously decaying nuclei of the heaviest existing atoms, which at first seemed so mysterious, is caused by an internal instability and that it occurs without any possibility of influencing it from the outside.

In general, the product of the decay of a radioactive nucleus is another unstable nucleus so that we find decay series whose end product is always a stable nucleus. Fig. 134 shows the decay series that are known today. The three long-known series end with the stable lead isotopes of mass numbers 206, 207, and 208. The mass numbers and atomic numbers of the sequences of nuclei which follow one another by the radioactive decay in the different families may be taken from Fig. 134. Three types of radioactive transformations are known, according to whether the nucleus emits a doubly positive helium nucleus (α-particle), a

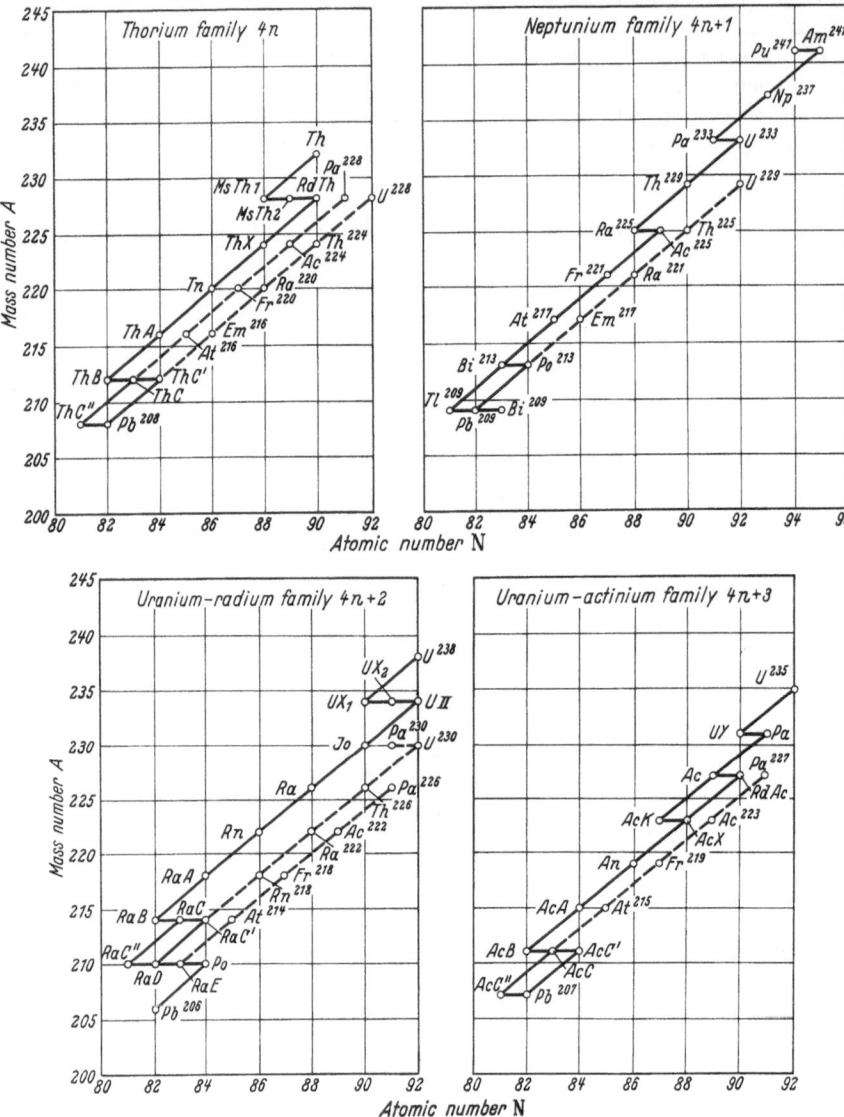

Fig. 134. The four radioactive decay families with their side branches

fast electron (β-particle), or a very energetic photon (γ-quantum). From these facts and from our knowledge of the structure of the Periodic Table, the displacement law of FAJANS and SODDY, though quite a discovery in its time, is self-evident: *The nuclear charge (i.e., the atomic number) of a decaying nucleus is reduced two units by α-decay, leading to an isotope of the element two places to the left in the Periodic Table. On the other hand, the positive nuclear charge is increased one unit by β-decay so that the resulting nucleus is an isotope of the element on the right-hand side of the decaying element. Since γ-radiation does not imply a change in the charge of the nucleus, it is not associated with a transformation of an element.* This displacement law is evident also from our illustration of the decay series. A line sloping to the left in Fig. 134 always corresponds to an α-transformation, whereas a horizontal line toward the right corresponds to a β-transformation (because it occurs without change of the mass number). One can see at once from Fig. 134 to which elements the different radioactive transformation products belong. Before their relation to the known elements became clear, each of these radioactive nuclides had been given its own name, which today can give rise to some misunderstanding. For example, we see from Fig. 134 that in the four decay series there are eight different *polonium isotopes* with mass numbers 210, 211, 212, 213, 214, 215, 216, and 218. Each of these has a different name, for example, the isotopes 210, 214, and 218 are called radium F, radium C′, and radium A, respectively. A similar situation holds for the other decay series, where, for example, mesothorium I and thorium X are radium isotopes, while mesothorium II is an actinium isotope.

In addition to the three decay series which have been known for half a century, some new decay series have been added in Fig. 134 as a result of research on artificially produced nuclei. Beside the branch series shown by the broken lines, all of which rejoin the original series after four or five steps, there is in particular one entirely new radioactive series which begins with the formerly unknown element plutonium (see V, 14). It is shown in Fig. 134 as the second series from the left. Since the element of this series which has the longest life time is the neptunium isotope $_{93}Np^{237}$ (also to be discussed later) with a mean life of 2.3×10^6 years, this new radioactive series is usually called the *neptunium series*.

It is of interest in connection with the special role which the α-particle with its four nucleons appears to play in the structure of the nucleus, that we now have exactly four radioactive series, whose members can all be represented by the following formulas:

$$
\begin{array}{llll}
\text{Thorium series:} & 4n & (n=58\to52) \\
\text{Neptunium series:} & 4n+1 & (n=60\to52) \\
\text{Uranium-radium series:} & 4n+2 & (n=59\to51) \\
\text{Uranium-actinium series:} & 4n+3 & (n=58\to51).
\end{array} \qquad (5.28)
$$

b) Type of Decay, Decay Constant, and Half Life

First, we will summarize the most important facts about natural radioactivity and then continue with their explanation.

Fig. 134 shows that in the majority of all cases a certain radioactive nucleus decays by emitting either α-particles or β-particles. Both types of decay, however, are often connected with the emission of γ-quanta. There are some cases, however, according to Fig. 134, where the same nucleus decays by emitting α-particles *or* β-particles (RaC, ThC, AcC, Bi^{213}). Here, the relative probabilities of α-decay and β-decay often differ greatly; its understanding, like that of all atomic transition probabilities, is one of the most important goals of theoretical nuclear studies.

We have already mentioned the important empirical fact that the naturally radioactive decay cannot be accelerated nor delayed from the outside, but that it occurs completely spontaneously and purely statistically. This can be seen from the observation that the number of nuclei of one kind decaying per second

(dN/dt) depends only on the number of nuclei which have not yet decayed and is proportional to this number,

$$\frac{dN}{dt} = - \lambda N.$$ (5.29)

By integrating we obtain the decay law, in analogy to the corresponding "decay" of excited atoms (see III, 23):

$$N_t = N_0 e^{-\lambda t}.$$ (5.30)

From this exponential decay law we conclude that the probability of decay of a nucleus does *not* depend on its age, in contrast to the probability of the death of any living organism. The occurrence of a radioactive decay process evidently is purely a probability matter. We have already discussed this in relation to the theory of causality in IV, 15.

The decay constant λ is a constant which is characteristic for each radioactive nuclide. Its value may be determined from Eq. (5.29) by measuring the radiation intensity of a radioactive sample as a function of time and plotting this intensity, which is proportional to the number of decays per second, on a logarithmic scale against time. Instead of the decay constant λ, the half life τ is often used, i.e., the time in which half of the presently existing nuclei of a certain kind have decayed. The relation between the decay constant λ and the half life τ is found if t and N_t in Eq. (5.30) are replaced by τ and $N_0/2$ respectively:

$$e^{\lambda \tau} = 2,$$ (5.31)

$$\tau = \frac{\ln 2}{\lambda} = \frac{0.693}{\lambda}.$$ (5.32)

The values for the half life of naturally radioactive nuclei vary between 10^{-7} sec and 10^{14} years or more. The relatively small half life of neptunium, by the way, is responsible for the fading-away of the neptunium series which, therefore, is not being found in nature any more.

c) Decay Energy and Half Life of Radioactive Nuclei

Exact energy values of the particles (α, β, γ) emitted in radioactive decay are of considerable theoretical interest since we may consider the different types of radioactive decay to be transitions from one discrete energy state of a certain nucleus to other states of the same or a neighboring nucleus. Therefore, the energy differences of these states (see the analogy in Chap. III) represent the most important empirical basis of a later theoretical analysis.

Formerly, the energy of α-particles was measured by their range in normal air. The Geiger relation, as a good approximation, states the connection between the range R (measured in cm) and the energy E_0 of the α-particles (measured in Mev)

$$E_0 = 2.12 R^{\frac{2}{3}}.$$ (5.33)

For determining the energy of α-particles as well as β-particles, i.e., fast electrons, more precisely, the different types of *magnetic spectrographs* (see page 216) are used. Quite a number of methods are available for measuring the energy of γ-quanta, i.e., high-energy photons. If the energy does not substantially exceed 1 Mev, the measurement of the wavelength is carried out by means of the crystal lattice spectrograph (page 51), a precision method developed by Du Mond. The energy then is computed from

$$E = h\nu = \frac{hc}{\lambda}.$$ (5.34)

All other exact methods, which are also usable for γ-quanta of higher energy than 1 Mev, are based on the transformation of γ-quanta into fast electrons whose energy is then measured with the magnetic spectrograph. This transformation may occur by absorption of the γ-quanta, e.g., in a lead screen. The kinetic energy of the photo-electrons produced by the absorbed γ-photons is equal to their energy minus the binding energy that is required for setting the photo-electrons free. Since the binding energy can be determined from the X-ray spectra of the absorbing material (cf. III, 10d), the measurement of the kinetic energy of the photo-electrons allows one to determine the energy of the absorbed γ-quanta. There is also a certain probability that the excitation energy of the nucleus is

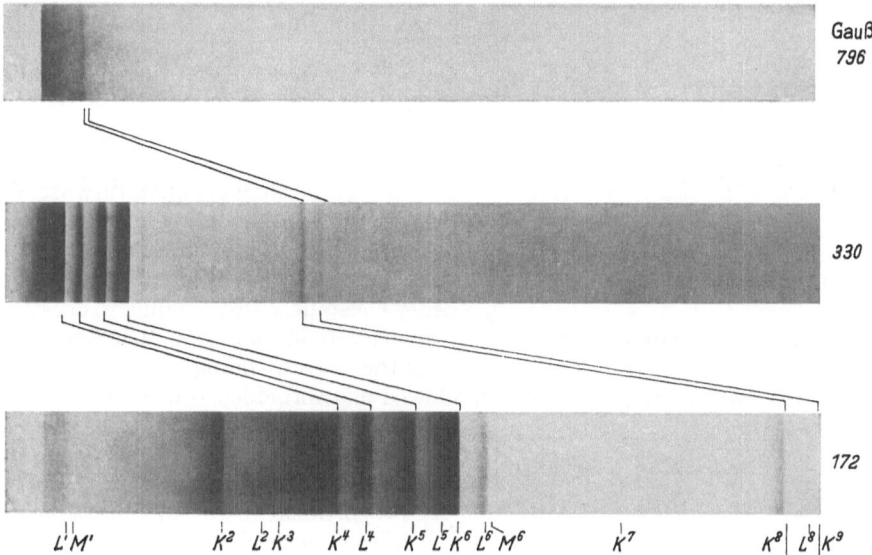

Fig. 135. Example of γ-β-conversion spectrum (Se⁷⁵). The kinetic energy of the freed electron is measured at different field strengths (indicated at the right) by means of a β-spectrograph (Fig. 116). The sharp edges of each "line" correspond to the binding energies in the different states. (After CORK and co-workers)

not used for emitting a photon but is transferred to the electron shell and results in the emission of a shell electron. These electrons originating from this *internal transformation* (often called *conversion*) again can be measured with the magnetic spectrograph; their energy equals the energy of the absorbed γ-quantum minus the binding energy of the shell electron. Since the absorption can occur in the K shell, L shell, or M shell etc., groups of sharply defined electron energies are observed, $E_e = h\nu - E_K$, $E_e = h\nu - E_L$, $E_e = h\nu - E_M$, etc. Here $h\nu$ is the energy of the γ-quantum and E_K, E_L, E_M etc. are the binding energies of an electron in the K shell, L shell, M shell, etc., respectively. Thus, the energy of the γ-quantum $h\nu$ corresponding to a nuclear transition may be easily determined by measuring the electron groups. Fig. 135 shows an example of a γ-β-conversion spectrum as it shows up on the plate of a β-spectrograph. Finally, in a process to be discussed in V, 21, γ-quanta of sufficiently high energy may produce electron pairs whose energy can also be used for determining the γ-energy.

From such energy measurements of the different particles we have learned that α-particles emitted by a certain α-decay process have one or a few discrete energies, comparable to those of a discrete line spectrum. The spectrum of γ-quanta consists of a number of sharp lines also. In contrast to these phenomena, high-energy electrons emitted by a β-decay process have a *continuous energy*

16*

spectrum. A typical case of this spectrum is given in Fig. 136. We shall come back to its explanation later.

An important theoretical relation exists between the kinetic energy of the α-particles or β-particles emitted by a radioactive decay and the half life of the emitting nuclei. GEIGER and NUTTALL found that the energy of α-particles, in a first approximation, is connected with the decay constant λ, which itself is inversely proportional to the half life τ [see Eq. (5.32)], by the equation

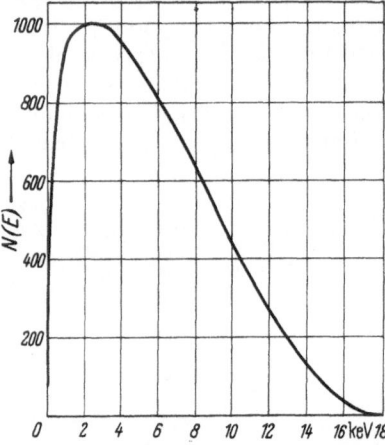

Fig. 136. The continuous β-spectrum of tritium, $_1H^3$

$$\ln E = A + B \ln \lambda. \qquad (5.35)$$

Fig. 137 shows a diagram of this relation; it will be explained later. If, in the same way, the logarithm of the maximal β-energies is plotted against the logarithm of the decay constants, according to SARGENT two straight lines (with considerable scattering of the measured points) are found (see Fig. 138). It means that two different values of the decay constant, corresponding to decay probabilities which differ by two orders of magnitude, exist for each value of the decay energy (β-energy). In analogy to the corresponding spectroscopic processes, we may distinguish between allowed and forbidden transitions.

We may conclude from these two empirical laws that the energy emitted by a disintegrating nucleus is inversely proportional to the stability or half life of the nucleus.

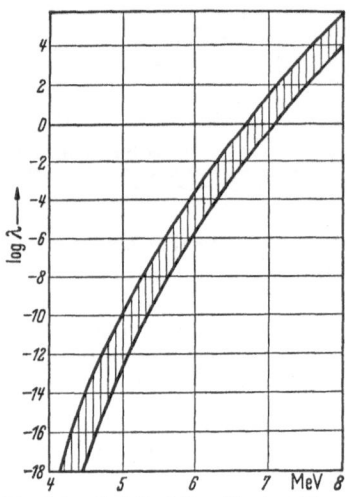

Fig. 137. Geiger-Nuttall relation. The logarithm of the decay constant is plotted versus the energy. The values of all known α-active nuclei lie within the hatched stripe

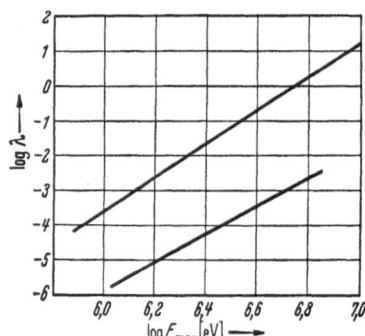

Fig. 138. Sargent diagram. Doubly logarithmic plot of decay constant versus energy for naturally radioactive β-emitters. The values lie on the two indicated curves

d) The Explanation of γ-radiation and the Mössbauer-Effect

The γ-radiation of atomic nuclei can be understood in complete analogy to the emission of radiation by atoms as discussed in Chap. III. It is interpreted as being emitted as a consequence of an energy change of the nucleus, i.e., upon a transition of the nucleus from an excited state to the ground state or in general to a state of lower energy. The mean life of an excited nucleus, however, is much smaller than that of an excited atom, obviously because of the much greater interaction of the nucleons compared to that of the shell electrons. In general, it is below 10^{-13} sec compared with the mean life of about 10^{-8} sec of an excited atom. However, special cases also exist where excited atomic *nuclei* have a mean life of 10^{-7} sec; this fact proved to be extremely important for the discovery and application of the Mössbauer effect, which we will now discuss.

Indeed, important details of the emission of photons can be studied with the help of γ-quanta because of their high energy $h\nu$, details that remain below the limit of observation for the emission of visible quanta. We know from II, 7 that a photon $h\nu$ has a momentum $p = h\nu/c$. The same momentum and the corresponding kinetic energy

$$E_R = \frac{p^2}{2M} = \frac{h^2\nu^2}{2Mc^2} \tag{5.36}$$

must be transferred by recoil in the emission process to the emitting nucleus and in the absorption process to the absorbing nucleus of mass M. According to the law of the conservation of energy, the energy of the re-absorbed photon therefore must be equal to the difference ΔE of the energy states of the emitting nucleus minus an energy twice the recoil energy $(2E_R)$. *Thus, in general, it is impossible that an emitted γ-quantum is re-absorbed and re-emitted by nuclei of the same kind. For nuclei, therefore, resonance absorption and resonance fluorescence in general are not possible.* For atoms, however, in contrast to nuclei, resonance absorption is observed; this is due to the fact that the recoil of optical photons (emitted or absorbed) does not play a role in comparison to the natural width of the optical energy levels because of the 10^6 times smaller energy of these photons.

If, however, according to MÖSSBAUER, the emitting as well as the absorbing nuclei are built into a crystal lattice, the recoil as a consequence of the emission as well as the absorption of the photon is taken up by the crystal as a whole, and the recoil energy (5.36), therefore, is zero because of $M \to \infty$. Thus, the energy $h\nu$ emitted by nuclei in a low-temperature lattice is directly equal to the difference ΔE of the energy states of the nuclei; therefore it can be absorbed and re-emitted by nuclei of the same kind. This is called *nuclear resonance fluorescence.*

It is possible, however, that quantized crystal vibrations are excited by the recoil according to VII, 8 if the recoil energy is larger than the lattice vibration quanta. What is the result? Since the nuclei bound in the lattice do not carry out any random thermal motion (which might result in a Doppler broadening), the absorption spectrum consists of a "Mössbauer line" of very small natural width (cf. III, 21) on a continuous background caused by the large number of possible lattice vibrations of different frequencies. This continuous background, which may make the observation difficult, is the less intense the lower the energy $h\nu$ of the γ-line and the lower the temperature of the absorbing lattice is. For that reason, low-temperature absorbers are frequently used.

Because of the dependence of the background on the energy of the photon, the low-energy 14 kev γ-line of Fe^{57} is often used. This has an other advantage. Since its upper state has the high mean life of 10^{-7} sec, this excited nuclear state, according to the uncertainty principle, has a half width of only 5×10^{-9} ev, and the corresponding frequency of the γ-quantum consequently may be measured with an accuracy up to 3×10^{-13}! *γ-radiation of this kind represents a frequency standard that is by orders of magnitude the most exact one known to-day.* The Mössbauer effect allows one to measure relatively easily Doppler-effects of very small magnitude as they are predicted, e.g., by the theory of relativity. Even a relative velocity of only 10^{-3} mm/sec (!) between a γ-emitter and a γ-absorber can be detected. The range of the possible applications of the Mössbauer effect, therefore, will be very wide and can scarcely be overestimated.

e) Energy Diagrams and Disintegration Possibilities of Radioactive Nuclei

By extending our concept of the energy level diagram, we get a good understanding of the different decay processes (α, β, γ) and the interrelations between these three types of radioactive transformation. The problem of energy level

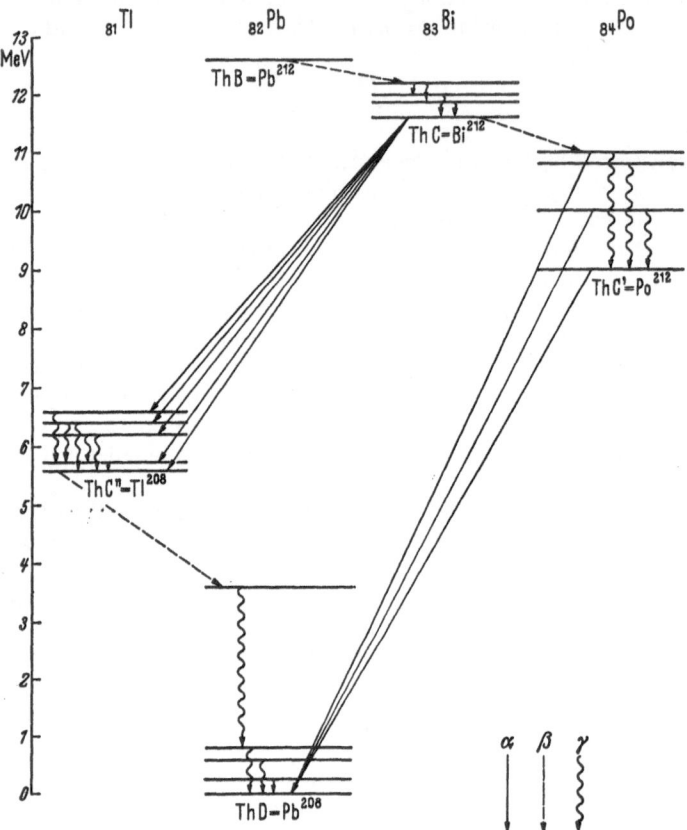

Fig. 139. Relations between the α-, β- and γ-transitions of 4 unstable nuclei at the end of the radioactive thorium family and its stable end nucleus $_{82}Pb^{208}$

diagrams of atomic nuclei, namely, is much more far-reaching than the relationship between the energy levels of each *individual* nucleus as inferred from γ-radiation. By the emission of *particles*, namely, transitions to the levels of *neighboring* nuclei are also possible. In anticipating the general case (to be discussed in V, 10), we show this with the aid of Fig. 139 where the somewhat simplified last steps of the radioactive thorium series with its branches are presented. The four columns next to each other give the energy states of the nuclei with the atomic numbers 81 through 84 respectively, while the energy, in units of Mev, is plotted as the ordinate. The two participating Pb isotopes with the mass numbers 208 and 212 are found under the atomic number 82. The distance of their two ground states (12.7 Mev) is equivalent to their mass difference, taking into account that Pb^{208} contains 4 neutrons less than Pb^{212}. Since γ-radiation does not result in transformations of one nucleus into another (charge and mass number remain unchanged), the corresponding changes of energy are indicated by vertical arrows, in analogy to the corresponding transitions of electrons (see Chap. III). In contrast to this, transition arrows that represent α-transformations or β-transformations must go obliquely from one nucleus to another. Evidently, arrows pointing to the right by one atomic number are associated with β-transformations; those pointing to the left by *two* charge numbers belong to α-decay processes. The vertical distances of the levels connected through arrows determine directly the energy of the emitted particles if these are α-particles or photons. For β-transitions, the distance measures the *maximal* energy of the emitted β-particles.

Fig. 139 demonstrates that there exist in general several competing transition probabilities. The relative probabilities for emitting the different particles (α, β, γ) are then inversely proportional to the half lives of the corresponding decay processes.

Furthermore, we learn from Fig. 139 an extremely important difference between the stability of atoms and of atomic nuclei. For atoms we distinguish (see Chap. III) the ground state and a sequence of excited states from which the atoms may be transferred to the ground state by emitting photons. The ground state of atoms is absolutely stable. Fig. 139 shows that atomic nuclei also have excited states from which the nucleus may be transferred to the ground state of the same nucleus by emitting γ-radiation. In contrast to the ground state of atoms (atomic shells), however, the ground states of nuclei often are not absolutely stable but only stable with respect to γ-radiation. Radioactive nuclei may still make transitions, by emission of an α-particle or a β-particle, from their ground state to excited states or the ground state of a *different*, more stable nucleus. In reality, *only* the ground state of thorium D, i.e., of the nucleus Pb^{208}, in Fig. 139 is really stable.

f) The Explanation of α-decay

We have already been concerned with the balance of energies, masses, and charges of the naturally radioactive decay processes without having dealt with the question how these α-particles or β-particles can leave the nucleus. We will now take up this problem. The mechanisms of the two types of decay are basically different and we treat the α-decay first.

The difficulty of an explanation obviously lies in the fact that α-active nuclei often have a very long half life, e.g., the one of the Ra nucleus is 1600 years. The

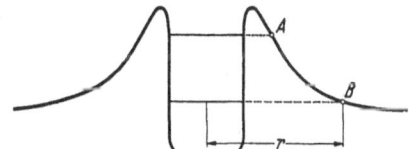

Fig. 140. α-decay from two nuclear states (or two nuclei) of different energy, used for explaining the Geiger-Nuttall relation

nucleons must be held together in the nucleus for a long time although, occasionally, two protons and two neutrons, in the form of one α-particle, may leave the nucleus spontaneously, i.e., without any outside stimulation. The relatively long cohesion of the nucleons in the nucleus means that the potential curve in and near the nucleus with reference to a doubly positive particle such as the α-particle must be approximately that shown in Fig. 140 since a potential minimum corresponds to the stable nucleus. The potential outside the nucleus, at least from a distance of 3×10^{-12} cm on, must fall off as Ze^2/r because of the electrostatic Coulomb repulsion. The normal state of the nucleus corresponds to the potential minimum. γ-radiation can be emitted by transition from an excited state of the nucleus (also indicated in Fig. 140) to the ground state. Classically, an α-particle can leave the nucleus only if it has sufficient kinetic energy for surmounting the enclosing potential wall. The quantum-mechanical tunnel effect explains the fact that α-particles with an energy of more than 10 Mev below that of the limiting potential wall have, nevertheless, a certain probability for leaving the nucleus.

The de Broglie waves which according to wave-mechanics correspond to the α-particles are not completely reflected by the potential wall (see IV, 12), but they can partially penetrate it, and the lower and narrower it is the easier

Fig. 141. Cloud-chamber photograph of a single high-energy α-particle of RaC′ among a large number of α-tracks of smaller energy. The high-energy α-particle is emitted by an excited energy state of the same nucleus. (After PHILIPP)

they do so. The absolute value of the penetration probability of a particle that hits the potential wall is very small according to Eq. (4.188). However, from the average velocity of an α-particle within a nucleus and the nuclear radius it follows that the α-particle hits the limiting potential wall about 10^{20} times per second so that obviously only very small penetration probabilities are necessary for explaining the observed mean lives of α-active nuclei. As soon as the α-particle has penetrated the potential wall, it is accelerated by the repulsive Coulomb force; its kinetic energy thus is inversely proportional to the distance from the center of the nucleus at which it leaves the potential wall. If we call this distance r, then the kinetic energy of a doubly positive particle, leaving a $(Z-2)$-fold charged rest nucleus behind, is

$$E = \frac{2(Z-2)e^2}{r}.$$

(5.37)

The Geiger-Nuttall relation is a consequence of this theory of the α-decay. To understand this we compare two α-particles, one of which may come from the ground state of the nucleus (Fig. 140) and the other from an excited state (or from a nucleus with correspondingly lower potential wall); both of them may emerge from the nucleus by the tunnel effect. The potential wall to be penetrated by the α-particle in the ground state is much higher and broader than that for the particle in the excited state. The decay probability of the normal nucleus, therefore, is much smaller, and thus its half life larger, than that of a nucleus in a state of higher energy. Conversely, the acceleration of the α-particle emerging from the higher state begins at A, that of the α-particle from the lower state begins at a greater distance r, at B. The energy of the α-particle originating from the higher state with the shorter half life thus is larger than that of the α-particle from the lower state with longer life (half life). The Geiger-Nuttall relation, which has been proved correct for the enormous range of half lives from 10^{10} years (Th^{232}) to 10^{-7} sec (Po^{212}) and the corresponding decay energies, thus is qualitatively understood. An excellent illustration is Fig. 141 where a single α-particle that is emitted from an excited state of RaC' exceeds the large number of α-particles emitted from the ground state of the same nucleus by its longer range and, therefore, higher energy. It is one of the finest achievements of quantum mechanics that in this way the radioactive α-decay is satisfactorily explained in agreement with experience.

We already mentioned in V, 4b that the distance r from the center of the nucleus at which the acceleration of the α-particles begins can be determined from the range of the α-rays (energy E) by Eq. (5.34). This distance thus represents an upper limit for the nuclear radius.

g) The Explanation of β-decay and the Existence of the Neutrino

In our attempt at explaining the radioactive β-decay we encounter fundamental difficulties of two types. First, it seems incomprehensible how electrons can come out of the nucleus since the nucleus consists exclusively of protons and neutrons. Secondly, the ejected electrons of the β-rays do not have, as do the α-particles, a definite energy or, if there are several nuclear states, a few discrete energies. On the contrary, Fig. 136 reveals a continuous energy distribution of the β-particles, which is called the *continuous energy spectrum of β-radiation*. If we take for granted that the primary nucleus which is to be transformed by β-emission has, of course, a well-defined energy, and that the same holds with the same certainty for the final nucleus, then we are forced to the conclusion that the energy set free in the transformation must also have a definite amount. Unless one assumes that in the case of β-emission the energy conservation law, which has been proved exactly valid for all phenomena of classical as well as atomic physics, does not hold here, one is forced to the only remaining conclusion (drawn in 1931 by PAULI) that in β-transformation an invisible second particle is emitted in addition to the electron (β-particle). In that case, the sum of its energy and that of the β-particle would be equal to the energy which is set free by the nuclear transformation. *This particle which can have no rest mass nor charge, because otherwise it could be observed in the cloud chamber, is called the neutrino.* The assumption of its existence is not only necessary for reasons of energy conservation but it removes also another difficulty associated with β-decay. As we have already mentioned in IV, 4c, nuclei of odd atomic weight have half-integral spin. Since the atomic mass number (number of nucleons) is not changed in a β-transformation, the spin must remain half-integral. On the other hand, the emitted electron has

a spin of $\frac{1}{2}$, so that here we have another contradiction which is solved if we ascribe
to the neutrino the same spin as that of all light elementary particles, i.e., $\frac{1}{2}$
(always measured in units of $h/2\pi$). Also the third of the mechanical conservation
laws, namely that for the momentum, requires the existence of the neutrino.
Experiments about the recoil of such β-active nuclei that disintegrate by emitting
a slow electron showed that the measured recoil can be understood only under the
assumption that a high-energy neutrino is emitted simultaneously with the electron.
Therefore, we have to assume the existence of the neutrino if we do not want to
give up the three fundamental laws of conversation of energy, momentum, and
angular momentum to which no exceptions were found so far, neither in classical
nor in atomic physics.

Now we discuss the second difficulty mentioned at the beginning, that in
β-decay an electron leaves the nucleus although, to our knowledge (cf. V, 4a),
the nucleus does not contain electrons. Fermi removed this difficulty in his theory
of 1934 by drawing a conclusion from the following analogy: In the case of the
"transformation" of an excited atom into a normal atom a photon is emitted,
although a photon as such did not previously "exist" in the atom. Similarly, the
electron with its neutrino, emitted by the nucleus in β-decay, is assumed to be
produced during the transformation of the nucleus. The process that leads to
β-decay is the transformation of a nuclear neutron into a proton combined with
the emission of an electron and a neutrino [or more exactly according to V 7a) of
an antineutrino $\bar{\nu}$]:

$$n \rightarrow p + e^- + \bar{\nu}_e. \tag{5.38}$$

It can be seen that this transformation does not change the mass number, while the
number of protons and, therefore, the atomic number is increased by one unit.
By the Dirac theory (see IV, 14), the emission and absorption of light (photons)
by an atom is considered to be the result of an interaction of the atom with the
electromagnetic high-frequency field. In a similar manner, just as the photon
corresponds to a change of the electromagnetic field, so, in the Fermi theory, the
electron and neutrino are associated with the nuclear field. The theory yields an
expression for the relation between the β-decay constant λ and the maximal
energy $E_{\max}$ of the emitted electrons.

$$\lambda = k E_{\max}^5. \tag{5.39}$$

The logarithmic form of Eq. (5.39) is indentical with the empirical relation of
SARGENT (Fig. 138). The constant k determines the probability of the β-transition
and, therefore, differs for the two (or more?) straight lines of the Sargent diagram.

This theory seems to explain the essential phenomena of the β-decay. Two
consequences of the theory ought to be mentioned as being of special interest. The
energy distribution of the continuous β-spectrum in Fig. 136 is *independent* of the
special individual structure of the nucleus and depends *only* on the interaction
of the nucleons in the nucleus. It is, therefore, one of the most interesting
empirical foundations of the quantitative theory of nuclear interactions. However,
the decay probability of a β-active nucleus or, conversely, its half life depends on
the individual structure of the disintegrating nucleus. Just as the transition
probabilities in the optical spectra are determined by quantum numbers, so is the
decay probability here determined by the quantum numbers of the nucleus
(see V, 12).

The existence of neutrino and antineutrino has been proved directly by the
detection of nuclear reactions caused by their absorption. These will be treated
farther below, together with the difference between both particles.

For examining the question, mentioned in V, 4f, whether the parity is conserved in β-decay or not, a correlation was looked for between the angular momenta of β-active nuclei and the direction of the electrons emitted by them. Contrary to all expectation, such a correlation was found. This discovery confirmed the theory of LEE and YANG about the non-conservation of parity in weak interactions. Only the future can show how significant this discovery may be for the theoretical understanding of the β-decay and the neutrino.

7. Artificially Radioactive Nuclei and their Transformations

Closely related to the decay processes of naturally radioactive nuclei are disintegration processes of the nearly 1000 different unstable atomic nuclei that do not occur in nature but are produced in certain induced nuclear reactions. They are called *artificially radioactive nuclei, radio-nuclides*, or *radio-isotopes*; the last expression indicates that these isotopes belong to elements most of which also have stable isotopes. Because of the close relationship between artifical and natural radioactivity we treat the former here, right after concluding our treatment of natural radioactivity, although the production of radio-nuclides by induced nuclear transformations will be treated later.

In contrast to so-called compound nuclei with an extremely short life time (to be treated in V, 8) which are produced in induced nuclear transformations, *nuclei with a measurable mean life are called artificially radioactive*. The determination of their life time enables us to separate the radio-nuclides from the inactive material and to identify them chemically unambiguously as isotopes of a certain element. Artificial radioactivity was discovered in 1934 by I. CURIE (daughter of the discoverers of natural radioactivity) and her husband JOLIOT; they found at the same time that quite a number of artificially radioactive nuclei, in contrast to natural radioactive ones, transform to more stable nuclei by emitting *positive* electrons. We know today that most of the artificial radio-nuclides disintegrate by emitting electrons (β^--decay) or positrons (β^+-decay), but often also by capturing electrons (K-effect, to be treated below). Only in rare exceptional cases the emission of α-particles or neutrons is chosen by nature in the transition to a more stable nucleus.

a) β^+-activity, Positrons, Neutrinos, and Antineutrinos

Since the β^- decay of artificially radioactive nuclei by emitting electrons and antineutrinos does not differ substantially from the β-decay of the natural radioactivity, we do not have to discuss it again. The β^+-decay however was a new and surprising phenomenon. First of all, we would like to mention that these positive electrons, or *positrons*, which occur frequently in artificial radioactivity, were found by ANDERSON shortly before the discovery of artificial radioactivity in cloud chamber studies of cosmic rays. Except for the positive sign of the charge, the positron corresponds in all its properties to the negative electron. The obvious question, why only negative electrons had been found in nature, found a surprising answer. The *positron cannot exist long in the presence of matter. It unites with a negative electron, whereby the electron pair completely vanishes and, according to the equivalence relation (2.32), its kinetic energy and its mass are emitted in the form of two γ-quanta with a combined energy of* 10^6 ev. The reverse process is also possible and has been studied: A γ-quantum of at least 10^6 ev energy is transformed into an electron-positron pair. Both processes are explained by DIRAC's "hole theory", which we shall discuss in V, 21.

*We shall learn in V, 8 that positrons are emitted by artificially induced radio-
active nuclei which have excess protons and, thereby, transform a nuclear proton into
a neutron, inversely to the normal β⁻-decay,*

$$p + 1 \text{ Mev} \rightarrow n + e^+ + \nu_e. \tag{5.40}$$

Since in this process the nuclear charge is reduced by one unit, the transforming atom simultaneously must eject a shell electron so that the atom remains
neutral. Because a positron *and* an electron thus leave the atom, only radioactive
atoms with an excess of two electron masses or an excess energy of 1 Mev
compared to the next stable nucleus can transform into this more stable nucleus.
This restriction obviously does not apply to the β⁻-decay functioning according
to Eq. (5.38). Finally, we shall learn in V, 11 that a few nuclei, such as $_{29}$Cu^{64},
are β⁻-active as well as β⁺-active; i.e., they may decay into more stable nuclei
either by emitting an electron or a positron; we shall also learn how this type of
transformation is possible.

By the way, the neutrino emitted with the positron in the transformation
process (5.40) is not identical with the antineutrino occurring in (5.38). This
fact was proved by experiments that unambiguously confirmed the existence
of the neutrino about a quarter of a century after the conception of the neutrino
hypothesis. These experiments dealt with the inverse β-effect [see Eq. (5.40)]:
Nuclear reactors (see V, 16) with their numerous β⁻-decaying fission products
represent huge antineutrino sources. COWAN and REINES observed that when
these antineutrinos react with protons, the protons are transformed into neutrons
and positrons, while the antineutrino is absorbed. The equation for this reaction is:

$$p + \bar{\nu}_e \rightarrow n + e^+. \tag{5.41}$$

Since the absorption probability for neutrinos is extremely small [the effective
cross section is only 10^{-44} cm^2 (V, 9b)], great experimental difficulties had to be
overcome in carrying out this important experiment.

Very similar experiments led to the decision whether actually neutrino and
antineutrino behave physically in a different way or not. If their behavior were the
same, energetic antineutrinos from a reactor should initiate not only the reaction
(5.41) but also the inverse reaction to (5.40). The corresponding reaction, e.g.,

$$_{17}\text{Cl}^{37} + \nu_e \rightarrow _{18}\text{Ar}^{37} + e^- \tag{5.42}$$

or

$$n + \nu_e \rightarrow p + e^- \tag{5.43}$$

that is expected to occur with neutrinos could not be confirmed with antineutrinos
from the reactor. From this result we conclude that neutrinos behave differently
physically from antineutrinos.

Experiments concerning the parity of neutrinos showed that this different
behavior is based on their different screw direction (helicity), i.e., the relation
between spin and the direction of neutrino emission. *While the antineutrino
emitted in β-decay corresponds to a right-hand screw where the vectors of momentum, $\vec{p}$,
and spin, $\vec{s}$, have the same direction, the two vectors for the neutrino are of opposite
direction, and the neutrino corresponds to a left-hand screw.*

An entirely new aspect of the neutrino problem was revealed in 1962 by the
discovery that, in addition to the neutrino and the antineutrino connected with
β-decay, there exists an other neutrino with its antineutrino which belongs to
the μ-meson (to be discussed in V, 23) just as the neutrino treated so far belongs

to the electron. This discovery resulted from an experiment similar to (5.43) in which neutrinos resulting from the decay of π-mesons (cf. V, 23) did not produce electrons but μ-mesons.

b) Nuclear Transformation by Capture of Shell Electrons

In 1936, YUKAWA and SAKATA pointed out the possibility that a nucleus which is unstable because of a relatively too high number of protons may, in certain cases, transform one of its protons into a neutron not by emitting a positron but, inversely, by capturing and absorbing a negative electron. In general, this electron would come from the shell closest to the nucleus, i.e., the K shell (see III, 10a), seldom from the L shell, and be drawn into the nucleus. The correctness of this prediction was confirmed by ALVAREZ two years later. The unstable vanadium isotope $_{23}V^{48}$ can be transformed into the stable titanium isotope $_{22}Ti^{48}$ either by β^+-emission or by acquiring an electron from the K shell. The isotope $_{23}V^{49}$ is transformed into stable $_{22}Ti^{49}$ exclusively by acquiring a shell electron. The hole, left in the K shell of the Ti atom by the electron capture, is filled by an electron transition from an outer shell, i.e., by the emission of radiation of the K series. It was the intensity of this X radiation fading away with the correct decay constant, by which the capture of shell electrons was proved; the process therefore is called K capture.

What are the relative probabilities for the emission of positrons (β^+-decay) and for the capture of shell electrons, which obviously are in competition with each other? First of all, it is evident that capturing a shell electron by a positive nucleus is the easier the higher the positive nuclear charge. Actually, we find only very few emitters of positrons among nuclei of high atomic number; *nearly all heavy unstable nuclei with excess protons are transformed into stable nuclei by capturing shell electrons, as is to be expected.* Both types of transformation, however, are found for lighter nuclei, a fact that can also be understood easily: A positron emission (β^+-decay) is possible only for nuclei that have an excess energy of at least 1 Mev [see Eq. (5.37)] compared with the more stable nucleus resulting from the transformation. This limiting condition obviously does not exist for the capture of shell electrons. *Therefore, unstable nuclei with excess protons will always transform by absorbing shell electrons if their energy exceeds that of the more stable final nucleus by less than 1 Mev, whereas positron emission starts only at an excess energy of at least 1 Mev and becomes more probable the higher the excess energy of the unstable nucleus.* For the lightest nuclei, capture of a shell electron is not possible because the shell electrons are too far away from the nucleus (too small nuclear charge), whereas for heavy nuclei, the innermost electrons are so close to the nuclei that decay always occurs by electron capture. Only for not too heavy nuclei, both forms of decay are possible.

The difficulties with regard to momentum and angular momentum for the β-decay that necessitate the introduction of neutrino or antineutrino also apply to electron capture by a nucleus. Therefore, it has to be concluded that simultaneously with capturing a shell electron either a neutrino is captured too (a very improbable case) or an antineutrino is emitted (a far more probable case).

c) Disintegration of Artificially Radioactive Nuclei by Emitting Neutrons or α-Particles

At first it was believed that artificially radioactive nuclei are transformed to more stable nuclei exclusively by emitting negative or positive electrons or capturing shell electrons. But why do unstable nuclei not compensate their too numerous excess neutrons by emitting neutrons instead of decaying under β^--emission? Again, this obvious question can be answered by energy considerations. *The emission of a neutron requires an energy of the unstable nucleus higher than the energy of the final products, i.e., of the stable nucleus and the free neutron,* whereas the β-transition (of course into a different final nucleus) has the advantage that energy in the amount of 0.76 Mev is released in the corresponding transformation of a neutron into a proton plus electron (with neutrino). Nevertheless, some examples of radio-isotopes emitting neutrons were found. Among the fission products of the uranium nucleus (see V, 14), there are several radio-isotopes disintegrating with a half life between 0.05 and 55.6 sec by emitting neutrons. We shall come back to the considerable significance of this "delayed emission of neutrons" for mastering the technique of nuclear energy.

Another well-known example is the nitrogen isotope $_7N^{17}$ which is transformed, by β^--emission, to a highly excited oxygen nucleus $_8O^{17}$ that itself changes to the stable $_8O^{16}$ by emitting neutrons:

$$\left.\begin{array}{l} _7N^{17} \rightarrow {}_8O^{17} + e^- \\ _8O^{17} \rightarrow {}_8O^{16} + n. \end{array}\right\} \tag{5.44}$$

The delay in the emission of neutrons is here always based on the long life time of the initial nucleus with regard to β-decay, while the unstable intermediate nucleus usually releases its neutron very quickly.

Finally, during the last years, some isotopes of rare earths, of gold, and of mercury were found that disintegrate by emitting α-*particles*.

d) Isomeric Nuclei and their Decay Processes

In V, 4c we defined nuclei as isomeric if they have above their ground state a metastable energy state with measurable life time. These nuclei may thus exist, for a measurable period of time, in two nuclear configurations differing in their energy values. If they have also an unstable ground state (radio-nuclides), stabilizing transitions (in most cases by β^--decay) are in principle possible from each of these two energy states, so that *the same nucleus is characterized by two different life times or decay constants.*

As early as in the twenties, HAHN discovered the first isomeric nucleus among the naturally radioactive nuclides, $_{91}Pa^{234}$. Today we know a great number of isomeric nuclei among the artificially radioactive nuclei as well as among the stable ones. In agreement with theoretical expectations, the following types of decay processes of isomeric nuclei with a stable or an unstable ground state were found.

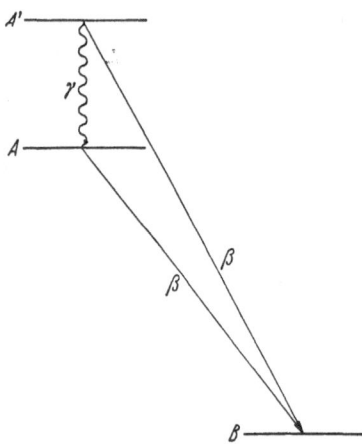

Fig. 142. Energy-level diagram for the possible transitions of an isomeric nuclear state A' to a different stable final nucleus B

The metastable excited energy states of isomeric nuclei with a ground state which is stable against β-decay emit γ-radiation with life times between 10^{-6} sec and several days for the following reason. If the ground state of a nucleus is stable, its excited states are also stable with respect to β-decay since the ratio of the number of protons to that of the neutrons is the same for all states of the same nucleus. Consequently, the *metastable* isomeric state of a *stable* isotope can lose its energy exclusively by γ-emission.

If, however, the isomeric nucleus is β-active because of its unfavorable ratio of the number of protons to that of neutrons, the situation for the simplest case may be explained by an energy diagram as shown in Fig. 142. In general, we have to distinguish between two different life times for the excited metastable nucleus A', namely one life time for the transition by γ-emission into the ground state of the β-active nucleus A, and another one for the direct β-decay into the stable nucleus B. If the probability for a γ-transition of A' is noticeably higher than that for a direct β-transition, a γ-radiation decaying with the γ-life time of A' will be observed simultaneously with a β-emission decaying with the life time of A, while the direct β-emission of A' will then not be detectable. Conversely, if the γ-transition from A' to A has an essentially smaller probability (forbidden transitions) than the β-decay from A' to B, a double β-decay with two different life times is observed, corresponding to the β-transitions from A' and A to B.

8. Artificially Induced Nuclear Transformations and their Mechanism

Although naturally radioactive transformations occur spontaneously and cannot be influenced by physical means, manifold and diverse transformations of normally stable nuclei can be *induced* by shooting high-energy particles at these stable nuclei. As a result of such experiments our knowledge of the behavior of the nucleus has been greatly increased. The rapid progress of modern nuclear physics actually began with the discovery of these induced transformations. The first induced transformation of this kind was discovered in 1919 by RUTHERFORD when he bombarded pure nitrogen by α-particles. A cloud chamber photograph of this transformation is reproduced in Fig. 143.

The arrows point to the place where a collision between an α-particle and a nitrogen nucleus occurred. A long track originates here which, according to the ionization strength, must be caused by a proton. Since the nitrogen contained no hydrogen, the proton must be the result of a nuclear transformation. The colliding α-particle, whose track ends at the place where the collision occurred, must have penetrated into the nitrogen nucleus, and as a result of this collision a proton was ejected by the nucleus. Thus, the charge of the nucleus is increased by one unit and the mass by three units: Consequently, an oxygen nucleus $_8O^{17}$, the rarest of the stable oxygen isotopes (cf. Table 3), must have been produced from the nitrogen nucleus $_7N^{14}$. We can write this first induced nuclear transformation in the form of a chemical equation,

$$_7N^{14} + {}_2He^4 \rightarrow {}_8O^{17} + {}_1H^1. \tag{5.45}$$

Here, because of the law of conservation of charge and mass, the sum of the units of charge (subscripts on the left) and the sum of the nucleons (superscripts on the right) must be the same on both sides of Eq. (5.45),

$$\left.\begin{array}{l} 7 + 2 = 8 + 1 \quad \text{units of charge} \\ 14 + 4 = 17 + 1 \quad \text{nucleons}. \end{array}\right\} \tag{5.46}$$

The momentum of the colliding particle in this case is shared by the ejected proton and the $_8O^{17}$ nucleus, whose trace is the short, thick line going toward the upper left from the point of collision. This artificially induced nuclear reaction already shows all the essential properties of induced transformations whose mechanism we now discuss.

Each transformation of a stable nucleus is initiated or induced by a nuclear particle being shot into the nucleus and attached to the nucleus to be transformed. As projectiles, mainly protons, neutrons, α-particles, and deuterons (nuclei of $_1H^2 \equiv d$) are used; for special investigations also tritons (nuclei of the β-active heaviest hydrogen isotope $_1H^3 \equiv t$), $_2He^3$ nuclei, and the nuclei of heavier atoms,

Fig. 143. Cloud-chamber photograph of the first artificial nuclear transformation, discovered by RUTHERFORD. The emission of a proton (proceeding from the point of disintegration to the lower right corner) by a nitrogen nucleus bombarded by α-particles. The nitrogen nucleus is transformed into an O^{17} nucleus which travels to the upper left due to the recoil. (After BLACKETT and LEES)

finally also electrons are employed. High-energy photons, i.e., γ-quanta, which can be produced at present by means of the electron synchrotron (cf. V, 3) up to at least 1000 Mev, play a special role because their absorption does not produce a *new* nucleus but in general only a highly excited or unstable state of the *same* absorbing nucleus. Since nucleons can be released from the nucleus by this absorption of γ-quanta in exactly the same manner as electrons are freed from the electron shell of atoms by the absorption of light (process of photo-ionization, see page 73), these nuclear processes are called *nuclear photo-effect*. Of great importance is the possibility of inducing nuclear transformations by neutrons. Since these have no charge, they are not affected by repulsive forces and can thus penetrate even into the heaviest nuclei with their high positive charge. We can show by a simple computation how much energy an impinging charged particle, e.g., a proton, must have in order to overcome the repulsive Coulomb force and penetrate into the nucleus. For enabling a proton of charge $+e$ to approach a nucleus of charge $+Ze$ to a distance r, an energy

$$E = \frac{Ze^2}{r} \text{ [erg]} \tag{5.47}$$

has to be expended. Substituting the value of e and converting to million electron volts by the relation $1 \text{ Mev} = 1.60 \times 10^{-6}$ erg, we obtain

$$E = (1.44 \times 10^{-13}) \frac{Z}{r} \text{ [Mev]}. \tag{5.48}$$

If we use the atomic number $Z = 40$ (zirconium) and for r the value 6×10^{-13} cm resulting from Eq. (5.9) for a nucleus of the corresponding mass, the energy

$$E = 9.6 \text{ Mev} \tag{5.49}$$

is obtained. In order to shoot a proton into a zirconium nucleus, it consequently must be accelerated by a potential of at least 10 million volts. The superiority of neutrons, which are not affected by repulsive forces, for use as nuclear projectiles is evident from this computation. The same advantage exists for the numerous nuclear reactions with high yield where deuterons and tritons are shot against nuclei of all kinds, always resulting in the emission of a proton. As we know, a deuteron consists of a proton and a neutron, a triton of a proton and two neutrons. In these processes, the relatively weakly bound deuteron (or triton) is "stripped" when hitting the nucleus (after OPPENHEIMER and PHILLIPS), i.e., it is *split* into a proton and a neutron (or a proton and a double neutron). While the proton is then reflected by the positive nucleus, the neutrons penetrate into the nucleus so that the bombardment by deuterons or tritons is actually equal to that by neutrons or double neutrons; thus for this excitation process a potential wall does not exist.

Finally, we have mentioned also electrons as projectiles for exciting nuclear reactions. It seems hard to decide whether the mechanism of this excitation consists of a direct transfer of the kinetic energy of the electron by impact or of the absorption of a γ-quantum produced by a stopped electron (cf. page 76).

The result of the bombardment of nuclei by projectiles of more or less energy is, in general, the emission of a nucleon, an α-particle, or a γ-quantum, depending on the amount of the absorbed energy. Sometimes, several nucleons or α-particles, rarely a deuteron $_1H^2$, a triton $_1H^3$, or a $_3He^3$ nucleus may be ejected. The reason why nuclei heavier than α-particles are very seldom ejected can be understood from the fact that the potential wall which has to be penetrated by these nuclei of relatively high charge is the higher the larger the charge, according to Eq. (5.47). Using the method of BOTHE and FLEISCHMANN, nuclear transformations in which one and in rare cases two or more particles are emitted from the nucleus, can be written in a manner which is shorter and just as clear as Eq. (5.45). In addition to the initial and final nucleus, we indicate the impinging particle and the particle ejected during the process by speaking of an (α, p) process or an (n, γ) process. If the original nucleus is written in front of the parenthesis and the final nucleus behind the parenthesis, the whole reaction is completely identified. Thus the transformation (5.45) discussed above can be written

$$_7N^{14} (\alpha, p) \, _8O^{17}. \tag{5.50}$$

Such reactions can be represented in a diagram by arrows from the initial nucleus to the final nucleus (see Fig. 144). Here the number of protons forming a nucleus (atomic number Z) is plotted versus the number of its neutrons $(A - Z)$. In Fig. 144, all known transformations of the fluorine nucleus $_9F^{19}$ are shown.

The new nucleus produced by the absorption of the projectile is responsible for the emission of particles in the transformation. It is called, after BOHR, *compound nucleus*. The hypothesis of its existence as a physical reality makes it easy to understand that the impinging particle not only can knock out a particle (nucleon or α-particle) which already exists in the nucleus but that, on occasion,

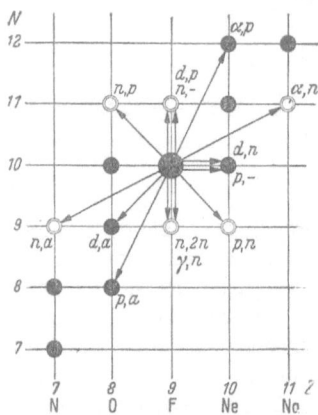

a rearrangement in the compound nucleus includes the absorbed colliding particle. This applies, apparently, to the bombardment of $_3Li^7$ nuclei by protons, by which the intermediate nucleus $_4Be^8$ is formed which, according to experiments, disintegrates into two $_2He^4$ nuclei, i.e., α-particles. In this case the proton shot into the Li nucleus must combine with two neutrons and one proton, already in the nucleus, to form the second α-particle. Then the large binding energy that is set free by this α-formation together with the original energy of the impinging proton causes the decay of the compound nucleus $_4Be^8$ into two α-particles. The nuclear reaction thus occurs in this case as in numerous other ones in two steps, which may be regarded as convincing evidence for the correctness of the assumption of a compound nucleus.

Fig. 144. Schematic representation of the presently known transformations of the fluorine nucleus. Dots = stable nuclei in the environment of the fluorine nucleus, circles = unstable, radioactive nuclei. Near the transformation arrows, the type of transformation is indicated by the abbreviated symbols. (After RIEZLER)

The same compound nucleus can be produced in different ways from stable nuclei, and it may disintegrate, depending on its state of excitation, in many different ways as well. In diagram (5.51), the compound nucleus $_{30}Zn^{65}$ is shown, for which at least two production possibilities and seven decay processes are known:

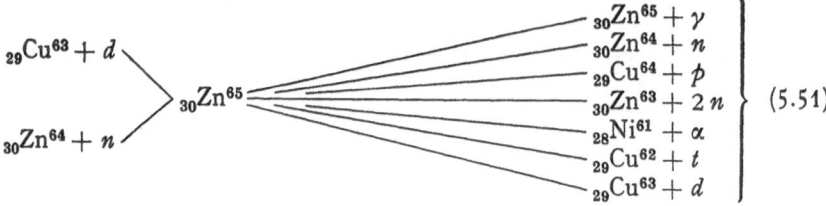

$$_{29}Cu^{63} + d$$
$$_{30}Zn^{64} + n$$
$$\left. \begin{array}{l} \searrow \\ \nearrow \end{array} \right\}_{30}Zn^{65}$$
$$\begin{array}{l} _{30}Zn^{65} + \gamma \\ _{30}Zn^{64} + n \\ _{29}Cu^{64} + p \\ _{30}Zn^{63} + 2n \\ _{28}Ni^{61} + \alpha \\ _{29}Cu^{62} + t \\ _{29}Cu^{63} + d \end{array} \right\} \quad (5.51)$$

The fact that both initial reactions (5.51) lead to the same final reactions with apparently the same probabilities may be taken as additional evidence for the correctness of the assumption about the compound nucleus.

We would like to point out that the compound-nucleus hypothesis corresponds exactly to the well-known hypothesis of unstable intermediate products in physical chemistry, which is used for explaining the mechanism of some complicated chemical reactions, e.g., the explosion of a $H_2 - Cl_2$ mixture. Just as in chemistry, the life-time of such an intermediate product (in nuclear physics: the compound nucleus) may be so large that it can be measured. On the other hand, in certain cases it may simply be equal to the collision time between inpinging particle and nucleus.

The nucleus was compared earlier with a fluid droplet held together by the surface tension (or the corresponding potential wall). If another molecule is shot into such a droplet, its binding energy (heat of condensation) *and* the kinetic energy carried along are released in the droplet as heat. In the same way, the energy of the nucleus that absorbs an impinging particle is increased by its kinetic energy (which may be measured) *and* its binding energy which is, according to Fig. 133, approximately 8 Mev for each nucleon. The nucleus is thus transformed into the compound nucleus which in general has a different charge and mass (except in the case of absorption of a γ-quantum). The theory of the compound nucleus maintains that in general, because of the large interaction between the closely packed nucleons in the nucleus, the absorbed energy is imparted to several nucleons so that we may speak of a heating-up process of the compound nucleus by the absorbed energy. It may occasionally occur that in the statistical play so much energy is concentrated on one molecule of a fluid droplet that this molecule evaporates from the droplet. Similarly it may occur that a single nucleon is separated from the nucleus instead of several nucleons just being excited. This is analogous to the process of autoionization (cf. III, 21). Since the average binding energy of a nucleon is 8 Mev, it is obvious that only a single nucleon can be ejected from the intermediate nucleus by an impinging particle of small kinetic energy. For each additional particle to be ejected, an additional kinetic energy of 8—10 Mev must be supplied by the colliding particle.

Such induced nuclear transformations, i.e., the disintegration of artificially produced compound nuclei, often lead to final nuclei with more protons or neutrons than in a stable nuclide (cf. Table 3), but with not enough energy to eject the excess nucleon directly. These nuclei, inspite of their instability, have a relatively long life, and they decay into stable nuclei by emitting negative or positive electrons and neutrinos or by capturing shell electrons. These processes of spontaneous transformation, which were already discussed on page 251, are called artificial radioactivity. *Consequently, artificially radioactive nuclei are nuclei which originate from induced nuclear transformations; they contain more or less protons than those corresponding to a stable configuration. After a measurable half-life they decay into stable nuclei by emitting electrons or positrons or by capturing shell electrons.*

We mentioned already that the maximum number of nucleons ejected in a nuclear transformation is given by the total available energy of the compound nucleus divided by the average nucleonic binding energy of 8 Mev. Only recently have we learned to accelerate impinging particles up to 100 Mev and more so that we are able to eject a relatively high number of nucleons out of heavy nuclei. For example, in Berkeley the bombardment of arsenic nuclei $_{33}As^{75}$ by α-particles of 400 Mev energy resulted in chlorine isotopes $_{17}Cl^{38}$; which means that the compound nucleus $_{35}Br^{79}$ must have emitted a total of 41 nucleons, 18 of which were protons. Extreme cases of such nuclear reactions (called "spallations") are observed in cosmic rays; undoubtedly, they will soon also be produced in the laboratory by means of the huge particle accelerators. There are two limiting cases for these collisions of 10^9 Mev and more kinetic energy. In a central collision with a heavy nucleus (one with many nucleons) the colliding particle imparts its energy by numerous internal collisions to the nucleons until its energy is expended and it is absorbed in the nucleus. Its absorption causes, therefore, an intense "heating" of the nucleus. As a result, many nucleons evaporate. In extreme cases the whole nucleus literally explodes. Fig. 167 shows one of the finest known examples of such a nuclear explosion in which the angular distribution of the nuclear fragments is isotropic, as is to be expected for an explosion of the heated-up compound nucleus. However, if the primary particle hits the nucleus well off-center and its initial energy is so high that the impinging particle is not absorbed, we cannot speak of an explosion, but of a direct ejection of nucleons by the impinging particle. The particle directions then show the expected preference for the direction of the colliding particle. Finally, for the very highest collision energies, which exceed the eigenenergy of the colliding particle, the concept of the compound nucleus fails completely, because the nucleus has to be regarded as a loose cluster of nucleons, their binding energy being negligible compared to the high collision energy. In this case we have to consider only collisions between the primary particle and the individual nucleons. The collision of a high-energy α-particle with a nucleus then has to be regarded as analogous to a simultaneous collision of four impinging nucleons with the nucleons of the target.

Since cosmic-ray processes are presently of outstanding interest to physicists, we shall come back to these problems in detail and we shall then learn more about collision processes of high energy and of the new elementary particles found in them. But, first of all, let us return to nuclear reactions initiated by colliding particles with energies of only a few Mev.

9. Energy Balance, Reaction Threshold, and Yield of Induced Nuclear Reactions

In the last section, we discussed the various types of induced nuclear transformations and their mechanism without regard to the energy balance of the reaction nor to its yield, i.e., the probability for the actual occurrence of a certain reaction as a function of the energy of the colliding particles. We shall now make up for this omission, especially since this question will be of interest in connection with the nuclear energy-level diagrams which are to be treated in the next section.

a) Energy Balance and Reaction Threshold

We begin with the problem of the energy balance and the associated question of the reaction threshold of some nuclear reactions. Just as for molecules, there exist also for nuclei exothermal and endothermal reactions, depending on whether the reaction energy is released or absorbed, respectively. Whereas exothermal reactions may occur spontaneously after the compound nucleus has been formed, the endothermal reactions need a certain minimum kinetic energy of the colliding particles; they consequently have a finite reaction threshold. It is easy to distinguish between the two types of reaction if the equivalence of energy and mass is taken into consideration. If, namely, the sum of the masses on the left-hand side of the reaction equation (initial nucleus plus colliding particle) is *higher* than the sum of the masses on the right-hand side (final nucleus plus emitted particles), the reaction is exothermal and, after the transformation, the mass difference appears as kinetic energy of the final nucleus and the ejected particle or particles, γ-quanta of course being considered here as particles. Conversely, if the sum of the masses on the left-hand side is *smaller* than that on the right-hand side, the exciting particle must have at least a kinetic energy corresponding to the mass difference in order to make the reaction possible.

A typical example of an exothermal nuclear reaction is the (p, α) reaction of the $_3\text{Li}^7$ nucleus, i.e., the explosion of the compound nucleus Be^8 (produced by the bombardment of Li^7 with protons) into two high-energy α-particles,

$$_3\text{Li}^7 + _1\text{H}^1 \rightarrow 2\,_2\text{He}^4 + Q \qquad (5.52)$$

with Q being the heat of reaction. From Table 3, page 34 it follows that the sum of the masses on the left-hand side of Eq. (5.52) equals 8.023857, that on the right-hand side only 8.005207. The energy of 17.2 Mev corresponding to the mass difference of 0.018650 [cf. Eq. (5.27)] thus has to appear as kinetic energy of the two α-particles; it is identical with the heat Q of the exothermal reaction. Since protons with a kinetic energy of 0.4 Mev were used for the experiment, the energy to be expected on the right-hand side of Eq. (5.52) was 17.6 Mev. In complete agreement with this expectation, the kinetic energy of each α-particle was found to be 8.8 Mev, computed from measurements of the α-range in a cloud chamber. Here we would like to add the remark that computations of masses have to use either the masses of stripped atoms without any electrons at all or, consistently, the *atomic masses* of Table 3, including that of the colliding particle, so that the same number of electrons appears on both sides of the reaction equation.

In principle, no minimum kinetic energy of the impinging particle is necessary for initiating an exothermal nuclear reaction as given in Eq. (5.52). Such a reaction may be started by neutrons of extremely small kinetic energy, and this even with a particularly high yield.

The situation is different for *endo*thermal nuclear reactions. Numerous reactions of the types (p, n), $(n, 2n)$, (γ, n), (γ, p), and $(d, 2n)$ require a sharply defined minimum energy below which the reaction cannot occur. In other words, the energy state of the compound nucleus that has to be reached in the collision is then *higher* than that of the initial nucleus. In all these cases, the sum of masses on the left-hand side of the reaction equation is *smaller* than that on the right-hand side. The experimental result that the reaction

$$_3\text{Li}^7 + _1\text{H}^1 + Q \rightarrow _4\text{Be}^7 + n \tag{5.53}$$

requires a minimum energy of the colliding proton of 1.88 Mev, whereas the reaction products at the right have a kinetic energy of only 0.23 Mev, leads to the conclusion that the left-hand sum of the masses is smaller than the right-hand sum by 0.0017 mass units, corresponding to the energy difference of 1.65 Mev. Since the mass of the neutron is exactly known, this reaction enables us to compute the accurate mass of the unstable $_4\text{Be}^7$ nucleus. In a similar way *the masses of stable and unstable nuclei today are determined from nuclear reactions with high accuracy without using the mass spectrograph. The mass spectrograph is predominantly applied only for a very accurate determination of a few standard nuclei.*

In V, 10 we shall deal with the question how to obtain the energy levels of different nuclei in their relation to each other so that we may read directly from the energy level diagrams whether a certain reaction is exothermal or endothermal. Practically, however, the reverse method is mostly applied. The relative energy levels of nuclei (in the ground state as well as in excited states) are determined from the energies released or absorbed in nuclear reactions, i.e., in chemical nomenclature, from the positive or negative reaction energies Q.

b) Yield and Excitation Function of Induced Nuclear Reactions

Even if the energy balance of a certain nuclear reaction is known, the question of the yield of this specific transformation is not yet answered. The yield, in general, depends on the kinetic energy of the exciting particle (or, for γ-quanta, on the energy $h\nu$), just as the excitation probability for atoms excited by electrons of various energies (page 64). This dependence is called *excitation function or yield function.*

The yield of a nuclear reaction may be characterized by the inverse value of the number of particles of a given velocity which must be shot into a thick layer of the target in order to produce *one* transformation. For those nuclear reactions which have the greatest yield, this number is of the order of 10,000; but for most reactions the number is 2 or 3 orders of magnitude higher.

In nuclear physics, however, it is preferred to characterize collision processes by the energy-dependent *effective cross section* of the nucleus to be transformed. By effective cross section we mean the circular area which a nucleus must have if each collision within this cross section would produce a transformation. Let us consider a thin foil (thickness d cm, N atoms per cm³); n impinging particles with an energy E may hit each cm² and produce $A\,(E)$ nuclear transformations. Then, the energy-dependent effective cross section of this nucleus for this specific reaction is

$$\sigma(E) = \frac{A\,(E)}{nNd}\ [\text{cm}^2].\qquad (5.54)$$

The effective cross section defined by Eq. (5.54) thus includes first the probability that a particle will penetrate into a nucleus; secondly the probability that the

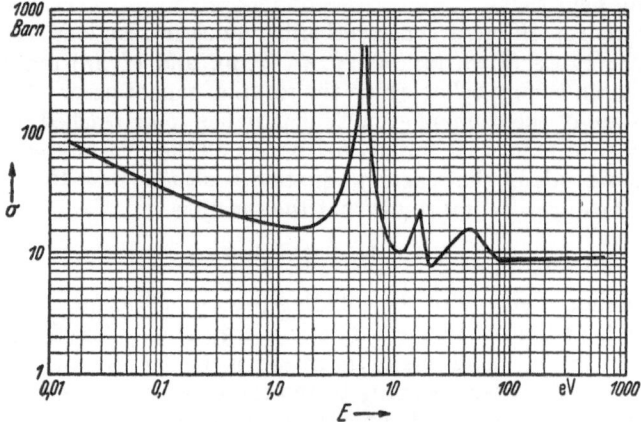

Fig. 145. Absorption cross section of neutrons versus their energy (with resonance peaks) for silver. (After measurements by GOLDSMITH, IBSER, and FELD)

transformation under consideration will actually occur. This second probability depends largely on the internal interaction of the nucleons. The effective cross section, therefore, is an *auxiliary concept* which is very practical for yield computations but does not have a direct relation to the actual nuclear radius discussed in V, 4b. Nevertheless, for most nuclear reactions it has at least the order of magnitude of the actual nuclear cross sections (10^{-24} cm²), although effective cross sections have been measured up to 10^{-20} cm² and down to 10^{-44} cm². Nuclear physicists have adopted their own unit for the effective cross section. They have given to the cross section 10^{-24} cm² the name "barn".

Besides this effective cross section for the occurrence of a certain nuclear transformation, we are often interested in the *total effective cross section* of a nucleus with regard to certain particles of a well-defined energy. This expression has also the form of Eq. (5.54); but one must substitute for A the nuclear processes by which impinging particles are deflected from their original course either through nuclear scattering or through absorption in combination with any kind of subsequent transformation. For shielding calculations, i.e., for computing the thickness or the most suitable type of matter necessary for shielding a given radiation source, it is obviously essential to know the total effective cross section of the specific nuclei with regard to the given radiation.

Figs. 145 and 146 show two frequently found types of excitation functions for neutrons and for charged particles. Since the neutron does not have to pass a potential barrier before penetrating into the nucleus, the effective nuclear cross section for a certain nuclear reaction induced by neutrons depends only on the probability for the subsequent disintegration of the compound nucleus. The probability for absorbing a neutron is, at least classically, determined by the interaction time. Because this time is inversely proportional to the velocity of the neutron, we also expect that the cross section for absorbing neutrons is in-inversely proportional to their velocity v, i.e., to the square root of the kinetic energy of the neutrons. For charged particles, conversely, the probability of penetrating into the nucleus first increases exponentially with increasing impact energy because of the repulsive Coulomb force, i.e., the necessity of penetrating the potential wall around the nucleus (see V, 8). At higher energies we expect the excitation function to decrease slowly because of the decreasing interaction time. This is exactly what we find in Fig. 146, which represents the effective cross section of the Cu^{63} nucleus for the (d, p) transformation as a function of the energy of the colliding deuterons. In Fig. 145 we see plotted, in a logarithmic scale, the absorption cross section of the Ag nucleus for neutrons versus their kinetic energy: For small energies, there is good agreement with the expectation; the cross section decreases with increasing energy. But we have to explain the spectacular maxima at energies of 5, 15, and 45 ev.

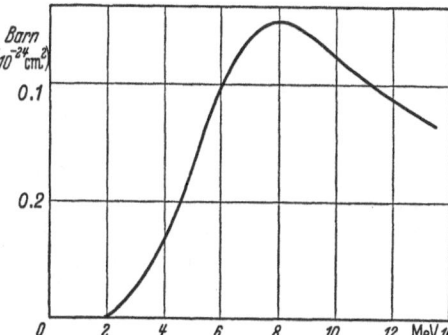

Fig. 146. Yield of the (d, p) reaction of Cu^{63} versus deuteron energy, used for illustrating the excitation by charged particles repelled by the nucleus. (After measurements by PEASLEE)

The absorption probability as discussed above would correspond directly to the effective cross section only if, after the particle has penetrated into the nucleus, the probability of a transformation would not depend on the kinetic energy transferred to the nucleus by the colliding particle. However, such a dependence does, in general, exist. We have discussed above (page 258) that nuclear transformations lead to the formation of compound nuclei whose heating up by the kinetic and potential energy carried along by the colliding particle results in the emission of nucleons or radiation. If the energy set free in the nucleus by the absorption of the colliding particle is exactly equal to one of the excitation energies of the compound nucleus, it is highly probable that a compound nucleus is formed, and only rarely will the colliding particle be reflected. In other words, if there is energy resonance between the colliding particle and the compound nucleus, the transformation probability and thus the effective cross section suddenly becomes very large; for these energy values steep maxima are superimposed over the smooth yield functions. These resonance energies are well known for a large number of nuclear reactions, especially for slow-neutron reactions. They are of great help in determining the energy-level diagrams of atomic nuclei, which will be discussed now.

10. Energy-level Diagrams of Nuclei and their Empirical Determination

In the last sections we dealt frequently (cf. Fig. 139) with excited energy states of nuclei and with the energy of their ground states relative to those of neighboring nuclei. Thus, our interest in knowing the complete energy-level diagrams is evident, especially since we are able to determine the exact masses of unstable nuclei by means of the energy mass relation if we know their relative ground-state energies. Also from the theoretical point of view we are interested in the energy-level diagrams since the energy states of the nuclei depend on the particular configuration of protons and neutrons and, therefore, allow conclusions about this configuration. Just as the exact knowledge of the atomic energy-level diagrams made possible the determination of the electron configuration in atomic shells (Chap. III), so will the knowledge of nuclear energy levels become a prerequisite for a quantitative theory of nuclear structure.

We begin by making a few general remarks about nuclear energy levels and the difference between energy levels of atoms and of nuclei.

It is clear that a nucleus, as a system consisting of a certain number of nucleons, is subject to quantum rules in the same way as an electron shell, which is a system of a certain number of electrons. Thus we expect a nucleus to have a series of discrete energy levels corresponding to the stationary energy states of the nucleus, as well as continuous energy regions corresponding to the decay processes of the nucleus, comparable to the ionization continuum of the atom (page 73) or the dissociation continua of a molecule (see VI, 7a). However, in contrast to the electron shells, the nucleus has no central force field. Each nucleon moves in the resulting force field of all the other nucleons or, at least, in the field of its neighbors. In contrast to the Rydberg series of the atom, which reflect the existence of a central force field, we do *not* expect the nucleus to have a regularly arranged system of energy states. Because of the strong coupling between the nucleons, the excitation energy of the nucleus can be easily imparted to many, and in the limiting case, to *all* nucleons of the nucleus ("heating" of the nucleus), so that there are corresponding stationary energy states of the nucleus whose energies are far larger than the separation energy of an individual nucleon or α-particle. Therefore, only in rare cases (ground states of stable nuclides) are nuclear energy states actually stable.

In general, an excited nuclear energy state can make a transition to a more stable state either by radiating (γ-emission) or without radiation (see auto-ionization, III, 21), by electron emission (β+ or β-), electron capture (see V, 7b), emission of an α-particle, proton, or neutron, in the last cases always to a state of a *different* nucleus. *In determining the energy-level diagram of an atom or a molecule we had to consider only that one atom or molecule. However, in determining the energy-level diagram of a nucleus, we must, as we shall soon see, consider the numerous possible transitions between the nucleus under consideration and its neighbors.*

It follows from the uncertainty relation (4.18) that the width of an energy level is inversely proportional to the life time τ of the system in this state. We learned in dealing with radiationless transitions within atomic electron shells (see III, 21) that the life time of an excited state can be very much reduced by such transitions. The same applies for nuclear states. According to our consideration above, the probability of radiationless transitions increases, in general, considerably with increasing nuclear excitation energy. We expect, therefore, *sharp* energy levels of excited nuclei only for comparatively small excitation energies, whereas *in the region of high excitation energy ($E \gg 10$ Mev) the width of the energy levels, in general, increases more and more until it becomes comparable to the distance of neighboring excited states, and the difference between discrete states and continuous energy regions gradually disappears.* This theoretical expectation has been confirmed very satisfactorily by the experimental results known so far.

The excitation probability measured as a function of the impact energy shows energy-level widths far below 0.1 ev for excitation by slow neutrons. In the region of high excitation energies, however, level widths of 1.0 Mev and more are found frequently. The correspondent life times of the nuclei in these excited states can be computed from the width with the aid of the uncertainty relation (4.18). Values from 10^{-14} sec to 10^{-20} sec, in special cases, however, up to 10^{-7} sec have been found.

An interesting application of these considerations concerns the compound nucleus B produced by the absorption of an impinging particle in a nucleus A. Its energy depends on the kinetic energy and the binding energy of the incorporated colliding particle. If the total energy of this compound nucleus corresponds to one of its quantized stationary energy states, the compound nucleus has a life time that corresponds to the width of this state. Otherwise, it cannot exist as a bound physical system. In this case, we have just a collision between an impinging particle and a nucleus A; the duration of the collision may be computed from the nuclear diameter of about 10^{-12} cm, traversed by the particle with a velocity of the order of magnitude of 10^{10} cm/sec. That results in a collision duration of about 10^{-22} sec.

So far, we spoke only of discrete energy states of nuclei or their nucleons without considering the fact that the nuclei are built up of two *different* although closely related particles, protons and neutrons, while the electron shells of atoms consist of elementary particles of *one* type only. There can be no doubt that the construction of nuclei follows a principle similar to that of the electron shell (see III, 19); i.e., the successive incorporation of new particles, here of protons or neutrons, will always take place into the lowest energy state that is not yet occupied. For the theoretical investigation of nuclear structure we thus have to distinguish between proton states and neutron states of the nucleus. The sequence of stable nuclides (Table 3, page 34) provides information about the relative energy of their ground states. Let us begin with the deuteron consisting of one proton and one neutron. If we want to construct the next heavier nucleus consisting of three nucleons, the question arises whether the lowest energy state not yet occupied by a nucleon is a proton state or a neutron state. Obviously, the occupation of the first state would produce the $_2\text{He}^3$ nucleus, the occupation of the latter, however, the isobaric nucleus $_1\text{H}^3$. Whereas the nucleus $_2\text{He}^3$ is stable, $_1\text{H}^3$ (the well-known triton) is radioactive. We conclude from this fact that the energy level of the second proton is lower than that of the second neutron. On the other hand, the small maximum β-energy of $_1\text{H}^3$ shows that the level of the second neutron is only very little (0.006 Mev) above that of the second proton. Let us discuss a second example, viz., the most stable of the sulfur nuclei, S^{32}, consisting of 16 protons and 16 neutrons. Table 3 shows that *both* lowest unoccupied levels here are neutron states which, when filled, lead to the stable sulfur nuclides S^{33} and S^{34}. The seventeenth proton state then is only a little lower than the nineteenth neutron state. This may be concluded from the fact that $_{17}\text{Cl}^{35}$ is the stable nucleus with mass 35, while its unstable isobar (nucleus with the same mass) $_{16}\text{S}^{35}$ emits β-radiation of only very small energy when transformed into $_{17}\text{Cl}^{35}$.

By referring to our empirical knowledge about the stability of certain nuclei and the β-energy of others we have already begun to deal with the various complementary empirical methods by which nuclear energy-level diagrams and the relative energies of their ground states are determined.

First of all, the most direct methods are those of a collision excitation of the higher nuclear states, in close analogy to the electron collision experiments of FRANCK and HERTZ (see III, 4). Artificial nuclear excitation was discovered in 1930 by BOTHE and BECKER who observed γ-radiation caused by the bombardment of various nuclei with α-particles. They noticed that the excited nuclear states responsible for the γ-radiation can be excited by other nuclear reactions as well.

Later, WIEDENBECK transformed cadmium into radioactive silver by bombarding it with electrons of 1 to 3 Mev energy. He found that the yield of β-radiation showed maxima at certain discrete energies of the exciting electrons. These maxima indicate that the radioactive Ag nucleus has excitation levels at the corresponding energies. In a similar way he succeeded in exciting seven different energy states of the silver nucleus by the absorption of continuous X-radiation of 1 to 3.5 Mev. He identified the excited states by their γ-radiation, which was delayed because of the existence of a metastable state. In both experiments,

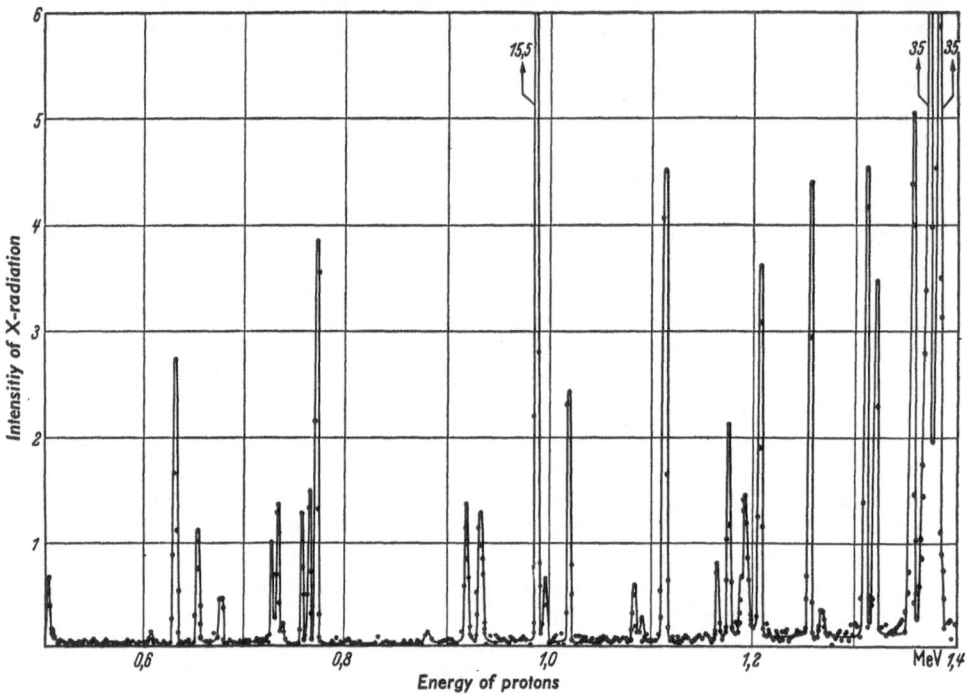

Fig. 147. Excitation of the compound nucleus Si²⁸ by bombardment of Al²⁷ with protons. The sharp peaks of the γ-radiation of Si²⁸ plotted against the energy of the protons bombarding the Al²⁷ nuclei show the well defined excitation energies of the compound nucleus Si²⁸. (After BROSTROM, HUUS, and TANGEN)

yield functions are observed whose maxima indicate the existence of discrete excitation levels of the compound nucleus. Excitation states of a great number of nuclei were measured, especially in the range of very small energies (i.e., of a few ev), in an analogous way by utilizing excitation functions of (n, γ) transformations (as shown in the example of Fig. 145). Fig. 147 gives another impressive example in which many discrete energy states of the Si²⁸ nucleus, ranging from 0.5 to 1.4 Mev, were excited. Si²⁸ nuclei in their different energy levels were here produced by bombarding Al²⁸ with "monochromatic" protons, and the intensity of the γ-quanta emitted in the (p, γ) transformation was measured as a function of the proton energy. Electrons, γ-quanta, neutrons, protons, deuterons and α-particles, i.e., in principle all available nuclear particles which can be accelerated to a predetermined energy can be used for exciting and measuring the energy levels of the corresponding (stable or radioactive) compound nuclei. Also, excitation processes were observed where the exciting particle itself left the excited nucleus with a correspondingly smaller energy, i.e., the particle was scattered and no transformation took place. These processes are obviously inelastic exciting collisions, fully analogous to the processes which were cleared up by the experiments of FRANCK and HERTZ (see III, 4).

Supplementing these direct methods for measuring excitation energies, accurately measured energies of particles emitted in induced nuclear reactions (i.e., reaction-heat values Q, see page 200) can also be used for computing the excited energy states of nuclei participating in the reaction and their relative ground state energies. The principle of all these methods may be seen from Fig. 139, treated in V, 6c. An unambiguous assignment of the different transitions is thus possible. These measurements thus lead to the energy-level diagrams of all the nuclei which participate in a certain reaction. If, in addition, the precise mass of *one* of the nuclei in its ground state is known (e.g., that of Pb^{208} in Fig. 139), the exact masses of the unstable nuclei participating in the reaction can be determined with the aid of the equivalence equation and the known mass of the α-particle.

What was shown in Fig. 139 for the example of a radioactive decay reaction, evidently applies in the same way for a combination of induced nuclear reactions. For instance, five proton groups of different but sharply defined energies were observed for the (α, p) transformation

$$_{13}Al^{27} + {}_2He^4 \rightarrow {}_{15}P^{31} \rightarrow {}_{14}Si^{30} + {}_1H^1. \tag{5.55}$$

We conclude that, in analogy to the α-decay processes of ThC or ThC' of Fig. 139, the P^{31} compound nucleus either is transformed (by proton emission) into different excited states of the Si^{30} nucleus, or that transitions occur from different excited levels of the compound nucleus P^{31} to the ground state of Si^{30}. In the first case, the transformation must be accompanied by γ-radiation of the final nucleus Si^{30}, in the second case, by γ-radiation of the compound nucleus P^{31} because the α-bombardment of Al^{27} produces a highly excited compound nucleus P^{31} of well-defined energy. Since the life time of a highly excited compound nucleus with respect to particle emission is generally small compared to that for γ-emission, we expect the excited states to belong to the final nucleus. If there is doubt about the assignment, more transformations must be investigated which include the two nuclei in question. A decision can always be made. Fig. 148 shows the level diagram of the $_4Be^8$ nucleus which is unstable with respect to disintegration into two α-particles. It serves to show how reliable information about nuclear energy-level diagrams can be obtained by combining the results of numerous nuclear transformations. Here, twenty different transformations of the nuclei He^4, Li^6, Li^7, Li^8, Be^9, B^8, B^{10}, B^{11}, and C^{12} were evaluated, all of which include the $_4Be^8$ nucleus as compound or final nucleus. One can see from Fig. 148 that the majority of the energy levels are fixed by two or three independent observations. Also some yield functions of the corresponding induced transformations are plotted versus the energy; they illustrate our discussion of V, 9b.

What is now the relative energy of corresponding proton and neutron levels? There is every reason for assuming that they would be very similar to each other if we would eliminate the energy shift which is due to the electrostatic contribution to the potential of the protons. Evidence of this is furnished by the so-called *mirror nuclei*, as we call pairs of neighboring isobars such as $_1H^3 - {}_2He^3$, $_6C^{13} - {}_7N^{13}$, $_7N^{15} - {}_8O^{15}$, $_{14}Si^{29} - {}_{15}P^{29}$, and so on. The only difference of the partners of such pairs is that in the first-named partner the unpaired nucleon is a neutron and in the other one, a proton. The complete energy-level diagrams of such mirror nuclei, in particular those excited states of the unpaired nucleon which normally are not occupied, are indeed very similar if they are corrected with respect to the electrostatic contribution of the proton. Since the first exited states are possible states for the neutron in the one nucleus of the pair and for the proton in the other nucleus, it follows that neutrons and protons are not basically different with respect to their energy levels and to their contribution to nuclear binding if we disregard the Coulomb forces.

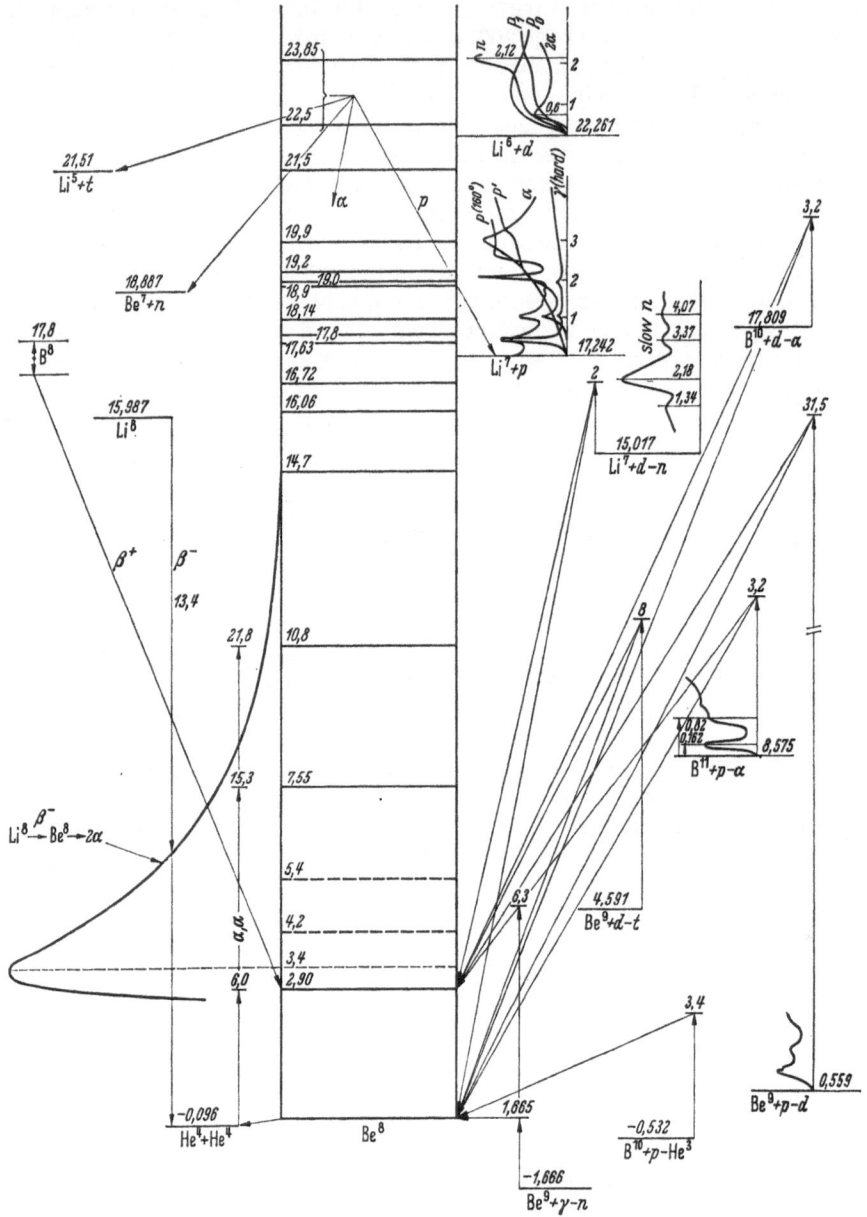

Fig. 148. Energy-level diagram and excitation functions of the $_4$Be^8 nucleus. (After T. LAURITSEN.) The curve at the left shows the energy distribution of the α-particles emitted in the decay process of the unstable Be^8 nuclei produced from the Li^8 nuclei by β-emission. Figures are energy values in Mev; colons are to be read as points

Our present knowledge of the excited states of nuclei still shows many gaps. But a considerable part of the research going on today in nuclear physics is devoted to the determination of complete nuclear energy-level diagrams, of level widths and the probabilities for the various possibilities of transitions or disintegration. Undoubtedly, this knowledge will become the most important prerequisite for a quantitative theory of the structure of nuclei. We shall learn in V, 12 that the energy states of nuclei, just like those of the electron shells of atoms and molecules, can be characterized by quantum numbers, that there exists a construction principle for nuclei that is strongly similar to that for atomic electron shells, and that closed nucleon shells account for the behavior of certain nuclei in a similar way as the closed electron shells explain the behavior of noble gases and palladium. Therefore, nuclear spectroscopy seems to be especially appropriate for illustrating the close relations between the various domains of atomic physics.

11. Droplet Model and Systematics of Nuclei

Several times in the preceding sections, we have made use of the fact that a nucleus can be compared, in a first approximation, to a drop of liquid. In support of this analogy we mention, first, that the density of all nuclei is approximately constant (see V, 4b) and, secondly, that according to Fig. 133 the binding energy per nucleon is approximately constant and lies between 8.5 and 7.5 Mev if we disregard the lightest nuclei. From these two results we conclude that, in contrast to electrostatic forces, the nuclear force has such a short range that it extends no farther than to the nearest neighbors. This conclusion can be experimentally verified by shooting protons or neutrons respectively against protons or neutrons in target nuclei. Then, by measuring the angular distribution of the scattered nucleons the nuclear forces may be computed. If we now plot the potential of these forces (just as in Fig. 91), as a function of distance

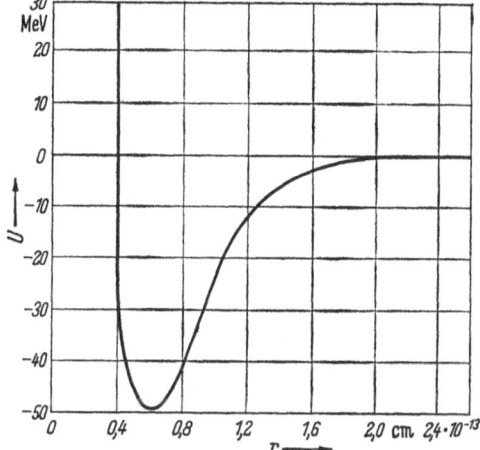

Fig. 149. Simplest interaction potential $U(r)$ between two nucleons as function of the distance of their centers

of the interacting nucleons, we get a $U(r)$ curve as in Fig. 149. This means, the nucleons behave in these collision experiments as if they had a "hard core" of about 0.5×10^{-13} cm diameter. The attractive nuclear forces between two nucleons are practically non-existent beyond 2.6×10^{-13} cm. Their nature shall be discussed in V, 23. The additional repulsive electrostatic force between two protons does not play any essential role as compared to the attracting nuclear forces. One more important point might be mentioned. At first one might think that the large energy corresponding to the minimum of the $U(r)$ curve would be released as binding energy when the two particles combine and thus, conversely, must be expended to separate them. Actually the binding energy of a proton and a neutron in the deuteron amounts to only about 5 per cent of the energy of the potential minimum, while the nucleons retain the remaining 95 per cent as kinetic energy in the bound state. This follows from the uncertainty principle (IV, 3). Since the particles are confined to the small volume of the nucleus, this small indeterminacy of their position accounts for a large indeterminacy of their momentum, i.e., velocity, from which follows the maximum kinetic energy of the nucleons, $mv^2/2$, in their bound state.

Let us find out now how far the presently known properties of the nuclei, especially their total binding energies (see Fig. 133) and the average binding energies per nucleon (cf. Fig. 150) can be understood from a detailed discussion of the droplet model of the nucleus. This theory, developed by VON WEIZSÄCKER in 1935, is based on the assumption that, because of the analogy to a liquid drop, the *binding energy per nucleon* is composed of five components:

(1) A contribution a_1 that is due to the nuclear forces between paired protons and neutrons. This constant term predominates for nuclei that are not too light, while the four other terms are, in general, only corrections.

(2) A contribution taking into account that the bond of the unpaired nucleons is weaker than the bond of the proton-neutron pairs. Term (2) has to be considered as an antibonding term because we had assumed for term (1) that *all* nucleons

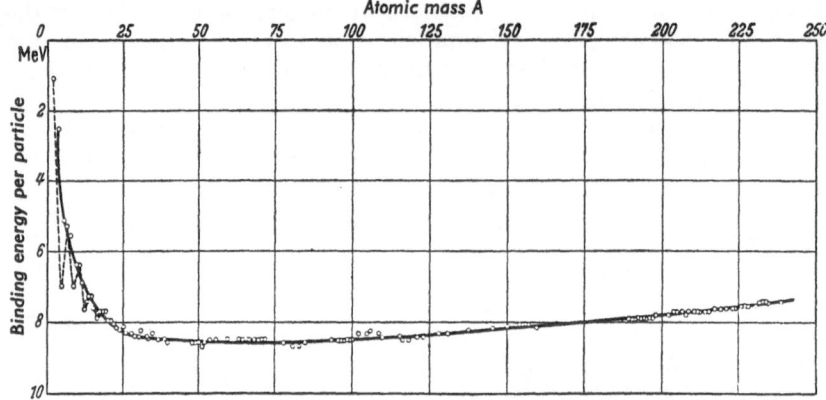

Fig. 150. Binding energy in Mev computed per nucleon (proton or neutron) versus the mass number of the nucleus

were bound with a strength that actually applies only to proton-neutron pairs. We expect this second term to be a function of the relative neutron excess; more exactly, to be proportional to $(N-Z)^2$ since it has a minimum für $N=Z$. For each excess neutron we thus have the contribution

$$a_2 \left(\frac{N-Z}{N+Z} \right)^2 . \tag{5.56}$$

(3) A contribution of a kind of surface tension with antibonding effect, being due to the unilateral binding of the surface nucleons from the inside. We expect this term to be proportional to the nuclear surface, i.e., proportional to $4\pi r^2$. Since, according to Eq. (5.9), r equals $r_0 (N+Z)^{\frac{1}{3}}$, and since the binding energy per nucleon is that of the nucleus divided by $(N+Z)$, the contribution of the surface tension is $a_3 (N+Z)^{-\frac{1}{3}}$.

(4) A mutual electrostatic repulsion of the protons in the nucleus. It is proportional to the square of the proton number Z and inversely proportional to the average distance of the protons which is, in a first approximation, equal to the nuclear radius $r_0 (N+Z)^{\frac{1}{3}}$.

The Coulomb term thus becomes

$$\frac{3}{5} \frac{Z^2 e^2}{r(Z+N)} = \frac{3}{5} \frac{Z^2 e^2}{r_0 (N+Z)^{\frac{1}{3}}} = a_4 \frac{Z^2}{(N+Z)^{\frac{1}{3}}} , \tag{5.57}$$

$\frac{3}{5}$ being a factor which follows from the quantitative theory.

(5) A contribution due to the spin compensation which takes into account the following fact: Nuclei with an even number of protons and neutrons (so-called gg-nuclei) have a somewhat larger binding energy than those with an even number of protons but an odd number of neutrons or vice versa (so-called *gu*-nuclei or *ug*-nuclei), while doubly-odd nuclei (*uu*-nuclei) have a smaller binding energy than the *gu*-nuclei or *ug*-nuclei. The amount of this contribution per nucleon is represented by a term $a_5(N+Z)^{-2}$ where a_5 is positive for even-even nuclei, negative for odd-odd nuclei, but zero for even-odd or odd-even nuclei. Numerical evaluations show that for all nuclei with a mass number higher than 40 this term is smaller than 1% of the total binding energy and, therefore, can be neglected for such nuclei. If we compare the analytic expression of the binding energy resulting from these 5 contributions with the empirical binding energies computed from mass defects of a number of well-known nuclei, the constants a_1, a_2, a_3, and a_5 (which cannot yet be computed theoretically) can be determined from the empirical mass defects. We thus obtain for the binding energy per nucleon the formula

$$E\,[\text{Mev}] = 14.0 - 19.3 \left(\frac{N-Z}{N+Z}\right)^2 - \frac{13.1}{(N+Z)^{\frac{1}{3}}} - 0.60 \frac{Z^2}{(N+Z)^{\frac{4}{3}}} \pm \frac{130}{(N+Z)^2}. \quad (5.58)$$

Fig. 150 shows a diagram of the binding energies computed from Eq. (5.58) versus the mass numbers of the nuclei. If we disregard the lightest nuclei, the curve follows closely the empirical values of the binding energy as indicated by small circles. It is not surprising that some of the lightest nuclei deviate from the general curve. The droplet model is only a poor approximation for these nuclei, and the individual binding forces between the few nucleons must be taken into consideration. But even in the range of light nuclei, the theoretical curve represents satisfactorily the general trend of the nucleonic binding energy. In order to get an indication of the relative significance of the various terms of Eq. (5.58), we compute the binding energy per nucleon for the $_{80}\text{Hg}^{200}$ nucleus. We find that the second term, which is due to the excess neutrons, has a value of only 0.77 Mev, while the unilateral binding of the surface nucleons (third term) reduces the binding energy per nucleon by 2.24 Mev. Similarly, the repulsion between the protons causes a reduction by 3.28 Mev; the last term however contributes only 0.003 Mev and can, therefore, be neglected as is always the case for nuclei with a large number of nucleons.

Formula (5.58), as derived from the droplet model, permits one to draw a number of simple and clear conclusions. In a first approximation, the fourth term, being proportional to Z^2, may be neglected for nuclei with only a few protons. According to Eq. (5.58), in this case we expect the highest stability if the second term disappears, i.e., for $N=Z$. Actually, the most stable isotopes of all relatively light elements have the same number of protons and neutrons. The small binding energy of the lightest nuclei is obviously due to the third term which represents the surface tension. Since the repulsive Coulomb force is proportional to Z^2, the fourth term becomes the more noticeable the heavier the nuclei are and it causes a slow decrease of the average binding energy per nucleon. It is also the cause of the α-emission of the heavy, naturally radioactive nuclei: α-emission occurs as soon as the energy necessary for ejecting two protons and two neutrons from the nucleus becomes smaller than the binding energy of 28 Mev that is released by the formation of an α-particle. However, in computing the binding energy of the last nucleons of a heavy nucleus, we cannot use the average binding energy per nucleon, (i.e., 7.5 Mev, as follows from Fig. 133). We must take into account instead that the last nucleons of a nucleus have a much smaller binding energy than those incorporated earlier. A comparison of the mass defects of the U²³⁸ nucleus and the Pb²⁰⁸ nucleus shows, with the aid of the mass values of Table 3, page 39, that the average binding energy of the *last* 30 nucleons is only 5.35 Mev per nucleon. Consequently, an energy of only a little more than 21 Mev is necessary for releasing four nucleons from the heaviest radioactive nuclei, while about 28 Mev are set free when the nucleons are combined into an α-particle outside of the nucleus. This rough calculation shows that the rest, an amount of 7 Mev, would be available for the α-particle in the form of kinetic energy, and this amount is actually of the same order of magnitude as the maximum α-energy found for radioactive nuclei. On the other hand, an analogous calculation with the mass values of Table 3 shows that an α-decay of nuclei with masses below 210 is not to be expected since here the binding energy even of the most weakly bound nucleons is too high.

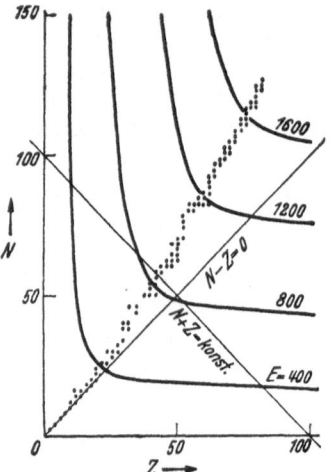

Fig. 151. Dependence of the nuclear binding energy (in 1/1000 mass units = 0.931 Mev) on the number of protons Z and neutrons N in the nucleus (so-called stability valley). "Contour lines" of equal binding energy are indicated

Finally, Eq. (5.58) makes us understand quite generally why nuclei which are not too light always have an excess of neutrons: Here, the relative increase of the second term is overcompensated by a comparatively larger decrease of the fourth term.

An interesting survey of the stability of all known nuclei, as determined from empirical mass defects, is shown in Fig. 151. This is actually a three-dimensional representation in which the number of protons is plotted as abscissa and the number of neutrons as ordinate, and the binding energy of each nucleus (in 1/1000 mass units = 0.931 Mev) is plotted perpendicularly to the plane of the paper, towards the back. The resulting surface is like a valley which falls off from the lower left to the upper right corner of the figure. The origin of the valley, whose bottom contains the points representing the stable nuclei, lies in the drawing plane in the left lower corner of Fig. 151, while the exit of the valley (upper right corner of the figure) is behind or below the drawing plane by the amount of the binding energy of the heaviest nuclei (1800/1000 mass units). Also some contour lines corresponding to specific binding energies are indicated in Fig. 151. Fig. 133 now may be understood as a cross section through the three-dimensional diagram of Fig. 151 along the bottom of the valley so that Fig. 133 shows the angle of inclination of the valley's bottom against the planes of constant energy. Isotopic nuclei lie in planes parallel to the NE plane of Fig. 151; isobaric nuclei (i.e., nuclei with equal mass), which can be transformed into each other by emission of electrons or positrons (or by capturing electrons), are located on planes ($N+Z$ = const) that intersect the planes ZE and NE of Fig. 151 at 45°.

We now discuss in more detail the stability of different nuclei. Theoretically it follows from the Pauli principle (see IV, 10) that two protons or neutrons with antiparallel spins can occupy one nuclear energy level, and we may expect, as in the case of electron shells, that nuclei which have only such doubly occupied energy states, so-called even-even nuclei, must be exceptionally stable. Furthermore, if the energy states of two protons and two neutrons are close together and those of the remaining nucleons are higher, which is apparently true for all light nuclei, we expect a particular stability of such α-complexes. Experience confirms this conclusion. *The α-particle, consisting of two neutrons and two protons, with its binding energy of 28 Mev (see V, 3), is the most stable of all known nuclei, and nuclei built up of even numbers of neutrons and protons exceed all others in stability and abundance.* This is particularly true for nuclides built up exclusively from α-complexes, such as He^4, C^{12}, O^{16}, Ne^{20}, Mg^{24}, Si^{28}, and S^{32}. 80% of the earth's crust consist of the even-even-nuclei O^{16}, Mg^{24}, Si^{28}, Ca^{40}, Ti^{48}, and Fe^{56}. Furthermore, 162 out of 276 known stable nuclides are even-even nuclei. 108 stable nuclides have an even number of protons and an odd number of neutrons or vice versa, whereas only six stable odd-odd nuclei exist. Nuclei with an even number of protons and an odd number of neutrons or vice versa are expected to be less stable than even-even nuclei since one state is only singly occupied, i.e., unsaturated, whereas we expect nuclei with an odd number of both protons and neutrons to be much less stable. Again experience is in agreement with these predictions. *The stability and abundance of the even-odd nuclei is considerably smaller than that of the doubly even nuclei, whereas all doubly odd nuclei with atomic numbers over 8, with the exception of the rare nuclides $_{23}V^{50}$ and $_{73}TA^{180}$, are unstable, i.e., they are β-active.* We shall explain the exceptions of the doubly-odd light nuclei $_1H^2$, $_3Li^6$, $_5B^{10}$, and $_7N^{14}$ below.

We get a better view of this stability problem if we discuss intersections of the stability surface Fig. 151 by isobaric planes, i.e., planes which cut the Z and N axes at 45° and are parallel to the E axis. We then see cross sections (Figs. 152

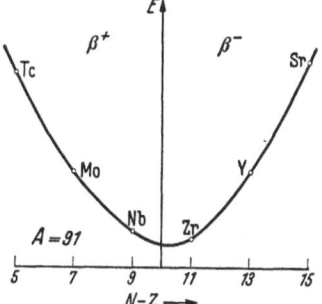

Fig. 152. Isobaric section through the stability valley at an odd mass number ($A = Z + N = 91$)

and 153) of the valley, and all isobaric nuclei with the same $(Z+N)$ value lie on the curve of intersection of the plane with the surface of the valley. The most stable nuclei lie on or near the bottom of the valley; on the left-hand side are those with an excess of protons and on the right-hand side those with an excess of neutrons. Those on the slope on the left of the bottom of the valley we expect to transform into stable nuclei by emitting *positrons*, those on the right by emitting *electrons*. As expected we find empirically that nuclides on the left-hand side of the valley's bottom are β⁺-emitters (and K radiators, cf. V, 7b), those on the right-hand side are β⁻-emitters. In such transformations, according to the theory, only *one* electron or positron can be emitted at a time, never two simultaneously.

Now, as an example, we consider an isobaric section through the odd mass number 91. It is shown in Fig. 152, where *decreasing* binding energy is plotted upward, and the neutron excess is plotted as abscissa. Here we expect only the lowest nucleus, $_{40}Zr^{91}$, to be stable. Actually there exists, according to MATTAUCH's first isobar law, *only one stable nucleus for each odd mass number*, whereas all nuclei lying to the left of it transform into more stable nuclei by β⁺-emission or electron capture, and all those lying to the right by β⁻-emission. The situation is more complicated if the isobaric plane has an even $A = Z + N$ value, e.g., (in Fig. 153) $A = 92$. Doubly even as well as doubly odd nuclei belong to this mass number. Since the latter ones are by far less stable than the former ones, the odd-odd curve must lie above the even-even curve. We have here, so to say, two layers on the bottom of the valley on which the doubly even and doubly odd nuclei are situated. Again the most stable nucleus is that with the lowest energy, $_{40}Zr^{92}$. In addition, however, also the $_{42}Mo^{92}$ nucleus is stable, because it could be transformed to $_{40}Zr^{92}$ only by a *double* positron emission, which is forbidden; whereas the transformation by way of $_{41}Nb^{92}$ is also impossible because this would require an expenditure of energy. Thus we see that stable doubly odd nuclei cannot exist; they should all be β-active. *On the other hand, there can exist several stable doubly even isobars* (second isobar law of MATTAUCH), but *they always differ by two units of nuclear charge.* The unexpected stability of the four lightest doubly odd nuclei $_1H^2$, $_3Li^6$, $_5B^{10}$, and $_7N^{14}$ is caused by the fact that for these light nuclei the valley of Fig. 153 is cut-in so deep and thus the slopes are so steep that the minimum of the upper odd-odd curve with the four nuclides in question is situated lower than the nuclei lying somewhat to the side on the even-even curve.

It might be of interest to point out, referring to Fig. 153, that the unstable odd-odd nucleus $_{41}Nb^{92}$ has *two* decay possibilities since it is situated between two stable even-even nuclei (isobars). It either may transform into the stable isobaric nucleus $_{42}Mo^{92}$ by emitting an electron or into the stable isobar $_{40}Zr^{92}$

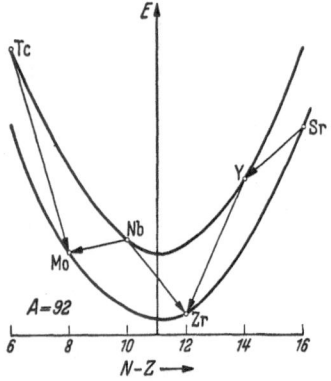

by positron emission. On page 252 we have already mentioned this possibility of a β^--activity *and* β^+-activity of certain odd-odd nuclei; its explanation here seems self-evident in the framework of nuclear theory.

In concluding, we have to discuss the problem how many stable isotopes an element with even or odd atomic number may possess. By plotting the stable nuclides in Fig. 151 we verify ASTON's *isotope rule according to which elements of odd atomic number have at most two stable isotopes, whereas elements with even atomic number often have many more than two.* This may be confirmed by checking Table 3 (page 34 ff.). ASTON's isotope rule is a direct result of the "structure" of the energy surfaces treated above. First of all, it is obvious that stable isotopes of odd atomic number (=number of protons) higher than 7 must always have an even number of neutrons since odd-odd nuclei above $_7N^{14}$ are not stable. Consequently, such isotopes, e.g., $_{29}Cu^{63}$ and $_{29}Cu^{65}$, never can have neighboring mass numbers. The curve of the minimal potential energy (bottom of the valley) therefore, in general, runs between the two nuclei, leaving them situated on opposite slopes of the valley. Other potentially stable isotopic nuclides, e.g., $_{29}Cu^{61}$ and $_{29}Cu^{67}$, must then be situated so high on their respective slopes that they can decay, according to Fig. 153, into more stable isobars, i.e., to isotopes of a *different* element, by emitting electrons or positrons (or capturing electrons): They are therefore radioactive. Thus, nuclei of odd atomic number actually cannot have more than two stable isotopes. Nuclei of even atomic number, however, have always one stable isotope with an odd number of neutrons (cf. Fig. 152); furthermore, they have, in general, *at least* two isotopes with an even number of neutrons, one of which will be situated somewhat to the right, the other one to the left of the minimum of the intersection curve. This can best be seen by actually plotting some isobaric intersections for the corresponding mass numbers, e.g., in the case of $_{28}Ni$ for the mass numbers 58, 60, 61, 62, and 64.

Analogous considerations lead to the result that we cannot expect the "new elements", $_{43}Tc$, $_{61}Pm$, $_{85}At$, and $_{87}Fr$, which fit into the former gaps of the Periodic Table, to have stable isotopes. Since all of them have an odd number of protons, MATTAUCH's first isobar law allows only *one* stable isobar for each mass number. It can be shown that for all mass numbers available for forming stable nuclides with 43, 61, 85, and 87 protons, stable isobars of neighboring elements already exist so that we do *not* expect stable isotopes of the four new elements.

Fig. 153. Isobaric section through the stability valley at an even mass number $(A = Z + N = 92)$

We may stress here that nuclear theory does *not* yet give a reason why the $_{42}Mo^{97}$ nucleus and not the $_{43}Tc^{97}$ nucleus is stable; it only shows *that Tc^{97} cannot be stable if Mo^{97} is known to be stable*. This line of thought is typical for the present status of nuclear theory. We are able to draw conclusions about the stability or the expected type of instability of almost any nucleus by combining certain general theoretical viewpoints with empirical knowledge. But we do not as yet have a nuclear theory that allows us to predict, without any empirical help, *which one* out of a number of isobars with odd mass number must be the stable one.

12. Shell Model and Collective Model of the Nucleus. Magic Numbers, Nucleon Quantum Numbers and the Influence of the Nuclear Core

It was shown in the preceding section that a detailed discussion of the droplet model for the nucleus yields a satisfactory understanding of the empirical results about the binding energy of nuclei, about the different stability of the different nuclei (even-even, even-odd, and odd-odd), about the instability of radionuclides with respect to β^--decay or β^+-decay or electron capture, and about MATTAUCH's isobar laws as well as ASTON's isotope rule.

We shall now show that a number of additional empirical results of nuclear physics may be explained, without any contradictions, from certain simple and reasonable assumptions concerning the interaction of the nucleons in the nucleus. At the same time, these assumptions illustrate the close relationship between our theoretical concepts of nuclear structure and electron shell structure. The empirical results that have to be explained are the following ones: The measured values of the nuclear angular momenta $\vec{I}$, the magnetic nuclear moments μ, the electric nuclear quadrupole moments Q, and the empirical relations between these observables. Furthermore, we want to explain certain empirical selection rules for the β-decay and, finally, the particular stability of some nuclei which are distinguished by certain "magic" numbers of neutrons and (or) protons.

We begin by discussing the empirical evidence for the distinguished role of the nuclei with the so-called magic numbers of protons and (or) neutrons 8, 14, 20, 28, and especially 50, 82, and 126, because this evidence provides the basis for the shell model of the nucleus. First of all, the nuclei $_2He^4$, $_8O^{16}$, $_{14}Si^{28}$, and $_{20}Ca^{40}$ whose numbers of protons *and* neutrons equal the first four magic numbers have a much higher cosmic abundance than their neighbors, and also the elements Zr ($N=50$), Sn ($Z=50$), Ba ($N=82$), and Pb ($Z=82$) show peaks of the relative abundance. Furthermore, we have a lot of evidence pointing to the exceptional role of the neutron numbers 28, 50, and 82. There is the extraordinarily high number of five stable nuclides with the neutron numbers $N=20$ and $N=28$, of six stable nuclides with $N=50$, and of seven stable nuclides with $N=82$. Among the only three isotopes of even-even nuclei which have more than 60% abundance in the isotope mixture of their elements, $_{38}Sr^{88}$ has the neutron number 50, whereas both $_{56}Ba^{138}$ and $_{58}Ce^{140}$ have 82 neutrons each. Among the five doubly even nuclei representing the lightest isotopes of their elements and occuring with an abundance of more than 2%, there are two nuclides with $N=50$ and two with $N=82$. The element Sn with 50 protons has the highest number (ten) of stable isotopes of all elements, a fact which indicates the particular stability of a configuration of 50 protons. Finally, the magic proton number 82 and the magic neutron number 126 stand out by the facts that Pb with 82 protons is the stable end product of all the three long-known radioactive families (cf. Fig. 134); and its most abundant isotope $_{82}Pb^{208}$ contains 126 neutrons, whereas the end product of the new neptunium family (cf Fig. 134) is the isotope $_{83}Bi^{209}$ with also 126 neutrons.

This surprising nuclear stability of certain numbers of protons and neutrons immediately calls to mind the special stability and behavior of the closed electron shells of the rare gases. Actually, there is general agreement today that these magic nucleon numbers must be explained by somehow closed and, therefore, especially stable shells of protons or neutrons ("shells" perhaps in the wider sense of energetically preferred configurations). In analogy to the situation of the electron shells of the alkali atoms, we would expect here that the neutron incorporated just after one nucleon shell is filled should have an especially low binding energy. This conclusion is confirmed by the fact that the only neutron emitters known so far, He^5, O^{17}, Kr^{37}, and Xe^{137}, have one neutron each in addition to the closed neutron shells of 2 or 8 or 50 or 82 neutrons. Furthermore, nuclei with nearly closed neutron shells have a relatively large effective cross section with respect to neutron capture; this is in full analogy to the electron affinity of the halogen atoms. Moreover, the neutron absorption cross section of nuclei with magic neutron numbers is especially small. All these facts excellently fit our concept.

The model which has proved very valuable for understanding the magic nucleon numbers and other properties of the nucleus, which will be discussed presently, is the so-called *shell model*. Just as the model for the many-electron atoms which was introduced in III, 8, it assumes the existence of a closed shell, here of nucleons, and of additional independent nucleons which are responsible for the most essential properties of the entire nucleus.

According to page 167, the energy states of the nucleons depend on the particular potential field in which a nucleon moves. For instance, the convergent sequence of energy states of the hydrogen electron corresponds to the Coulomb potential, while the equidistant energy levels (4.93) correspond to the parabolic potential of the harmonic oscillator. We do not yet have a detailed knowledge about the potential of nuclear forces: As a model, a more or less rectangular potential trough is assumed (see Fig. 140) which degenerates to a parabolic trough for light nuclei with very few nucleons. GOEPPERT-MAYER and, in particular, HAXEL, JENSEN, and SUESS proved theoretically that the magic nucleon numbers can satisfactorily be understood by means of the shell model if the sequence of energy states as derived quantum-mechanically is combined with the assumption that spin and orbital momentum of each nucleon are very closely coupled so that jj-coupling (cf. III, 13) exists in the nuclei. This additional assumption, although deviating from the relations in the electronic shell, is confirmed by measurements. The ls-coupling for each nucleon leads to a large doublet splitting where the state with the higher total momentum (here designated by $\vec{I}$ instead of $\vec{j}$) is always the lower one. Without further assumptions, we thus obtain an energy level diagram (cf. Fig. 154) in which each energy state occupied by a magic number of protons or neutrons (2, 8, ... or 126) is always separated from the next higher one by a relatively large energy gap. It thus is possible to understand the exceptional role, e.g., the particular stability, of the "magic" nuclei. To avoid misunderstandings, we wish to point out again, in supplementing page 265, that the energy-level diagram of each nucleus consists of one level diagram for protons and one for neutrons. It is obvious that the relative energy of the levels of both diagrams determines the number of protons and neutrons forming a certain stable nucleus, since both types of levels must be filled with protons or neutrons to about the same height because of the possible transformation into each other by β-emission.

By using the terminology known from atomic electrons, we now may characterize a nucleon by three quantum numbers or their symbols. We explain this for
the example of a $4f_{\frac{7}{2}}$-nucleon. The first number, 4, corresponds to the principal
quantum number. The symbol "f" characterizes the orbital momentum $|\vec{l}| = 3\,h/2\pi$
of the nucleon (generally: s, p, d, f, for $l = 0$, 1, 2, 3, ... respectively), whereas the
subscript, $\frac{7}{2}$, gives the total
momentum $\vec{I}$ ($\vec{j}$ for electrons),
which is the vector sum of
orbital momentum $\vec{l}$ and spin
$|\vec{s}| = \hbar/2$, again in full analogy to the electron case. An
atomic shell electron of the
total momentum number j
can be oriented in an electric
or magnetic field in $2j + 1$ different directions (see III, 16)
which are characterized by
different values of the directional quantum number m.
We have, therefore, $2j + 1$
different electrons of quantum number j, and the same
is true for nucleons of quantum number I. Moreover,
the Pauli principle which
proved so essential for the
structure of the electron shell,
also applies to the nucleus
because of the universal validity of the quantum laws.
Thus, each nuclear state described by an angular-mo-
mentum quantum number I
can be occupied by $2I + 1$
protons or neutrons. This
number we write as a right-
hand superscript of the symbol of the nucleon, which for
clarity is put into parentheses. We therefore designate
five nucleons in a $4f_{\frac{7}{2}}$ energy
state by $(4f_{\frac{7}{2}})^5$. With these

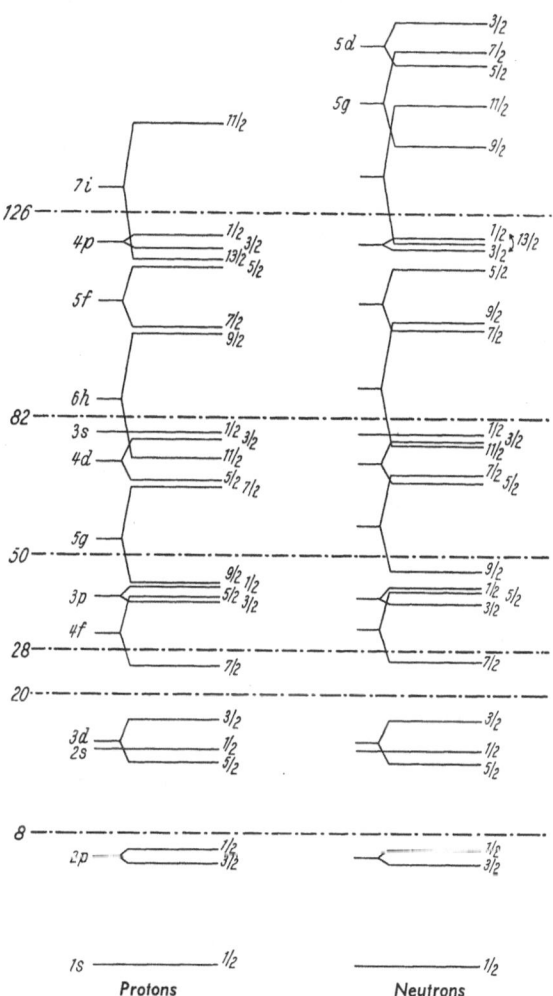

Fig. 154. Term diagrams for protons and neutrons in nuclei. On the left-
hand side the symbols for the principal quantum numbers, on the right-
hand side the symbols for the total angular momentum. On the far left
side the "magic nucleon numbers"

definitions we can describe the energy arrangement of the protons and neutrons
of each nucleus by writing down the corresponding sequence of nucleon symbols
just as in the analogous case of the shell electrons. For the proton arrangement
of the "magic" nucleus $_{50}$Sn we thus obtain

$$(1\,s_{\frac{1}{2}})^2\,(2\,p_{\frac{3}{2}})^4\,(2\,p_{\frac{1}{2}})^2\,(3\,d_{\frac{5}{2}})^6\,(2\,s_{\frac{1}{2}})^2\,(3\,d_{\frac{3}{2}})^4\,(4\,f_{\frac{7}{2}})^8\,((3\,p_{\frac{3}{2}})^4\,(4\,f_{\frac{5}{2}})^6\,(3\,p_{\frac{1}{2}})^2\,(5\,g_{\frac{9}{2}})^{10}.$$

We see from this example that the *construction principle of the atomic nucleus*
deviates in one essential point from that of the electronic shell, which is represented in Table 10. While the electronic energy states with smallest orbital

momentum quantum numbers are the lowest (viz., the construction sequence is $1s, 2s, 2p, 3s, 3p, 3d, 4s, 4p, 4d \ldots$), the opposite is true for the nucleons, where the highest orbital momenta with $l=n-1$ are preferred. Thus, the sequence of construction is $1s, 2p, 3d \ldots$ with occasional inclusion of nucleons with smaller orbital momenta ($3s, 4p \ldots$). Another difference between the nuclear and electronic shell construction principles is the fact that, for nuclei, parallel spin and orbital momentum give lower energy states than antiparallel orientation (cf. Fig. 154).

     The shell model is also able to explain the empirical relation between the mechanical angular momentum and the magnetic moment of the nuclei, both properties of the nucleus which are attributed to the one unpaired external nucleon. The implied assumption that always two protons as well as two neutrons compensate each other with respect to their mechanical angular momenta and to their magnetic moments does not only seem reasonable from energy considerations, but it is a necessary consequence of the empirical result that all known nuclei with an even number of protons *and* neutrons have the nuclear angular momentum $\vec{I}=0$ and the magnetic moment $\mu=0$. *The*

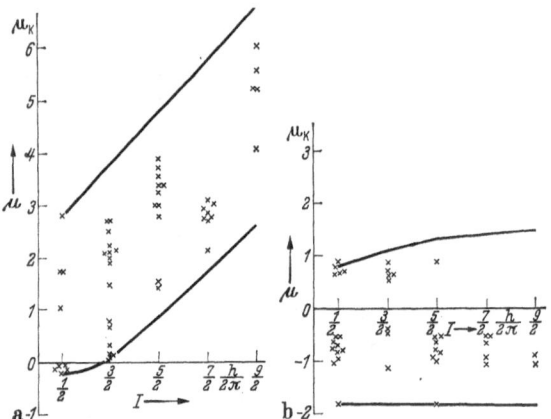

Fig. 155. Dependence of the magnetic moments of atomic nuclei on their mechanical nuclear angular momenta (so-called Schmidt-curves):a) for nuclei with an unpaired excess proton. b) for nuclei with unpaired excess neutron. ¡The values of the magnetic moments plotted against the values I of the mechanical angular momentum are situated between the two Schmidt-curves

values found for I and μ differ from zero only if the nucleus contains an unpaired neutron or proton whose existence is thus responsible for these moments.

     Since the angular momentum $\vec{I}$ of the unpaired nucleon is composed vectorially of its spin $\vec{s}$ of value $h/4\pi$ and its orbital angular momentum $\vec{l}$ of value $0, 1, 2, 3, \ldots, h/2\pi$, the magnetic moment μ also must be composed of an eigenmoment μ_s resulting from the spin and a component μ_l due to the orbital momentum (which of course is zero for neutrons because of their zero charge).

     Now we ask how the magnetic moments of the nuclei are expected to depend on their angular momenta. We begin with an unpaired neutron whose magnetic moment equals -1.91 nuclear magnetons (μ_n). Since the magnetic component of the orbital momentum is $\mu_l=0$, we expect the magnetic moments $-1.91\,\mu_n$ or $+1.91\,\mu_n$ for all nuclei with an unpaired neutron, independent of the value of their mechanical momentum $\vec{I}$. The negative or positive value corresponds, respectively, to the parallel or antiparallel orientation of mechanical spin and orbital momentum. According to THEODOR SCHMIDT, the μ values of all nuclei with an unpaired neutron are indeed situated between the two parallels $\mu=\pm1.91\,\mu_n$ in Fig. 155b. Also the magnetic moments of nuclei with an unpaired *proton* can easily be computed. We know that the magnetic spin moment of the proton is $+2.79$ nuclear magnetons, whereas the contribution from the orbital momentum $\vec{l}$ of the proton is l nuclear magnetons according to Eqs. (3.92) and (3.93). As we know, no anomaly exists for the orbital momentum (see III, 15). Just as in the case of the electron shell (cf. III, 9b), we have the two values $(l \pm s)\,h/2\pi$ for the resulting nuclear momentum $\vec{I}$. Since $s = \frac{1}{2}$, there are two l-values for each quantum number I of the nuclear momentum:

$$\left.\begin{array}{l} l_1 = I - \frac{1}{2} \\ l_2 = I + \frac{1}{2} \end{array}\right\} \tag{5.59}$$

and, since $\mu = \mu_l + 2.79$ nuclear magnetons, only two values of the nuclear magnetic moment are possible for each I-value, viz.

$$\left.\begin{array}{l} \mu_1 = (I - \frac{1}{2} + 2.79) \quad \text{nuclear magnetons} \\ \mu_2 = (I + \frac{1}{2} - 2.79) \quad \text{nuclear magnetons.} \end{array}\right\} \tag{5.60}$$

μ_1 and μ_2 correspond to parallel and antiparallel orientation of spin and orbital momentum, respectively. If the magnetic moments of all nuclei with an unpaired proton are plotted versus their nuclear momenta, the resulting points are situated *between* the two Schmidt-lines of Fig. 155a.

It is a strong argument in favor of our independent-nucleon model that, in principle, the relation between nuclear momentum $\vec{I}$ and nuclear magnetic moment μ can be understood as discussed. We learned in V, 11 that a great number of empirical facts about properties and behavior of nuclei could be understood with the aid of the nuclear-droplet model. This model assumes, as its name implies, a very close packing of the nucleons in the nucleus so that a free motion of single nucleons would scarcely seem possible. On the other hand, the empirical results discussed in this section are as satisfactorily explained by the independent-nucleon model, in which the individual nucleons are characterized by individual orbital and spin momenta. This model, therefore, seems to require a practically free and ordered motion (orbital revolution!) of at least the unpaired nucleon in the average field of all the others.

It is characteristic for the present state of nuclear theory that no single nuclear model is sufficient for describing all the empirically known properties of nuclei. By means of the shell model the quadrupole moments cannot be explained since the empirical values, in particular those of nuclei with partially filled sub-shells, are considerably larger than was to be expected according to the shell model for the one external nucleon. For that reason RAINWATER has introduced a hypothesis, in full analogy to the interpretation of the alkali atoms (cf. III, 8) by the core model, according to which the nuclear core with its nucleons in closed shells is deformed by the external nucleon and thus contributes to the nuclear quadrupole moment. This hypothesis seems to describe the properties of the nucleons surprisingly well.

The quantum theory of a nuclear model with a deformed core is much more complicated than that of the shell model. In particular the energy state of a nucleon which moves in the deformed potential field of the core cannot anymore be described correctly by the quantum numbers l and j, but only by the component Ω of its nuclear momentum in the direction of the axis of symmetry. This quantum number and not the total nuclear momentum j is a "good" quantum number of the external nucleon. Its wave function thus is a linear combination of eigenfunctions which correspond to different j-values.

The nuclear core is deformed all the time or only part of the time, depending on the strength of the interaction between core and external nucleon. In the second case we expect vibrations by which the core alternatingly is dilated and compressed, whereas an always deformed core should be able to carry out rotations similar to those of an ellipsoidal diatomic molecule according to VI, 9.

In carrying through this theory, BOHR and MOTTELSO Ntreated the nuclear core as a droplet of an incompressible liquid without friction and computed the vibration states of the nucleus to be expected according to these assumptions. These states should be equidistant like those of the harmonic oscillator (cf. IV, 7c), whereas the rotation states of the always deformed nucleus correspond to those of the free rotator (cf. IV, 7b). Comparison with the empirical energy states resulted in moments of inertia of the nuclei which where considerably smaller than to be expected for solid rotating cores. This seems to indicate that only layers close to the surface of the model droplets participate in the rotation.

The computed vibration and rotation spectra of the nuclear core can be observed according to the theory only for such nuclei for which the motion of the external nucleon is very fast compared to the motion of the core, i.e. for those nuclei in which the polarization of the core by the external nucleon is only a relatively small perturbation of the strong bond of the external nucleon to its core. The distances between the excitation levels of the external nucleon in this case are large compared to the vibration and rotation levels of the core. The comparison of the measured energy levels of nuclei with the expected rotation levels shows that the latter are observed only for those nuclei for which large quadrupole moments aree vidence for the perpetual deformation of the core. Surprisingly, the moments of inertia derived from the observed levels of the nucleus prove to be larger than was expected for a rotating liquid droplet, whereas they are small in comparison to those expected for the rotation of a solid deformed nucleus.

Vibration levels of the core are found in the spectra of gg-nuclei. In this case, however, the collective model describes the results in a very rough approximation only. This may be caused by the fact that the complicated behavior of nuclear matter cannot be described sufficiently well by a simple harmonic oscillator according to IV, 7c, in particular not for strong vibrations. Nevertheless we may derive from the vibration spectra of the nuclear core that its behavior is between that of the liquid droplet and a solid core.

For the explanation of the magnetic moments of the nuclei in their ground states the collective model is an improvement in so far as for strong interaction between core and external nucleon the values are expected to move into the range between the Schmidt curves of Fig. 155, as was actually observed. A further improvement of the theory by NILSSON leads to an understanding of the core deformations as derived from the measured quadrupole moments.

We see that the collective model of the nucleus, which in a certain sense may be regarded as a combination of the old droplet model and the shell model, allows us to understand the empirical behavior of nuclei to a surprising extent.

13. Discovery, Properties, and Effects of the Neutron

We referred already, in the preceding sections of this chapter, to the neutron as the nuclear particle with charge zero whose mass is approximately equal to that of the proton. But we did not go into the details of its discovery, its properties, its detection, and its production. We shall now take up this question after we have assembled the necessary information to do so.

a) Discovery, Mass, and Radioactivity of the Neutron

The existence of the neutron was inferred by CHADWICK in 1932 in a bold but physically logical analysis from experiments of CURIE and JOLIOT as well as from his own cloud-chamber experiments carried out with various gases. CURIE and JOLIOT had repeated the excitation experiments of BOTHE and BECKER (mentioned in V, 10) but, in contrast to them, they happened to use an arrangement sensitive also to the yet undiscovered neutron. After the α-bombardment of Be, an unknown radiation appeared to have been emitted besides the γ-radiation; for the ionizing effects of the total radiation increased if hydrogen-containing substances were brought into the ionization chamber. At first, CURIE and JOLIOT assumed that energy of BOTHE's γ-quanta was being transferred to the scattering nuclei by COMPTON scattering (see IV, 2). CHADWICK, however, deduced that the phenomenon must be the effect of a new uncharged particle which he called the neutron, a particle searched for in vain since 1920 by RUTHERFORD. By applying the laws of conservation of energy and momentum to a central collision of the hypothetical neutron with a proton and with a nitrogen nucleus, respectively, CHADWICK obtained two equations for the unknown mass and velocity of the neutron. From these he found that the mass of the neutron approximately equals that of the proton. The presently most accurate determination of the mass of the neutron uses reactions with nuclei whose masses are mass-spectroscopically well known, e.g., the photo-fission of the deuteron into a proton and a neutron, which requires an energy of 2.2247 Mev. The resulting atomic weight of the neutron is 1.0086654, i.e., about 0.00084 mass units (corresponding to 0.78 Mev) higher than the sum of the masses of proton and electron. *The neutron therefore may disintegrate spontaneously into a proton and an electron (plus an antineutrino):*

$$n \rightarrow p + e^- + \bar{\nu}_e + 0.78 \text{ Mev.} \tag{5.61}$$

It must thus be called radioactive or, more exactly, β-active. If we insert the maximum energy of 0.78 Mev set free in the disintegration process into the Sargent diagram Fig. 141, we obtain a half life of the neutron of about 20 minutes, which is in satisfactory agreement with the best experimental value of 12.8 ± 2.5 minutes. Because of the strong interaction between neutrons and nearly every type of nuclei, i.e., because of the neutron's large absorption cross section, most nuclear experiments do not show evidence for this spontaneous disintegration of the neutron. Nevertheless, the process is significant for the understanding of the neutron. Certainly, it cannot "consist" of a proton plus an electron (the original assumption of RUTHERFORD's) since the combination of these two particles into a neutron would imply the release of binding energy so that the mass of the neutron would then be smaller than the sum of the masses of proton and electron by a corresponding mass defect: Actually, however, the neutron's mass is larger than this sum. Consequently, the neutron must be a new elementary particle, although, in contrast to the proton, an unstable (β^--active) one.

b) Neutron Sources

The nuclear transformation which led to the discovery of the neutron can be represented in the terminology introduced in V, 8 as

$$_4\text{Be}^9 \, (\alpha, \, n) \, _6\text{C}^{12}. \tag{5.62}$$

This reaction is still used for producing neutrons; we have already mentioned a neutron source consisting of a mixture of beryllium and an α-emitting radioactive substance. The energy of the fastest neutrons emitted in this reaction is 13.7 Mev; unfortunately the scattering of the energy values is considerable. Neutrons of much lower energy (24 to 830 kev) can be released from deuterium or beryllium by γ-radiation of artificially radioactive isotopes through a (γ, n)

process. Such *photoneutron sources* therefore consist of a mixture of a deuterium compound or of beryllium powder with a radioisotope. They are comparatively cheap and easy to handle, but they have the disadvantage that the neutron yield decreases in time because of the decay of the radionuclides. The photo-neutron sources thus have a relatively small life time. Table 11 shows several of these sources with their half lives and neutron energies. The neutrons of the source mentioned last are very nearly monochromatic.

With artificially accelerated particles, considerably higher neutron intensities can be produced. Of special interest is the induced reaction which produces neutrons in abundance through collisions of deuterons of less than 0.5 Mev energy against deuterons:

$$_1H^2(d, n)\,_2He^3. \tag{5.63}$$

In addition to this reaction, the bombardment of tritium, lithium, beryllium, and carbon by fast deuterons, and the bombardment of tritium and lithium by fast protons also produces neutrons with high yield.

Table 11. *Photo-neutron Sources with Half Life Times and Neutron Energies*

Neutron Source	Half Life	Neutron Energy in Mev
beryllium with Sb^{124} .	60 days	0.024
deuterium with Ga^{72} .	14 hours	0.13
deuterium with Na^{24} .	15 hours	0.22
beryllium with La^{140} .	40 hours	0.62
beryllium with Na^{24} . .	15 hours	0.83

Neutrons of very high energy in the range of 100 Mev can be obtained from deuterons accelerated in large cyclotrons. The deuterons disintegrate into protons and neutrons by peripheral collisions with the nuclei of a target inside a cyclotron since they have a binding energy of only 2.2 Mev. The protons then are deflected into the interior of the cyclotron by its magnetic field, whereas a well-collimated neutron beam leaves the machine. By far the most powerful sources for neutrons of intermediate and thermal energy are the nuclear fission reactors which will be treated in detail in V, 16. A large part of today's neutron research is therefore concentrated in laboratories equipped with high-power reactors. The neutron beams that emerge from windows of these reactors presently have a flux density up to more than 10^{14} neutrons/cm² sec.

c) Production of Thermal and Monochromatic Neutrons

Because of their missing electric charge, neutrons are not deflected by positive nuclei, so that even very slow neutrons can penetrate into nuclei where they are absorbed and where they thereby initiate nuclear reactions through their released binding energy of about 8 Mev. The smaller their kinetic energy the greater is their time of interaction and, therefore, the reaction yield. For this reason, the production of slow and extremely slow neutrons is of special interest.

Thermal neutrons are those neutrons whose kinetic energy equals the most probable kinetic energy kT of their environment, i.e., at room temperature $kT = 0.025$ ev, so that $v_0 = \sqrt{2kT/M} = 2.2 \times 10^5$ cm/sec. Such thermal neutrons are obtained when fast neutrons lose their kinetic energy by elastic collisions while they diffuse through sufficiently thick layers of light material. This energy transfer is the more effective the smaller the mass difference between the two collision partners is. For instance, 18 collisions are necessary in the average to stop 1 Mev-neutrons in light water, 25 collisions in heavy water, 90 in beryllium, and 114 in carbon. According to the law of conservation of momentum, the energy transferred by elastic collisions decreases with increasing mass difference of the collision partners. This explains the seemingly paradox observation that neutrons, in contrast to charged particles and photons, are easily stopped by paraffin and water, whereas they can penetrate thick layers of lead. For producing thermal neutrons, not only the "moderator" effect but also the absorption cross section of the nuclei for slow neutrons is important, since it is desirable to moderate as many neutrons as possible but with the smallest possible absorption loss. Some typical examples of absorption cross sections, in units of $10^{-24}\,\mathrm{cm^2}=1$ barn, are given in Table 12.

Table 12. *Absorption Cross Sections of Several Nuclei for Thermal Neutrons*

H	D	He	Be	C	O	U²³⁸	U²³⁵
0.33	0.00046	0	0.01	0.003	<0.0002	2.75	687 barn

Since gaseous helium has too small a density, heavy water is obviously a particularly good moderating substance; next comes pure carbon (graphite) as well as beryllium and its oxide.

Graphite plays an important role in the production of extremely slow neutrons of 0.0018 ev, corresponding to a temperature of only 20 °K: If neutrons diffuse through a large graphite block, they are irregularly scattered into all directions by the numerous microcrystals of the material, until they finally are absorbed if their de Broglie wavelength h/Mv is smaller or equal to twice the maximal lattice constant of graphite, which is 3,4 Å. From BRAGG's relation [see (7.11)] it follows, namely, that waves with $\lambda > 7$ Å, equivalent to neutrons with a kinetic energy smaller than 0.0018 ev, cannot be diffracted by the graphite lattice. Consequently, these and only these extremely slow ("cold") neutrons are able to leave a reactor through a huge graphite block.

While the thermal neutrons, including these coldest ones, have a statistical velocity distribution, it is of interest for the investigation of resonance effects (see below) to produce neutrons with well-defined and known energies. These we call *monochromatic neutrons* in analogy to light of exactly defined wavelength. The best method for their production is the application of a crystal spectrograph for neutrons which corresponds completely to the X-ray crystal spectrograph described in III, 1 a, although for intensity reasons it requires a very large flux density, of the primary neutron beam. If neutrons strike a crystal under grazing incidence they are diffracted according to their de Broglie wavelength. Between the lattice constant a, the angle φ of the first diffraction maximum, and the wavelength λ of the neutron we have the relation

$$\lambda = 2a \sin \varphi \qquad (5.64)$$

which yields, under consideration of the de Broglie wavelength (4.9), for the velocity v of the neutrons diffracted by the angle φ the relation

$$v = \frac{h}{2Ma \sin \varphi}. \qquad (5.65)$$

The neutron diffraction by a crystal can thus be utilized for monochromatizing neutrons in the energy range of 0.01 to 100 ev.

The other methods for obtaining monoenergetic neutrons make use of mechanical choppers or pulsed neutron sources for producing sharply separated groups of neutrons. After a certain traveling time, these groups extend in length because of the different velocities of the neutrons contained in each group so that neutrons of definite velocity can be singled out by correctly synchronized choppers. Unfortunately, these methods are restricted to relatively slow neutrons ($E < 1000$ Mev). The production of approximately monochromatic neutrons of high energy requires special nuclear processes by which such neutrons are obtained directly.

d) Detection and Measurement of Neutrons

Indirect methods are necessary for detecting neutrons because uncharged particles do not ionize matter. Neutrons thus are not directly detectable by cloud chambers, photographic plates, or counters.

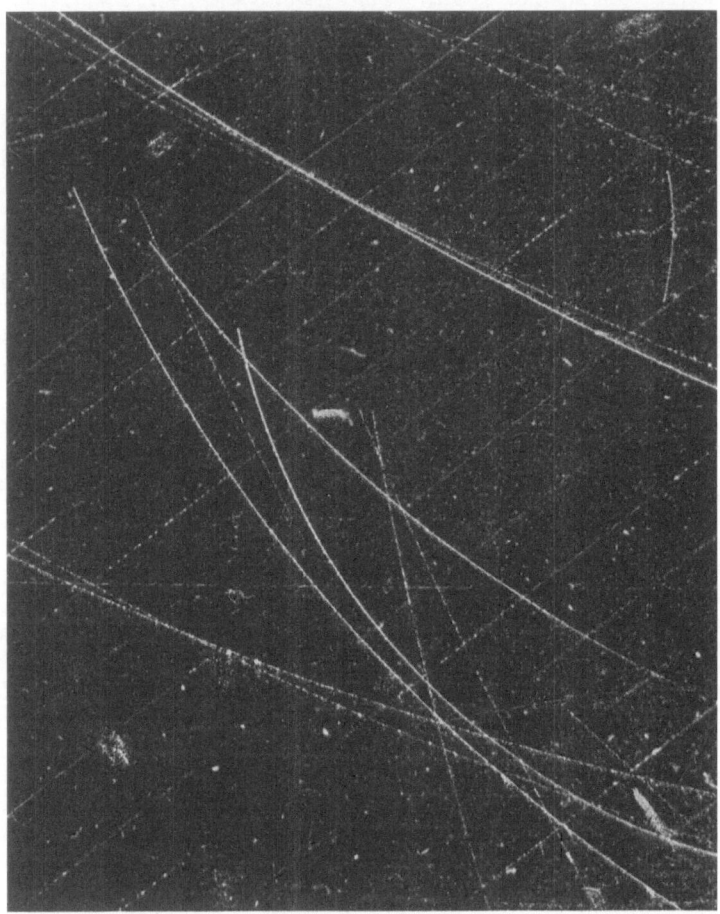

Fig. 156. Cloud-chamber photograph of protons, knocked out of hydrogen gas by impact of fast neutrons (not visible), which enter from the upper left corner. Proton tracks are curved due to magnetic field. (Courtesy of Radiation Laboratory, University of California)

For detecting and measuring *fast* neutrons, we make use of the fact that they can transfer their total energy by central collisions to protons of hydrogen-containing substances (Fig. 156). These fast protons then are able to ionize and thus can be detected (and their energy measured) in all devices mentioned in V, 2. For that purpose, cloud chambers, ionization chambers, or the counters are either filled with hydrogen or some hydrocarbon vapor of sufficient pressure; or the walls of these devices are coated by paraffin. Hydrogen-containing luminescent materials (cf. VII, 23) in scintillation counters are used for measuring fast

neutrons also. For the detection of slow and thermal neutrons which cannot transfer to protons an energy sufficient for ionization, the β-activity of (n, γ)-produced
radionuclides is used. Their β-radiation is measured by a counter and serves as a
measure of the number of slow neutrons absorbed. Indium and gold are examples
for such *neutron indicators* of high sensitivity.

A special Geiger counter has proved very valuable for measuring neutrons if
filled with the trifluoride BF_3 of the boron isotope with mass number 10. Collisions
with neutrons transform $_5B^{10}$ nuclei with a high yield according to the reaction

$$_5B^{10} + n \rightarrow {_5B^{11}} \rightarrow {_3Li^7} + {_2He^4}. \tag{5.66}$$

The α-particles produced in this transformation cause ionization in the Geiger
counter and, therefore, can be used for indirectly counting the neutrons. A related
method for counting slow neutrons uses B^{10}-containing phosphors (cf. VII, 23)
or LiI-crystals. The α-particles produced by (n, α)-reactions with B^{10} or Li^7
excite the phosphor to scintillations which thus permit counting of the neutrons
indirectly.

A final detection method for neutrons is based on the fission process that is
caused by them. We shall learn in V, 14 that slow neutrons cause the nuclei of
the uranium isotope 235 to split, whereas fast neutrons have the same effect,
although to a smaller degree, for uranium nuclei U^{238}. Ionization chambers are
therefore coated with uranium or a uranium compound, and the ionization of
the high-energy fission products produced in the fission process is used for
detecting the neutrons. It is obvious that a U^{235} coating makes the ionization
chamber sensitive to slow neutrons while the common U^{238} (or bismuth) is used
in this *fission chamber* for the detection of fast neutrons.

e) Specific Nuclear Reactions Initiated by Neutrons

We end our treatment of neutrons with a short discussion of nuclear reactions
initiated by them. First of all, it is evident that the effects of a bombardment by
fast neutrons are not very different from those by fast protons; both may cause
the emission of one or more neutrons, protons, and α-particles from the target
nucleus.

It is, however, a characteristic property of neutrons that they are able to
penetrate into light nuclei as well as heavy ones even when they have an extremely
low kinetic energy. Their binding energy of $7-8$ Mev set free in their absorption
by a nucleus leads always to an excitation of the compound nucleus, which then
returns to its ground state by emitting a γ-quantum. These (n, γ)-reactions thus
are typical for slow neutrons. They are also of great practical significance because
most neutron-absorbing nuclei can free themselves of the neutron excess acquired
by this absorption only by a subsequent β^--decay. *The majority of (n, γ)-reactions
therefore leads to radionuclides, and the irradiation of appropriate substances by slow
(thermal) neutrons is the most common method for producing radionuclides.*

We have mentioned on page 262 that the neutron-capture cross section of
practically all nuclei depends very strongly on the neutron energy and has sharp
resonance peaks for certain energy values. A typical case is the (n, γ)-process of
the uranium isotope U^{238} which has a cross section peak of several thousand
barns for a neutron energy of 6.7 ev. We shall come back to this fact later
since the U^{239} produced in this process is finally transformed into plutonium. The
problem of resonance capture was treated theoretically by BREIT and WIGNER
in analogy to the theory of optical dispersion where a similar resonance problem
exists between an arriving photon and an absorbing or scattering atom.

The technically most important reaction initiated by neutrons will now be treated in detail; it is the fission of the heaviest nuclei with its far-reaching consequences.

14. Nuclear Fission

In the nuclear transformations discussed so far we have found that single nucleons or α-particles are emitted. Only very high collision energies lead to multiple particle emission or disintegration of the entire nucleus (see Fig. 167).

A fundamentally different type of nuclear transformation, whose possible existence had been suggested in 1934 by I. NODDACK, was discovered by HAHN and STRASSMANN in 1938. At first, the radioactive products resulting from the irradiation of uranium by slow neutrons were believed to be transuranium isotopes, i.e., nuclides of atomic number larger than 92. But instead, HAHN and STRASSMANN found them to be nuclear fragments of intermediate atomic weight originating from the fission of the uranium compound nucleus produced by the absorption of a neutron (cf. the cloud chamber photograph in Fig. 157). Further investigation, especially by MEITNER and FRISCH, soon revealed that the fission process occurs in very different ways, depending on its initiation by thermal neutrons or by fast neutrons with $E_k \geq 1$ Mev. Fig. 158 shows the fission cross section plotted versus the neutron energy for the most important nuclide, U^{235}, which is fissionable by thermal neutrons, and for U^{238}, which is fissionable only by fast neutrons and even then only with a cross section which is by three orders of magnitude smaller.

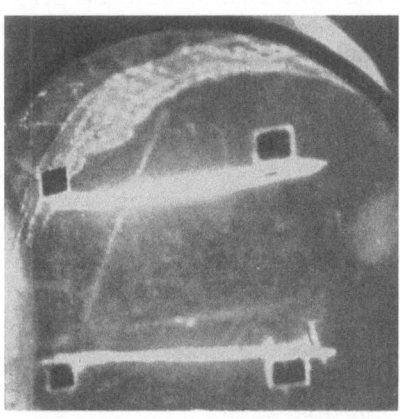

Fig. 157. Cloud-chamber photograph of HAHN and STRASSMANN's fission of a uranium nucleus upon absorption of a neutron. The two fission products leave the upper horizontal uranium plate in opposite directions with a combined energy of 160 Mev. (Photograph by CORSON and THORNTON)

Since the two fission products repel each other because of their positive charges, they gain kinetic energy which then is transferred to the surrounding matter as frictional heat. Each fission process releases the high energy of about 200 Mev. The compound nucleus U^{236} produced in the absorption of a thermal neutron by a U^{235} nucleus has 52 excess neutrons and thus a much larger neutron excess than the stable nuclei of the fission products of medium atomic weight. Therefore, the fission products must either transform some of their neutrons into protons by multiple β^--emission, thus decaying step by step into more stable nuclei, or they must eject the excess neutrons directly. Actually, both transformations occur. The fission products (see Fig. 159) loose the major part of their excess neutrons by β^--emission. In addition, however, two to three neutrons are set free in each fission process. The observations indicate that this emission of neutrons predominantly occurs directly during the fission process; but a small percentage is ejected later by highly excited fragments which cannot get rid of their excess neutrons fast enough by β^--transformation. This spontaneous and delayed emission of neutrons in the nuclear fission process is of utmost importance since these neutrons can induce fission in further uranium nuclei; a *chain reaction* is thus possible. In the next sections we shall come back to the possibility of producing useful energy in this way.

In the fission process as well as in the subsequent β^--emission, excited radio-active fission fragments are produced which release their excitation energy at least partly by emission of γ-quanta so that nuclear fission is always accompanied by strong γ-radiation. Such γ-radiation is also emitted by the radioactive

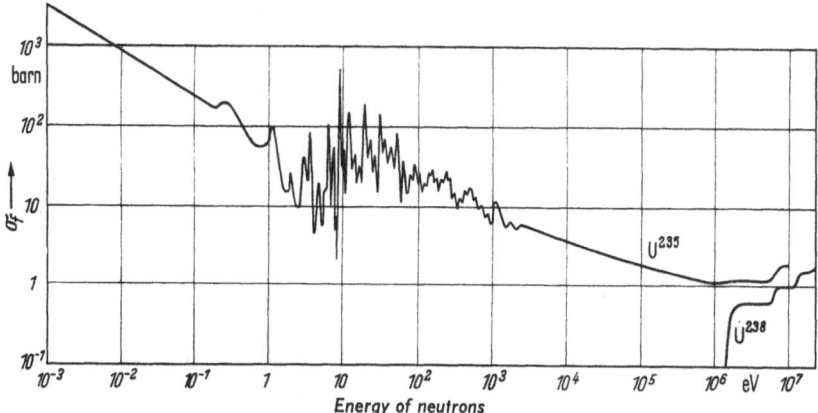

Fig. 158. Fission cross section of U^{235} and U^{238} versus neutron energy

fission products during their often considerable life time. The total energy set free in fissioning U^{235} by slow neutrons therefore consists of four different parts. During the fission process, 162 Mev are released as kinetic energy of the fragments, 6 Mev as kinetic energy of free neutrons, and additional 6 Mev in the form of γ-radiation. Finally, 21 Mev are emitted as more or less delayed β- and γ-radiation by the radioactive fissions products. Consequently, the total energy per fission process is 195 Mev, of which only 174 Mev are released at the moment of fission.

Beside the uranium isotope U^{235} mentioned before, two new fissionable nuclides can be produced artificially. These are the uranium isotope U^{233} and the isotope 239 of the new element plutonium, $_{94}Pu^{239}$. Both absorb thermal neutrons and then split with cross sections roughly comparable to U^{235}. We mentioned already that U^{238}, which is the most abundant uranium isotope, is stable with respect to slow neutrons but has a very large capture cross section (several thousand barns) for neutrons of 6.7 ev energy [(n, γ)-process]. The equally stable thorium nuclide Th^{232} also

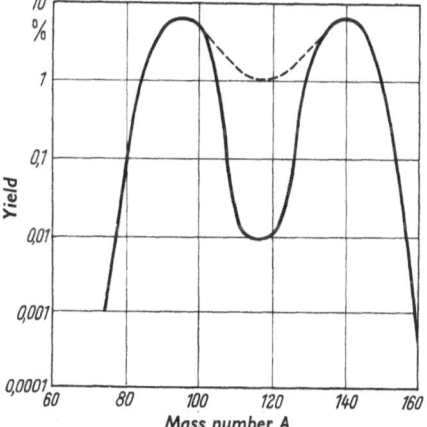

Fig. 159. Yield curve of the fission products of the U^{235} nucleus. Fission induced by thermal neutrons and neutrons of 14 Mev energy (dashed line). Percentage of fission products versus mass number.

absorbs slow neutrons with a large cross section. The compound nuclei U^{239} and Th^{233} produced by this neutron capture are β-active and decay by two successive electron emissions [always combined with neutrino emission (see V, 6g)] to easily fissionable nuclides of mass 239 and 233, respectively. Whereas the nuclides of the thorium reaction sequence

$$_{90}Th^{232} (n, \gamma)\,_{90}Th^{233} \rightarrow\,_{91}Pa^{233} + e^- \rightarrow\,_{92}U^{233} + e^- \qquad (5.67)$$

belong to well-known elements, nuclei of two new elements are produced by the β-decay of U^{239}. These are neptunium with the atomic number 93 and plutonium with the atomic number 94:

$$_{92}U^{238}\,(n,\,\gamma)\,_{92}U^{239}\rightarrow\,_{93}Np^{239}+e^{-}\rightarrow\,_{94}Pu^{239}+e^{-}.\tag{5.68}$$

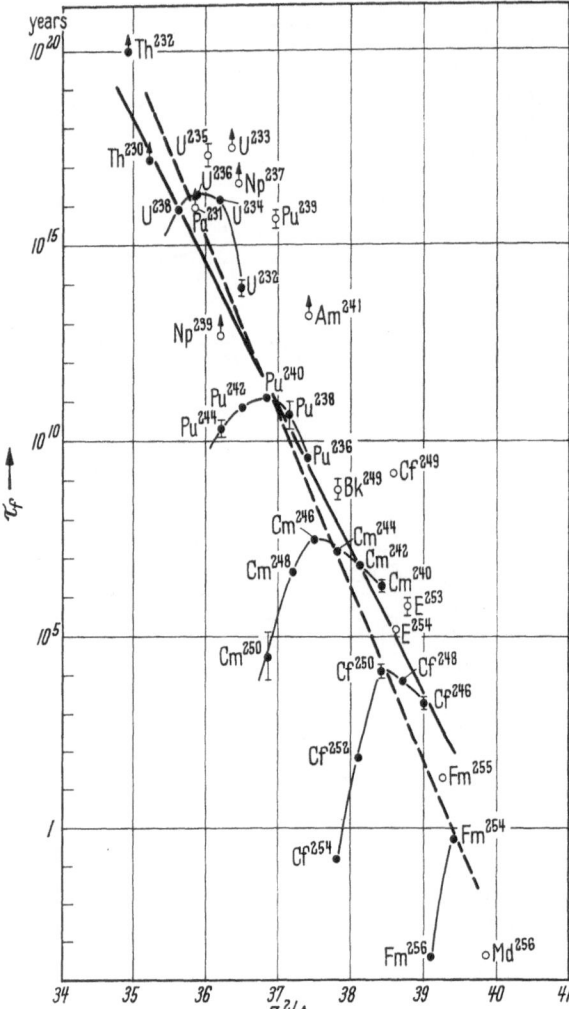

Fig. 160. Half lives of the heaviest nuclides for spontaneous fission versus Z^2/A. (After Kraut)

Also the new nuclei U^{233} and Pu^{239} are fissionable by thermal neutrons, but their cross sections differ somewhat from those of U^{235}.

We see that the process of nuclear fission is not restricted to the uranium nucleus U^{235}. Neither can it be initiated only by neutrons. Practically all nuclei can be split by appropriate experimental means. But the necessary energy varies considerably from nucleus to nucleus and depends on their different internal stability with regard to fission. First of all, all the heavy nuclei beyond uranium split spontaneously, i.e., without activation energy, though with a probability which is very small for uranium but increases with mass number (see Fig. 160). The compound nuclei produced by absorption of a neutron have very different stabilities also. Whereas the even-odd nuclides U^{233}, U^{235}, and Pu^{239} have fission cross sections of many hundreds of barns for thermal neutrons, their neighboring even-even nuclei U^{238} and Th^{232} require neutrons of higher energy. The fission of lighter elements is possible only by neutrons or α-particles of many hundreds of Mev from the large particle accelerators. This last remark implies that nuclear fission can be initiated not only by neutrons. So far, bombardment by deuterons, α-particles, and high-energy γ-radiation proved to be effective besides that by neutrons. In analogy to photoionization (page 72) and photodissociation (see VI, 7a), this last-mentioned form of nuclear fission is called *photofission*. Furthermore, nuclear fission does not exclusively produce two fragments (and some neutrons) but with a probability of 1:400, three fragments. The third fragment is frequently an α-particle, but occasionally also one with a mass comparable to that of the two other fragments.

The result that fission by thermal neutrons is asymmetric in so far as the mass ratio of the fragments is about 2:3 was at first very surprising. Fig. 159 illustrates the fact that the fission product masses scatter widely and that all elements of the Periodic Table with masses in the range from 70 to 165 occur as fission products. The frequency curve shows, however, distinct peaks at the mass numbers 95 and 140. Fission with the mass ratio 2:3 is much more probable than a disintegration into two fragments of equal mass. This asymmetry of nuclear fission is characteristic for slow neutrons only; it disappears, according to Fig. 159, for higher energies of the initiating particles. For fission of uranium by α-particles of 380 Mev, we obtain a yield curve with only one peak at 115, and for fission of bismuth by neutrons of 200 Mev or deuterons, a similar curve with a peak at about 100.

The fact that the heaviest nuclei do split is not too surprising from the theoretical point of view because the phenomenon of natural radioactivity is evidence of an instability of the nuclei near the end of the Periodic Table. If we consider the deviation from a spherical shape, which is verified for the heaviest nuclei (cf. V, 4b), it seems understandable that an appropriate energy supply leads to internal vibrations between proton groups which repel each other until, finally, the nucleus splits. The theory shows, furthermore, that the empirical value of approximately 200 Mev for the fission energy was to be expected. We arrive at this result in three different ways. First, it follows from Fig. 133 that the average binding energy per nucleon is about 7.5 Mev for uranium, but for the fission products of medium mass numbers it is 8.4 Mev. Indeed, the difference of 0.9 Mev per nucleon multiplied by the number 236 of nucleons in the uranium compound nucleus gives the correct order of magnitude of 200 Mev. The same value follows from the plausible assumption that the fission energy originates from the electrostatic repulsion of the fission fragments:

$$E = \frac{Z_1 Z_2 e^2}{(r_1 + r_2)}. \tag{5.69}$$

Here, the values 38 and 54 of the peaks in Fig. 159 have to be inserted for the charges Z_1 and Z_2 (whose sum represents the number 92 of the uranium protons); the radii r_1 and r_2 of the fragments are computed from Eq. (5.9). Finally, the fission energy can be computed from the mass balance of the reaction. The mass of the compound nucleus U^{236} produced by the absorption of one neutron is 236.052 mass units (see Table 3, page 39). If we assume it to fission into the most frequently occuring fragments of masses 95 and 139, the stable end products are the nuclei Mo^{95} and La^{139}. In order to obtain the total number 236 of the nucleons, two more neutrons have to be added. The sum of the masses of the fission products and the two neutrons is 235.829, according to Table 3. By subtracting this value from the mass value of the compound nucleus U^{236}, we find a total mass defect of 0.223 mass units, corresponding to a theoretical fission energy of 207 Mev.

We mentioned above that for nuclei with medium mass numbers the average binding energy per nucleon is 0.9 Mev higher than that of the heaviest nuclei near the end of the Periodic Table. It seems plausible to conclude from this fact that *all* heavy nuclei should be unstable with respect to fission. This leads to the question why among all the nuclei which we find in nature only U^{235} is so easily fissionable. The reason is that the fission process, though always exothermal, requires an activation energy in exactly the same way as an energetically unstable molecule of an explosive which does not decay under normal conditions before it has been activated. The binding energy of a neutron of about 7 Mev is sufficient as activation energy for the three nuclei U^{235}, U^{233}, and Pu^{239}, fissionable by slow neutrons. A higher excitation energy, however, is required for more stable nuclei; it can be provided by the kinetic energy of the impinging particle. Consistent with this picture, the activation energy varies from nucleus to nucleus, but it is always smaller for even-odd nuclei with an even number of protons and an odd number of neutrons (cf. page 273) than for even-even nuclei.

The electrostatic computation of the fission energy is evidence that fission is caused by the repulsion of groups among the 92 protons in the uranium nucleus. By applying the theory of the droplet model of the nucleus, BOHR and WHEELER were able to show that the inner "electrostatic pressure" of the protons beyond a certain critical number Z_{cr} of protons exceeds the surface tension caused by the nuclear binding forces so that the nucleus then splits spontaneously. The critical number of protons Z_{cr} follows from the relation

$$Z_{cr} > \sqrt{45\,A} \qquad\qquad (5.70)$$

where A is the mass number of the nucleus under consideration. According to (5.70), spontaneous fission of spherical nuclei should be expected beyond the atomic number 100. Deviations from the spherical shape, however, which are known to exist from the nuclear quadrupole moments, increase strongly the fission probability. From model experiments with soap bubbles whose surface tension was nearly compensated by charging them electrically, DÄNZER could show that fission of normally stable nuclei is caused by internal mechanical vibrations when excited by an activation energy. Nuclear fission as a consequence of internal vibrations, therefore, is closely related to the dissociation of diatomic molecules (see VI, 7a) caused by overexcited vibrations of the atoms oscillating against each other. Fission of nuclei and dissociation of molecules thus can be treated theoretically in an analogous way. By making use of the droplet model of the atomic nucleus we are thus able to explain nuclear fission satisfactorily.

The spontaneous fission of the heaviest nuclei (cf. Fig. 160) is undoubtedly responsible for the end of the Periodic Table. One of the long discussed fundamental problems of atomic physics was thus solved. But the breaking-off of the Periodic Table is not a sudden one since the stability of different isotopes even of the same element can vary considerably. It is therefore not surprising that nuclides of the eleven *transuranium* elements with atomic numbers 93 to 103 were discovered (cf. Table 3) among the products resulting from irradiating uranium with neutrons and heavier particles. They were isolated, just as the previously unknown elements $_{43}$Tc, $_{61}$Pm, $_{85}$At, and $_{87}$Fr, by SEABORG and co-workers with the help of microchemical methods. Their properties were often investigated from only a few micrograms of the substances, an admirable achievement of chemical experimental art.

In returning to nuclear fission, we must briefly deal with the problem of asymmetrical fission by thermal neutrons, in contrast to the apparently symmetrical fission by fast neutrons, deuterons, α-particles, or protons. MEITNER considers the particular stability of the closed nucleon shells (treated in V, 12) at least partly responsible for this asymmetrical fission. Preceding the actual fission process, a regrouping of the fragments-to-be leads to vibrations against each other. The tendency of forming closed neutron shells of 50 and 82 should begin to show its effect. Let us assume that the twelve neutrons of $_{92}$U^{236}, remaining after the directly emitted ones have left the nucleus, are distributed among the closed subgroups of 50 or 82 neutrons in equal parts and that the 92 protons are divided as 2:3 between the two fragments. Then we obtain for the mass numbers of these fragments the values 91 and 142. This is in good agreement with the experimental values 95 and 140 to be taken from Fig. 159. A quantitative statistical theory of fission by FONG, who applied Eq. (5.58) to the fragments vibrating against each other, proved this assumption to be essentially correct. Thus, at least the basic facts of the nuclear fission follow satisfactorily from general nuclear theory as treated in V, 11.

15. The Nuclear Fission Bomb and its Effects

We mentioned above that the discovery of neutron emission in nuclear fission has made it possible to liberate nuclear energy, either in the catastrophic form of the exploding nuclear bomb or controlled in nuclear reactors which can continuously produce energy. Because of this fact, nuclear and atomic physics suddenly have become the center of general interest as well as of political discussion. Unfortunately the term "atomic energy", instead of the correct term "nuclear energy", has come into common use and has caused an apparently irreparable confusion among laymen.

We begin by discussing the nuclear fission bomb, indicating by this notation the fundamental process for the energy production and thus stressing the difference to the hydrogen bomb or nuclear fusion bomb, which we shall discuss later.

Let us consider a sphere composed of pure fissionable material U^{235}, U^{233}, or Pu^{239}, unstable with respect to the capture of neutrons. If the diameter of this sphere is larger than the mean free path of the neutrons set free in these metals by nuclear fission, a single slow neutron or one spontaneous fission process will initiate an explosion. Namely, two or three neutrons ($\nu \geq 2$) are released in the average by each fission process of a U^{235} nucleus, and each of these neutrons may activate another nucleus to fission which again releases ν neutrons, and so on. Such a sequence of reactions is called a *chain reaction*. The reaction products (here neutrons) initiate further reactions so that the fission reaction spreads through the total fissionable mass by an avalanche-like multiplication. We emphasize that fission in the uranium bomb occurs by *fast* neutrons in contrast to reactors (to be treated below) where controlled nuclear fission in general is caused by thermal neutrons. For each split U^{235} nucleus, an energy of 180 Mev is directly released. Thus, if every nucleus in one kg of U^{235} could be split (which is practically impossible), an energy of about 7.5×10^{20} erg or 20 millions kwh would be freed within an extremely short period of time in a very small volume.

The first task in developing the nuclear bomb was the computation of the important nuclear data and the free path of the neutrons in the "bomb material" in order to obtain the correct dimensions for a bomb made from uranium 235 or plutonium. The bomb material, of course, is not pure at all, and the computations were based on experimental data which were sparse and obtained from very small quantities of material. The critical data are still kept secret, but HEISENBERG estimated the critical radius in U^{235} to be 8.4 cm in his "Theorie des Atomkerns" cited below. Without the application of neutron reflectors (see later), this value would require a minimal mass of about 50 kg U^{235} for a nuclear bomb. The critical mass (in any case, with reflector) actually seems to be considerably smaller.

The second condition for the production of an uranium bomb was the isolation of the fissionable U^{235}, previously achieved only in microscopic quantities. This U^{235}, which represents only 0.72% of natural uranium, had to be separated from the other isotope U^{238}. The expenditure in equipment, development work, and money necessary for this task was enormous. The official report by SMYTH (see the bibliography of this chapter) gives an idea of the problems and their actual solution in the USA where the first bomb was exploded by OPPENHEIMER,

BACHER and co-workers on July 16, 1945 near Alamogordo in the desert of New Mexico.

A further problem was how to prevent with certainty the self-ignition of the bomb before the planned moment. *For an explosion of the total fission mass must occur automatically as soon as the critical mass necessary for an explosion is united at any spot*, since uranium nuclei can split spontaneously and a sufficient number of neutrons is always in existence, for instance from cosmic radiation. Thus, self-ignition can only be prevented if the fissionable material in the bomb before its ignition is kept separately in the form of several parts of subcritical size. The ignition is then initiated by the sudden mechanical union of the subcritical parts to one piece of supercritical size. This mechanical union must be done so fast and complete that as many nuclear fissions occur as possible before the bomb explodes mechanically due to its large internal energy production (heating-up), which process would interrupt the chain reaction. American newspapers unveiled that present bombs consist of a relatively large number of subcritical masses, which are brought together in the ignition process by the explosion (actually implosion) of appropriately shaped explosive "lenses" acting concentrically toward the center. Since the continuously multiplying number of fission processes in the bomb material is stopped by the mechanical explosion of the bomb, its effect is the greater the faster the fission processes follow one another. In order to avoid the capture of neutrons by non-fissionable nuclei, extremely pure material is used and the bomb proper is surrounded by a shield of suitable material of high density. This reflector is made to scatter back at least part of the neutrons that would normally leave the fissionable material toward the outside. This shield, furthermore, retards the mechanical explosion of the bomb due to its high inertial mass.

It is not known which fraction of the fissionable material used in a nuclear bomb actually does split and therefore contributes its energy to the effects of the explosion. Probably, this fraction is far below 100%. The official report on the effects of nuclear fission bombs, cited in the bibliography, states only that the bombs dropped on Japan were equal to the effect of 20,000 tons TNT and equivalent to the complete fissioning of about one kg U^{235}. This bomb is called the *nominal bomb*, and the discussion in the report is based on this nominal bomb. More recent statements, however, indicate that the effect of modern nuclear fission bombs is a multiple of that of the nominal bomb.

The application of neutron reflectors around the fissionable material of a uranium bomb makes it possible to construct "baby" bombs. Without the neutron reflector, the radius of the uranium or plutonium mass must be larger than the mean free path of the neutrons in the fissionable material. This restriction is obviously not necessary if neutrons can be reflected back into the fissionable material. The minimal mass applicable in nuclear weapons then depends on the reflectivity or the scattering power of the neutron reflector which surrounds the fissionable material.

We should not omit to discuss the non-mechanical effects of a nuclear explosion because of their general physical interest: According to official reports, during the explosion proper about 3 % of the total energy released by a bomb are emitted as γ-radiation and another 3 % as fast neutrons. This nuclear radiation emitted by the nominal bomb would kill the majority of people exposed to it at any distance less than 1 km. But its effect declines rapidly with increasing distance from the center of the explosion (assumed to occur in free atmosphere) so that the primary nuclear radiation would not be an essential danger at a distance of more than 2 km. An additional 83 % of the total energy of the bomb appear as kinetic energy of the fission products and thus serve to heat up the central vapor mass that originally represented the bomb. The temperature obtained in this way is said to be of the order of 10^7 °K. This means that atomic physicists actually have "made" a real although short-lived small star with a temperature characteristic for the center of fixed stars. After the end of the explosion proper, this fireball, very small in the beginning, expands very rapidly so that the radiating surface increases quickly and at the same time cool off. The maximum of the heat radiation of the bomb will therefore be reached after a few tenths of a second, when the surface temperature of the fireball, now having a diameter of over 100 m, has decreased to 7000 °K, comparable to the surface temperature of the sun. Depending on the transmittance of the atmosphere, this radiation may cause extremely dangerous burns at distances up to several kilometers, i.e., far beyond the range of the direct nuclear radiation. The absorption of γ-radiation and neutrons as well as that of the short-wavelength part of the heat radiation may initiate in the surrounding atmosphere a large number of photochemical effects such as dissociation of molecules (cf. Chap. 6) and ionization of gases.

The remaining 11% of the energy released by the explosion of a nominal fission bomb become free some time after the explosion proper in the form of β-radiation or γ-radiation of radioactive fission products. Together with the radioactive decay of the radionuclides produced by (n, γ)-processes in the surroundings of the explosion center, this radiation causes the dangerous after-effects of a nuclear bomb explosion that rightly is being dreaded by mankind.

Totally different is the mechanism of the hydrogen bomb. This bomb utilizes the high temperature that is produced for a short time during the explosion of a nuclear fission bomb in order to obtain thermonuclear fusion reactions. It is not possible to simply "reproduce" the nuclear processes occurring in fixed stars (see V, 19); for these reactions are much too slow. According to the reports, not protons but nuclei richer in neutrons, such as deuterons, $_1H^2$, and tritons, $_1H^3$, are used since these unite much more easily to α-particles, be it directly or after a partial decay into protons and neutrons. The hydrogen bomb probably consists, therefore, of a conventional uranium or plutonium bomb surrounded by a large mass of heavy-hydrogen compounds and of certain elements like Li^6 which can produce tritium by (n, t)-processes. The use of an outer shell of neutron-producing matter is even more efficient in increasing the energy release of the bomb than a beryllium reflector. Quite a number of fissionable nuclei are activated to split by these additional neutrons while the bomb is still expanding, i.e., at a time when the probability of colliding with a neutron would already be small if the neutron-producing envelope were missing.

We conclude this section with a few remarks on the scientific results of the atomic-bomb experiments. First, we mention the so-called *Godiva* experiments in which nearly critical fission bombs were used. These could be made just critical for a very short time without having the energy liberation become sufficient for an actual explosion. Experiments with this bomb-like reactor arrangement led to important conclusions with regard to the fission mechanism and its temperature dependance. The extremely short pulses of 10^{16} neutrons produced in these experiments have been used for experiments on neutron physics.

Not only novel knowledge about the high atmosphere, its turbulence, and about geophysical problems was obtained from the atomic-bomb experiments, particularly by recording the shock waves. Also the elements einsteinium and fermium were discovered in bomb residues, and the mechanism of the build-up of the heaviest elements by rapidly following neutron absorptions (see V, 18) was cleared up. The subterranean American atomic-bomb explosions, in which no radioactivity reached the surface of the earth, showed how it is possible to produce unheard-of effects *without* endangering the surroundings on the surface. The extremely high energy of 7×10^{19} erg $= 2 \times 10^6$ kwh was released in one microsecond at a depth of 240 m beneath the surface. This energy produced a pressure of 7 million atmospheres and a temperature of one million degrees in a volume of 9 m³, which melted 800 tons of rock in one second into a glass-like mass containing nearly all the radioactivity. No wonder scientists as well as engineers are seriously pondering how these enormous effects and the 10^{24} neutrons produced in a microsecond might be utilized for peaceful scientific and technical purposes.

16. The Liberation of Useful Nuclear Energy in Fission Reactors

The first experimental fission reactor was put in to operation by FERMI on December 2, 1942, in Chicago. In contrast to a bomb, the reactor must operate under stationary conditions and produce an adjustable power; the number of nuclear fissions per second must therefore be carefully kept constant. The reactor fuel is a more or less concentrated, in most cases heterogeneous mixture of an easily fissionable material (U^{235}, U^{233}, or Pu^{239}) with U^{238} or Th^{232} whose nuclei cannot be split by slow neutrons. The fission process is controlled by means to be discussed below so that the number of neutrons desired per unit time initiates new fission processes. A certain percentage of the remaining neutrons is absorbed by U^{238} or Th^{232} (which may be located in a shield around the reactor core) and transforms them to Pu^{239} or U^{233} according to Eqs. (5.67) and (5.68); the rest is either captured by absorbing materials necessary in the reactor or escapes to the outside.

If the *multiplication factor k* is defined in the usual way as the ratio of the neutron flux densities at the end and at the beginning of a "generation", it is obvious that only a reactor with a multiplication factor $k \geq 1$ can be operated steadily. On the other hand, k must be kept slightly above 1 during adjustment to the desired power and then must be kept exactly equal to 1. This is in contrast to the bomb for which a multiplication factor as high as possible is desired. Since, in the average, each fission sets free 2.5 neutrons, the value of the multiplication factor depends on the fraction of these neutrons that, in the average, is absorbed in the uranium or lost in the other reactor materials or, finally, diffuses out of the reactor without initiating new fissions. Even in an infinitely large quantity of pure natural uranium, for instance, k always remains below one so that a chain reaction cannot be maintained because the large majority of primary fast neutrons is absorbed by U^{238}, forming Pu before these neutrons are able to initiate fission of further U^{235} nuclei and thus produce new neutrons.

Yet there are two ways to obtain a "critical" reactor with $k \geq 1$. First, a reactor can be operated with *slow or fast* neutrons if the absorption of neutrons in U^{238} is essentially diminished by using uranium highly enriched with U^{235}. Secondly, a natural-uranium reactor with $k \geq 1$ can be obtained if its uranium rods are imbedded in a slowing-down substance called the "moderator". Here the fast neutrons which are produced in nuclear fissions are slowed down by elastic impacts to energies below 6.7 ev (see page 287), i.e., below the energy region in which they are subject to strong absorption by U^{238}. This slowing-down should occur before the neutrons have a chance of colliding with U^{238} nuclei. We see that in such a "heterogeneous thermal reactor" the fission processes are induced by slow thermal neutrons with their large fission cross section. The slowing-down substance of the reactor must be a scattering material (see V, 13 c) of small atomic weight so that as high a percentage of the neutron energy as possible is transferred in every impact; i.e., the material must have a high scattering power. Its absorption of neutrons, on the other hand, must be as small as possible so that valuable neutrons do not unnecessarily get lost in the scattering process. Particularly suitable moderator materials therefore are deuterium in the form of heavy water (D_2O), carbon (graphite) of highest purity, and finally highly pure beryllium or its oxide. Whereas the fast neutrons produced in the fission process are slowed down to thermal velocity of some 10^5 cm/sec by about 25 collisions in D_2O, they need about 100 nuclear collisions in graphite or beryllium. The largest reactors in operation today use graphite moderators because of their smaller cost; other reactors use heavy or light water. We shall show below that the latter ones have the advantage of smaller size and higher neutron flux density. However, if inexpensive light water is utilized as slowing-down substance, the loss of neutrons by absorption has to be compensated by using expensive enriched fuel.

A multiplication factor $k > 1$ can thus be obtained for a thermal reactor by appropriate arrangement and sufficient size. This factor is given, according to a very simple theory, by

$$k = \varepsilon p f \eta L. \tag{5.71}$$

Here ε is the so-called *fast-fission factor*. It states by what fraction the number of primary fast fission neutrons (first-generation neutrons) is increased due to the fact that a few of these fast neutrons initiate new fission processes of U^{235} and U^{238} nuclei. Reactors with natural uranium have an ε of about 1.03. The factor p is the so-called *resonance escape probability*; it is the fraction of the primary fast neutrons which is slowed down to thermal energy without getting lost by resonance absorption in uranium. For well-designed natural-uranium reactors, p has a value of about 0.9. f measures the percentage of neutrons slowed down to thermal energy which is absorbed in uranium and not in the moderator or other reactor materials; its value is also approximately 0.9. On the average, each of the $\varepsilon p f$ thermal neutrons absorbed in uranium produces, by thermal fission, η new fast neutrons, with η being 1.32 for natural uranium but 2.08 for pure U^{235}. Although $\nu = 2.5$ neutrons are primarily produced in each fission process, η is much lower. One reason is that thermal neutrons in natural uranium are also captured by U^{238}. Furthermore, not all thermal neutrons absorbed by U^{235} nuclei cause fission processes, but 20% of them initiate (n, γ) processes and thus do not produce new neutrons. An infinitely large reactor would thus have a multiplication factor

$$k_\infty = \varepsilon p f \eta \quad \text{(``four-factor formula'')}. \tag{5.72}$$

For the actual reactor of finite size, however, k_∞ must be multiplied with the "non-leakage factor" L,

$$k_{\text{eff}} = k_\infty L, \tag{5.73}$$

since the neutron fraction $(1-L)$ that leaks out of the reactor surface is lost for carrying on the chain reaction.

Thus we see: *With natural uranium, because of its small η-value of 1.32, high enough values of p and f for k to be larger than one can be reached only by the most favorable arrangement of fuel elements and moderator and by using reactor materials with small neutron absorption.* Furthermore, since the reactor surface responsible for the leakage loss increases with increasing size less than its volume, there exists a "critical" reactor size (depending on its shape) below which the reactor cannot become critical because of too high neutron leakage. If enriched uranium with its higher η value is used, the critical reactor size is correspondingly smaller than that of reactors using natural uranium.

The factors in the expression (5.71) for k depend on temperature in a complicated manner. Furthermore, during operation of a reactor not only fissionable material is consumed but neutron-absorbing fission products are formed also and are accumulated in the reactor. Therefore, the values of k_{eff} and $(k_{\text{eff}}-1)/k_{\text{eff}}$, called "reactivity", depend on temperature and the burn-up rate of the fuel elements. In general, k decreases with increasing temperature and operation time (a fact that is important for reactor power plants with high operating temperature!). It is therefore necessary to plan for a new cold reactor a considerable excess reactivity (up to 20%, corresponding to $k_{\text{eff}}=1.20$). The value $k_{\text{eff}}=1$ for stationary operation can then be reached by varying the insertion depth of rods or strips of neutron-absorbing cadmium or boron which absorb the undesired excess neutrons in the reactor. By further insertion of control rods, k can of course be decreased below 1, and the reactor can thus be shut down. This can be done, for instance in an emergency, by an automatic release of boron steel rods suspended electromagnetically above openings in the reactor.

There is one important point left to be discussed. The control mechanism as described does not seem fast enough for the rapid sequence of fission processes so that we could expect the energy increase to destroy the plant within fractions of a second if nature itself had not provided the required delay. Namely, about 0.75% of the neutrons emitted in fission of U^{235} are not "immediately" emitted during the fission process proper, but with an average delay of about 10 seconds because some fission products are not able to get rid of their excess neutrons by β-transformations fast enough (cf. page 286). Since the multiplication factor of every reactor is one or very little above one, each increase of this factor, and thus of the energy production, is ultimately based upon the effect of these *delayed neutrons*. Reactor power thus increases, after a first rapid growth, slowly enough so that it can be compensated by automatically controlled insertion of absorbing rods into the reactor.

From the energy of about 180 Mev immediately released in each fission process we can easily compute with the help of the energy relation (3.10) that the production of 1 watt of power requires the fission of 3×10^{10} nuclei per second. The energy production per sec, i.e., the power of a nuclear reactor, is proportional to its volume V, to the average density N of the fissionable nuclei per cm³, to their fission cross section σ for neutrons, and to the mean flux of neutrons nv per cm² and sec in the reactor, with n and v designating the average density and velocity of the neutrons. Consequently, the power of a nuclear reactor measured in watts is given by

$$L \text{ [watt]} = \frac{nvN\sigma V}{3 \times 10^{10}}. \tag{5.74}$$

The fission cross section of U^{235} for thermal neutrons is 580 barns, but for fast neutrons it is smaller by a factor up to 10^3.

Eq. (5.74) represents the power of a critical reactor as a function of its volume, of the concentration of the fissionable material, and of the neutron flux density; but it does not indicate anything about the critical size of the reactor. This size depends on k_∞ and the so-called *migration length l*, which is the average path of a neutron in a specific reactor between its origin and its absorption. For the radius R_c of a just critical spherical reactor we have the relation

$$R_c = \frac{\pi l}{\sqrt{k_\infty - 1}}. \tag{5.75}$$

Actually, a reactor is always built somewhat larger in order to have excess reactivity available for control purposes and in order to compensate for the decrease of the fissionable material and the production of neutron-absorbing fission products during the reactor core life.

If natural uranium is used as fuel, the multiplication factor k_∞ is so slightly above one, particularly with graphite as moderator, that only extreme purity of the materials used in the reactor provides sufficient excess reactivity. Natural uranium graphite reactors therefore require a much larger volume (and more uranium) than those with D_2O as moderator, while reactors with nearly pure U^{235} are the smallest ones.

In order to reduce the neutron loss due to radial leakage, the reactor core is surrounded, similar to the fission bomb, by a *neutron reflector*. It consists of graphite, light or heavy water, beryllium, or its oxide. This reflector offers also a possibility for controlling the "fast" reactor, which operates on fast neutrons without any moderator. It is possible to regulate, by a controlled movement of a reflector section, the radial neutron loss and thus indirectly the number of fissions per second, i.e., the reactor power. As a second possibility for controlling a fast reactor, a fixed reflector may be used and the reactivity may be varied by controlled insertion of a fuel rod without which the reactor would not be critical.

Difficulties and details of design vary considerably for research reactors and reactors for plutonium production or energy production. While the produced heat is a useless byproduct for the first two reactor groups and therefore is removed at temperatures of only 50° to 100 °C, we want heat of a temperature as high as possible for a power reactor. This heat is used for producing steam for a turbine-generator. In developing power reactors, it is a very difficult task to master technologically the behavior of structural materials with respect to corrosion and radiation damage at temperatures of 300° to 800 °C and to safeguard the surroundings against the high radioactivity of the reactor.

Research reactors are required for all experiments with neutrons of high flux density, e.g., for neutron diffraction, for the investigation of the behavior of materials under neutron bombardment or γ-irradiation, and also for the production of radionuclides, which are becoming increasingly important for science and technology. These radionuclides accumulate in the reactor as fission products, or they may be produced by neutron irradiation. They are chemically separated and shipped to research laboratories all over the world. The thermal power of research reactors lies between fractions of a watt and about 10,000 kilowatts, materials-test reactors have a power between 10,000 and 200,000 kilowatts. They consist mostly of rod-shaped or, because of the better heat removal, of strip-shaped subdivided fuel elements of natural or enriched uranium. The fuel is enclosed in vacuum-tight capsules (in order to avoid corrosion and to contain the radioactive fission products) and imbedded in graphite or heavy or light water. In addition, the reactors have rods of neutron-absorbing material which serve for controlling and shutting-down the reactor; finally they have a cooling system. Fig. 161 shows schematically the layout of a small research reactor. All reactors are surrounded by the neutron reflector and by the so-called biological shield of about two meters concrete for radiation protection, which consists mostly of heavy concrete. Openings in this shield, some of which reach into the core of the reactor, allow neutron beams to leave the reactor. Many types of irradiations and measurements with neutrons can thus be carried out outside of the reactor (cf. Fig. 162). On the other hand, materials to be irradiated can be inserted into the reactor through the beam holes and there be exposed to the full radiation flux. Whereas about 30 tons of natural uranium are required for a graphite research reactor, only a few tons of natural uranium are needed for a heavy-water reactor and less than one kilogram of U^{235} for the smallest training reactors. The neutron flux density nv, which is decisive for research reactors, lies between 10^7 and 10^{13} per cm² sec for standard research reactors, but it reaches values up to 10^{15}. These flux densities are important especially for testing materials where the effects of slow and fast neutrons as well as of γ-radiation on the structure of solids and liquids are to be investigated. Interesting cross relations

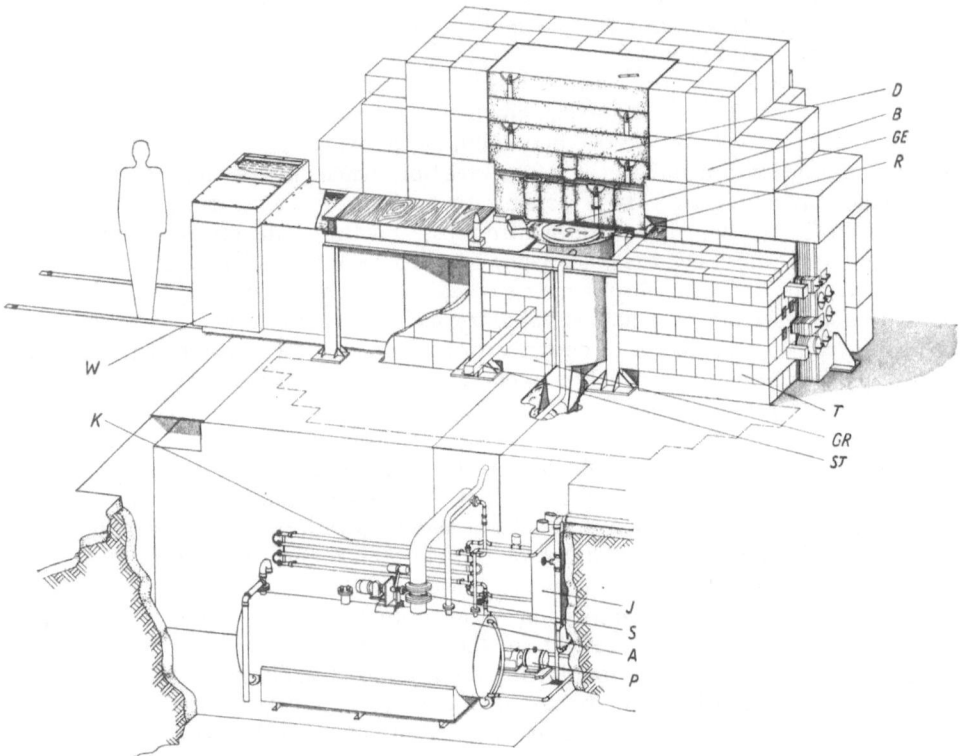

Fig. 161. The Siemens-Argonaut research reactor. $R=$ ring core with sheet elements of 20% enriched uranium; $GE=$ internal graphite column; $B=$ concrete-block shield; $D=$ removable cover; $T=$ thermal graphite column; $GR=$ graphite reflector; $St=$ beam tube for neutron experiments; $J=$ ion exchanger; $S=$ dump valve; $A=$ dump tank; $P=$ cooling-water pump

between nuclear physics and solid-state physics (see Chap. VII) come to bearing in this field.

The development of reactors for power plants, which are characterized by a high temperature of the cooling substance and are used for the production of electricity or for ship propulsion, is still in its infancy. So far, only heterogeneous thermal reactors have stood the test. They are operated either with light water as moderator and coolant, but with enriched uranium as fuel because of the neutron absorption of water, or with graphite or D_2O as moderators, CO_2 or D_2O as coolants, and the less expensive natural uranium as nuclear fuel. Also organic liquids have become of interest as moderators and cooling substances; finally, sodium is being tested as a cooling substance for a number of reactors. It is obvious that in *all* reactors operating with natural or slightly enriched uranium U^{238} is transformed into plutonium, which later may be chemically separated and may be utilized for manufacturing new fuel elements.

Beside these reactors operating with thermal neutrons, the "fast reactor" mentioned above may, in the long run, play an important role as a so-called *breeder reactor*. Since the absorption cross sections of most construction materials of a reactor are very small for fast neutrons, it is possible by an appropriate arrangement to have the majority of the neutrons which are not consumed in fission processes leave the reactor core and produce fissionable Pu^{239} or U^{233} in a shield of U^{238} or Th^{232} surrounding the reactor. In the average, 2.5 neutrons are released per uranium fission, but only one of these is required for carrying on the reaction,

Fig. 162. General view of the large Canadian Chalk-River research reactor NRX operating with heavy water. With experimental arrangements. (Courtesy Atomic Energy of Canada Limited)

i.e., for a subsequent fission process. Therefore, the possibility exists in principle to produce, by means of a fast reactor, *more fissionable material from unfissionable U^{238} or Th^{232} than is simultaneously consumed for power production*. In this case, we speak of a breeding process in a *breeder reactor*, while a reactor is called a *converter* if less fissionable material is produced than is simultaneously consumed by fission. The first fast breeder reactor, built in 1951 in U.S.A., succeeded in proving this breeding process feasable with a breeding factor above one. Very probably it will also be possible for certain types of thermal reactors to produce more fissionable material than is consumed simultaneously. But the conversion factor of the *thermal breeder reactor* will be only slightly above one because of the relatively high absorption cross section of most materials for slow neutrons.

The slow and fast reactor types treated above are called "heterogeneous" reactors because fuel, coolant, and moderator (if it is used) are separated from each other in the reactor. In contrast to this, the *homogeneous reactor* uses fuel and moderator as a homogeneous, in most cases liquid, mixture. This mixture may be pumped through a heat exchanger for removing the heat produced in the reactor. In the simplest case, such a homogeneous reactor consists of a sphere of steel sheet with a diameter of only 30 cm, filled with an aqueous solution of nearly pure fissionable uranyl sulphate. The dissolved uranium atoms, whose total mass may be less than 800 g for this reactor, represent the fuel elements, whereas the hydrogen and oxygen atoms surrounding them in the solution act as moderator. In small research reactors of this type (so-called *water boilers*), the produced heat is removed by an inserted cooling coil. But in power reactors, the whole solution, which of course is highly radioactive, is pumped through the heat exchanger. Instead of a solution, an aqueous suspension of UO_2 dust may be used.

Aside from its great difficulties due to the highly radioactive coolant as well as due to corrosion and erosion, this reactor type has also great advantages. By using a liquid fuel, the technically complicated and expensive manufacture of fuel elements is eliminated, and so is the neutron absorption due to the cladding material. The second decisive advantage of an aqueous homogeneous reactor is its excellent *self-stabilization*. If, namely, the power of the reactor should increase because of a failure in the control system (which latter is not even necessary), the density of the fuel solution would decrease because of the thermal expansion and the formation of vapor bubbles. This would cause a decrease of the power production so that such a reactor never can "run away". Finally it should be possible with this reactor type to continuously eliminate the radioactive fission substances produced and also the fissionable material bred during the operation, whereas it is difficult and expensive to exchange the fuel elements of heterogeneous reactors and have them later chemically reprocessed.

We cannot discuss here the detailed technical problems connected with the development of entire nuclear power plants, e.g., the spatial separation of radioactively contaminated and of uncontaminated regions of the plant, and the safe storage, refining, or removal of large quantities of radioactive waste. But it may be pointed out that the many technical problems of the various types of power-plant reactors are often very different from one another so that experiences with one type can be transferred to the operation of another type to a very limited extent only.

The liberation of atomic energy for technical purposes, as treated so far, was based exclusively upon the nuclear fission process. Less than one thousandth of the mass of the fissionable nuclei is transformed into energy. In realizing this, we ask at once whether we do not have more efficient processes for the liberation of nuclear energy, i.e., whether it is not possible to transform a larger part, if not the total mass of nuclei, into energy. This latter question has obsessed physicists and engineers since the beginning of nuclear physics. For a while, the question was seriously discussed whether the huge energy of the cosmic ray particles (see V, 20) might originate from such annihilation processes of protons, helium nuclei, and heavier nuclei. This assumption, however, proved to be wrong. We shall learn in V, 21 that the complete annihilation does occur, in agreement with theoretical expectations, for pairs of positive and negative electrons or other elementary particles whose combined mass can actually be completely transformed into radiation energy. But it is hardly to be expected that such processes can be technically utilized.

Fig. 150 shows, however, that besides the energy release by fission of the heaviest elements the reverse process to fission, i.e., the fusion of relatively light nuclei to heavier ones which have a higher average binding energy should be possible also. This too is an exothermal process, equivalent to the chemical "fusion" of carbon and oxygen to CO_2 with release of heat in our furnaces. We shall learn in V, 18 that the energy production in our sun and in most of the other stars is due to this reaction, and, in particular, due to the fusion of hydrogen into helium. Therefore, we shall come back to the interesting problem how to utilize these nuclear fusion reactions after we have treated those nuclear reactions that are important for astrophysics (see V, 19).

17. Applications of Stable and Radioactive Isotopes

The development of mass-spectroscopic and other methods of isotope separation (cf. II, 6d) made the production of usable quantities of pure or highly enriched stable isotopes possible. As a consequence of the development of nuclear reactors, hundreds of radioactive nuclides have become available in quantities that are sufficient for many types of application. Some of these radioactive nuclides can be separated from radioactive fission products by chemical methods, others are produced by neutron irradiation of appropriate elements in reactors. The application of isotopes in science, medicine, and industry solved such a great number of formerly unsolvable problems that it seems necessary to present at least a brief survey of this most recent application of nuclear physics. With respect to all details we refer to the literature cited at the end of this chapter.

The basis of all isotope applications is the fact that a certain atom can be distinguished from others of the same element by its mass (for stable isotopes), or by its β-radiation and often γ-radiation (for radioisotopes). If, for example, heavy water D_2O is added to the drinking water of an animal, deuterium is later found in its body fat. This serves to prove a hydrogen exchange between the water drunk by the animal and the fat of its body, a proof that could hardly have been established otherwise. Another example may show the application of radio-nuclides: A thin layer of radioactive iron is evaporated upon the surface of an iron block and, after a certain period of time, the surface is abraded layer by layer. If we then measure the activity from deeper and deeper layers of the iron block, we may determine the self-diffusion of iron in iron by this so-called *tracer method*, a measurement that would obviously be very difficult without such a tracer.

The advantage of using stable isotopes is the fact that they do not decay and thus exist for an unlimited period of time. But they have the disadvantage that their detection requires the complicated method of mass spectroscopy. The stable isotopes D^2, C^{13}, N^{15}, and O^{18} play nevertheless an important role in biological investigations. One reason is that radioactive isotopes of the important elements oxygen and nitrogen with a life time long enough to carry out precise experiments do not exist. The advantage of using radionuclides is obviously their easy detection by counters (treated in V, 21) or by the photographic plate, which allows, by means of the autoradiography, to obtain directly the distribution of the radioisotope, e.g., in a leaf, or a bone, etc. The disadvantages of using radioisotopes are, however, in many cases their limited life time, the difficulty and sometimes also danger in handling them, and the possible falsification of the results by decay products. The last-mentioned difficulties are alleviated by the extremely small quantity of radioisotopes which is required for a measurement.

According to V, 6b, the number of β-particles emitted per second by N atoms of half life τ is given by

$$\frac{dN}{dt} = -\lambda N = \frac{0.693\, N}{\tau}.$$

(5.76)

Since an activity of four emissions per second can easily be detected by a Geiger counter, the number of radioactive atoms necessary for a measurement is

$$N = \frac{4\tau}{0.693} \approx 6\tau. \qquad (5.77)$$

From Eq. (5.77) it follows that only 10^{12} atoms or 2×10^{-11} grams of radiocarbon C^{14} with a half life of 5360 years [produced from N^{14} in a (n, p) reaction] are required for its detection, whereas even 10^7 atoms or 4×10^{-16} grams suffice for detecting radiophosphorus P^{32} with a half life of 14.1 days, an isotope that is equally important for biological and medical investigations.

Only a few particularly interesting or important examples can be given from the enormous number of results which have been brought to light within a few years by means of the isotope methods. The calibration of the meter in wavelengths of a sharp spectral line was previously accomplished with the red Cd line: This was unsatisfactory because of the hyperfine structure of the line. Today a red Kr^{86} line is used since the necessary quantity of Kr^{86} for filling Kr lamps could be produced by isotope separation. The meter is thus defined, independent of the further fate of the original meter in Paris, as $1,650,763.73$ wavelengths of this Kr^{86} line.

Radiocarbon C^{14} with its convenient half life of 5360 years offers an interesting possibility for the age determination of organic matter after LIBBY. Neutrons from cosmic rays continuously transform a small number of N^{14} atoms in the atmosphere by (n, p) reactions into radioactive C^{14}. These carbon atoms are absorbed by plants in the form of CO_2 and, when the plants are eaten by animals, the C^{14} is embodied by the animal. An equilibrium is thus established so that, in the average, 1 gram of living substance has an activity of 12.5 β-particles per minute due to its radiocarbon content. As soon as the carbon exchange with the atmosphere comes to an end, e.g., in dead wood or bones, this C^{14} activity decreases according to the C^{14} half life. From measuring the C^{14} activity we thus are able to determine the time since the specific organic matter left the cycle of life. With this method the age of wood, for instance, from an Egyptian Pharaonic tomb was determined to be 4500 years, a result which was in best agreement with archeological data.

Tracer methods have also proved useful for the elucidation of the mechanism of chemical reactions. In photosynthesis, for instance, green plants absorb CO_2 from the atmosphere and H_2O from the soil and transform these molecules to starch and free oxygen under the influence of the sunlight that is absorbed by the chlorophyll. The unstable intermediate products of this most important biological reaction were determined with radioactive tracers, whereas the application of O^{18} helped in proving that the released oxygen, which is exhaled by the plants, originates from absorbed H_2O and not from decomposed CO_2. It would have been hard to find this result by any other method.

The speed of metabolism in living organisms was also determined by tracer methods. Examples are the speed of the hydrogen exchange between body water and body fat or that of the nitrogen exchange between the amino acids of proteins incorporated with food and the body proteins. These results are of highest physiological interest. Similar investigations have clarified to a large extent the specific role of nitrogen and iron in and for our blood.

Isotope research may well have found its most sensational applications in the field of medicine, for which it is of particular interest that radiosodium, radiophosphorus, and numerous other radioisotopes, are not, in contrast to radium, elements foreign to the body and therefore are assumed to cause less dangerous side-effects. The fact that the injection of radioactive sodium chloride into a vein of one arm causes the appearance of radiosodium in the perspiration of the other arm only 75 seconds later proves again the enormous velocity of the metabolism of the body. The Geiger counter enables us to follow, point by point, the spreading of the injected radiosodium in our body. Inhibitions of the circulation can thus be determined and localized. Certain elements, in stable or radioactive form, are deposited with preference at specific places or organs of the body, an observation which is of great significance for medical diagnosis and therapy. For instance, the preferred deposition of iodine in the thyroid gland allows the treatment of the superactivity of this organ. In certain cases, even cancer of the thyroid gland can be treated by the γ-radiation of radioiodine received with the food. A brain tumor can hardly be localized from the outside; it is even difficult to determine its boundary toward the healthy brain tissue during an operation. Therefore the preferred deposition of fluorescein in the tumor is used when a fluorescein compound containing radioiodine is injected and the γ-radiation of the radioiodine that penetrates the scull is used for localizing the tumor. Later, during the operation, the short range of the β-radiation of radiophosphorus, which is also deposited with preference in the tumor, is used for determining its exact boundaries. Furthermore, it is possible to influence the superproduction of red blood particles in a certain blood disease by depositing radiophosphorus in the spinal cord. Finally, malignant tumors are treated today more and more with radioactive cobalt Co^{60} instead of the expensive radium which has also other disadvantages. Co^{60} can be directly inserted, in needle-form, into the tumor.

If we consider that the isotope methods gain a steadily widening field of application such as in the study of technological processes of all types, in chemical industry and in metallurgy as well as in agriculture (e.g., the study of the atom exchange between fertilizer and plant), we see what a powerful instrument for science, medicine, and technology has grown out of atomic physics.

18. Thermonuclear Reactions at Very High Temperatures in the Interior of Stars. The Origin of the Elements

Apart from the spontaneous radioactive decay of the heaviest elements, which cannot be influenced at all, the nuclear reactions considered so far were initiated by collisions of the nuclei with artificially accelerated nucleons. What is, on the other hand, the mechanism of the usual chemical reactions? These too can be initiated by accelerated particles (i.e. atoms, ions, electrons); but mostly they occur in "thermal equilibrium". If, e.g., oxyhydrogen gas is heated, the molecules are gaining, according to the Maxwellian energy distribution, so much energy that they may dissociate other molecules by collisions until the reaction (in this example, the explosion) begins. This comparison justifies the question whether there exist also *thermonuclear reactions* in which the fast initiating particles have their kinetic energy due to the sufficiently high temperature of the gas.

There are indeed such thermonuclear reactions. But it is easily seen that the necessary temperatures must be *very* much higher than those that, so far, can be produced on earth. Because of the much higher "activation energy" necessary for the intrusion of the colliding particles into the nucleus and the initiation of nuclear reactions than that required for chemical reactions, we expect temperatures of the order of 10^7 to 10^8 degrees Kelvin.

Such temperatures occur in the interior of the sun and of stars, as astrophysicists have found. ATKINSON and HOUTERMANS (in 1929) and, in more detail, v. WEIZSÄCKER (in 1936) were the first physicists to point out that exothermal thermonuclear reactions must occur in stars. In these reactions they found large amounts of energy to be set free, and it seemed possible to solve, by means of such reactions, the problem of the origin of the continuously emitted energy of the sun. This assumption was indeed confirmed. Moreover, it is believed today that nuclear reactions in the interior of very hot stars may be the source of continual heavy-element build-up.

From the equation

$$\frac{m}{2} v^2 = \frac{3}{2} k T \tag{5.78}$$

we see that the average kinetic energy $m v^2/2$ of a particle at the temperature of 1.4×10^7 degrees in the center of the sun is equivalent to only 2,000 ev. In spite of this low value of the mean particle energy, compared with the millions of ev of particles accelerated in our laboratories, it is possible that they initiate a sufficient number of nuclear reactions per second for two reasons: Because of the Maxwellian velocity distribution, there is always a small number of particles with a kinetic energy far higher than the mean thermal energy. Furthermore, the very large volume of the stars leads to such a high number of collisions that even reactions with a small probability occur with sufficient frequency. Since the number of particles of sufficiently high velocity as well as the probability of entering the nucleus increase rapidly with temperature, the number of nuclear reactions per sec and cm³ increases rapidly too. This means, in the chemist's language, that the thermonuclear reactions have a high temperature coefficient.

The mechanism of the thermonuclear reactions at the sun's temperature of 1.4×10^7 degrees, or at the even higher temperatures of many giant stars, can be computed with reasonable accuracy from the experimentally known yields of most nuclear reactions and from the half lives of radioactive nuclides. Starting with the plausible assumption that the first stars originated from condensation of hydrogen, we may conclude that occasionally deuterons $_1\text{H}^2$ were being formed in collisions between protons at temperatures of the order of 10^7 degrees. The excess energy and charge is emitted in the form of a positron and a neutrino. The deuterons then react with other protons and form He^3. From here on, we have two possibilities according to (5.79). Either two $_2\text{He}^3$ nuclei collide and form a compound nucleus which immediately decays into He^4 and two protons, or the He^3 nuclei build up Be^7 nuclei in collisions with previously formed He^4 nuclei. These Be^7 nuclei decay by positron emission into Li^7 nuclei, and the Li^7 nuclei react with protons and form the unstable Be^8 nuclei which decay into two He^4 nuclei:

$$_1\text{H}^1 + _1\text{H}^1 \rightarrow _1\text{H}^2 + e^+ + \nu$$
$$_1\text{H}^2 + _1\text{H}^1 \rightarrow _2\text{He}^3$$

$$\left. \begin{array}{l|l} _2\text{He}^3 + _2\text{He}^3 \rightarrow _2\text{He}^4 + 2\,_1\text{H}^1 & \begin{array}{l} _2\text{He}^3 + _2\text{He}^4 \rightarrow _4\text{Be}^7 \\ _4\text{Be}^7 \qquad\quad \rightarrow _3\text{Li}^7 + e^+ + \nu \\ _3\text{Li}^7 + _1\text{H}^1 \quad \rightarrow 2\,_2\text{He}^4 . \end{array} \end{array} \right\} \tag{5.79}$$

The balance of this reaction cycle results in the fusion of four protons to a $_2\text{He}^4$ nucleus with the simultaneous emission of two positrons and 2 neutrinos.

$$4\,_1\text{H}^1 \rightarrow _2\text{He}^4 + 2\,e^+ + 2\,\nu \tag{5.80}$$

or

$$2p \rightarrow 2n + 2\,e^+ + 2\,\nu . \tag{5.81}$$

As we know, the reaction (5.80) is strongly exothermal. Energy is set free since the mass of the four protons is 4 times 1.00723, which is larger by 0.02741 mass units than the originating He^4 nucleus whose mass is 4.00151. Therefore, in this reaction the large energy of 25.5 Mev per He nucleus is liberated which corresponds to 1.5×10^8 kcal/gram or 6×10^8 kcal/mole. It is furthermore of interest that the neutrinos set free during the reaction (5.80) are responsible for about 10% of the energy flux from the sun. We do not notice, however, the high neutrino flux of about 10^{11}/cm^{2}sec at the surface of the earth because of the small absorption cross section for neutrinos.

In the sun and most of those stars in which the existence of C^{12} nuclei was proved spectroscopically, another reaction cycle is possible which was first studied by BETHE. Again, He^4 nuclei are formed from four protons:

$$\left.\begin{array}{l} {}_6C^{12}+p \rightarrow {}_7^*N^{13} \\ {}_7^*N^{13} \quad \rightarrow {}_6C^{13}+e^++\nu \\ {}_6C^{13}+p \rightarrow {}_7N^{14} \\ {}_7N^{14}+p \rightarrow {}_8^*O^{15} \\ {}_8^*O^{15} \quad \rightarrow {}_7N^{15}+e^++\nu \\ {}_7N^{15}+p \rightarrow {}_6C^{12}+{}_2He^4 . \end{array}\right\} \qquad (5.82)$$

Here the normal ${}_6C^{12}$ nucleus reacts with a proton to produce the positron-active ${}_7^*N^{13}$ nucleus, which decays under positron emission to ${}_6C^{13}$. This nucleus again reacts with a proton and the stable ${}_7N^{14}$ is formed which, in its turn, reacts with another proton forming a radioactive ${}_8^*O^{15}$ nucleus that is transformed to ${}_7N^{15}$ under positron emission. Finally, ${}_7N^{15}$ reacts with a fourth proton and decays, by a (p, α)-reaction, into a stable nucleus ${}_6C^{12}$ and an α-particle ${}_2He^4$. *It is the surprising and important result of this reaction cycle that the initial nuclei ${}_6C^{12}$ are not used up but are set free again.* In the chemist's language, we have a reaction catalyzed by carbon nuclei since these, as a catalyst, come out of the reaction unchanged. Reaction cycles similar to (5.82) can be started with the nuclei O^{16} and Ne^{20}. *The result of computations based on the reaction cycles (5.79) and (5.82) is that the transformation of four protons into one helium nucleus produces the total energy which is continually emitted by the sun,* and that the hydrogen content suffices to guarantee the life of the sun for about 10^{11} years.

It seems to be certain that also in the other normal stars, i.e., in those of the main branch of the Hertzsprung-Russell diagram, the energy production is essentially maintained by the reaction sequences (5.79) or (5.82) or by both together. In the cooler stars, reaction sequence (5.79) seems to prevail, in the hotter ones, sequence (5.82) predominates. In the sun, both seem to participate to about equal parts.

According to recent investigations by FOWLER, the production of C^{12} nuclei necessary for the sequence (5.82) is supposed to happen in the interior of very hot giant stars at a temperature of about 10^9 degrees. Here, two He^4 nuclei react to produce unstable Be^8. But from experimentally confirmed considerations it is inferred that the equilibrium concentration of Be^8 is high enough to produce C^{12} in collisions of Be^8 with a third α-particle in an (α, γ) reaction. Also the other nuclei that are exclusively composed of α-particles, such as O^{16}, Ne^{20}, ... up to Ca^{40}, can be formed at temperatures of about 10^9 °K. Since the nuclei C^{13}, O^{17}, and Ne^{21}, which are rich in neutrons, are produced in the reaction cycle (5.82) and in the corresponding ones with O^{16} and Ne^{20}, (α, n) reactions with these nuclei are possible after GREENSTEIN and BURBRIDGE. In these reactions,

$$
\left.
\begin{aligned}
&C^{13} (\alpha, n) \ O^{16}, \\
&O^{17} (\alpha, n) \ Ne^{20}, \\
&Ne^{21} (\alpha, n) \ Mg^{24},
\end{aligned}
\right\}
\tag{5.83}
$$

free neutrons are produced. Thus, one more puzzle of nuclear astrophysics apparently has been solved, since free neutrons are indispensable for the build-up of higher nuclei by (n, γ)-reactions.

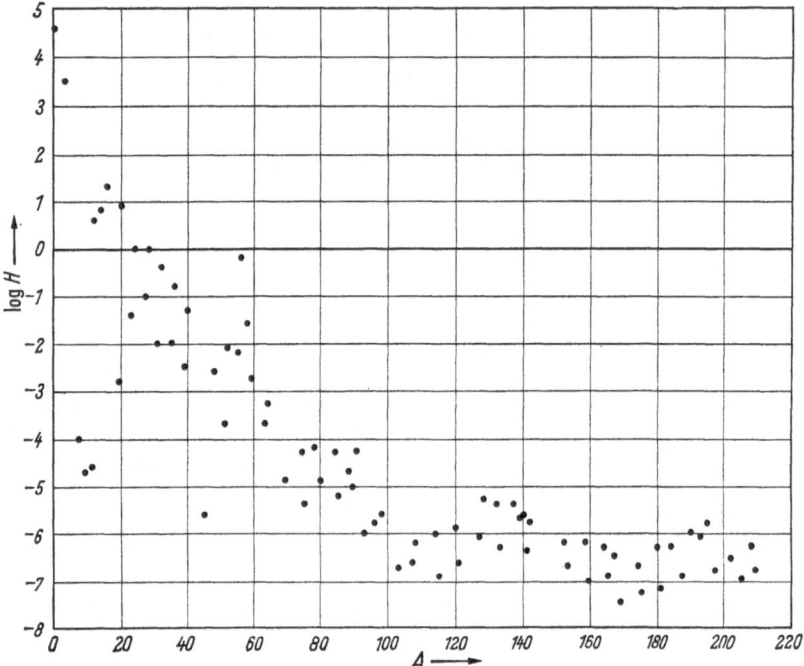

Fig. 163. Logarithm of relative cosmic abundance of stable nuclides (referred to Si=1) versus mass number. (After SUESS and UREY)

The considerations described above about thermonuclear reactions in the interior of stars throw some light on the cosmologically significant problem of the origin of the elements in their present frequency distribution, which agrees for the earth, the sun, and most of the stars. A conspicuous fact of this distribution (see Fig. 163) is the high percentage of hydrogen and helium, which together comprise 99% of the matter of the universe (hydrogen alone about 80%), while the heavy elements above the iron group contribute only 10^{-6}% to the total number of atoms. The problem is to understand how this distribution is compatible with the build-up of the elements from protons and neutrons (and, of course, the electrons of the atomic shells).

All previous theories were based on the assumption that the composition of heavy nuclei is not possible any more in our present universe. Accordingly, this build-up was supposed to have occurred at a moment shortly after what could be considered as the moment of "birth" of our present universe. This birth could be dated back about 7×10^9 years by measuring the age of radioactive elements, and by astrophysical observations such as the red-shift in the spectra of spiral nebulae and the expansion of the universe inferred from it, or from the dynamics of spherical star clusters. Three different theories have been discussed. The equilibrium theory assumes that the elements were formed from protons, neutrons, and electrons according to the laws of chemical equilibrium at a very high initial temperature of the order of 10^{10} °K. The equilibrium attained at that temperature was then assumed to have "frozen" by the rapid temperature decrease caused by the expansion of the original plasma cloud. Such an equilibrium theory is apparently able to explain by and large the abundance distribution of the light elements up to the minimum in Fig. 150. But it proved impossible to find any suitable temperature and pressure that would yield at the same time the correct distribution of the heavy elements of low abundance. Finer details of the abundance distribution of the lighter elements (cf. Fig. 163) cannot be explained either. MAYER and TELLER therefore assumed that the heavier elements originate from fission of a cold primary neutron fluid with subsequent β-decay and neutron evaporation. This is by no means a satisfactory assumption. GAMOW instead assumed a very dense neutron cloud of high temperature as the primary state of the universe. The formation of the elements in their present distribution was supposed to have occurred within a few minutes by decay of neutrons into protons and electrons, and by multiple absorption of neutrons with subsequent β-decay. Certain consequences of this hypothesis seem to agree remarkably well with experience, for instance the fact that the cosmic abundance of neighboring nuclear species depends not so much on their intrinsic stability than on their effective cross section for neutron absorption. However, the non-existence of elements with masses 5 and 8 is a serious difficulty for this theory; and the same is true for the build-up of the heaviest nuclei.

The fundamental assumption of all these theories, namely that of the production of all the elements in their present abundance about 7×10^9 years ago, was decisively shattered by MERILL's observation that certain stars emit spectral lines of the unstable element *technetium* whose isotope of longest life has a half life of little above 2×10^5 years only. This fact proves that *even today heavy elements are formed in certain stars.*

A new theory of the origin of the elements was developed by HOYLE and FOWLER who take into account the new results mentioned above about thermonuclear reactions in giantstars in connection with recent astronomical discoveries. The fundamental ideas of this theory are as follows: The initial state of the universe is assumed to have been a homogeneously distributed gas of hydrogen atoms or of protons and electrons. Statistical fluctuations of the density lead to star-like condensations by means of gravitational forces. These become gradually stars because of the gravitational energy liberated during the contraction, and the interior of the stars heats up until thermonuclear reactions begin. Helium is then first produced from protons by the mechanism (5.79). As soon as a sufficiently large inner sphere of helium has been created that is free of hydrogen, the energy production ceases. Renewed gravitational contraction then leads, for sufficiently large stars, to a central temperature of the order of 10^9 degrees. At this temperature, carbon nuclei are formed from helium nuclei and finally, under further contraction and at temperatures of 4×10^9 degrees, also the higher elements up to the middle of the Periodic Table (Fe, Co, Ni, i.e., the minimum of the curve of Fig. 150) are produced in thermal equilibrium. By neutron production from (5.83) and by neutron attachment with subsequent β-decay, even the elements up to lead may be created. The heaviest nuclei of the radioactive families, however, cannot be built up in this way because their step-by-step creation would be prevented by the short half life of some of these nuclei. Therefore, radium and uranium can have been formed only in regions with neutron densities high enough so that the *stepwise absorption of numerous neutrons could occur in a period of time short compared even to the shortest half lives of its radioactive decay products*. The discovery of the californium isotope $_{98}Cf^{254}$ in the fall-out of the hydrogen bomb on Bikini is an interesting proof for the possibility of such a process. Its nuclei can have been formed only by an extremely rapid attachment of not less than 16 neutrons to an uranium nucleus of mass 238 with a subsequent sixfold β-decay. There is much evidence for the assumption of very high neutron densities in the explosive stages of stars known as supernovae just as in the explosion of atomic bombs. In these regions of high neutron density even the heaviest nuclei of the Periodic Table can have been formed and may still be formed. The scarcity of such events would explain the low cosmic abundance of those nuclei.

The reaction cycle (5.82) has not been mentioned in FOWLER's theory of the origin of the elements because we discussed so far only the formation of stars from pure hydrogen. An example of a "pure hydrogen star" was recently studied extensively by UNSÖLD. It is known that during the relatively frequent nova explosions of such "stars of the first generation" large masses of stellar matter in which the normal distribution of the elements has already been established, are hurled into interstellar space where they may serve for the formation of "stars of the second generation". In these stars of the second generation, nuclei of carbon, oxygen, and neon already exist so that the fusion of hydrogen into helium may now occur also via the carbon-nitrogen cycle and other similar cycles which were not yet possible in the stars of the first generation.

We cannot deal here with all the details of these astronuclear theories and the experimental investigations confirming them. But it is of interest to recognize how microcosm and macrocosm here come into contact and how the elucidation of a single nuclear reaction such as that of the formation of carbon from helium permits us to understand the origin of the distribution of the elements, which is of such utmost importance for our whole universe.

19. Problems of Future Energy Liberation by Nuclear Fusion

Since we know what enormous energies are continuously liberated in the sun and the stars by fusion of hydrogen into helium, it is tempting to physicists to try to imitate this nuclear process in *fusion reactors*. They thus hope not only to get rid of all raw-material problems but also to develop power plants which will yield only harmless helium as waste instead of the radioactive fission products of the normal fission reactors. A more detailed investigation shows, however, that the problems encountered in a terrestrial fusion reactor differ from those in the sun in two very essential points. Since there do not exist any material walls that could hold together a plasma of many hundreds of millions degrees, we have to make use of non-material walls for containing a fusion plasma. For this purpose, appropriate magnetic fields are being used which contain the hot plasma so that it cannot reach the relatively cool walls (so-called magnetic bottle). This means that the magnetic pressure of such fields, directed towards the interior, must be larger than the plasma pressure which is very high at the extraordinarily high plasma temperature, even for a small gas density. Calculation shows that it is possible, with fields from 20,000 to 50,000 gauss, to reach magnetic pressures from 15 to 100 atmospheres. The gas density that corresponds to this pressure at the plasma temperature is $10^{14}-10^{15}$ particles/cm^3. This is, however, many orders of magnitude below the gas density in the sun. If we, nevertheless, want to obtain a power output of some 100 megawatts from a power plant of reasonable size (corresponding to an energy density of the order of at least 1 w/cm³) in spite of this small plasma density and the correspondingly small number of collisions per second that will lead to fusion reactions, the plasma temperature must be much higher than that of the sun.

Calculations for such a low plasma density show that fusion of protons into helium requires a plasma temperature far above 10^9 °K for the reaction

$$4p \rightarrow {}_2\text{He}^4 + 2e^+ + 2\nu + 25 \text{ Mev} \tag{5.84}$$

that occurs in the sun. Such a temperature seems to be unattainable. Terrestrial nuclear fusion experiments, therefore, start from heavy hydrogen $_1\text{H}^2$, which exists, as a practically unlimited supply, in the oceans. It would be a very inexpensive raw material inspite of the high cost of its production by isotope separation because the energy liberated in fusion is so very large. The nuclear reactions leading to fusion of heavy hydrogen would essentially be the following ones, taking the place of (5.79) and (5.82),

$$
\left.
\begin{aligned}
\text{H}^2 + \text{H}^2 &\nearrow \text{He}^3 + n + 3.25 \text{ Mev} \\
&\searrow \text{H}^3 + p + 4.0 \text{ Mev} \\
\text{H}^2 + \text{He}^3 &\rightarrow \text{He}^4 + p + 18.3 \text{ Mev} \\
\text{H}^2 + \text{H}^3 &\rightarrow \text{He}^4 + n + 17.6 \text{ Mev}.
\end{aligned}
\right\} \tag{5.85}
$$

The gross reaction is the transformation of three deuterons into one He⁴ nucleus, one free proton, and one free neutron:

$$3\,\text{H}^2 \rightarrow \text{He}^4 + p + n + 21.6 \text{ Mev}. \tag{5.86}$$

The neutrons thus produced serve as indicators for the occurrence of such reactions in heavy hydrogen. They may also be utilized in future large industrial plants.

In principle, two different procedures for operating fusion reactors might be visualized. First, it might be possible to pulse an electric discharge in heavy hydrogen of appropriate density for short periods of time so that a temperature might be reached which would be sufficient for fusion reactions. This was already experimentally achieved to a small extent in the laboratory. The power thus liberated in the form of neutrons, protons, and γ-radiation might then be utilized. Among other possibilities, the direct inductive transformation of the plasma currents occurring in such machines into electric energy has been discussed. But it seems doubtful whether economical power (i.e., a sufficiently large excess of power output over the power used by the plant itself) can be obtained by such pulsed operation. The second possibility corresponds to the oxyhydrogen gas burner, in which the heat of combustion can be controlled by the supply of hydrogen and oxygen. Such a continuous nuclear fusion burner would have to be supplied continuously with as much heavy hydrogen as would be required for the production of the desired power by fusion to helium. It would seem difficult, however, to control such a plant against a "run-away", which would lead to self-destruction of the plant with its grave consequences. A prerequisite for such a continuous operation of a fusion reactor would be, of course, that the operating temperature be above the *critical temperature* at which the energy production just compensates the radiation losses of the plasma. Otherwise, the fusion reactor would shut down automatically.

Unfortunately, the operating temperature of such a continuous fusion reactor with heavy-hydrogen fuel must be 10^9 °K, as is shown by computation.

Small additions of the expensive tritium H^3 to the heavy hydrogen H^2 would lower the operating temperature considerably, however, and a mixture of equal parts of D_2 and T_2 as fuel would lead to a critical temperature of only 50×10^6 °K. Apart from all technical difficulties, it seems very doubtful whether an economical operation can be expected with such an expensive fuel in competition with that of the then highly developed fission reactors.

In spite of much effort and impressive scientific results, nuclear fusion for the present time still belongs definitely to the domain of the research physicist. His efforts are concentrated on the study of the methods for heating up hydrogen plasmas of low gas density and on the stationarity problems of such plasmas under the simultaneous action of magnetic and electric fields and the plasma flow caused by them (magnetohydrodynamics). Since after the heating-up we are dealing with completely ionized plasmas of temperatures comparable to that in the center of stars (20×10^6 degrees seem to have been reached for short periods), this research work is of high interest for astrophysics as well as for the yet little explored plasma state of matter.

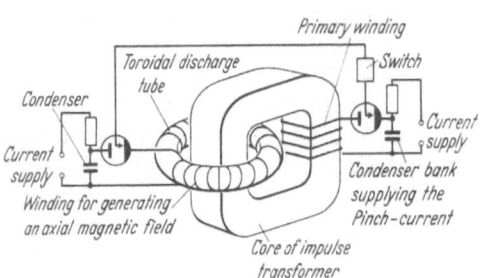

Fig. 164. Schematic arrangement for producing and heating-up a toroidal plasma column for fusion experiments. (After Riezler and Walcher)

So far, there are two basically different methods by which one tries to obtain the high temperature required for fusion reactions. Both methods start from magnetically constricted discharges, either between electrodes or in electrodeless ring discharges. The linear discharges, which in the beginning were predominantly used, are primarily of scientific interest. Temperature increases up to several million degrees are obtained by current pulses which cause the discharge to be compressed by its own magnetic field (pinch effect). These compressions often have the character of shock waves. Technologically these pinch discharges have the disadvantage of a strong cooling effect by the electrodes. Ring discharges don't have electrodes and permit a heating-up by electric plasma currents inductively generated or by their Joule heat (cf. Fig. 164). By excitation of plasma oscillations or by adiabatic compression, one tries to obtain increasingly higher temperatures of the plasma. Finally, energy may be fed in by periodically varied magnetic fields, again causing a heating-up. The second method mentioned above is different in principle (see Fig. 165). Here, powerful ion beams of high kinetic energy are injected into the discharge plasma. Their energy originates from an external accelerator and is transferred to the plasma, thus heating it up. For all details, including also the heating-up by linear shock waves in "shock tubes", we refer to the literature quoted on page 341.

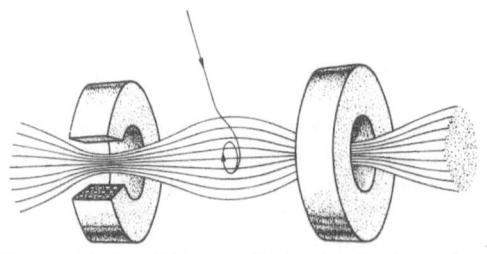

Fig. 165. Injection of high-energy D_2^+-ions into the plasma of an arc, for heating it up to fusion temperature. (DCX experiment, Oak Ridge)

Because of its fundamental interest, we finally mention the so-called cold fusion by means of the μ-mesons which will be treated in V, 23. These μ-mesons form, together with deuterons, hydrogen-like atoms which have a 206 times smaller diameter than normal hydrogen because of the μ-meson's mass of 206 electron masses [cf. Eq. (3.18)]. In a mixture with light hydrogen, these very small mesonium atoms of heavy hydrogen then form HD molecules with normal H atoms. The very small distance of proton and deuteron caused by the μ-meson in these peculiar molecules often leads to fusion under formation of a $_2\text{He}^3$ nucleus. An energy of 5.4 Mev is set free in this process. This energy may lead to the emission of another μ-meson by the He^3 nucleus; a chain reaction would thus become possible. A technical exploitation of this reaction seems, nevertheless, to be improbable because of the difficulty of the production and the short life time of the μ-mesons.

20. Collision Processes of Highest Energy

a) The Primary Particles of the Cosmic Radiation

The processes in and near the nucleus which we discussed so far were initiated by collisions in which the colliding particle had energies between 10^6 and 10^8 ev. Entirely new and fundamentally significant information about the transformation and creation of matter and radiation was obtained from experiments on radiation coming from outer space, the so-called *cosmic radiation*. Measurements have proved that primary cosmic-ray particles with a kinetic energy up to 10^{20} ev corresponding to 10^{11} times the rest mass of the nucleon do occur.

Only slowly, by cooperation of observation, experiment, and theory has the abundance of complicated primary and secondary cosmic-ray phenomena been analyzed, phenomena first observed by HESS and KOHLHÖRSTER in the years after 1911. We cannot go into any details and want to restrict ourselves here to the results pertaining to primary cosmic radiation and to those elementary processes occurring between particles of highest energy which are definitely proved today. By "highest energy" we mean particle energies well above 10^8 ev so that in most cases the collision energies even of the heaviest primary particles exceed their rest energies $m_0 c^2$ by orders of magnitude. We are thus really dealing with the most extreme physical processes which we know.

Only in recent years has conclusive evidence been obtained about the primary particles which fall into the earth's atmosphere from outer space and which cause the abundance of complicated cosmic-ray phenomena. Observations with rockets and artificial satellites have considerably contributed to the results known today. The energy density of the primary radiation is practically identical with that of the optical radiation from the stars. It seems to be certain today that electrons are not contained in the "primary component" of the cosmic radiation, and high-energy photons do not play an essential role either. It was, moreover, definitely proved that the main constituents of this primary component are high-energy protons, whose energy distribution over the range from 10^9 to 10^{17} ev is shown in Fig. 166. Fig. 166 gives also information about the absolute numbers of primary protons. Furthermore, recent observations leave no doubt that also α-particles and, with much lower abundance, even heavier nuclei with atomic numbers up to 30 and more occur in the primary cosmic radiation. It is interesting to note that the composition of the primary component agrees rather well with the abundance distribution of the elements in the universe, except for a relatively too high abundance of the rare elements Li, Be, and B, which probably is caused by secondary processes in interstellar space.

These results are of great significance for the question about the origin of the cosmic radiation, i.e., about the source of the tremendous energy of the primary particles. Only a short time ago, the opinion seemed to prevail that the immense acceleration of the primary particles was impossible in our universe as known today so that the primary component of the cosmic radiation had to be regarded as a remnant from an earlier stage of the universe. But this hypothesis has become untenable through the result mentioned above that the primary component contains composite nuclei, since the high temperature of such an original explosion would exclude the existence of composite nuclei. Recently, therefore, the formerly discussed idea has found increasing support that the acceleration is due to vortical magnetic fields, comparable to the acceleration in a betatron (cf. V, 3), and that it is still occurring. This hypothesis is supported by BABCOCK's disco-very that some stars have a magnetic field varying periodically in strength and sign and they thus resemble huge betatrons.

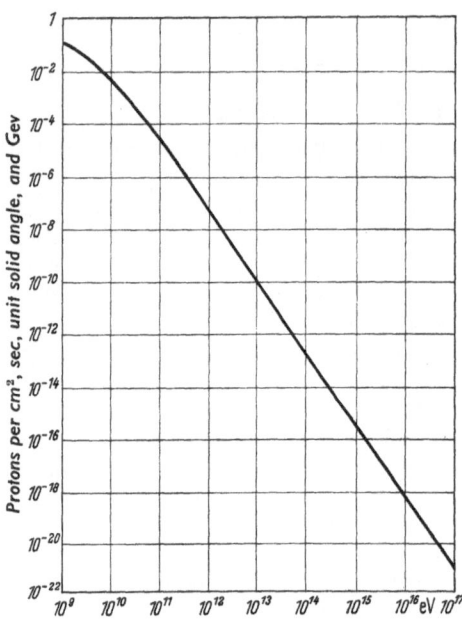

Fig. 166. Energy distribution of primary cosmic-ray protons (so-called differential spectrum), extrapolated to the surface of the earth's atmosphere. (After LOHRMANN and SCHOPPER)

But in the sun and, probably, in most of the stars there exist also *local* magnetic fields by which charged particles may be accelerated. They originate, for example, in the large plasma eruptions from the interior of the sun, known as "flares", since in these flares electric currents result from the different mobilities of elec-trons and ions in the vortical plasma motion. The magnetic fields on the sun, whose existence was proved by measurements of the Zeeman effect (see III, 16c), can apparently acceler-ate protons only up to about 10^9 ev. But according to UNSÖLD, there is a sufficient number of stars in our galaxy with a turbulence larger by so many orders of magnitude that it does seem possible to cause an acceleration of the observed number of primary particles to the highest energies mentioned. According to a theory of FERMI, primary cosmic-ray particles may also be accelerated by the immensely extended magnetic fields of low field strength in interstellar space, which are connected with the motion in ionized gas clouds. According to this theory, the particles passing through these clouds gain, in the statistical average, more energy than they lose. Apparently, there is no lack of possible acceleration mechanisms, but we cannot yet decide between them.

One other surprising observation about the primary cosmic radiation is the apparent non-existence of high-energy electrons, although electrons are pre-sent all over the universe for space-charge compensation and, therefore, should also be accelerated to high energies. Thus we must probably assume that they are quantitatively eliminated before they reach the earth or the upper boundary of its atmosphere. If this does not happen by the terrestrial magnetic field itself, whose effect will be discussed below, collisions with solar photons of relatively low energy may come into play. In these collisions, the electrons may transfer energy and momentum to the photons in an inverse Compton effect (cf. IV, 2).

It is difficult to draw clear conclusions from the distribution of the primary cosmic radiation *observed* at high altitudes to that actually *occurring* in outer space around the earth because of the action of the terrestrial magnetic field. This field has hardly any influence on particles with a kinetic energy above about 10^{10} ev ($=10$ Gev). The much more numerous primary particles of lower energy, however, are deflected by the terrestrial magnetic field perpendicular to their direction of motion and to the direction of the field lines so that they may be completely prevented from reaching the earth. This deflecting effect, which falsifies the energy spectrum of the primary particles, is of course weakest near the poles and strongest near the equator of the earth because of the directions of the magnetic field lines at different latitudes. This action of the terrestrial magnetic field may have quite unexpected effects, such as the broad belts of very intense cosmic radiation recently discovered by VAN ALLEN with the aid of the American satellites in altitudes of 1000 and 20,000 km. Evidently, its particles are captured and held by the focussing action of the terrestrial field in a similar way as the electrons or protons on their magnetically determined orbits in betatrons and synchotrons. Although our knowledge about the primary cosmic radiation made great progress in recent years, there are many of its problems still to be solved.

b) The Secondary Processes of the Cosmic Radiation

It is very difficult to analyze quantitatively the observed cosmic radiation. This is due first to the action of the terrestrial magnetic field, which changes the radiation. Furthermore, the composition of the cosmic radiation is fundamentally changed in the transition from higher to lower layers of our atmosphere by a multitude of secondary particles, produced in collisions of primary particles with atomic nuclei in the uppermost atmosphere, and by their subsequent processes. In the highest layers, protons predominate, as we mentioned before, and to a small degree there exist also heavier nuclei which have lost their shell electrons by stripping collisions. But even in these highest layers, the kinetic energy of each primary particle is subdivided among a large number of secondary cosmic-ray particles by collisions of the high-energy primary particles with nuclei of atmospheric molecules. We shall discuss these interesting processes in somewhat more detail.

The binding energy of 8 Mev of the nucleons in a nucleus is negligible compared to the kinetic energy of the colliding primary protons so that the nucleus can be considered, in a first approximation, as a spherical cluster of independent nucleons. The bombarding proton, therefore, effectively collides only with the few nucleons in its direction of impact. Thus, the nucleons which are hit directly and, through these, a small number of secondary nucleons are thrown out of the nucleus, gaining high values of energy and momentum. At the same time a considerable number of free π-mesons is produced by the interaction of the colliding particle with the π-meson clouds (cf. V, 25) of the nucleons. To a smaller degree, also heavier mesons, hyperons, and antiparticles are produced, which we will discuss later. If the bombarding particle is not a proton but a heavier nucleus, it may be regarded as a group of bombarding nucleons of equal velocity, whereby the number of collisions and thus the number of secondary ejected nucleons and mesons is correspondingly increased. Figs. 167 and 168 show photographs of the tracks of such collisions. Narrow showers of nucleons and mesons can be seen, which are caused by the transfer of momentum from the bombarding to the ejected particles. In addition, a relatively large number of heavier tracks with isotropic angular distribution can be recognized. We interpret them as follows: After the ejection of a number of nucleons and mesons, the bombarded nucleus remains in a highly disordered, i.e., highly excited, state; it can be regarded as a highly heated-up compound nucleus. As a consequence, a number of protons and neutrons, occasionally even larger nuclear fragments, evaporate from the target nucleus. The ejection of these subsequently evaporating nuclear disintegration products occurs with an isotropic angular distribution.

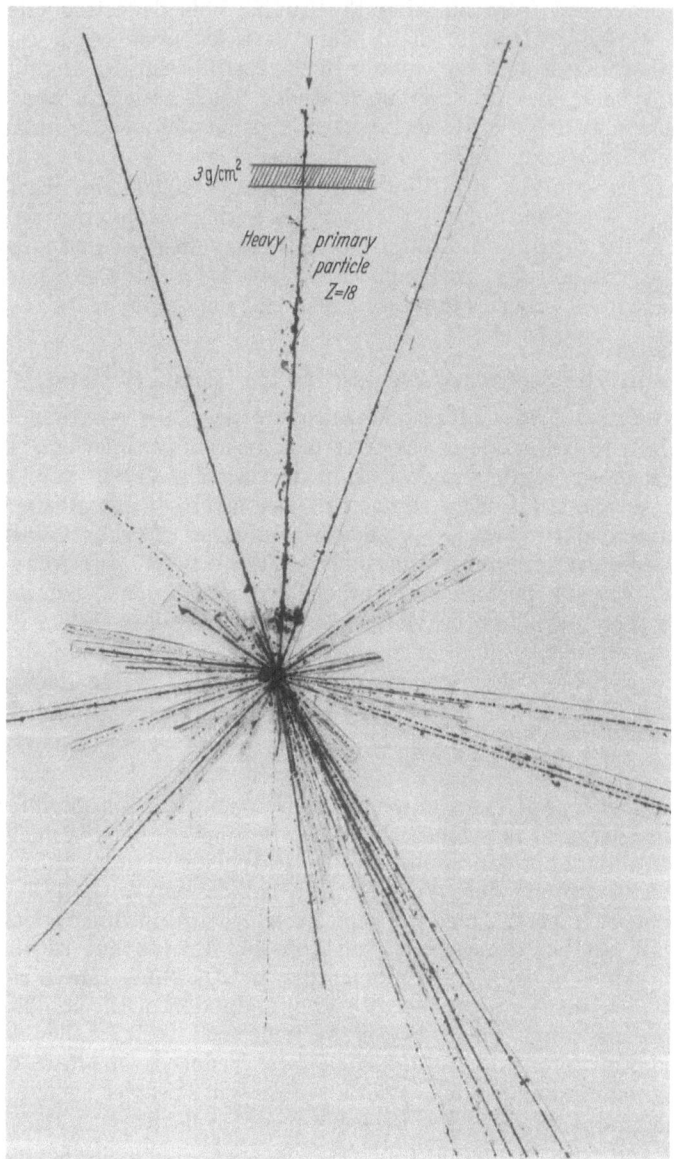

Fig. 167. Nuclear explosion caused by an extremely energetic primary cosmic-ray particle of atomic number 18. Photograph courtesy of L. Leprince-Ringuet

The ejection of all these particles from the primarily hit nuclei is not yet the end of the secondary processes. In particular, the secondary nucleons and mesons ejected with high momentum are able to produce additional free nucleons and mesons in collisions with further atmospheric molecules as long as their kinetic ener-

gy is above approximately 10^9 ev. Thus the primary cosmic-ray particle generates
an entire *cascade of secondary nucleons and mesons.* The protons of the cascade

lose further energy by ionizing collisions until they come to rest, while the neutrons are finally attached to N^{14} nuclei where they produce N^{15} by (n, γ) processes or by (n, p) processes C^{14}, i.e. radiocarbon. We discussed already in V, 17 the significance of this radioisotope for age determination problems.

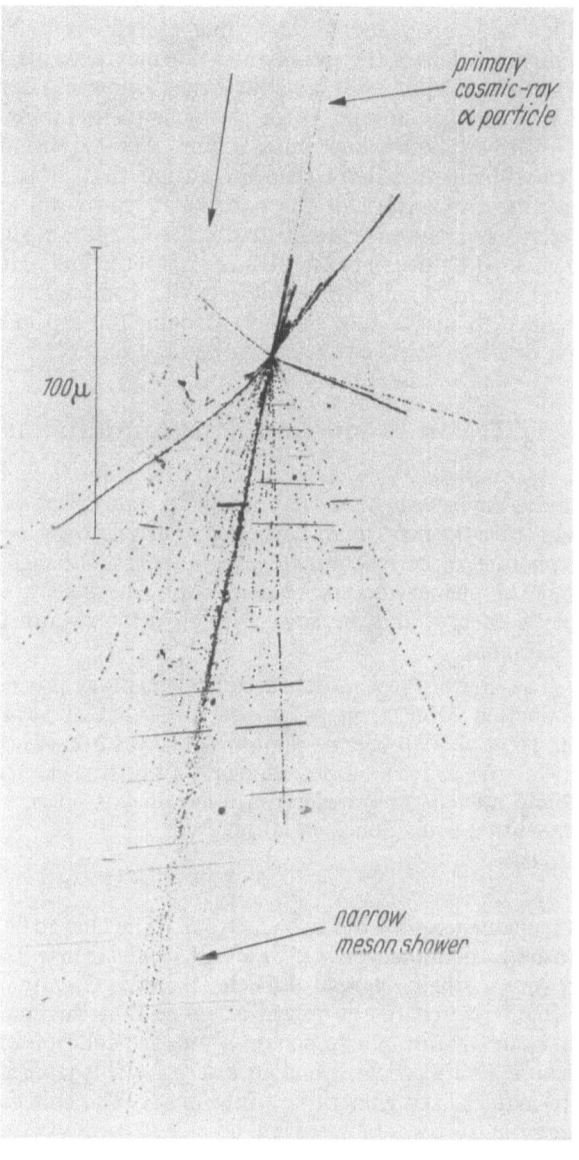

The nucleon and meson component of the secondary cosmic radiation is finally supplemented by a large number of lighter secondary particles.

The neutral π^0-mesons disintegrate, according to V, 23, very rapidly into two extremely energetic photons (γ-quanta). These in turn generate electrons and positrons of high energy by processes to be dealt with below. Moreover, the charged π-mesons decay to a considerable extent into the somewhat lighter μ-mesons and neutrinos. Therefore, we find at atmospheric altitudes below 20 km predominantly electrons of both signs, photons, and μ-mesons. The μ-mesons, which have a very high energy and little interaction with atomic nuclei, can penetrate large layers of matter without an essential loss of energy. They reach, as the *penetrating component* of cosmic radiation, the surface of the earth nearly without absorption, in contrast

Fig. 168. Microphotograph of the explosion of an Ag or Br nucleus in a photographic emulsion at high atmospheric altitude, caused by a primary cosmic-ray α-particle with an energy of 10^{13} ev. At least 18 heavy and 53 light particles (mesons and electrons) can be noticed in the nearest neighborhood of the star. Among them there is a number of extremely energetic electron pairs, whose generating photons (see p. 316) probably come from the decay of neutral π-mesons (cf. p. 323). Photograph courtesy of M. F. KAPLAN, B. PETERS, and H. L. BRADT

to the electrons. Here they comprise about 80% of all observed cosmic particles. Even at depths of several hundreds of meters below the surface of the earth or below the water level of lakes can they be detected.

The multiplicity of interactions, still to be treated, between the elementary particles has an enormous effect on the cosmic radiation observed in the lower layers of the atmosphere. This is best demonstrated by the example of the so-called *large air showers*. According to present knowledge, a single high-energy primary nucleus of the cosmic radiation may produce, by a complicated interchange of subsequent processes, an avalanche ("shower") of up to 10^{10} secondary particles of low energy, most of them electrons. The major part of these is reabsorbed at altitudes between 10 and 20 km. The extension of such a particle shower is clearly illustrated by the observation that, in extreme cases, their last traces reaching the surface of the earth may cover an area of several square kilometers. For the measurement of these large air showers, the entire area has to be covered by instruments with central recording. However, before we can understand the mechanism of the electronic component of the cosmic-ray showers, we have to discuss yet another fundamental aspect of matter, the anti-particles, first detected for electrons in cosmic-ray observations.

21. Pair Production, Pair Annihilation, and Anti-Matter

A completely new aspect of matter came to light in particle collisions of a kinetic energy so high that its mass equivalent $m = E/c^2$ corresponds to more than twice the mass of an elementary particle, or as a consequence of the absorption of photons of correspondingly high energy: Elementary particles together with their complementary *anti-particles* may be generated from radiation energy or kinetic energy, and, inversely, they can be re-transformed ("annihilated") into radiation.

The first of these most startling elementary processes of modern physics is the production of electron pairs, first observed in 1934 in the cloud chamber (see Fig. 169). A high-energy photon with $h\nu > 1.02$ Mev $= 2m_e/c^2$ is transformed into an electron pair, i.e., a negative and a positive electron, in the electric field of an atomic nucleus or an electron. The excess energy appears here, according to EINSTEIN's equivalence equation

$$h\nu = 2m_e c^2 + E_k, \tag{5.87}$$

as kinetic energy of the two electrons. The charge balance shows that the photon can be transformed only into a pair of electrons: The uncharged photon cannot produce a single charged particle; it may generate, however, a *pair* of particles having equal charge of opposite sign so that the charges compensate each other. This *pair production* can occur only in the collision of the photon with a particle because a strong field seems necessary for this transformation of radiation energy into materialized energy, i.e., into mass. The collision partner takes over excess energy and excess momentum.

The discovery of pair production is fundamentally significant in many respects. *It proves for the first time that production of matter from energy actually occurs in nature; the theory of relativity had predicted this as a possibility.* The equivalence equation is hereby quantitatively verified. Since the inverse process of the pair production, i.e., the transformation of an electron and a positron into radiation energy, viz., into 2 or 3 photons of an energy according to Eq. (5.87), was discovered a little later, quite generally mass particles can be produced from radiation energy and vice versa. We must therefore get accustomed to the idea of *mass as a specific form of (materialized) energy, beside the mechanical and electrical energy. The factor c^2 of* EINSTEIN's *equivalence equation is, in this sense, the mass-energy equi-*

valent, which allows the conversion of mass into energy or vice versa in the same way as the mechanical heat equivalent relates mechanical energy to thermal energy.

The positron which appears here as partner of the normal electron has the same mass but the opposite charge and, with reference to the vector of the mechanical spin, also the opposite magnetic moment as the electron. Therefore it is called the *anti-particle* of the electron. Similar anti-particles were found during the last years to all important elementary particles, i.e., to the neutrino, the mesons, the nucleons proton and neutron, as well as to the hyperons. *The general and fundamentally important phenomenon of anti-matter thus exists beyond any doubt.*

An energy of more than 2 Gev is required for the production of nucleon-antinucleon pairs. In 1955/56, the Segrè group actually discovered the negative proton (antiproton) by means of the bevatron (mentioned on page 228), which was designed and built for just this purpose. The negative proton as the anti-particle of the proton has the opposite charge as the proton and therefore, with respect to the vector of the mechanical spin, the opposite magnetic moment. The antineutron was first produced from an antiproton which gave up its charge in colliding with a nucleus. It has no charge, just like the neutron, but must disintegrate into a negative proton and a positron and, therefore, has the opposite magnetic moment as the neutron, again with reference to their spin directions.

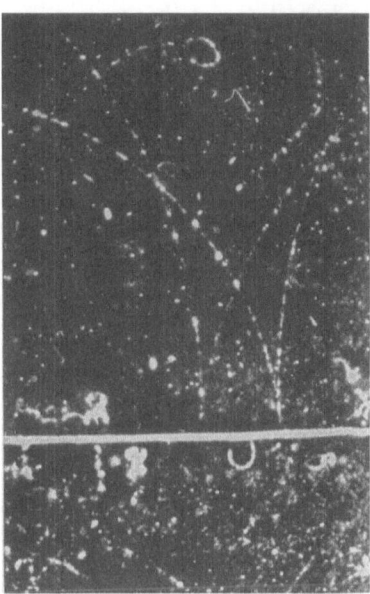

Fig. 169. Cloud-chamber photograph of the production of two electron pairs by high-energy photons (hard γ-radiation of 17.6 Mev). Photographed in a magnetic field of 2500 gauss by Fowler and Lauritsen

We come back to the problem of annihilation of particle-antiparticle pairs and will treat first the electron-positron pairs. As most other recombination processes, the annihilation of electron and positron occurs much easier between slow particles than between fast ones since these pass each other without sufficient interaction. A fast positron of whatever origin, which penetrates matter, therefore first loses its kinetic energy by collisions, and only then it interacts strongly with an electron. Because the tow particles have equal charges of opposite sign they "capture each other" and revolve around their common center of gravity for a short time. They thus form a stationary system comparable to that of the hydrogen atom; it is called *positronium atom*. Its binding energy follows from Bohr's theory to be 6.76 ev. Even compounds such as $H^- e^+$ and $Cl^- e^+$, in which a positron circles around a negative ion, seem to possess some stability.

The positronium atom generated by the mutual capturing of electron and positron can exist in two different states. One of them is a singlet state $^1S_0$ with anti-parallel spin directions of electron and positron, the other is a triplet state $^3S_1$ with parallel spin directions. After a life time of only 8×10^{-9} sec, singlet positronium decays into two photons, each of the energy $h\nu = m_e c^2$, which are emitted into opposite directions. Triplet positronium has a thousand times longer life time, viz. 7×10^{-6} sec; it then annihilates its total energy of $2 m_e c^2$ by emitting three photons for reasons of conservation of angular momentum. Since triplet positronium has an angular momentum of $h/2\pi$ and since the same is true for each photon, the annihilation of positronium can produce only three photons; two of them with opposite spins compensate their momenta, whereas the third one carries along the angular momentum of the electron pair.

Whereas the annihilation of electron pairs occurs into two or three photons, *observations show that the annihilation of protons or neutrons with their respective anti-particles creates in genreal π-mesons* whose average number is approximately 5. Occasionally also one or the other of the heavy K-mesons, to be discussed later, occurs (see Fig. 170). The often-mentioned relation between proton and neutron as two states of the nucleon has the consequence that *a proton can annihilate with an anti-proton or an anti-neutron and, conversely, the same is possible for a neutron with an anti-neutron or an anti-proton.*

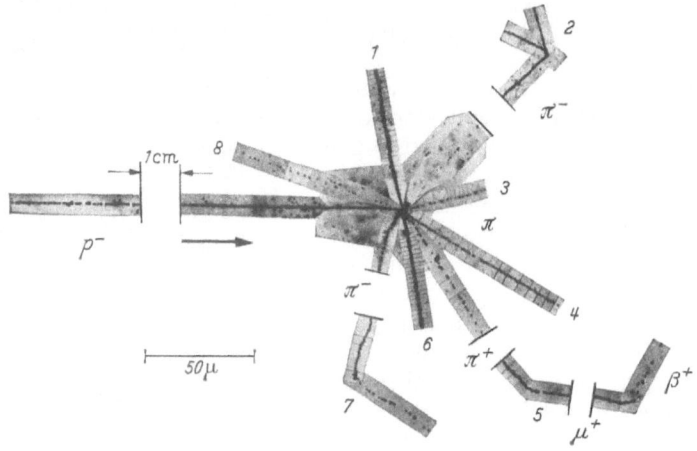

Fig. 170. Photograph of the annihilation of an anti-proton (incident from the left) with a nucleon of an atomic nucleus of the photographic layer. In addition to nuclear fragments, five charged π-mesons are emitted. (After Segrè)

The annihilation of electron pairs occurs predominantly at rest. In contrast to this, experiments of AMALDI show that anti-nucleons can be annihilated in collisions with nucleons with a high probability, i.e., with an effective cross section of nearly 100 mb $= 10^{-25}$ cm^2, even at kinetic energies of 100—300 Mev. On the other hand, it is not yet clear how far bound states of the type of positronium play a role in the annihilation of nucleon-antinucleon pairs. Such a bound state would be the *nucleonium* or protonium where a positive and a negative proton revolve around their common center of gravity, or also an atom-like structure in which a negative proton, instead of an electron, circles around a positive atomic nucleus, the anti-proton afterwards annihilating with one of the nuclear nucleons into π-mesons.

From all investigations known so far, one result seems to be certain: *Matter in all its forms has its counterpart in anti-matter which is completely equivalent to it, except for the opposite sign of its charge, spin direction, and parity. Thus, a world of anti-matter might very well exist. However, matter and anti-matter don't go along with each other. When meeting, they rapidly transform themselves by annihilation into photons and mesons.*

The discovery of the positron and of the other anti-particles as well as the production and annihilation of pairs found a lively response among the theoretical physicists since DIRAC in his attempt at a relativistic theory of the electron had already published strange results which now suddenly seemed to get physical meaning by these discoveries. The fundamental ideas of DIRAC's much discussed *hole theory* are as follows:

If we draw an energy diagram of all possible energy states of the "common" negative electron and take into account the eigen-energy of the electron, which corresponds to its rest mass, i.e.,

$$E_e = m_e c^2 = 0.511 \text{ Mev}, \tag{5.88}$$

we notice that the energy of the free electron at rest is about 0.5 Mev above zero energy (see Fig. 171). Somewhat below, we find the discrete energy states of the electrons bound in an atom and above these there is the continuum of the free electron with kinetic energy. DIRAC now concludes that there should exist, in addition to these already verified energy states of the electron, also states with negative mass energy. According to Fig. 171, these should be *below* zero energy by more than 0.5 Mev. One might object that such negative energy states should be noticed by electron transitions from the known positive energy states to these lower states so that atoms should not be stable. DIRAC, however, meets this objection by assuming that the negative energy states are normally completely occupied so that transitions can not occur because of the Pauli principle. If this ad-hoc assumption is accepted in spite of all possible doubts, a number of remarkable conclusions follows automatically. In order to lift an electron from one of these un-detectable states to the "upper world", an energy amount of more than 1 Mev must be supplied (cf. Fig. 171), for instance by absorption of a high-energy photon. By this absorption process, a negative electron and a "hole" in the negative energy states are produced simultaneously. This hole behaves like a particle of positive charge, and it can be shown that it has all qualities known so far of the positron as the anti-particle of the electron. The absorption process mentioned thus is equivalent to the "production" of an electron pair or, more generally, of a pair of particle and anti-particle. By using this model, the transformation of an electron pair into radiation energy by an annihilation process has to be interpreted as the transition of a particle from a positive energy state into a hole in the continuum of the negative energy states. The hole theory explains without any difficulty why anti-particles are relatively rarely observed: Since our world is more or less densely filled with matter, i.e., with particles in positive energy states, there exists always a particle close to an anti-particle created before. The particle will therefore very soon jump into this hole, i.e., it recombines with an anti-particle under annihilation.

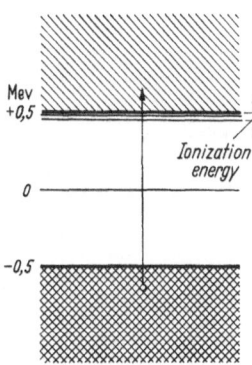

Fig. 171. Complete energy diagram of an electron according to DIRAC's theory. Arrow: "production" of a positive and a negative electron by absorption of a photon with an energy above 1.02 Mev. (See process of Fig. 169)

We see that the hole theory as briefly sketched here is able to explain the known observations about anti-particles in a satisfying way. Apparently we must therefore conclude, in spite of certain misgivings that the world of normal matter is confronted with the world of anti-matter and that both are related to each other like world and underworld according to DIRAC's hole theory.

22. Collision Processes of High-Energy Electrons and Photons

We now return to the role which the electrons play in cosmic radiation, a role which we understand fully only after we have learned about pair production and annihilation. High-energy electrons occur in great numbers in the cosmic radiation as secondary particles of the primary cosmic radiation and of meson decay, which will be dealt with further below. They are being produced in high as well as low layers of our atmosphere.

These electrons may lose energy, while passing through matter, by ionizing electron shells of atoms and molecules; but this process is the less probable the higher the energy of the electron. Fast electrons may furthermore be slowed down in collisions with atomic nuclei by the electric field of the nuclei and thus produce bremsstrahlung of extremely short wavelength (cf. III, 6e). If their energies are higher than 10^8 ev, the loss by bremsstrahlung becomes larger than that by ionization. In this process, the kinetic energy of the electron is used for the production of a quantum of bremsstrahlung, i.e., for producing a high-energy photon. The direct inverse of this process is the absorption of a high-energy photon by an electron in an atom, with the electron acquiring the total energy as kinetic energy. But this process happens rarely since the absorption probability of a photon is the smaller the more the energy of the photon exceeds the binding energy of the absorbing electron in the atom.

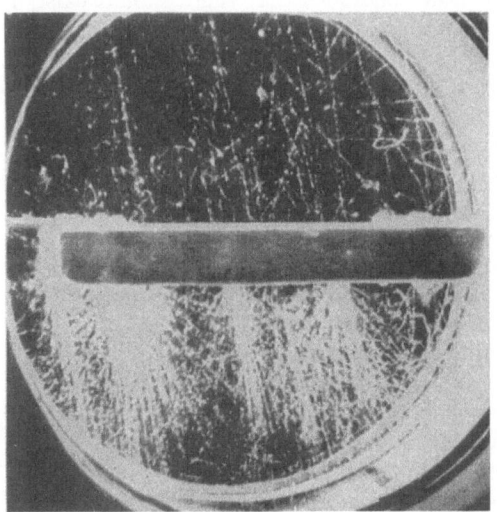

Fig. 172. Cloud-chamber photograph of five showers of electrons and positrons, produced in a lead plate by parallel incident cosmic-ray particles (cascade showers). Photograph by BRODE and STARR

For the processes of the *multiplication or cascade showers of the cosmic radiation* it is essential that the bremsstrahlung of fast electrons or positrons and the production of electron pairs follow each other very rapidly. These cascade showers are showers of positive and negative electrons extending through the atmosphere over hundreds or even thousands of meters of altitude. When produced in the densely packed matter of a lead plate on the other hand, they may be observed in a cloud chamber. If the one thick plate of Fig. 172 is exchanged by a number of thin lead sheets (cf. Fig. 173), it is possible to recognize that these showers are not produced in *one* collision process but in a sequence of individual processes. That is why they are called multiplication showers or cascade showers.

We explain them theoretically in the following way: If a very fast electron, i.e., a high-energy electron, penetrates the lead plate, it soon produces in a nuclear collision a bremsstrahlung quantum of correspondingly high energy. This photon in its turn, after a short flight, produces an electron pair in which each of the two particles gets approximately half of the energy. Shortly thereafter electron and positron will produce one bremsstrahlung quantum each, these two photons produce again two electron pairs, and so on. In the end, the energy of the primary electron has thus be distributed among a very large number of electrons and positrons produced by continued multiplication. A primary electron of 10^{11} ev energy may thus be able to generate a multiplication shower of approximately 1000 particles in a layer of lead only 5 cm thick. The lead with its high nuclear charge and high density serves only to compress the entire multiplication process into a limited space so that it can be observed in the small cloud chamber, whereas in our atmosphere it is extended over a very much thicker layer. It is obvious that such multiplication showers can be initiated by electrons as well as by high-energy (i.e. hard) photons, and such high-energy photons are produced not only by bremsstrahlung but also by decay of π°-mesons, to be treated in V, 23.

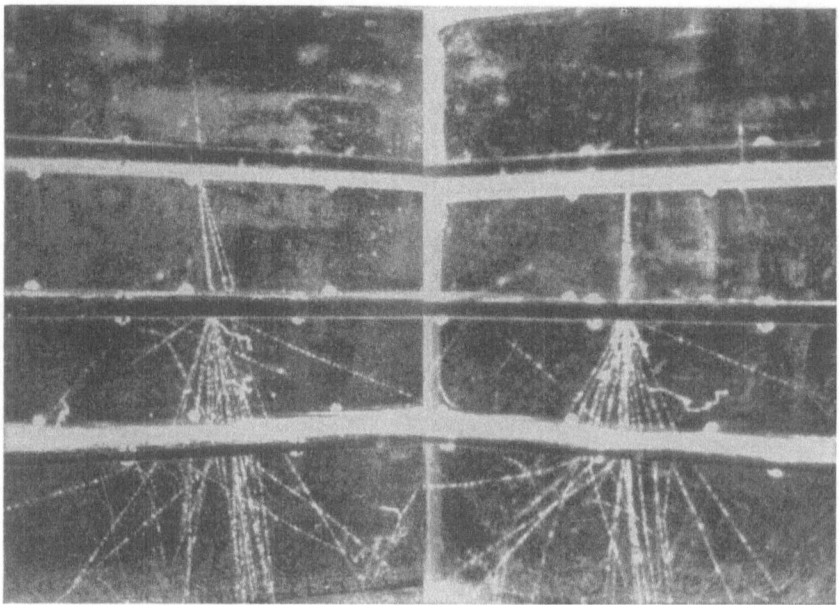

Fig. 173. Cloud-chamber photograph of a cascade shower originating from a number of thin lead plates, providing evidence for the multiplication mechanism of cascade showers. (After Fussell)

23. Mesons and Hyperons

It is not uncommon in research that a discovery is at first misleading. So, the first elementary particle that was discovered with a mass between that of the electron and that of the nucleon proved to be a very atypical representative of this group of unstable particles, called mesons. The existence of this μ-meson or muon was definitely proved in 1935 by Anderson and Neddermeyer, though some years before particles with a positive or negative elementary charge and a mass of some hundred electron masses had been assumed as being responsible for the so-called penetrating component of the cosmic radiation. Only cloud-chamber observations, however, proved unambiguously that a formerly undetected particle actually exists whose charge is equal to that of the electron and whose mass has a value between those of electron and proton. For that reason, it was called *meson*.

Its discovery was not entirely unexpected. Some years before, in working out his theory of nuclear forces (cf. V, 25), Yukawa had postulated the existence of charged elementary particles with masses of the correct order of magnitude. Gradually, however, it became apparent against all expectations that the cosmic-ray meson or μ-meson could not be identical with Yukawa's meson. Its great ability of penetrating matter shows clearly that it has very little interaction with nuclei and that its weak interaction is caused by its charge (equal to that of the electron) and by its magnetic moment. In its interaction with matter the μ-meson behaves in every respect as the electron and *may be regarded as its heavy brother*, whereas YUKAWA expected for "his" meson an extremely strong interaction with nuclei. The μ-meson occurs (just as the electron) with positive or negative charge and has the same spin $(h/4\pi)$ as the electron. Its mass, however, is 206.77 electron masses and its magnetic moment is only $eh/4\pi M c$ with $M = 206.77 \, m_e$. As all heavier elementary particles except the proton, the μ-meson is not stable but decays after a mean life of 2.21×10^{-6} sec into an electron and a neutrino and antineutrino (see Fig. 174):

$$\mu^+ \rightarrow e^+ + \nu_e + \bar{\nu}_\mu, \tag{5.89}$$

whereby electrons of a maximum kinetic energy of 55 Mev are produced. This decay, which corresponds in every way to the radioactive β-decay, happens to all μ^+-mesons since they are repelled by the nuclei because of their positive charge. Occasionally, a μ^+-meson forms with a negative electron a positronium-like "atom" in which the electron circles around the μ^+-meson which then represents the positive nucleus. This strange "atom", however, lasts only for a very short time until the μ^+-meson decays.

In contrast to the behavior of the μ^+-mesons, the negative μ^--mesons usually are being captured by a positive nucleus after they have been slowed down; they then revolve around this nucleus in Bohr orbits. With a proton as nucleus and a meson in the ground state, we obtain for the radius of the orbit the small value of 2.5×10^{-11} cm [see Eq. (3.18)] because of the approximately 200 times

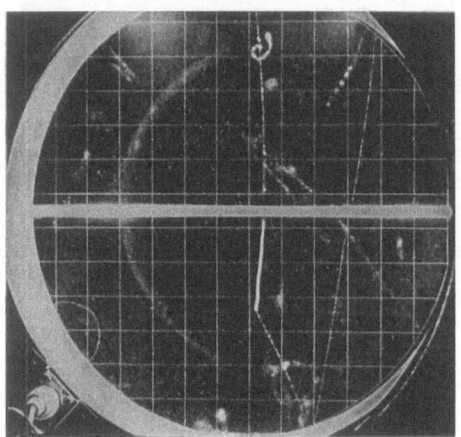

Fig. 174. Cloud-chamber photograph of a μ-meson. The μ-meson is decelerated in the lead plate and therefore ionizes heavily below the plate until it decays into an electron and two non-ionizing neutrinos. The electron track extends from the end of the meson track toward the lower right corner. (Courtesy of R. W. THOMPSON)

larger mass of the μ-meson. In this orbit, the muon is so close to the nucleus that the latter cannot be regarded as a point charge any more. This is one of the reasons for the theoretical significance of these *meson atoms*: In general, the mesons are not captured in the ground orbit but in excited outer orbits from which they jump to inner orbits with emission of X-rays. From the analysis of this radiation, we obtain the energy states of the meson atoms in the same way as those of regular atoms from their spectra. The energies of the different states may be computed from BOHR's theory and the known meson mass if we assume the nucleus to act like a positive point charge. The deviations between computed and observed frequencies of the X-ray lines emitted in such meson transitions thus furnish evidence on the actual forces between the meson and the spatially extended nucleus.

A μ^--meson thus bound to a positive nucleus may either decay, just as a free μ-meson, into an electron and two neutrinos, or it may react with a nuclear proton according to the reaction

$$\mu^- + p \rightarrow n + \bar{\nu}_\mu. \tag{5.90}$$

This is called *bound meson decay*. If the nuclear charge of the mesonic atom is small, the free decay predominates. For large nuclear charge Z, the bound decay predominates since the distance of the μ-meson from the nucleus decreases with increasing Z.

The first and most important representative of the "real" mesons was found in 1947 by POWELL and OCCHIALINI in cosmic-ray experiments. It was called *primary* or π-meson, today often also *pion*. Like all real mesons it has, in contrast to the muon, the spin zero. A little later the π-meson was also observed in experiments with the Berkeley cyclotron, when nuclei were bombarded by

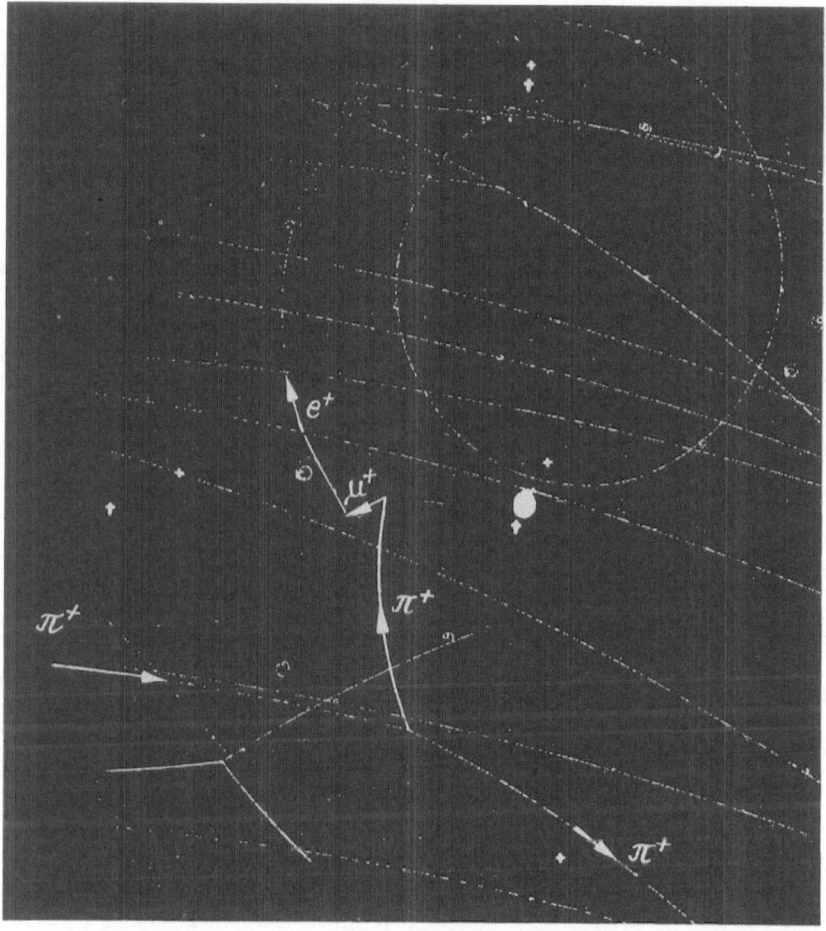

Fig. 175. CERN bubble-chamber photograph of the decay of a secondary π^+-meson, produced by collision of a fast primary π^+-meson with a proton, into a μ-meson and a neutrino. The μ^+-meson in its turn decays after a short way into two neutrinos (again invisible) and an electron, which can be well recognized by the strong curvature of its track in the magnetic field and its small bubble density. (Courtesy CERN)

380 Mev α-particles. The π-meson occurs with a positive or negative charge and a mass of 273.23 m_e, and in addition as a neutral π^0-meson with a smaller mass of 264.4 m_e. The charged π-mesons decay, after a mean life of 2.5 × 10⁻⁸ sec, nearly always into a μ-meson of 34 Mev and a neutrino (see Fig. 175), very rarely also directly into an electron and a neutrino. The neutral π^0-meson, however, has a mean life of only 2.3 × 10⁻¹⁶ sec and decays nearly always into two high-energy photons (γ-quanta), very rarely into an electron pair and a neutrino. The spin of the π-mesons is zero, and so is therefore their magnetic moment. The value zero of the spin follows from a detailed investigation of the reaction $p + p \rightarrow d + \pi^+$.

The π-mesons show the strong interaction with nuclei that was expected for the Yukawa meson; practically in every collision with nuclei they are absorbed. Again, the negative π^--mesons are mostly first captured by positive nuclei and form *π-mesonic atoms*. Then they either decay or they react with a nuclear proton to produce a neutron. In contrast to the corresponding μ-process, the excess energy (mass energy of the π-meson of 140 Mev) is here not taken over by the neutrino but is transferred to the nucleons because of the strong interaction with them. The excess energy serves to heat up the nucleus, which subsequently may emit nucleons, mesons, or larger nuclear fragments. Because of the strong interaction, a positive π-meson of sufficient energy, in contrast to the μ-meson, is able to initiate also nuclear processes in a direct collision, e.g., to transform neutrons into protons. Fig. 176 shows this process as well as the electron pair into which the simultaneously produced π^0-meson decays. One group of π-meson reactions is particularly important for the understanding of the interrelation of all elementary particles. These are collision processes in which π-mesons of sufficiently high energy in the so-called associated production with nucleons generate *K-mesons* and *hyperons*. We shall come back to this process in more detail.

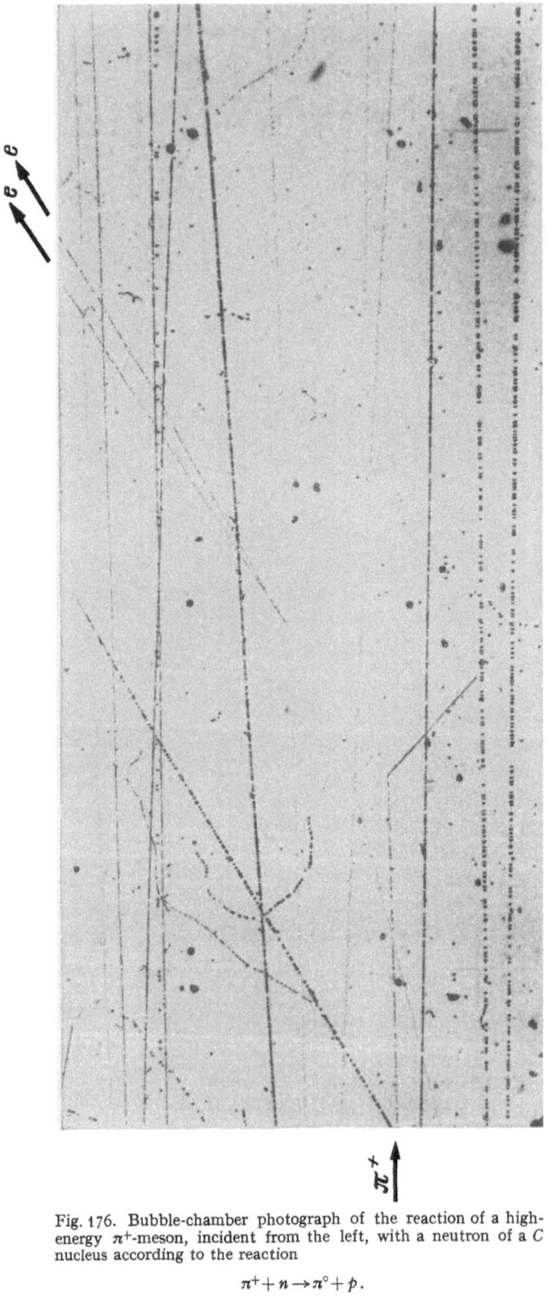

π-mesons may be produced either as photomesons by photoabsorption of γ-quanta of sufficiently high energy by nucleons, for instance

$$\left.\begin{array}{l} \gamma+p\to n+\pi^+ \\ \gamma+p\to p+\pi^\circ \end{array}\right\} \quad (5.91)$$

and in similar reactions with neutrons, or in collisions of two nucleons of sufficient kinetic energy. In this case, single π-mesons may be produced according to the reactions

Fig. 176. Bubble-chamber photograph of the reaction of a high-energy π^+-meson, incident from the left, with a neutron of a C nucleus according to the reaction

$$\pi^+ + n \to \pi^\circ + p .$$

The proton travels to the lower right, the invisible π°-meson to the upper right, decaying into two γ-quanta, one of which produces a high-energy electron pair still in the chamber. (After GLASER)

$$p+n\to n+n+\pi^{+}$$
$$p+n\to p+n+\pi^{\circ} \quad (5.92)$$
$$p+n\to p+p+\pi^{-}$$

or π-meson *pairs*, according to the reactions

$$p+n\to p+n+\pi^{+}+\pi^{-}$$
$$p+n\to p+p+\pi^{0}+\pi^{-}. \quad (5.93)$$

Shortly after the discovery of the π-mesons which proved particularly interesting because of their significance for the nuclear forces, as predicted by YUKAWA, ROCHESTER and BUTLER discovered an additional group of mesons. The mass of these so-called K-*mesons* was determined to be $966\ m_e$; for the neutral K°-meson, a value of $973\ m_e$ has recently been quoted. Therefore they are also called *heavy mesons*. They occur positive, negative, or neutral, and they have zero spin and thus zero magnetic moment. They are produced in collisions of high-energy π-mesons with nucleons, and it is surprising that never single K-mesons are observed but always K-mesons together with hyperons, which we shall discuss below. The mean life of the charged K-mesons is 1.2×10^{-8} sec; whereas there are two different K°-mesons with the mean lives 10^{-10} and 10^{-7} sec. They seem to be a kind of "mixture" of K°-mesons with their antiparticles, so far a unique phenomenon. The charged K-mesons may decay, according to Table 13, either into three π-mesons plus an energy of 75 Mev, or into a neutral and a charged π-meson plus 219 Mev, or into charged μ-mesons and finally even into electrons and neutral π°-mesons and neutrinos. For at least *one* of the

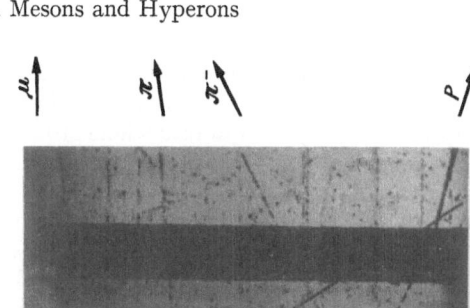

Fig. 177. Bubble-chamber photograph of the precuction of a K_1°-meson and a Λ°-hyperon by a 1.1 Gev π^{-}-meson, incident from the left, according to the reaction

$$\pi^{-}+p\to K_1^{\circ}+\Lambda^{\circ}.$$

At the lower right, the Λ° decays into a proton and a π^{-}; at the upper right, the K_1°-meson into two π-mesons, one of which suffers a subsequent μ decay. (After GLASER)

neutral K°-mesons, a decay into a positive and negative π-meson with 215 Mev excess energy was proved to exist. Because of these multiple decay possibilities, the K-mesons were first believed to be a whole group of different heavy mesons. They were also called τ-mesons and Θ-mesons according to their decay into three or two π-mesons, respectively. More recent investigations showed, however, such an agreement of their masses that one is now more and more inclined to assume that there is only one charged K-meson, but with different decay possibilities. Fig. 177 shows a bubble-chamber photograph of a K^+-meson (τ^+-meson) which decays into two positive π^+-mesons and one negative π^--meson; the latter is

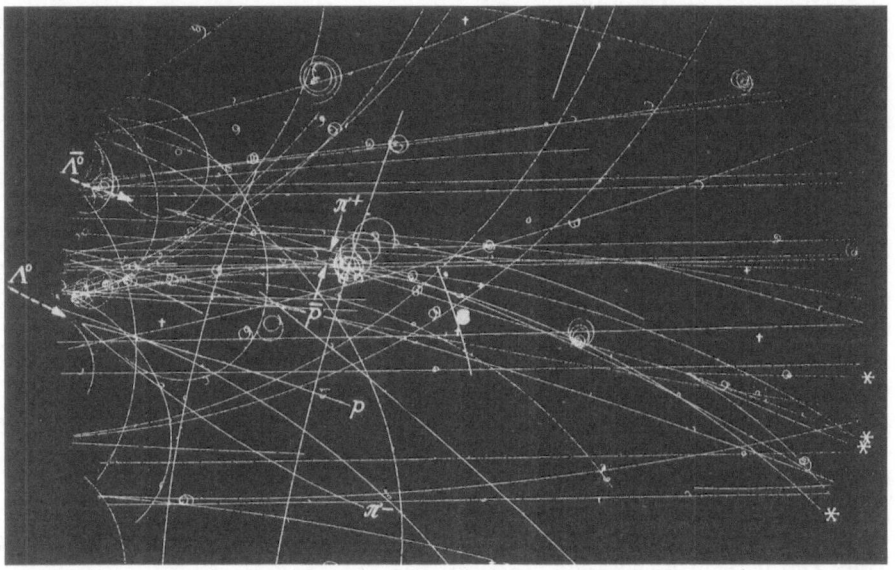

Fig. 178. CERN bubble-chamber photograph of two Λ° decays. The non-ionizing Λ°-hyperon, incident from the left, decays into a proton and a π^--meson. Above these tracks an even more interesting $\bar\Lambda^\circ$ decay is visible. The anti-lambda-hyperon $\bar\Lambda^\circ$ decays into a π^+ and an anti-proton p^-, which is annihilated together with a nucleon with emission of four charged π-mesons (see lower right-hand corner where the end-points of the four π-meson tracks are marked by asterisks). This photograph was the first one to show the decay of an anti-hyperon. (Courtesy CERN)

invisible, of course. In general, the K^+-meson decays, after it has been slowed and stopped, as a free meson because of its repulsion by the nucleus. The K^--meson, however, may form a K-mesonic atom. Because of its small radius and the strong interaction of the K^--meson with the nucleus, it is then soon absorbed by the nucleus and may cause explosive nuclear reactions.

At about the same time when the heavy K-mesons were discovered, also the *hyperons* were found. They are produced together with the K-mesons in collisions of high-energy π-mesons with nucleons. The hyperons are a group of unstable elementary particles which differ from mesons by their mass which is larger than the mass of the nucleons. Their spin is half-integer, $h/4\pi$, since all of them decay directly or indirectly into a nucleon and π-mesons. Their life times are about 10^{-10} sec; contrary to the theoretical expectation for such high-energy particles, for which a much lower value was expected. The only exception is the neutral Σ°, which is the only hyperon known so far which does not decay by emitting a meson, but decays into a Λ°-hyperon and a high-energy γ-quantum after a mean life of only 10^{-18} sec. Table 13 presents the masses, decay possibilities, and other properties of the hyperons. Except for the just mentioned Σ° and the Ξ-hyperons, which have the largest mass, all of them decay into a nucleon and one π-meson each[1]. The heaviest $\varXi$-hyperons decay cascade-like into nucleons by double emission of a π-meson. Anti-particles have been found for all hyperons. As an example, Fig. 178 shows a $\varLambda$-decay and an anti-$\varLambda$-decay; the anti-proton produced in the latter event later collides in the bubble chamber with a proton or a neutron and then disappears under emission of a number of π-mesons of which only the charged ones are visible. Just as the π- and K-mesons, all hyperons seem to interact strongly with nucleons. This is understandable from their mutual relation. It is interesting to note that the neutral hyperon $\varLambda^\circ$ can be bound to a nucleus. This means that it can be a substitute for a neutron. Such nuclei are called *hyperfragments*.

We mentioned already that hyperons and K-mesons are produced together in π-meson collisions: A K-meson either together with an anti-K-meson, or a $\varLambda^\circ$-hyperon, or a $\varSigma$-hyperon, while a $\varXi$-hyperon is always observed together with *two K-mesons*. In the next section, we shall return to this observation which proved particularly important for the understanding of the hyperons.

In reviewing the multiplicity of mesons and hyperons discussed, we realize that all of them eventually decay into protons, electrons, and neutrinos. The prominence of these three particles among the confusing abundance of "elementary particles" seems thus emphasized.

The experimental investigations concerning the behavior of mesons and their interaction with atomic nuclei have thrown new light upon one property of all nuclei and elementary particles, namely its even or odd parity, mentioned already on pages 237 and 252. For all atomic systems composed of several particles, parity is defined as the property of their wave-function to remain unchanged upon reflection at the origin of the coordinates (even or positive parity) or to change the sign (odd or negative parity). As an example, from the behavior of the deuteron in the ground state there follows unambiguously that its parity is even. From the interaction of a π-meson with a deuteron we must conclude that parity has surprisingly to be ascribed also to a single particle such as the π-meson ("intrinsic parity"). According to the experiments, the parity must be odd in this case. The intrinsic parity therefore belongs to the properties of a particle just as its spin and may be characterized, again like the spin, by a quantum number ± 1. The parity of a single elementary particle may be determined from its behavior with respect to strong interaction with other particles of known parity. Vice versa, the parity is responsible, together with the other quantum numbers, for the behavior of the particle in interaction with other particles. Until quite recently, physicists had been convinced that parity satisfies a conservation law as do energy, momentum, and total angular momentum since parity was always found conserved in all interactions previously investigated. LEE and YANG, however, were stimulated by theoretical consequences of the conservation of parity to consider theoretically also the consequences of a possible non-conservation of parity for weak interactions between elementary particles. This theoretical investigation led them to conclusions that could be tested experimentally. These experiments, taken up independently by several laboratories, proved indeed, as we mentioned before, that in weak interactions parity is *not* being conserved.

The theory showed, for instance, that in the decay of a π^+-meson into a μ^+-meson and a neutrino the spin of the μ^+-meson should be oriented in such a way that the spin predominantly coincides with the direction of motion of the μ^+-meson if parity is not conserved. This polarization of μ^+-mesons in turn would lead, again according to the theory, to an asymmetric distribution of the

---

[1] *Note added in proof-reading:* We have added to Table 13 the Ω^- particle of strangeness 3, which was discovered in February 1964. It was not possible, however, to discuss its far-reaching consequences for the elementary-particle theory at this late date.

direction of the positrons produced in the μ^+-meson decay. And this asymmetry was indeed found in agreement with LEE and YANG's theory. If we assume non-conservation of parity, we furthermore expect an asymmetric angular distribution of the electrons emitted from β-active nuclei with oriented spin directions. We mentioned on page 251 that also this asymmetry and some further consequences with respect to the β-decay were confirmed in the meantime by experiments.

We cannot yet fully realize the significance of the sensational discovery of non-conservation of parity in weak interactions. But one statement can be made already with certainty: Particles of opposite spin orientation with respect to their direction of motion may be regarded as right-hand or left-hand screws. So far, *in nature both screw orientations were considered to be equivalent. Parity investigations have shown that this is not so.* In V, 7a we learned that the neutrino behaves as a left-hand screw and the anti-neutrino as a right-hand screw. All parity investigations known today indicate that this result is a very general one. *Matter and anti-matter are apparently distinguished by the sense of their respective screw orientation. In other words, matter and anti-matter are mirror-symmetrical to each other.* This means that there exists a generally valid conservation law, although not for the parity operator P (reversal of sign for all coordinates) but for the operator PC, with C signifying the operator of charge reversal. *PC-invariance* states that all processes in nature are invariant with respect to a simultaneous change of the sign of all coordinates *and* of the charge, since by the operation PC every particle is transformed into its anti-particle which again is a particle occurring in nature.

Recently a new group has been added to the large number of elementary particles, a group which has caused much discussion whether it is reasonable from the physics point of view to talk of short-lived particles with a life time of the order of 10^{-22} or 10^{-23} sec rather than of excitation or resonance states of systems of baryons or mesons (see Table 13a).

In studying the scattering of π- or K-mesons by nucleons, maxima of the scattering cross section were found at definite energies, and from the widths of these maxima the life time of these resonance states was determined by means of the uncertainty relation.

If we take into consideration the role which the mesons play as quanta of the nuclear force field and the photons as quanta of the electromagnetic field, the idea offers itself to compare these new resonance states with the excited states of an atom which are reached by absorption of a photon. If we plot the cross section for the absorption of a photon by an atom against the energy of the photon, it is self-evident that we find sharp maxima of the absorption cross section at those photon energies which correspond to the excited states of the atom. Such an excited "resonance state of the system" can make a transition to the ground state only by emission of one or several photons. In analogy to this, the new resonance states of elementary particles decay either into several π-mesons or into a nucleon and a π-meson, a hyperon and a π-meson etc.

For quite a number of these excited or resonance states of elementary particles, the energies or masses of which follow from the maxima of the cross section of the exciting reaction, we have not yet any detailed knowledge of the decay reaction. Research aims at determining the quantum numbers which characterize all these excited states because this would increase our empirical knowledge on the elementary particles and thus the possibilities of testing the theories of elementary particles to be discussed presently.

24. The Theoretical Interpretation of the Elementary Particles

Reviewing the elementary particles and their properties as they are compiled in Table 13, we recognize four distinctly different classes. First there are the nucleons and their excited states which decay partly by emitting π-mesons, partly by emitting K-mesons and which have the spin $\frac{1}{2}$, $\frac{3}{2}$, $\frac{5}{2}$ etc. The best-known ones among these excited states are the hyperons. The nucleons and their excited states are called heavy elementary particles or *baryons*. Closely related to them are the *mesons* with the spin zero, being the quanta of the nuclear force field. More precisely, we are talking here about the π- and K-mesons, whereas the μ-meson cannot be a true meson because it has a spin $\frac{1}{2}$ and quite generally exhibits a very close relation to the electron. Electrons, μ-mesons (muons), and the two types of neutrinos (ν_e, ν_μ) are alled the light elementary particles or *leptons*; they all have the spin $\frac{1}{2}$. Finally there is a certain relation between the mesons, being the quanta of the nuclear force field, and the photon being the quantum of the electromagnetic field. It has the spin 1.

This classification of the elementary particles in four classes corresponds to a classification with respect to their interactions, i.e. the forces acting between them. Between the baryons we have the very strong nuclear forces which, however, act over distances of a few 10^{-13} centimeters only. Their quanta are the π- and K-mesons. We infer from the short range of the nuclear forces and from their strength reaction times of this *strong interaction* of only 10^{-23} sec.

Since these reaction times apply for all reactions among baryons and mesons, we should expect the decay of the hyperons into nucleons and π-mesons to take place also with this extremely short life time so that hyperons as well as K-mesons decaying into π-mesons should not occur with measurable life times. Actually, these particles (K-mesons and hyperons) have a life time which is at least by the factor 10^{12} larger than expected. Since this fact could not be understood theoretically, the particles were called "strange" particles. We shall discuss their theory below.

The *electromagnetic interaction* among the charged nuclear particles which leads to the emission of γ-quanta (photons), is considerably weaker than that due to nuclear forces. It manifests itself in the somewhat longer life times of 10^{-18} to 10^{-16} sec known from the γ-decay of excited nuclei, of the Σ°-hyperon, and the π°-meson. Still weaker by a factor of at least 10^{10} are the forces among leptons on the one hand and baryons and mesons on the other hand *(weak interactions)*. They manifest themselves in life times of baryons and mesons longer by many orders of magnitude with respect to μ-decay or β-decay, i.e., *with respect to every decay connected with the emission of neutrinos.*

The classification of the different interactions among elementary particles permits to treat physics of elementary particles theoretically in three steps of increasingly better approximation. In a first approximation, we would consider only the strong interactions. No photon exists in this approximation, the masses of proton and neutron would be equal, the same would apply for the masses of the charged and the neutral meson since all these mass differences are of electromagnetic origin. Finally, all particles decaying into leptons or γ-quanta, as the neutron and all π-mesons, would be stable in this approximation. A second approximation to the theory of elementary particles would include the electromagnetic interactions. It would be able to describe the γ-decay of π°-mesons and the mass differences of elementary particles of the same type but different charge. Only by taking into account the weak interactions in a third approximation, the theory would give the complete multiplicity of possible transformations among

Table 13. *The Presently Known Elementary Particles*

	Particle	Anti-Particle	Mass m_e	Mass Mev	Charge	Spin $h/2\pi$	Life Time sec	Decay
Leptons	Electron e^-	e^+	1	0.510976	$\mp e$	$\tfrac{1}{2}$	∞	—
	e-Neutrino ν_e	$\bar{\nu}_e$	0	0	0	$\tfrac{1}{2}$	∞	—
	Muon μ^-	μ^+	206.77	105.66	$\mp e$	$\tfrac{1}{2}$	2.212×10^{-6}	$e^- + \nu_\mu + \bar{\nu}_e + 105$ Mev
	μ-Neutrino ν_μ	$\bar{\nu}_\mu$	0	0	0	$\tfrac{1}{2}$	∞	—
Photon	Photon γ	γ	0	0	0	1	∞	—
Mesons	π^-	π^+	273.18	139.59	$\mp e$	0	2.55×10^{-8}	$\mu^- + \bar{\nu}_\mu + 33.9$ Mev
	π°	π°	264.20	135.00	0	0	2.3×10^{-16}	$2\gamma + 135$ Mev
	π^+	π^-	273.18	139.59	$\pm e$	0	2.55×10^{-8}	$\mu^+ + \nu_\mu + 33.9$ Mev
	K^+	K^-	966.6	493.9	$\pm e$	0	1.224×10^{-8}	$\begin{cases}\mu^+ + \nu_\mu \text{ or } \pi^+ + \pi^\circ \\ \pi^\circ + e^+ + \nu_e \text{ or } \pi^\circ + \mu^+ + \nu_\mu \text{ or } 3\pi\end{cases}$
	$K^\circ \Big\langle \begin{smallmatrix} K_1^\circ(\Theta^\circ) \\ K_2^\circ(\tau^\circ) \end{smallmatrix} \Big\rangle \bar{K}^\circ$		974.2	497.8	0	0	1×10^{-10}	$\pi^+ + \pi^- + 218.6$ Mev / $2\pi^\circ + 227.8$ Mev
			974.2	497.8	0	0	6.1×10^{-8}	$\pi^+ + \pi^- + \pi^\circ + 83.6$ Mev / $3\pi^\circ + 92.8$ Mev
Baryons — Nucleons	p^+	p^-	1836.12	938.21	$\pm e$	$\tfrac{1}{2}$	∞	—
	n	$\bar{n}$	1838.65	939.51	0	$\tfrac{1}{2}$	10^{13}	$p + e^- + \bar{\nu}_e + 0.8$ Mev
Baryons — Hyperons	Λ°	$\bar{\Lambda}^\circ$	2182.8	1115.36	0	$\tfrac{1}{2}$	2.36×10^{-10}	$\begin{cases}p + \pi^- + 37.6 \text{ Mev} \\ n + \pi^\circ + 40.8 \text{ Mev}\end{cases}$
	Σ^-	$\bar{\Sigma}^-$	2340.6	1196.0	$\mp e$	$\tfrac{1}{2}$	1.61×10^{-10}	$n + \pi^- + 117$ Mev
	Σ°	$\bar{\Sigma}^\circ$	2331.8	1191.5	0	$\tfrac{1}{2}$	10^{-18}	$\Lambda^\circ + \gamma + 76$ Mev
	Σ^+	$\bar{\Sigma}^+$	2327.7	1189.4	$\pm e$	$\tfrac{1}{2}$	0.81×10^{-10}	$\begin{cases}p + \pi^\circ + 116 \text{ Mev} \\ n + \pi^+ + 110 \text{ Mev}\end{cases}$
	Ξ^-	$\bar{\Xi}^-$	2580.2	1318.4	$\mp e$	$\tfrac{1}{2}$?	1.3×10^{-10}	$\Lambda^\circ + \pi^- + 63$ Mev
	Ξ°	$\bar{\Xi}^\circ$	2566	1311	0	$\tfrac{1}{2}$?	1.5×10^{-10}?	$\Lambda^\circ + \pi^\circ + 61$ Mev
	Ω^-	$\bar{\Omega}^-$	3297	1686	$-e$	$\tfrac{3}{2}$?	10^{10}	$\Xi + \pi + \downarrow$? Mev

Table 13a. *Excited States of Baryons and Mesons* (as known by 1963)

		Mass m_e	Mass Mev	Spin $h/2\pi$	Isospin T	Parity	Strangeness S	Life Time sec
Nucleon Resonances		2422	1240	$\frac{3}{2}$	$\frac{3}{2}$	$+$	0	0.45×10^{-24}
		2958	1510	$\frac{3}{2}$	$\frac{1}{2}$	$-$	0	0.50×10^{-24}
		3305	1690	$\frac{5}{2}$	$\frac{1}{2}$	$+$	0	0.46×10^{-24}
		3760	1920	$>\frac{3}{2}$	$\frac{3}{2}$	?	0	0.35×10^{-24}
Hyperon Resonances	Y_0^*	2750	1405	?	0	?	-1	?
	Y_0^*	2970	1520	$\frac{3}{2}$	0	$-$	-1	4.3×10^{-24}
	Y_0^*	3550	1815	$<\frac{3}{2}$	0	?	-1	?
	Y_1^*	2710	1385	?	1	?	-1	1.3×10^{-24}
	Ξ^*	3000	1535	?	$\frac{1}{2}$	?	-2	?
Meson Resonances	η	1072	548	0	0	$-$	0	6.5×10^{-24}
	ω	1526	780	1	0	$-$	0	4.4×10^{-24}
	ϱ	1463	748	1	1	$-$	0	0.65×10^{-24}
	K^*	1735	885	$\frac{1}{2}$?	$\frac{1}{2}$	?	1	1.3×10^{-24}

all elementary particles including the leptons. It is not surprising that in these processes due to the comparitively weak interactions, such as the β-decay, many selection rules and conservation laws are violated which normally are valid, e.g., the conservation of parity. These processes may be designated as "forbidden" in a certain sense, since they are less probable by many orders of magnitude than processes due to strong and electromagnetic interactions.

We gain a still deeper insight into the foundations of a future theory by considering those conservation laws that are always valid, so far as we know from experiments. Here are first the laws of conservation of the total charge, of the number of leptons, and of the number of baryons. There is ample evidence that in all collision processes and other transformation processes of elementary particles the sum of the positive charges minus the sum of the negative charges remains constant. The same is true for the sum of all leptons minus that of all anti-leptons, and finally the sum of all baryons minus that of the anti-baryons. The conservation law for leptons requires, however, an agreement about the question which particles we define as leptons and which ones as anti-leptons. The conservation law is valid if we consider electrons, μ^--mesons, and neutrinos (ν) as leptons, but positrons, μ^+-mesons, and anti-neutrinos ($\bar{\nu}$) as anti-leptons. According to the lepton conservation law, β-decay always occurs into a lepton and an anti-lepton, i.e., an electron and an anti-neutrino or a positron and a neutrino. In the π-decay we get a μ^--meson together with an anti-neutrino or a μ^+-meson and a neutrino; the μ^--decay yields an electron, a neutrino, and an anti-neutrino.

A comprehensive theoretical description of our knowledge about mesons and baryons, which agrees surprisingly well with observations, is obtained if two new quantum numbers are introduced (after GELL-MANN and NISHIJIMA) in addition to the value of the charge Q ($+1$, or 0, or -1), which also has to be regarded as a kind of quantum number, and to the mass number A, also called baryon quantum number ($+1$ for baryons, -1 for anti-baryons, 0 for mesons and leptons). These two new quantum numbers are the *isotopic spin* T, already mentioned in V, 4a, and the "strangeness" S. The different components of the isotopic spin T with respect to some direction z of a configuration isospace distinguish particles of different charge but equal mass (isobars); T therefore should better be designated as *isobaric spin*. Today it is often called *isospin*. According to III, 9, the mechanical spin $s=\frac{1}{2}$ has two quantized components $+\frac{1}{2}$ and $-\frac{1}{2}$ with respect to a given direction, and these components describe two states of the system with different spin orientation. In an analogous manner, the two states of different charge of the nucleon, the proton and the neutron, may be formally distinguished by the isospin components $T_z = +\frac{1}{2}$ and $T_z = -\frac{1}{2}$ of the total isospin $T = \frac{1}{2}$ which describes the nucleon. This way of description ascribes to the π-meson the isospin $T = 1$ so that the π-mesons π^+, π°, and π^- represent an isospin triplet with the components $T_z = +1, 0, -1$, respectively.

From these examples, we find the general relation

$$Q = A/2 + T_z \qquad (5.94)$$

between charge number Q, mass number A, and isospin component T_z.

In order to describe also the observations about K-mesons and hyperons, GELL-MANN extended this relation by introducing the strangeness quantum number S. Its physical meaning is not yet understood, but it is postulated that *it shall be conserved in collisions or processes with strong or electromagnetic interaction. The same we assume for the isospin component T_z.* S is supposed to change, however, by unity, as other known quantum numbers do, in the decay of the K-mesons and hyperons, which are conspicuous by their unusually long life time. S never changes by more than unity, and this change of S during the decay is supposed to be the reason for the long life time of the strange particles. Let us ascribe to the normally decaying π-mesons and nucleons the strangeness quantum number $S = 0$; to the K^+-mesons and K°-mesons $S = +1$; to their anti-particles K^- and $\overline{K}^\circ$, to the hyperons Λ°; Σ^+, Σ°, Σ^- the strangeness $S = -1$; and finally to the heaviest hyperons Ξ, which decay by double π-emission into nucleons, the value $S = -2$. We then obtain the surprisingly successful extension of Eq. (5.94) in the form

$$Q = T_z + A/2 + S/2. \qquad (5.95)$$

In taking into account that elementary particles with charge or baryon numbers larger than one have never been observed so that the additional conditions

$$A = 0 \quad \text{or} \quad \pm 1 \quad \text{and} \quad Q = 0 \quad \text{or} \quad \pm 1 \qquad (5.96)$$

seem to hold, Eq. (5.95) leads to combinations of quantum numbers summarized in Table 14. The mesons and baryons listed in the last column of Table 14 correspond to the sets of quantum numbers given in the first five columns. The particles appear as doublets (such as p and n) or as triplets (such as the three π-mesons) depending on their value of the isospin of $\frac{1}{2}$ or 1, respectively. The anti-particles are listed only for the K-mesons, for there is a principal difference between K^- and K^+, as we shall see below. For the anti-particles of the baryons, however, only the signs of A, T_z, S, and Q have to be changed if they are not zero.

We see that all known mesons and baryons can easily be deduced from Eq. (5.95) if the quantum number S is assigned to mesons and hyperons as indicated above: $S = \pm 1$ to K-mesons and normal hyperons; $S = 2$ to the heaviest hyperons (Ξ) which decay into nucleons by double π-meson emission. Only a very few combinations of quantum numbers follow from Eq. (5.95) for which corresponding elementary particles have not yet been found. The particle determined by $S = A = T = T_z = Q = 0$ should be a neutral π°-meson. It is not yet quite sure whether it is distinguished from the normal neutral π°-meson with $T = 1$ and $T_z = 0$. According to HEISENBERG, it might have so large a mass and be so unstable that it could decay into common π-mesons. Another unknown particle seems to be the positive baryon with $S = +1$; its decay should be similar to that of Σ^+. Two mesons with $S = \pm 2$ and the isospin $T = 0$ would also be

Table 14. *The Mesons and Baryons (without Anti-particles) and their Quantum Numbers according to the Theory of Gell-Mann*

S	A	T	T_z	Q	Particles
0	0	0	0	0	π°-isosinglet (unknown)
0	0	1	$\begin{cases} +1 \\ 0 \\ -1 \end{cases}$	$\begin{matrix} +1 \\ 0 \\ -1 \end{matrix}$	$\left.\begin{matrix} \pi^+ \\ \pi^\circ \\ \pi^- \end{matrix}\right\}$ isotriplet
0	1	$\frac{1}{2}$	$\begin{cases} +\frac{1}{2} \\ -\frac{1}{2} \end{cases}$	$\begin{matrix} +1 \\ 0 \end{matrix}$	proton neutron
+1	0	$\frac{1}{2}$	$\begin{cases} +\frac{1}{2} \\ -\frac{1}{2} \end{cases}$	$\begin{matrix} 1 \\ 0 \end{matrix}$	K^+ K°
+1	1	0	0	+1	unknown baryon
−1	0	$\frac{1}{2}$	$\begin{cases} +\frac{1}{2} \\ -\frac{1}{2} \end{cases}$	$\begin{matrix} 0 \\ -1 \end{matrix}$	K^- $\bar{K}^\circ$
−1	1	1	$\begin{cases} +1 \\ 0 \\ -1 \end{cases}$	$\begin{matrix} +1 \\ 0 \\ -1 \end{matrix}$	$\begin{matrix} \Sigma^+ \\ \Sigma^\circ \\ \Sigma^- \end{matrix}$
−1	1	0	0	0	Λ°
−2	1	$\frac{1}{2}$	$\begin{cases} +\frac{1}{2} \\ -\frac{1}{2} \end{cases}$	$\begin{matrix} 0 \\ -1 \end{matrix}$	Ξ° Ξ^-
−2 +2	0 0	0 0	0 0	−1 +1	$\left.\begin{matrix} \\ \end{matrix}\right\}$ unknown: Apparently, mesons with $S = \pm 2$ do not exist.

compatible with Eq. (5.95) and the restricting conditions. This may point to the non-existence of mesons with strangeness 2.

All known mesons and baryons thus fit into the Gell-Mann theory, and only one of the expected baryons has not yet been found. This is strong support for the strangeness theory.

Still more conspicuous is its success in interpreting the observations on reactions among K-mesons, hyperons, nucleons, and π-mesons. Since the strangeness S is supposed to be conserved in collisions with strong interaction, *elementary particles with $|S| > 0$ cannot be produced individually in collisions, but only two or three together so that the sum of their S-values remains zero.* According to Table 14, a K-meson with $S = +1$ can be produced only together with a particle with $S = -1$, i.e., with an anti-K-meson, a Λ°-hyperon, or a Σ-hyperon. A Ξ-hyperon can be produced exclusively together with *two* K-mesons. These postulates of GELL-MANN's theory of the "associated production" of elementary particles with $|S| > 0$ are fully confirmed by experiments; no exception is known so far. The same is true for the conservation of the quantum number S in collisions of K-mesons or hyperons with nucleons or π-mesons. In particular, another requirement of the theory was confirmed here, viz., that K^+-mesons behave completely different from K^--mesons because of their different strangeness. Besides simple scattering processes, K^+-mesons may initiate only processes of the type

$$K^+ + n \to K^\circ + p, \tag{5.97}$$

while the strangeness S remains constant in collisions of K^--mesons or K°-mesons also if in collisions these mesons are transformed into hyperons Λ° or Σ. The actual observation of the different behavior of the two kinds of K-mesons is further evidence for the correctness of the result of GELL-MANN's theory that the K-mesons K^+, K°, K^- are not a triplet analogous to π^+, π°, π^-. The theoretically expected result that *two* Λ°-hyperons ($S = -1$ each) should originate from a collision of a Ξ^--hyperon ($S = -2$) with a proton has also been confirmed.

Furthermore, the free decay of hyperons under π-emission fits into the picture because the strangeness is changed by unity:

$$\varDelta S = \pm 1. \tag{5.98}$$

In this case, the $\varXi$-hyperon decays in two steps, first into a $\varLambda°$-hyperon, then, by again emitting a π-meson, into a proton. Undoubtedly, therefore, the Gell-Mann theory points into the right direction and will be an integral part of any future consistent theory of all elementary particles.

In Fig. 179 the presently known mesons and baryons are plotted with their S-values versus their mass in units of m_e and Mev as ordinates. It must be

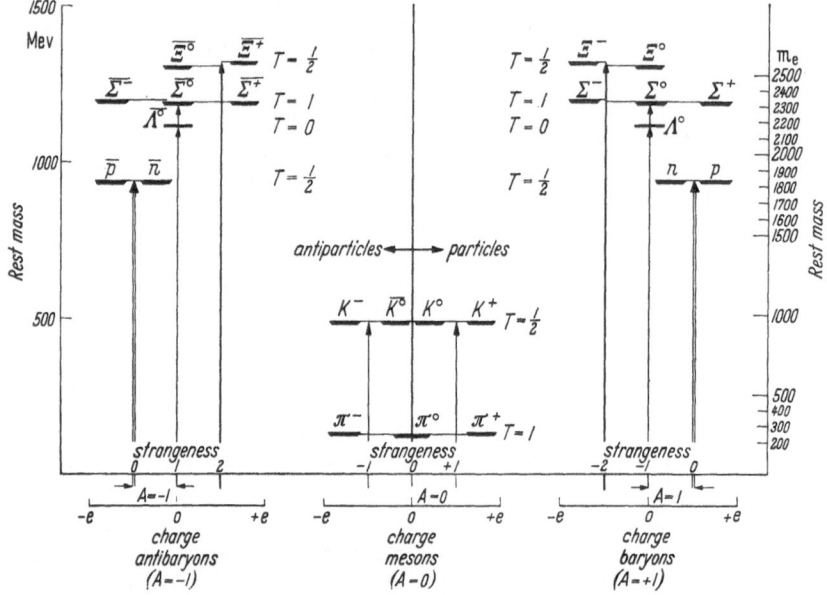

Fig. 179. Mass spectrum of all known mesons and baryons as well as of their anti-particles with their corresponding quantum numbers for charge, strangeness, and isospin. (Gell-Mann scheme)

emphasized that these mass values are empirical, since the Gell-Mann theory cannot yield mass values beyond the assignment to $A = 0$ (mesons) or $A = 1$ (baryons) since it is concerned with quantum numbers only.

In recent years, HEISENBERG has developed fundamental ideas of a theory of the elementary particles and has begun with its mathematical elaboration. He hopes his theory to be able to describe all essential features and properties of all elementary particles known today and yet to be found. It may open the door to a new epoch of atomic physics. In contrast to earlier attempts at theories of individual elementary particles or classes of such, the fundamental idea of this theory is to search for a *"world formula"* of matter. This formula contains the three fundamental empirical constants h (PLANCK's quantum of action), c (velocity of light, as representative of every field theory), and l_0 (universal smallest length). Its mathematical form represents those general symmetry properties which are expected to apply to all elementary particles as well as to every fundamental physical theory. HEISENBERG may be quoted: "By these requirements everything else seems to be determined ... Similarly as in PLATO's concept, it looks as if our seemingly so complicated world of elementary particles and force fields is based on a simple and translucent mathematical structure. All those relations which we are accustomed to call laws of nature in the different fields of physics should be deductible from this one structure." If HEISENBERG's world formula can be solved, its solution should furnish the masses of all elementary particles (their *mass spectrum*) and their other properties in the same way as we can deduce the abundance of atomic and molecular energy states with their transition probabilities and quantum numbers from the solution of the *one* Schrödinger equation describing an atom or molecule. And again, in the same way as all energy states of an atomic system must differ, according to PAULI, by at least one of their four quantum numbers, also the system of the elementary particles can be described by a system of "qunatum numbers" in such a way that no two particles are described by identical quantum numbers. The physical significance of the quantum numbers occurring in HEISENBERG's theory is not yet quite clear. It can be shown, however, that 5 quantum numbers are sufficient for describing all elementary particles. These are the baryon quantum number (mass number) A, which is 0 for all leptons and mesons, 1 for baryons and -1 for antibaryons; the charge quantum number Q (0 or ± 1); the isospin T and its component T_z in some prescribed direction of isospace (0, $\pm\frac{1}{2}$, or ± 1), and finally the strangeness quantum number S, also introduced already, with its possible values 0, ± 1, or ± 2. Q is here an independent quantum number, not related to the other quantum numbers, because Eq. (5.95) which describes Q as a function of A, T_z, and S is valid only for baryons (see page 332), and not for leptons, whereas HEISENBERG's theory assigns quantum numbers T, T_z, and S to leptons also. Originally, the strangeness quantum number S too was introduced by GELL-MANN for baryons only, whereas HEISENBERG distinguishes μ-mesons and electrons as well by different quantum numbers S ($S \neq 0$ for μ-mesons). In view of the fact that all observable properties of μ-meson and electron seem to be equal so far, this distinction still seems problematic.

HEISENBERG was able to show in the development of his theory that his equation

$$\gamma_\nu \frac{\partial \psi}{\partial x_\nu} \pm l_0 \gamma_\mu \gamma_5 \psi (\overline{\psi} \gamma_\mu \gamma_5 \psi) = 0, \qquad (5.99)$$

which we quote here without explanation, does indeed possess all the symmetry properties required by the theory of relativity and by quantum theory as well as those which follow from the empirical properties of the elementary particles. Furthermore, it seems possible to deduce from this equation the quantum numbers for all different elementary particles. Up to the present time, the mass values have been computed by means of rather raw approximation methods only. Although the numerical values are not in any way quantitatively correct, there is agreement with the observation in so far as the theory predicts the existence of nucleons and π-mesons with very different mass values. We mentioned already the interesting result that the mass of the not yet observed iso-singlet π_0^0-meson, the existence of which follows from the Gell-Mann theory, according to HEISENBERG's theory is supposed to be only slightly smaller than the nucleon mass. It should be unstable therefore with respect to decay into "normal" π-mesons. We may look forward to the further development of the theory as well as to the results of further experimental research with the big accelerators which will show, for instance, which hyperons with $S > 2$ exist. The Ω^- hyperon with $S = 3$ has just been discovered.

25. Nucleons, Meson Clouds, and Nuclear Forces

Our preceding presentation is not yet satisfying in so far as we could give only hints about the nature of the nuclear forces which are responsible for holding protons and neutrons together in the nucleus. Their explanation is a central problem of nuclear physics, and we shall now look somewhat deeper into its present state.

First of all, we realize that the lack of a satisfactory theory of nuclear forces is indeed very characteristic for the present state of nuclear physics: Physics of the electron shells, of molecules, and of solids can be described, in principle, correctly and quantitatively by quantum mechanics. But not even the basic phenomena of nuclear physics, although we may be able to understand them qualitatively, can be described quantitatively. This is exemplified by the fact that we cannot compute any binding energy, not even that of the simplest composite nucleus, the deuteron. Neither can we calculate the life times of β-active nuclei or the magnetic moments of proton and neutron. The ultimate reason for having a quantitative theory of the processes in the electron shells lies in the fact that electrons and nucleus are bound by the electric or, more generally, the electromagnetic field. Its properties we know very well, together with the quantum of this field, the photon. Evidently, the lack not only of a theory of nuclear forces but, quite generally, of a theory of the nucleus is due to the fact that the interaction between the nucleons is due to a field whose details are not yet known. π-mesons and K-mesons are the quanta of this field; but we do not yet know them sufficiently well in their interaction with the other elementary particles. *We notice here the close relationship between the theories of nuclear forces and of elementary particles.*

What can we presently state about the forces between nucleons in analogy to those between nucleus and electrons? In the atom, the electric field is instrumental in binding nucleus and electron shell. According to this concept, a certain amount of electrostatic energy is "located" in an excited atom, and this energy will be emitted as a photon, i.e., an electromagnetic wave, when the atom returns to its ground state. Since the electrostatic field is only a special case of the electromagnetic field, we have two equivalent statements about the bond between atomic nucleus and electron shell: "The bond is due to the electromagnetic field", or "the bond is due to the possibility of emission and absorption of photons". We learned already on page 203 (see IV, 14) that DIRAC has elaborated these concepts in his quantum electrodynamics to a satisfactorily consistent theory of the emission and absorption of radiation by atoms. In contrast to that of the electromagnetic field, the particle corresponding to the nuclear field has a considerable rest mass. This follows, according to YUKAWA, from the small range of nuclear forces, which is connected with the mass m by the relation

$$R = \frac{h}{2\pi m c}.$$
(5.100)

The particle mass computed from Eq. (5.100) with empirical data for the range of nuclear forces is about 270 electron masses; the particle is the π-meson. That it actually is the quantum of the nuclear field is confirmed by numerous observations about the strong interaction between nucleons and π-mesons. The π-mesons, therefore, are produced in great numbers in all collisions between nucleons of sufficient high energy. It is therefore believed today, in agreement with YUKAWA, that proton and neutron have a "core", which is the same for both particles, and a virtual π-meson cloud. The difference of neutron and proton is caused by the different composition of this cloud of charged and uncharged π-mesons. We call the cloud a *virtual* π-meson cloud since it is impossible that the nucleon "consists" of a nucleonic core and a meson cloud in the same way as the atom consists of nucleus and electron shell. From the uncertainty principle [Eq. (4.18)] for energy and time it follows that a proton may very well "dissociate", in spite of the mass difference, into a neutron and a π^+-meson for a period of some 10^{-24} sec. This time is sufficient for the meson to fly some 10^{-13} cm, i.e., the range of nuclear forces. We therefore may assume that the bond of proton and neutron, for instance in the deuteron, is due to a mutual emission and absorption of π-mesons by the nucleons. By this emission and absorption, the momentum of the mesons is exchanged between the nucleons. The change of momentum of the interacting particles, however, corresponds to forces which in this case are the nuclear forces. The bond between two protons or two neutrons in a nucleus, according to this concept, would require the exchange of neutral π°-mesons. Looking back to IV, 11, we recognize that the reduction of nuclear forces to an exchange of mesons between nucleons is quite analogous to the explanation of the bond of two hydrogen atoms in the H_2-molecule as being due to an exchange of binding electrons.

Obviously, the problem of nuclear forces is much more complicated than that of the chemical bond, not to mention the simple electrostatic bond. The two nucleons in the deuteron, for instance, have the same spin orientation. This is unexpected but possible according to the Pauli principle, since the two nucleons have different charge, i.e., they possess different isospins. There do not exist, however, bound double neutrons or double protons ($_2He^2$) although nuclear forces must act between protons as well as between neutrons. But the Pauli principle in this case prohibits the parallel spin orientation which seems to be preferred by nuclear forces.

The interrelation between the nuclear forces and the exchange of π-mesons by the nucleons is supported by quite a number of experimental results. Among these are the β-decay of radioactive nuclei, the anomalous magnetic moments of protons and neutrons, the evidence on "excited nucleons", and finally HOF-STADTER's results about a "structure" of the nucleons.

The theory of nuclear forces is based on the concept of a charge exchange among nucleons, and this idea is well supported by the phenomenon of β-decay. According to YUKAWA, the π-mesons primarily emitted and absorbed by the nuclei decay directly or indirectly into electrons and neutrinos. It seems therefore reasonable to bring the β-decay into relation with the meson theory of nuclear forces.

The fact that the magnetic moments of protons and neutrons are not multiples of the nuclear moment μ_n (cf. V, 4e) is also a consequence of the meson theory of nuclear forces. A proton can be transformed into a neutron by emission of a positive meson, the neutron into a proton by emitting a negative meson. Therefore, we have to add to the nuclear magnetic moment belonging to the "proton alone" a contribution from the π^--meson which occassionally revolves around the proton; this contribution is relatively large because of the small mass of the meson [cf. Eq. (5.10)]. The neutron as an uncharged particle does not have any magnetic moment belonging to "itself alone". Its moment as empirically found is exclusively due to the occasional rotation of a π^--meson around the proton. This moment has an orientation opposite to that of the mechanical spin of the neutron and therefore in a first approximation the same value (but negative sign) as the positive additional moment of the proton. The magnetic moments of proton and neutron thus are indirect evidence of the "ball game" which protons and neutrons play with π-mesons. Direct evidence for the reality of the neutron-proton exchange was found in measuring the scattering of 90 Mev neutrons by nuclei. The number of protons scattered into the direction of impact proved to be much higher than was to be expected theoretically by taking into account the binding energy of the scattering nuclear protons. The only explanation for this proton excess seems to be the assumption that a certain percentage of the primary neutrons is being transformed into protons by meson exchange during the collision with the nucleus.

Surveying again the relation between nuclear forces and elementary particles, we may conclude that the fermions p, n, e, μ, and ν with their respective anti-particles, all having the spin $\hbar/2$, can be regarded as "real" elementary particles, whereas the bosons with spin zero, i.e., the π-mesons and K-mesons, are responsible for the forces between the fermions. We might even say that the nucleons, at least partially, "consist of" these bosons, although in a way that is not easily to be visualized. If the nucleons act in collision processes as "soft" spheres with a small "hard" core ,the soft shell is certainly due to their ability to emit π-mesons (and to a smaller radial distance also K-mesons), i.e., due to their *meson cloud*. Recent experiments point to the reality of this concept. First of all, measurements of the scattering cross section of protons with respect to π-mesons and data on the photo-production of π°-mesons from bombarding protons by γ-quanta have provided evidence for the excited states of the proton or its meson cloud which were mentioned above. Secondly, Hofstadter bombarded protons and neutrons by high-energy electrons from the Stanford linear accelerator. From the angular distribution of these scattered electrons he drew conclusions about the radial distribution of the electric charge and magnetic moment, which in principle seem to agree with the above indicated ideas about the "structure" of the nucleons. The "hard" core of the nucleons then might be caused by their ability to emit also hyperons and nucleon-antinucleon pairs to radial distances of some 10^{-14} cm, again according to the uncertainty principle.

The nucleons thus seem to have a much more complicated structure than we previously assumed. This statement is the very end of our present knowledge about nuclei and elementary particles.

26. The Problem of the Universal Constants in Nature

In concluding this chapter, we briefly deal with one of the most fascinating problems of modern physics, that of the fundamental constants of nature. Taking into account all our present knowledge of atomic physics, we ask to what ultimate fundamental data all physics can be reduced and what we know at present about them.

If we disregard for the moment the mesons and consider matter only in its normal state, our entire material world consists of only three types of elementary particles: Protons, neutrons, and electrons. In addition to their masses and the elementary charge e, we have furthermore the universal constants contained in the fundamental laws of physics: PLANCK's quantum of action h, the velocity of light c, and the gravitational constant f. If we remember that the masses of proton and neutron are equal within about 0.1% and if we neglect this small difference in a first approximation, the remaining six constants should be sufficient to develop the structure of our whole world:

$$\left.\begin{aligned} M_P &= 1.672 \times 10^{24} \text{ g} \\ m_e &= 9.108 \times 10^{-28} \text{ g} \\ e &= 4.803 \times 10^{-28} \text{ g}^{\frac{1}{2}} \text{ cm}^{\frac{3}{2}} \text{ sec}^{-1} \\ h &= 6.625 \times 10^{-27} \text{ g cm}^2 \text{ sec}^{-1} \\ c &= 2.998 \times 10^{10} \text{ cm sec}^{-1} \\ f &= 6.670 \times 10^{-8} \text{ g}^{-1} \text{ cm}^3 \text{ sec}^{-2}. \end{aligned}\right\} \tag{5.101}$$

Most atomic physicists agree that we need, as an additional constant, a fundamental smallest length l_0 as mentioned before, which must have the order of 10^{-13} cm. The question thus arises whether this new constant is an independent seventh fundamental constant or whether it can be derived from the constants (5.101). It is interesting to note that from the constants (5.101) two combinations can be derived which have the dimension of length and the correct order of magnitude. One of them is the classical radius of the electron[1], known already from page 25

$$r_e = \frac{e^2}{2m_e c^2} = 1.41 \times 10^{-13} \text{ cm},\tag{5.102}$$

the other is the so-called Compton wavelength of the proton

$$l_0 = \frac{h}{M_P c} = 1.32 \times 10^{-13} \text{ cm}.\tag{5.103}$$

Both values agree surprisingly well with the value of the nuclear radius for the mass number $A=1$, according to Eq. (5.9), and also with the range of nuclear forces as computed from YUKAWA's formula (5.100) making use of the π-meson mass. Not only is it remarkable that the four magnitudes: electron radius, limit of nuclear radius, range of nuclear forces, and Compton wavelength of the proton agree so well numerically, but also that Eq. (5.102) is a formula of classical physics, while only Eq. (5.103) contains the quantum constant h. This makes the assumption look at least interesting that the quantity l_0 might play a particular role in describing nature since, according to Eq. (5.103), l_0 contains the two fundamental constants (h and c) of every field theory and quantum theory in addition to the mass of the nucleon.

From the six fundamental constants (5.101), three dimensionless constants may be formed so that it should be possible, in principle at least, to reduce the number of independent fundamental constants of physics to three. In the first place, we have the mass relation of proton and electron

$$\frac{M_P}{m_e} = 1836.12 = \beta.\tag{5.104}$$

In the second place, there is the dimensionless quantity e^2/hc which, multiplied by 2π,

$$\frac{2\pi e^2}{hc} = \frac{1}{137.037} = \alpha,\tag{5.105}$$

is known as SOMMERFELD'S fine-structure constant (see page 85). It is the ratio of the velocity of BOHR's hydrogen electron on its first orbit to the velocity of light. A third dimensionless quantity contains the gravitational constant f and may be written as

$$\frac{e^2}{m_e M_P f} = 2.28 \times 10^{39} = \gamma.\tag{5.106}$$

It is the ratio of the electrostatic to the gravitational attraction between electron and proton and thus also has a clear physical significance. If we regard, for lack of any better relation, the expression (5.103) as the universal smallest length, which is certainly correct with respect to the order of magnitude, then we have as a fourth dimensionless constant

$$\frac{h}{M_P l_0 c} = 1.\tag{5.107}$$

---

[1] The factor $\frac{1}{2}$ is obtained by assuming that the charge e is brought onto the sphere of radius r_e step by step in smallest quantities.

By means of the four relations (5.104) to (5.107) we may, at least formally, express four of the seven constants (5.101) and (5.103) by the three remaining ones. This means that we can transfer from the cgs-system with its arbitrarily defined units to a naturally defined system of physical units of h, c, l_0.

The decisive problem, still to be solved, is of course how to derive in a physically reasonable way the dimensionless constants (5.104) to (5.106) which should be founded in nature and, consequently, should follow *automatically* from a complete theory of the elementary particles. In spite of a number of interesting suggestions, a satisfactory theory of the physical significance of the dimensionless constants is still missing. Moreover, there exists not even an agreement on the most promising approach. Particular difficulties may be expected for the constant (5.106) because it seems extremely improbable that a constant of the order of 10^{39} follows from normal atomic or generally physical relations. This has led to the attempt to regard this quantity as evidence for a relation between atomic and astronomical data. It is noteworthy that the diameter of our universe measured in elementary units of length l_0 and its age in elementary units of time l_0/c have practically the same order of magnitude, viz. 10^{40}. The pertinent considerations by EDDINGTON, ERTEL, DIRAC, JORDAN, and others will be of great interest to any unbiased reader and will sharpen his view for the possibilities of such relations. It has to be stressed, however, that up to the present time all these combinations are no more than interesting but *very vague hypotheses*. These considerations show clearly on the other hand that, besides the extension of quantum mechanics (cf. IV, 15) and the theory of elementary particles, the deduction and the explanation of the dimensionless constants of physics presents the third fundamental problem of modern physics. There is little doubt that these three problems of present physics are closely interrelated and will find their solution by *one* new idea.

Problems

1. Deuterons circulate within the dees of a (fixed frequency) cyclotron in circular orbits. The largest orbit has a radius of 45 cm where the magnetic field is 13 kilogauss. Calculate (a) the maximum kinetic energy of deuterons in Mev; (b) the frequency of the applied electric field. If it is required to use the cyclotron to accelerate alpha particles using the same magnetic field, what frequency of the electric field would now have to be used and what would be the maximum kinetic energy of the alpha particles? Assume that an alpha particle has twice the mass of a deuteron.

2. Calculate the number of beta particles emitted per second by 10 g of Na^{24} whose half life is 15 hours. What would be the activity of this sample after 20 hours?

3. A gamma-ray photon of 2.22 Mev materializes in the neighbourhood of a heavy nucleus into a positron-electron pair. Assume that the electron and the positron move in opposite directions with equal speed. Compute the kinetic energy of each particle in Mev and the mass of each particle at this speed.

4. A radon nucleus at rest decays to a polonium nucleus with the emission of an α-particle. Using the table of atomic masses compute (a) the disintegration energy; (b) the energy and the momentum of the α-particle; (c) the recoil velocity of the daughter nucleus.

5. A stationary deuteron nucleus, when struck by a 2.62 Mev photon, disintegrates into a proton and a neutron. The total kinetic energy of the two particles is found experimentally to be 0.45 Mev. Determine from these data the mass of the neutron. The masses of deuteron and proton are given in the table of atomic masses.

6. Calculate (a) the energy release in the following reaction:

$$_7N^{13} \rightarrow {}_6C^{13} + \beta^+;$$

(b) the energy necessary to separate one neutron from $_8O^{17}$.

7. Assuming 200 Mev as the energy released in the fission of a single $_{92}U^{235}$ nucleus compute (a) the number of fission processes taking place each second in a nuclear reactor operating at a power level of 50 MW; (b) how much U^{235} is consumed each day in the reactor operating at this power level.

8. In the thermonuclear reactions taking place constantly in the deep interior of the sun, four protons combine to form a He^4 nucleus and two positrons. If about 4000 kg of mass is annihilated per second, calculate (a) the power in watts radiated from the surface of the sun and (b) the number of protons that must be converted per second into helium to produce energy at this rate.

Literature

General nuclear physics:

BITTER, F.: Nuclear Physics. Cambridge, Mass.: Addison-Wesley Press 1949.
CORK, J. M.: Radioactivity and Nuclear Physics. New York: Van Nostrand 1950.
EVANS, R. D.: The Atomic Nucleus. New York: McGraw-Hill 1955.
FERMI, E.: Nuclear Physics. Chicago: University Press 1950.
GENTNER, W., H. MAIER-LEIBNITZ u. W. BOTHE: Atlas of Typical Expansion Chamber Photographs. London: Pergamon Press 1954.
HALLIDAY, D.: Introductory Nuclear Physics. 2nd Ed. New York: Wiley 1955.
HANLE, W.: Künstliche Radioaktivität. 2nd Ed. Stuttgart: Piscator 1952.
HEISENBERG, W.: Die Physik der Atomkerne. 3rd Ed. Braunschweig: Vieweg 1949.
HEISENBERG, W.: Theorie des Atomkernes. Göttingen: Max-Planck-Institut 1951.
International Bibliography on Atomic Energy. Vol. II: Scientific aspects. New York: Columbia University Press 1951.
KAPLAN, I.: Nuclear Physics. 2nd Ed. Reading: Addison-Wesley 1963.
KORSUNSKI, M. I.: Isomerie der Atomkerne. Berlin: Deutscher Verlag der Wiss. 1957.
LAPP, R. E., and H. L. ANDREWS: Nuclear Radiation Physics. 2nd Ed. New York: Prentice Hall 1955.
MATTAUCH, J., u. S. FLÜGGE: Kernphysikalische Tabellen. Berlin: Springer 1942.
Nuclear Data. National Bureau of Standards. Circular 499ff. Washington D. C.: US Govt. Printing Office 1950.
POLLARD, E. C., and W. L. DAVIDSON: Applied Nuclear Physics. 2nd Ed. New York: Wiley 1951.
PRESTON, M. A.: Physics of the nucleus. Reading: Addison-Wesley 1962.
RIEZLER, W.: Einführung in die Kernphysik. 6th Ed. München: Oldenbourg 1959.
RUTHERFORD, E., T. CHADWICK and D. C. ELLIS: Radiations from Radioactive Substances. Cambridge: University Press 1930.
SEGRÉ, E. (Editor): Experimental Nuclear Physics. 2 Vols. New York: Wiley 1953.
SIEGBAHN, K.: Beta and Gamma Spectroscopy. Amsterdam: North Holland Publ. Co. 1955.
SULLIVAN, W. H.: Trilinear Chart of Nuclides. 2nd Ed. Washington D. C.: US Govt. Printing Office 1957.
WEIZSACKER, C. F. v.: Die Atomkerne. Leipzig: Akademische Verlagsgesellschaft 1937.
WIGNER, E. P.: Nuclear Physics. Philadelphia: University of Pennsylvania Press 1940.

Special fields of nuclear physics:

ALLEN, J. S.: The Neutrino. Princeton: University Press 1958.
BARKAS, W. H.: Nuclear Research Emulsions. New York: Academic Press 1963.
BUTLER, S. T., and D. HITTMAIR: Nuclear Stripping Reactions. New York: Wiley 1957.
CURRAN, S. C.: Luminescence and the Scintillation Counter. London: Butterworths 1953.
CURRAN, S. C., and T. D. CRAGGS: Counting Tubes. London: Butterworths 1949.
CURTISS, L. F.: Introduction to Neutron Physics. New York: Van Nostrand 1959.
DEVONS, S.: Excited States of Nuclei. Cambridge: University Press 1949.
FEATHER, N.: Nuclear Stability Rules. Cambridge: University Press 1952.
FRAUENFELDER, H.: The Mössbauer Effect. New York: Benjamin 1962.
FÜNFER, E., u. H. NEUERT: Zählrohre und Szintillationszähler. 2nd. Ed. Karlsruhe: Braun 1959.
HARTMANN, W., u. F. BERNHARD: Photovervielfacher und ihre Anwendung in der Kernphysik. Berlin: Akademie-Verlag 1957.
HUGHES, D. J.: Neutron Optics. New York: Interscience 1954.
KOLLATH, R.: Teilchenbeschleuniger. 2nd. Ed. Braunschweig: Vieweg 1963.

KOPFERMANN, H.: Kernmomente, 2nd Ed. Frankfurt: Akademische Verlagsgesellschaft 1956.
LIBBY, W. F.: Radioactive Dating. Chicago: University Press 1952.
LIVINGOOD, J. J.: Principles of Cyclic Particle Accelerators. Princeton: Van Nostrand 1961.
LIVINGSTON, M. S.: High Energy Accelerators. New York: Interscience 1954.
LIVINGSTON, M. S., and J. P. BLEWETT: Particle Accelerators. New York: McGraw-Hill 1962.
LÖSCHE, A.: Kerninduktion. Berlin: Deutscher Verlag der Wiss. 1957.
TOLANSKY, S.: Hyperfine Structure in Line Spectra and Nuclear Spin. London: Methuen 1948.
WILKINSON, D. H.: Ionization Chambers and Counters. Cambridge: University Press 1950.
WILLIAMS, J. G.: Principles of Cloud Chamber Technique. Cambridge: University Press 1951.
WIRTZ, K., u. K. H. BECKURTS: Elementare Neutronenphysik. Heidelberg: Springer 1958.
WLASSOV, N. A.: Neutronen. Köln: Hoffmann 1959.

Cosmic radiation and physics of elementary particles:
DALITZ, R. H.: Strange Particles and Strong Interactions. London: Oxford University Press 1962.
HEISENBERG, W.: Kosmische Strahlung. 2nd Ed. Berlin-Göttingen-Heidelberg: Springer 1953.
JANOSSY, L.: Cosmic Rays and Nuclear Physics. 1953.
LEPRINCE-RINGUET, L.: The Cosmic Rays. New York: Prentice Hall 1950.
MARSHAK, R. E.: Meson Physics. New York: McGraw-Hill 1952.
MARSHAK, R. E., and E. C. G. SUDARSHAN: Introduction to Elementary-Particle Physics. New York: Interscience 1961.
MONTGOMERY, D. J. X.: Cosmic Ray Physics. Princeton: University Press 1949.
POWELL, C. F., P. H. FOWLER, and D. H. PERKINS: The Study of Elementary Particles by the Photographic Method. London: Pergamon Press 1959.
ROCHESTER, G. D., and J. G. WILSON: Cloud Chamber Photographs of the Cosmic Radiation. London: Pergamon Press 1952.
ROMAN, P.: Theory of Elementary Particles. 2nd Ed. Amsterdam: North Holland 1961.
ROSSI, B.: High-Energy Particles. New York: Prentice Hall 1952.
THORNDIKE, A. M.: Mesons: A Summary of Experimental Facts. New York: McGraw-Hill 1952.
WILLIAMS, W. S. C.: An Introduction to Elementary Particles. New York: Academic Press 1961.

Nuclear Theory:
BETHE, H. A., and P. MORRISON: Elementary Nuclear Theory. 2nd Ed. New York: Wiley 1956.
BLATT, J. M., and V. M. WEISSKOPF: Theoretical Nuclear Physics. New York: Wiley 1952.
DÄNZER, H.: Einführung in die theoretische Kernphysik. Karlsruhe: Braun 1948.
FEENBERG, E.: Shell Theory of the Nucleus. Princeton: University Press 1955.
FERMI, E.: Elementary Particles. New Haven: Yale University Press 1951.
FRENKEL, J. I.: Prinzipien der Theorie der Atomkerne. 2nd Ed. Berlin: Akademie-Verlag 1957.
GAMOW, G., and C. L. CRITCHFIELD: Theory of the Atomic Nucleus and Nuclear Energy Sources. Oxford: Clarendon Press 1949.
GOEPPERT-MAYER, M., and J. H. D. JENSEN: Elementary Theory of Nuclear Shell Structure. New York: Wiley 1955.
NEMIROWSKII, P. E.: Contemporary Models of the Atomic Nucleus. New York: Pergamon Press 1963.
ROSENFELD, L.: Nuclear Forces. Amsterdam: North-Holland Publishing Co. 1948.
ROSENFELD, L.: Theory of Electrons. Amsterdam: North-Holland Publ. Co. 1951.
DE-SHALIT, A., and I. TALMI: Nuclear Shell Theory. New York: Academic Press 1963.

Radioactive isotopes and their applications:
BRODA, E.: Advances in Radiochemistry. Cambridge: University Press 1950.
FRIEDLÄNDER, G., and J. W. KENNEDY: Nuclear and Radiochemistry. New York: Wiley 1955.

HEVESY, G. v.: Radioactive Indicators. New York: Interscience 1948.
KAMEN, M. D.: Isotopic Tracers in Biology. 3rd Ed. New York: Academic Press 1957.
LIBBY, W. F.: Radioactive Dating. Chicago: University Press 1952.
MATTAUCH, J., u. A. FLAMMERSFELD: Isotopenbericht. Tübingen: Verlag der Zeit-
    schrift für Naturforschung 1949.
SCHUBERT, G.: Kernphysik und Medizin. 2nd Ed. Göttingen: Muster-Schmidt
    1948.
SCHWEITZER, G. K., and J. B. WHITNEY: Radioactive Tracer Techniques. New
    York: Van Nostrand 1949.
SCHWIEGK, H. (Editor): Künstliche radioaktive Isotope in Physiologie, Diagnostik
    und Therapie. Berlin-Göttingen-Heidelberg: Springer 1953.
SIRI, W. E.: Isotopic Tracers and Nuclear Radiations. New York: McGraw-Hill 1949.
WILLIAMS, R. R.: Principles of Nuclear Chemistry. New York: Van Nostrand 1950.
YAGODA, H.: Radioactive Measurements with Nuclear Emulsions. New York: Wiley
    1949.
ZIMEN, K.: Angewandte Radioaktivität. Berlin-Göttingen-Heidelberg: Springer 1952.

Nuclear energy and its application:

REACTOR Handbook. 6 Vols. Washington: US Atomic Energy Commission 1955.
BONILLA, C. F. (Editor): Nuclear Engineering. New York: McGraw-Hill 1957.
CAP, F.: Physik und Technik der Atomreaktoren. Wien: Springer 1957.
ETHERINGTON, H.: Nuclear Engineering Handbook. New York: McGraw-Hill 1958.
GLASSTONE, S.: Principles of Nuclear Reactor Engineering. New York: Van Nostrand
    1955.
GLASSTONE, S., and M. C. EDLUND: The Elements of Nuclear Reactor Theory. New
    York: Van Nostrand 1952.
HUGHES, D. J.: Pile Neutron Research. Cambridge, Mass.: Addison-Wesley 1953.
LITTLER, J. J., and J. F. RAFFLE: An Introduction to Reactor Physics. 2nd Ed.
    London: Pergamon Press 1957.
MURRAY, R. L.: Introduction to Nuclear Engineering New York: Prentice Hall 1954.
MURRAY, R. L.: Nuclear Reactor Physics. Englewood Cliffs, N. J.: Prentice-Hall 1957.
RIEZLER, W., u. W. WALCHER (Editors): Kerntechnik. Stuttgart: Teubner 1958.
SMYTH, H. D.: Atomic Energy for Military Purposes (Smyth Report). Princeton:
    University Press 1946.
STEPHENSON, R.: Introduction to Nuclear Engineering. 2nd Ed. New York: McGraw-
    Hill 1958.

VI. Molecular Physics

1. Molecular Physics and its Relation to Chemistry

In Chap. III we acquired a fairly thorough understanding of the structure, properties, and behavior of the atoms. In this chapter we shall discuss the structure of aggregates of atoms and their properties and processes, i.e., primarily physics of molecules. The natural extension of molecular physics, solid-state physics, will be discussed in Chap. VII.

Molecular physics is the science of the structure and properties of molecules so far as they are determined by physical methods. In this sense it is a logical continuation and extension of atomic physics proper. From what has been said it is obvious that molecular physics is closely related to chemistry. Chemistry attempts to determine the composition of a compound and the formula of its structure (e.g., of CH_3Cl). From such chemical formulae, the chemist derives conclusions as to the behavior of the molecule, i.e., the possibility of its reaction with other molecules or atoms. The correctness of the chemical formulae is checked by chemical analysis, and finally methods are developed for synthesizing the molecules from their constituent elements. Chemical methods are also used for determining molecular properties and certain characteristic quantities, among them the heat of formation of the molecule, e.g., the energy freed in forming

HCl from $\frac{1}{2}$ mole H_2 and $\frac{1}{2}$ mole Cl_2. Chemistry is not able however, as we mentioned already when discussing the Periodic Table, to *explain* the chemical valency of the atoms forming molecules, nor the molecular structures found, nor finally the different stabilities of different bonds in molecules.

It is therefore the most important task of molecular physics to determine how the atoms are bound in a particular molecule. The young student of chemistry often asks himself the question why there is a stable molecule NH_3 but not NH, a stable CO and CO_2 but no CO_4. It is an achievement of atomic physics with its quantum-mechanical atomic theory that we have now a basis for answering this fundamental question by a theory of the chemical bond. It is, therefore, not an overstatement when we say that only atomic and molecular physics has furnished a solid theoretical foundation for its older sister science, chemistry.

Molecular physics, on the other hand, attacks the molecule by *physical* methods. It determines the spatial arrangement of the atoms and the distances between them, its possibilities of rotation, and the moments of inertia of the molecule referred to the different axes of rotation, its modes of vibration, and the dissociation energies required to split a diatomic molecule or to separate an atom or group of atoms from a polyatomic molecule. Last but not least it determines the arrangement of the electrons in the shells of the molecule, the possibilities of exciting electrons, and the effects of excitation and ionization on the properties of the molecule. There is actually no known property of a molecule which is not tested and studied carefully by the molecular physicist.

The foundations of molecular physics rest, as we hardly need to say, on atomic physics. The properties of the atoms which were described in Chaps. III and IV are prerequisite for understanding the structure of molecules. Therefore, the molecular physicist makes use of modern theoretical physics, especially quantum physics, to an increasing extent. He also utilizes in his experimental studies such specific physical methods as spectroscopy, X-ray, neutron and electron diffraction, the determination of dipole moments, or anisotropy measurements by means of the electro-optical Kerr effect. In general, chemical results are not used in molecular physics. The chemical behavior of a molecule should follow from the correct physical picture of the molecule, and it is used to verify the correctness of this picture.

The results of three decades of research in molecular physics are so striking that modern inorganic and physical chemistry to an increasing extent not only has taken over many individual methods of molecular physics, but is actually doing molecular-physics research just as it was done before by physicists. There is scarcely a border line any more between physics and chemistry, and this is especially welcome and necessary in the study of the polyatomic molecules of organic chemistry, since few physicists have that extent of chemical knowledge which is necessary for investigations in the field of organic molecular physics.

In our presentation we shall be concerned chiefly with the diatomic molecules because their behavior is comparatively simple and easy to understand. Furthermore, fundamental research on diatomic molecules seems to have reached a certain state of completion. The polyatomic molecules will not be considered in detail because this field belongs more distinctly to the realm of chemistry. This treatment seems the more justified as the behavior of polyatomic molecules, though much more complicated, does not differ fundamentally from that of the diatomic molecules.

2. General Properties of Molecules and their Determination

a) Size and Nuclear Arrangement of Molecules

In discussing the special methods of molecular research, we limit ourselves to purely physical methods. We thus disregard entirely the methods used in chemistry for determining structure formulae.

The diameter of a molecule may be determined by methods discussed for the atom on page 14 from the covolume [the constant b in the van der Waals equation of state (2.8)], from the density in the liquid or solid state, and for gases, from the measurement of viscosity. The statements made about the problem of defining the radius or diameter of the atom, page 14, hold for molecules also. Therefore, different methods produce different values, and the viscosity method in particular yields a molecular diameter which is temperature-dependent. The diameter of most diatomic molecules is about 3 to 4 Å. However, we should not think of the molecule as spherical in shape but regard the diatomic molecules as ellipsoids, whereas the shape of the polyatomic molecules, depending on their particular structure, has to be regarded in a rough approximation as rod-like, tetrahedral, etc.

X-ray, neutron and electron diffraction methods have been developed into important methods for investigating the sizes, internuclear distances, and binding strengths especially of polyatomic molecules. In a manner which is not simple nor easy to understand theoretically, the type and spatial arrangement of the diffracting centers may be determined by measuring the angular distribution of X-rays, neutrons or electrons scattered by molecules. By a refinement of the experimental methods as well as of the methods of computation, it has been possible to determine not only the internuclear distances and thus the atomic structure of the molecule but also the electron density distributions (cf. Fig. 227) and, from it, the type and strength of the binding forces between the different molecular groups or atoms of a complicated molecule. According to ABBE's theory, each optical image can be interpreted as a diffraction phenomenon of light waves originating from the object. By comparing this theory with X-ray diffraction by atomic systems, BRAGG came to the conclusion that the pictorial representation of a molecule or crystal (Figs. 180 and 227), obtained from the so-called Fourier analysis, corresponds to what one would actually see by means of a hypothetical X-ray microscope. An X-ray diffraction pattern is the result of the fact that monochromatic X-rays are scattered by the lattice points (atoms) only into certain directions of space and with characteristic intensities (cf. VII, 4). Since there is a definite relation between the diffraction pattern and the arrangement of the scattering atoms in a molecule or elementary crystal cell, this arrangement can be determined by working backwards from the angles and intensities of the X-ray diffraction pattern. The path of the rays is then sort of reversed: From the observed interference pattern, the spatial arrangement of the atoms that produces this pattern by diffracting the X-rays of known wavelength can be computed. It is even possible to determine the atomic distribution in the diffracting objects (molecules or crystals) directly by photographic means instead of mathematical ones. Fig. 180 shows a result of this method for the hexamethylbenzene molecule, $C_6(CH_3)_6$. The hexagonal ring of the benzene molecule as well as the carbon atoms of the CH_3 groups attached to its corners are clearly recognized, whereas the hydrogen atoms, because of their small scattering power which is proportional to the square of the atomic number Z, do not show up. In contrast to X-ray diffraction, the diffraction of neutrons by atoms is nearly independent of the atomic mass. Neutron diffraction is therefore successfully utilized for determining the structure of such molecules in which hydrogen atoms occur beside heavy atoms.

DEBYE has developed methods for measuring the size and approximate shape of very large molecules containing thousands of atoms whose detailed structure cannot be determined by other methods because they are too complicated. He utilized either the scattering of light by these molecules in solutions or their behavior when they move in solutions through capillary tubes (change of behavior as a consequence of orientation).

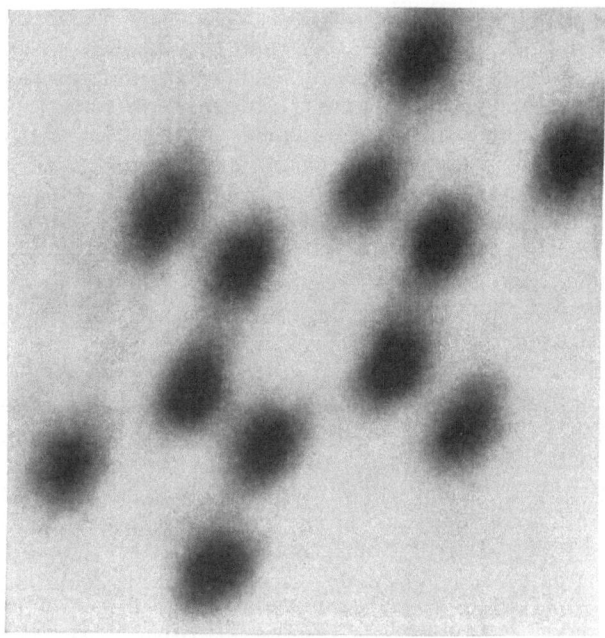

Fig. 180. Photographic construction of a picture of the hexamethyl-benzene molecule, $C_6(CH_3)_6$, based on X-ray data by BROCKWAY and ROBERTSON. (Courtesy of M. L. HUGGINS)

By making use of one or several of the methods mentioned it is possible to obtain a fairly accurate picture of the structure of a molecule; these methods are particularly valuable for investigating the structure of complicated polyatomic molecules.

b) Permanent Dipole Moments of Molecules

If the spatial arrangement of the atoms in a molecule is known, we still have to find out whether the molecule consists of neutral atoms or, in a first approximation, of positive and negative ions. In the *atomic molecules* such as H_2, O_2, etc., the centers of the positive charges and the negative charges coincide; this may be seen from the observation that such molecules do not have a resulting electric dipole moment. However, if a molecule such as NaCl consists of an electropositive and an electronegative partner, the more electronegative atom (Cl) attracts the valence electron of the more electropositive atom (Na) so that the NaCl molecule may be written, in a first approximation, as an ionic molecule Na^+Cl^-. Here, the centers of the positive and the negative charges do *not* coincide; such molecules have a *permanent dipole moment* $\mathfrak{M}_p$ which equals the product of the displaced charge (here the charge of the electron) and the distance l

between the two charge centers,

$$\mathfrak{M}_p = e l. \tag{6.1}$$

We expect all molecules which consist of *different* partners because of their different electronegativities to exhibit some displacement of their positive and negative charges against each other. Such molecules we expect to have permanent dipoles. If it is possible to determine these electric moments, Eq. (6.1) allows to compute the relative charge displacement l. For $l = 0$ we have an ideal atomic molecule; for l equaling the distance r_0 of the atomic centers, we have (for diatomic molecules) an ideal *ionic molecule*. Most of the real molecules are intermediate cases. Determination of their dipole moments represents the easiest method for finding their actual charge distribution.

The method is based on the measurement of the dielectric constant ε of the molecular gas or vapor in a condenser. If the dielectric polarization of the gas is designated by $\mathfrak{P}$, which is the resulting dipole moment per cm³, and if $\mathfrak{E}$ is the electric field strength, their relation is

$$\varepsilon = 1 + \frac{4\pi\mathfrak{P}}{\mathfrak{E}}. \tag{6.2}$$

If, furthermore, $\overline{\mathfrak{M}}_p$ is the average contribution of each molecule to the resulting moment $\mathfrak{P}$ and N_0 the number of molecules per cm³, then we have

$$\mathfrak{P} = \overline{\mathfrak{M}}_p N_0. \tag{6.3}$$

In a gas-filled condenser, the electric field attempts to orient the molecular dipoles $\mathfrak{M}_p$ parallel to the field direction, while the thermal motion counteracts this orientation. The actual average electric moment per molecule is therefore directly proportional to the field strength and inversely proportional to the thermal energy kT,

$$\overline{\mathfrak{M}}_p = \frac{\mathfrak{E}\mathfrak{M}_p^2}{3kT}. \tag{6.4}$$

The factor 3 in the denominator originates from the averaging over all angular orientations of the individual dipoles. Except for this factor, the formula may be understood from dimensional reasoning. By combining Eqs. (6.2) to (6.4), we obtain the relation between measured dielectric constant ε and unknown molecular dipole moment $\mathfrak{M}_p$:

$$\varepsilon = 1 + \frac{4\pi N_0 \mathfrak{M}_p^2}{3kT}. \tag{6.5}$$

Eq. (6.5) is incomplete in so far as the electric field used in the experiment shifts the electrons in the molecule and thus induces an additional dipole moment which is proportional to the field. Its amount, according to Eq. (6.6), is *independent* of the temperature, contrary to Eq. (6.5). If we therefore measure the dielectric constant ε as a function of the absolute temperature T, and if we then plot ε versus $1/T$, we obtain a straight line the gradient of which allows to compute the dipole moment of the molecule according to Eq. (6.5).

For several diatomic molecules, Table 15 shows the permanent dipole moments determined in this way and the distances l of their charge centers computed from Eq. (6.1). Since the internuclear distances in these molecules are of the order of 1 to 3×10^{-8} cm (cf. VI, 9a), the distance of the charge centers is in general small compared to the internuclear distances of the atoms which form the molecules.

Table 15. *Dipole Moments* ($\mathfrak{M}_p$) *and Distances* (l) *of the Charge Centers for Some Diatomic Molecules*

	CO	NO	HI	HBr	HCl	NaI	KCl	CsI
$\mathfrak{M}_p \times 10^{18}$ esu	0.1	0.13	0.38	0.78	1.03	4.9	6.3	10.2
$l \times 10^8$ cm	0.02	0.03	0.08	0.16	0.21	1.0	1.3	2.1

Even for the alkali halides, l does not quite reach the value of the internuclear distance. Table 16 lists some dipole moments of molecules with several atoms; in many cases they allow conclusions about the molecular structure or they can be used to distinguish between different possible atomic arrangements obtained from

Table 16. *Dipole Moments of Some Polyatomic Molecules* (in units of 10^{-18} esu)

N_2O	PH_3	AsH_3	H_2S	$CHCl_3$	NH_3	SO_2	H_2O	CH_3Cl	NCN
0.14	0.55	0.15	0.93	0.95	1.46	1.61	1.79	1.97	2.8

other experiments. Thus we know, for example, from the dipole moment of N_2O that the symmetric shape of the molecule with the oxygen atom in the center is not possible since in this case the dipole moment should be zero because of the symmetrical charge distribution. A more detailed discussion actually shows that the N_2O molecule has the asymmetric form $N\equiv N=O$. In the physics of polyatomic molecules, the so-called physical stereochemistry, it is in many cases possible to assign electric moments to certain definite bonds in the molecule. The expected resulting electric moment for different possible molecular models can then be estimated from the contributions of the different molecular groups if the mutual disturbance of different bonds affecting the same atom is taken into account. This resulting electric moment, computed for the different possible atomic arrangements, when compared with the actually measured dipole moment, often allows one to determine the correct structure of the molecule.

c) Polarizability and Induced Dipole Moments of Molecules

Besides orienting permanent dipoles (if these exist), an electric field has still another effect on molecules. It may induce in every atom or molecule an additional dipole moment $\mathfrak{M}_i$ by displacing the negative and positive charges in opposite directions, $\mathfrak{M}_i$ being proportional to $\mathfrak{E}$,

$$\mathfrak{M}_i = \alpha\,\mathfrak{E}. \tag{6.6}$$

The constant α is characteristic for a specific atom, ion, or molecule. In the cgs system it has the dimension cm^3 and is called *polarizability*; it is a measure of the deformability of the electron shells. Since $\mathfrak{M}_i$ is created by the field $\mathfrak{E}$ and therefore, in isotropic media, always has the same direction as the field, it is *not* subject to the desorienting effect of the thermal motion. The induced dipole moment per cm^3 for N_0 molecules per cm^3 thus is simply

$$\mathfrak{P}_i = N_0\,\mathfrak{M}_i = N_0\,\alpha\,\mathfrak{E}. \tag{6.7}$$

By using Eq. (6.2), we obtain instead of Eq. (6.5) the most general expression for the dielectric constant ε:

$$\varepsilon = 1 + 4\pi N_0\left(\frac{\mathfrak{M}_p^2}{3kT} + \alpha\right). \tag{6.8}$$

The first term in the parenthesis is zero if the molecules do not have permanent dipoles.

According to Eq. (6.8), the determination of the polarizability α is, in principle, possible by an absolute measurement of the dielectric constant ε. In practice it is done by measuring the refractive index n, the square of wich equals ε for sufficiently long wavelengths. The refractive index is determined for long waves, i.e., for red or infrared light, so that the inertia of the electron shells does not yet play a role during the change of the field strength in the lightwave. If the measurements are not carried out in gases but in liquids or solids, the interaction between atoms or between molecules has to be taken into consideration and the formula for α that follows from Eq. (6.8),

$$\alpha = \frac{\varepsilon - 1}{4 \pi N_0},\tag{6.9}$$

has to be replaced by the so-called Lorentz-Lorenz formula

$$\alpha = \frac{3}{4 \pi N_0} \frac{\varepsilon - 1}{\varepsilon + 2}.\tag{6.10}$$

In Table 17, α-values for several atoms, atomic ions, and molecules are listed. The ions are included because they play an essential role in atomic physics. We mentioned already on page 83 that the character of the spectra of the alkali

Table 17. *Values of the Polarizability of Some Atoms, Ions, and Molecules* (in units of 10^{-24} cm^3)

H	He	Xe	Na+	K+	Cs+
0.56	0.21	4.0	0.17	0.8	2.4

H_2	Cl_2	NO	CCl_4	CS_2	NH_3	C_6H_6
0.61	3.2	1.8	10.5	5.5	2.2	6.7
0.85	6.6	5.3		15.1	2.4	12.8

atoms is mainly determined by the polarizing influence of the valence electron on the atomic core; and the polarizability of the atomic ions plays a great role as well for ionic crystals (cf. VII, 5).

d) Anisotropy of the Polarizability.
Kerr Effect, Rayleigh Scattering, and Raman Effect

We listed in Table 17 two different values for the polarizability of all molecules except for the carbon tetrachloride molecule with its spherical symmetry. The reason is evidently that only systems with spherical symmetry, such as CCl_4, atoms, and atomic ions, have an isotropic polarizability, i.e., a polarizability which is the same for all angular orientations toward the electric field, whereas molecules without spherical symmetry show different values for the polarizability along different molecular axes. *Knowledge of the anisotropy of the polarizability thus allows direct conclusions concerning the molecular shape.* In general, this anisotropy can be measured directly if the molecules are oriented in a molecular crystal (cf. VII, 3). In that case the dielectric constant ε may have very different values for different directions of the crystal.

Measurements of the *electro-optical Kerr effect* allow conclusions about the anisotropy of the polarizability of the electron shells of molecules. According to this effect, which was discovered in 1875, certain molecular gases and liquids show birefringence in a strong electric field. This means that they propagate light that is polarized parallel to the electric field (x-direction) with a different velocity than light polarized parallel to the y-direction. By measuring the refractive indices in both directions of polarization and under consideration of $n^2 = \varepsilon$ and Eqs. (6.9) or (6.10), it becomes possible to determine the anisotropy of the polarizability.

We may mention that the absorption spectrum is different for the two directions of polarization also. This effect, called *dichroism*, follows from the relation between refractive index and absorption given by Eq. (3.149) of the dispersion theory, applied here to birefringent substances.

The anisotropy of the polarizability of gaseous substances may be determined roughly also by measuring the degree of polarization of scattered light. Incident light is scattered by atoms and molecules to a small extent into all directions (so-called Rayleigh scattering), and the intensity of the scattered radiation depends on the average polarizability $\bar{\alpha}$ of the scattering molecules. For anisotropic molecules, the electric moment induced in the molecule by the incident light wave does in general not coincide with the direction of the electric field vector. Consequently, the incident polarized light is the more depolarized by the scattering process the more asymmetric the electron shell of the scattering molecules is. *A high degree of depolarization (10% to 25%) thus indicates relatively long or flat molecules.*

Another phenomenon related to the molecular effects discussed so far is the *optical activity*. This is the fact that certain molecules have different refractive indices for left-handed and for right-handed circularly polarized light. From this phenomenon conclusions can again be drawn about the electron arrangement and its polarizability. Molecules with optical activity also have different absorption spectra for left-handed and right-handed circularly polarized light, an effect called *circular dichroism*. The optical activity is caused by such an asymmetric arrangement and coupling of atom groups within the molecules (details cannot be discussed here) that a dipole moment induced in the x-direction by an incident light wave [Eq .(6.11)] leads to the excitation of a dipole moment in the y-direction in another atom group of the same molecule. This asymmetry, in combination with the fact that the finite size of the light wavelength excites the various groups of the molecule with a different phase, explains the optical activity. Conversely, conclusions may be drawn about the asymmetry of the polarizable electron shells of large molecules from measuring the optical activity.

Finally, we have to discuss the *Raman effect*, which has to do with the polarizability of the electron shells of molecules also. As was mentioned before, a light wave with the frequency ν_0 incident on an isotropic molecule induces in this molecule a dipole moment

$$\mathfrak{M}_i = \alpha\,\mathfrak{E}_0 \sin 2\pi\nu_0 t, \tag{6.11}$$

the amount of which is proportional to the polarizability α. Let us investigate now the special case in which α is not constant but changes linearly with the elongation if the atoms in the molecules vibrate against each other with the frequency ν_s or if optically anisotropic molecules rotate around their centers of gravity. By setting

$$\alpha = \alpha_0 + \alpha_1 \sin 2\pi\nu_s t \tag{6.12}$$

and inserting this expression into (6.11), we obtain

$$\mathfrak{M}_i = \alpha_0\,\mathfrak{E}_0 \sin 2\pi\,\nu_0 t + \frac{1}{2}\,\alpha_1\,\mathfrak{E}_0 \left(\cos 2\pi\,(\nu_0 - \nu_s)\,t - \cos 2\pi\,(\nu_0 + \nu_s)\,t\right). \tag{6.13}$$

While the first term represents an induced dipole moment that vibrates with the exciting frequency ν_0 and thus corresponds to classical Rayleigh scattering, the two terms in brackets obviously indicate the scattering of two light waves the frequencies of which are displaced against that of the exciting wave by $\pm\nu_s$. This is the Raman effect, which is very important for molecular physics: If molecules are irradiated by monochromatic light of frequency v_0, we observe in the spectrum of the laterally scattered light some faint so-called *Raman lines* (cf. Fig. 181) symmetrically on both sides of the spectral line of the exciting frequency v_0. Their frequencies are $v_0 \pm v_s$, *where v_s is equal to some vibration or rotation frequency of the scattering molecule, which thus can be determined by measuring the wavelength difference between the primary line, v_0, and the Raman lines v.*

The frequency change $\pm v_s$ of the scattered light is easily understood quantum-theoretically. Because of $E = hv$, it means that either the scattered photon transfers part of its energy to the molecule and thus excites a higher rotational or vibrational state, whereas the rest of the energy is scattered as a Raman line of lower frequency; or vice versa that the scattering molecule transfers some part of its rotational or vibrational energy to the photon. There is no space to discuss here the wave-mechanical theory of the effect. It is sufficient to know that, according to

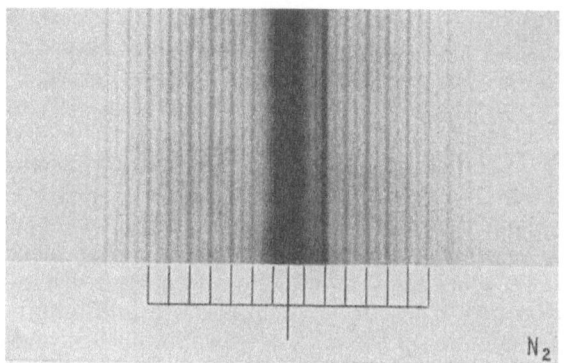

Fig. 181. Raman spectrum (rotational Raman effect) of the N_2 molecule. (After RASETTI)

Eq. (6.13), *only those vibrations (and rotations) are "Raman-active" the excitation of which changes the polarizability of the molecule.* This is different from the case of the normal optical spectra where, according to IV, 9, only those changes of the state of a system can be observed in absorption or in emission in which the *electric moment* of the system changes.

We conclude our discussion by adding a few remarks about the relation between classical scattering (Rayleigh scattering), quantum scattering (Raman effect), and molecular fluorescence. Eq. (6.13) contains implicitly the assumption that the incident frequency v_0 differs from all the resonance frequencies of the molecule, in other words, that its energy hv_0 does not coincide with the excitation energy of any of the stationary molecular energy levels. If v_0 approaches such a resonance frequency, however, the electric moment (6.11) induced in the molecule by the incident light wave and, with it, the intensity of the scattered light of this wavelength becomes larger and larger, until the scattering of the incident frequency, in the case of resonance, is identical with its absorption and re-emission frequency. The molecule then emits fluorescent radiation. If merely the incident frequency is re-emitted, i.e., if a transition of the excited molecule into its original state occurs, then we have the analogue to the classical Rayleigh scattering, which also occurs without any change of frequency. However, if the fluorescence emission is due to a transition into another vibrational or rotational state of the molecule, i.e., if the fluorescence spectrum is not identical with the absorption spectrum, we recognize the relation to quantum scattering, i.e., to the Raman effect which also involves an energy change.

3. Spectroscopic Methods of Determining Molecular Constants

The Raman effect leads to the spectroscopic methods of molecular physics to which we owe the decisive part of our knowledge of structure, properties, and dimensions of the simpler molecules. Beside the Raman spectroscopy, these methods comprise band spectroscopy, including infrared spectroscopy, and the high-frequency spectroscopy which supplements the other methods to an increasing extent. Spectroscopic research has yielded a basic understanding of even the finest details of the behavior of the diatomic molecules, e.g., the interaction between electronic motion, vibration, and rotation. It has made possible the determination of all interesting molecular data with high precision. In the case of polyatomic molecules, spectroscopic research has provided information about the possibilities of vibration and dissociation and the influence of the excitation of electrons. Spectroscopic methods also permit us to determine the type and stability of different chemical bonds, the geometrical arrangement of excitable electrons, and other details. In all these studies, the investigation of the mutually complementary absorption and emission spectra is rounded out in a fortunate way by the Raman effect, which permits us to determine the optically inactive vibration and rotation frequencies which cannot be determined from the other spectra.

Investigation of the longwave rotation spectra, of the rotational structure of rotation-vibration and band spectra as well as the rotational Raman effect permits the exact determination of the moments of inertia of a molecule about its different axes of rotation. In the case of diatomic molecules with known masses of the constituent atoms, the internuclear distances may be computed from the moments of inertia for each electron and vibration state of the molecule. The investigation of the infrared vibration spectra and the vibrational structure of the band spectra as well as the vibrational Raman effect enables us to compute for diatomic molecules the constants of the quasi-elastic binding forces for each electron configuration from the fundamental vibration frequencies. In the case of polyatomic molecules with many possibilities of vibration, however, the modes of vibration must be known in order to determine the constants of the binding forces from the measured vibration quanta. Investigation of the vibrational structure of band spectra and especially the study of continuous molecular spectra furnishes the decisive data of the "breaking strength", the dissociation energy of the normal molecule as well as of the molecule in its different excited electron states. These data, as well as the details of the dissociation as a consequence of absorption of radiation, form the basis of photochemistry. Important conclusions may also be drawn from continuous spectra about the process of formation of molecules from their constituent atoms. For all these studies, the spectroscopist has available not only the stable molecules known to the chemist but also normally unstable molecules. It has been especially valuable for the explanation of molecular structure and for the kinetics of reactions that radicals and unstable intermediate products such as OH, NH, ClO, CN, C_2, diatomic metal hydrides, etc. can be produced in sufficient concentrations in electric discharges so that their spectra can be studied.

The same spectroscopic equipment which was mentioned in III, 1 may be used for recording molecular spectra. Emission and absorption spectra complement each other in a very convenient way. Fluorescence, i.e., the emission of light by molecules upon absorption of exciting radiation, usually of monochromatic light, often is of great help for investigating the behavior of molecules.

The conditions for intense excitation of molecular emission differ considerably from those for the excitation of atomic spectra. For atomic spectra, strong excitation is favorable. In order to produce band spectra, the excitation must be "soft" enough to prevent dissociation, because otherwise the molecules which emit the band spectra may dissociate into atoms or groups of atoms. The same conditions hold for thermal excitation. Line spectra of *atoms* appear in the axis of the electric arc with its temperature of 7000 °K and more, whereas the

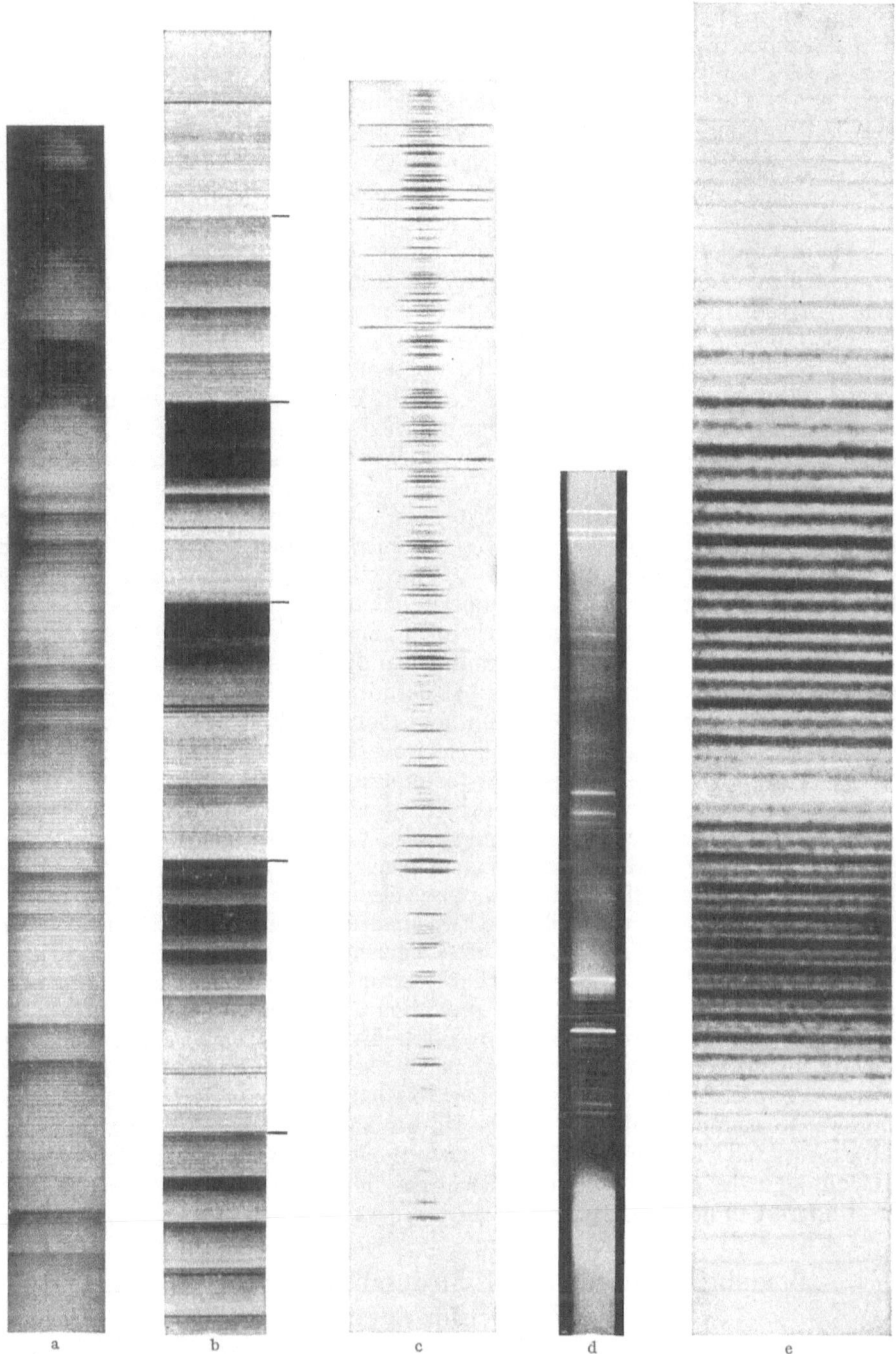

Fig. 182a–e. Examples of molecular band spectra: (a) absorption band spectrum of the iodine molecule I_2 photographed by MECKE with high dispersion (Rowland grating); (b) band spectrum of the PN molecule after CURRIE and HERZBERG; (c) widely separated and therefore scarcely recognizable bands of the so-called many-line emission spectrum of the H_2 molecule (large Rowland grating), photographed by the author; (d) emission spectra (diffuse bands and continua) of the Hg_2 van der Waals molecule (see VI, 8), photographed with low dispersion by MROZOWSKI; (e) absorption band of the C_2H_2 molecule, photographed with high dispersion (Rowland grating) by MECKE. Fig. (d) is, in contrast to the other photographs, a positive

cooler envelope of the arc serves as one of the most intensive sources of *band spectra*. In an electric arc in air, spectra of the oxides and nitrides of most elements may be excited by introducing a trace of the element into the arc. From a hydrogen atmosphere, similarly, a large number of hydride spectra may be obtained. The various forms of glow discharges with or without electrodes (high-frequency discharges) are well suited for exciting molecules of gases or vapors. The band spectra of neutral molecules are predominantly found in the positive column of the discharge, whereas in the negative glow the molecular ions such as CO^+, N_2^+, O_2^+, etc. are excited.

The excitation of band spectra by fluorescence, in which the molecule is irradiated by suitable shortwave (mostly monochromatic) light, is of interest for many investigations. Fluorescent radiation is observed perpendicularly to the direction of the exciting radiation. By varying the wavelength and thus the energy of the exciting radiation, one part of a molecular spectrum after the other may be excited. The absolute position of the corresponding levels in the term diagram thus can be established. Our knowledge of the molecules I_2, S_2, Hg_2, Cd_2, and Zn_2, for example, depends almost entirely on such fluorescence studies. Fig. 182 gives a survey of the various forms of band spectra of molecules.

Microwave spectroscopy (see page 50), applied to molecular studies, has been a great success. With its help, very small energy differences can be measured with high precision. According to VI, 9a, for example, the quantized changes of the rotation energy are inversely proportional to the moment of inertia of the rotating molecule. Microwave methods are a means of determining the moments of inertia and in simple cases the internuclear distances and the atomic arrangement of such molecules the rotational structure of which cannot be resolved by normal spectroscopic methods. STARK and ZEEMAN splitting of molecular lines can be measured precisely with weak electrostatic and magnetic fields, respectively. Their knowledge is important for molecular theory. Also, electric dipole moments of normal as well as excited atoms may be determined from Stark-effect measurements by means of microwaves. A further field of application concerns the energy splitting of rotation levels which is caused by the interaction of the molecular field with that of the electrons in motion as well as by the interaction of the molecular field and the quadrupole moment of the nucleus or the nuclei (cf. V, 4b). The results of investigating such interactions seem to give information also about changes of the electron arrangement due to chemical valence saturation (see VI, 14), thus providing a new experimental approach to this fundamental field of molecular physics. Determination of exact values of nuclear quadrupole moments as well as of the nuclear spins from molecular microwave spectroscopy establishes a new relation between molecular and nuclear physics. Measurement of small energy differences (and thus frequencies) such as those associated with the pendular motion of the N atom in the ammonia molecule NH_3 through the plane of the three H atoms (inversion spectrum of NH_3) is of great interest to molecular physics itself.

4. General Survey of the Structure and Significance of Molecular Spectra

If the emission and absorption spectra of any molecule built up of two dissimilar atoms is investigated by means of the spectroscopic equipment described in III, 1, three distinct groups of spectra are distinguished more or less clearly, which differ with respect to their locations in different spectral regions and by their different degrees of complexity.

For diatomic molecules, a series of equidistant lines is found in the far infrared. Polyatomic molecules show in this region some relatively simple sequences of lines. For reasons to be discussed below, these spectra are called the rotation spectra of the molecules.

In the near infrared at several μ, a number of line sequences is found which are clearly arranged in a regular manner. The sequences with the shortest wavelengths extend into the photographic infrared, e.g., Fig. 182e. These sequences of lines are called the rotation-vibration spectrum.

Finally, we find in the photographic infrared, in the visible, and ultraviolet regions' of the spectrum a larger or smaller number of regularly arranged but sometimes very complicated groups of lines (so-called bands). In addition to these bands (see Figs. 182a—d) continuous spectra are frequently found also. The line sequences which appear here as in most molecular spectra, and which are "shaded toward one side", are called *bands* because they appear as structureless band-like phenomena (cf. Fig. 182b) in low-dispersion spectroscopes. Each such band spectrum in general shows a threefold structure. It consists of a number of clearly separated groups of bands which, according to their general appearance, belong together, the so-called band systems. Each band system is built up of a number of bands which are occasionally arranged in sequences such as in Fig. 182a, b. Finally, each band consists of a number of regularly arranged band lines.

This threefold structure of a band spectrum corresponds to a partition of the total energy of the molecule into three parts, the excitation energy of the electrons (electron energy), the energy of vibration of the atom or of the nuclei of the molecule with respect to each other (vibration energy), and the kinetic energy of rotation of the molecule about an axis perpendicular to the line joining the nuclei (or in the case of polyatomic molecules, about the three principal axes of inertia), the rotation energy. These three parts of the total energy can change individually or together upon emission or absorption of radiation and thus give rise to the different spectra.

The theory which we shall discuss later shows that the relatively small amounts of energy which correspond to the emission or absorption of the lines in the far infrared are due solely to changes of the rotation energy of the molecule. The spectrum associated with such changes of the rotation energy is called the rotation spectrum. The spectrum which we find in the near infrared corresponds to changes of the vibration *and* rotation of the molecule. The superposition of these two types of energy changes produces the numerous lines and the complexity of the rotation-vibration spectra. The band spectra which are found in the shortwave region, i.e., in the visible and ultraviolet, correspond to changes in the configuration of the electrons, the vibration and the rotation of the molecule. These energy changes account for the often very complicated structure of the electron spectra (band spectra in the more restricted sense) of the molecules.

The largest change of the total energy of a molecule occurs as a result of a change of the configuration of the electrons, i.e., by an electron transition. Thus the position of the band system in the spectrum is determined by the energy of the electron transition. All bands of a band system, therefore, belong to the same electron transition. The change in molecular energy due to a change of the electron configuration is by an order of magnitude larger than that due to quantized changes of the vibration. The position of a band in the band system thus is determined by the vibrational transition, i.e., the difference of the vibration energy in the initial and final state. The different lines of a band are the result of quantized

changes of the rotation of the molecule. The change of the electron configuration, i.e., the electron transition, as well as the change of the vibration state
are the same for all lines of the same band. The change of the total molecular
energy as a result of a rotational transition is again by an order of magnitude
smaller than the change in energy corresponding to a vibrational transition.

We can treat this same problem of the structure of a band spectrum from a
somewhat different point of view by considering the simplest model of a diatomic
molecule. This model consists of two positively charged nuclei which are separated
by a distance r_0, and of an electron shell which in some way yet to be determined
holds the two nuclei together in spite of the electric repulsion between them. In
the old pictorial model of a molecule, the outer electrons revolve around both
nuclei. This idea of a common electron shell is in good agreement with the kinetic

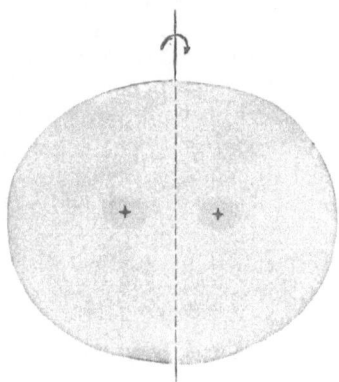

theory, according to which a diatomic molecule
acts in a collision as if it were an ellipsoid of
rotational symmetry with a relatively small
eccentricity (Fig. 183). If one imagines the internuclear distance in a diatomic molecule to
approach zero with the two nuclei fusing into
one, one obtains an atom. From this concept it
follows that the electron shells, and with them
the energy-level diagram and spectrum of a
diatomic molecule, must be closely related to
the electron shells, the energy-level diagram, and
the spectrum of the atom which is formed by
fusion of the two atoms of the molecule. *The* H_2
molecule thus must be similar to the He *atom*[1].

Fig. 183. Schematic representation of nuclei
and electron cloud of a diatomic molecule
(with rotation axis)

Now if we assume that in our model of the
molecule, Fig. 183, changes of the energy state
occur only by electron transitions as they do
in the atom, then the spectrum would consist of individual lines corresponding to
the different electron transitions, the so-called fundamental lines, which would
appear in the wavelength regions of the actually observed band systems. If we
then include the changes of vibration of the nuclei within the molecule as possible
changes of energy states, we superimpose on each electron transition a sequence
of possible changes of the vibration energy. Each fundamental line thus would
split up into a system of so-called zero lines which would appear at the same wavelengths in the spectrum as the actual bands in the band system. As a last approximation to the actual state of affairs, we must include the quantized changes of
the rotation of the molecule about its rotation axis, shown as a dotted line in
Fig. 183, as possible changes of the total energy. Then each zero line splits up
into a whole band, i.e., a series of band lines, because to each quantized change
of the electron and vibration energy the multiplicity of the possible changes of
the rotation energy has to be added. In this way we obtain the actually observed
threefold structure of a band spectrum, though this is not always immediately
apparent in a band spectrum (compare the spectra, Fig. 182).

We shall discuss in the following first the motion of the electrons in the molecule,
i.e., the systematics of their electron states, then the molecular vibration and the

---

[1] That this picture of the He atom which originates from the fusion of the two H atoms
of the H_2 molecule, pertains only to its electron shells and nuclear charge but not to
the nuclear mass is unimportant since the two missing nuclear neutrons have almost
no influence on the electron shells and their energy states and spectra, which alone are
of interest here.

phenomena related to it, and thirdly the rotation of the molecule and its influence on the spectra. This then leads to the discussion of the interaction of electronic motion, vibration, and rotation, from which the complete band spectra result.

5. Theory of the Electron Energy Levels of Diatomic Molecules

In this discussion of the electron states of diatomic molecules and their term symbols, we start from the corresponding treatment of the atomic states in III, 12, where we find the detailed reasons for the following facts.

First we investigate the behavior of electrons in the so-called two-center molecular model, i.e., in the electric field of two positive nuclei rigidly held at some finite distance r (not necessarily identical with the equilibrium distance r_0 of the nuclei). Then we determine what influence a change of the internuclear distance, resulting from vibration or dissociation, has on the electron arrangement. If we imagine the internuclear distance to be reduced until the nuclei coincide, we have a conversion of the electron shell and thus of the electron states of the molecule to those of an atom with a nuclear charge equal to the sum of the charges of the nuclei of the molecule. On the other hand, if the internuclear distance is increased to infinity, the electron levels of the molecule go over continuously into those of the two separated atoms which actually are produced by dissociation of the molecule. Between these two limiting cases, in which we have atoms in one case produced by fusing the nuclei and in the other by completely separating them, we can insert a case which is already known to us. The electron configuration of a molecule is influenced by the axial electric field, the direction of which is that of the molecular axis joining the two positive nuclei. The effect of this field on the configuration of the electrons and thus on the electron states must be similar to that of an electric field on the electron shell of the atom produced by fusing the two nuclei. Its electron terms in the Stark effect, consequently, must be very similar to those of the molecule. Thus, we have the following similarity sequence of electron levels: combined-nuclei atom → atomic Stark effect → molecule → separated atoms. In this sequence, the energetic order and arrangement of the electron levels of the strongly bound normal molecules with a relatively small internuclear distance must be similar to that of the Stark effect of the atom formed by fusing the nuclei. On the other hand, weakly bound molecules with a large internuclear distance (molecules that are nearly dissociated, or van der Waals molecules, see VI, 8) can be described in a better approximation by comparing their energy levels with those of the completely separated atoms. A development of this method of approach actually allows an accurate prediction of the electron term diagram of the molecule, as has been shown by HUND.

The behavior of molecular electrons, just as that of atomic electrons, can be quantitatively described by wave-mechanical eigenfunctions or, instead, by four quantum numbers. As in the case of atomic electrons, the principal quantum number n represents the number of the electron shell to which the molecular electron under consideration belongs. The orbital quantum number l indicates the wave-mechanical "form" of the electron or, in the Bohr theory, the eccentricity of the orbit (see page 85) which, however, is here of less importance because of the influence of the axial nuclear field. The orientation of the orbital momentum $\vec{l}$ which we have studied in the case of the Zeeman and Stark effects in atomic spectra (see III, 16c) is in the case of the molecule exclusively due to the axial field of the two positively charged nuclei. Thus the electron orbital momentum $\vec{l}$

precesses about the axis joining the nuclei with a quantized component $\vec{\lambda}$ in the direction of this axis. The quantum number λ which corresponds to this orbital momentum component $\vec{\lambda}$ is given by the relation

$$|\vec{\lambda}| = \lambda \frac{h}{2\pi}. \tag{6.14}$$

λ corresponds to the orientation quantum number m of the atomic electrons, introduced in III, 16a. Therefore, λ can have the $2l+1$ different values

$$\lambda = l, l-1, l-2, \ldots, 0, -1, \ldots, -l, \tag{6.15}$$

in which, in general, positive and negative values have the same energy. These states thus are degenerate if the degeneracy is not removed by a perturbation such as by the rotation of the molecule. The fourth electron quantum number, s, designates the spin direction and, as in the case of the atomic electrons, it can only have the values $\pm\frac{1}{2}$. The field with respect to which the spin is oriented is again that along the line joining the nuclei of the molecule.

Table 18. *Symbols of Molecular Electrons according to Eq. (6.17)*

l \ λ	0	1	2	3
0	$s\sigma$			
1	$p\sigma$	$p\pi$		
2	$d\sigma$	$d\pi$	$d\delta$	
3	$f\sigma$	$f\pi$	$f\delta$	$f\varphi$

Just as for atomic electrons, the following relations hold

$$l \le n-1, \tag{6.16}$$

$$|\lambda| \le l. \tag{6.17}$$

Instead of the quantum numbers $l=0, 1, 2, 3, \ldots$, the symbols s, p, d, f are used for designating the electrons, again as in the case of the atom. In addition, instead of the quantum numbers $\lambda=0, 1, 2, 3, \ldots$, the symbols $\sigma, \pi, \delta, \varphi, \ldots$ are used. Just as for atomic electrons, the principal quantum number n is written in front of the electron symbol (cf. Table 18). Thus a $3 d\pi$ electron is one which has the quantum numbers $n=3, l=2$, and $\lambda=1$. Here, too, small letters are used to characterize individual electrons.

For molecules with several external electrons the behavior of the entire electron shell of the molecule follows, in a manner similar to that of atomic electrons (see III, 12), from the vectorial addition of the momenta which correspond to the individual electrons. Because of the strength of the axial field in the direction of the line joining the nuclei, the coupling within the electron shell of a molecule is similar to the Paschen-Back effect of an atom in a strong magnetic field (see III, 16c). The interaction between the orbital momenta $\vec{l_i}$ of the individual electrons which combine to form the resulting total angular orbital momentum $\vec{L}$ is smaller in the molecule than the coupling of each individual electron to the axial field. The $\vec{l_i}$ of the outermost electrons (the inner, closed-shell electrons can be neglected in the first approximation just as in the case of atoms) precess individually about the internuclear axis with quantized integral components $\vec{\lambda_i}$. These $\vec{\lambda_i}$ combine additively or subtractively, according to their direction, to give the resultant quantized orbital momentum $\vec{\Lambda}$ about the internuclear axis, which is characteristic for the electron states of the molecule,

$$|\vec{\Lambda}| = \Lambda \frac{h}{2\pi}. \tag{6.18}$$

Instead of the Λ-values 0, 1, 2, 3, ..., the symbols Σ, Π, Δ, and Φ, ... are used for designating the corresponding electron configuration of the molecule.

The spin momenta $\vec{s_i}$ of the electrons combine vectorially, as they do for atoms, to form the resultant spin $\vec{S}$ of the electron shell, which then precesses about the direction determined by the axis joining the nuclei so that its components in the direction of $\vec{\Lambda}$, designated by $\vec{\Sigma}$, may have all integral values between $+S$ and $-S$. As a result of the magnetic coupling between $\vec{\Lambda}$ and the resultant spin $\vec{S}$, similar to the result of the coupling between $\vec{L}$ and $\vec{S}$ in the atom, each term belonging to a definite value of Λ splits into a term multiplet of $2S+1$ terms which are distinguished by the quantum number $\Omega = \Lambda + \Sigma$ of the resultant angular momentum of the electron shell about the molecular axis.

However, in contrast to the corresponding total momentum $\vec{J}$ of the electron shells of the atom, $\vec{\Omega}$ is not the total momentum of the molecule [even if we dis-

regard the small nuclear spin (cf. III, 20)], because the rotation of the entire molecule contributes appreciably to the total angular momentum of the molecule. Details about the vectorial combination of $\vec{\Omega}$ and molecular rotation follow in VI, 9d.

Just as for the atom, the value of the multiplicity of the term $(2S+1)$ is written as a superscript to the left of the Λ-symbol, and the value of

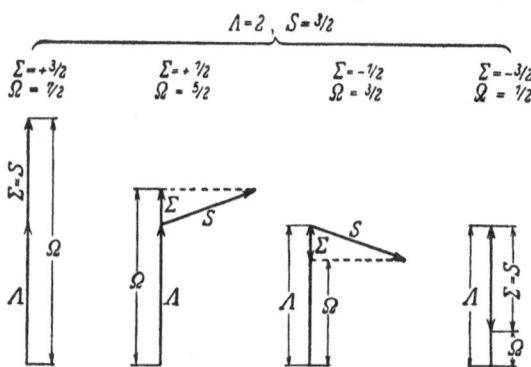

Fig. 184. The different possibilities of vector combination of the angular momenta belonging to the molecular quantum numbers S and Λ in the case of a $^4\Delta$ term

the resultant angular momentum Ω about the line joining the nuclei as a subscript on the right. The four components of an electron state with $\Lambda = 2$ of a three-electron molecule with $S = \frac{3}{2}$ thus are designated by $^4\Delta_{\frac{1}{2}}$, $^4\Delta_{\frac{3}{2}}$, $^4\Delta_{\frac{5}{2}}$, and $^4\Delta_{\frac{7}{2}}$. Fig. 184 shows the vector combinations for this case. To completely characterize an electron term of a molecule, the symbols of the individual electrons are written in front of the symbol of the total molecular term. The ground state of the H_2 molecule with two $1s\sigma$ electrons thus is written $(1s\sigma)^2\,^1\Sigma_0$. *The multiplicity law* (see III, 11) holds for the molecule as it does for the atom. It states that *a molecule with an even electron number has an odd multiplicity and vice versa*. This law may be used to distinguish between spectra of molecules and their ions (e.g., N_2 and N_2^+), which otherwise may be difficult to separate, since the ions and their neutral molecules differ by one electron and thus their spectra must have different multiplicity.

The selection rules for electron transitions in molecules correspond to a certain extent, but not entirely, to those of the atom. The difference is caused by the influence of the axial field. The principal quantum number n can change by an arbitrary integer, whereas the selection rule for Λ is

$$\Delta\Lambda = 0 \quad \text{or} \quad \pm 1. \qquad (6.19)$$

The selection rule for the spin is the same as in the atom: Spin changes in optical transitions, i.e., intercombinations between term systems of different multiplicity, are strongly forbidden for light molecules such as H_2 (just as for light atoms, cf. III, 13.) But they do occur for heavy molecules just as for heavy atoms, though with relatively small intensity. The so-called atmospheric oxygen bands, a $^3\Sigma \rightarrow {}^1\Sigma$ transition of O_2, and the Cameron bands of CO, a $^3\Pi \rightarrow {}^1\Sigma$ transition, are well-known examples of such *intercombination bands*.

The quantum numbers Λ and Ω of a molecular term as well as its multiplicity can be determined empirically from the band spectra. In doubtful cases the Zeeman effect plays a decisive role as a criterion, just as it does for atomic spectra.

Again as for atoms, Rydberg series of band systems are found in the spectra of diatomic molecules with not too many outer electrons. From this we conclude

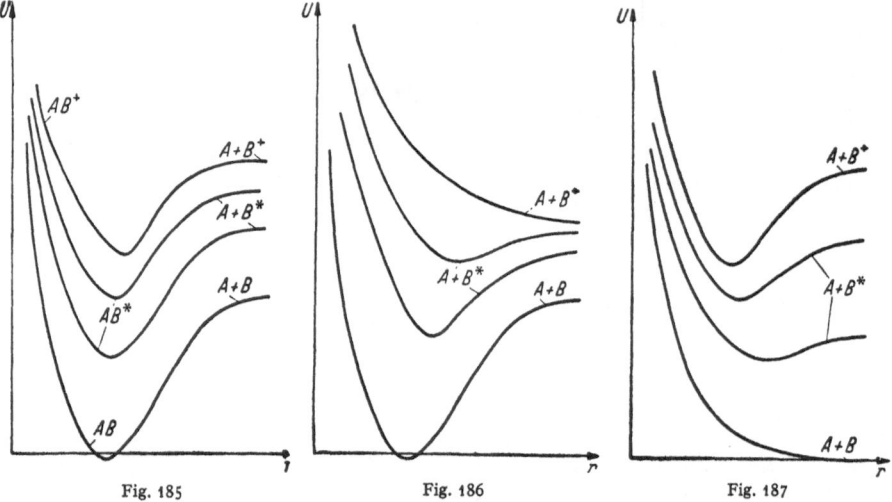

Fig. 185   Fig. 186   Fig. 187

Fig. 185. Potential curves of a normal molecule, an excited molecule (two different excitation states), and the molecular ion for the case of very small influence of the valence electron on the molecular bond. On the right-hand side, the dissociation products for the dissociation from the corresponding molecular states are indicated

Fig. 186. Potential curves of a normal molecule, excited molecule (two different excited states), and ionized molecule for the case of a strongly binding influence of the valence electron

Fig. 187. Potential curves of an unstable normal molecule, excited molecule (two different states), and ionized molecule for the comparatively rare case of a strongly antibonding effect of the valence electron on the molecular bond (for example, He_2)

that a large amount of excitation energy in most cases is used for the higher excitation of one electron and not for exciting several electrons. Such Rydberg series do not occur, however, in all molecular spectra. The reason for this becomes clear when we discuss the dependence of the potential energy of the electron configuration of a molecule on its internuclear distance and the influence of the valence electron on the molecular bond. Some typical potential curves of molecules are plotted in Figs. 185 to 187. They show that for a given electron arrangement the potential energy U of the molecule is a function of the internuclear distance. We shall return to the details of this potential-curve concept later. These potential curves are plotted in Fig. 185 for the ground state of the molecule, $A\,B$, for two excited states, $A\,B^*$, and for the ground state of the molecular ion, $A\,B^+$. Only curves with a pronounced potential minimum correspond to a stable molecular state. Thus for the case shown in Fig. 185, the molecule and the molecular ion (just as the intermediate states of the excited molecule) are stable. In this case the excitation and finally the separation of an electron has no essential influence on the binding of the atoms, and Rydberg series can occur.

Two other cases are found and are important for the understanding of molecular structure. They are those cases for which the normal molecule or the molecular ion are unstable. Instability, i.e., a potential curve such as that of the ground state in Fig. 187, means that for any separation between the atoms the repulsive forces predominate. Consequently, potential curves of this type of unstable molecular states are called *repulsion curves*. It is possible also that a stable molecule loses its stability by the separation of its valence electron so that the potential curve of the molecular ion has no minimum. Fig. 186 shows the potential curves for this case. We conclude from it that the valence electron was essentially responsible for the bond. Typical for the third case are the noble gases which do not form normal molecules in the ground state but which do have stable ions. This is shown in Fig. 187. When the valence electron approaches the molecular ion (the molecular core), the ion loses its stability so that a stable molecule does not exist in the ground state: We call this valence electron a *antibonding electron*. Since the excited molecular states to some extent represent intermediate steps between the normal molecule and its ion (Figs. 185 to 187), it is possible to extrapolate values of molecular constants of unstable molecules or ions by studying the change of the spectroscopically determined molecular constants with increasing excitation (principal quantum number n) of the valence electron.

6. Vibrations and Vibration Spectra of Diatomic Molecules

Diatomic molecules consist of two atoms that are held together by valence electrons in a way to be discussed later. According to classical concepts, these electrons are orbiting around both atoms. Fig. 185 shows that the potential energy of the system then has a minimum at a certain internuclear distance r_0. If this distance is momentarily reduced by outside effects, i.e., by impact with neighboring molecules, restoring forces become effective. The atoms of the molecule begin to vibrate against each other. Many important phenomena of molecular physics are related to these vibrations in molecules. According to general usage in molecular physics, r or r_0 represents the *distance* of the atomic nuclei in the molecule, not half this distance (i.e., the distance from the rotational axis). This is in contrast to the usage of r in the discussion of the rotator in IV, 7a, b.

a) Vibration Levels and Potential Curves

We begin with the theory of molecular vibrations. As always, we start from the simplest model and gradually refine it. Thus we assume we have two point masses, the nuclei, which can move only along a straight line joining them. Furthermore, the forces between the nuclei may be a function only of the distance r between them. For the case $r=r_0$ the repulsive and attractive forces may be in equilibrium and the potential $U(r)$ thus have a minimum.

If we choose the origin of the coordinates to be in one of the nuclei (the one at the left), we can treat the vibration of both nuclei about the center of mass as if the second nucleus alone would vibrate with respect to the origin located in the first nucleus. We introduce as a variable the relative internuclear distance, measured in units of the equilibrium distance r_0,

$$\varrho = \frac{r - r_0}{r_0}.$$
(6.20)

The reduced mass μ and the moment of inertia I of the molecule can be computed from the masses m_1 and m_2 of the two atoms and their equilibrium distance according to

$$I = \mu r_0^2 = \frac{m_1 m_2}{m_1 + m_2} r_0^2.$$
(6.21)

As a reasonable first approximation, we assume that the restoring force which becomes effective upon displacement from the equilibrium position is proportional to the displacement $r - r_0$, so that we can use the linear harmonic oscillator (see IV, 7c) as a wave-mechanically well-known model of the vibrating molecule. Taking into account a ϱ-independant contribution of the electron shell, we write the potential U:

$$U(\varrho) = E_{el} + \frac{k r_0^2 \varrho^2}{2} \tag{6.22}$$

and we obtain, instead of Eq. (4.86), the Schrödinger equation of the linear harmonic oscillator in a notation suited to our particular problem,

$$\frac{d^2 \psi}{d \varrho^2} + \frac{8 \pi^2 I}{h^2} \left(E_v - \frac{k r_0^2}{2} \varrho^2 \right) \psi = 0. \tag{6.23}$$

Here E_v is the vibration energy of the molecule obtained by subtracting from the total energy E a possible excitation energy of an electron. k is the constant of the binding force, which is related to the eigenfrequency v_0 of the vibrating molecule by the relation (see IV, 7c)

$$v_0 = \frac{1}{2} \sqrt{\frac{k}{\mu}} = \frac{1}{2} \sqrt{\frac{k r_0^2}{I}}. \tag{6.24}$$

Since experimental band spectroscopists prefer to work with wave numbers, measured in cm^{-1}, the eigenfrequency v_0 is replaced by $\omega_0 = v_0/c$. By using these units, the discrete energy eigenvalues that belong to the Schrödinger equation (6.32) according to Eq. (4.93) become

$$E_v = h c \omega_0 (v + \tfrac{1}{2}); \qquad v = 0, 1, 2, 3, \ldots \tag{6.25}$$

The energy level diagram of the linear harmonic oscillator thus consists of equidistant steps of magnitude $h v_0 = h c \omega_0$. However, the lowest energy value is not zero but, according to Eq. (6.25), $h c \omega_0 / 2$. v is called *the vibration quantum number*. This quantization of the vibration energy has a very obvious significance. Just as a violin string, the harmonic oscillator substituted here for our molecule can vibrate only with its eigenfrequency v_0 (fundamental vibration) and its integer multiples (higher harmonics). From this quantization of the frequency follows the quantization of the vibrational energy under consideration of the general quantum equation $E = h v$.

For illustration and further study of the molecular vibrations, we use the potential curve of Fig. 188. It shows the potential of the forces acting between the two nuclei as a function of the internuclear distance. The quantized vibrational energy states E_v are indicated also. The potential curve of the linear harmonic oscillator, according to Eq. (6.22), is a parabola. This potential curve can be interpreted by a mechanical analogy. If one nucleus is imagined to be fixed at the zero point, then the other nucleus oscillates back and forth, just as a ball would roll up and down in the potential trough. The point of reversal of the vibration is the intersection of the discrete level of the vibration energy with the potential curve. At this point, all the energy of the molecule is potential energy.

There are two points in which the model of the harmonic oscillator does not agree with the actual molecule. First, the internuclear distance cannot be reduced to zero because the electrostatic repulsion between the positive nuclei rapidly becomes very large as the distance between the nuclei is reduced. Furthermore, the inner electron shells of the atoms in the molecule offer much resistance to any deformation. Conversely, with increasing internuclear distance, the potential curve $U(r)$ cannot rise to infinity, since the binding forces due to the electrons which hold the nuclei together must decrease with increasing distance until finally the atoms are torn apart, i.e., the molecule is dissociated. The real case thus corresponds to the asymmetric curve in Fig. 189, i.e., we must go from the harmonic oscillator to the molecular model of an anharmonic oscillator. The potential curve $U(\varrho)$ or $U(r)$

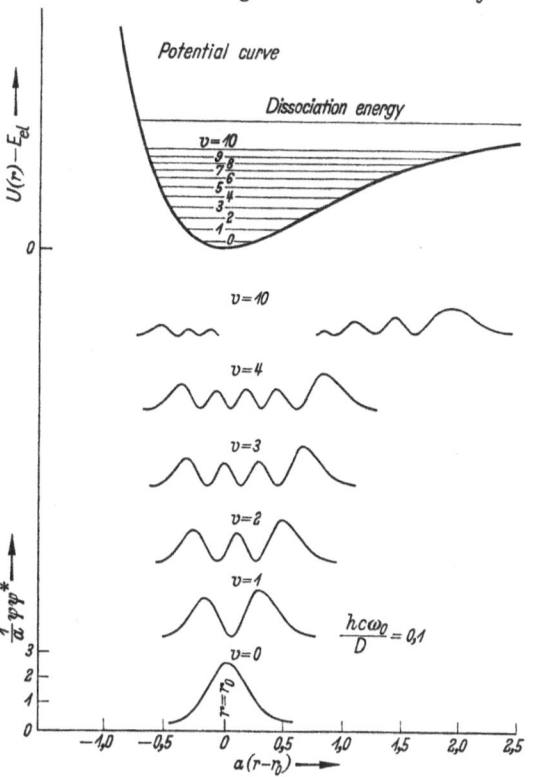

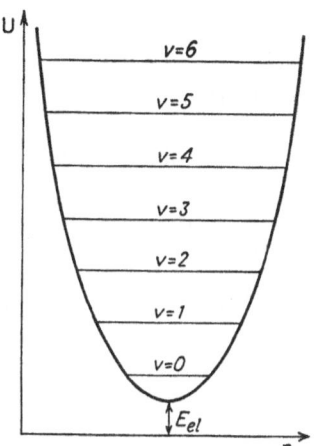

Fig. 188. Potential curve and vibration states for the harmonic oscillator

Fig. 189. Potential curve of an anharmonic oscillator (corresponding to approximation for diatomic molecules) and $\psi\psi^*$ functions for a number of vibration states. Colons are to be read as points

can be best represented by the expression

$$U(r) = E_{el} + D\left(1 - e^{-a(r-r_0)}\right)^2,\qquad(6.26)$$

where a is a constant measuring the curvature of the potential curve in the neighborhood of the minimum. Its value is

$$a = 2\pi\nu_0\sqrt{\frac{\mu}{2D}} = \sqrt{\frac{k}{2D}},\qquad(6.27)$$

where μ is the reduced mass of the molecule according to Eq. (6.21). The significance of D will be discussed presently. If we insert instead of Eq. (6.22) the potential (6.26) into Eq. (6.23), we obtain as eigenvalues of the Schrödinger equation, instead of (6.25), the vibrational energy states E_v of the anharmonic oscillator:

$$E_v = hc\omega_0\left(v + \tfrac{1}{2}\right) - \frac{h^2 c^2 \omega_0^2}{4D}\left(v + \tfrac{1}{2}\right)^2.\qquad(6.28)$$

By going to the anharmonic oscillator, we correct automatically for the second point in which the first model, the harmonic oscillator, did not agree with experience. According to the spectra, the difference between consecutive vibration levels is not constant ($hc\omega_0$) but decreases with increasing v value and, in agreement with Eq. (6.28), converges toward a limit which is approached asymptotically by the right-hand branch of the potential curve. The value of this energy limit turns out to be identical with the constant D of Eq. (6.26) for $r \to \infty$. *Thus D is nothing else than the dissociation energy of the molecule.* Fig. 189 shows a potential curve with characteristic vibration levels computed by means of MORSE's Eqs. (6.26) and (6.27). The norm $\psi\psi^*$ of the vibrational eigenfunctions, as computed from the solutions of the Schrödinger equation, is plotted here for the vibrational states $v = 0, 1, 2, 3, 4,$ and 10. These energy levels and eigenfunctions have to be compared with those of the harmonic oscillator as shown in Fig. 101. We see that, as a result of the anharmonicity of the molecular vibration, the nuclei are more likely to be found at large internuclear distance than close together (left-hand side of the curve).

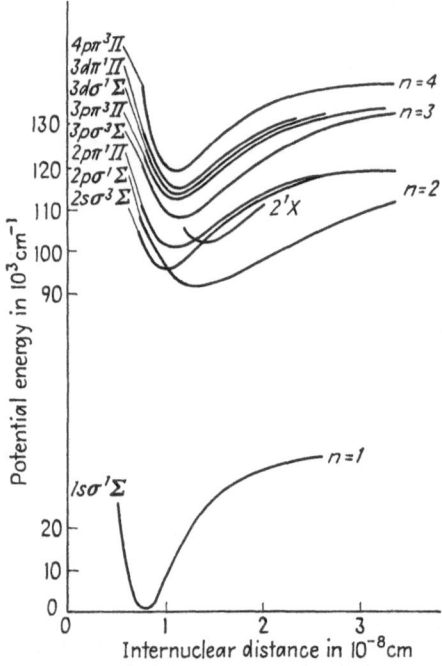

Fig. 190. Potential curves of the most important electron states of the H_2 molecule. (Computed by F. HUND from the band-spectroscopic analysis by MECKE and the author, and by RICHARDSON)

Since the shape and position of the potential curve is determined by the bond between the nuclei and since this depends upon the electron configuration, it is clear that for each configuration of the electrons, i.e., for each of the excited electron states of the molecule, we have a different potential curve. According to page 367, the valence electron in general participates in the bond, and its excitation tends to loosen the bond. The equilibrium distance r_0, consequently, will be greater and the dissociation energy smaller in an excited state than in the ground state. As an actual example, the potential curves of the H_2 molecule are shown in Fig. 190. These curves are computed from Eqs. (6.26) and (6.27) by making use of the values of the dissociation energy D and the fundamental vibration quantum $hc\omega_0$ taken from the spectrum of the H_2 molecule.

We may mention that there is direct evidence for the quantization of molecular vibrations besides the spectroscopic results. This evidence is the fact that molecular vibrations in general do not contribute to the specific heat of diatomic gases. In IV, 10, we shall come back to this important point.

b) Vibrational Transitions and Infrared Vibration Bands

We now inquire about the transitions between the vibrational states which are responsible for the emission and absorption spectra (bands). We have to distinguish between transitions within the same vibration level diagram, i.e., *without* change of the electron configuration, and vibrational transitions with simultaneous electron transitions. We begin with the first case. It accounts for the infrared rotation-vibration bands if we take into consideration the molecular rotation which will be discussed later. For the harmonic oscillator, the selection rule for the vibration quantum number is $\Delta v = \pm 1$ (cf. page 141). This means that transitions with emission or absorption of radiation can occur only between neighboring vibration levels. For the anharmonic oscillator, which corresponds closely to the actual molecular case, the selection rule is less stringent in so far as with decreasing probability (intensity) the transitions $\Delta v = \pm 2, \pm 3, \ldots$ are possible also.

The energy difference of neighboring vibration levels of diatomic molecules is of the order of magnitude of 0.1 to 0.3 ev, i.e., it is considerably larger than the average thermal energy kT. Consequently, higher energy levels are normally not excited so that the rotation-vibration bands are observed almost exclusively in absorption. According to Fig. 191 only the bands starting from the ground state and, at higher temperatures, those from the first excited vibration state are being observed. Very hot incandescent gases and vapors may also *emit* rotation-vibration bands, of course. The fundamental bands, corresponding to the energy $\Delta E = 0.1$ to 0.3 ev, are found in the region from 4 to 12 μ, whereas higher-energy transitions are found in the region up to the photographic infrared at about 1 μ.

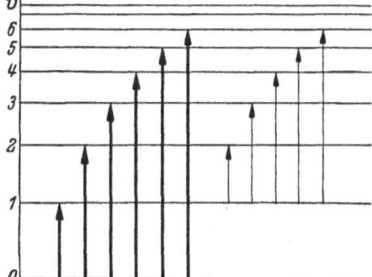

Fig. 191. Absorption transitions within a vibration level diagram of the electron ground state of a molecule, leading to the vibration bands (though here without rotation!) of the molecule

The emission and absorption of rotation-vibration bands is restricted by an important rule. According to IV, 9, emission or absorption of radiation is possible only if the electric moment of the system changes as a result of the transition. This is nearly always the case for electron transitions because here the distance between the electron and the nucleus changes. However, when identical atoms vibrate with respect to each other, e.g., in H_2, O_2, N_2 etc., the electric moment does not change. From this follows *the important rule, which is in agreement with our empirical knowledge, that diatomic molecules consisting of identical atoms do not have rotation-vibration bands of normal intensity.* The corresponding vibrational frequencies are therefore called *infrared-inactive* or *optically inactive* frequencies. From the fact that in classical as well as in atomic physics the emission or absorption of radiation is connected with a change of the electrical moment, there follow the selection rules for permitted or forbidden transitions as well as the non-appearance of entire groups of spectra, e.g., of vibration and rotation spectra of symmetric molecules.

The non-appearance of these spectra is no essential disadvantage for *practical* molecular physics. These optically inactive frequencies namely are Raman-active according to VI, 2d since the polarizability of the molecule changes during these vibrations. They thus occur in the Raman spectrum and can be determined. Furthermore, the optical inactivity of the vibration of symmetric molecules is caused by their missing electric dipole moments. Since these molecules mostly possess electric moments of higher order, however, the vibration bands may be observed nevertheless, though with an intensity that is by several orders of magnitude smaller than that of optically active bands. For this purpose, extremely long tubes filled with the absorbing gas must be used. In this way HERZBERG observed the first lines of the fundamental vibration band of H_2. Higher electric moments of considerable magnitude may also be produced at high pressure by collisions between molecules, i.e., by deformation of their electron shells. The corresponding absorption of the fundamental infrared vibration band, observed in H_2, O_2, N_2,, etc., is then called *collision-induced absorption.*

c) The Franck-Condon Principle as a Transition Rule
for Simultaneous Electron and Vibrational Transitions

We now discuss changes of the vibration state of molecules when coupled with electron transitions. There is no strict selection rule for such transitions

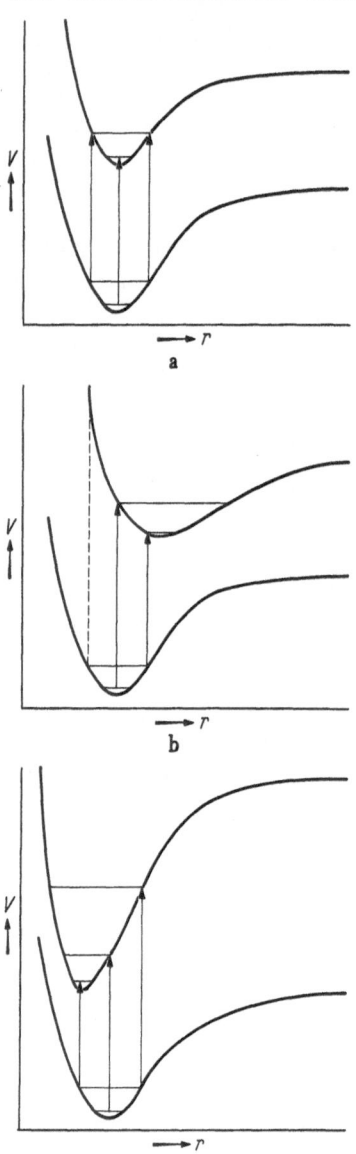

between different vibration term systems, i.e., between potential curves of two *different* electron states of the molecule, Fig. 192. These are governed by the Franck-Condon principle which indicates the preferred, most intensive transitions and thus the general appearance of all the electron band systems and the continuous spectra of a molecule. In discussing these transitions, FRANCK started from the plausible assumption that, because of the small mass of the electron compared to that of the nuclei, *the change of the electron configuration during an electron transition occurs so fast that the position and velocity of the nuclei does not change appreciably during the transition.* Therefore, the equilibrium internuclear distance for the new electron arrangement is reached only *after* the electron re-arrangement has been finished. This means that, in the potential curves shown in Fig. 192, transitions occur without change of the internuclear distance, i.e., vertically (with little deviation). Furthermore, because the velocity is preserved, transitions occur always between states for which the nuclei have the *same* velocity. Since in the classical picture the nuclei remain longest at the reversal points of their vibration, there is the highest probability for transitions to occur between two such reversal points. Transitions thus occur vertically from and to the points where the energy levels intersect the potential curves. This is shown in Fig. 192. CONDON verified this classical principle wave-mechanically. According to wave mechanics (see IV, 9) the transition probability is given by the integral

$$\int \psi_i r \psi_f d\tau, \tag{6.29}$$

in which ψ_i and ψ_f are the eigenfunctions of the initial and final states of the transition under consideration. From this it follows that the transition probability is large only for those combining states the eigenfunctions of which have maxima at the *same* internuclear distance. If it is assumed, as a first approximation, that that part of the transition probability which is due to the electron transition is constant

Fig. 192. Schematic representation of the transitions between ground state and excited electron state of a diatomic molecule, to be expected according to the Franck-Condon principle: (a) for identical internuclear distances in both electron configurations; (b) for loosening of the bond by electron excitation; (c) for tightening of the bond by electron excitation

and independent of the internuclear distance, it follows from the wave-mechanical Franck-Condon principle that the preferred transitions occur between a maximum of the vibrational eigenfunction of one potential curve to a maximum of the other potential curve. However, since, according to Figs. 101 and 189, the maxima of the vibrational eigenfunctions almost coincide with the reversal points of the classical vibration, the wave-mechanical computations lead to the same results as the classical picture. Only for the lower vibration levels is there a departure from the rule of vertical transitions between the reversal points, since we know that here the maxima of $\psi\psi^*$ do not closely coincide with the classical reversal points but that they are somewhat displaced toward the center. For the lowest level, the maximum is almost exactly in the center. The fact that the molecular spectra show an intensity distribution which is in excellent agreement with the form of the principle derived from wave mechanics and not with the classical picture, is further unambiguous evidence for the correctness of wave mechanics and the necessity to use it in treating atomic and molecular physics.

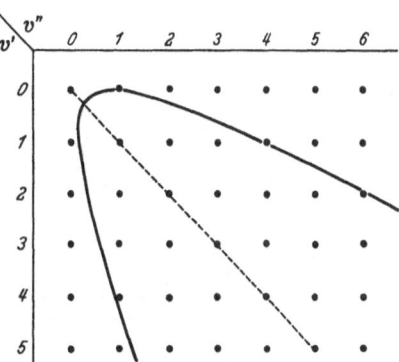

Fig. 193. Representation of the most intense bands in the edge diagram as expected according to the Franck-Condon principle (cf. Table 19). The dotted line shows the most intense bands for the case of Fig. 192a; the solid parabola the most intense bands for electron configurations corresponding to Fig. 192b, c

It is very important, especially for the theory of continuous molecular spectra, that transitions do not occur exactly vertically but in a narrow region of internuclear distances with a width determined by the finite width of the $\psi\psi^*$ maxima. From Fig. 192 we see immediately that it depends on the relative position of the potential curves which of the vibrational transitions are preferred, i.e., which appear with large intensity. If the equilibrium internuclear distances for two electron states are the same, as in Fig. 192a, and if v' is the vibration quantum number of the upper and v'' of the lower electron state, then the bands with $v' \to v'' = 0 \to 0 ; 1 \to 1 ; 2 \to 2$; etc. appear with high intensity. If, however, r_0 is larger for the upper state than for the lower one (the usual case of weaker binding as a result of electron excitation, Fig. 192b), then there are, for each v'', two different preferred transitions for absorption to the upper state. In the rare case of stronger binding in the upper state (Fig. 192c, where r_0 is smaller in the upper state than in the lower one), there are two other preferred transitions.

This intensity distribution in a band system, which follows from the Franck-Condon principle, becomes clearer if a two-dimensional diagram is used for plotting the transitions between two vibrational term systems. In the so-called *edge diagram* of a band system, the vibration quantum numbers v'' of the lower state are plotted as abscissas, and the v' of the upper state are plotted as ordinates. At the point corresponding to the quantum numbers v', v'' the wavelength, wave number, or intensity of the band corresponding to the transition $v'' \to v'$ is written (cf. Fig. 193). From the Franck-Condon principle it then follows that for the case shown in Fig. 192a bands of highest intensity lie on the diagonal of the diagram. In the two other cases they lie on a more or less open parabola. It is evident that the bands of the "vertical" branch of this parabola appear preferably in absorption, those of the "horizontal" branch in emission. The Franck-Condon principle thus explains why absorption and emission band spectra complement each other so well. *From an investigation of the intensity distribution in a band system by means of the Franck-Condon principle, the relative positions of the potential curves of the two combining states can be determined. This, in turn, permits us to determine the influence of the valence electron on the bond strength (see page 630).*

d) The Structure of an Electron Band System.
Edge Diagram and Band Edge Formulae

Because of the Franck-Condon principle we have, at least in the two extreme cases (when the change in the internuclear distance is zero or very large), an entirely different appearance of a band system. The two different cases are called "group spectra" and "sequence spectra", respectively.

Fig. 194 shows the transitions which occur if the potential curves (Fig. 192a) have almost the same position. The spectrum then consists of the bands on the principal diagonal $\Delta v = 0$ and those on the two neighboring diagonals $\Delta v = \pm 1$ that occur with smaller intensity. It can be seen from Fig. 194 that there is

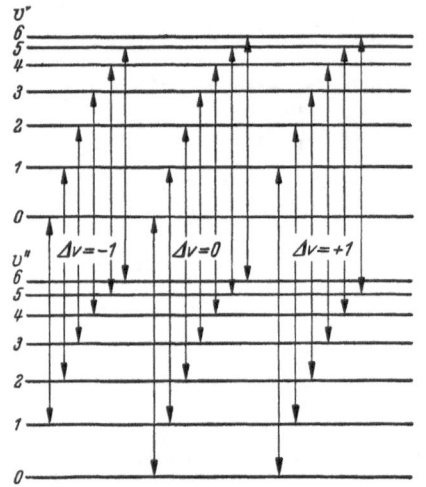

Fig. 194. The most intensive vibration bands in a "group spectrum" corresponding to the potential curves in Fig. 192a

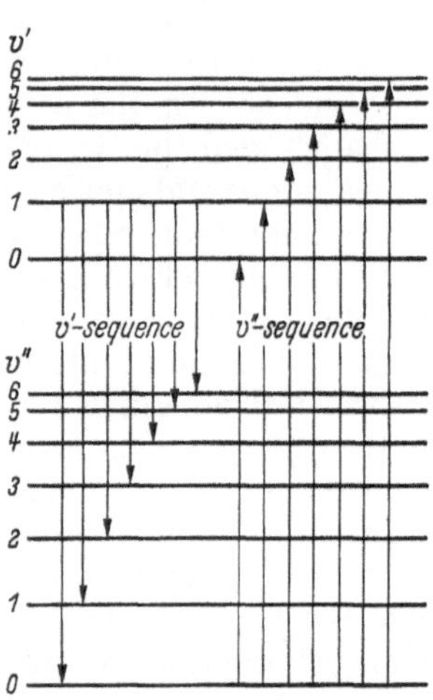

Fig. 195. Most intensive vibrational transitions (different in emission and absorption, see directions of the arrows!) in a "sequence spectrum", corresponding to the potential curves in Fig. 192b or c

little difference in the length, in reciprocal centimeters, of the transition arrows for each of the three groups, and thus very little difference in the wavelengths of the corresponding bands. Consequently, the spectrum consists of few groups of narrow bands lying close together. For this reason it is called a group spectrum. Typical group spectra are the band spectra of CN and C_2 which are emitted by every carbon arc. Conditions are different for the cases of Fig. 192b and 192c, i.e., for a large change of the internuclear distance. Here, as can be seen from the parabola in the band edge diagram, Fig. 193, the *emission* spectrum consists mainly of a band sequence starting from the same *upper* state (Fig. 195, left side). In *absorption*, a similar band sequence is observed the bands of which have the same *lower* state (Fig. 195, right-hand side). Compared to the group spectra, the wavelength differences of neighboring bands are large; we have a *band-sequence spectrum* such as that of the I_2 molecule (see Fig. 182).

We now take up the question how the vibration level differences in the upper and lower states can be determined from an observed band system. We answer the question with the aid of the band edge diagram in which the wave numbers of the bands of the system are noted. Table 19 shows the first bands of a band system of H_2 in which the differences between two neighboring bands are noted in italics between the wave numbers of the two bands. According to the structure of the edge diagram (see Figs. 193 and 195), all bands written in a horizontal line have the same initial state v', those beneath have the next higher initial state, etc. The wave-number differences of each two bands which belong to $v'=0$ and $v'=1$ respectively therefore must be constant and equal to the difference of the term values $v'=0$ and $v'=1$ in the upper state. Table 19 shows that these wave-number differences actually are constant up to a few thousandths of one per cent. Correspondingly, all bands in the same vertical column have the same end state. The differences between consecutive bands in the vertical columns are equal to the differences of the vibration terms in the lower state. The decrease of consecutive term differences ($1,312 \rightarrow 1,276 \rightarrow 1,242$ or $2,105 \rightarrow 1,944 \rightarrow 1,766$) is a measure of the anharmonicity of the vibration.

Table 19. *Wave Numbers of a Band System of the* H_2 *Molecule (incomplete), arranged in an Edge Diagram (after* FINKELNBURG *and* MECKE*). Wave-number Differences (in Italics)* $=$ *Term Differences*

v' \ v''	0		1		2		3
0	21 827.99	*1312.51*	20 515.48	*1276.63*	19 238.85	*1242.39*	17 996.46
	2 105.23		*2 115.22*		*2 105.26*		*2 105.28*
1	23 933.22	*1312.52*	22 620.70	*1276.59*	21 344.11	*1242.37*	20 101.74
	1 944.75		*1 944.63*		*1 944.74*		*1 944.64*
2	25 877.97	*1312.63*	24 565.34	*1276.49*	23 288.85	*1242.47*	22 046.38
	1 766.52		*1 766.64*		*1 766.53*		*1 766.58*
3	27 644.50	*1312.52*	26 331.98	*1276.60*	25 055.38	*1242.42*	23 812.96

We now try to describe the empirically determined vibrational structure by an analytical formula. We can then compare these empirical term formulae with the theoretical formulae for the vibration terms and thus determine important molecular constants. According to Eq. (6.28), the wave number of a band which corresponds to the vibrational transition $v' \rightarrow v''$ is

$$\bar{v}\,(v',\,v'') = \frac{E_{v'} - E_{v''}}{hc} = v_{el} + \omega'\,(v' + \tfrac{1}{2}) - \frac{hc\omega'^2}{4D'}\,(v' + \tfrac{1}{2})^2 - \left.\begin{array}{c}\\\\\end{array}\right\}$$
$$- \omega''\,(v'' + \tfrac{1}{2}) + \frac{hc\omega''^2}{4D''}\,(v'' + \tfrac{1}{2})^2. \qquad (6.30)$$

For brevity we introduce

$$x = \frac{hc\omega}{4D}. \qquad (6.31)$$

If the actual anharmonicity is not represented with sufficient accuracy by the quadratic formula of our model, we supplement it by a cubic term in $(v + \tfrac{1}{2})$ including a new constant y and obtain

$$\bar{v}\,(v',\,v'') = \bar{v}_{el} + \omega'\,(v' + \tfrac{1}{2}) - \omega'\,x'\,(v' + \tfrac{1}{2})^2 + \omega'\,y'\,(v' + \tfrac{1}{2})^3 - \left.\begin{array}{c}\\\\\end{array}\right\}$$
$$- \omega''\,(v'' + \tfrac{1}{2}) + \omega''\,x''\,(v'' + \tfrac{1}{2})^2 - \omega''\,y''\,(v'' + \tfrac{1}{2})^3. \qquad (6.32)$$

Here $\bar{v}_{el}$ is the wave number of the pure electron transition, i.e., the perpendicular distance between the potential minima of the combining states. However, the pure electron term never really appears, but only the vibrational ground level $v=0$ which lies, according to Eq. (6.25), above the pure electron state by the zero point vibration $hc\omega/2$. Empirical band spectroscopy therefore uses the term difference of the two lowest vibrational levels; it is denoted by $\bar{v}\,(0, 0)$ instead of $\bar{v}_{el}$. Furthermore, band spectroscopists are accustomed to work with *integral* term number v. The empirical band formula for representing the vibrational structure of a molecule thus is written

$$\bar{\nu}(v', v'') = \bar{\nu}(0, 0) + (\omega_0' v' - \omega_0' x' v'^2 + \omega_0' y' v'^3) - \\ - (\omega_0'' v'' - \omega_0'' x'' v''^2 + \omega_0'' y'' v''^3). \Bigg\} \tag{6.33}$$

Its constants can easily be expressed in terms of the theoretical formula above. As we have not yet discussed the rotation of a molecule, the band formula holds only for the case of no rotation, i.e., for the so-called zero lines of the bands. However, the formula is used mostly for representing the band edges which alone can be measured with small-dispersion spectrographs. Eq. (6.33) as an edge formula describes the actual vibrational structure very well and it permits the determination of the vibration quanta and thus the binding force constants of the two combining molecular states.

7. Dissociation and Recombination of Diatomic Molecules and their Relation to the Continuous Molecular Spectra

a) Molecular Dissociation and Determination of Dissociation Energies

In introducing the concept of the potential curve, we mentioned the dissociation energy D and the possibility of a molecule dissociating as a result of excessive vibration. Now we treat in some detail this process and the problem of determining the dissociation energy which is one of the most important molecular constants. By dissociation energy we mean the energy necessary for completely separating the atoms of a diatomic molecule. In addition to the dissociation energy of a normal molecule, the physicist is interested in the dissociation energies of the excited molecule. The dissociation from an excited state usually produces a normal and an excited atom, although according to VI, 8c not always.

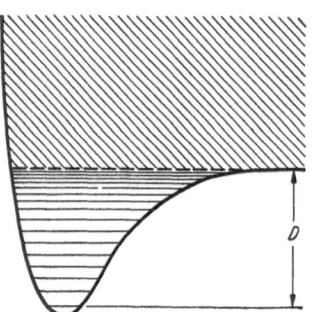

Fig. 196. Potential curve of a diatomic molecule with discrete energy states and continuous energy range corresponding to molecular dissociation. D indicates the dissociation energy

For discussing the dissociation process, we must extend our vibration level diagram. This, namely, extends not only to the dissociation limit but beyond toward higher energies (Fig. 196). Here we have a continuous energy region which corresponds to the states where the molecule is decomposed into its constituent atoms, which may have varying amounts of kinetic energy. This continuous energy range beyond the dissociation energy of a molecule is analogous to that beyond the series limit of an atom (cf. Fig. 45), which corresponds to the ionization of the atom into an ion and electron with varying amounts of kinetic energy.

One might be inclined to think that an optical dissociation should be possible by absorption of the dissociation energy (i.e., by overexciting the nuclear vibrations), while the electron remains in the ground state, i.e., without exciting the electron, according to the following equation

$$AB + hc\bar{\nu}_c \rightarrow A + B. \tag{6.34}$$

This, however, is not the case because, according to the selection rule mentioned on page 364, such large changes of the vibration quantum number, as they correspond to the transition from the vibrational ground state into the dissociation continuum, occur only with infinitely small probability. The same conclusion may be drawn from the Franck-Condon principle according to which the transition probabilities for such transitions are extremely small. As can be seen from Fig. 197, the maximum of the eigenfunction of the dissociated state under consideration does not lie above the maximum of the eigenfunction of the vibrational ground state at all. Of the two transition arrows of Fig. 197, the left one corresponds to a forbidden transition because it contradicts the conservation of the internuclear distance in the classical picture. The right-hand vertical transition is forbidden since it would contradict the conservation of momentum of the vibrating nuclei, because their velocity is much larger at the upper end of the arrow than at the lower one (as for a pendulum passing the rest position). *Thus an optical dissociation or photodissociation by a change of the vibrational state alone without a simultaneous electron excitation is not possible.*

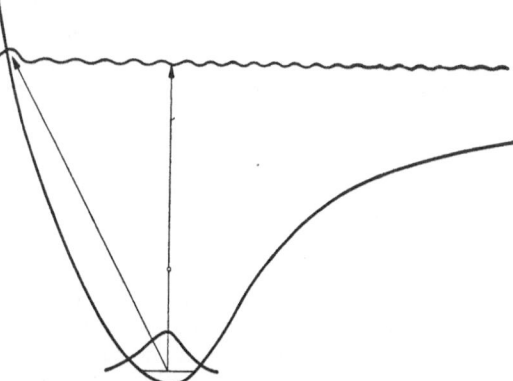

Fig. 197. Potential curve with vibration eigenfunctions for the vibrational ground state and a dissociated state, indicating the impossibility of optical dissociation of a molecule without simultaneous electron excitation. Left-hand transition forbidden because of change of internuclear distance; perpendicular transition impossible because of change of velocity

The case is different for *thermal* dissociation. If we gradually heat up a diatomic gas, increasingly higher vibration levels are gradually excited by the collisions until the molecules are finally "thermally" dissociated.

However, a photodissociation *is* possible if an electron is simultaneously excited according to the equation

$$A B + h c \bar{\nu}_c \rightarrow A + B^*. \qquad (6.35)$$

We explain this process and the determination of the dissociation energies from the spectrum with the aid of Fig. 198. By absorption of photons from the vibrational ground state $v'' = 0$ of the molecule, transitions can occur (for the case illustrated) into the discrete states and into the continuous region of the upper electron state. Thus we find a band sequence which converges toward a limit, with a dissociation continuum beyond

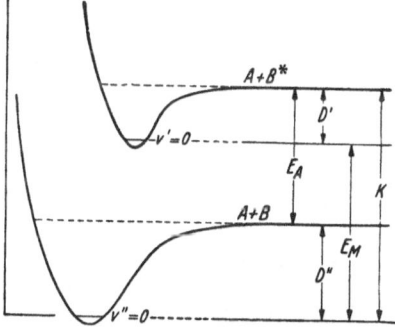

Fig. 198. Determination of the dissociation energy of a diatomic molecule in its ground state (D'') and an excited electron state (D'). E_M = excitation energy of the molecule; E_A = excitation energy of the excited atom originating from the dissociation of the excited molecule; K = energy of the limit of the band sequence to be taken from the band spectrum

this shortwave limit. The wave number of the long-wave limit of the continuum corresponds to the energy K. It seems reasonable to assume, and is in agreement with experience, that in a photodissociation of the type described the excited molecule decomposes not into two normal atoms but, in general, into a normal and an excited atom (e.g., $A + B^*$), since in the process of separation of the nuclei the excited electron remains in its excited state. According to Fig. 198 we can arrive at the final state (normal + excited atom, $A + B^*$) in two ways from the initial state (normal molecule $A B$). First, we can excite the molecule (electron transition) by transferring the excitation energy E_M of the molecule and then, by adding vibration energy of the amount D', dissociate the excited molecule into a normal and an excited atom ($K=E_M+D'$). Here D' is the dissociation energy of the excited electron state of the molecule. Since the molecular excitation energy E_M is equal to hc times the wave number of the $0\rightarrow0$ band and thus is obtained from the spectrum, and since K is the observable convergence limit of the band sequence, $D'=K-E_M$ can be determined directly. By the second way we can, theoretically, first add the vibration energy D'' to the molecule, thus dissociating it into normal atoms, and then, by adding the atom's excitation energy E_A, excite one atom. We then have $K=D''+E_A$ and arrive at the same final state, one normal and one excited atom ($A+B^*$).

From this consideration it follows that we obtain the dissociation energy of the ground state, D'', if we subtract from the energy value of the band convergence limit K the excitation energy E_A of the excited dissociation product. The quantity E_A can be determined in many cases. If the excited atom returns to the ground state, it emits a line corresponding to the energy E_A and we get E_A directly from wavelength measurements. In the case of the oxygen molecule, for example, the value of the dissociation energy D'' of O_2 was long in doubt because E_A was unknown. Finally the analysis of the atomic spectrum of oxygen allowed E_A and thus the dissociation energy of O_2 to be established.

Only in a few cases (e.g., in the molecular spectra of the halogens and the ultraviolet spectra of H_2 and O_2) do the potential curves have such favorable relative positions that the band limits K themselves appear in the spectra. Usually we find only a shorter or longer band sequence the convergence limit of which has to be determined by extrapolation. From the band formulae (6.30) and (6.31) we get for the dissociation energy D

$$D = \frac{hc\omega_0}{4x_0}. \tag{4.36}$$

Thus we need only to take the value of the energy of the vibration quantum $hc\omega_0$ of the ground state and that of the anharmonicity constant x_0 from the empirical band formula (6.33) in order to be able to approximately compute the dissociation energy.

b) Predissociation

Closely related to the dissociation process is the interesting phenomenon of predissociation which, in addition to its basic interest, is of importance for photochemistry. In several spectra of diatomic molecules (e.g., S_2) and especially in those of numerous polyatomic molecules, regions of diffuse bands without rotational structure are observed. In a sequence of bands, we often observe first sharp bands, then diffuse ones which often merge toward shorter wavelengths into the dissociation continuum. This phenomenon is called *predissociation*. Chemical investigations proved the existence of dissociation products when the molecules were irradiated by wavelengths of the diffuse bands. Dissociation thus occurs *before* the convergence limit of the bands, i.e., by light of longer wavelength than corresponds to the dissociation continuum.

As in the case of the phenomenon of preionization discussed in III, 21, predissociation is caused by nonradiative transitions from quantized energy states into a region of continuous energy, which in the case of predissociation corresponds to that of the dissociated molecule. To explain predissociation, Fig. 199 shows the potential curves and vibration level diagrams of two excited molecular states. We assume that a transition from a lower state, not shown, to the state a' is possible. Then in the absence of state b' we would observe a sequence of absorption bands which correspond to transitions from a lower state to all a' states, beyond which the dissociation continuum of a' would follow after the dissociation limit of a' has been reached. Now we consider the influence of the perturbing molecular state b' with a dissociation limit below that of a'. If certain selection rules between the electron quantum numbers of a' and b' are fulfilled, there exists for the discrete a' levels, which lie above the dissociation limit of b', the possibility of nonradiative transitions into the continuous energy region of b', and thus of dissociation *before* the dissociation limit of a' itself has been reached. Since the Franck-Condon principle applies for non-radiative transitions also, these transitions from a' into the continuum of b' occur with highest probability in the region of the intersection of a' and b', whereas for higher states the probability of predissociation decreases. We see that at the intersection of the two potential curves an

uncertainty exists with respect to the mutual association of the two left-hand branches to the two right-hand branches of the curves a' and b'. Therefore, only the relative probabilities can be given for a crossing of the potential curves or a mergence of curve a' into the lower one of the two right-hand branches.

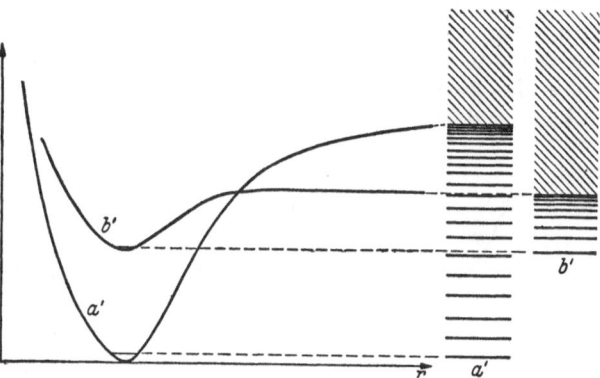

Fig. 199. Potential curves and corresponding vibration-term sequences leading to predissociation of a diatomic molecule

As in the case of pre-ionization, the possibility of transitions into the continuum implies a reduction of the life time of the discrete states of a'. Thus, according to the uncertainty relation, an increase of the width of these energy states and the lines result from their combination. In the region of predissociation, this line width usually becomes so large that the rotation lines of a band merge into one another, thus accounting for the diffuse appearance of the predissociation bands. Predissociation is observed only in absorption because, as a result of nonradiative transitions into the continuous energy region, the radiation probability of predissociating states is extremely small.

For the photochemist, this section contains an important result: *Irradiation by those wavelengths which correspond to sharp absorption bands of the irradiated molecule produce excited and thus sometimes reactive molecules. Irradiation by those wavelengths, however, that correspond to continuous absorption spectra or diffuse absorption bands (which appear diffuse not only because of the small dispersion of the spectroscope) produces dissociation of the irradiated molecules.* Thus the relation between the continuous absorption spectra of molecules and their decomposition into atoms is explained.

c) The Formation of Molecules from Atoms

What do we know about the reverse process, the formation of (diatomic) molecules from their atoms? In this case the situation is similar to the recombination of an ion and an electron in forming a neutral atom as discussed in III, 6. If two atoms which can form a molecule do collide, they form a molecule only if the binding energy released in the collision (which of course is numerically equal to the dissociation energy) is somehow carried away. This can be done by a third atom (or a molecule). In that case we have the formation of a molecule in a triple collision of atoms and can write the process as an equation, in analogy to Eq. (3.41),

$$A + B + C \rightarrow A B + C_{fast}. \tag{6.37}$$

If, by coincidence, there is resonance between the energy released in the formation of the molecule $A B$ and an excited state of the atom or molecule C, then the latter may take over the binding energy as excitation energy, so that Eq. (6.37) may be replaced by

$$A + B + C \rightarrow A B + C^*. \tag{6.38}$$

The question whether recombination can occur not only by this triple collision, but also when two atoms collide and the binding energy is emitted as a photon, has to be answered in different ways for atomic molecules and ionic molecules. *By atomic molecules we mean those homopolar molecules in which mainly uncharged atoms vibrate against each other. Therefore they decompose into normal neutral atoms when dissociated in the ground state. Examples are* H_2, O_2, *or* CO. *However, in ionic or heteropolar molecules such as* NaCl, *ions are vibrating against each other, in our example* Na+ *against* Cl-. *A dissociation of the ionic molecule into neutral atoms is nevertheless quite possible and for energy reasons even probable because of the uncertainty of associating the four branches of potential curves at the intersection point (see Fig. 200).*

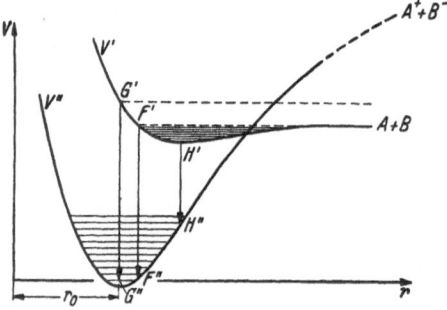

Fig. 200. Set of potential curves for explaining the formation of an ionic molecule in a collision of two atoms upon emission of the recombination continuum

In the usual case of atomic molecules, the colliding atoms A and B "run along" the potential curve of the normal molecule (see Fig. 198). The formation of a molecule by emission of the binding energy and the relative kinetic energy of A and B in this case is just as impossible as the reverse process, the optical dissociation without electron excitation. The reason is that the probability for transitions from the continuous energy region to the vibrational ground state of the *same* electron state is practically zero (Fig. 197). *Two normal atoms which in principle might be able to form a molecule thus cannot recombine in a two-particle collision but only in a triple collision in which the third partner, which may be the wall of a glass tube, for example, takes over the excess energy.*

The situation is different in the case of an ionic molecule (e.g., NaCl). Here the colliding atoms are in a *different* electron state (curve V' of Fig. 200) than the ground state V'' of the ionic molecule. Consequently, a recombination in a two-particle collision is possible if the colliding atoms change from the state of the atomic molecule $A B$ into the more stable ground state of the ionic molecule $A^+ B^-$ by a change of their electron arrangement and emission of the binding energy. In this case, the recombination probability is determined by the probability of the electron transition $A B \rightarrow A^+ B^-$ and by the Franck-Condon principle. In the case of the alkali halides, for example, transitions to the ground state of the normal ionic molecule $A^+ B^-$ can occur, according to the Franck-Condon principle, from the reversal points between F' and G' of the colliding normal atoms A and B. Here the energy is radiated as an emission continuum the wavelength limits of which are given by the transition arrows $F'F''$ and $G'G''$ so that we have the two-particle collision recombination process

$$A + B \rightarrow A^+ B^- + hc\,\bar{\nu}_c. \tag{6.39}$$

Consequently, just as *photodissociation with continuous absorption is possible only if there is a simultaneous change of the electron configuration, the formation of molecules in a two-particle collision with continuous emission of the binding and kinetic energies occurs only in connection with a simultaneous change of the electron configuration.* Such an electron re-arrangement is also connected with the radiative combination of normal and *excited* atoms in two-particle collisions, according to the equation

$$A + B^* \rightarrow A B + hc\,\bar{\nu}_c. \tag{6.40}$$

In a highly dissociated and excited gas, for example, in a discharge plasma, collisions between normal and excited atoms are sufficiently frequent. These take place "along" the curve of the excited molecular state $A + B^*$, Fig. 198. As in the case mentioned above, in collisions a region above the dissociation limit may be reached on the left-hand branch of the upper curve. By transitions from this branch of the potential curve to the ground state, the colliding atoms A and B^* may recombine with emission of continuous radiation.

Thus, the observation of such continua of molecules furnishes evidence for the process of radiative recombination, and its probability can be determined as a function of the velocity of the colliding atoms. Emission continua which are to be interpreted as recombination processes (6.39) are found in the reaction of alkali vapors with halides, whereas continua corresponding to radiative recombination (6.40) have been observed for the halogens and tellurium.

8. Limitations of the Concept of a Molecule. Van der Waals Molecules and Collision Pairs

We have so far used the term molecule as if it were a self-evident concept. Now we must refine our knowledge and proceed to a more sophisticated point of view. By a molecule we usually mean, without giving it too much consideration, a system of two or more atoms or groups of atoms, if the potential energy for a definite internuclear distance is a minimum and if this system is, to a certain extent, independent of its environment. The latter means that (at least at the instant of observation) the interaction within the system is large compared to that between the system and its environment. As we know, such a molecule is physically characterized by the forces acting between the atoms, i.e., by the change of potential with the internuclear distance (potential curve) and by the value of its total energy. The energy is either always taken positive from the ground state of the normal molecule or (as we do here) from the free atoms without kinetic energy at the zero point. In this case the binding energy of the stable molecular states is negative, whereas the kinetic energy of the free dissociated states of the molecule is positive.

We now can distinguish between three types of molecules, depending on the shape of their potential curves, i.e., their binding forces (lower part of Fig. 201). We consider first the genuine molecules bound by valence forces in the chemical sense, which were almost exclusively treated in the last section. Their characteristic features are a pronounced potential minimum at a relatively small internuclear distance ($r_0 \approx 1$ Å), and a relatively large dissociation energy of 1 to 10 ev

(Fig. 201, curve a). The second type of molecules is called "van der Waals molecule" for reasons which will become clear below. It is characterized by the potential curve b with a shallow minimum at a large internuclear distance ($r_0 \approx 3$ to 5 Å) and by a usually very small dissociation energy (order of magnitude of 0.01 to 0.1 ev in the ground state). These molecules consist of atoms which are not able to form a genuine molecule by re-arranging their electron shells. However, there exists between them some attraction that is due to interaction forces of the second order (cf. VI, 15). These second-order forces imply a possibility of formation of loosely bound molecules. The same types of interatomic and intermolecular attractive forces which produce these weak bonds are also responsible for the deviation in the behavior of a real gas from an ideal one. Since these deviations are described by the van der Waals equation of state, we call those molecules which are bound by these forces van der Waals molecules. If we now, consider as the third type of molecules an atomic system which is characterized by the potential curve c, we see that it differs from a van der Waals molecule only in that it cannot have negative energy values at all. The repulsive forces always exceed the attractive forces; in every other respect such a system of atoms behaves just as any real molecule in a free state of positive energy. It is represented by two atoms at the instant of a collision if only repulsive forces act between them. Such a system is sometimes called a collision pair. We introduce it here as the

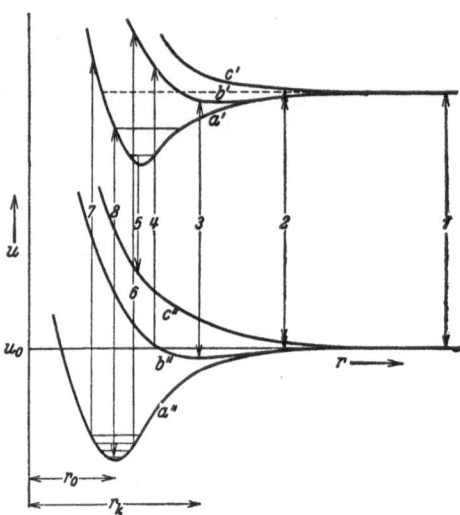

Fig. 201. Potential-curve diagram for the electron ground state and an excited electron state of a diatomic molecule with transitions indicating all possible discrete and continuous molecular spectra

limiting case of the molecule, although it lacks a potential minimum and thus does not fit the customary definition of a molecule given above. Evidently no sharp line can be drawn between a van der Waals molecule and a collision pair. Furthermore, a collision pair can emit and absorb spectra just as other molecules. Finally, a purely repulsive potential curve as it characterizes a collision pair not seldom also occurs in the case of excited states of regular molecules, as we know from several examples above.

The best known example of a vander Waals-type molecule is the Hg_2 molecule, which consists of two Hg atoms. All types of binding which we mentioned above occur among its excited energy states. Fig. 202 shows its potential-curve diagram which was determined from its mostly continuous spectra. The He_2 molecule, which consists of two helium atoms and was established spectroscopically, is another example of a van der Waals molecule, but only in the ground state. Pronounced potential minima occur for the excited states of He_2 and allow the emission of numerous He_2 bands, while an emission continuum lying in the far ultraviolet is emitted by transitions to the unstable ground state. A noticeable van der Waals attraction also exists between two O_2 molecules and is responsible for the double molecule $O_2-O_2 = (O_2)_2$. It consists of two normal oxygen molecules and, beside other spectra, absorbs the continuous bands which are responsible for the blue color of liquid oxygen.

After this extension of our knowledge about the concept of molecules and the potential curves of the different types of molecules, we return to the discrete and to the continuous spectra which result from the shapes and mutual positions of the potential curves of the combining molecular states and, conversely, may be used for determining them. For this purpose we consider the schematic diagram Fig. 201 in which at the right-hand side the ground state and an excited state of an atom are shown, which represent the asymptotes of the two families of curves. They originate from bringing an additional normal atom into interaction with the first one. The different possible types of transitions are represented by arrows.

However, we have to consider that the transitions occur also from all neighboring points of the potential curve under consideration.

Whereas the transition number 1 between the undisturbed atomic states produces a sharp spectral line, a broadened atomic line results from transition number 2. It is evident that the amount and character of the broadening depends upon the average interatomic distance (and thus on the density of the gas), on the temperature, and on the shape of the potential curves. With these few remarks, we have considered the process of collision broadening, already discussed in III, 21, from an entirely new point of view. We now regard the emitting or absorbing and the colliding atom together as a molecule (or collision pair) and thus conceive of collision broadening as a limiting case of a molecular spectrum. It is now evident that it is only a small step from the simple broadened atomic lines to the spectra of the weakly bound van der Waals molecules as represented by transitions 3 and 4. They consist of narrow continuous or seemingly continuous bands which are closely related to the corresponding atomic lines. The emission and absorption of

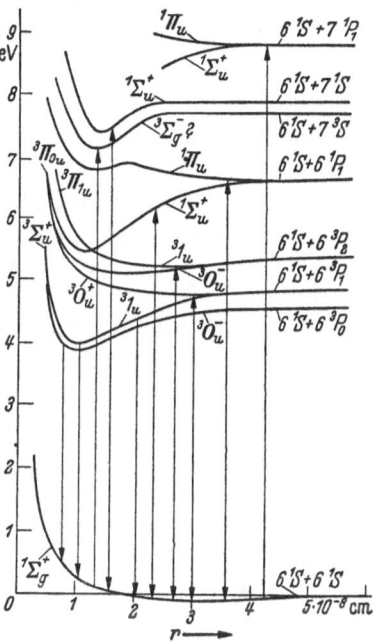

Fig. 202. Potential-curve diagram of the Hg$_2$ van der Waals molecule, determined from the study of the diffuse bands and continua. (By Mrozowski and the author)

mercury vapor at not too low pressures provide numerous examples of this type of band associated with atomic lines. We have already referred to the identification of these bands as spectra of a Hg$_2$ van der Waals molecule. The transitions 5 to 8 occur in the region of internuclear distances in which the electron clouds of the two atoms penetrate each other considerably and by re-arrangement thus cause major energy changes compared to the unperturbed atomic states. They represent, therefore, molecular spectra in the more restricted sense. Whereas the normal electron band spectra correspond to transitions like 8 between the discrete states of the two potential curves, an extended emission continuum originates from transition 5 from the minimum of an excited molecular state to the repulsion curve of a lower state. The best-known examples of this are the continuum of the helium molecule in the extreme ultraviolet and the hydrogen molecular continuum which extends from the green to the far ultraviolet. The latter is emitted by any glow discharge in dry hydrogen at several millimeters pressure. In conclusion, transition 7 from the discrete states of the lower potential curve to the continuous energy region of an upper curve accounts for an extended absorption continuum. The best-known examples are the absorption continua of the halogen molecules I_2, Br_2, Cl_2, IBr, etc., which lie in the visible or near ultraviolet. This discussion reveals how worthwhile it is to become thoroughly familiar with the concept of potential curves. They often represent a tool for drawing far-reaching conclusions from the molecular emission or absorption spectra upon the structure of the molecule under consideration.

9. Rotation of Molecules and the Determination of Moments of Inertia and Internuclear Distances from the Rotational Structure of the Band Spectra of Diatomic Molecules

After we have become acquainted with the numerous phenomena which are related to molecular vibration and dissociation, we discuss now molecular rotation and the conclusions which may be drawn from investigations of the rotational phenomena in molecular spectra.

a) Rotation Levels and Infrared Rotation Spectrum

The simplest model of a rigid molecule rotating about the axis of its largest moment of inertia is the dumbbell model of the rotator shown in Fig. 100. The quantization of the rotation energy, and thus the discrete energy states of the rotator, follow from the general rule, derived quantum-mechanically in IV, 8, that atomic angular momenta are quantized and can be written:

$$|\vec{J}| = I\omega = \sqrt{J(J+1)}\, h/2\pi \qquad J = 0, 1, 2, 3, \ldots \qquad (4.125)$$

where the moment of inertia I may be computed from Eq. (6.21), ω is the angular velocity and J the quantum number of the angular momentum. From this we obtain the discrete states of the rotation energy of a rotator with its axis free to move in space,

$$E_{\text{rot}} = \frac{1}{2} I\omega^2 = \frac{h^2}{8\pi^2 I}\, J(J+1), \qquad (6.41)$$

in agreement with the solution of the Schrödinger equation for this problem (see page 172). Molecular spectroscopists are used to write this result in the form

$$E_{\text{rot}} = hc\,BJ(J+1) \qquad J = 0, 1, 2, 3, \ldots, \qquad (6.42)$$

since the so-called *rotation constant*

$$B = h/8\pi^2 cI, \qquad (6.43)$$

which has the dimension of a wave number (cm^{-1}), can be taken directly from the rotation spectrum and thus allows to determine the moment of inertia I.

Since the term values measured in wave numbers (cf. page 58) are equal to the energy values divided by hc, Eq. (6.42) gives the rotation level diagram which is shown in Fig. 203. On the right-hand side, the quantum numbers of the angular momentum, here called rotational quantum numbers J, are indicated, while on the left-hand side the term values in units of B are given. For optical transitions between these rotation terms, the selection rule

$$\Delta J = \pm 1 \qquad (6.44)$$

applies, which follows from the correspondence principle (cf. III, 22) or from wave mechanics since rotation is a purely harmonic motion. Thus each rotational state can combine by emission or absorption of radiation only with its two neighboring states. From this selection rule and Fig. 203 it follows that the rotation spectrum of a diatomic molecule due to a change of the rotation energy is to be looked for in the far infrared and consists of a sequence of equidistant spectral lines (rotation lines) with the wave numbers $2B$, $4B$, $6B$, $8B$, etc. The distance between two consecutive lines consequently is $2B$. Pure rotation spectra in good agreement with this theory were first found by CZERNY in the spectra of the hydrogen halides. Just as in the case of the vibration spectrum (cf. VI, 6b), the rotation spectrum is absorbed or emitted with normal intensity only *if a change of the component of the electric moment in the direction of observation is associated with the rotation, i.e., if the centers of the electric charges do not coincide with the molecule's center of mass. This condition is fulfilled only for the asymmetric molecules such as HCl but not for H_2, Cl_2, O_2, etc. Consequently, the latter do not show easily observable rotation or rotation-vibration bands.* These bands can be observed, however, although with an intensity lower by a factor 10^8, if very thick absorption layers and extremely sensitive measuring methods are applied. These weak bands are then due to electric quadrupole moments of the molecules. A sufficiently sensitive method is microwave spectroscopy (cf. page 50); with its help Gordy observed the $0 \rightarrow 1$ transition of the rotation spectrum of O_2 at the wavelength of 2.5 mm.

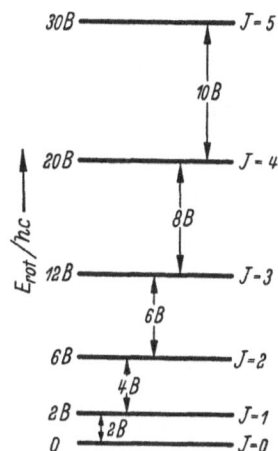

Fig. 203. Possible transitions in the rotation-level diagram of the electron and vibration ground state of a diatomic molecule (rotation band)

If the rotation spectrum is observable, then the rotation constant B can be taken from it at once. The moment of inertia of the molecule, one of its most important constants, can be computed from B according to Eq. (6.43). Moreover, if the masses m_1 and m_2 of the constituent atoms of a molecule are known (and this is always the case if we know the molecule responsible for the spectrum), then the internuclear distance of the molecule follows immediately from the moment of inertia by means of Eq. (6.21). The internuclear distances determined in this way are of the order of 1 Å, in agreement with our knowledge of the dimensions of atoms and molecules. Table 20 presents the most important data for a number of diatomic molecules, as determined from their spectra. If the rotation and vibration are optically inactive, as in the case of molecules consisting of identical atoms, then B, and with it I and r_0, can be determined by means of microwave spectroscopy or, though not quite as easily, from the rotational structure of the electronic bands (see below) or sometimes from the rotational Raman effect.

Table 20. *Internuclear Distances, Moments of Inertia, Fundamental Vibrational Quanta, and Dissociation Energies of Some Important Molecules in the Ground State*

Molecule	Internuclear distance, r_0 Å	Moment of inertia I, 10^{-40} g cm²	Fundamental vibrational quantum ω_0 cm⁻¹	Dissociation energy	
				Electron volt	kcal/mole
H_2	0.77	0.47	4390	4.46	103
O_2	1.20	19.1	1580	5.11	118
N_2	1.09	13.8	2360	9.76	225
S_2	1.60	67.7	727	4.45	103
Cl_2	1.98	113.5	565	2.47	57
Br_2	2.28	342	324	1.96	45
I_2	2.66	741	214	1.53	35
CO	1.13	15.0	2169	9.6	220
NO	1.15	16.3	1907	5.3	122
HCl	1.27	2.60	2989	4.40	102

b) The Rotation-Vibration Spectrum

From the discussion of the pure rotation spectrum we proceed to the rotation-vibration spectrum which is produced by changes of the rotational and vibrational states of the molecule (without a change in the configuration of the electrons). Two sequences of rotational terms belonging to the vibration levels $v=0$ and $v=1$ are shown in Fig. 204. According to the selection rule (6.44), the resulting

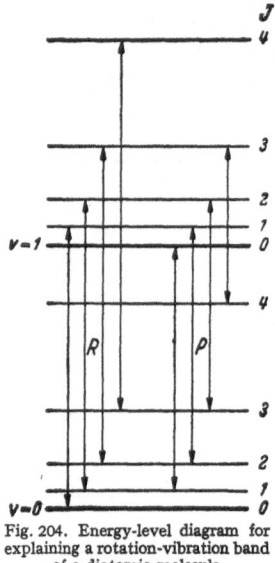

rotation-vibration band consists of two "branches" which belong to the transitions $\Delta J=+1$ and $\Delta J=-1$. They are shown at the left and right in Fig. 204. The line sequence corresponding to $\Delta J=+1$ extends with increasing number J toward shorter wavelengths and is called the positive or R branch. The sequence corresponding to $\Delta J=-1$ extends toward longer wavelengths and is called the negative or P branch. If we assume, in a rough approximation, that the moment of inertia of the molecule is the same for the upper and for the lower vibrational state in spite of the difference in vibration, then the separations of the rotation levels are equal in both vibrational states. This case is shown in Fig. 204. Once more we find equidistant lines separated by the constant distance $2B$ (Fig. 205). The zero line, which cannot appear because the transition $\Delta J=0$ is forbidden, can be easily determined. Its frequency v_0 (position indicated in Fig. 205) is the energy of the pure vibrational transition. For every vibrational transition which is possible according to Figs. 194 and 195, we expect a band of the type described. This corre-

Fig. 204. Energy-level diagram for explaining a rotation-vibration band of a diatomic molecule

sponds roughly to the spectroscopic results. A small deviation from the assumed constancy of the distances between consecutive lines, however, is evidence that, first, the moment of inertia is not constant and independent of the vibrational state and that, second, rotation and vibration of the molecule are not quite in-

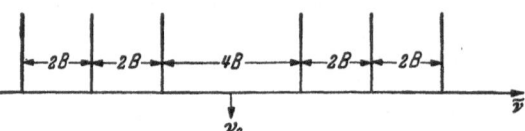

Fig. 205. Schematic representation of a rotation-vibration band

dependent of each other. The energy values of the pure rotation and vibration thus cannot simply be added, but the interaction of vibration and rotation must be taken into account by introducing a mixed term containing v and J in the expression for the molecular energy. If this is done, complete agreement between theory and spectroscopic results is achieved.

c) The Rotational Structure of a Normal Electron Band

We now take up the most complicated case, that of band spectra resulting from simultaneous changes in the electron configuration, vibration, and rotation of a molecule. These bands occur in the visible and ultraviolet spectral regions. *The participation of an electron transition in the absorption and emission in general results in a change of the electron configuration and, therefore, of the electric moment which is essential for emission and absorption.* This fact has two consequences. First, electronic band spectra can be observed also for symmetric molecules, such as H_2, O_2, etc., whereas their pure rotation spectra and vibration spectra do *not* show up with normal intensity because of the missing electric moment. Secondly, the electron transitions make it possible that transitions occur without a change of the rotation quantum number. Thus, instead of Eq. (6.44), the selection rule for the rotation in electronic band spectra is

$$\Delta J = 0 \quad \text{or} \quad \pm 1. \tag{6.45}$$

Therefore, a zero or Q branch $(\Delta J = 0)$ has to be added to the branches of a band discussed before, which corresponded to $\Delta J = \pm 1$ (positive or R branch, negative or P branch). This zero branch cannot appear, however, for the $\Sigma \rightarrow \Sigma$ transitions which occur without a change of the orbital quantum number, since no resulting electronic moment Λ in the direction of the axis joining the nuclei exists for Σ states and, consequently, the electric moment cannot change without a change of the rotational state. Furthermore, in all electronic bands the transition between the two states with the total angular momentum zero of the molecule, the so-called $0 \leftrightarrow 0$ line, is missing.

In addition to the transition (6.45), with much smaller intensity transitions are observed which correspond to quadrupole radiation and to a change of the rotational quantum number by two,

$$\Delta J = \pm 2. \tag{6.46}$$

Thus we have to add to the P, Q, and R branches an O branch and an S branch.

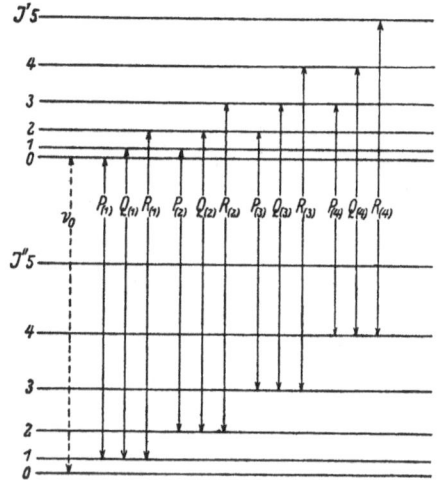

Fig. 206. Transitions between the rotational term sequences of two different electron states of a diatomic molecule. This explains the P, Q, and R branch of a band connected with an electron transition

We know from the discussion of the potential curves that the internuclear distance, the moment of inertia, and the binding forces of a molecule may be greatly changed by an electron transition. In general, therefore, the sequence of the vibrational states as well as that of the rotational states are different in the upper and lower electron states. Accordingly, two different rotational term sequences are shown in Fig. 206, which may be thought of as belonging to any two vibration levels of any two different electron states. The energy difference of the two zero-rotation levels is the wave number of the zero line (forbidden) for which the vibration formula (6.33) applies. It is equal to the sum of the electron and vibration transitions $v' - v''$ and is written $\bar{v}_0(v', v'')$. The individual rotation lines of a band are designated, as in Fig. 206, by the J number of the *lower* state J''. Because of Eq. (4.80) we then have the formula for the wave numbers of the rotation lines of a band

$$\bar{v} = \bar{v}_0(v', v'') + B_{v'} J'(J'+1) - B_{v''} J''(J''+1) \tag{6.47}$$

in which, according to the selection rule (6.45), J'' can only be equal to J' or $J' \pm 1$. These three possible changes in the rotational quantum numbers correspond to the three branches of a band,

$$J' = \begin{cases} J'' - 1: & P \text{ branch} \\ J'' & Q \text{ branch} \\ J'' + 1 & R \text{ branch}, \end{cases} \tag{6.48}$$

which are shown in Fig. 206.

The best representation of the structure of an electron band as determined by Eqs. (6.47) and (6.48) is its so-called Fortrat diagram. If, as in Fig. 207, the wave numbers of the band lines, or their wavelengths, are plotted on the abscissa, and the rotational quantum numbers J'' are plotted on the ordinate, the result is a parabola for each of the three branches of the band, since the wave numbers of the band lines depend quadratically on J, according to Eq. (6.47). The Fortrat diagram thus provides a means of separating the lines of the different branches and rotational quantum numbers which, in the observed spectrum, appear superimposed in a sometimes confusing way. Conversely, we may derive from the Fortrat diagram a picture

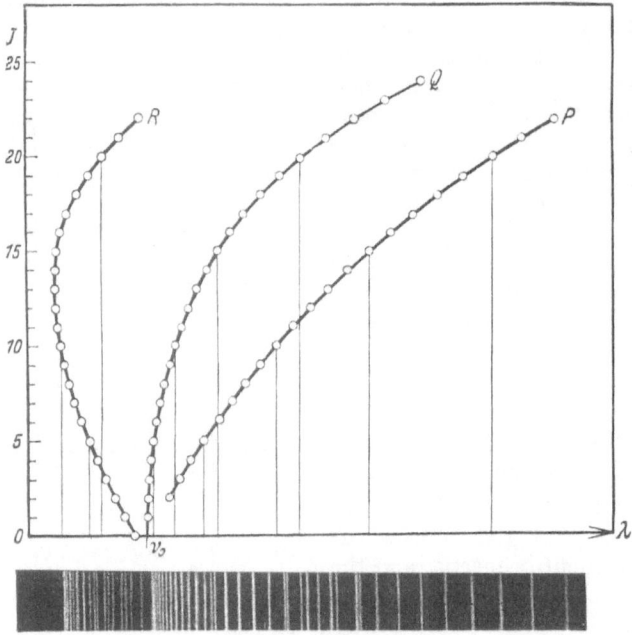

Fig. 207. Fortrat diagram of a band with three branches. Below is the complete spectral band originating from the superposition of the lines of the three branches. (After FERMI)

of the spectrum by projecting on the ν axis or λ axis (abscissa) the points in the diagram which correspond to the different lines. This has been done in Fig. 207 where the picture of the entire band is shown underneath the diagram. It is evident from the diagram that the band edge is actually not due to a convergence of the band lines, as is the series limit in an atomic spectrum, but is a more or less accidental phenomenon. It depends, just as the interval between it and the physically significant zero line (for the pure electron and vibrational transition), on the intervals between the rotation levels of the upper and lower states. If the P branch forms an edge, as in Fig. 210, we say that the band is shaded toward the violet; in the case of an edge formed by the R branch (Fig. 207) it is shaded toward the red. It follows from the relation between the Fortrat diagram and the transitions in the term diagram that the shading depends on whether the internuclear distance (and with it the moment of inertia) is larger or smaller in the upper state than it is in the lower one. We thus have the following relations:

$$\text{Red-shaded band } (R \text{ edge}): \quad r_0' > r_0''; \quad I' > I''; \quad B' < B''.$$
$$\text{Violet-shaded band } (P \text{ edge}): \quad r_0' < r_0''; \quad I' < I''; \quad B' > B''.$$

For equal internuclear distances and moments of inertia in the two combining states, it follows from Eq. (6.47) that, because of $B_{v'} = B_{v''}$, the parabolas of the P and R branch degenerate into straight lines and that all lines of the Q branch merge into a single line as shown approximately in Fig. 208, which is an example of an actually measured band.

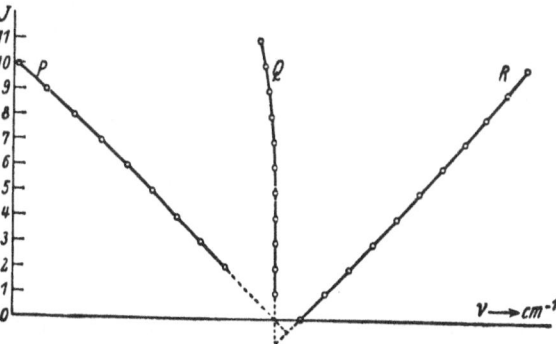

Fig. 208. Fortrat diagram of a band without edge, due to identical internuclear distances in the upper and lower electron state. (After WEIZEL)

Thus a short glance at the shading of the bands of a band spectrum allows us to draw important conclusions about the combining molecular states, whereas an exact analysis provides us with the absolute values of the rotation constants, moments of inertia, and internuclear distances.

d) The Influence of Electron Transitions on the Rotational Structure

In our treatment of the rotational structure of the electron bands we have, so far, not considered the fact that the total momentum $\vec{J}$ of the molecule which controls the spacing of the rotation terms results from the vectorial addition of the angular momentum of the molecular rotation and the angular momentum of the electron shell. This last one itself is composed, according to VI, 5, of the angular momentum of the electron shell about the internuclear axis $\vec{\Lambda}$ and the resultant spin $\vec{S}$, and both may change by the electron transition also. These relationships are complicated by the fact that the rotation of the entire molecule produces a new magnetic field which, with respect to the orientation of the spin of the electron shell, competes with the field in the direction of the internuclear axis. Therefore, the coupling between the angular momenta of the molecular rotation $\vec{K}$, the resulting orbital momentum in the direction of the internuclear axis $\vec{\Lambda}$, and of the resultant spin of the electron shells $\vec{S}$ changes with increasing rotation and depends on the strength of the internuclear field. The different possible cases of coupling for this interaction of electronic motion and rotation have been analyzed and clarified by HUND. These different coupling cases cause differences in the rotational structure of the electron bands, such as the failure of the zero line and of certain neighboring lines to show up, as well as deviations from the parabolic curves in the Fortrat diagram. We must pass over these details of the band structure and can only mention that, as a result of the higher multiplicity of many-electron molecules, multiple branches of bands can also appear. An investigation of all these details, however, is of interest because it permits us to determine unambiguously the change of the electron state that causes a band, i.e., to determine empirically the quantum numbers Λ or Ω of the electron configurations in the upper and lower states (cf. VI, 5).

The complete band analysis thus furnishes nearly all information which we can get about a molecule. The consistent agreement of even the finest details of the often very complicated spectra with the theory is the best evidence for the correctness of our theoretical concept of the molecules and their behavior.

e) The Influence of the Nuclear Spin on the Rotational Structure of Symmetric Diatomic Molecules. Ortho- and Parahydrogen

In conclusion we discuss the influence of the nuclear spin on the rotation of diatomic molecules with *identical* nuclei, upon which an interesting effect of molecular hydrogen depends. We saw in discussing the two-electron system, such as the He atom and the H_2 molecule, that because of the exchange possibility of the two identical electrons under consideration of their electron spin, these two-electron systems have two term systems, a singlet system and a triplet system which do not intercombine. In the case of H_2 and the other molecules consisting of identical atoms we now have, in addition to the possibility of exchange of electrons, the exchange possibility of the identical nuclei. This again causes the occurrance of two term systems which do not combine with each other. The wave-mechanical treatment showed that in the electron singlet system as in the electron triplet system of H_2 the rotation terms belong alternately to one and to the other of the two term systems which in this case are called even and odd because of their different parity. The two nuclear spin momenta of $h/4\pi$ for each of the two protons can have the same direction (analogous to the triplet system) or opposite directions (analogous to the singlet system; see III, 13). We thus expect that every other of the consecutive rotation terms should consist of three closely spaced term components because of the three orientation possibilities of the spin $\hbar$ of the two nuclei. Actually, these three components do not show up since the interaction between the nuclear spin and the molecular motion is too small. They only cause a threefold statistical weight for every other rotational term, by which *alternating intensities* of consecutive band lines of all molecules consisting of identical atoms are produced. This phenomenon was discovered by MECKE. From this explanation follows that the effect of alternating intensities can not appear for molecules which are even slightly asymmetric with respect to their nuclei, such as $N^{14} N^{15}$; and it is here missing indeed. The intensity ratio of a strong rotation line of a symmetric molecule to the mean intensity of its two weaker neighbors is theoretically given by

$$\text{Intensity ratio} = \frac{I+1}{I} \qquad (6.49)$$

with I being the nuclear spin of the identical atoms in the molecule, measured in units of $h/2\pi$. The derivation of Eq. (6.49) would lead us too far. Since the proton has the spin $h/4\pi$, we expect for H_2 the intensity ratio $3:1$, which is in agreement with the experiment. The molecules O_2^{16} and He_2^4, however, because of their resulting nuclear spin zero, have an infinite intensity ratio; i.e., every second rotation line is missing. *A measurement of the intensity ratio of consecutive rotation lines of diatomic molecules formed from identical atoms thus allows us to determine the nuclear spin of the atoms forming the molecule. It complements the measurements of hyperfine structure in atomic lines* (see III, 20). This is an excellent example of the interrelationship of the various fields of atomic physics.

In the case of the hydrogen molecule, with its widely separated rotation states because of its small moment of inertia [Eq. (6.42)], the existence of the two rotational term systems, which do not combine with one another, leads to an interesting effect. Because this effect contradicts classical theory, it can be regarded as further evidence for the correctness of quantum physics. By gradually decreasing the temperature, not all H_2 molecules can enter the rotationless state $J=0$, as is expected classically. Only the molecules with antiparallel nuclear spins can enter this state while those with parallel spins, because of the Pauli principle, must remain in the next higher state $J=1$, which cannot combine with the state $J=0$, regardless of how far we cool down the hydrogen. This fact becomes evident in a well-known anomaly of the specific heat of hydrogen at low temperatures. *Thus molecular hydrogen behaves as if it consisted of two different modifications which* BONHOEFFER *and* HARTECK *called ortho- and parahydrogen, in analogy to ortho- and parhelium. It must be realized, however, that in contrast to helium, where the phenomenon depends on the effect of the electron spin, the difference between ortho- and parahydrogen depends on the nuclear spin of the atoms which form the molecule, or, more accurately, on the exchange possibility of the two identical nuclei.* A transformation of orthohydrogen (parallel nuclear spin momenta, $J=1, 3, 5, \ldots$) into parahydrogen (antiparallel nuclear spin momenta, $J=0, 2, 4, \ldots$) would require a flip-flop process of one of the nuclear spin momenta, which normally is as forbidden for nuclei as it is for electrons. By special treatment of H_2 under high pressure and by adsorption at cooled charcoal, one has succeeded, nevertheless, in producing pure parahydrogen. It was found, as was expected, that the rotation lines corresponding to $J=1, 3, 5, \ldots$ were then missing.

10. The Effect of Quantization of Vibration and Rotation on the Specific Heat of Molecular Gases

As empirical evidence for the quantization of the vibration and rotation energy of molecules (see VI, 6 and VI, 9), we discussed so far only the discrete band spectra. But there is other and just as clear evidence of this quantization and of the impossibility of applying classical physics to molecular motions; this is the specific heat of gases.

According to the equipartition theorem of classical physics, the specific heat per degree of freedom of a molecule is equal to $k/2$ where k is, as usual, the Boltzmann constant. A diatomic molecule according to Fig. 183 has classically three translational degrees of freedom which correspond to the motion of the center of gravity along the three coordinate axes. In addition, there are two rotational degrees of freedom according to the two angles that describe the position of the internuclear axis in space and one vibrational degree of freedom corresponding to the variation of the internuclear distance. This latter has to be counted twice since potential and kinetic energy contribute $k/2$ each to the specific heat. We therefore expect classically a specific heat of $7k/2$ per molecule or $7R/2=7$ cal per mole of diatomic molecules (R=general gas constant) because of the seven degrees of freedom.

Actually, this total specific heat of $7k/2$ per molecule is found only at very high temperatures of several thousand degrees, whereas at normal temperatures the specific heat of diatomic gases is only $5k/2$, and at lowest temperatures it even decreases to $3k/2$. This, however, can be observed only for H_2. The reason for this behavior is easy to understand: The rotational as well as the vibrational energy of molecules is quantized. This is equivalent to saying that the energies can increase only by discrete steps. The equipartition theorem, therefore, can be valid for rotational and vibrational degrees of freedom only if the mean thermal energy kT is at least of the same order of magnitude as the energy steps, i.e. the fundamental vibrational energy quantum $hc\omega_0$ or the first rotational energy quantum $2hcB_0$. If, namely, kT is smaller than the first vibrational energy quantum, a small temperature increase obviously *cannot* cause an increase of the vibration energy, contrary to classical expectation. The vibration then does *not* contribute to the specific heat. Considering nitrogen as an example, we can infer from the molecular data of Table 20 that the *vibration* energy begins to contribute to the specific heat essentially only above 3400 °K, whereas the *rotation* participates fully in the specific heat for all temperatures for which N_2 exists as a gas. The specific heat of $5k/2$ observed for nitrogen and most other diatomic gases is therefore caused by the fact that the two thermal degrees of freedom that are due to the vibration do not contribute to the specific heat at normal temperatures because of the quantization of the vibrational energy. This is often expressed by stating that *the vibrational degrees of freedom at room temperature are already "frozen".* In hydrogen which has particularly large rotational energy quanta because of its small moment of inertia [cf. Eqs. (6.41) and (6.42)] and which remains a gas even at very low temperatures, also the freezing of the rotation has been observed, in agreement with theory.

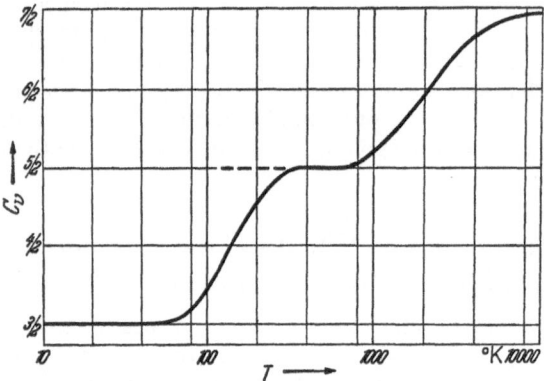

For H_2, Fig. 209 shows this "freezing" of vibration as well as rotation when the temperature is decreased. *The temperature dependence of the specific heat, therefore, is quite as unambiguous evidence of the quantization of rotation and vibration of molecules as are their discrete rotation and vibration spectra.* In some cases of polyatomic mole-

Fig. 209. Stepwise "freezing" of vibrational and rotational degrees of freedom for the H_2 molecule with decreasing temperature

cules, it was even possible to determine vibration quanta of certain optically inactive vibrations directly by measuring the temperature dependence of the specific heat. The same problem, viz., the relation of quantized vibrations to specific heat, will be treated in VII, 9 for the more complicated case of the solid state.

What we discussed here for the case of molecules applies for atoms also. If these could be considered as small spheres obeying classical laws of physics, they would have, in addition to their three translational degrees of freedom, three rotational ones because of the possibility of rotation about the three coordinate axes. Actually, the specific heat of monatomic gases is, however, $3k/2$ per atom instead of $3k$. Obviously, only the three translational degrees of freedom, not the rotational ones, contribute to the specific heat, and this is just what we should expect from quantum theory: The rotation of an atom, according to Chap. III, can be considered as being due to the orbital motion of the electrons, and the quantization of the orbital momentum of these light particles leads to energy steps of the order of several electron volts because of the smallness of the moment of inertia of the atoms. *The three rotational degrees of freedom are therefore practically always frozen, and we may consider the specific heat of $3k/2$ per atom of the monatomic gases to be direct evidence of the quantization of the orbital momenta of the electrons* [see Eq. (6.75)].

11. Band Intensities and Band-Spectroscopic Temperature Determination

Knowing about the structure of the spectra of diatomic molecules and its relation to molecular structure, we are able to consider the problem of the intensity of band spectra. This is important, first, because line structure *and* intensity distribution are necessary for a complete description of a spectrum and, second, because the measurement of the intensity distribution in band spectra has led to an important method for determining high temperatures.

According to III, 23, the intensity of a spectrum is determined by the transition probability between the two combining states and by the number of molecules occupying the initial state. The transition probability, and with it the intensity of the entire band system, is partly determined by the probability of the electron transition, computed according to IV, 9. The transition probability for the individual bands, on the other hand, follows from the Franck-Condon principle (cf. VI, 6c). Within each branch of a band, finally, the transition probability is constant, whereas the relative intensity of the different branches can be computed from J and the electronic quantum numbers according to formulae by HÖNL and LONDON. The occupation number of the initial state follows from MAXWELL's distribution law, but only for the case of thermal equilibrium, i.e., in general for the *absorption* of molecules. The case of *emission*, however, can be computed only if the excitation is exclusively due to thermal excitation. For the intensity of a spectral line resulting from a transition between two states of energies E_1 and E_2 we have

$$I_{E_{1s}} = C_{1,2}\, g_1 e^{-\frac{E_1}{kT}} \quad (6.50)$$

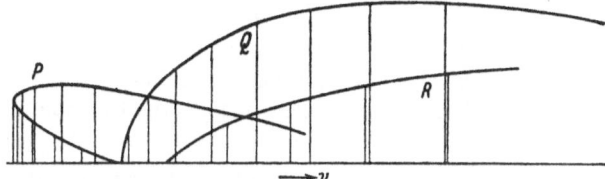

where $C_{1,2}$ is a constant containing the transition probability between

Fig. 210. Intensity distribution in the three branches of a normal band. (After JEVONS.) The intensity of the lines is indicated by their length

states 1 and 2, g_1 is the statistical weight of state E_1 (cf. III, 23), and T is the absolute temperature of the absorbing or emitting molecules.

From the general formula (6.50) we obtain the intensity distribution in a rotation band if we substitute for E_1 the rotation energy

$$E_1 = h c\, B J\, (J+1) \qquad (6.51)$$

and for g_i the statistical weight of a rotational state

$$g_1 = 2J + 1. \qquad (6.52)$$

Eq. (6.52) follows from the fact that each rotational state in a diatomic molecule is $2J$-fold degenerate because of the two equal rotation axes which are perpendicular to the internuclear axis. With this we have

$$I(J) = C(2J+1)e^{-\frac{h c B J (J+1)}{kT}}. \qquad (6.53)$$

Eq. (6.53) describes the intensities of the rotation lines of a band as a function of the J values of the initial state. Fig. 210 shows the intensity distribution of the rotation lines of a band computed according to Eq. (6.53).

According to Eq. (6.53), the intensity distribution in a band is temperature-dependent. Thus by measuring the maximum of the intensity in the branches of a band, i.e., the J value of the line of highest intensity, the temperature of the emitting or absorbing gas can be determined. This method has been applied successfully for measuring the temperatures in electric arcs, especially in the outer zones, because it was possible there to prove the existence of thermal excitation. However, in discharges resulting from nonthermal exciting collisions, Eq. (6.50) is not applicable; and attempts at determining the temperature with its help lead to gross errors.

If the condition of thermal equilibrium is fulfilled, the method described above may be used for computing how molecules at a given temperature are distributed among the vibrational states of the initial electronic state. By applying the Franck-Condon principle, the intensity distribution of the bands within a band system can be determined. In this case we have to substitute in Eq. (6.50)

$$E_1 = hc\omega\,(v + \tfrac{1}{2}).\tag{6.54}$$

Since the vibrational states of diatomic molecules are not degenerate, the statistical weight of all vibrational states is unity, i.e., $g=1$. Thus we have

$$I(v) = C\,e^{-\frac{hc\omega(v+\frac{1}{2})}{kT}}.\tag{6.55}$$

Because of the difficulty of numerically computing the transition probability contained in C from the Franck-Condon principle, Eq. (6.55) is less suited for the absolute determination of temperature from the relative intensities of the different bands of a system. However, the determination of changes of the temperature from the corresponding changes of the relative intensities of the various bands is easily possible since in this case C remains constant. This vibration method has been used also with success in investigating temperatures of electric arcs and of flames.

12. Isotope Effects in Molecular Spectra

We have already mentioned in discussing the phenomenon of isotopes (see II, 6c) that the detection of isotopes as well as the determination of their masses and relative abundances is possible by optical spectroscopic methods just as by mass spectroscopy. We treated in III, 20 the influence of isotopes on the hyperfine structure of line spectra. Now we discuss briefly the isotope effect in band spectra.

Since the isotopes of an atom differ only in their masses, the incorporation of different isotopes in the same molecule (e.g., Li^{6}H and Li^{7}H) affects the band spectra through the corresponding change in the moment of inertia (6.21) of the molecule. According to Eq. (6.41), the separation of the rotational levels is inversely proportional to the moment of inertia I of the molecule, while according to Eqs. (6.24) and (6.25) the spacing of the vibrational levels is inversely proportional to the square root of the moment of inertia I. The results of these two influences on the spectra are the rotational and the vibrational isotope effect.

Let us consider two diatomic molecules, one of which may have the atomic masses m_1 and m_2; the other one $m_1 + \Delta m$ and m_2. By making use of the theory presented in VI, 9 for the rotation and in VI, 6 for the nuclear vibration, the separation Δv_r of two equivalent rotation lines of the two isotopic molecules and the separation Δv_s of two equivalent band edges of these two molecules, may be computed. We thus find

$$\Delta v_r = \frac{m_2\,\Delta m}{(m_1 + m_2)\,(m_1 + \Delta m)}\,v_r\,,\tag{6.56}$$

and

$$\Delta v_s = \frac{m_2\,\Delta m}{2\,(m_1 + m_2)\,(m_1 + \Delta m)}\,v_s\,,\tag{6.57}$$

where v_r is the wave-number distance of the rotation lines from the zero line of the band and v_s represents the distance of the band edges from the 0,0 band of the band system.

Thus the isotopic splitting increases linearly with increasing distance from the zero line of a band (cf. Fig. 211) and from the 0,0 band in a band system. By measuring the spacings $\Delta\nu_r$ or $\Delta\nu_s$ and the distances ν_r or ν_s, respectively, the mass ratio $m_1/(m_1 + \Delta m)$ of the isotopic atoms in the molecule can thus be determined. Because of the high accuracy of spectroscopic wavelength measurements this method of determining isotope masses is very accurate. By measuring the intensity ratios of the lines belonging to the isotopic molecules, the relative abundance of the isotopic atoms in the molecule can be determined.

By means of this isotope effect in band spectra, new isotopes of the elements C, N, and O were discovered for the first time. A large number of isotopes were confirmed after they had been detected by mass-spectroscopic experiments. Mass ratios and relative abundances have been measured for many new isotopes. The application of high-frequency spectroscopy to this field has increased considerably the attainable accuracy.

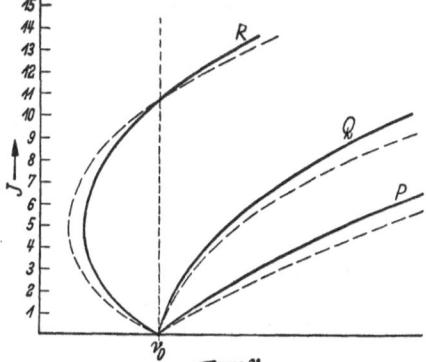

Fig. 211. Rotational isotope effect in a band. The branches belonging to the two isotopic molecules are indicated as solid and dotted curves, respectively. (After MECKE)

13. Survey of the Spectra and Structure of Polyatomic Molecules

The results of molecular structure deduced from the spectra of diatomic molecules as they were presented in this chapter can be applied, at least in their fundamentals, to polyatomic molecules. However, because of the large number of possibilities for excitation and ionization, for vibration and dissociation, as well as for the rotation about the three axes associated with the principal moments of inertia, the spectra of these molecules are very complicated. A complete analysis of the electronic, vibrational, and rotational structure, and the corresponding complete knowledge of all molecular data, therefore, is available only for some simpler polyatomic molecules (such as H_2O). In most cases we have to be content with incomplete information. Therefore, we shall not discuss in all detail the electronic, vibrational, and rotational structures of the spectra of polyatomic molecules but rather limit ourselves to a survey of the more important phenomena which are typical of such molecules.

a) Electron Excitation and Ionization of Polyatomic Molecules

The systematics of the electronic states of polyatomic molecules is closely related to that of diatomic molecules (cf. VI, 5). The difference compared to diatomic molecules consists, in the first place, in the larger number of electrons which may be excited and ionized. In the second place, these electrons can play very different roles in holding the molecule together. We may distinguish three limiting cases:

First, the light absorption by non-binding molecular electrons, which leads to an excitation of stable molecular states; second, the absorption by localized electrons of the so-called chromophoric groups; and finally, the absorption by binding electrons, which often produces dissociation.

The absorption spectra of quite a number of polyatomic molecules, e.g., of benzene, show a clear vibrational structure even in the liquid state, or for molecules in solution. We thus are compelled to conclude that absorption in these molecules occurs by electrons which participate to such a small extent in the binding that stable electronic states of the molecule are excited by the absorption. In this case of photo-excitation, as it is presented in Fig. 185 for diatomic molecules, even ionization may occur as a limiting case. The molecule remains stable in the excited or ionized state; neither the nuclear arrangement nor the vibration are basically changed. Spectroscopically, Rydberg series of bands are observed in the vacuum ultraviolet onto which is joined in many cases an ionization continuum the long wavelength limit of which corresponds to the ionization energy. However, it must be remembered that, for polyatomic molecules, one can no longer speak strictly of *one* ionization energy because the ionization energy varies greatly depending on the binding state of the particular separated electron. Only because, for some reason not yet known to us, one electron is preferably excited and ionized, its ionization energy can be called that of the molecule. Our molecular theory, here as so often in atomic physics, has not yet arrived at its final state. We can determine what electron is excited and in what manner if the molecule absorbs a certain spectrum. However, we cannot yet answer the question why, in a given electron configuration of a large molecule, one particular electron and only that one electron absorbs the incident light and becomes excited.

A systematic investigation of the second group of electronic excitations, the study of chromophoric groups, can perhaps contribute to the solution of this question. In a polyatomic molecule, radiation may be absorbed by an electron which does not belong to the shell of the entire molecule but which is localized in a definite group of atoms. This may be a benzene ring or any similar atomic group belonging to a large molecular complex. In this case the resulting spectrum in a first approximation is independent of the large complex to which the atomic group (e.g., $-N=N-$, $-HC=CH-$, $>CO$ or Cr^+) is attached, and only finer differences (wavelength displacements) permit us to draw conclusions on the state of binding of the particular group to the whole molecule. Since the absorption spectra of this type of electrons lie mostly in the visible spectral region and thus determine the color of the substance which consists of these molecules, these absorbing groups are called *chromophoric groups*.

In the third limiting case, the absorption is due to electrons which are responsible also for holding the molecule together and which therefore are called *binding electrons*. In analogy to the case of the diatomic molecule (see Fig. 186), their excitation can reduce the bond to such an extent that the molecule dissociates, as a consequence of the absorption, into two atomic groups. According to VI, 7, the corresponding absorption spectrum must be continuous. This photodissociation as a consequence of the absorption by binding electrons explains why the ultraviolet absorption spectra of numerous polyatomic molecules are continuous. But we must point out here that the absorption spectra of polyatomic molecules that appear to be continuous are by no means always genuine dissociation or predissociation spectra. On the one hand, discrete band spectra appear sometimes to be continuous at low resolution, in particular because the distance of neighboring band lines is inversely proportional to the molecular moment of inertia (cf. VI, 9a) so that it often becomes extremely small for polyatomic molecules. On the other hand, there exists quite a number of apparently genuine continuous absorption spectra of molecules which do not seem to lead directly to dissociation because small structural changes in the molecule again produce a vibrational structure. According to KORTÜM, this is true for molecules in which large groups carry out torsional vibrations against one another, e.g., the two phenol rings in diphenyl ($\langle \bigcirc - \bigcirc \rangle$). If the amplitude of this torsional vibration is sufficiently large, the structure of the originally rigid and plane molecule is changed in such a way that radiationless transitions from the double-ring structure to the single-ring structure become probable. The conjugation of the two rings forming the entire molecule thus disappears. The life time of the excited state of the molecule is correspondingly shortened by many orders of magnitude so that, according to the uncertainty principle (see IV, 18), the absorption spectrum of the whole molecule looses its structure without dissociation or predissociation occurring. If torsional vibrations are made impossible in such molecules by suitable substitutions or by freezing them in through cooling, the structure in the spectrum re-appears. The molecules become rigid with respect to torsion, and the spectra of rigid molecules show clear vibrational structure. We thus see an interesting interaction between electron arrangement and torsional vibrations.

This knowledge of the possibility of exciting particular electrons, which is based exclusively on the investigation of absorption, was supplemented in some cases by investigating the fluorescence. The investigation of the emission spectra of polyatomic molecules for a long time failed to produce reliable results because these complicated molecules in most cases dissociated in gas discharges as a consequence of exciting electron collisions. What could be observed, therefore, was a confusing superposition of the continuous spectra of the molecule and its components. However, by using a low-current glow discharge with special precautions, SCHÜLER was able to excite discrete as well as continuous emission spectra of very complicated and easily dissociable polyatomic molecules. By investigating whole series of such molecules in which either different groups were attached to the same molecule or, conversely, the same groups were attached to different molecular complexes, and by comparing the emission spectra with the absorption and fluorescence spectra of the same molecule, he was able to establish a large number of rules and relations which seem to be of importance for understanding the interaction in polyatomic molecules. Interesting differences have been ascertained concerning molecular excitation by light (absorption) and by electron impact (excitation in a discharge). These differences will aid in an attempt at answering the above-mentioned question why and how a particular electron is preferentially excited in a certain process. In the case of formaldehyde, acetone, and diacetyle, for example, fluorescent light is emitted from a lower energy state of the molecule than that which is excited by absorption of light. This initial state of fluorescence, which is normally reached by collisions of the second kind (page 71) from the state excited by absorption can be excited directly from the ground state by electron impact. In other cases, as in the case of benzene and its derivatives, light absorption always leads to an excitation of the non-localized π electrons of the benzene ring (see VI, 14d). Recent investigations proved, however, that this emission of the C_6 ring (excited by electron impact), which corresponds to a spectrum between 2000 and 3000 Å, is excited only if the mass number of the individual substituents, or that of two substituents coupled by a double bond, is smaller than 27. Since this value is of the order of magnitude of the mass of two C atoms in the ring (mass number 24), we seem to have a theoretically interesting relation: Electron excitation can apparently be blocked by externally attached atom groups with a mass larger than that of the absorbing group. Similarly, though not quite as pronounced, the mass influence seems to be effective in the case of excitation of the molecules to fluorescence (decreasing intensity of fluorescence with increasing mass of the substituents in the series F-, Cl-, Br-benzene). Many more interesting results may be expected from this type of research concerning the detailed behavior of different outer electrons of larger molecules.

b) Rotational Structure and Moments of Inertia

Rotational structure, the investigation of which provided so simple a method for determining moments of inertia and internuclear distances of diatomic molecules (cf. VI, 9), is (except for the special case of linear molecules which we shall discuss soon) of less value for studying polyatomic molecules. First, the moments of inertia of even relatively simple polyatomic molecules are so large that the rotation bands can be resolved only with high-frequency methods because the distance of consecutive band lines is inversely proportional to the corresponding moment of inertia of the molecule. Furthermore, in the most general case, a polyatomic molecule has three principal moments of inertia (corresponding to the model of the asymmetric gyroscope), and the rotational structure rapidly becomes extremely complicated for this model. Only in the case of the H_2O molecule, the details of the rotational structure were almost completely analyzed and correspondingly accurate molecular data were derived. Even in the apparently simple

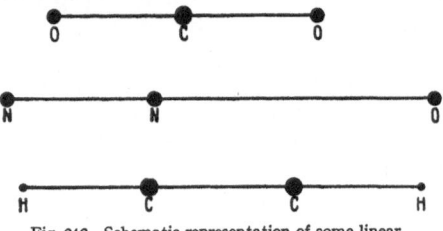

Fig. 212. Schematic representation of some linear polyatomic molecules, CO_2, N_2O, C_2H_2

cases in which two or even three of the moments of inertia are equal, the rotational structure is so complicated because of the interaction between rotation and vibration that only a few molecules were thoroughly investigated. Only linear molecules such as CO_2, N_2O, C_2H_2, etc. (Fig. 212) have a simple and clear rotational structure. In these molecules, as in the diatomic molecules, the moment of inertia about the molecular axis is zero in the first approximation (if one neglects the deformation vibrations and the resulting orbital momentum of the molecular electrons), and the other two moments of inertia are equal. In this case the rotation bands are exactly equal to those of the diatomic molecules so that the formulae of VI,9 can be applied. Thus the moment of inertia can easily be deduced from the spectra. However, even if the masses of the atoms forming the molecule are known, in this case the internuclear distances cannot be computed without additional information, because different nuclear configurations may have the same moment of inertia. One of the non-spectroscopic methods, e.g., electron diffraction, thus must be used for determining the internuclear distances.

c) Vibration and Dissociation of Polyatomic Molecules

The problem of the vibration of a polyatomic molecule formed of N atoms is rather complicated also. In order to determine the number of possible vibrations, we have to subtract from the $3N$ degrees of freedom of the N atoms the three degrees of freedom of translation and in general three degrees of freedom for the rotation of the entire molecule so that there are $3N-6$ degrees of freedom of vibration. Consequently, $3N-6$ fundamental vibration frequencies have to be derived from the infrared and Raman spectra. From these, the actual vibrations of the molecule originate (just as Lissajous figures of mechanical vibrations) by superposition of the vibrational amplitudes associated with the individual frequencies. These $3N-6$ independent harmonic vibrations from which all actual vibrations of the molecule evolve are called the *normal vibrations*. The six normal vibrations of a plane four-atomic molecule such as formaldehyde H_2CO are shown, as an example, in Fig. 213. The sixth normal vibration is perpendicular to the plane of the molecule in which the other five vibrations happen.

The fundamental difficulty in investigating the vibration of polyatomic molecules lies in the fact that, in contrast to the diatomic molecules, the spectroscopic knowledge of the fundamental frequencies of the normal vibrations is not sufficient for determining the forces between the different atoms and atomic groups. One can easily see, for example, from Fig. 213 that the number of spiral and leaf springs one must think of so that by their action the normal vibrations become possible is much larger than the number 6 of the normal vibrations the frequencies of which can be determined from the spectrum. The number of force constants for f normal vibrations is equal to $f (f+1)/2$. For the H_2CO molecule this number thus is 21. Consequently, determining the normal vibrations from the spectrum is *not* sufficient to establish a molecular model. In addition, the geometry of the vibrations must be known, and the correct spectroscopically determined fundamental vibration frequencies must be assigned to the different *geometrically determined normal modes*. Determining the vibration modes can be done by geometrical considerations in the case of simple molecules as soon as the nuclear configuration has been established, e.g., by electron diffraction. The vibrational isotope effect (cf. VI, 12) may be used to aid in establishing the modes of vibration of complicated molecules. If an isotope of different mass is substituted

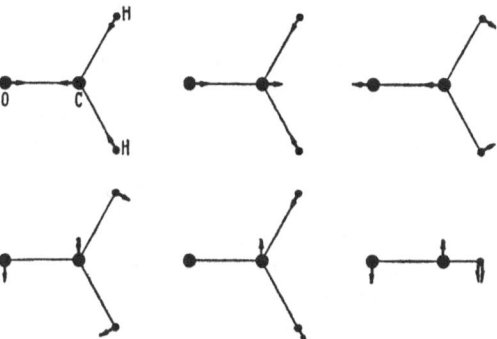

Fig. 213. The normal vibrations of the formaldehyde molecule H_2CO. (After Mecke)

for any atom of a molecule, the change in the vibration spectrum produced by this substitution is the larger the more the substituted atom participates in the vibration. By systematic substitutions, the different forms of vibrations and their associated frequencies may thus be determined. Another aid is the fact established by Mecke that the vibrations along the line joining two nuclei, the "valence vibrations" (e.g., the first normal vibration of Fig. 213), always have considerably higher frequencies than those associated with the angular changes, the "deformation vibrations", such as the fourth and sixth normal vibration shown in Fig. 213. Thus the difficulty in the vibrational analysis of polyatomic molecules lies not so much in determining the fundamental vibration frequencies than in assigning these frequencies to the normal vibrations of the vibrational modes.

It is self-evident that the vibrations of polyatomic molecules are anharmonic, and thus each valence vibration can be described by a potential curve such as Fig. 189. This representation is of value in explaining dissociation processes. From the anharmonicity of the vibration it follows, as in the case of diatomic molecules, that, in addition to transitions between neighboring states, also larger vibrational transitions occur in the spectrum. However, these appear with smaller intensity. Moreover, it follows from the anharmonicity of the vibrations that there is a mutual interaction of the different vibrations which is manifest in the appearance of combination vibrations, i.e., the superposition of different normal vibrations. Obviously, this leads to an increased complication of the vibration spectrum. This situation has an analogue in acoustics where the analysis of a composite sound phenomenon, as from an organ pipe, offers many difficulties. Just as in the spectra of diatomic molecules, the frequencies of all those vibrations which cause a change of the electric moment of the molecule occur in the emission and absorption spectra of polyatomic molecules. The frequencies of those vibrations which change the polarizability of the molecule may be taken from the Raman spectrum. Again infrared and Raman spectra in most cases complement each other excellently so that almost all molecular vibrations can be determined experimentally. In the exceptional cases where dipole moment *and* polarizability remain constant during a vibration, so that the vibration is optically *and* Raman inactive, the frequency can at least approximately be determined for not too complicated molecules (see VI, 10) from the temperature dependence of the specific heat.

In discussing the motion of electrons in polyatomic molecules, we distinguished between localized electrons associated with a definite atom or atomic group, and non-localized electrons associated with the shell of the entire molecule. In an analogous way, we can distinguish between *localized* and *non-localized vibrations* according to whether only two neighboring atoms or atomic groups participate in the vibration under consideration and, therefore, one bond only is affected, so that the energy of the vibration quantum is independent of the nature and the state of motion of its environment in the molecule, or not. For example, all atoms of the formaldehyde molecule participate in the vibrations presented in Fig. 213; these vibrations therefore are *non*-localized. The vibration of two doubly bound C atoms, however, has the same fundamental vibration quantum ω_0 of about 1600 cm^{-1} in all hydrocarbons and thus is strongly localized.

Of much higher importance for polyatomic molecules than for diatomic ones is the interaction between vibration and rotation. It is evident from the vibrational modes such as those in Fig. 213 that these vibrations will be more or less distorted by molecular rotation, depending on the particular type of vibration. This is analogous to the perturbation of the vibration of a Foucault pendulum by the Coriolis force due to the earth's rotation. Finally there are certain cases where a molecular vibration can change directly into a rotation. This is possible if, for example, a molecular group such as CH_3 is held in its stable position with respect to the rest of the molecule by weak forces of such a type that it can carry out slow rotatory vibrations about its rest position, which at higher energy may change into a real rotation. Thus the clear distinction between vibration and rotation, which is characteristic for the case of diatomic molecules, may get completely lost for some polyatomic molecules. We cannot discuss here the corresponding complicated spectral phenomena. It is obvious, however, that vibrational and rotational structure must disappear if the free vibration and rotation of individual molecules are prevented by bonds which may be formed between the molecules. According to SCHÜLER, this happens if phenol is added to quinine so that hydrogen bridges form (see VI, 15).

We finish our discussion by briefly considering the dissociation processes. As in the case of diatomic molecules, dissociation of polyatomic molecules by absorption of continuous radiation is possible as a consequence of excessive molecular vibration, but it is limited to the overexcitation of valence vibrations (breaking of bonds). This clearly emphasizes the significance of valence vibrations as compared to deformative vibrations. Because of the large number of possible vibrations, there is also a large number of possibilities of dissociation of a molecule. For this same reason, there appears much more frequently than for diatomic molecules that superposition of discrete and continuous vibrational levels which, according to VI, 7b, may lead to predissociation. In both cases the large number of possibilities makes it difficult to attribute a particular absorption continuum or a region of predissociation in the spectrum of a polyatomic molecule to a definite dissociation process. We may note also that the dissociation energies of polyatomic molecules that are derived from the long-wave limit of absorption continua (or predissociation wavelengths) can represent only their *upper limit*. A certain amount of the absorbed energy may be used, namely, not for the dissociation but also for the excitation of other vibrations because of the above-mentioned coupling between different vibrations.

Because of this ambiguity of the spectroscopic results, every information gained with different methods about the dissociation processes is of high interest, in particular for polyatomic molecules. Such information may be derived from photochemistry which allows sometimes to prove photodissociation by secondary reactions. In certain cases dissociation products may be detected optically. It is possible, finally, to investigate the dissociation of organic gases and vapors after ionization by electron impact with subsequent mass-spectroscopic analysis of the ionized fragments. In general, the dissociation here follows immediately after the ionization. But occasionally an ionized molecule may pass the entire acceleration space of the mass spectrograph before it disintegrates.

We summarize: The investigation of the spectra of polyatomic molecules may provide answers about the excitation, ionization, and dissociation processes and the amounts of energy required for these processes. In the simpler cases it also permits to determine the possibilities of vibration, the force constants, and the moments of inertia. For all complicated molecules, however, the determination of the nuclear configuration and the binding forces is possible only if the spectroscopic methods are supplemented by the other methods of molecular physics mentioned in VI, 2, and especially by electron diffraction.

14. The Physical Interpretation of the Chemical Bond

In this chapter we discussed so far the most important facts about the structure and properties of molecules. In doing this, we took the molecules as facts and did not consider the difficult problem, which kind of force causes the binding of a number of atoms, either like or unlike, in a molecule. This is really the fundamental problem of chemistry, because otherwise chemistry has to accept as a sort of miracle the fact that there is an H_2 but no H_3 molecule, a CO and a CO_2 but no CO_3, an H_2SO_4 but no HSO, etc. The theoretical explanation of the Periodic Table (cf. III, 19) was the first major achievement of atomic physics for chemistry. The physical explanation of chemical bond in homopolar molecules, achieved by the methods of quantum mechanics in 1927, completed the successful attempts at understanding the basic features of chemistry from the properties of the atoms.

a) Pre-quantum-mechanical Attempts. Heteropolar Bond and Octet Theory

Partial success had been obtained, as early as 1916, by KOSSEL who succeeded in explaining the bond in heteropolar or ionic molecules, such as NaCl, as an effect of electrostatic forces. According to III, 19, the unique behavior of the chemically inactive noble gases is due to their completed (saturated) electron shells. Atoms with one or a few external electrons outside of closed shells (such as the alkali or alkaline-earth atoms) are electropositive because they can attain noble-gas configuration by releasing these external electrons. The halogen atoms, on the other hand, are electronegative because they can complete their electron shells to a noble-gas shell by accepting an additional electron. The bond of an alkali-halide molecule such as NaCl, according to this very simple concept, is based on the following: An Na atom gives up its outermost electron to a Cl atom and thus produces an Na^+ and a Cl^- ion. These ions have closed electron shells as noble gases and attract each other electrostatically.

Polyatomic molecules can be explained by this effect of ionic attraction also. Of atoms which belong to the same period of the Periodic Table, those from the left side always appear positively charged, those from the right side negatively charged. Atoms from the central groups of the table may appear as positive or negative ions depending on whether they combine with atoms which are farther to the right or farther to the left in the Periodic Table. The molecules PH_3 and PCl_5 are examples of this. In PH_3 the electron shell of the P atom is *completed* by inclusion of the three hydrogen electrons and thus forms the argon noble-gas shell. The P atom thus becomes a threefold negatively charged P ion which binds the three positive H^+ ions. In the PCl_5 molecule, on the other hand, each of the five Cl atoms attracts an electron of the P atom so that its electron shell is *reduced* to the noble-gas shell of neon and we get a fivefold positive P^{5+} ion which binds electrostatically the five Cl^- ions. In an analogous way, we may regard the C atom in CH_4 as a fourfold negatively charged, and in CCl_4 as a fourfold positively charged ion. ABEGG's rule, known since 1906, according to which the sum of the highest positive and negative valency of an atom is always 8, thus follows automatically from the possibility of completing or reducing the outermost noble-gas shells which contain eight electrons.

By applying such electrostatical considerations, KOSSEL was even able to explain the complicated phenomena of A. WERNER's complex chemistry: A usually highly charged central atom binds a larger number of outer atoms than should be possible according to its chemical valency. Not only the spatial extension of the participating ions must here be taken into account but also the "penetration" of the central field through the "envelope" of the nearest neighbors.

This explanation of the ionic bond seemed so evident and its success so convincing that one was inclined for a long time to overlook the inherent difficulties. The energy necessary for ionizing the metal atom is smaller than the energy released in attaching the electron to the halide atom only for the CsF molecule, so that only in this case two oppositely charged ions actually can be produced, which then may form an ionic molecule. However, numerous other molecules also seem to have this heteropolar bond. This is possible because the excess of the ionization energy over the electron-affinity energy is overcompensated by the energy gained in forming the molecule. Accordingly, diatomic ionic molecules have considerable dipole moments (cf. page 348) which nevertheless are much smaller than would be expected if these molecules were actually formed from ions. The dipole moment of KCl, for instance, should be equal to 15×10^{-18} esu., viz. one electronic charge multiplied by the internuclear distance of 3.1 Å, whereas the measured value is only 8×10^{-18} esu. For most molecules, the discrepancy is even larger. We see that the theory of the heteropolar bond cannot be as simple as it first seemed to be, in spite of its success in explaining a wide variety of facts. We shall show below how quantum mechanics has modified and incorporated this simple theory into the framework of the valence theory.

Beside these complications, the theory of the heteropolar bond is obviously restricted to such atoms of different electronegativity with electron shells which complement each other in forming noble-gas shells. The binding of the simplest existing molecules, such as H_2, N_2, O_2, and many others, i.e., of the so-called *homopolar* molecules, cannot be explained on this basis.

An attempt of an explanation was made somewhat later by G. N. LEWIS with his *octet theory* which is still popular with some chemists. LEWIS's theory also is based on the fact that all electron shells tend to complete the noble-gas shell of eight electrons, i.e., to form so-called octets. By indicating each valence electron (i.e., electron of the outermost non-closed shell) by a dot, LEWIS writes the ethane molecule C_2H_6 in the following way:

$$
\begin{array}{cc}
\text{H} & \text{H} \\
\!\cdot\cdot & \cdot\cdot \\
\text{H:C:C:H} \\
\cdot\cdot & \cdot\cdot \\
\text{H} & \text{H}
\end{array}
$$

By making use of the H electrons, the C atoms thus succeed in surrounding themselves by complete octets. Each electron pair, however, belongs to two atoms at the same time and thus is supposed to cause the bond between them. For many well-known molecules the octets cannot be complete. NO_2, for instance, may be written as

$$:\!\ddot{O}\!:\!\ddot{N}\!:\!\ddot{O}\!:$$

However, LEWIS was able to show that such molecules are the less saturated, i.e., the more chemically active the less complete their octet shells are. It is an important feature of LEWIS' concept that he recognized homopolar chemical binding as an effect of electron pairs common to the two bound atoms. He could not yet give, however, an explanation for the re-arrangement of the electron shells that leads to the bond.

b) Quantum Theory of the Chemical Bond

The quantum theory of the homopolar chemical bond has been presented in IV, 11 in the discussion of the H_2 molecule. In spite of the significance of the electron *pair* bond, the physical reason for the bond is best illustrated by the example of two protons bound by *a single* electron, i.e., by the example of the H_2^+ ion. Because of the identity of the H_2^+ nuclei 1 and 2, its electron may be found near the first nucleus (wave function ψ_1) just as well as near the other one (wave function ψ_2). According to IV, 11, the actual behavior of the electron has to be described by the wavefunctions $\psi_1 \pm \psi_2$. Fig. 214 shows the symmetric wavefunction with even parity $\psi_s = \psi_1 + \psi_2$ and the antisymmetric wavefunction with odd parity $\psi_a = \psi_1 - \psi_2$ for a finite internuclear distance. Considering the fact that the square of the wavefunction

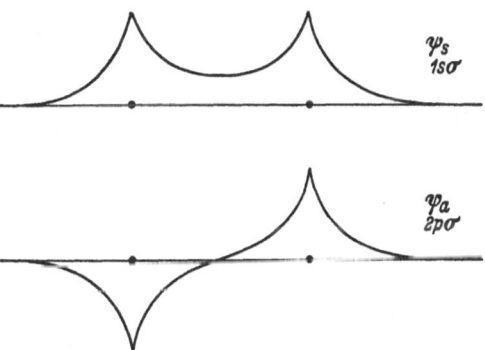

Fig. 214. Wave functions of H_2^+, symmetric and antisymmetric with respect to mirror reflection at the origin. The symmetric wave function corresponds to a considerable electron density between the nuclei and is responsible for their binding. For the antisymmetric wave function, the electron density has a minimum between the nuclei so that binding is impossible

represents the time average of the spatial electron density distribution, we see that the electron density between the two nuclei is large for ψ_s but very small for ψ_a. Since the symmetric wavefunction does not have nodes on the axis joining the nuclei, the electron described by ψ_s is a $1s\sigma$ electron, whereas the electron characterized by ψ_a with a node on the axis is a $2p\sigma$ electron. According to Fig. 214, the $1s\sigma$ electron stays predominantly between the protons of the H_2^+ so that it binds them, whereas the probability for the $2p\sigma$ electron to stay between the nuclei is very small. This electron thus is not able to compensate the mutual electrostatic repulsion of the nuclei so that in this case repulsion occurs for every internuclear distance. The discussion shows clearly *that also according to quantum mechanics the chemical bond is caused by electrostatic forces, namely the compensation of the mutual repulsion of the positive nuclei by the negative charge density that exists between the nuclei of a molecule which is described by a symmetric wavefunction.*

The situation is similar for two electrons participating in the bond, i.e., in the H_2 molecule. Here, in addition to the considerations above, the Pauli principle becomes instrumental, according to which all electrons of a molecule must differ with respect to at least one quantum number. Therefore, the two $1s\sigma$ electrons which cause the bond must differ in the direction of their spins. *This spin compensation of the two binding electrons in H_2, which is necessary according to PAULI's principle, thus does not cause the bond as, mistakenly, it was occasionally assumed. It only makes possible the presence of two $1s\sigma$ electrons in the molecule, which cause the electrostatic bond between the two nuclei by the large probability for their staying between them.*

It was shown in IV, 11 that the behavior of the two H_2 electrons may also be described by a symmetric eigenfunction $\psi_s = \psi(1, 2) + \psi(2, 1)$ and an antisymmetric one $\psi_a = \psi(1, 2) - \psi(2, 1)$, with $\psi(1, 2)$ and $\psi(2, 1)$ being distinguished by the association of the electrons 1 and 2 to the nuclei a and b, respectively. If now the previously neglected interaction of the two atoms of the molecule is introduced as a perturbation, the two states of the total system described by ψ_s and ψ_a, which were at first degenerate, i.e., had the same energy, split up into two different energy states, one of which always has a smaller energy than *each* of the two original ones which belonged to the wave functions $\psi(1, 2)$ and $\psi(2, 1)$. The most stable state of the system corresponds to the lowest possible potential energy of the nuclei; and evidently its stability is the larger the larger the interaction of its two atoms. The interaction of the two atoms, and thus the stability of the molecular bond formed by them, increases with decreasing distance the more their electron eigenfunctions overlap. Since every physical system left to itself attempts to reach the state of lowest potential energy, *the atoms in a molecule arrange themselves, as far as the Pauli principle permits, in such a way that their electron eigenfunctions overlap as much as possible.* However, the internuclear distance cannot decrease infinitely, because too strong a penetration of the atoms is prevented by the mutual repulsion of the nuclei and the stability of closed inner electron shells. From these attracting and repelling forces results the well-known potential curve of the interatomic forces, Fig. 189. It is a consequence of the Pauli principle that, according to Fig. 107, the approach of one H atom to another H atom leads to one energy state with a minimum of the potential energy and to a second one with continuously increasing potential energy. According to PAULI, both electrons may be in the K shell only if their spins are antiparallel, whereas for parallel spin directions one of them must move into the higher L shell and thus, being a $2p\sigma$ electron, has an antibonding effect.

The quantitative wave-mechanical theory of the chemical bond is so difficult to comprehend because we do not have exact solutions of the Schrödinger equation for a complicated system of several interacting atoms. Even the approximative solutions are not very good. The reason for this is that the re-arrangement of the electrons during the formation of the molecule from the atoms as well as, vice versa, during the dissociation of a molecule into the atoms is too deep-reaching to be considered as a perturbation that may be introduced subsequently.

From this discussion we understand that there are two different approximative methods for computing the wavefunctions that describe the behavior of the electrons in a molecule. The first method starts from the separated atoms and their electron eigenfunctions ("orbitals"); the second one, conversely, begins with the complete nuclear skeleton of the molecule in which its electrons must move. For the hydrogen molecule, the first method is identical with the Heitler-London theory which we have treated in detail. It starts from the "atomic orbitals" of the electrons and then takes into account, as a perturbation, the re-arrangement of the electron shell as a consequence of the atoms' interaction which leads to the chemical bond. The method works well for large internuclear distances for which the re-arrangement is indeed a perturbation only. But the numerical results are *not* satisfactory for the small internuclear distances of real molecules.

The second method which is called the method of "molecular orbitals" was developed by HUND and MULLIKEN. Here the arrangement of the nuclei of the molecule is assumed to be known, and the behavior of each electron, i.e., its wavefunction, is investigated in the so-called self-consistent field of the positive nuclei and of all other electrons, which are considered to be distributed uniformly over the entire nuclear skeleton. In the case of H_2, each electron is assumed to move in the field of two positive nuclei with the charge $+e/2$ each, since half of the charge is supposed to be compensated by the other electron distributed among both nuclei. By this method, binding energies are obtained that are closer to the observed ones than those from the Heitler-London method, so that the molecular-orbitals method proves to be more appropriate for strongly bound molecules. But the results are poor or even wrong if this Hund-Mulliken method is applied to losely bound or nearly dissociated molecules. Obviously the assumption that each electron moves in the field of both nuclei *and* in that of the other electrons distributed over the nuclei must be the less correct the larger the interatomic distance is. For in the limiting case of dissociation, one electron must stay with one nucleus and the other with the second nucleus. While the method of atomic orbitals thus automatically leads to the dissociation into those atoms whose eigenfunctions were used in the first approximative description of the molecule, the method of molecular orbitals does not yield a clear assignment between molecule and dissociation products.

The treatment of the H_2 problem by means of the Hund-Mulliken method leads, however, to an important conclusion: The behavior of each of the two electrons in the field of the two nuclei a and b is approximately represented by a wavefunction $\psi_a + \psi_b$. We thus obtain for the total wavefunction of the ground state of the H_2 molecule, if the normalizing factor is neglected,

$$\psi = [\psi_a(1) + \psi_b(1)] [\psi_a(2) + \psi_b(2)]. \tag{6.58}$$

The number of the electron being found close to nucleus a or b is indicated in parantheses. Eq. (6.58) may be written as

$$\psi = [\psi_a(1)\psi_a(2) + \psi_b(1)\psi_b(2)] + [\psi_a(1)\psi_b(2) + \psi_a(2)\psi_b(1)], \tag{6.59}$$

In this form we may compare it with Eq. (4.164). We see that the right-hand bracket is identical with the wave-mechanical description of the H_2 ground state in the Heitler-London theory. It describes a state of the molecule in which one electron is always with one nucleus and the other one with the other nucleus. The square of the left-hand bracket, however, indicates the probability for both electrons being either with nucleus a or with nucleus b. This expression describes an ionic molecule H^+H^-. We thus find a surprising result: *The wave function which describes the ground state of the H_2 molecule is, as a purely mathematical consequence of the successful formalism of wave mechanics, a linear combination of polar and non-polar terms.* Carrying through the computation shows that this description is a better approximation of the empirical results than the Heitler-London method which neglects the polar terms.

This fact indicates in a satisfying way *that the homopolar and heteropolar bond are not different in principle but that they are limiting cases of a multiplicity of transitional binding cases.* If, after the re-arrangement of the electrons leading to the bond, in the time average one electron is always with nucleus a, the other one with nucleus b, then we deal with a pure atomic molecule, i.e., with an ideal homopolar bond. If in the time average, however, both electrons responsible for the bond are predominantly with atom a and none of them with atom b, nucleus a appears to be negatively charged and nucleus b positively charged. The molecule is then an ionic molecule with a heteropolar or ionic bond. We mentioned already in VI, 2b that only a few such molecules exist, since most measured dipole moments are considerably smaller than those expected for ideal ionic molecules. The real molecules are intermediate cases between the two limiting ones: Even for symmetric molecules such as H_2 there exists a small probability for both electrons to be found with one atom, whereas for asymmetric molecules such as CO the probability is somewhat larger for finding both electrons with the electronegative partner so that the molecule has a more polar character. Only for a few alkali halides, the polar character is clearly predominant and thus determines the bond in these molecules. Since heteropolar and homopolar bonds are not different in principle but borderline cases of the electron re-arrangement responsible for the bond, they find a uniform description in the quantum theory of the chemical bond. This follows automatically from the Hund-Mulliken approximation, as we showed before when considering wave function (6.59), whereas no polar terms can occur in the Heitler-London approximation that starts from neutral atoms.

c) The Binding of Atoms with Several Valence Electrons

What do we know about the bond between atoms with several valence electrons, such as in molecules like N_2, O_2, Cl_2, and CO? This problem may be approached from the two standpoints discussed before too. We may, with HEITLER-LONDON, start from the atoms forming the molecule and arrive automatically at an explanation of the chemical valency of the atoms. We shall discuss this in detail below. Or we may start again, according to HUND-MULLIKEN-HERZBERG, from the complete nuclear configuration of the molecule but with the electrons still in their atomic energy states. We may then assume that successively one electron each of the two original atoms after the other is being built into the completed configuration of the nuclei. For the existence of a bond it is essential whether energy is used or released when the two spin-compensated electrons which form a bond are transferred from the separated atoms into the molecule. In the first case, the electron pair is designated as *anti-bonding*, in the latter one as *binding*. Only the outermost electrons, i.e., the valence electrons of the atoms which form the molecule, have to be taken into account, of course, since the closed inner electron shells each have compensated spins and stay with "their" atoms so that they cannot participate in the chemical bond. In a first approximation, the binding state may then be characterized by the difference in the number of binding and anti-bonding electron pairs. For the O_2 molecule, for instance, the four valence electrons of each of the two oxygen atoms may form three binding electron pairs in the energetically low states $2p\pi^4$ and $3p\sigma^2$, and one anti-bonding pair that has to go into the higher state $3p\pi^2$ by using up energy. Thus the anti-bonding pair compensates one binding pair, and two binding electron pairs remain that are responsible for the chemically well-known double bond of the

O_2 molecule. In a similar way, it can easily be shown that two atoms with two *s*-valence electrons each (e.g., atoms of the second group of the Periodic Table) should not be able to form a molecule. In integrating these electrons into the molecule, namely, only one of the electron pairs can be built into a lower state, thus gaining energy, whereas the other one has to move into a higher state while using up energy so that the effects of the two electron pairs would compensate each other. Therefore, an Hg_2 molecule bound by valence electrons cannot exist, in contrast to the molecule discussed in VI, 8 that is very loosely bound by van der Waals forces. It is evident that this method of studying the bond type requires an exact knowledge of the electronic states of the atoms and molecules and thus demonstrates their significance.

Let us now conversely investigate the separated atoms and their *chemical valency*. We thus ask whether it is possible that hydrogen atoms are bound by a certain atom, and how many of them. It is evident that all outer electrons with uncompensated spins of a certain atom may form valence bonds with corresponding electrons of the other atom. In principle, this valence bond is similar to that of H_2. We see from Table 10, page 131, for instance, that the nitrogen atoms in N_2 each have 5 electrons above the closed K shell; two of them are $2s$ electrons, three are $2p$ electrons. The $2s$ electrons have opposite spin directions so that they apparently cannot act as valence electrons. The three $2p$ electrons, however, have parallel, un compensated spins as we see from the quartet ground state of the atom (cf. column 3 of Table 10). They may therefore form three spin-compensated electron pairs (valence bonds) with the three corresponding $2p$ electrons of another N atom. The resulting high density of binding electrons between the nitrogen nuclei is responsible for the large binding energy (i.e., dissociation energy) of the nitrogen molecule, namely 9.76 ev. According to this simplest form of the valence theory, *the chemical valency of an atom is equal to the number of its unpaired outer electrons, i.e., equal to the multiplicity of the ground state of the atom minus one*. However, this rule does not hold throughout. It explains why, for instance, the alkali and the halide atoms are monovalent, whereas oxygen is bivalent and nitrogen trivalent. It does *not* explain, however, why the elements of the second group of the Periodic Table which have two outer spin-compensated *s* electrons according to their singlet ground state are nevertheless bivalent (for example in $BeCl_2$). The simple theory does not explain, furthermore, why carbon is tetravalent (which is the basis of the entire organic chemistry!) and why the atoms of the third group such as boron are trivalent in spite of their single non-paired outer electron.

The reasons for these deviations from the simple valence rule is the fact mentioned before that each genuine chemical bond represents a deep-reaching re-arrangement of the electrons and that the participating outer electrons of the reacting atoms may change their properties very basically in this process. The beryllium atom, for instance, was expected to be chemically inactive because of its two spin-compensated outer $2s$ electrons. It is able, however, by using 2.7 ev energy, to change into an excited state in which its two outer electrons, being a $2s$ and a $2p$ electron, have parallel spin directions so that the Be atom in this state is bivalent. But the Be atom is able to form chemical bonds only if the electronic excitation energy just mentioned is overcompensated by the gain of binding energy. In the same way, the normal carbon atom which has the outer electrons $2s^2\,2p^2$ with only two unpaired p electrons, by using 4.2 ev energy may change into an excited state with the electron arrangement $2s\,2p^3$ in which, according to the spectrum, all four outer electrons have unpaired spins. *This is the tetravalent state of the C atom which is so important for chemistry*. The deep-reaching rearrangement of the electrons in the chemical bond here and in many similar cases has another consequence: The difference between $2s$ electrons and $2p$ electrons, being very marked in isolated atoms, tends to disappear in some molecules. In methane (CH_4), for instance, we have practically four equivalent valence electrons which bind four hydrogen atoms symmetrically and with the same strength.

d) Multiple Bonds, Directed Valences of Stereochemistry, and the Effect of Nonlocalized Valence Electrons

All finer details of the chemical bond, including the double bond which is rigid with respect to torsion, and the angular valence bonds of stereochemistry, follow from the overlap rule by SLATER and PAULING which we explained on page 398.

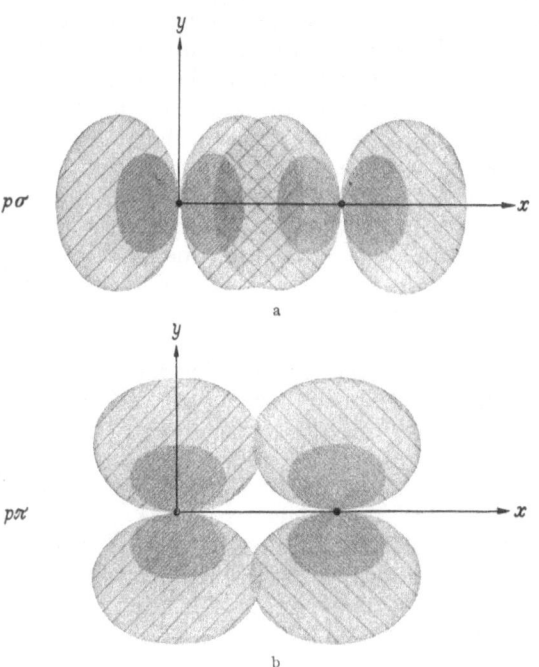

It states that the energy decrease due to the atoms approaching each other, i.e., *the stability of the molecular bond, is largest if the eigenfunctions of the valence electrons of the interacting atoms which form electron pairs overlap to a maximum extent.* The time-averaged illustrations of the spatial distribution of atomic electrons in different quantum states (Fig. 103) are particularly suitable for a qualitative discussion of this rule. Here the p electrons play a decisive role. Since they have an axial symmetry in contrast to the spherical symmetry of the s electrons, the three possible p electrons with the same spin orientation of the N atom, for instance, orient themselves so that their symmetry axes are perpendicular to each other, similar to the three coordinate axes. We distinguish them by

Fig. 215. Schematic representation of binding possibilities by two p electrons. (a) stronger $p\sigma$ bond, without resistance against rotation about x-axis; (b) weaker $p\pi$ bond, rigid with respect to rotation about x-axis

denoting them as p_x, p_y, and p_z electrons. If we now consider the two nitrogen atoms of the N_2 molecule with their p valences, and if we choose the internuclear axis as the x axis, Fig. 215 shows that the p_x electrons of the two atoms overlap the most and thus form a very stable bond (cf. Fig. 215a), whereas *for the same internuclear distance* the overlapping of the $2p_y$ electrons is markedly smaller (cf. Fig. 215b). We thus see that the three bonds of the N_2 molecule are by no means equivalent. We have a very strong $p_x - p_x$ bond and two much weaker bonds $p_y - p_y$ and $p_z - p_z$.

Since the p_x electrons extend in the direction of the molecular axis and thus do not have nodes on it, their angular momentum $|\vec{l}| = h/2\pi$ does not have a component in this direction. Therefore, the p_x electrons become σ electrons with the quantum number $\lambda = 0$ (cf. VI, 5) in the molecule. The bond formed by two σ electrons (which in the atom may be p_x or s electrons) is designated as σ bond.

In contrast to this, the p_y and p_z electrons have nodes on the molecular axis (see Fig. 215 b) and a component of their angular momentum in the direction of the molecular axis of $|\vec{\lambda}| = h/2\pi$ so that these electrons become π electrons in the molecule. For this reason the bonds $p_y - p_y$ and $p_z - p_z$ are called π bonds. We thus have in the nitrogen molecule a strong σ bond and two weaker π bonds. This distinction between σ and π bonds has principal significance for the understanding of the free or hindered rotation of molecular groups against each other. *For a σ bond, the electron distribution between the bound nuclei has rotational symmetry with respect to the axis joining the nuclei.* Let us consider the two groups CH_3 of ethane as an example; they are bound by a σ bond. Because of the rotational symmetry of the σ bond of the two C atoms, these groups may freely rotate against each other, in agreement with experience. *For a π bond, however, the two p_y electrons have rotational symmetry with respect to the y direction (cf. Fig. 215 b) but not to the x axis which is identical with the molecular axis.* We thus do not have a maximum of the electron density on the axis joining the nuclei (x axis), but, according to Fig. 215, there is one in the x y plane. *This electron distribution within a plane, caused by the π bond, resists an axial torsion of the molecule, as may be intuitively recognized, so that torsional vibrations are possible between molecular groups that are bound by a π bond but no free rotation.* Since most of these molecules have also a σ bond, we observe this case only for $\sigma\pi$ *double bonds.* It explains

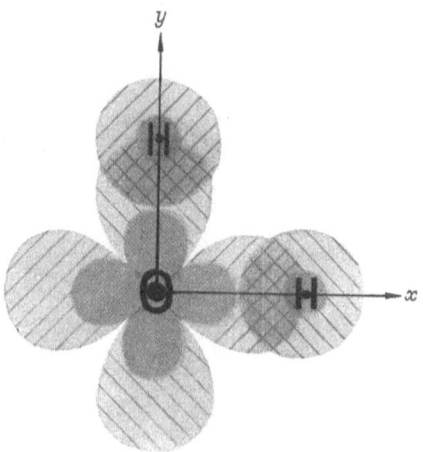

Fig. 216. Schematic representation of the bonds of the two H atoms in H_2O caused by p electrons of O atom

the rigidity of the C=C double bond as it is found in the plane ethylene molecule $H_2C = CH_2$, the rigidity with respect to torsion of which is well known to organic chemists.

The spatial structure of the eigenfunctions of the p electrons is also responsible for the angular valences of stereochemistry. The best-known example is the H_2O molecule in which the two s electrons of the hydrogen atoms are bound by the two p electrons of the oxygen atom. It is evident that a linear H_2O molecule is impossible, since the p_x electron oriented along the x axis can bind only *one* s electron, i.e., one hydrogen atom, by a σ bond (cf. Fig. 216), whereas the p_y electron of the O atom may bind with the same strength only an H atom on the y axis. *The spatial ψ^2 density of p electrons* gives thus direct and pictorial evidence of the angle between the valences of the H_2O molecule and, quite generally, the angular valences of stereochemistry. In all these cases we should expect right angles. Analysis of the band spectrum of H_2O, however, yielded an angle of somewhat over $100°$; and in other molecules deviations from a right angle were found also. We may explain them by assuming that either the atoms on the legs of the angle are mutually influencing each other because of their size and charge, thus increasing the valence angle by "steric hindrance", or that the molecular rotation causes a centrifugal stretching of the molecule or, finally, that other interatomic forces are effective.

The NH_3 molecule is another example for σ bonds between three s electrons and the three p electrons of another atom with mutually perpendicular axes. We now understand its pyramidal shape as a consequence of the binding of one

H atom on each of the three coordinate axes by the three unsaturated p electrons of the N atom, which forms the pyramidal top if we take into account a certain deformation by interatomic forces.

The tetravalent carbon with its four equivalent valences was treated above. Symmetry considerations indicate that these four valences of the C atom are spatially arranged in such a way that they form equal angles with each other so that the CH_4 molecule is a regular tetrahedron of H atoms with the C atom in its center. This is in agreement with experience. Corresponding considerations lead to the result that the long CH_2 chains well known to organic chemists have angles of about $110°$ between the C atoms (cf. Fig. 217b) and not the plane stretched form of Fig. 217a. The H atoms must lie above and below the plane determined by the C atoms. The $120°$ angles between the C atoms of the C_6 benzene ring

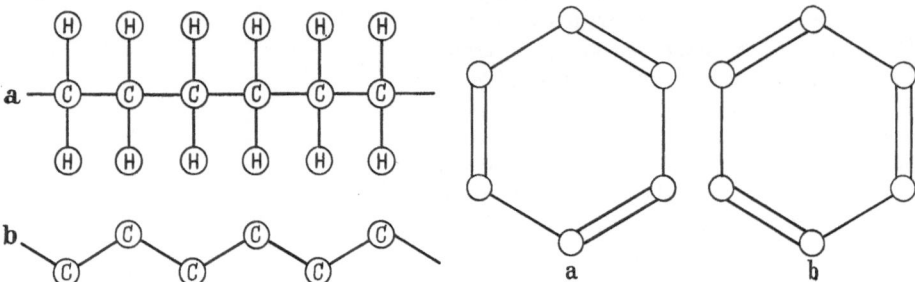

Fig. 217. Linear (a) and angular (b) configuration of C atoms in a hydrocarbon molecular chain

Fig. 218. The two possible configurations of single and double bonds in a benzene ring; their superposition produces the actual binding character

(see Fig. 218) can be explained in a similar way. All major phenomena of the angular valences may thus be understood from the spatial structure of the p electrons or, for the C atom, of the sp^3 electrons.

There remains a final problem to be discussed which becomes evident if we consider the benzene ring. The bonds treated so far are designated as *localized bonds*, since every electron pair can be assigned unambiguously to a certain bond between two atoms and thus causes this bond. The most important organic molecule, however, the benzene ring C_6H_6, cannot be understood along the lines of this bond theory that so far seemed to be the only one possible. If one of the four C valences each binds an H atom, three valences per C atom remain so that (according to KEKULÉ's hypothesis) single and double bonds should alternate between the C atoms. But this concept does not agree with the empirical chemical results nor with the findings of molecular physics, according to which the ring has to be completely symmetrical. E. HÜCKEL investigated this problem quantum-mechanically and showed that, as expected, every C atom is bound to its neighbors and to its H atom by localized σ bonds. The remaining six valence electrons of the six C atoms, however, form π bonds and act as if they were *uniformly* distributed over the entire benzene ring so that *one might speak of a one-and-one-half bond between two C atoms each of the benzene ring. Such π bonds are designated as non-localized*, since there is empirical evidence that the π electrons move freely all over the ring. This behavior is a consequence of the resonance phenomenon, quantum mechanically treated in IV, 11, between two electron configurations of equal energy. Since from every C atom three σ bonds extend to the two neighbors and to its H atom, the fourth valence electron can form nothing but a π bond with a similar valence electron of a neighboring C atom. Because of the complete symmetry, there exists no reason for preferring a bond towards the right-hand side as compared to that towards the left-hand side. We thus have energy resonance between the two possibilities of binding represented in Fig. 218. As in all such cases of energy resonance, the eigenfunction which describes the behavior of the electrons (here that of the last six π electrons of the C_6H_6 molecule) is a linear combination of the wave functions describing the two possibilities 218a and 218b. This means that the π electrons oscillate between the configurations indicated in Fig. 218 and that they are distributed over all six bonds with the same probability. The benzene ring is only the best-known example of such *conjugated double bonds*, a phenomenon that is as frequent as it is important in organic chemistry. In the usual presentation of conjugated double bonds, consecutive C atoms of a ring or molecular chain are bound alternately by σ single bonds and $\sigma\pi$ double bonds. It is obvious that in such a chain we have resonance between different equivalent configurations of the π bonds and, therefore, a relatively free mobility of these π electrons along the entire C chain. An interruption of this bond sequence by two successive single bonds or two double bonds evidently disturbs the resonance so that it has the effect of blocking the free mobility of the π electrons.

There exists impressive evidence *for the free mobility of the π electrons along conjugated double bond chains and rings*, a phenomenon closely related to metallic conductivity (cf. VII, 13). This free mobility of the π electrons causes the unusually large polarizability in the molecular plane or along the molecular axis of such molecules in an electric field, whereas the polarizability does not exceed normal values in directions perpendicular to the molecular chain. Furthermore, according to III, 15, the phenomenon of diamagnetism is caused by an inductive influence on the atomic or molecular electrons by the varying magnetic flux, in other words, is a consequence of the production of atomic or molecular eddy currents. The magnetic susceptibility therefore must be the larger the larger the electronic mobility is on closed orbits perpendicular to the magnetic field. This effect has been observed for ring molecules with free mobility of the π electrons (for example, naphtalene), where the diamagnetic susceptibility for a magnetic field perpendicular to the ring's plane is three times larger than for a field in the plane of the ring. Finally, if we connect numerous C_6 rings in a plane with each other and thus form the lattice planes of graphite, then the electronic mobility of the π electrons within these planes causes a nearly metallic electronic conductivity which does not exist in a direction perpendicular to these planes. This observation leads us from the problems of the molecular bond to the phenomena of the solid state, which we shall discuss in the next chapter.

The present valence theory based upon quantum mechanics is in principle able to explain satisfactorily all details of the manifold phenomena of the chemical bond. It enables us also to predict the bond character of unknown, so far not yet investigated or not yet produced molecules. Although the mathematical difficulties of the perturbation calculations are so immense for more complicated molecules that we have to restrict ourselves to qualitative statements, this theory of the chemical bond is of utmost significance for theoretical and practical chemistry. It is particularly suitable for illustrating to the chemist the necessity of mastering atomic physics with its electron states, including the necessary foundations of quantum mechanics.

15. Van der Waals Forces

Valence theory as treated so far is concerned with the binding of atoms at internuclear distances of the order of an atomic diameter or that of an atomic group (measured from center to center). At distances of atoms in molecules which are considerably larger than one atomic diameter, such deep-reaching re-arrangements of the electron shells as we realized as essential for genuine chemical bonds do not occur any more. In V, 8 we mentioned, however, the loosely bound van der Waals molecules with large equilibrium distances which are due to weaker interactions between atoms or groups of atoms. Such forces are also known from the existence of cohesion and adhesion. Forces of this type are collectively called *van der Waals forces*. We may obtain information about them from scattering experiments with crossed atomic or molecular beams, from the compressibility of van der Waals crystals (cf. VII, 3), and from the behavior of the real gases, i.e., from the van der Waals equation. The quantum-mechanical treatment of these forces was first undertaken by LONDON. We can, however, best understand the essential features of these *forces of second order* if we consider the classical interaction of atomic or molecular dipole moments, may they be permanent or produced by polarization. We then obtain (in the sense of the correspondence principle) not only the correct dependence on the distance but also the correct order of magnitude of the binding energies. This classical procedure, which starts from the properties of isolated atoms or molecules, fails of course if the electron shells begin to become deformed due to a closer approach of the atoms or molecules. We then need the quantum theory of the chemical bond.

We start with the interaction energy of two permanent dipoles lying anti-parallel side by side, neglecting in this qualitative discussion the angular dependence of the forces. The energy of a dipole with the moment $\mathfrak{M} = e\vec{l}$ in an electric field of the strength $\mathfrak{E}$ is $\mathfrak{M}\mathfrak{E}$. A dipole of the moment $\mathfrak{M}_1$ produces in its neighborhood an electric field strength

$$\mathfrak{E}_1 = \mathfrak{M}_1/r^3. \tag{6.60}$$

The interaction energy of two antiparallel dipoles at rest which may have the moments $\mathfrak{M}_1$ and $\mathfrak{M}_2$ is, therefore,

$$U(r) = \mathfrak{M}_1\mathfrak{M}_2/r^3. \tag{6.61}$$

An example will prove that Eq. (6.61) gives reasonable values of the binding energy for two dipole molecules in rest. According to Table 15, page 348, the dipole of an HCl molecule is 1.03×10^{-18} esu. At a distance of 3 Å, which often occurs for van der Waals bonds, the binding energy of the two molecules, therefore, would be 3.9×10^{-14} erg $= 0.024$ ev, according to Eq. (6.61). This small value of the binding energy is reasonable since we know that HCl at room temperature still consists of individual molecules, i.e., in the vapor state.

Since, in general, thermal motion prevents a permanent parallel orientation of the dipole moments of the interacting molecules, we have to consider the binding energy of two dipoles for the case that the orientations of $\mathfrak{M}_2$ in the field of the dipole $\mathfrak{M}_1$ as described by Eq. (6.60) varies statistically. The computation yields for $\mathfrak{M}\mathfrak{E} \ll kT$ the following value of the average dipole moment

$$\overline{\mathfrak{M}}_2 = \mathfrak{M}_1\mathfrak{M}_2^2/3\,kT\,r^3. \tag{6.62}$$

Multiplied by the value of $\mathfrak{E}$ from Eq. (6.60), it yields the binding energy $U(r)$. It is evident that $\overline{\mathfrak{M}}_2$ is the larger, the larger the dipole moments $\mathfrak{M}_1$ and $\mathfrak{M}_2$ are that strive for parallel orientation and the smaller the mean thermal energy kT is. Multiplication of Eq. (6.60) by Eq. (6.62) gives the interaction energy of two dipole molecules

$$U(r) = \frac{\mathfrak{M}_1\overline{\mathfrak{M}}_2}{r^3} = \frac{\mathfrak{M}_1^2\mathfrak{M}_2^2}{3\,kT\,r^6} \tag{6.63}$$

for sufficiently high temperature. This is the well-known r^{-6} law for the inter-action potential of two dipole molecules for not too low a temperature. If we compute the energy for the example of two HCl molecules at an assumed distance of 3 Å for room temperature, we obtain the same energy as from Eq.(6.61). The influence of the temperature, therefore, becomes essential only for still higher temperatures.

We consider now the attraction between a polar molecule and an atom or molecule *without* a permanent dipole. According to V, 2c, it is clear that the electric field (6.60) of the dipole molecule induces polarization of the non-polar atom or molecule, and this polarization is the larger the larger its polarizability α is. Therefore we have

$$\mathfrak{M}_2 = \alpha \, \mathfrak{E}_1 = \alpha \, \mathfrak{M}_1 / r^3 \tag{6.64}$$

and, consequently,

$$U(r) = \frac{\mathfrak{M}_1 \mathfrak{M}_2}{r^3} = \frac{\alpha \, \mathfrak{M}_1^2}{r^6}. \tag{6.65}$$

In an equal manner, we obtain for the attractive interaction energy of *two* polarizable molecules with permanent dipoles which are field-strength dependent,

$$U(r) = \frac{\alpha_1 \, \mathfrak{M}_2^2 + \alpha_2 \, \mathfrak{M}_1^2}{r^6}. \tag{6.66}$$

In all these cases, the attractive potential of the van der Waals forces follows the r^{-6} law.

What do we know about the attraction of two atoms or molecules if *both* are *without* permanent dipoles? Here we expect and find a small mutual polarization and, therefore, an attraction which again is proportional to r^{-6}. Taking two inter-acting hydrogen atoms as an example, we may consider them as very rapidly rotating dipoles each consisting of a proton and an electron. This yields an electric moment equal to the elementary charge e multiplied by the atomic radius R (in this case 0.5 Å). The trend towards antiparallel orientation of these two dipoles depends on the magnitude of the dipole moments and is inversely proportional to the rotational energy, which is related to the ionization energy E_i. We obtain the right order of magnitude of the resulting attracting force if we take as the dipole moment of the atom the value

$$|\mathfrak{M}| = eR \tag{6.67}$$

and if we assume that $3kT$ corresponds to the ionization energy E_i of the atom. Substituting these values into Eq. (6.63), we obtain

$$U(r) = \frac{e^4 R^4}{E_i \, r^6}. \tag{6.68}$$

The binding energy of two H atoms at a distance of 3 Å then follows from Eq. (6.68) as 2.6×10^{-15} erg $= 1.6 \times 10^{-3}$ ev, which is not unreasonable.

Defining the polarizability α as the mean dipole moment $\mathfrak{M}$ produced per unit field strength by polarization in the electric field, we obtain from Eqs. (6.4) and (6.67), with E_i instead of $3kT$,

$$\alpha = \frac{e^2 R^2}{E_i}, \tag{6.69}$$

and instead of Eq. (6.68), we may also write

$$U(r) = \frac{\alpha^2 E_i}{r^6}. \tag{6.70}$$

This result agrees, except for a factor of the order 1, with the result of the strictly quantum-mechanical computation. The attraction energy between helium atoms ($\alpha = 2 \times 10^{-25}$ cm^3; $E_i = 24.6$ evolts) at a distance of 3 Å follows from Eq. (6.70) to be 6×10^{-4} evolts, in good agreement with the extremely low boiling point of helium.

Although there is only a loose relation to the van der Waals forces, we finally mention a bond between external groups of larger molecules, which has become familiar to chemists by the name of *hydrogen bond*. Also in this bond, polarization effects are essential. In molecules with external OH or NH groups, the electron of the hydrogen atom is often attracted by the more electronegative partner to such an extent that, in a first approximation, we may speak of a proton instead of an H atom attached to the outside of a molecule. This proton then may polarize the end groups of neighboring molecules, and this leads to a bond between the molecules in which the hydrogen may be regarded as a bridge between the two groups. Often, it cannot be decided to which of two molecules or groups of molecules the proton belongs. Empirical evidence of the existence of such hydrogen bridges follows from Fourier analysis (cf. V, 2a) of electron or neutron diffraction diagrams.

16. Molecular Biology

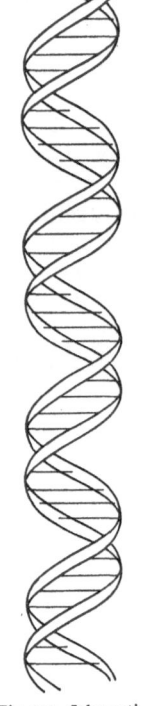

We cannot conclude our presentation of molecular physics without at least briefly mentioning the significance which research in the field of molecular structure has gained for biology and in particular for genetics. It had been believed for a long time that the genes, the carriers of hereditary characteristics of the biological cells, are individual although very large protein molecules, and that the immense numbers of genes are not different from each other in principle in spite of minute differences of their molecular structure. But only during recent years it became possible to elucidate the structure of at least some of these giant molecules which are so essential to biology, and which contain several hundred thousands of atoms each, but atoms of only the five elements H, N, O, C, and P. Even an understanding was gained of the basic features of the mechanism of reproduction of identical gene molecules in cell division as well as of the production of the enzyme molecules which are necessary for the metabolism of the cell. According to results first obtained by WATSON and CRICK, the genes are huge deoxyribonucleic-acid molecules (abbreviated to DNA), several microns in length, which are built up from only four different chain-like "nucleotides". There is hardly any doubt possible that the sequence in the arrangement of these four different nucleotides in the DNA chain represents all the information required for the fulfilment of the

Fig. 219. Schematic drawing of the helical structure of the deoxyribonucleic acid molecule

task of a gene, e.g., for the synthesis of a certain enzyme, in the same manner as the magnetic tape of an electronic computer contains information in its sequence of magnetic north and south poles. Each of these nucleotide links of the chain consists of only three molecular groups, a phosphoric acid, a sugar molecule (deoxyribose), and one of four nitrogen-containing bases, viz. adenine, thymine, guanine, or cytosine. The entire DNA molecule resembles a helically twisted ladder, like winding stairs (see Fig. 219) whose sidepieces are formed by alternating sugar and phosphoric acid links, whereas the crosspieces, which always are fastened to the sugar links, consist of two bases each, connected by hydrogen bonds (cf. VI, 15). Of these four bases, adenine is always connected to thymine and guanine to cytosine only. Fig. 220 shows a section of a DNA ladder spread out in a plane. It is very essential that the DNA sidepieces with their adjoined bases are connected only by the relatively weak hydrogen bonds in the center of each cross piece. During cell division, these hydrogen bonds are severed, and by this longitudinal division two half-molecules are produced, which then complement each other from the surrounding free nucleotides (the "chain links") and thus form complete DNA molecules again. The afore-mentioned fact that adenine is always connected with thymine only, and cytosine only with guanine is essential for always obtaining the correct sequence of the nucleotides, so that the DNA molecules are reproduced exactly with their entire structure and thus with the information contained in them.

However, the genes must do more than just reproduce themselves by division, for we know that they have to pass on the inherited properties of the living organisms, which develop from the cells. This is done by the genes through protein enzymes produced by them, and each of these enzymes consists of polypeptide chains which in turn are built up from 20 different amino acids in an arrangement determined by the gene. Many of the details are still unknown. But we see already a fascinating outline of how a longitudinally split DNA molecule (i.e., a gene), instead of

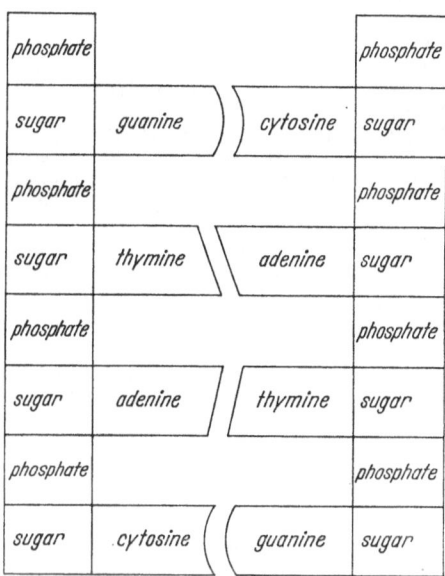

Fig. 220. Section of the DNA molecule

reproducing itself, may produce ribonucleic-acid molecules by utilizing still another base, urazile. These molecules then carry the gene information into the cell plasma in the form of ordered molecular groups, and there the enzymes controlling the chemistry of the cell are being built up from amino acids which are in turn produced in just the right manner from ribosomes in the order (molecular structure) dictated by the gene.

We cannot go into details here, and research is not yet in its final state. But we want to show how molecular structure as treated in this chapter, including the fine details of the different bonds between atoms and molecular groups, determines the mechanism of cell reproduction and even the building up of biological individuals from the primitive cell, far beyond its significance for chemistry proper.

Problems

1. Calculate the reduced mass and the moment of inertia of the hydrogen molecule whose equilibrium distance is 0.77 Å.

2. The equilibrium distance of the HCl molecule is 1.27 Å. Calculate the energy in ev and the wavelength of the photon which, when absorbed by this molecule, would induce pure rotational transitions from the state $J=0$ to the state $J=1$.

3. The spacing between the lines of a rotation band of the HCl molecule is experimentally observed to be 20.68 cm^{-1}. Using this information calculate the moment of inertia and the equilibrium distance of the HCl molecule.

4. Find the temperature at which the ratio of the number of NO molecules in the pure rotational state $J=1$ to that in the state $J=0$ is 0.1. The moment of inertia of the NO molecule is 16.30×10^{-40} g cm².

5. Find the force constant, k, and the lowest energy level, E_0, of a hydrogen molecule vibrating with natural frequency 1.32×10^{14} sec⁻¹. The moment of inertia of the molecule is 4.7×10^{-41} g cm² and the equilibrium distance is 0.77 Å [cf. Eq. (6.24)].

Literature

General Molecular Physics:

ARKEL, A. E. VAN: Molecules and Crystals. Den Haag: Van Stockun 1949.
DAUDEL, R., R. LEFEBRE and C. MOSER: Quantum Chemistry. New York: Inter-science 1959.
DEBYE, P.: Polare Molekeln. Leipzig: Hirzel 1929.
FRÖHLICH, H.: Theory of Dielectrics. Oxford: University Press 1949.
GLASSTONE, S.: Theoretical Chemistry. New York: Van Nostrand 1944.
HÜCKEL, W.: Anorganische Strukturchemie. Stuttgart: Enke 1948.
HUND, F.: Theorie des Aufbaues der Materie. Stuttgart: Teubner 1961.
NACHOD, F. C., and W. D. PHILLIPS: Determination of Organic Structures by Physical Methods. New York: Academic Press 1962.
POLANYI, M.: Atomic Reactions. London: Williams and Norgate 1932.
PREUSS, H.: Grundriß der Quantenchemie. Mannheim: Bibliographisches Institut 1962.
RICE, F. O., and E. TELLER: The Structure of Matter. New York: Wiley 1949.
SETLOW, R. B., and E. C. POLLARD: Molecular Biophysics. Reading: Addison-Wesley 1962.
SIMPSON, W. T.: Theories of Electrons in Molecules. Englewood Cliffs: Prentice-Hall 1962.
SLATER, J. C.: Electronic Structure of Molecules. New York: McGraw-Hill 1963.
STUART, H. A.: Die Struktur des freien Moleküls. Berlin-Göttingen-Heidelberg: Springer 1952.
WAGNER, K. W.: Das Molekül und der Aufbau der Materie. Braunschweig: Vieweg 1949.
WOLF, K. L.: Theoretische Chemie. 3. Aufl. Leipzig: Barth 1954.

Molecular Spectra:

BELAMY, L. J.: The infrared Spectra of complex molecules. 2nd Ed. New York: Wiley 1958.
BRANDMÜLLER, J., and H. MOSER: Einführung in die Ramanspektroskopie. Darmstadt: Steinkopff 1962.
FINKELNBURG, W.: Kontinuierliche Spektren. Berlin: Springer 1938.
FINKELNBURG, W., u. R. MECKE: Bandenspektren. Vol. 9/II of „Hand- und Jahrbuch der chemischen Physik". Leipzig: Akademische Verlagsgesellschaft 1934.
GAYDON, A. G.: Dissociation Energies and Spectra of Diatomic Molecules. New York: Wiley 1947.
HERZBERG, G.: Molekülspektren und Molekülstruktur I. Dresden u. Leipzig: Steinkopff 1939. Improved 2nd English Ed.: New York: Van Nostrand 1950.
HERZBERG, G.: Infrared and Raman Spectra of Polyatomic Molecules. New York: Van Nostrand 1945.
JOHNSON, E. C.: Introduction to Molecular Spectra. New York: Pitman Publ. Co. 1949.
KOHLRAUSCH, K. W. F.: Der Smekal-Raman-Effekt (with „Ergänzungsband"). Berlin: Springer 1931/38.
KOHLRAUSCH, K. W. F.: Raman-Spektren. Leipzig: Akademische Verlagsgesellschaft 1943.
PEARSE, R. W. B., and A. G. GAYDON: The Identification of Molecular Spectra. 2nd Ed. New York: Wiley 1950.
PRINGSHEIM, P.: Fluoreszenz und Phosphoreszenz. 3rd Ed. Berlin: Springer 1928. Improved English Ed.: New York: Interscience 1949.
SCHAEFER, CL., u. F. MATOSSI: Das ultrarote Spektrum. Berlin: Springer 1930.
SPONER, H.: Molekülspektren und ihre Anwendung auf chemische Probleme. 2 Vols. Berlin: Springer 1935/36.
WEIZEL, W.: Bandenspektren. Ergänzungsband I of „Handbuch der Experimentalphysik". Leipzig: Akademische Verlagsgesellschaft 1931.

WILSON, E. B., J. C. DECIUS and P. C. CROSS: Molecular Vibrations. New York: McGraw-Hill 1955.
WU, TA YOU: Vibrational Spectra and Structure of Polyatomic Molecules. Peking: National University Press 1939.

The Chemical Bond:
BRIEGLEB, G.: Elektronen-Donator-Akzeptor-Komplexe. Berlin: Springer 1961.
COULSON, C. A.: Valence. Oxford: University Press 1952.
DEWAR, M. J. S.: The Electronic Theory of Organic Chemistry. Oxford: University Press 1949.
HARTMANN, H.: Theorie der chemischen Bindung auf quantenmechanischer Grundlage. Berlin-Göttingen-Heidelberg: Springer 1953.
JAWSON, M. A.: The Theorie of Cohesion. London: Pergamon Press 1954.
KOSSEL, W.: Valenzkräfte und Röntgenspektren. 2nd Ed. Berlin: Springer 1928.
KRONIG, R. DE L.: The Optical Basis of the Theory of Valency. New York: McMillan 1935.
PALMER, W. C.: Valency Classical and Modern. Cambridge: University Press 1944.
PAULING, L.: The Nature of the Chemical Bond. 3rd. Ed. New York: Cornell University Press 1959.
RICE, O. K.: Electronic Structure and Chemical Binding. New York: McGraw-Hill 1940.
SIRKIN, Y. K., and M. E. DYATKINA: The Structure of Molecules and the Chemical Bond. London: Butterworth 1949.

VII. Solid-State Physics from the Atomistic Point of View

1. General Survey of the Structure of the Liquid and Solid State of Matter

We now turn from the molecules to very large assemblies of atoms, ions, or molecules, i.e., to the liquid and solid state of matter. The binding forces which are acting between the individual particles of large aggregates of atoms have the effect that in the equilibrium state, defined by a minimum of the potential energy, the atoms or molecules arrange themselves in a symmetrical geometric pattern or lattice which we call a crystal. For weaker binding forces or at higher temperatures, the thermal motion of the atoms prevents them from attaining this state of equilibrium. The atoms or molecules then have some mobility, and the regularity of the atomic pattern is disturbed: This is the liquid state of matter.

The solid state, consequently, is identical with the crystalline state. In the apparently noncrystalline or amorphous structures we have either a large number of coalesced microcrystals or, as in the case of glass, an undercooled liquid which is not a genuine solid at all. The very few solids which are regarded as "really amorphous" at the present time, such as the explosive modification of antimony and the gray selenium, apparently are not thermodynamically stable modifications of the solid state, as is evidenced by the explosiveness of amorphous antimony. The rest of this chapter will be devoted to a thorough discussion of the genuinely solid crystalline state. In this section, however, we treat very briefly the structure of the liquid state, of which not too much is known yet.

From the above interpretation it is clear that in liquids, in which the average interatomic distances are essentially the same as in crystals because of their equivalent densities, the forces between atoms or molecules also tend to establish a crystal-like, regular arrangement of the atoms. However, the crystalline structure is disturbed by the thermal motion of the atoms, molecules, or ions. This explains that in liquids we find phenomena of order as well as of disorder. For the same reason, a completely random distribution of atoms is not to be expected even in compressed gases, and evidence of some degree of order in such gases has been found by the method of X-ray diffraction. Although it had been assumed for a long time that lack of a geometrically regular molecular arrangement was a characteristic of liquids, more and more evidence has been found for so-called *short-range order* in liquids. This means that we have a crystalline order in regularly arranged atomic or molecular groups. These groups, however, are not arranged regularly with respect to each other and have some mobility. What is missing in the liquid state of matter thus is *the characteristic of the solid crystal, its so-called long-range order, which extends over macroscopic distances and does not change with time.*

What at first appeared to be a radical difference between the solid crystalline and the liquid state, now is recognized to be rather a difference in the degree of order and mobility of the constituents of the crystal lattice or in groups of them. We shall learn in VII, 8 from the discussion of diffusion in crystals that even in the most ideal solid crystals the constituents have a certain mobility, and that a number of very important effects depends upon this mobility. This mobility of individual lattice constituents (and with it the general disorder of the crystal) increases with increasing temperature until the melting point is reached above which some mobility is a characteristic property of *all* particles. It is not quite clear yet why the melting point of pure liquids is so sharp. In any case, the increase of the volume during the melting process is a consequence of the increasing number of lattice vacancies in the melt, and the melting heat is nothing but the energy necessary for forming these vacancies. With temperature increasing beyond the melting point, the mobility and the density of defects in the fluid increase more and more so that we expect the quasi-crystalline order gradually to disappear. Because the assertion of a quasi-crystalline short-range order in liquids is so surprising, we briefly discuss the most important evidence in favor of it.

DEBYE's method of scattering of X-rays by liquids provides one direct proof. If the distribution of molecules and the intermolecular spacing were completely random, the scattered intensity would decrease uniformly with increasing scattering angle. However, if there is a regular molecular arrangement, then and only then maxima and minima of the scattered intensity (plotted against the scattering angle) are to be expected as a result of the interference of the X-rays scattered by the regularly spaced molecules. This was actually observed. If for a number of possible geometrical arrangements (i.e., structures of the liquid) the theoretically expected scattering curves are derived and are compared with those obtained empirically, the molecular structure in a liquid may be determined with considerable accuracy in those cases which are not too complicated. The atomic arrangement in liquid mercury has been studied in this way; at 18 °C the arrangement agreed basically with that of crystalline mercury and thus deviated considerably from the closest-spherical packing which was formerly assumed for the liquid state of mercury.

According to SAUTER, the short-range order in liquids is evident also from the behavior of the electric resistance of pure metals, which shows a surprisingly small increase when the melting point is passed. Measurements seem to be in agreement with the assumption of crystal-like groups of from 50 to 150 atoms in the liquid.

Even more unambiguous evidence for the quasi-cristalline structure of liquids follows from the value of their atomic heats. Whereas for monatomic liquids the value $R/2$ is to be expected for each of the three translational degrees of freedom of the atoms, i.e., an atomic heat of $3R/2$ per mole, twice this value is found, i.e., 6 cal/mole, for liquid mercury and for liquid argon. This is just the amount to be expected for *crystals*, since their atomic heat is composed of the kinetic vibrational energy *and* the potential energy of the constituents which vibrate about fixed positions. For harmonic vibrations, the average potential energy is equal to the vibrational energy so that the atomic heat of solids (with all degrees of freedom excited) is $2 \times 3 R/2 = 6$ cal per mole and degree. The fact that this value was found for monatomic liquids can be explained only by the assumption that also here the atoms carry out threedimensional vibrations about their rest positions. In contrast to the solid crystals, however, these centers of vibration are not fixed in space but move with a temperature-dependent velocity. A number of further results of optical and electrical measurements which we cannot discuss here indicate just as clearly the crystal-like arrangement of the molecules in liquids. STUART and KAST have investigated the dependence of liquid structures on the type of atoms or molecules, on the temperature, and density by model experiments. Their model liquids consisted of large numbers of little magnets which were shaken continuously and in which the intermolecular forces were replaced by the forces between the small magnets which substitute the molecules. The different structures of dipole and quadrupole liquids and quite a number of finer details can be shown very nicely in this manner.

Fig. 221. The quasi-crystalline tridymite structure existing in liquid water due to association of the H_2O molecules. Black spheres = oxygen atoms, white spheres = hydrogen atoms. Each O atom is tetrahedrally surrounded by four H atoms. (After KORTÜM)

Quite generally the existence of strong dipole or quadrupole moments introduces a complication in the normal liquid structure due to the formation of molecular chains (e.g., in alcohols) or molecular clusters as in the case of the so-called associated liquids, the anomalous behavior of which depends on the formation of these clusters. By far the most important associated liquid is water; its anomalous behavior has long been attributed to association. It had been assumed for a long time that water consisted of associated molecules of the type $(H_2O)_n$ with a constant degree of association n. However, all attempts at determining n have failed so that the conviction grew that water consists of molecular clusters of indefinite size. The structure of these molecular clusters can be determined by comparing the experimental X-ray scattering curves with the theoretically deduced patterns. By this method a tridymite structure has been found in which, according to Fig. 221, each O atom is tetrahedrally surrounded by four H atoms. As to be expected according to valence considerations (cf. VI, 14d), always two of the four H atoms are bound somewhat more strongly and, therefore, are closer to the O atom than the two remaining ones which form hydrogen bonds (cf. page 408) between these H_2O groups. This particular geometric arrangement of the H_2O molecules, i.e., *this semicrystalline structure of water, is responsible for its well-known abnormal properties*. From this theory, the change in the structure and properties of water as a result of adding relatively few ions (or a little alcohol) may also be understood: The tridymite structure of water which depends on secondary valence forces is greatly distorted by the electrostatic forces of even a few added ions or by the inclusion of a few large foreign molecules which, at the very least, influence the size of the molecular clusters. Conversely, the addition of a few H_2O molecules cannot change the chain structure of pure alcohol; hardly any change of the properties of alcohol consequently is found when it is mixed with small amounts of water. Even empirically known minor properties of liquids can be understood in this way from their crystal structure.

The strong electrolytes, which are dissociated to a high percentage into ions, exhibit a number of properties, such as the unexpectedly small mobility of the ions, which are accounted for by the theory of DEBYE and HÜCKEL and may be considered as direct evidence for the quasi-crystalline short-range order. Similar to the case of ionic crystals to be discussed later, in strong electrolytes more negative ions are found close to a positive ion than should be expected from a purely statistical distribution. Therefore, electrostatic forces between the ions which arrange themselves approximately as in a lattice are responsible for the characteristic behavior of strong electrolytes.

So far as can be concluded from the incomplete investigations, short-range order and thus a quasi-crystalline structure of liquids and solutions is the more pronounced the larger the electric charges or moments (dipole or quadrupole moments) or the larger and the more complicated the molecules are which form the liquid. Evidence for this can be derived from many remarkable phenomena which LEHMANN found in his studies of large and complicated organic molecules, described in his books on liquid crystals. Most surprising among his results is the birefringence of some liquid droplets and, in many cases, a very pronounced dichroism (cf. VI, 2d). Both of these phenomena are clear evidence for an anisotropy which is characteristic of a crystal. On the other hand, the internal mobility, i.e., the unambiguously liquid character of the particles investigated, was just as evident a property. Therefore, LEHMANN's designation "liquid crystals" seems a good one in spite of the objections of many a crystallographer of his time. We thus have to distinguish between liquid crystals and the much more general phenomenon of solid crystals. A discussion of LEHMANN's results from the standpoint of modern atomic theory is unfortunately still lacking in spite of the increasing interest in the structure of liquids.

Whenever in the following the term crystal is used, we mean exclusively the normal solid crystals.

2. Ideal and Real Crystals.
Structure-sensitive and Structure-insensitive Properties of Crystals

Physics of the solid state is more or less identical with crystal physics. We begin our discussion by stressing the important difference between ideal and real crystals. The ideal rock salt crystal, for example, consists of negative chlorine ions and positive sodium ions arranged in a regular geometric lattice in which each sodium ion is surrounded by six chlorine ions, and conversely. However, this ideal crystal is only a model and does not actually exist, first because crystals do not grow with absolute regularity so that there is quite a number of lattice defects in each cm^3 and, secondly, the purity of the crystal material is not so high that not here and there an impurity atom or ion is incorporated in the crystal lattice. Actually, even the best real crystals have at leat one defect among 10^9 constituents, i.e., at least 10^{13} lattice defects per cubic centimeter.

SMEKAL was the first to point out that an important difference in the properties of crystals is related to this distinction between ideal and real crystals. He designated those properties of crystals which are not, or at least not essentially, influenced by lattice defects or impurity atoms as *structure-insensitive*. They are almost exclusively determined by the *basic lattice* of the crystal. Among these structure-insensitive properties we may name the lattice structure, the specific heat, elasticity, thermal expansion and compressibility, the energy of formation, the principal features of optical absorption (color) and dispersion, as well as, for metallic crystals, the normal electronic conductivity, and finally dia- and paramagnetism. Those properties of crystals, on the other hand, which depend essentially on the number, arrangement, and type of lattice defects and impurities are called structure-sensitive. Among these are the diffusion phenomena, the ionic and electronic conductivity of non-metals, and the multiplicity of semi-conductor effects, including their optical properties and the phenomenon of phosphorescence, not to forget the crystal strength.

Crystal defects consist either of built-in impurity atoms, which distort the lattice more or less depending on their size, or of structural defects in the regular lattice. The latter ones are missing atoms or interstitial atoms which are located outside of their regular geometric positions and thus distort their lattice environment. There are, finally, the important *dislocations* in which ten constituents, for example, take the place that nine would take in the ideal lattice. Furthermore, real crystals often consist of small, almost ideal crystal blocs of several microns diameter with an accumulation of defects and impurity atoms near and on the boundaries of the individual blocs (layer or mosaic structure of crystals).

In the following we shall treat first the ideal crystals and the structure-insensitive properties of solid matter. In the second part of this chapter, we shall then discuss the real crystals and their structure-sensitive phenomena.

3. The Crystal as a Macromolecule.
Ionic Lattice, Valence Lattice, and Molecular Lattice

We have already interpreted the crystal as a limiting case of a very large but unusually regular and symmetrical molecule. This point of view presents only *one* side of the picture and thus has its limitations. However, it allows us to understand many crystal phenomena without further explanation. It seems clear, for instance, that there may be non-localized electrons in crystals, just as in benzene ring molecules (cf. VI, 14d), which are only part of the crystal as a whole. Conversely, there are other electrons, just as the absorbing electron in a chromophoric group of a complicated polyatomic molecule (cf. VI, 13a), which we can attribute to a definite group of atoms or even to a definite atom or ion of a crystal. In discussing the optical properties of crystals, this difference will prove to be very important.

A similarity also exists between molecules and crystals with respect to the nature of their bonds. Thus we may classify crystals according to the type of binding, which is quite analogous to what in molecules we called polar, homopolar, and van der Waals bonds, including all intermediate cases. The *ionic crystal lattice* is built up from positive and negative atomic ions which are mainly bound to each other by electrostatic forces. They correspond to those in predominantly polar molecules such as Na^+Cl^-. The alkali halide crystals are typical representatives of this lattice type. The atomic crystal or *valence crystal* has as its best-known representative the diamond in which each of the four valence electrons of every C atom forms a strongly localized bond with one valence electron of one of the four neighboring C atoms. This valence crystal corresponds closely to the homopolar molecules such as H_2, N_2 with strictly localized electron pairs between the atoms. Finally, we have atomic crystals with non-localized bonds (cf. page 404); these comprise the metals and most of the alloys. In VII, 7 we shall recognize their most important properties as a consequence of this bond type.

We have to stress, however, as we did for molecules (see page 400), that ionic lattices and atomic lattices are only two limiting theoretical cases with all kinds of transitional types between them. Let us consider the example of the quasi-crystalline structure of water as represented by the model of Fig. 221. It is doubtful whether it is more correct to say that each O atom is tetrahedrally surrounded by four hydrogen atoms, or that always a doubly charged negative O^{--} ion is surrounded by four protons (H^+). The electric neutrality is of course maintained by the fact that in the average only two hydrogen atoms "belong" to each oxygen atom. The truth is in between: The O atom which is electronegative corresponding to its position in the Periodic Table (Table 1) attempts to attract the electrons of at least two of its neighbors. This accounts for the electric dipole moment of the H_2O molecule (cf. VI, 2b). The size of the dipole moment indicates that this attraction does not go so far that we should speak of an ionic structure, whereas the assumption of an ice crystal consisting of atoms would not explain the actual displacement of the electrons and the observed dipole moment. We thus note that molecules and crystals which consist of *different* constituents can in general neither be considered as being built up from neutral atoms nor from ions. Actually,

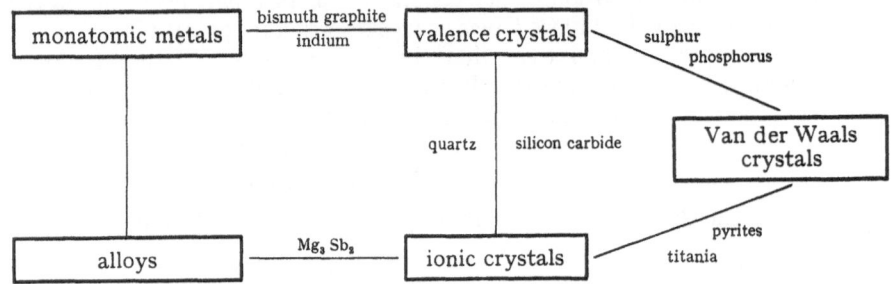

Fig. 222. Schematic diagram of the relation between the different types of solids. (After SEITZ)

the more electronegative constituent carries some negative charge and, conversely, the more electropositive constituent has the corresponding positive charge. However, this charge, caused by the displacement of valence electrons, in most cases is much smaller than a full electron charge.

The last lattice type, that of the molecular crystals, corresponds to the loosely bound van der Waals molecules (cf. VI, 8). This type should be called *van der Waals crystal*, because not only molecules but also atoms may be the constituents, for instance, in the solidified noble gases. If the lattice constituents are molecules (e.g., H_2, Cl_2, CO_2, CH_4), their atoms may be bound by heteropolar or homopolar forces, whereas the molecules themselves are held in their lattice positions by the weaker van der Waals forces (see VI, 15). Accordingly, molecular crystals melt at relatively low temperatures. All organic compounds, for example, crystallize in molecular lattices.

Finally, we should not forget to mention solids with a structure in which several of the bond types mentioned above participate. First, there is the "layer crystal" type in which (chosen at random) LiOH, CdI_2, and $CrCl_3$ crystallize. In these crystals we have genuine atomic or ionic bonds within the crystal layers, whereas the individual crystal layers are held together by van der Waals forces. Graphite (see page 405) belongs to these crystals also. The bonds within each layer are due to localized and non-localized valence electrons, while the layers among themselves are bound by van der Waals forces. Since non-localized electrons participate in the bond *within* each layer, nearly metallic electronic conductivity is found in the layers but not perpendicular to them, as was mentioned before. Asbestos is an example for a solid in which only fibers or needles are kept together by chemical bonds proper, whereas van der Waals forces are responsible for holding together the molecular chains.

A diagram according to SEITZ (cf. Fig. 222) presents the relationship between the different types of solids. In the upper row, we find the crystals with predominantly homopolar bonds, in the lower one those with essentially heteropolar bonds. The first column contains the crystals with predominantly non-localized bonds, the second one those with strongly localized bonds. Naturally, the van der Waals crystals are outside of this double scheme since their bond type is quite different. In this sense, the monatomic metals exhibit some relationship to the valence crystals, whereas the alloys are more related to the heteropolar ionic crystals. Between the pure metals and the non-conducting valence crystals, the crystals of bismuth, indium, and graphite may be regarded as transitional cases. The transition between alloys and ionic crystals is represented by certain intermetallic compounds with fairly well localized ionic-like bonds, such as Mg_3Sb_2. Furthermore, there are transitions between the valence and ionic crystals, just as in the analogous case of molecules. Thus we may regard the crystals quartz, SiO_2, and carborundum, SiC, as valence crystals with some contribution of ionic binding. Nor are the van der Waals crystals completely isolated, since solid sulphur and phosphorus represent transitional cases between them and the valence crystals and, on the other hand, ice, pyrite, and TiO_2 form transitional cases to the ionic crystals.

Before we consider further the question of lattice binding, at least for the more important cases of ionic crystals and metals, we must briefly review the geometric arrangement and the determination of the lattice structure of a crystal.

4. Crystal Systems and Structure Analysis

We have already explained that because of the forces acting between the atoms or ions of a solid (crystal) a stable equilibrium is possible only if the lattice constituents are arranged in a regular geometric pattern. Group-theoretical methods show that there is a finite but large number (230) of crystal systems which are possible arrangements of the lattice constituents in a crystal. In which one of these systems a particular material crystallizes, depends upon the size of the lattice constituents and upon the magnitude and orientation of the forces acting between them. The same material may crystallize in different systems depending upon the type of binding forces. Carbon, for example, can crystallize in the graphite or diamond lattice. The determination of the structure of a crystal, the so-called *structure analysis*, is often not simple; it is the task of the crystallographer. It is achieved almost exclusively by means of X-ray or neutron diffraction following the procedures of VON LAUE or DEBYE-SCHERRER and variations of these methods (rotating-crystal and goniometer methods).

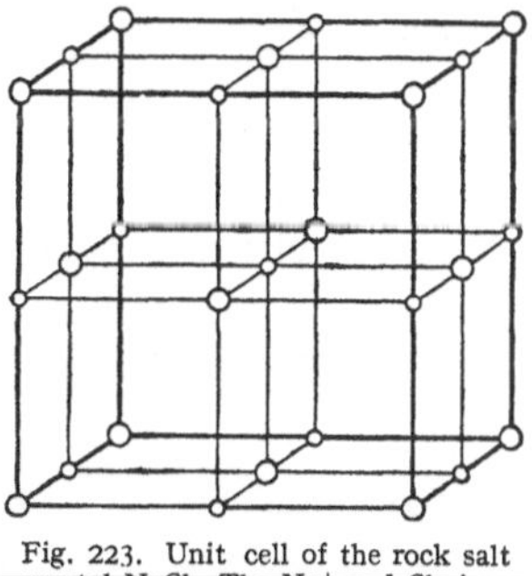

Fig. 223. Unit cell of the rock salt crystal NaCl. The Na+ and Cl− ions are indicated by their different sizes. The NaCl lattice results from intertwining the two face-centered cubic lattices of the Na+ and Cl− ions

It follows from the periodicity of the lattice structure that for each crystal there is a smallest element, the *elementary cell*, from which the entire crystal can be built by translational repetition along the crystal axes. In the cubic NaCl crystal, for instance, the elementary cell (Fig. 223) is a small cube. In most other crystals, the elementary cell is more complicated. For a cubic lattice, the smallest distance of two equivalent constituents is designated as *lattice constant a*; the distance of two neighboring lattice planes as *lattice-plane distance d*. If a coordinate system is introduced with its axes parallel to the edges of the crystal, each crystal plane (and thus the whole crystal as determined by the crystal planes) may be characterized by specifying the lengths from the origin to the intersections

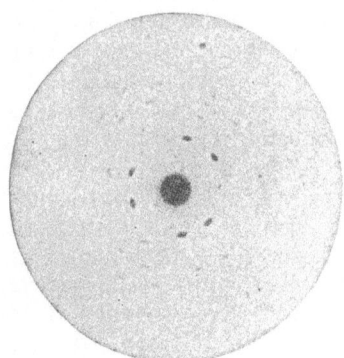

of every crystal plane with the axes. These lengths are measured in units of the corresponding edges of the elementary cell. The system which is generally applied for characterizing crystal planes uses not the values from the origin to the intersections of the planes with the axes themselves but their reciprocals which, when reduced to smallest integral numbers, are called the Miller indices. Thus, for example, the (123) plane corresponds to the intercepts $1, \frac{1}{2}, \frac{1}{3}$ or to the smallest integral numbers 6, 3, 2. The (100) face, on the other hand, is a crystal plane parallel to the yz plane which is shifted in the x direction by one lattice constant (the right-hand surface of the cube in Fig. 223),

Fig. 224. Laue diagram of a crystal with threefold symmetry

whereas (110) represents a plane parallel to the z axis, on which the right-hand front edge and the left-hand back edge of the cube in Fig. 223 are located.

For determining the system of a given crystal, one uses the fact that when X-rays of wavelength λ traverse a crystal, they are reflected by the atoms of any

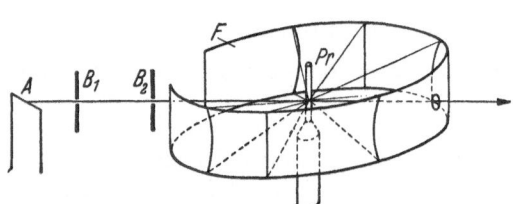

crystal plane under the angle α (or, in a different representation, diffracted by the lattice atoms about the angle α), if BRAGG's relation

$$2d \sin \alpha = n\lambda, \quad n = 1, 2, 3, \ldots \quad (7.1)$$

holds between λ, α, and the distance d of the lattice planes. Eq. (7.1) is the basis of all X-ray or neutron diffraction methods.

Fig. 225. Debye-Scherrer arrangement for photographing X-ray diffraction by a crystal powder. A = anticathode of the X-ray tube; B_1, B_2 = diaphragms; Pr = sample of crystal powder; F = photographic film

Different methods make use of monochromatic light (λ = const) or of continuous X-ray spectra with λ being variable.

In the Laue method, a monocrystal is irradiated by a narrow beam of continuous X-radiation. The different spots of a Laue pattern (see Fig. 224) therefore

Fig. 226. Debye-Scherrer diagram of cerium dioxide

belong to different lattice planes *and* to different wavelengths, a fact which makes the evaluation somewhat complicated. However, we see immediately from the Laue pattern what type of symmetry the crystal has with respect to the direction of the incident X-rays. In Fig. 224 it is a threefold symmetry, i.e., periodicity with respect to rotation about 120°. By taking Laue diagrams with directions of incidence parallel to the different crystal axes and to the diagonals, a clear picture of the symmetry and the relative distances of the lattice planes may be obtained. We thus may assign indices to the crystal planes and arrive at the complete geometrical structure of the crystal if *one* distance between two lattice planes can be determined by an X-ray photograph with known wavelength.

Determining lattice plane distances d is usually done by means of the Debye-Scherrer method, which has the advantage of not needing single crystals (which are not available for many solids) but just crystal powder. Fig. 225 indicates how a *monochromatic* X-ray is scattered by the microcrystals of the powder sample Pr. Since these microcrystals are oriented in all possible directions, the pattern on a film circularly surrounding the sample shows sharp lines (see Fig. 226) from which the scattering angles α can be determined. For a given wavelength λ, such as the Cu line $K_\alpha = 1.54$ Å, the lattice-plane distances which are responsible for the scattering then may be computed from Eq. (7.1).

An important extension of these methods is the so-called *Fourier analysis*, a theoretically difficult and rather cumbersome quantitative evaluation of X-ray diffraction photographs which takes into account the intensity of the radiation scattered into different directions. This Fourier analysis yields not only the arrangement of the crystal constituents in the crystal but also the electron distribution and thus the type of bond between the atoms or ions. This method corresponds to X-ray and electron

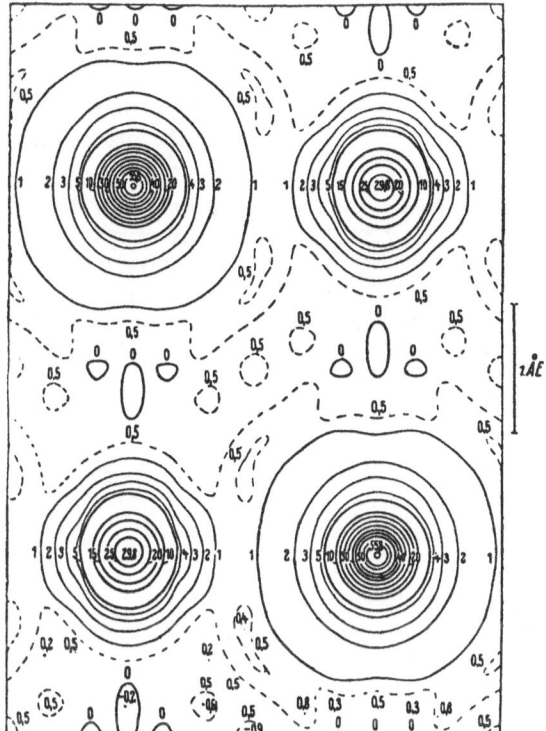

Fig. 227. Distribution of the electron density in a NaCl crystal, determined by Fourier analysis of X-ray data. The numerals indicate the electron density per cube of 1Å at an edge. (After GRIMM, BRILL, and co-workers)

diffraction experiments treated in molecular physics in VI, 2a. Fig. 227, as an example, shows the electron density in an NaCl crystal. The lines indicate equal electron density, the numerals the number of electrons per cube of 1 Å edge length. This method is especially important for investigating the bond type in certain transitional bond cases.

5. Lattice Energy, Elasticity, Crystal Growth, and Atomistic Interpretation of the Properties of Ionic Crystals

After this survey of the methods for determining crystal structures, we shall consider crystal binding and the phenomena which depend upon it more carefully for ionic crystals. If we choose the model of the ideal rock salt crystal, then the binding depends on the electrostatic attraction between the singly charged ions Na^+ and Cl^-. Consequently, the potential of this force of attraction between two isolated ions is

$$u\left(r\right) = -\frac{e^2}{r}. \tag{7.2}$$

However, the two ions are not isolated in space but are in an environment of identical positive sodium and negative chlorine ions. One can easily see that the nearest neighbors of the ion pair under consideration, because of their opposing charges, tend to reduce the bond. Quite generally, all ions of the environment will contribute to the binding energy positive or negative amounts which decrease

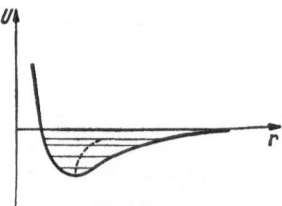

with increasing distance from the ion pair under consideration. The exact computation as carried out by E. MADELUNG showed that the above expression for the attractive potential must be multiplied by a factor 0.29 in order to take into account the contributions of the other lattice constituents.

Fig. 228. Interaction potential $u(r)$ between two crystal constituents. The average internuclear distance is indicated as a function of the vibration energy

Since the two ions cannot penetrate each other, there must be a repulsive force counteracting the attractive force. Since the deformation of the electron shells requires a considerable amount of energy, the potential of the repulsive forces increases with a high power of $1/r$. The experimental results can best be described by the ninth power of $1/r$. Thus we have for the potential of the forces acting between two ions

$$u\left(r\right) = -0.29\,\frac{'e^2}{r} + \frac{c}{r^9} \tag{7.3}$$

where c is a constant. In order to determine its value, we make use of the fact that for the equilibrium internuclear distance r_0 the potential energy must be a minimum,

$$\left(\frac{\partial u}{\partial r}\right)_{r=r_0} = 0. \tag{7.4}$$

By performing the differentiation we get

$$c = \frac{0.29}{9}\,e^2 r_0^8 \tag{7.5}$$

and thus

$$u\left(r\right) = -0.29\,e^2\left(\frac{1}{r} - \frac{1}{9}\,\frac{r_0^8}{r^9}\right). \tag{7.6}$$

By plotting the potential $u(r)$, just as we have done in molecular physics (see Fig. 189), we get the diagram Fig. 228. So far we have only considered the bond between an Na^+ and one of its neighboring Cl^- ions. By the term lattice potential or lattice energy we mean the total energy per mole of the crystal, i.e., the energy required per mole to form the crystal from widely separated Na^+ and Cl^- ions. To compute the lattice energy, we have to consider that each Na^+ ion in the lattice has six neighboring Cl^- ions, each of which is bound to it by the same $u(r)$ computed above. Furthermore, 1 mole of an NaCl crystal has L Na^+ ions from each of which binding forces extend to each of its six neighbors. Thus the lattice energy (referred to a normal crystal with an equilibrium distance r_0 between two consecutive constituents) is

$$U = 6L\,u\left(r_0\right) = -1.74\,L\,e^2\,\frac{8}{9}\,\frac{1}{r_0} \tag{7.7}$$

with L representing Avogadro's number.

This lattice energy (7.7) is negative because it is released if the crystal is formed from its ions. If we include the numerical values which refer to the specific case of an NaCl crystal in a specific constant C, and if we take into consideration that r_0 is equal to one half of the lattice constant a, then the lattice energy of *all* ionic crystals can be expressed by the general formula

$$U = C \frac{e^2}{a}. \tag{7.8}$$

Lattice energies computed according to this theoretical formula cannot be compared directly with experimental data, because the experimentally measured heat of formation E_B of the NaCl crystal refers not to the ions which form the crystal but to metallic sodium and Cl_2 molecules. However, BORN has shown that the lattice energy U can be computed from the measured heat of formation E_B by a hypothetical cyclic process, provided some atomic data are known. We denote the measured heat of formation of the crystal by E_B, the heat of vaporization of sodium metal (always referred to one mole) by V, the dissociation energy of the chlorine molecules by D, the ionization energy of one mole of sodium atoms by E_i, and by E_a the electron affinity energy which is set free when electrons are attached to one mole of Cl atoms.

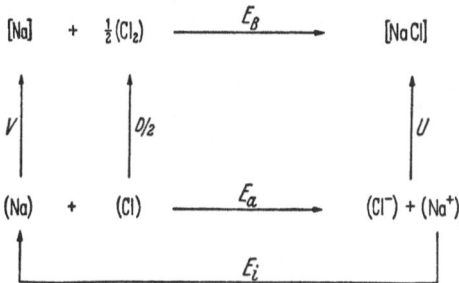

Fig. 229. BORN's cycle: Symbols in parentheses indicate gaseous substances; those in brackets, solids. Arrows give the direction of processes leading to the release of indicated energies. Denotations explained in the text

BORN's cycle is then represented in the scheme of Fig. 229 where the arrows indicate the direction of the spontaneously occuring processes by which energy of the corresponding amount is released. On the one hand, we may build one mole of the crystal from one mole each of Cl^- ions and Na^+ ions thereby setting free the lattice energy U. On the other hand, by using up the electron affinity E_a and setting free the ionization energy E_i, we can go from the free Na^+ ions and Cl^- ions to a mole each of free neutral atoms. The Na atoms may then be transformed into a mole of Na metal, thus releasing the heat of evaporation V, and the one mole of Cl atoms may be transformed into one half of a mole of Cl_2 molecules by setting free half of the dissociation energy, viz. $D/2$. Finally, we may form one mole of solid NaCl from the solid metal and the Cl_2 gas, thus releasing the heat of formation E_B. Consequently, the lattice energy V may be computed from the constants of the materials, according to the scheme of Fig. 229,

$$U = E_B + V + D/2 + E_i - E_a. \tag{7.9}$$

These "experimental" values of U may be compared with the theoretical results from Eq. (7.7). For the alkali halide crystals,

Table 21. *Lattice Energies of Several Alkali Halide Crystals in Cal/Mole. Comparison of the Measured and Computed Values* (after BORN)

Alkali halide crystal	Experimental value	Computed value
NaCl	183	182
NaBr	170	171
NaI	159	158
KCl	165	162
KBr	154	155
KI	144	144
RbCl	161	155
RbBr	151	148
RbI	141	138

Table 21 shows an agreement of both values that is as good as can be expected from the accuracy of the constants of the equation.

The lattice energy as computed from Eq. (7.7) relates to lattice constituents bound from all sides and thus indicates the amount of energy which would be set free if a mole of lattice constituents would be integrated into the interior of the crystal. The actual growth of a real crystal from the melt as well as from the vapor phase, however, occurs by attaching constituents at suitable places of the crystal surface. The energy thus released is smaller since the binding does not occur from all sides. According to KOSSEL, its amount depends on the specific position where the constituent is attached. A maximum bond energy is released if the constituent is attached to an incomplete atomic row in an incomplete lattice plane. If the constituent is attached to a more or less complete row of an incomplete lattice plane, the binding energy is about 20% of the maximal bond energy. If, however, the constituent is attached on a complete lattice plane, then it is bound from one of the six planes of a cube only and this attachement leads to the release of only 7.6% of the maximal binding energy. In thermal equilibrium, the energetically most favorable possibilities for the attachement are of course preferred. We thus can understand the well-known regular growth of crystals. KOSSEL and STRANSKI succeeded, by detailed considerations of this kind, in making understandable a large variety of empirical rules about the *growth of crystals*. Beyond this it seems that the growth of many real crystals is essentially influenced by lattice defects and in particular by dislocations. We may thus understand that the spiral growth of crystals which has often been observed is energetically preferred to that by ideal consecutive planes.

From the curve of the mutual potential $u(r)$ of two lattice constituents of an ionic crystal, derived on page 420, a number of important macroscopic properties of crystals can be deduced or explained.

The *compressibility* is the relative decrease in volume per unit of pressure. That energy must be supplied to reduce the volume, i.e., to decrease the equilibrium distance r_0, can be seen from the potential curve. Moreover, it can be seen immediately that the compressibility of a crystal is the smaller the steeper the potential curve rises from the minimum toward smaller internuclear distances. The force F necessary for reducing the distance can be computed by differentiating $u(r)$ with respect to r. Thus the work per mole required to reduce all distances from r_0 to r is

$$A = 6L \int_{r_0}^{r} F\, dr = \int_{v_0}^{v} p\, dv = \frac{P}{2}\, \Delta v. \qquad (7.10)$$

Here p is the pressure which has to be increased with the compression and P is the final pressure which corresponds to the reduction of the volume, Δv. The function $p(v)$ has been approximated by a linear function. Carrying through the computation results in an expression for the compressibility,

$$K = c' \frac{a^4}{e^2}. \qquad (7.11)$$

The compressibility thus increases with the fourth power of the lattice constant a, a plausible result since the compressibility must be the smaller the more densely the lattice components are packed in equilibrium.

If the compressing force is removed, the equilibrium distance r_0 is restored, since without external forces equilibrium can exist only if all distances correspond to the minimum of the potential curve. Consequently, the crystal is elastic. From this fact it follows that the crystal can vibrate, i.e., that the distance between the lattice points can periodically increase and decrease. We shall consider this important property in some detail in VII, 8.

The last important conclusion from the potential curve which we have to mention concerns the thermal expansion of crystals and, in general, of all solids. The vibration energy of the crystal and thus the amplitude of the crystal vibrations increases with increasing temperature. Because of the asymmetry of the potential curve (anharmonicity of the lattice vibration, cf. Fig. 228), the average distance between two lattice points, shown in Fig. 228, increases with increasing amplitude of the vibration. Consequently, the volume of the crystal increases with increasing temperature. Thus the fundamental thermal property of all known solids, their expansion with increasing temperature, finds a very simple explanation from atomic theory. Ultimately it depends upon the asymmetry of the potential curve $u(r)$.

We have limited ourselves in this section to a discussion of ionic crystals because their properties can be accounted for quantitatively. Qualitatively, the behavior of valence crystals with respect to the potential curve and the conclusions drawn from it, such as compressibility, thermal expansion, etc., is very similar to that of the ionic crystals. Phenomenologically, valence crystals such as diamond are characterized by hardness and excellent electrical insulating properties. The latter is due to the fact that electrolytic conductivity, which is characteristic for ionic crystals and which will be considered in detail in VII, 18, is not possible because the constituents are neutral atoms. Electronic conductivity, on the other hand, is impossible since all electrons are strongly bound in localized valence bonds. The hardness of the valence crystals is related to this localized valence bond too.

6. Piezoelectricity, Pyroelectricity, and Related Phenomena

A number of conspicuous and in some cases also practically important phenomena of asymmetric crystals are related to the elastic binding of ions in the crystal lattice, which was discussed in the last section. These are piezoelectricity and its reversal, electrostriction, as well as the less common pyroelectricity, well-known particularly from tourmaline, with its reversal, the electrocaloric effect.

Piezoelectricity is the phenomenon that quartz and those other crystals in which not each lattice constituent is a center of symmetry of the lattice exhibit electric surface charges if elastically compressed in certain directions. The reverse process, i.e., the mechanical deformation of a crystal by the action of an electric field, is called *electrostriction*. The much rarer phenomenon of *pyroelectricity* refers to surface charges which are observed if crystals are submitted to temperature changes. Conversely, in the *electrocaloric effect* the temperature of the crystal changes if it is exposed to an electric field.

Whereas in *every* dielectric substance dipole moments are induced in an electric field, piezoelectricity and pyroelectricity occur only in *ionic* crystals, i.e., in crystals in which the constituents have at least some ionic character. If these ions are arranged so symmetrically that each ion is surrounded by the same number of oppositely charged ions in equal distances as is the case in NaCl according to Fig. 223, neither a compression nor a change of the vibrational energy is able to perturb this symmetrical arrangement and thus the mutual charge compensation.

The situation is different for the zincblende modification of ZnS (cf. Fig. 230), where a tetrahedron of four sulfur atoms is built into the face-centered cubic lattice of the Zn atoms. If we now cut a (110) section from the upper right edge of the cube to the lower left one, this plane contains two space diagonals (see Fig. 231). On each of them a sulfur atom is asymmetrically located so that the diagonal is divided in the ratio 1:3. It is obvious that pressure applied in the direction of a space diagonal influences the larger distance Zn-S more than the smaller one and thus causes an asymmetric displacement of the atoms which carry small amounts of excess charge. Consequently, this displacement causes the appearance of charges on the two opposite end faces of the crystal. This explains

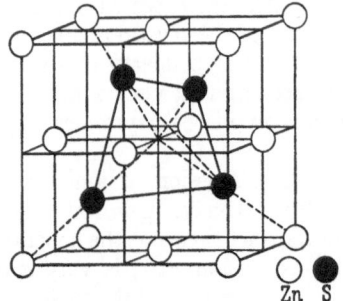

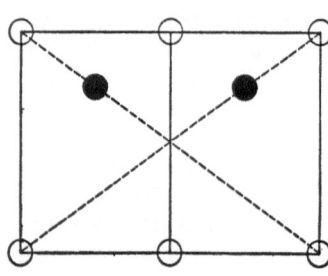

Fig. 230. Arrangement of constituents in a zincblende lattice, used for explaining piezoelectricity

Fig. 231. (110) section through zincblende lattice shows asymmetric position of sulphur atoms

the phenomenon of piezoelectricity. Piezoelectric crystals show, of course, electrostriction if exposed to an external electric field. This is of high practical importance in the case of quartz. Properly cut quartz plates exposed to a cor-

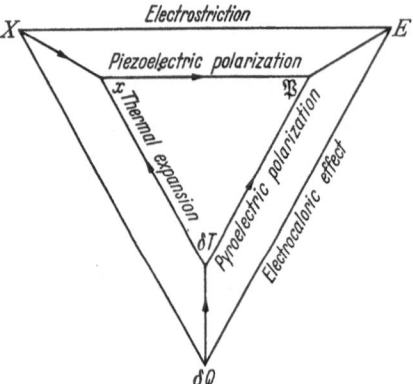

Fig. 232. Relation between the mechanical, thermal, and electrical effects of a pyroelectric crystal (after HECKMANN). E = electric field strength, $\mathfrak{P}$ = dielectric excitation (polarization), X = mechanical stress, x = mechanical deformation, δT = temperature change, δQ = change of heat energy

rectly tuned high-frequency electric field can be excited to their characteristic mechanical eigenvibrations by periodic electrostriction. These eigenvibrations, which are extremely constant with regard to their frequency at constant temperature, are being utilized for controlling high-frequency transmitters and quartz clocks.

Whereas the dipole moment of a piezoelectric crystal is caused by external forces, certain crystals with a so-called polar axis (e.g., tourmaline) have a permanent dipole moment even without external forces acting. Normally, this is compensated by accumulated surface charges and therefore is not in evidence. But temperature variations change the dipole moment of these crystals because

of the increase or decrease of the distance between the lattice ions so that opposite charges occur on the corresponding end faces of the crystal. This is the explanation of pyroelectricity.

We see that only piezoelectric crystals can be pyroelectric. But not all piezoelectric crystals are also pyroelectric, because piezoelectricity requires only that there are ions which are not centers of symmetry of the crystal. Pyroelectricity, in addition requires the existence of permanent electric dipoles in the direction of the crystal axis of highest thermal expansion.

Related to these phenomena is *ferroelectricity*, which is caused by a spontaneous parallel orientation of the electric dipoles in entire crystal domains. We shall come back to this phenomenon, which is of considerable scientific and technical interest, in VII, 16.

The interrelations between elastic, electric, and thermal phenomena and effects in piezoelectric and pyroelectric crystals may be seen from the diagram Fig. 232, for instance, by starting from the upper left corner: A mechanical pressure X produces a lattice deformation which causes a piezoelectric polarization $\mathfrak{P}$ that becomes evident as an electric field. Vice versa, an external electric field produces a mechanical stress in the crystal (electrostriction), and this in turn produces a lattice deformation. If, on the other hand, we increase the heat capacity of the crystal by δQ, we increase its temperature by an amount δT which depends on its specific heat. This temperature increase causes a mechanical deformation (expansion) x of the crystal, which can produce a piezoelectric polarization $\mathfrak{P}$. This production of an electric polarization via $\delta T \rightarrow x \rightarrow \mathfrak{P}$ has sometimes been called *secondary pyroelectricity*, whereas the direct production of a polarization $\delta T \rightarrow \mathfrak{P}$ represents the primary pyroelectricity. Fig. 232 shows furthermore that the heat capacity of the crystal is changed by elastic distortion of the crystal via $X \rightarrow \delta Q$, so that an external electric field E changes the heat capacity of a pyroelectric crystal either directly (electrocaloric effect $E \rightarrow \delta Q$) or indirectly by electrostriction and deformation heat via $E \rightarrow X \rightarrow \delta Q$. From Fig. 232 we see, finally, that the result of the change of *one* property of the crystal (we have to include here an external pressure and an electric field with the variables which determine the "state" of a piezoelectric crystal) depends essentially on which of the non-varied parameters of Fig. 232 are kept constant.

7. Survey of Binding and Properties of the Metallic State

Having understood the essential properties of ionic and valence crystals from their bond characteristics, we now discuss briefly the relation between binding and properties of the other extreme among the solids, the metals, which are characterized by a lack of localized bonds. Such a survey appears to be the more appropriate as the large number of typical properties of metals, their modifications (different forms of lattices in different temperature regions), and their alloys can be understood only from atomistic concepts which in turn also throw new light on the technological treatment of metals such as forging and annealing.

The most important properties which distinguish metals and alloys from other solids, viz. their large electric and thermal conductivity as well as their plasticity that is so important for their workability, are direct consequences of the type of bond in metals. As we indicated already in VII, 3, in monatomic metals the bond must be purely homopolar since there is no polarity, in contrast to the alloys. However, contrary to what we found in valence crystals, the metals do not have localized electron-pair bonds with compensated spins, but rather a type of bond which is closely related to the non-localized π-electron bonds in benzene rings and in molecular chains with conjugated double bonds (page 405). Just as in these cases, the alkali metals, silver, and copper, have *one* valence electron per atom which is responsible for the bonds to *all* neighbors in the lattice. This electron causes the binding as a non-localized and therefore mobile electron. It is valence electron and conduction electron at the same time. This becomes clear if we compare the valence crystal diamond and the metal silver. In diamond, each carbon atom is symmetrically surrounded by four nearest neighbors and contributes one of its four valence electrons to each of the four bonds. Each of these electrons forms, together with the valence electron contributed by the neighboring carbon atom, one of the strongly localized spin-compensated electron-pair bonds which we discussed in VI, 14b. In metallic silver, on the other hand, each Ag atom is symmetrically surrounded by twelve nearest neighbors, but it has only *one* valence electron. Therefore, taking into account the valence electrons of the neighbors, in the time average we find only one sixth of one electron charge between two Ag atoms. This one sixth of one electron is responsible for the bond, in contrast to the two full electron charges between each two C atoms in diamond.

From this difference in bond type and bond strength, the essential differences of the properties of diamond and monovalent metals may be derived. The much lower melting temperature and the smaller heat of evaporation of all metals compared to valence crystals follows from the twelve times smaller electronic charge between each two atoms in silver than in diamond. The hardness of the valence crystals and the plasticity of metals follow from the localized bonds in the former and the non-localized bonds in the latter case: For in order to shift two carbon atoms of diamond with respect to each other, localized electron-pair bonds have to be severed, whereas the non-localized bonds remain unchanged by a displacement of metal atoms. In this case, the resistance against shifting is mainly due to mutual steric hindrance of the metal atoms. The complete saturation of the four valence electrons of each C atom of diamond in four localized electron pairs which bind four neighbors has the effect that it is impossible for any carbon atom, according to the Pauli principle, to bind more than four C atoms. *The coordination number (number of nearest neighbors in the lattice) of valence crystals* (in diamond, four) *is therefore determined by the valency of the atoms which form these crystals. In contrast, there is no valence saturation in non-localized metal bonds, and the generally larger coordination number of metals* (8 for alkali-metals, 12 in silver and copper) *is determined by the available volume rather than by specific properties of the atoms.*

Metallic bonds are thus characterized by a lack of valence saturation and therefore by a high symmetry, i.e., equally strong bonds to all equally distant neighbors independent of the valence of the metal's constituents. Quantum-mechanically, the wave function describing the behavior of the valence electrons is a linear combination of the wave functions describing the twelve possible bonds to the twelve partners. It is irrelevant whether one speaks of a distribution among the twelve bonds or, with DEHLINGER and PAULING, of a rotation of *one* "real" electron-pair bond, caused by the valence electron successively forming binding, spin-compensated electron pairs with the valence electron of each of its neighbors. What counts in the final analysis, is only the average electron density between the atoms to be bound. The bond may also contain a polar contribution since both valence electrons of the atoms may occassionally be found near the same atom.

In general it can be said that in metals the nonselective metallic bond predominates and determines the metallic character. Nevertheless, the individual properties of the atoms, especially their spin, orbital momentum, and the valency determined by them play a certain role. Thus, in addition to the general metallic bond, we find a small amount of heteropolar or homopolar binding. Consequently, a metal always crystallizes in the lattice of highest symmetry which is compatible with the specific properties of the lattice constituents. For example, in the case of the alkali metals, the body-centered lattice in which each atom is symmetrically surrounded by eight atoms of opposite spin is more stable than the lattice that would be expected for *purely* metallic bonds, in which each atom would be surrounded by 12 others at equal distances, half of which have one spin direction and half the other. The preferred antiparallel spin directions of the alkali electrons thus are responsible for the fact that not the lattice corresponding to pure metallic binding with a coordination number 12 but the body-centered lattice with coordination number 8 has the lower potential energy. This body-centered lattice has actually been found for the alkali metals by X-ray experiments.

We know from experience that the elements on the left side of the Periodic Table exhibit metallic character, whereas localized binding predominates in crystals which are formed from elements on the right-hand side of the table, and it is here that we find the insulating crystals. According to HUME-ROTHERY, the reason for this distinction is that metals generally crystallize in a body-centered or face-centered cubic lattice or in a cubic or hexagonal close-packed structure. On the other hand, the elements on the right-hand side of the table belong to very different lattice types in which the coordination number is always equal to the valency of the atom under consideration. According to FRÖHLICH this difference depends upon the repulsive Coulomb forces between the outer electrons. Sodium and chlorine, for example, have both the valency 1. Sodium has only one outer electron, whereas chlorine, which lies on the right-hand side of the Periodic Table, has seven. The binding effect of the one electron pair in the Na—Na and Cl—Cl pairs therefore is different because in Cl—Cl we have an additional repulsion between the two times six excess electrons. Such a repulsion is not present in the case of Na—Na. Nevertheless, in the case of the diatomic Cl_2 molecule, we find a strong homopolar bond because the exchange forces (cf. IV, 11) produce a unilateral charge displacement (and thus an attraction between the atoms) by forming a spin-saturated electron pair; this compensates for the repulsion due to the other electrons. However, such an asymmetry of the electron configuration is incompatible with the high symmetry (coordination number) of the metal lattice. Thus, because there is no Coulomb repulsion between excess electrons of the outermost shell, the small average electron density between the atoms is sufficient for binding them metallically, whereas in chlorine saturated Cl_2 molecules are produced. These molecules may then be bound in a crystal lattice by the relatively small van der Waals forces. This contrast, which was explained here as an example for Na and Cl, holds in general for all elements on the left or right-hand side of the Periodic Table respectively: On the left side we have metallic binding with high symmetry and coordination number; on the right, complicated lattices strongly influenced by the valency of the constituent atoms and always without electrical conductivity.

It is of further interest to ask about the binding of atoms in the middle region of the Periodic Table. There we have a competition between the metallic bond and the formation of valence-saturated electron pairs, and we cannot decide without further considerations which of the possible lattice types is the more stable one in a particular case. In general, metallic binding is more stable at higher temperatures since its degree of order is lower. The strictly localized, high-order valence bond, however, may be stable at lower temperatures. It is known indeed that some elements may crystallize in different lattices at different temperatures. Tin, for example, above 18 °C is a genuine metal with practically closest-packed layers of the atoms. Below this temperature, it exists in a non-metallic valence-crystal modification (so-called *gray tin*) which is related to diamond. Another example of technical importance are the intermetallic compounds, e.g., compounds of metals from the third and fifth column of the Periodic Table, such as indium antimonide. Although consisting of two metals, crystals of this type do not show metallic behavior but are semiconductors (after WELKER) with predominant valence bonds just as gray tin.

Let us now have a brief look at the *alloys*, which are of such technological importance. Alloys are homogeneous atomic mixtures, also called solid solutions, of one metallic element with another, sometimes also of some other element with a metal. The proportions of the mixture are in general variable within rather wide ranges, in sharp contrast to the behavior of chemical compounds with valence saturation. We distinguish *substitutional alloys* and *interstitial alloys*. The former are genuine alloys of two metals in which all atoms occupy regular lattice positions. In forming the alloy, the atoms of one kind of metal may substitute for those of the other metal. Interstitial alloys, however, are solid solutions of non-metallic elements in a metal. Here, the interstitial atoms which mostly are of small diameter (such as H, B, C, or N) occupy interstitial sites (cf. VII, 18, Fig. 258). Examples are the systems palladium-hydrogen and iron-carbon.

While a small atomic diameter of the added element is favorable for the formation of interstitial alloys, it is evident that the formation of substitutional alloys is the easier the less different the diameters and chemical valencies of the alloying metals are. The broad range of mixing ratios of the alloys compared to that of normal chemical compounds, combined with the fact that the bond strength does not change appreciably, can be explained from their binding by non-localized valence electrons. Every impurity atom that has a valence electron of low binding energy may take part in the general metallic bond. But it is evident too that in alloys of metals with different ionization energies polar bonds will participate. In Fig. 222 we therefore have indicated the relationship of the alloys to the ionic crystals. Many metals can be alloyed in nearly arbitrary proportions. However, we conclude from changes of the electric resistance as well as of the magnetic properties that for certain rational proportions of the components (corresponding, for example, to the compounds $AuCu$, $AuCu_3$, $FeCo$, or Ni_3Fe) especially stable so-called *superstructures* occur, which provide further evidence that the general metallic bond is superposed by some localized binding which exhibits saturation characteristics. These localized valence forces which cause an ordered structure of alloys (e.g., Cu ions in the center and Zn ions at the corners of the elementary cubes in β-brass) are very weak forces. Evidence for this is that at temperatures of only a few hundred degrees this ordered structure disappears and is replaced by a disordered lattice with a probability distribution of Cu and Zn ions.

We conclude our discussion of the behavior of metals with a few remarks on the ductility of metals and its dependence on technological treatment. In metallic single crystals, which are rather an exception, the lattice planes in general extend throughout the whole crystal in a well-ordered fashion. Even in these, we often find lattice defects of many types, and in particular *lattice dislocations*. These are extended one-dimensional or two-dimensional regions where the ideal lattice structure is disturbed. It is evident that the energy necessary for the mutual displacement of lattice planes during plastic deformation of a solid is considerably smaller than in an ideal crystal if such dislocations exist. Metal single crystals are exceedingly easy to deform, but they are surprisingly hardened by one single deformation. This is so because the first deformation causes angular disorientations of the lattice planes, and the internal corners produced in this way make further sliding of lattice planes much more difficult. That technical metals are much less ductile than single crystals is due to the fact that in the technical metals we have always a texture of mutually dislocated and indented micro-crystals. By hammering or other cold-working, this internal indentation is increased due to compression of the texture and internal stress. Sudden quenching of a heated metal has a similar effect. The thermally produced disorder of the lattice is "frozen in" and the formation of extended glide planes is prevented. On the other hand, slow heating and cooling, i.e., thermal annealing, relieves internal stresses and, in general, produces an approximation to a higher state of order in the metal. In agreement with experience, it thus increases the ductility.

8. Crystal Vibrations and the Determination of their Frequencies from Infrared and Raman Spectra

Because there is a quasi-elastic binding force effective between the lattice constituents in a crystal, vibrations are possible between them just as they are in a molecule (cf. VI, 6). We now discuss these vibrations in detail for the example of the ionic crystals. The asymmetric potential curve $u(r)$ (see Fig. 228) is evidence that the vibrations are anharmonic. We have used this fact in explaining the thermal expansion of solids.

We distinguish between *internal vibrations* and *lattice vibrations*. Internal vibrations occur particularly in molecular crystals in which vibrations are possible within the molecules forming the lattice. These molecular vibrations may be disturbed more or less by the crystal lattice but have nothing else to do with the

crystal as such. In the lattice vibrations proper, on the other hand, the whole crystal participates. To illustrate these two types of vibrations, we consider the calcite crystal $CaCO_3$. In this crystal the lattice components are the ions Ca^{++} and $(CO_3)^{--}$. The vibrations of all ions of one kind with respect to the ions of the other kind are called lattice vibrations because the lattice itself participates in this vibration. On the other hand, there are internal vibrations within the CO_3 group (e.g., the O atoms vibrate with respect to the C atom) which have little to do with the calcite

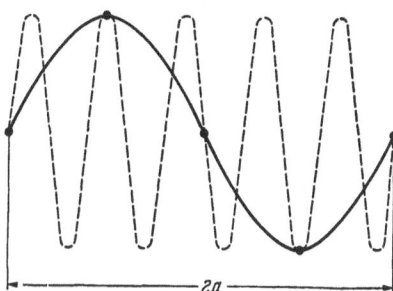

Fig. 233. Representation of a vibration by the largest possible wavelength (solid line) and a smaller wavelength (dashed line)

crystal as a whole. Therefore they occur with nearly the same frequencies in other crystals containing the CO_3 group, such as $FeCO_3$ or $MgCO_3$. The lattice vibrations proper have a large variety of vibrational possibilities, as may be illustrated by a three-dimensional arrangement of mass points elastically bound to each other: In the *stationary* case to which we shall restrict ourselves, we have *standing* waves with a wide range of wavelengths. If we consider an ionic lattice such as NaCl with the lattice constant a indicating the smallest distance of two identical ions, we infer from the one-dimensional example of Fig. 233 that the smallest physically meaningful wavelength is $\lambda = 2a$. Every smaller wavelength describes the same vibrational state in a more complicated way, as can be seen for a wavelength $\lambda = 2a/5$ in Fig. 233 (dashed line). However, *all* wavelengths between $2a$ and the length of the crystal in the corresponding direction are physically meaningful and may occur if the crystal length can be considered infinite compared to a.

Just as in the case of a vibrating string or membrane, only such waves can produce a *stationary vibrational state* in the form of standing waves which have nodes at the external boundaries of the crystal. This means that for every eigenvibration state the crystal length must be an integral multiple of the corresponding half wavelength. Let us consider the one-dimensional example of an NaCl chain with N ions and the lattice constant a. Its length is $N \times a/2$ and we obtain for the possible wavelengths of the standing waves or *eigenvibrations* of this crystal the condition

$$N\frac{a}{2} = n\frac{\lambda}{2}, \qquad (7.12)$$

or

$$\lambda = \frac{Na}{n} \quad \text{with} \quad n = 1, 2, 3, \ldots N/2. \qquad (7.13)$$

We thus have only $N/2$ possible discrete wavelength values because the smallest physically meaningful wavelength is $\lambda = 2a$ to which the value $n = N/2$ corresponds according to Eq. (7.13). Since two different vibrational modes belong to each of these $N/2$ discrete wavelengths (as we shall show immediately), *the total number of discrete eigenvibrations of the one-dimensional lattice is equal to N; i.e., it is equal to the number of lattice constituents.*

It is essential that we have two different vibrational modes of different frequency for every wavelength of the standing waves in question. For reasons to be explained presently, these are called optical and acoustical (or elastic) vibrations. On the one hand, a crystal lattice may perform eigenvibrations which are completely equivalent to those of a continuum insofar as regions large compared to the lattice constant undergo a *uniform* motion just as in a string or membrane. Since, in a first approximation, the relative distances between the positive and negative charges here remain constant, these vibrations are *not* connected with a change of the electric dipole moment

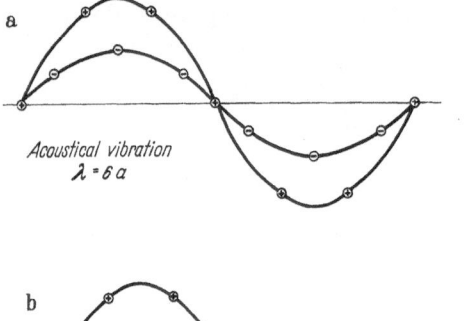

Acoustical vibration
$\lambda = 6a$

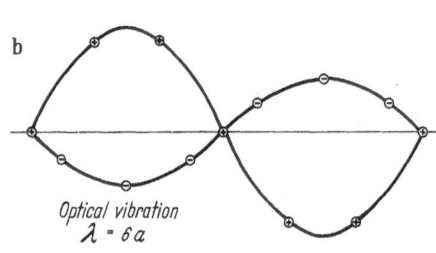

Optical vibration
$\lambda = 6a$

Fig. 234. Acoustical vibration (a) and optical vibration (b) of an ionic crystal with different masses of the constituents

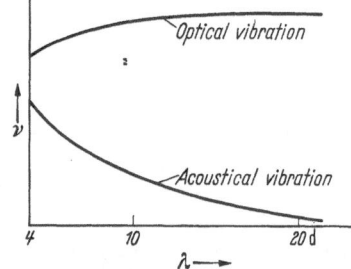

Fig. 235. Vibrational frequency versus wavelength for optical and acoustical vibrations (so-called optical branch and acoustical branch)

so that they do not appear in the absorption spectrum of the crystal. These vibrations are normal elastic vibrations and therefore are called *acoustical* vibrations. On the other hand, in every ionic crystal (and according to VII, 3 nearly all real crystals have some ionic character) eigenvibrations of the same wavelength are possible in which neighboring lattice ions are moving in *opposite* directions. Obviously, these vibrations produce electric dipole moments which vary with the vibration frequency. Since the corresponding frequencies are therefore able to absorb and to emit radiation, this vibrational mode is called *optical* vibration.

Comparing Fig. 234a and b, we see that the quasi-elastic force between neighboring lattice ions is changed very little by acoustical vibrations but very much by the optical vibrations due to neighboring ions moving in opposite directions. Consequently, *for the same wavelength the frequency of the optical vibration is always larger than that of the acoustic one* (see Fig. 235). If the ions have equal masses, then for an infinite wavelength only the optical vibration is connected with a periodical change of the electric dipole moment, while there is no such change for the acoustical vibration. For the general case of different ion masses of opposite sign and for smaller wavelengths, we see from Fig. 234 that also the acoustical vibration brings forth a change of the dipole moment. This change, however, is *smaller* than that corresponding to the optical vibration of the same wavelength. Optical and acoustical vibrations coincide only for the limiting wavelength $\lambda = 2a$; for equal masses of the ions they then have also the same frequency. Conversely, the frequency difference of the two vibrational types becomes largest for $\lambda \rightarrow \infty$ because then the bonds of the atoms are practically not under stress in the acoustical vibration but under maximum stress in the optical vibration. By plotting the frequencies of the two types of vibration against the wavelength, we thus obtain for the general case of different ion masses the diagram of Fig. 235.

So far we discussed only *stationary* vibrations. But lattice vibrations may also be excited locally by a force acting at a particular point. This happens, for instance, when a valence electron of a lattice constituent is excited by light absorption or is separated from its ion. Since this electronic process produces a sudden change of the local bonds nearby (cf. the Franck Condon principle and Figs. 192b and 238), the ions begin to vibrate. Such locally excited vibrations then propagate through the crystal as waves. Using the particle language we may speak of a *production (and migration) of lattice vibrational quanta or "phonons". Thus phonons correspond to the lattice vibrations propagating in the crystal in the same manner as do photons to light waves.* According to VII, 10d, an excited electron can lose its energy either by emission of a photon or by emission of one or several phonons, i.e., in the wave language, by excitation of lattice vibrations.

How about our experimental knowledge on lattice eigenvibrations? We know mainly three different effects of lattice vibrations from which they may be determined: They are, first, modifications of X-ray diffraction; secondly, changes of the optical behavior, in particular of the absorption and reflection of solids and, thirdly, the temperature dependence of the specific heat of solids. For metals we know a fourth effect, namely the temperature dependence of the electric conductivity, which is caused by lattice vibrations (cf. VII, 13).

Since the diffraction of X-rays depends on the spatial arrangement of the diffracting atoms in the crystal lattice, periodic changes of this arrangement due to lattice vibrations cause changes of the diffraction pattern. Investigations of this effect are under way. It is evident that *optical* studies of lattice vibrations must make use of the optical eigenvibrations which are connected with a periodic change of the electric dipole moment, even if this may be only due to the polarization of the vibrating atoms. This periodic modification of the resulting dipole moment is the larger the larger the wavelength is and reaches its maximum for the optical vibration with $\lambda \rightarrow \infty$, which is called *fundamental vibration* of the lattice. In the fundamental vibration of rock salt, the entire lattice of the Na^+ ions vibrates in phase, i.e., synchronously, against the entire lattice of the Cl^- ions. This vibration thus has the most pronounced optical effect. For NaCl it has a wavelength of $61\,\mu$. The absorption coefficient of the corresponding optical wavelengths is large because all lattice ions participate in this vibration. Absorption measurements therefore are made with very thin crystal layers. A more elegant and simpler method, although not so exact, is the reststrahlen method of RUBENS. Because of the relation between absorption and reflection, the crystal reflects strongly those wavelengths which correspond to the fundamental vibration, whereas the other wavelengths are mainly transmitted by the crystal. If a light ray is reflected back and forth several times between two plates of the material under investigation, then all wavelengths except those of the fundamental vibration are gradually eliminated by transmission. The residual rays or "reststrahlen" which remain after many reflections thus consist of the fundamental vibrations so that their wavelengths can be measured.

Occasionally it happens that, as a result of special symmetry relations which exist, for example, in fluorspar CaF_2, the fundamental vibration is not associated with a change of the dipole moment and thus it is optically inactive. In that case it cannot be found in the absorption spectrum. If the polarizability of the lattice constituents is changed by such optically inactive vibrations, as is almost always the case, the frequencies of the vibrations under consideration can be determined by Raman effect measurements as in the corresponding case in molecular physics (see VI, 2d). Infrared and Raman investigations again complement each other so that there is no basic difficulty in determining the frequencies of the fundamental lattice vibrations.

9. The Atomistic Theory of the Specific Heat of Solids

One of the most striking effects of lattice vibrations is found in the specific heat of solids. It is of particular interest because it played an important role in the historical development of quantum physics. In VI, 10 we discussed the corresponding relations with respect to molecules.

According to the equipartition law of classical physics, the thermal energy transferred to a solid should be partitioned equally among all degrees of freedom of the system; and even near the absolute temperature $T=0$ the thermal energy per degree of freedom was expected to be $kT/2$.

Since every crystal constituent which is able to vibrate possesses three degrees of freedom each for the kinetic and the potential energy, we expect a thermal energy per atom of $3kT$ and, therefore, a thermal energy per mole of L atoms of $3LkT=3RT$, where R is the general gas constant with the value 1.987 cal per degree and mole. According to classical theory, the specific heat at constant volume is therefore expected to have the constant value

$$c_v = 3R = 5.96 \text{ cal/degree} \cdot \text{mole}. \tag{7.14}$$

This is the Dulong-Petit law, which follows directly from the classical theory according to which the atomic heat, defined as the specific heat per mole of atoms, is expected to be close to 6 cal per degree and mole for all solids.

It has been known, however, for a long time that the Dulong-Petit law is only a very rough approximation and even this only at temperatures which must be the higher the smaller the atomic weight of the element in question is. For example, the atomic heat of diamond at room temperature is a little less than 1.5 cal/degree instead of 6 cal/degree. In approaching absolute zero of temperature, the atomic heat of *every* crystal even goes down towards zero, in sharp disagreement with the classically expected independence of the temperature.

EINSTEIN has given an interpretation of this discrepancy in 1907. He realized that the gradual disappearance of the atomic heat in approaching absolute zero is a typical quantum effect. It is due to the fact that, in contrast to the fundamental assumptions of the classical law of equipartition, the oscillating lattice constituents cannot take up arbitrarily small amounts of thermal energy but, according to quantum mechanics, only integral energy quanta $h\nu_0$ just as *every* oscillator of the eigenfrequency ν_0. With decreasing temperature, the mean thermal energy kT finally becomes smaller than one vibrational quantum $h\nu_0$. Below that temperature, less and less acceptable energy quanta are offered to the crystal. In VI, 10, we discussed already the corresponding effect for molecules, and we called it a successive "freezing" of the degrees of freedom. With decreasing number of the degrees of freedom which are able to absorb thermal energy, the specific heat decreases towards zero. The lighter the atoms of a specific crystal are, the higher is the temperature at which this "starving" of the degrees of freedom occurs. This is due to the fact that, according to IV, 7c, the vibrational energy quanta are the larger the smaller the oscillating masses are.

The statistical computation is very similar to that of the cavity oscillations which lead to PLANCK's radiation formula (2.44), and the result reflects this similarity. If we assume a single eigenfrequency ν_0 of the crystal atoms, EINSTEIN's theory leads to the atomic heat

$$c_v = 3R\left(\frac{h\nu_0}{kT}\right)^2 \frac{e^{h\nu_0/kT}}{(e^{h\nu_0/kT}-1)^2},\qquad(7.15)$$

instead of Eq. (7.14). At a sufficiently high temperature, this expression approaches $3R$, in accordance with experience, whereas for $T \to 0$ the atomic heat correctly goes to zero. The characteristic eigenfrequency ν_0 of Eq. (7.15) was determined by EINSTEIN from a comparison of his formula with the empirical temperature dependence of the specific heats. For alkali halide crystals, he obtained values of ν_0 that agreed within about 20% with the frequencies of the optical fundamental vibrations as determined by the reststrahlen method (cf. page 431).

In spite of this success, it was clear that the basic assumption of EINSTEIN's theory, the existence of only one eigenfrequency of the crystal, is too much of a simplification. Thus it was not surprising that discrepancies were found between EINSTEIN's theory and the empirical values, especially for low temperatures, because the higher frequencies should freeze at a higher temperature than the lower ones which were not taken into account by EINSTEIN. Therefore, DEBYE in 1912 first carried out the quantization of the eigenfrequencies of the entire lattice which we discussed in VII, 8. He then substituted for the statistical energy distribution among the vibrating atoms that among all possible eigenvibrations of the crystal. The crystal in this theory is treated as *one* atomic system the states of which are characterized by the different eigenfrequencies which are possible according to VII, 8. The n-th eigenfrequency ν_n then contributes the temperature-dependent energy

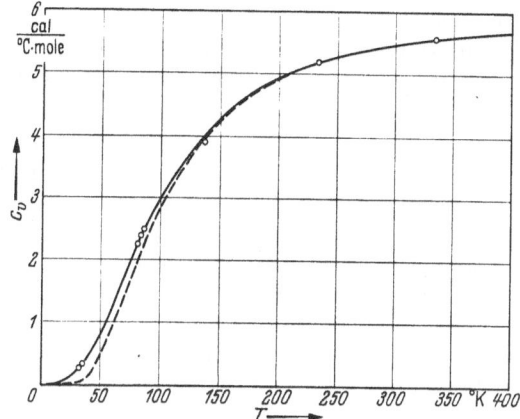

Fig. 236. Specific heat of aluminum versus temperature according to EINSTEIN's theory (dashed line) and to improved theory of Debye (solid curve) with measured values by RICHTMYER and KENNARD

$$E(\nu_n) = \frac{h\nu_n}{e^{h\nu_n/kT}-1}\qquad(7.16)$$

to the thermal energy of the crystal. The final formula of DEBYE's for the specific heat is much more complicated than Eq. (7.15), but it describes much better the measured specific heats, especially at low temperatures. In this Debye theory, all theoretically expected properties of the lattice vibrations play a role, for instance, that optical vibrations "freeze" at higher temperatures than the acoustical ones since the former have higher frequencies. Fig. 236 shows the temperature dependence of the specific heat of aluminum according to the theories by EINSTEIN and DEBYE, respectively, together with some experimental values. Certain deviations from the Debye theory that are still to be observed at low temperatures as well as at very high temperatures are due to the still insufficient treatment of details of the vibrational spectrum. These include the anharmonicity of the vibrations which causes a decrease of the higher vibrational quanta. For metals, a small contribution of free electrons (cf. IV, 13) to the observed specific heat has also to be taken into account.

10. General Survey of the Electronic Processes in Solids and their Relation to Optical and Electrical Properties

An essential part of the optical and electric properties of all solids is caused by the behavior of their electrons and by their interaction with the lattice vibrations and with external particles or photons. Before discussing in detail the theoretical concepts about the arrangement and energy states of the electrons in solids and the manifold phenomena which depend on them, it seems appropriate to give a brief survey of the action of electrons in solids and to introduce some pictorial concepts which will aid in understanding the following sections.

a) The Physical Significance of Excitation and Internal or External Release of Electrons from Solids

We begin by discussing the differences between excitation and ionization of an individual atom or molecule on the one hand, and an atom or molecule in the solid on the other hand. In principle, a valence electron of an atom, ion, or molecule in a crystal may be excited or completely separated from "its" lattice constituent by a colliding particle or by photon absorption. These processes correspond to the excitation and ionization of an individual atom (cf. III, 6). In addition, however, an electron of a solid may pass through the surface and emerge to the outside, i.e., it may leave the solid. This process, which will be treated in detail in VII, 14 and VII, 22, may be considered as *ionization of the crystal as a whole*. For a crystal, in contrast to an individual atom or molecule, we thus have to distinguish *three* electron processes: First, the excitation of electrons; second, the separation of electrons from "their" ions or atoms by which process more or less freely mobile electrons are produced in the crystal; and finally, the emission of electrons from the solid.

In further contrast to an individual atom or molecule, *the solid has neither firmly bound nor completely free outer electrons*. That the electrons are not *firmly* bound to one constituent is due to the possibility of their moving from one lattice constituent to the next by tunnel effect (see IV, 12) since the atoms or molecules in a lattice are closely coupled. Vice versa, even an electron which is released from its original lattice atom by energy absorption is not at all completely free because it moves within the potential fields of atoms or ions, i.e., it is subject to attracting and repelling forces. In certain directions and for certain kinetic energy values it is even reflected by the lattice constituents in such a way that it becomes unable to move at all.

b) The Relation between Spectrum (Color) and Conductivity in Solids

The fact that in a solid there are neither completely bound nor completely free electrons agrees with the optical result that in general neither series of sharp spectral lines are known nor distinct series-limit continua (cf. III, 6c). In most cases, only wide bands are found, the assignment of which to distinct electron transitions has been possible in relatively few cases only. We shall discuss this problem in the following section.

The investigation of *photoconductivity* offers a possibility for experimentally distinguishing between electron excitation and electron separation and for attributing observed absorption bands to these two processes. For this purpose, the crystal is connected by two electrodes, according to Fig. 237, with a battery and a measuring device. The crystal is then illuminated by light of wavelengths which correspond to its different absorption bands. In this way, it was found that for the case of alkali halides, illumination by wavelengths of the longest-wave absorption band (located in the ultraviolet) does not cause any changes in the crystal which are electrically detectable. The crystal remains an insulator. Obviously, this absorption leads to electron *excitation* only. However, the illumination of the crystal by light of the absorption bands of shorter wavelengths causes an electric conductivity of the crystal which increases with increasing illumination. Obviously, absorption of these bands leads to the production of mobile electrons in the crystal, i.e., to their separation from their bonds. Details will be discussed later.

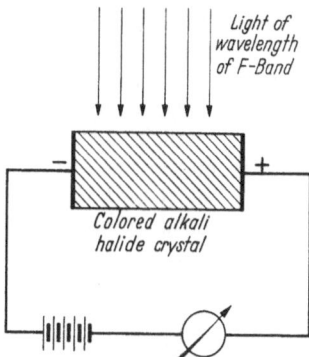

Fig. 237. Arrangement for the detection of photoconductivity of an alkali halide crystal colored by irradiation with the wavelength of its F band (cf. p. 473)

Without going into the details of the theory, we may understand the important empirical relation between the spectra and the electric conductivity of solids. In the simplest form it states: *In wide spectral ranges, especially in the visible range, insulators do not absorb; they are transparent. Metals and semiconductors (cf. VII, 20) on the other hand, in general do absorb strongly within the entire spectral range from the infrared to the far ultraviolet; they are non-transparent.* We know from page 431 that the absorption of solids due to lattice vibrations is limited to the infrared spectral range. The absorption leading to electron excitation and electron separation, however, requires wavelengths that are the shorter the stronger the absorbing valence electrons are bound to their lattice constituents, i.e., the better the crystal (without external excitation) insulates electrically because of missing mobile electrons. *The transparency of inorganic insulators thus is causally related to their insulating property*[1]. In metals, on the other hand, the absorbing quasi-free electrons are also responsible for the electric conductivity, whereas in semiconductors the absorbing electrons are so loosely bound that they are released by a small temperature increase and then are able to conduct the current. In both cases we expect and find absorption spectra that consist of broad overlapping continua. *Good electronic conductors thus are non-transparent because of the weak or even missing binding of their electrons.* In this respect they are related to a highly ionized plasma in which the electrons are not bound any more to their ions as a consequence of the high interatomic field strength (cf. III, 21). In the ideal case, such a plasma emits that continuous spectrum which is known from discharges of high current density.

c) Energy and Charge Transport in Solids.
Electrons, Positive Holes, Excitons, Phonons, and their Significance

We shall now discuss an important property of the electron excitation in insulating crystals which is due to the fact that the crystal always has a large

---

[1] The fact that the synthetic materials, widely used today for insulating purposes, are *not* transparent does not contradict our proposition. These insulating materials are not solids proper but compressed molecular materials in which the light is predominantly scattered and not absorbed.

number of identical constituents. If an electron of one of these constituents is excited, we have an exchange degeneracy as treated in IV, 11 since the energy state of the crystal is independent of *which one* of the numerous identical lattice constituents is excited. Therefore the electron excitation may be handed over from the originally excited constituent to a neighboring one etc. through the entire crystal. This is the more probable the stronger the coupling is between the equivalent atoms, i.e., the more their electron eigenfunctions overlap. This migration of excitation energy over long distances is a well-confirmed empirical phenomenon. It may be compared to the transfer of excitation energy in a collision of the second kind (cf. III, 6a) from one atom to another where the collision couples the electrons of the atoms which exchange their energy. In the closely packed solids, the interaction is always present.

Before we consider this migration of excitation energy through a crystal from a somewhat different standpoint, we introduce a new concept which is essential for solid-state electronics but involves a conceptual difficulty. It is the *positive electron hole* or *defect electron*. In order to bring out the essential point, let us assume a large volume filled with electrons in such a way that they form a solid lattice in which each electron has the same distance from its neighbors. We now assume that an electron is missing at a certain place in this electron lattice so that an "electron hole" exists. If we bring this electronic crystal into an electric field, the electron on the negative side of the hole will follow the field and fill the hole, the next neighbor will follow, and so on. This consecutive motion of the electrons obviously causes a step-like displacement of the hole itself toward the negative electrode. *The electron hole thus moves in an electric field as if it had a positive charge.* The concept of the moving *positive* hole thus represents a simplified way for describing by means of the hole the actual movement of the *negative* electrons that follow one another step by step. A positive hole has, of course, also an apparent effective mass which depends on the forces that act upon the hole due to the surrounding lattice with its ions and electrons. *Since the electrons in their normal states always occupy the lowest available energy states, they push the defect electrons (positive holes) into the highest possible energy states.*

Let us now assume that an electron hole was produced when the electron which occupied this position before was removed from it by external action, and that it is now migrating in the neighborhood of its original position via interstitial lattice places. Since this electron is repelled from all lattice places occupied by electrons except for the electron *hole* which does not carry any negative charge, the hole affects the electron in this respect as if it had a positive charge. If the electron succeeds in moving into the hole, this free interstitial electron disappears simultaneously with the positive hole. This process is comparable to the simultaneous disappearance of an electron and a positron in the pair annihilation by radiation (cf. V, 21). We speak of the *recombination of an electron and a positive hole*.

The recombination of electrons and positive holes requires, just as the analogous process between electrons and positrons or positive ions, that the released energy and the excess momentum are somehow dissipated. Since this is not always possible, it often happens that in a solid an electron orbits about a positive hole without being able to recombine. Such a bound electron-hole system is somewhat similar to the positronium treated in V, 21. It is called an exciton.

If we now turn from our model lattice of electrons to the real crystal, nothing is changed in principle since the lattice of positive ions which was disregarded in our previous approximation merely compensates the charge of the electrons. If now, for instance by light absorption, a valence electron is freed from its bond so that it may move around in the lattice, a positive hole (defect electron) is created at its original position. According to our description above, this hole migrates when in an electric field neighboring valence electrons successively move into the hole. The positive holes thus participate in the charge transport. If, however, the valence electron is not completely separated from its ion by light absorption but is only excited, this process obviously means, in our new way of speaking, *the production of an exciton*. The above-described migration of excitation through a crystal from one atom to a neighboring identical one may then be described as the *migration or diffusion of excitons*. Since these consist of coupled negative and positive charges, their movement is independent of any electric field, and it does not contribute to the charge transport.

An exciton may, moreover, dissociate. It may split into an electron and a positive hole (which both follow the field and contribute to the charge transport) if the energy required for disrupting their bond is available. The energy may be supplied as thermal energy by lattice vibrations or, in the particle language, by absorption of phonons (cf. page 431). Vice versa, in the recombination of electrons and positive holes energy and momentum may be conserved by *emission* of phonons or, in the conventional manner of speaking, by excitation of lattice vibrations. In thermal equilibrium, a temperature-dependent fraction of the produced excitons will thus always be dissociated.

The above-mentioned mobility of excitons and defect electrons in the crystal is subject to an important restriction. This migration, because it is due to energy resonance and exchange degeneracy (cf. IV, 11), can occur only if the excited electron which is responsible for the production of the exciton, or the separated electron which is responsible for the positive hole, belong to a normal lattice constituent with numerous identical neighbors. If, however, we have an impurity atom in a diamond lattice, its excitation or ionization leads to an exciton or positive hole of such an energy that energy resonance with the neighboring C atoms is not possible. In this case, the exciton or positive hole cannot migrate. These "particles" then are localized; they have fixed positions. This consideration about the conditions for mobile or localized excitons and positive holes will prove to be important for the understanding of electronic semiconductivity.

d) The Interaction between Crystal Lattice and Electronic Processes. Electron Traps

In our previous discussion, we took into consideration the reaction of the crystal lattice to electronic processes only in so far as the recombination of electrons and positive holes by emission of phonons was concerned. Now we must discuss the problem more thoroughly. A crystal lattice, like a molecule, is in a dynamic equilibrium in so far as the positive ions are located in the potential minima given by the arrangement of the negative valence electrons, and vice versa. Just as in general the bond and with it the internuclear distance of a molecule in equilibrium is changed by the excitation or separation of a molecular electron (cf. VI, 5), excitation or separation of a valence electron in the crystal lattice disturbs the equilibrium arrangement of the lattice constituents in its environment and with it the lattice symmetry. The situation is described by the Franck-Condon principle introduced in VI, 6c. According to this principle, *the lattice re-arrangement to a new equilibrium after electron excitation or electron separation occurs after the electronic process has been completed.* The electronic process itself therefore occurs without a change of the internuclear distance or of the velocity of the heavy constituents. In general, the excitation or separation of a valence electron causes a loosening of the bond between the two constituents in question, just as in molecules, so that their potential energies for the normal state and the state with excited or separated valence electrons may be represented, in analogy to Fig. 192b, by a diagram like Fig. 238. The optical energy required for exciting or separating an electron here is indicated by the vertical arrow. We see that this energy is considerably larger than the energy difference of the two potential minima which correspond to the states of equilibrium. This latter energy difference is called *thermal* excitation or separation energy because it represents the energy necessary for exciting or separating an electron in thermal equilibrium. We thus learn from the Franck-Condon principle *that the optical excitation or separation energies to be obtained from the absorption spectra are considerably larger*

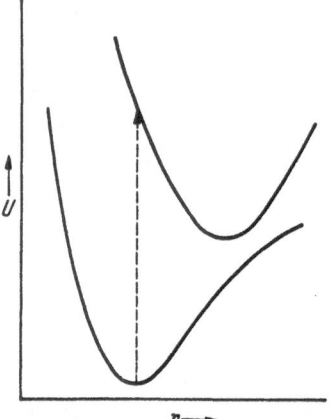

Fig. 238. Potential curves for explaining the difference between optical and thermal excitation of lattic electrons

than the thermal or equilibrium excitation energies which play a decisive role in statistical computations. In optical excitation, the energy excess is transferred to the lattice as vibrational energy (or, in the particle language, is used for the production of phonons).

Just as these electronic processes are influenced by the lattice, also freely moving electrons, positive holes, and even excitons exert polarizing forces upon their lattice environment. The resulting polarization, in its turn, reduces the mobility of the polarizing charge carriers. This retardation effect is the larger the slower the particles move. It disappears if the polarizing particle travels the distance between two lattice constituents within a time small compared to their vibrational period. An electron or hole, which migrates through a lattice while it polarizes its environment and thus carries along its polarization, is called a *polaron*. The significance of the polaron for the optical and electric properties of ionic crystals is not yet clear. The same is true for a process called *self-trapping* in which an electron or positive hole which moves not too fast polarizes its environment and thus creates a potential well from which it is unable to escape without energy being supplied from the outside.

In contrast to this possible but apparently not very important trapping process, there exist in all real crystals and especially on their surfaces numerous potential wells which may act as traps for electrons or defect electrons of not too high velocity. Lattice vacancies and impurity atoms in the lattice form traps for electrons or positive holes. Lattice dislocations, boundary surfaces between microcrystals as well as external surfaces and structural defects also are places of asymmetrical potential distribution by which electrons and holes may be captured. The life time of free electrons and holes in an insulating crystal therefore is less limited by free recombination (see page 436) than by being captured in traps with subsequent recombination. This is much more probable with *bound* partners because excess energy and momentum may then be removed very easily. We shall come back later to the decisive role of these traps for semiconductor physics and many phenomena connected with it.

11. The Energy Distribution of Electrons in a Crystal. Energy Band Diagrams and Electron Spectra of Solids

After this discussion of the general behavior of electrons in a solid we turn to details of their energy distribution. For this purpose we have to get acquainted with the energy band model with its wide and far-reaching applications. This model may be approached from very different points of view, all leading to the same result.

We consider first the transition from an undisturbed atom to one which is strongly perturbed by its environment and finally to one which is incorporated in a crystal as a greatly disturbed lattice atom. By this very pictorial approach, we arrive at an essentially correct picture. According to III, 6c, undisturbed atoms have the sharp energy levels of the bound electrons and the continuous energy regions of the free electrons. In the spectra of atoms which are disturbed by the microfields of their environment, the broadening of the higher energy states of the electrons as a result of the perturbation is already very noticeable (cf. III, 21). For the highest energy states just below the ionization limit, it can no longer be decided whether these are to be considered as discrete energy states of bound electrons or as continuous energy regions of free electrons. Finally, if we go to the lattice constituents in a crystal, we expect a different behavior, aside from the increasing magnitude of the perturbation and broadening of the energy states, only insofar as the perturbing centers are now arranged in a regular lattice.

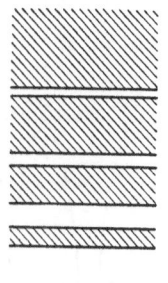

An electron which was released from its original atom now moves "quasi-freely" throughout the crystal in the periodic potential field of all lattice ions. We have thus arrived, in a pictorial way, at the correct picture of the electrons in a crystal. It is evident that by incorporating the atom in a crystal lattice we do not influence the inner electrons which are responsible for X-rays. These innermost electrons remain strongly bound to their respective nuclei; their energy states are practically undisturbed and therefore

Fig. 239. Energy band diagram of the electrons in a crystal

remain sharp. Evidence for this is the sharpness of the X-ray lines emitted by metallic anticathodes (see Fig. 60). The perturbation and thus the width of the energy levels of electrons (cf. Fig. 239) increases greatly with increasing principal quantum number n. For the optical energy levels, which in general are not occupied by electrons, this width reaches several electron volts so that one can readily speak of energy bands of the electrons which are now quasi-free (see Fig. 239). We shall see that these quasi-free electrons are responsible for the electronic conductivity of the metals (cf. VII, 13). For the highest energy levels, the band width occasionally becomes so large that there is an overlapping of the different bands. We shall discuss this when we take up the theory of metallic conductivity.

It is interesting to show how one arrives at the same result by an entirely different approach. For that reason we mention two other concepts. They are less pictorial but they have the advantage that they permit an exact computation of the behavior of crystal electrons.

One approach is based on the concept of resonance or exchange splitting which was discussed in IV, 11. Quantum mechanics shows that in the case of two coupled atomic systems of equal energy their common energy state (disregarding a shift) splits into two energy states as a result of this energy resonance. The separation of these new states is the larger the stronger the coupling between the two systems is. This case occurs in crystals where in principle an exchange of electrons by any two lattice constituents is possible because of their complete identity. Thus, if the crystal consists of N atoms, as a result of the possibility of exchange of each electron with the $N-1$ analogous electrons each energy state of the atoms in the crystal splits up into N energy levels. We distinguish these N energy levels of every energy band by quantum numbers k. Each k state can be occupied by two electrons of opposite spin. *The magnitude of the splitting, and thus the width of the band resulting from the N levels, depends on the degree of coupling, i.e., on the exchange probability of the electrons.*

From IV, 11 we know how the splitting of the originally degenerate energy states depends on the coupling of the interacting atoms and the corresponding exchange frequency. The energy splitting is equal to the exchange frequency multiplied by h, and this frequency may be computed in principle from the theory of the tunnel effect (see IV, 12). In the solid, the width of the energy band originating from the N states takes the place of the splitting of two energy states.

Now we discuss in detail the possibility and probability of an electron exchange between the atoms of a crystal lattice by means of the potential curve diagram of Fig. 240. In Fig. 91, page 137, we had represented the potential near a hydrogen nucleus together with the energy states of the electron which moves in this potential field. In a solid, because of the geometrically regular arrangement of the atoms, we have a three-dimensional periodic repetition of potential wells and potential walls, which are indicated in Fig. 240. The innermost electrons of the crystal atoms lie

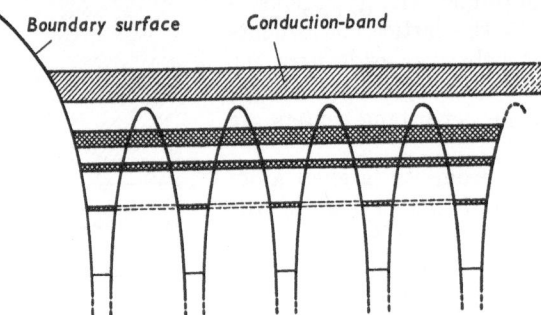

Fig. 240. Periodical potential field in a crystal with electron states, excitation bands, and ionization (conduction) band

almost completely within the potential wells of their ions. An exchange of these electrons with corresponding electrons of other atoms is impossible classically. Because of the tunnel effect, it is not quite impossible according to quantum mechanics, but it is very improbable because of the height of the potential wall which must be penetrated. Because of the small exchange probability, the splitting of the energy levels and thus the width of the energy bands of the innermost electrons is very small, in agreement with our first picture and with experience. Conversely, the exchange probability and the width of the energy bands resulting from the energy levels is very large for the quasi-free valence electrons. From these considerations we understand that the band width is the larger the more the lattice atoms interact. This interaction depends on the average distance of the lattice constituents (measured in units of their electron shell diameters); it thus depends on temperature and pressure. Fig. 241 shows schematically this dependence of the band width on the distance of the lattice constituents for several bands of different energy. This is of significance for the understanding of numerous solid-state phenomena. The picture of the energy band model which follows from these more refined considerations thus agrees well with our first merely pictorial derivation. But in addition it teaches us that the energy bands theoretically consist of discrete energy levels and appear only continuous because of the immense number of these levels, which is equal to the number of lattice constituents, i.e., 6×10^{23} per mole of the crystal.

We understand the consequences of the important energy band model even better by starting from the motion of *free* electrons in the crystal. This third approach, therefore, is a good approximation for metal electrons only. Here we try to show the significance of the quantization of the electron states within an energy band and of their characterization by a number k. Furthermore, we shall understand why there are energy ranges between the energy bands that are "forbidden" for the actually not completely free electrons. If we think of the entire crystal not as of a huge number of coupled systems (atoms) but wave-mechanically (see VII, 3) as *one* system, then the eigenfunctions of this system, as BLOCH has shown, are propagating plane electron waves which may be represented as sine waves by the formula

$$\psi_k = e^{2\pi i k r},\qquad (7.17)$$

which is identical with Eq. (4.47). The influence of the potential field of the ionic lattice may at first be neglected. By combining the expression (7.17) with the Schrödinger equation (4.46) for a plane wave of *free* electrons we obtain

$$k^2 = \frac{2m}{h^2} E \qquad (7.18)$$

for the relation between k and the electron energy E. If we take into account that the electron momentum p is related to the energy E by

$$E = \frac{p^2}{2m} \qquad (7.19)$$

and that the electron momentum is connected with the electron wavelength through the de Broglie relation (4.27), we find

$$k = p/h = 1/\lambda. \qquad (7.20)$$

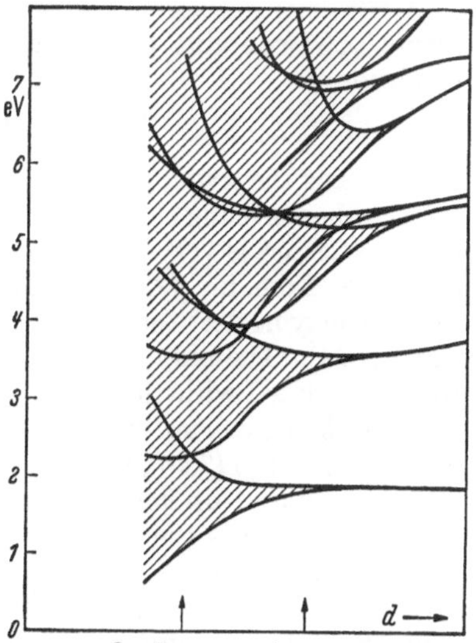

k thus is proportional to the momentum (or the velocity) of the electrons; it is equal to the wave number of the corresponding de Broglie wave of the electrons. Since the electron wavelength depends on the lattice

Fig. 241. Transition from sharp energy states of atoms to increasingly broadened energy bands and, finally, to a uniform energy continuum due to increasing interaction of the atoms with decreasing interatomic distance. The two vertical arrows indicate the relatively small interaction and, therefore, small band width of the lower bands of insulator crystals, and the large interaction with corresponding overlapping of energy bands of metals. Curves according to model calculations by SLATER

symmetry and, therefore, in general on the *direction* of propagation, k is, exactly spoken, a vector as is p; it is called *wave number vector*.

In order to understand the quantization of the k states of an energy band, we consider a crystal cube with the edge length $N^{\frac{1}{3}} d$, where N is the number of atoms in the crystal and d the distance between two consecutive lattice planes. It is evident that a stationary electron state of the crystal is possible only if every electron wave has the same phase on both crystal boundaries, exactly as if beginning and end of the crystal (in a one-dimensional description) would be connected to a circle. This is possible only if the length of the edge of the crystal cube is equal to an integral multiple of the wavelength,

$$N^{\frac{1}{3}} d = n\lambda = n/k; \qquad n = 1, 2, 3, 4, \dots . \qquad (7.21)$$

The wave number

$$k = \frac{n}{N^{\frac{1}{3}} d} \qquad (7.22)$$

therefore must have discrete values, and it can then be used as a kind of quantum number for characterizing the N states of every energy band.

Actually, the electrons are not free in the crystal. They move in the periodic potential field of the ions (see Fig. 240) so that the relation (7.18) between k and E cannot be valid exactly since it was derived from the potential-*free* Schrödinger equation. The electron waves cannot be represented any more by Eq. (7.17); they are described by the more complicated expression

$$\psi_k = \psi_k^0(r)\, e^{2\pi i k r},\qquad (7.23)$$

where $\psi_k^0(r)$ contains the lattice periodicity, which means that it satisfies the condition

$$\psi_k^0(r) = \psi_k^0(r + na);\qquad n = 1, 2, 3, \ldots,\qquad (7.24)$$

with a being the lattice constant. What is the influence of such a periodic potential field on the traveling electron waves? As long as the velocity v or the momentum p of the electrons is very small, the de Broglie wavelength of the electron waves, as computed from Eq. (7.20), is large compared to the lattice constant a. In this case, we may neglect the specific influence of the lattice periodicity on the electron wave (7.23). The only effect of the periodic potential field then is that the electrons associated with the waves (7.23), at a given energy E move with a *different* velocity than was expected according to Eq. (7.19). If in spite of this we want to use Eqs. (7.17) to (7.19), we have to ascribe to the electrons an *effective mass* which differs from that of free electrons. For metals for which this approximation is valid, m_{eff} becomes the larger the smaller the probability is for surmounting the potential walls between the minima, or the smaller the width ΔE of the energy band of the electrons and the lattice constant a are, i.e., the shorter the distance is in which the potential minima follow each other. We then have, as always for electrons of small kinetic energy according to BLOCH, for dimensional reasons

$$m_{\text{eff}} = \frac{h^2}{8 a^2 \Delta E}.\qquad (7.25)$$

Making use of this effective electron mass is a way of taking into account in a simple manner the effect of the periodic crystal field. In insulators and semiconductors, m_{eff} may be smaller than m_0 and may even be negative, and this for electrons as well as for holes. The reason is that considerable deviations from the simple relation (7.19) between momentum and energy of the electrons occur if the wavelength of the traveling electron waves approaches $2a$, i.e., twice the lattice constant. This may be understood by considering the critical case $\lambda = 2a$. For this wavelength, the waves are reflected by all lattice points with a common phase shift against the incident wave so that a standing wave instead of a traveling one results. The critical wavelengths for which Bragg reflection occurs [cf. Eq. (7.1)] and therefore standing waves result are

$$\lambda_{\text{crit}} = \frac{2a}{n}\quad \text{with}\quad n = 1, 2, 3, 4, \ldots.\qquad (7.26)$$

An electron motion in the crystal, therefore, is *impossible* for the electron momenta

$$p_{\text{crit}} = h/\lambda_{\text{crit}} = h n/2a\qquad (7.27)$$

or the wave numbers

$$k_{\text{crit}} = 1/\lambda_{\text{crit}} = n/2a.\qquad (7.28)$$

The surprising result thus is: *The electron eigenfunctions for all non-critical k quantum numbers represent traveling waves; but those for the discrete critical k numbers (7.28) or electron momenta (7.27) represent standing waves.*

The situation becomes even more remarkable if we ask for the relation between electron momentum and energy for the critical momenta (7.27). It is evident that two standing waves of *different* energies belong to the same electron momentum p_{crit}. If for instance, $\cos x$ represents a wave the nodes of which coincide with the maxima of the potential, then the nodes of the wave $\sin x$ which belongs to the same value λ_{crit}, coincide with the minima of the potential. According to IV, 6, the squares of the ψ amplitude correspond to the probability that an electron occupies a certain volume element. We therefore conclude that for the cosine wave the electrons predominantly occupy regions of small potential energy, whereas for the sine wave we find them in regions of high potential energy. The energy E of the cosine wave, therefore, is the smaller (compared to that of the sine wave) the larger the potential energy is in comparison to the total energy.

This may be seen clearly by plotting the energy E of the lattice electrons against their momentum p or their wave number k. For the unperturbed traveling electron waves without any potential field, we have the dashed parabola of Fig. 242. But as a consequence of the periodic potential field of the lattice ions, discontinuities occur at the critical momenta (7.27). This means that forbidden energy ranges exist which are the broader the more marked the potential maxima are, or the larger the difference is between the sine waves and the cosine waves which belong to the same electron wavelength. Because of these discontinuities, the continuous sequence of possible energy states of the free electrons in a periodic potential field changes into a number of "allowed" energy bands which

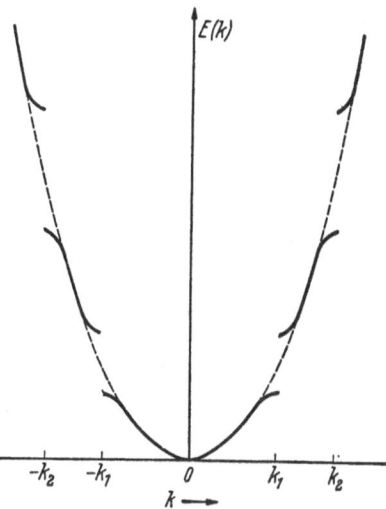

Fig. 242. Relation of electron energy E with wave number vector $\vec{k} = \vec{p}/h$ for undisturbed electron waves (dashed line) and for electron waves in a crystal lattice (solid lines), with discontinuities at wave numbers for which Bragg reflexion occurs

are separated by gaps or forbidden energy ranges. The k ranges which correspond to the allowed energy bands are called *Brillouin zones* for reasons to be discussed below. In order to get from one energy band to another one with different eigenfunctions, the electrons have to undergo transitions that are completey analogous to the transitions from one electron state of an atom to another one.

The critical electron momenta at which the discontinuities occur depend on the lattice constant a, according to Eq. (7.27). Since the distance of two consecutive equivalent atoms in a crystal in general depends on the direction of propagation of the electron waves in the crystal, the energy band diagram is different for different directions of the electron motion in the crystal and depends on its specific lattice structure. These relations are graphically represented by the so-called Brillouin zone diagrams of which Fig. 243 gives an example for the simplest possible case of a two-dimensional square lattice. These zone diagrams are representations in k space or, in our two-dimensional example, in the k plane. Since k, according to Eq. (7.20), is equal to the inverse electron wavelength, i.e., to the electron momentum p divided by h, every point of the k space corresponds to an electron momentum well defined with respect to magnitude and direction from the origin of the coordinates to the point in question. Lines in the k plane of Fig. 243 mark those k values for which the corresponding electron wavelength

$\lambda = 1/k$ has one of the critical values, viz., $k=1/2a$, $2/2a$, $3/2a$, etc., for which values the electron waves are reflected by the lattice. If electrons in a crystal are being steadily accelerated from zero, for instance, by an electric field, then higher momenta than those corresponding to the lines which separate the central white square of Fig. 243 from the outside cannot be reached because the electrons are reflected by the lattice planes as soon as they reach this critical value of the momentum. This inner white square is called the *first Brioullin zone. It corresponds to the first energy band of the quasi-free electrons, which the electrons, as mentioned above, can leave only by electron transitions and not by steady acceleration.* The boundary of this Brillouin zone therefore indicates the k value (or p value) for the energy gap beginning at the upper edge of the first energy band. The zone diagram shows just as clearly that by acceleration in the directions x or y the electrons may reach the maximum momentum

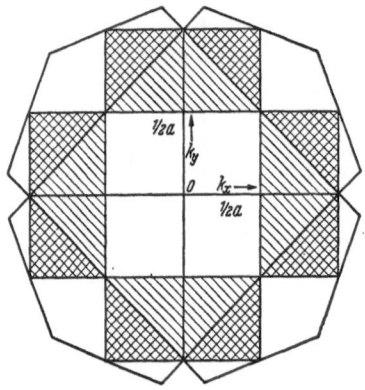

$$p_{max}(x) = h\,k_{max}(x) = \frac{h}{2a}, \qquad (7.29)$$

but by acceleration in a direction of $45°$ against the x axis the higher value

$$p_{max}(45°) = h\,k_{max}(45°) = \frac{h\sqrt{2}}{2a}. \qquad (7.30)$$

Fig. 243. The first four Brillouin zones of a quadratic two-dimensional lattice. (After SEITZ)

Similarly, critical momenta may be obtained from zone diagrams for more complicated lattices. These diagrams therefore are as important for the computation of electron diffraction patterns as for the understanding of the energy band diagrams of electrons in more complicated crystals. This becomes clear if we restrict ourselves to the x axis or to the y axis since here the single-hatched parts of the zone comprise the k values between $1/2a$ and $1/a$, positive as well as negative. BRILLOUIN has shown, as may be seen geometrically from Fig. 243, that the area of the k plane of every zone is the same. This is evident for the third Brillouin zone, indicated in Fig. 243 as a double-hatched area. It is obvious that for three-dimensional crystals also the Brillouin zone diagrams have to be extended to three dimensions, although they thus lose their pictorial clearness.

After this review of the energy bands and the different possibilities of representing them, we ask how the $2N$ k-states that build up an energy band of a crystal with N atoms, each of which may be occupied by one electron, are distributed over the width of the band. Since in our first method of representation (see page 439) we looked at the energy bands as a consequence of the broadening of the sharp energy states of the separate crystal constituents, we expect a more or less bell-shaped distribution of the k states with a maximum energy-state density in or near the band center. The intensity distribution in a spectrum due to a transition from such an energy band to a sharp inner level should therefore be similar to that of a broadened spectral line (cf. Fig. 89), and indeed is. Just as the splitting of terms which lead to line broadening, also the splitting of the k states is not necessarily symmetrical to the undisturbed term. Therefore the distribution of the k states over the band width is not always symmetrical to the band center. Near the lower or upper band edge, the energy state density $N(E)$ depends parabolically on the distance from the band edge, as can be seen from the electron distribution in a solid according to Fermi statistics (cf. Fig. 110). The resulting energy-state distribution is indicated in Fig. 244.

As the selection rule for optical transitions between energy bands, theory predicts that transitions between *all* the different energy bands (different n values) are allowed, but with the important restriction that in optical transitions the wave number vector k and with it the electron momentum p must be conserved with respect to amount and direction. The enumeration of k in the energy bands is not always the same. If k increases always from the lower to the upper edge of the energy band (as is the case for energy bands formed by s electrons), optical transitions of shortest and longest wavelength are indicated by the arrows in Fig. 245 a. The width of the corresponding spectral band is equal to the *difference* in the widths of the two combining energy bands. However, if k increases in one band from the lower to the upper edge of the band and in the other band from the upper to the lower edge (as it is the case in certain lattice directions, for instance, for p-electron bands), then the transitions of shortest and longest waves are indicated by the arrows in Fig. 245 b. Here

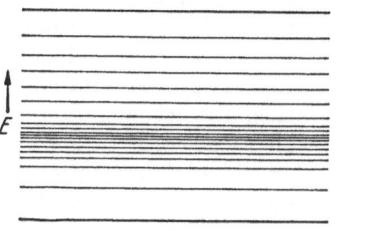

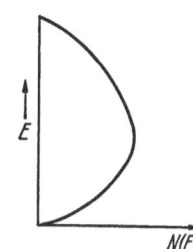

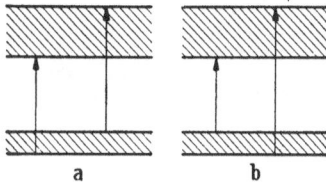

Fig. 244. Schematic representation of the distribution of the k-energy levels over an energy band (left-hand side) and the corresponding distribution function $N(E)$ (right-hand side)

Fig. 245. Different cases of allowed transitions between two energy bands

the width of the spectral band is given by the *sum* of the widths of the energy bands. For metals, for example, both cases may occur.

Conclusive evidence for the width and arrangement of the non-occupied optical energy bands may be derived from X-ray absorption spectra. According to III, 10d, the longwave limit of the X-ray absorption edges originates from raising an electron to an unoccupied energy level near the ionization energy of the atom. Its structure thus reflects directly the energy-band structure in the crystal.

Optical absorption spectra of crystals result from transitions of electrons from the highest occupied energy band into one of the higher unoccupied bands. Since the interaction of the electrons is different for different types of solids, we find broad continuous absorption bands in the ultraviolet and partly in the visible spectral regions for the metals, and very narrow bands, almost lines, for the crystals of the rare earths. This last result agrees well with our pictorial concept: The spectra of the rare earths result from transitions of electrons within the inner $4f$ shell which is shielded by the outer electrons (see Fig. 85) against external influences. These energy bands, consequently, are broadened only slightly. The large number of lines is due to the fact that the selection rules which are valid for isolated atoms become invalid by the influence of the interatomic electric fields in the crystal.

Emission spectra of pure crystals might result from transitions of excited electrons from a normally unoccupied band to holes of a lower band, but these are improbable since the requirement of equal k-quantum numbers for the excited electron and the hole in the nearly filled lower band is almost never fulfilled. We shall come back to the details of this in VII, 23.

12. Fully Occupied and Not Completely Occupied Energy Bands in Solids. Insulator and Metallic Conductor in the Energy-Band Model

We may use the energy-band diagram of a crystal for drawing some conclusions which are as pictorially clear as they are important. Our aim is to give an atomistic explanation of a fundamental property of crystals: Either a crystal is a good electronic conductor or it is, as far as we consider only ideal crystals at low temperatures, an insulator.

In order for electric charge to be transported through a crystal by its electrons, it is necessary that, as a result of the motion of electrons in the electric field to the positive pole, there be an excess of electrons on the positive side and a deficit on the negative side. Now, if the highest band occupied by electrons is completely filled and if we disregard the possibility of an electron jumping to a higher band (since this would require *very* high temperature or optical excitation), then the accumulation of an excess of electrons on one side of the crystal is not possible. The migration of electrons toward one side of a crystal through a fully occupied band is

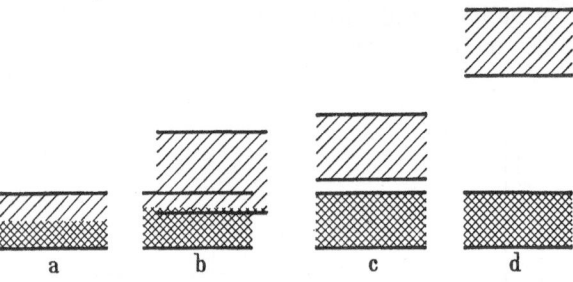

Fig. 246. Arrangement of energy bands for (a) a one-electron metal such as sodium, (b) a two-electron metal such as beryllium, (c) an intrinsic semiconductor, (d) an insulator. Cross-hatched regions of energy bands are completely filled with electrons

possible only if simultaneously an equal number of electrons migrates toward the other side. It is evident that a unidirectional migration of electrons in a filled band is impossible because the electrons have to take energy from the field if they are accelerated by it. However, there are no energy states of higher energy available for these electrons in a filled band. Thus, electron conduction is not possible in crystals with a completely filled highest energy band. These crystals therefore are called insulators. On the other hand, if in a crystal the highest energy band which contains electrons is *not* completely filled, then, according to the above statements, unidirectional migration of electrons in an electric field is possible. Consequently, metallic conductors are characterized by the property that the highest of the occupied energy bands is incompletely filled.

What do we actually know about the occupation of the highest energy bands of the metals? According to our discussion on page 440, each energy band of a crystal consisting of N atoms has room for $2N$ electrons, i.e., for two electrons per atom. According to the Pauli principle, each k state may be occupied by two electrons with oppositely directed spin. We also know that the highest energy band of a crystal is occupied by the valence electrons of the constituent atoms. Consequently, it is clear that the monovalent metals are electric conductors because they have only one electron per atom in the highest energy band, which thus is only half filled (cf. Fig. 246a). However, according to this simplest concept, bivalent metals should be insulators, in contrast to experience. The interaction of the electrons which determines the width of their energy bands (and which, according to page 425, also causes the metallic bond between the atoms in the crystal), however, in metals is often so large that the upper energy bands of the metals partially overlap. Since the stable state to which the system tends is always that of the smallest potential energy, the $2N$ valence electrons distribute themselves in the manner shown in Fig. 246b instead of completely filling one band and leaving the other unfilled. For the sake of clarity, the one band in this figure is shifted against the other. *As a result of this overlapping of the highest bands in a crystal with large interaction of the electrons, the highest occupied bands of the bivalent metals are not completely filled, and these solids thus are electric conductors.*

That the situation presented here is not only a hypothesis but that it corresponds exactly to reality may be shown by investigating the long-wave X-ray emission spectra of metals. These are emitted by transitions from the very broad, highest filled energy band to the next lower electron level, which is scarcely broadened because of the small disturbance of the innermost electrons (cf. page 440) by the lattice. The width and intensity distribution of the emitted X-ray band thus provides direct evidence for the width of the upper energy band and its electron population. Densitometer curves of these X-ray emission bands are shown in Fig. 247 for the monovalent lithium metal and the bivalent magnesium metal. Only half of a bell-shaped curve is clearly visible in the first case, and this is pictorial evidence that the upper band of monovalent lithium is only half filled with electrons. In the case of the bivalent magnesium, on the other hand, it is easily recognized

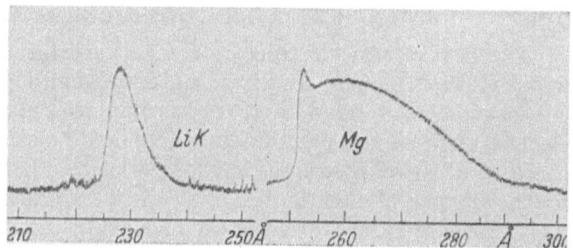

Fig. 247. Densitometer recordings of the X-ray emission bands of a monovalent (Li) and a bivalent metal (Mg), as evidence for the correctness of the band arrangement in Mg according to Fig. 246 b. (After SKINNER)

that the X-ray emission band originates from the transitions from *two* partially occupied, overlapping energy bands to the same lower state. *This is in complete agreement with our discussion given above.* In a similar way the energy bands of insulator crystals may be investigated by means of X-ray emission, and sometimes the energy bands may be attributed to certain electron configurations of the atoms which form the lattice.

We have explained the metallic character of the alkali metals by the fact that the uppermost energy band of the crystal is only half occupied since these metals have only *one* valence electron. This brings up the interesting question why hydrogen atoms do not form a metallic lattice. The answer lies in their tendency to build diatomic spin-saturated molecules H_2 as evidenced by the large dissociation energy of H_2 of more than 4.4 ev. These molecules are bound to each other, at sufficiently low temperature, by van der Waals forces (see VI, 15) in a molecular lattice. Here the lattice constituents are H_2 molecules, and their two valence electrons completely fill the only existing energy band of the crystal. This explains why solid hydrogen has no metallic conductivity and is optically transparent. We may, however, infer from Fig. 241 that *at a sufficiently high pressure (of some 10^5 atm) the molecular lattice of H_2 should change into a cubic closely packed atomic lattice which then should show metallic conductivity and, contrary to the molecular lattice, should not be transparent.*

One more relation about the fundamental difference between metallic conductors and insulators follows from the concept of energy bands. Metallic electronic conduction is always found if the highest occupied energy band of the crystal is not completely filled by electrons (cf. Fig. 246a). If, however, this band *is* completely filled because of a sufficient number of valence electrons of the crystal constituents, then we have to distinguish three basically different cases:

If the highest occupied band and the lowest unoccupied band which corresponds to an electron excitation overlap as in Fig. 246 b, we have metallic conductivity. Conversely, if the completely filled energy band and the first unoccupied band are separated from each other by a gap of several ev (see Fig. 246d) as in diamond, we have a good insulator if the crystal is pure. Most interesting, however, is the case that the two energy bands don't overlap but are separated by a very small gap (cf. Fig. 246c). In this case, electrons may be thermally excited from the completely filled band into the higher empty band if the temperature is not too low. This results in an electronic conductivity of the crystal, which *increases* with temperature. Such crystals are called *electronic semiconductors* and more specifically *intrinsic semiconductors*. Before we discuss them in detail in VII, 20, we first want to treat metallic conductivity due to electrons in half occupied or overlapping energy bands.

13. The Theory of Metallic Conductivity

The behavior of electrons in a metal crystal, which was described in VII, 7 and VII, 11/12, leads directly to an understanding of metallic conductivity and the phenomena related to it. According to FERMI's theory of the degenerate electron gas (see IV, 13), the quasi-free metal electrons have a very large kinetic energy and therefore a large random velocity. The latter is responsible not only for the transfer of electric charge by these "conduction electrons" but also for the transfer of kinetic energy from the high-temperature side of a metal crystal to the low-temperature side, i.e., for the high heat conductivity of metals. The fact that the electric conductivity and the thermal conductivity both depend on the velocity of the electrons explains WIEDEMANN-FRANZ's law according to which the ratio of the thermal to the electrical conductivity is the same for all metals and depends only on the absolute temperature. We have already mentioned the result of Fermi statistics that metal electrons contribute very little to the specific heat of metals, in contrast to classical expectations, but in excellent agreement with experience.

We now discuss the electric conductivity of metals in more detail. If by applying a voltage we produce an electric field E in a metal, the quasi-free metal electrons which according to FERMI's theory possess a high random thermal velocity are accelerated by the field. At the same time, however, a retarding frictional force acts upon them since they move through the lattice of the metal ions. This force, to be discussed later, increases with the electron velocity. The conduction electrons of the metal therefore are accelerated by the electric force eE acting on them only until they have reached a drift velocity v_E in the direction of the field, for which electric and friction force are in equilibrium. The electric current density j in the metal then equals the product of the volume density of the conduction electrons, their drift velocity v_E in the direction of the field, and the elementary charge e:

$$j = n e v_E. \tag{7.31}$$

Since the electric conductivity σ is defined as the current density produced per unit field strength,

$$\sigma = j/E, \tag{7.32}$$

and since the field velocity v_E produced by the unit field strength is called the *electron mobility μ*,

$$\mu = v_E/E, \tag{7.33}$$

the electric conductivity is given by

$$\sigma = e n \mu. \tag{7.34}$$

Measuring the conductivity σ, therefore, yields only the product of the density n of the conduction electrons per cm³ of the metal by the electron mobility μ. For obtaining each of these important figures separately, we make use of the *Hall effect*.

If an electric field is applied in the y direction (see Fig. 248) to a metal plate placed in the xy plane while a magnetic field acts in the z direction, the electrons moving in the y direction are deflected by the magnetic field into the x direction. Thus an excess of electrons is produced at the left-hand edge of the plate carrying the current, which together with the corresponding electron deficit at the right-hand edge causes a potential difference between the two edges of the plate. This is the so-called Hall voltage which can be measured as indicated in Fig. 248.

The electric Hall field strength which corresponds to this electric voltage in the x direction counteracts the magnetic deflection of the electrons and thus leads to an equilibrium in which the magnetic force acting on the electrons and the transverse electric counter force compensate each other. The Hall voltage can be measured only for conducting *electrons*, since for conducting ions the effect remains below the limit of accuracy of measurement because of the low ion mobility. The Hall voltage, therefore, can be used for determining the type of conduction (by

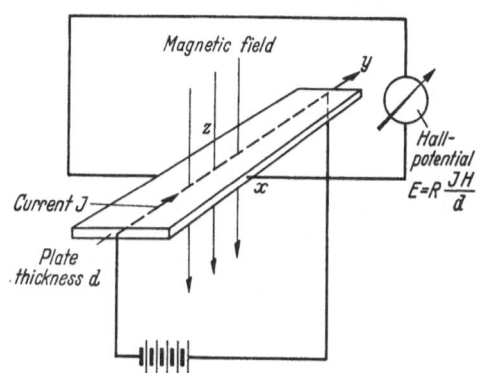

Fig. 248. Arrangement for measuring the Hall voltage of electronic conductors

electrons or by ions, cf. VII, 18) and in addition the charge-carrier density n. The value of the transverse force acting upon an electron of charge e and velocity v in a magnetic field of strength H is

$$K_m = evH,\tag{7.35}$$

which may also be written, because of Eqs. (7.32) and (7.33),

$$K_m = \frac{e\mu jH}{v}.\tag{7.36}$$

In stationary equilibrium, this magnetic transverse force (7.36) is compensated by the electric force (7.37) which results from the Hall field strength E_H:

$$K_e = eE_H.\tag{7.37}$$

Equalizing (7.36) and (7.37) yields

$$E_H = \frac{\mu}{\sigma}jH = RjH.\tag{7.38}$$

All magnitudes in Eq. (7.38), except the so-called Hall constant R, are known or may be measured. The Hall constant, however, may be written, by using Eqs. (7.31) to (7.33),

$$R = \mu/\sigma = v/j = v/nev = 1/ne.\tag{7.39}$$

The Hall constant thus yields directly the number n of the conduction electrons per cm³ and with the measured conductivity σ also the mobility μ. For most metals, the number of conduction electrons lies between 10% and 100% of the number of metal ions in the lattice. This means that in the metals of highest conductivity in the average each atom gives up to the lattice *one* valence electron as a conduction electron. In metals of low conductivity, however, only every fifth or tenth of all atoms contributes a conduction electron. In transition cases between metals and non-metals, e.g., in bismuth, the density of conduction electrons may be still smaller by two or three orders of magnitude.

How can we interpret these results in the light of the theoretical concepts developed in the previous sections? According to VII, 11, the valence electrons in metals are not entirely free. They can oscillate between the potential wells of neighboring lattice ions and may thus travel from one ion to the next one and so forth. The frequency of this electron exchange among the ions is proportional to the width of the energy band occupied by these electrons. We may express this fact by saying that the fraction of electrons which at a certain moment participate in the electron motion is the larger the wider the valence electron band of the metal in question is. The average conduction electron density as determined from the Hall effect is therefore the density of those electrons which, in the time average, participate in the electron drift in an electric field.

The real problem of the theory of metallic conductivity is the explanation of the frictional force which, as mentioned above, acts on the conduction electrons, and its dependance on temperature and lattice structure. If this frictional force would not exist, the electrons would be accelerated by an electric field without limit, and there would be no OHM's law according to which the drift velocity of the electrons in a field is proportional to the field strength [see Eq. (7.33)]. In the particle picture we may interpret the friction of the electrons in the ion lattice as the consequence of their collisions with lattice ions, by which their velocity is reduced. In the equivalent wave concept, the friction corresponds to a scattering of the electron waves by the lattice. As is to be expected, both concepts lead to the same result: In an ideal lattice without any motion of the lattice constituents (i.e., at absolute zero of temperature), there should be no friction. In the particle picture, we can visualize electron trajectories without collisions with lattice ions. In the wave concept, scattering would disappear just as for the propagation of light in an ideal optical medium. To be sure, theoretically there is scattering by the lattice points, but because of the perfect order of the lattice the secondary waves extinguish each other by interference. It is known that the conductivity of metals actually increases with decreasing temperature as T^{-5} so that their resistance decreases while approaching zero temperature proportional to T^5. Nevertheless, even at lowest temperature there is always a *residual resistivity* (disregarding here the special phenomenon of superconductivity, which will be discussed in VII, 17a). This residual resistivity is caused by the always existing lattice defects, by crystal distortions due to internal stress, and by built-in impurity atoms. All these defects produce perturbations of the periodic potential field, indicated in Fig. 240, through which the electrons are moving. These perturbations of the ideal lattice periodicity are also responsible for the residual resistivity of alloys, which is much larger than that of the pure metals because of the different size of the two types of atoms and increases very strongly with increasing degree of disorder of the lattice. In the wave picture, this residual resistivity is due to the scattering of the electron waves by inhomogeneities of the lattice, analogous to the scattering of light by small air bubbles, inclusions, or other inhomogeneities in glass. At higher temperatures, even ideal metal crystals show an electric resistivity which strongly increases with T. This is due to temperature excitation of acoustical lattice vibrations (cf. VII, 8) which cause density fluctuations in the metal, periodic in space and time, by which the electron waves are scattered. The optical analogy to this phenomenon is the Debye-Sears effect, the scattering of light by periodic density fluctuations produced in a homogeneous optical medium by ultrasonic waves. The size of the perturbation of the electron drift by the lattice vibrations may be seen from the change of the mean free path of the electrons. While for non-superconducting metals these are of the order of 1000 Å at the lowest attainable temperatures, at room temperature the mean free path amounts to only a few Å, i.e., to 1 to 2 lattice spacings.

14. The Potential-Well Model of a Metal.
Work Function, Photoemission, Thermal Emission, Field Emission, Contact Potentials

In our discussion of solids we have so far neglected all surface effects. We shall now be interested in these phenomena, in particular in the passage of electrons through the surface of a crystal into free space or into other solids.

In this respect we must think of the conduction electrons as bound not to the individual lattice ions but to the crystal as a whole so that for an electron to escape from the metal energy must be expended against this binding force.

This energy is called the work function W. The potential energy of an electron in free space thus is larger by the amount W than it is in the crystal. We may represent this by a potential curve in the customary manner and thus obtain the potential-well model of a metal (Fig. 249). In this representation we have neglected the periodic fluctuations of the potential in the crystal (represented in Fig. 240) as well as the energy states of the bound inner electrons. Only the energy states of the conduction electrons are indicated in the potential well. An electron thus needs at least the kinetic energy W_0 for leaving the metal.

Fig. 249. Potential-well model of a metal. W = effective work function, W_0 = work function for electrons in the lowest energy state of the conduction band, E_F = zero-point energy of the electrons in the highest occupied electron levels (energy of Fermi surface)

According to the Fermi theory of the electrons in a metal (cf. IV, 13), the highest occupied energy band of a metal is filled with electrons up to the level E_F, the *Fermi surface*, so that the *effective work function*

$$W = W_0 - E_F \tag{7.40}$$

is necessary for releasing from the metal one of the most energetic conduction electrons which possess the kinetic energy E_F even at absolute zero of temperature. We know three different methods for measuring the effective work function. These are based on three effects of theoretical and practical importance, namely the photoelectric emission of electrons, the thermal emission of electrons, and the contact potential between different metals.

If photons with an energy $h\nu$ higher than the effective work function W hit a clean metal surface, they may liberate electrons which then leave the surface with a velocity v which is given by the relation (4.1):

$$h\nu = W + \frac{m}{2} v^2. \tag{7.41}$$

If we use photons of increasingly lower energy, i.e., longer wavelength λ, we reach a long-wave limit at which electrons are just barely released from the metal surface. For this limit we have the relation

$$h\nu_l = W, \tag{7.42}$$

which may be used for determining the *photoelectric work function*. Table 22 presents values for some important metals.

Table 22. *Work Functions (ev) of Several Metals Determined by Different Methods*

	Cs	Ba	Th	W	Pt
Thermal Emission	1,8	2,1	3,35	4,52	5,32
Photoemission	1,9	2,5	3,5	4,57	6,35
Contact Potentials	—	2,4	3,46	4,38	5,36

It is also evident that metal electrons which due to sufficiently high temperature have a kinetic energy higher than the work function W may leave the metal if conditions are favorable, i.e., if the electrons are close to the surface and move toward it. This is analogous to the ability of H_2O molecules to evaporate from the surface of sufficiently warm water. This *evaporation of electrons from incandescent metals* (or semiconductors, see VII, 21) explains the electron emission from hot cathodes as used in a large variety of electronic devices. The electron saturation current, which we measure if all released electrons are drawn off by a sufficiently high voltage, is given by RICHARDSON's equation

$$j = A\,T^2 e^{-W/kT}. \tag{7.43}$$

The theoretical value of the factor

$$A = \frac{4\pi e m k^2}{h^3} \tag{7.44}$$

is 120 if the electron current density is measured in amp/cm^2. It is plausible that the electron emission is proportional to $e^{-W/kT}$ since this term measures the fraction of electrons that have the energy W at temperature T. It is still *a* question, however, why the empirical value of A deviates from the theoretical result if the normal effective work function is substituted for W. The explanation may partially lie in the fact that the W values of different metals depend in different ways on the temperature. Customarily, measurements are represented by formula (7.43) with a *constant* W. If now the logarithm of the measured saturation current j is plotted versus $1/T$, we obtain, according to Eq. (7.43), a straight line, the gradient of which is a measure of the *thermal* electron work function W (cf. Table 22). Not much can be said about the deviations from the values obtained from the photoelectric method. While the longwave limit of the photoemission and the W values determined from it may be falsified in the same way as those from contact measurements (see below) by surface layers or electrons in surface states, the determination of W from RICHARDSON's equation may be disturbed by the above-mentioned temperature dependence of W.

We see from Table 22 that the variation in the work function among the metals corresponds to that of the ionization potentials of the atoms which form the metal (see Table 2, page 21). Cesium, for example, has the smallest work function of all pure metals and likewise the smallest ionization potential of all atoms. This relation might be expected from atomic physics since *the work function may be regarded as the ionization energy of the metal atoms that are perturbed by the integration into the crystal lattice.* Since this perturbation is different in different lattice planes of a single crystal because of the different interatomic distances, *the work function W is not constant but is different for different crystal surfaces.* This effect has been investigated in particular for single crystals of tungsten, where the effective work function of different crystal surfaces varies between 4.2 and 5.6 ev.

The strong dependence of the thermal emission on the work function W which follows from RICHARDSON's equation is used widely in technical applications. Since hot cathodes cannot be made of pure cesium, barium, or thorium, tungsten is being used, which can withstand high temperatures. However, its work function is reduced by depositing on it a layer of cesium or thorium, monatomic if possible, which results in a large increase of the electron emission. If the hot tungsten cathode is operated in an extremely rarefied atmosphere of about 10^{-6} mm of cesium vapor, the desired thin cesium layer on the tungsten surface results from the statistical interplay of condensation and evaporation. For producing a thorium surface, tungsten is mixed in the manufacturing process with thorium oxide and this is reduced by heating so that thorium diffuses toward the surface, thus producing a monatomic thorium layer on the tungsten base. This *thoriated tungsten cathode* has the great advantage that it can be used in high vacuum. Surprisingly, by properly coating tungsten with Cs or Th atoms, a work function is obtained which is only slightly higher than 1.3 or 2.6 ev, respectively, and thus is noticeably smaller than that of pure metallic cesium or thorium, respectively.

This is due to the fact that the Cs and Th atoms adsorbed on the tungsten surface are *polarized* with the positive pole directed away from the tungsten surface. The electric double layer produced in this way deforms the edge of the potential well in such a way that the effective work function is lowered. The evaporation of cesium on tungsten oxide results in a surface with a work function of only 0.71 ev. This is due to a semiconductor effect which will be treated in VII, 21 a.

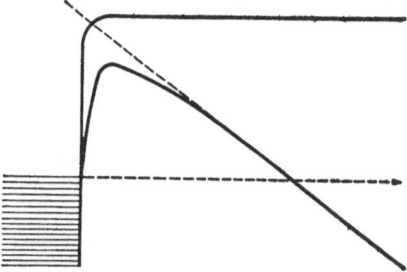

Fig. 250. "Bending" of the edge of a potential well by a strong electric field, leading to field emission of electrons from a metal surface by tunnel effect in the direction of the arrow

Other applications of the potential-well model concern the increase of the thermal electron emission of metals in electric fields (so-called Schottky effect) and the field emission of electrons from non-heated metals in very high electric fields.

We learned in III, 21 that a strong external electric field causes a reduction of the ionization energy of an atom because the potential of the external electric field changes the Coulomb potential of the electrostatic force between electron and nucleus. The result is the "distorted" potential of Fig. 91. The same effect occurs if a sufficiently strong electric field is produced between a metal surface and an electrode outside of it so that the electrons are torn out of the metal electrostatically. According to SCHOTTKY, the work function W then is reduced by the external field strength E to

$$W_{\text{eff}} = W - \sqrt{e^3 E}. \qquad (7.45)$$

For a field of 10^7 volts/cm, which can easily be obtained today, this field correction has the considerable amount of 1.2 ev and thus produces a very large increase of the thermal electron emission since W_{eff} occurs in the exponential of RICHARDSON's equation.

A pure field emission from *non*-heated metals is often observed for very high electric fields. It plays a role in the starting mechanism of certain discharges since sufficiently high field strengths may occur near sharp microscopic metal points. This field emission of electrons according to Fig. 250 is due to electrons of the conduction band tunneling through the potential wall and thus escaping into free space. This explanation of the field emission corresponds closely to the explanation of the pre-ionization of atoms (cf. III, 21) by a distortion of the potential curve and by tunnel effect. According to IV, 12, the penetration probability increases exponentially with decreasing height and width of the potential wall to be penetrated. Therefore, the field emission current increases exponentially with increasing field strength, since the width and the height of the potential wall to be penetrated decrease with increasing field strength because of the increasing distortion of the potential wall (see Fig. 250). Since its height depends on the original work function W (for field zero), provided the other conditions are kept constant, field emission depends exponentially on W also. Actually the somewhat complicated theory of the tunnel effect leads to the following formula for the field emission density j:

$$j = 1.55 \times 10^{-6} \frac{E^2}{W} e^{-\frac{6.9 \times 10^7 \, W^{\frac{3}{2}}}{E}} \ [\text{amp/cm}^2], \qquad (7.46)$$

with E being the field strength measured in volt/cm, and W the normal effective work function of the metal, measured in ev. There seems to be satisfactory agree-

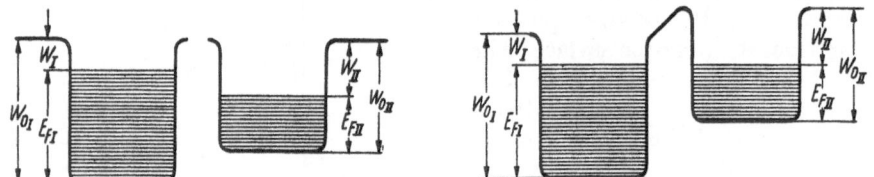

Fig. 251. Potential wells of two different metals before contact

Fig. 252. Explanation of the contact potential between two different metals I and II

ment between this formula and experimental results, provided carefully purified and degassed metal surfaces are used and the increase of the effective field strength above the measured field strength at sharp microscopic points is taken into account. Evidence for the latter influence was provided by measuring the size of emitting spots by means of the electron microscope.

This field emission method was used for obtaining the above-mentioned result that the work function for different faces of tungsten microcrystals differs by as much as 1.4 ev because of the different bond strengths of the electrons. It is evident that adsorbed surface layers influence the field emission since they change the effective work function of the electrons. E. W. Müller investigated this effect thoroughly; he showed that it is possible to utilize the change of the field electron emission by single molecules adsorbed on a pure metal surface for projecting an image of these molecules linearly enlarged by the factor 10^6 on a luminescent screen surrounding the emitting metal point. The further development of this *field emission electron microscope* will be of high interest.

As a last example of the application of the potential-well model, we discuss the contact potential between two different metals which are characterized by their different values of W_0, W, and E_F. If both metals are without excess charges and not in contact with each other, their relative energetic position (cf. Fig. 251) is given by the potential of the outside which we choose as zero. If they now are brought into contact, electrons flow to metal II, because of the higher Fermi level in metal I, until the Fermi surfaces have the same level. This exchange, however, does *not* occur by a flow of very many conduction electrons from I to II but by metal II becoming increasingly negative with regard to metal I due to the few incoming electrons. This negative charge corresponds in the potential-well model to a lifting of the potential well II relative to I until both Fermi surfaces are equal, so that there is no reason for any further electron flow. The process thus does not appreciably change the depth of the Fermi seas filled with electrons (and thus does not change the effective work functions W_I and W_{II}). It only shifts the two potential wells against each other until both Fermi surfaces have the same level. If the two metals are now seperated, we find the situation represented in Fig. 252: Between the metals or two points of their surfaces a measurable potential difference exists, the so-called *contact potential*. Since the potential of a point just outside of the metal surface I is equal to W_I and that of a point close to metal II is equal to W_{II}, this contact potential obviously equals the difference of the two effective work functions divided by the electronic charge e,

$$\Delta U = \frac{1}{e}(W_{II} - W_I). \qquad (7.47)$$

This theoretical result is in good agreement with the experiment, again provided that extremely clean and degassed surfaces are applied. If the effective work function of *one* metal is known from photoelectric or thermal measurements, the measurement of the contact potential allows us to determine the effective work function of every metal that is brought into contact with the known one. Some W values thus determined are presented in Table 22. For the sake of completeness we want to mention that the so-called *Galvani potential*, the potential difference $\frac{1}{e}(E_{F_I} - E_{F_{II}})$ *cannot* be measured directly.

15. The Magnetic Properties of Solids and their Interpretation

In dealing with the magnetic properties of atoms (cf. III, 15), we learnt that their magnetic moments result from vectorial addition of two different individual magnetic moments. These are due to the orbital angular momentum and the spin of the valence electrons of the atom. In a gas, without magnetic field, the magnetic moments of the atoms have random directions because of the random thermal motion of the atoms. In an orienting magnetic field, however, a component of all moments, though in general a small one, gets oriented into the direction of the field. This component according to Eq. (3.103) is directly proportional to the orienting field $\mathfrak{H}$ and inversely proportional to the absolute temperature T. The magnetic susceptibility χ is defined by the relation

$$\mathfrak{P} = \chi \mathfrak{H} \qquad (7.48)$$

between the orienting magnetic field $\mathfrak{H}$ and the polarization $\mathfrak{P}$, which is the resulting magnetic moment $\mathfrak{M}$ per cm³. The value of χ thus depends strongly on the temperature T. Atoms with closed electron shells do not possess resulting magnetic moments. In an external magnetic field they behave diamagnetic because of the inductive action of the field on the electrons. Their susceptibility is negative.

a) Electron Binding and Magnetism of Solids

How does the association of atoms to molecules or crystals influence the magnetic properties of the atoms? The answer to this question follows from the fact that magnetism is caused by the *un*compensated spin and orbital momenta of the electrons. Since, according to VI, 14, the formation of molecules and crystals from atoms is in general connected with a saturation of spin and orbital momenta, the magnetic moments of the atoms in general disappear as a consequence of their association to molecules or solids. We discuss hydrogen and sulfur as examples. The hydrogen atom with its single electron (ground state $2S_{\frac{1}{2}}$) has a magnetic moment of one Bohr magneton. In an H_2 molecule, built up from two

H atoms, the spin momenta of the two electrons, however, compensate each other by forming an electron-pair bond, and the H_2 molecule does *not* have a resulting magnetic moment. Hydrogen thus is diamagnetic. The situation is more complicated for sulfur. According to its $^3P_2$ ground state (cf. Table 10), the S atom has a magnetic moment of three Bohr magnetons. In forming the S_2 molecule, only the orbital momenta compensate each other in this case. The S_2 molecule, according to its $^3\Sigma$ ground state, thus has two uncompensated spin momenta and, therefore, a magnetic moment of two magnetons. Experiments show, however, that solid sulphur is diamagnetic. Obviously, the spin momenta which remained uncompensated in the molecule saturate each other in forming a crystal so that the paramagnetism disappears completely.

Therefore we do not expect to find paramagnetism in solids as a rule, but only in the exceptional cases where the crystal constituents have unfilled electron shells in the solid state, or where the valence electrons do not saturate each other by forming pairs. The best-known examples of the first group are the compounds of the rare earths with their unfilled f shells and of the transition metals with their incomplete d shells, cf. Table 10, page 131, whereas magnetism of the second type is found in nearly all metals. The compounds of the rare earths are of little interest from our viewpoint of solid-state physics, since the f shells which are responsible for the magnetism are situated deep within the electron core so that they are not influenced by forming a solid. The magnetism of this group of solids, therefore, is due to pure atomic magnetism. The perturbation by the surrounding lattice, however, is considerable in the compounds of the transition elements with incomplete d shells. There we find, as a typical solid-state effect, the disappearance of the orbital magnetism, which we will treat in more detail below in connection with the magnetism of metals. In discussing the metallic bond in VII, 7, we noticed already that there is in general *no* compensation of the spins of the valence electrons so that we expect a paramagnetism caused by the spin momenta of the valence electrons. This paramagnetism, however, may be more or less compensated by diamagnetic contributions of the valence electrons and the electrons in closed shells. This happens quite frequently: Whereas the alkali metals, some bivalent metals, and also some heavy metals like molybdenum, tungsten, and uranium are paramagnetic, copper, silver, gold, and the majority of metals of the second and third column of the Periodic Table are slightly diamagnetic. Finally, iron, cobalt, and nickel show the phenomenon of ferromagnetism. What does theory tell us about these problems?

b) Paramagnetism and Diamagnetism of the Metals

If we begin with the simple case of the monovalent metals, we have to distinguish the lattice of the positive ions and approximately *one* quasi-free conduction electron per ion. The electrons of the ions are all in closed shells and therefore, because of their diamagnetism, yield a negative contribution to the magnetic susceptibility. The conduction electrons do not perform stationary orbits around "their" ions anymore, in contrast to their original behavior as atomic electrons. Thus, not only the orbital magnetism disappears in the metal, but there is even a diamagnetic contribution to the susceptibility because of the effect of the field on the random motion of the free conduction electrons. This diamagnetism competes with the paramagnetic (positive) contribution of the magnetic spin of the free electrons. *The observed magnetic behavior of the metals therefore is determined by three different effects: (1) a diamagnetism of the closed electron shells of the positive ions; (2) a diamagnetism caused by the influence of the magnetic field on the motion of the free electrons;* (3) *a paramagnetic contribution of the spin momenta of the free electrons.*

According to PAULI, the theory of paramagnetism of the conduction electrons may be described as follows: Without an external magnetic field, the N conduction electrons of a monovalent metal, according to VII, 12, occupy the lower half of their energy band in which there is room for $2N$ electrons. Each quantum state is occupied by two electrons of opposite spin so that the resultant magnetic moment is zero. If a magnetic field H is applied, primarily one half of all spins is oriented *in the direction* of the field, the other half in the *opposite direction* (directional quantization). The difference of the energy of an electron with a magnetic moment μ_0 in the direction of the field H and one with opposite spin direction is, according to (3.110),

$$\Delta E = 2\mu_0 H. \tag{7.49}$$

This may be represented in the energy band model by Fig. 253, where the energy band is regarded as consisting of two half-bands A and B, each of which contains electrons of *one* spin direction only. According to Fig. 253 b, these half-bands are shifted against each other by the amount ΔE. Since the Fermi energy E_F (cf. IV, 13) of the highest occupied states of both A and B in this case would be different, electrons flow from the half-band A to the lower one B until the Fermi levels of both half-bands are equal. The entire band according to Fig. 253 c then contains more electrons with spins oriented parallel to the field than with

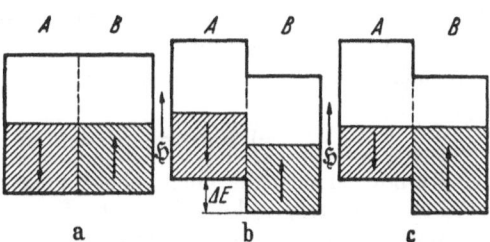

Fig. 253. Variation of occupation of a half-filled band by electrons of different spin directions, theoretically resolved into three steps, viz. (a) orientation of the spin moments parallel to and opposite to the field direction, (b) energy shift of the two half-bands by the energy amount given by Eq. (7.49), (c) equalization of the Fermi surfaces. See explanation in the text

opposite orientation, and the paramagnetic moment $\mathfrak{M}$ of the metal is obtained by multiplying the difference of the two numbers of electrons by μ_0.

In contrast to the temperature-dependent paramagnetism of gases [cf. Eq. (3.103)], *this paramagnetism of the metals obviously is independent of the temperature*, since the orientation of the spin momenta of the degenerate electrons (see VI, 13) is due to the directional quantization in the field and is not affected by temperature.

The value of this paramagnetic moment $\mathfrak{M}$ per cm³ and thus the magnetic susceptibility (see 7.48) depends on the number of excess electrons of one spin direction in the energy band B, i.e., of the number of k states in the energy range ΔE of Fig. 253. This has two important consequences. First, we expect a maximum of the total magnetic moment $\mathfrak{M}$ if the shift ΔE of the two half-bands becomes equal to their widths since then *all* electrons of this band have parallel spins. This case corresponds to *ferromagnetism*, which will be discussed below. Secondly, we realize that quite generally *the number of k-states in the range ΔE, and thus also the number of excess electrons of one spin direction, is the larger the smaller the band width of the conduction band is.*

The theory of diamagnetism of the conduction electrons as a consequence of the inductive effect of the magnetic field on their random orbital motion, first developed by LANDAU and PEIERLS, yields the result that in a first approximation the negative diamagnetic contribution of the free electrons is one third of their positive paramagnetic spin contribution. Observed deviations from this result are caused by the fact that this *diamagnetic contribution is the larger the more freely the electrons move in the metal, i.e., the larger the width of their energy band is.*

We thus expect the following behavior of the three contributions to the resultant magnetism of the metals: The diamagnetic contribution of the lattice ions depends on the number of electrons in closed shells; it therefore increases in going downward in the Periodic Table. The contribution of the free conduction electrons to the resultant metal magnetism is always positive and amounts to approximately two thirds of the paramagnetic spin contribution; it further depends on the effective mass of the electrons (see page 442) and thus may differ from the theoretically expected value in particular if different energy bands overlap.

The fact that the magnetic behavior of the atoms in metals is determined by additive contributions of diamagnetic ions and, in general, paramagnetic conduction electrons explains that paramagnetic atoms (such as Bi) frequently form diamagnetic metals, whereas diamagnetic metals may form paramagnetic alloys. This algebraic addition of comparably large positive and negative contributions to the susceptibility also explains why, with few exceptions, paramagnetic as well as diamagnetic susceptibilities of metals are very small (10^{-6}—10^{-7}). There is good qualitative agreement between theory and observation; but the quantitative agreement is by no means satisfactory.

c) Ferromagnetism as a Crystal Property

We now turn to the phenomenon of ferromagnetism. That this actually is a *solid-state property* follows from the fact that, for example, iron vapor or iron compounds do not exhibit ferromagnetism, whereas the Heusler alloys formed from the non-ferrous metals copper, manganese, and aluminum (Cu_2MnAl) are ferromagnetic. Moreover, certain non-magnetic crystals can be made ferromagnetic just by changing their lattice structures. The most remarkable property of ferromagnetic materials is that even a very small external magnetic field produces a large magnetization. Moreover, the magnetization does not increase steadily by increasing the field. An exact large-scale picture of the hysteresis curve (Fig. 254) shows that the magnetization increases in discontinuous steps, so-called Barkhausen steps. These facts can be explained only by the assumption that even without an external field entire crystal domains are magnetized (spontaneous magnetization) and that the magnetic field only serves to overcome the resistance which prevents these magnetic domains from turning into the direction of the external field.

Since strong local fields must exist near the boundaries of these domains of spontaneous magnetization, their structure can be made visible by fine iron powder which accumulates at these spots and may be photographed. This method yielded the result that frequently not the entire magnetic domains orienting themselves in an external field undergo flip-flop processes, but that those with a spontaneous magnetization in the direction of the external field grow at the expense of neighboring "wrongly" oriented domains. The occurrence of domains is easily understood from the energetic point of view: The magnetic field energy per unit volume is considerably larger for spontaneous magnetization of an entire single crystal than for a crystal consisting of numerous small domains of different directions of magnetization which compensate each other. The state with many domains of different magnetization is therefore thermodynamically more stable than that with a uniform magnetization of a whole single crystal. However, since the formation of every domain requires some kind of surface energy because of the fields existing between neighboring domains, the average size of a ferromagnetic domain is energetically determined by the condition that the sum of the magnetic volume field energy and the surface energy of the domains per unit volume is a minimum.

It is a problem of atomic physics to explain *this spontaneous magnetization of large regions of ferromagnetic substances even without an external field. It can result only from the parallel orientation of the magnetic moments of all or almost all electrons of the highest occupied band or of the highest partially occupied bands in each domain of the metal.*

In treating atomic electron shells, we learnt that energetically it seems to be more favorable if the spin momenta of the valence electrons are parallel instead of compensating each other, since the ground state of atoms according to HUND has the highest possible multiplicity. According to Table 10 on page 131, the spin momenta of the three $2p$ valence electrons of the N atom are parallel; Cr and Mo have six electrons with non-compensated spin momenta and the Gd atom even has eight. As long as different orbital momenta or orientation possibilities of the orbital momentum are available for the valence electrons, as it is the case for the half-occupied p, d, and f shells, parallel orientation of the spins is preferred rather than mutual compensation. However, spin saturation, i.e., the non-magnetic state, is energetically preferred by the two electrons of the H_2 molecule since the electron accumulation between the nuclei compensates for the nuclear repulsion, and the Pauli principle requires that for parallel spins one of the

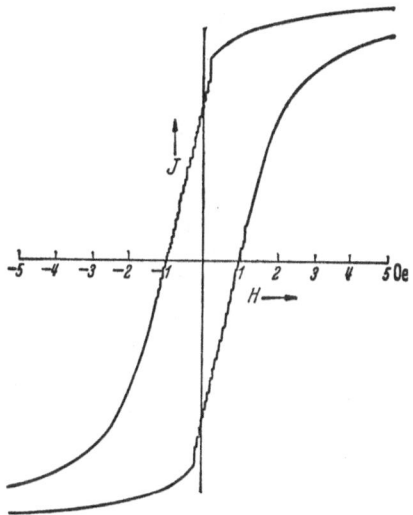

Fig. 254. Magnetization curve with Barkhausen steps. (After BECKER and DÖRING)

two electrons has to go to an excited, much higher state. We thus conclude that parallel orientation of the spin momenta seems to yield a lower potential-energy state than mutual compensation if the latter is not required by the Pauli principle. How about the orientation of the electron spin momenta of ferromagnetic metals? The conduction electrons of iron, cobalt, and nickel, which amount to about 0.6 per atom, do not seem to contribute essentially to ferromagnetism because of their predominantly antiparallel orientation. Ferromagnetism then must be due to parallel orientation of the spin momenta of the majority of d electrons which occupy a non-closed shell. Such a parallel orientation means that each k state, which is normally occupied by two electrons of antiparallel spin orientation, is occupied by only *one* electron so that twice as many k states are necessary for housing *all* electrons. The non-closed d shell does have free k states available, and the amount of magnetization per atom of the various ferromagnetic substances agrees well with the value to be expected if *all* k states are at least singly occupied. If the d electrons occupy *higher* k levels in the ferromagnetic state, their zero point energy is increased. A parallel spin orientation therefore can occur spontaneously only if more energy is gained by this orientation than is lost by bringing the electrons into higher k levels. The energy gained is the quantum mechanical exchange energy. Whether a parallel or antiparallel orientation of the spin momenta of the electrons is energetically preferred for a given configuration of nuclei and electrons, depends on the value of the exchange integral, which was treated in IV, 11,

$$A = \int \psi_a(1)\,\psi_b(2)\,\psi_a(2)\,\psi_b(1)\left[\frac{1}{r_{ab}} - \frac{1}{r_{a2}} - \frac{1}{r_{b1}} + \frac{1}{r_{12}}\right] d\tau. \qquad (7.50)$$

Here a and b indicate two nuclei, 1 and 2 the two electrons of two neighboring lattice constituents. If this exchange integral for eigenfunctions of electrons with parallel spins is *positive* and if it is larger than the above-mentioned increase of the Fermi energy by magnetization, then the state of the crystal with *parallel spins of the electrons is energetically favored compared to the state with compensated magnetic electron moments.* According to Eq. (7.50), A is positive if the average distance between the two interacting electrons, r_{12}, is small and the distances between the nuclei and electrons, r_{a2} and r_{b1}, are large. Accordingly we expect, after SLATER, a spontaneous orientation of the spin momenta of the quasifree electrons in a crystal if two conditions are fulfilled:

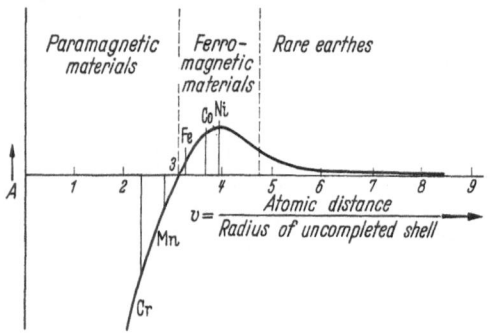

1. The atoms forming the crystal must have uncomplete shells so that the spin momenta of the external electrons are not mutually compensated.

2. The radius of the d or f electron shell in question must be small compared to the lattice distance, since only then the integral over the product of the electron eigenfunctions divided by r_{12} and the exchange integral (7.50) has a large

Fig. 255. Value of the exchange integral (which determines ferromagnetism and Curie point) for a number of ferromagnetic or nearly ferromagnetic metals as a function of the ratio of interatomic distance to the radius of the incomplete electron shell. (After BECKER and DÖRING)

value. At the same time, the small mutual perturbation of the just slightly overlapping electrons causes their energy band width to be small (cf. IV, 11) so that the increase of the Fermi energy by magnetization is small also. Accordingly, ferromagnetism is found only if the distance between neighboring lattice atoms is at least three times as large as the radius of the corresponding electron shell.

Condition 1 is fulfilled for all transition metals in the Periodic Table with partially filled d or f shells, whereas condition 2 applies only to the rare earths and to iron, cobalt, and nickel. Though the exchange integral is positive for the rare earths, its value is so small that even a slight thermal excitation is able to offset the spontaneous orientation of the spin magnets in the crystal. The crystals of the rare-earth elements thus are ferromagnetic only in the neighborhood of absolute zero. Their Curie points, above which their ferromagnetism disappears, is far below room temperature. Only for the metals iron, cobalt, and nickel is the excess of the exchange integral over the increase of the Fermi energy large enough to cause a stable spontaneous orientation of the spin momenta. Only these metals therefore are ferromagnetic up to their fairly high Curie points of 360 to 1000° C. Fig. 255 shows the value of the exchange integral A (7.50) with its dependence on the ratio of the interatomic distance to the radius of the incomplete electron shell. This atomic explanation of ferromagnetism is confirmed by the fact that ferromagnetism can be produced by increasing the lattice distance of a non-magnetic metal, e.g., of manganese by addition of nitrogen atoms.

The spontaneous magnetization (without an external field) of certain crystals can thus be understood on the basis of the atomic theory. That a piece of iron is in general not magnetic without an external field, is due to the fact that this spontaneous magnetization occurs only in small crystal domains of 100 to 10,000 atoms in diameter, and these elementary domains are usually oriented randomly in the metal. An external magnetic field is thus required for orienting the domains and for overcoming the internal resistance against this orientation. This peculiar crystal property has been studied carefully by measuring the hysteresis curves of single crystals of iron. For these studies the external magnetic field was directed along an edge, along a surface diagonal, or the volume diagonal of the cubic iron crystal. Fig. 256 shows the results. If the magnetic field acts along a cube edge (100), the magnetization increases rapidly with increasing orienting field up to the saturation value $\mathfrak{M}_s$. Evidently a small orienting force is sufficient for turning the spontaneously magnetized domains into the direction of the cube edge. If the direction of the field coincides with a surface diagonal (110), the magnetization increases rapidly up to the value $\mathfrak{M}_s/\sqrt{2}$; and with the field in the direction of the cube diagonal (111), up to the value $\mathfrak{M}_s/\sqrt{3}$. It then rises slowly to the saturation value $\mathfrak{M}_s$. This behavior is understandable if we assume that the spontaneously magnetized domains offer little resistance against being rotated into the direction of the cube edge closest to the field direction, but that it requires more force to turn the spins from this preferred direction into the actual field direction. After the external field has been removed, the domains re-orient themselves in the direction of the

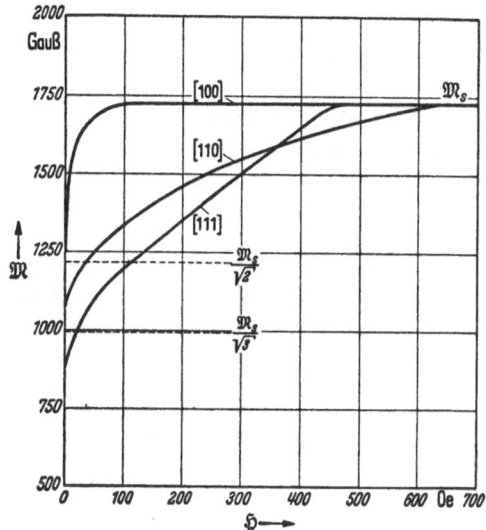

Fig. 256. Magnetization of an iron single crystal for orientation of the field in different crystal directions. (After HONDA and KAYA)

nearest cube edge and remain there if the thermal agitation is not too large. This is the explanation of the well-known remanent magnetism. Surprisingly, the directions of easiest magnetization are not the same for the crystals of Fe, Co, and Ni. For Ni it is the volume diagonal (111), for the hexagonal Co it is the (0001) axis.

For the usual polycrystalline iron, and particularly for forged or rolled iron, the behavior is much more complicated than for single crystals. This is chiefly because the inner stresses produced in working the iron provide preferred directions for the orientation of the domains.

Results of experiments on smallest ferromagnetic crystals indicate that an Fe microcrystal must have at least 64 elementary cells in order to show ferromagnetism. By disregarding the surface atoms, which supposedly are unessential for the spin orientation, this number reduces to 8, i.e., to a body which also seems to be significant as a crystal seed. Here we seem to see the borderline between a mere aggregate of some atoms and a real well-ordered crystal.

A spontaneous orientation of spins in large crystal domains is responsible for *antiferromagnetism* also. If we consider a cubic body-centered crystal lattice in which we assume the exchange integral (7.50) for each atom with its nearest neighbors to be negative, contrary to the case of ferromagnetism, then each two neighbors in the lattice have *opposite* electron spin directions for sufficiently low temperature, just as binding electron pairs in a molecule. The crystal as a whole therefore appears diamagnetic, although it is built up from paramagnetic atoms. We may think of such a cubic body-centered lattice as consisting of two cubic lattices with parallel but opposite spin momenta, which are shifted with respect to each other. We then have two ferromagnetic lattices the effects of which compensate each other because of their opposite directions. For this reason the phenomenon is called antiferromagnetism. Just as for ferromagnetism, above a temperature characteristic for each antiferromagnetic substance, the thermal motion prevents the antiparallel orientation of neighboring spin momenta. *Above this Curie point, antiferromagnetic substances show normal paramagnetism.* The best-known anti-ferromagnetic substances are MnO and $\alpha\text{-}Fe_2O_3$ with Curie points of 122 °K and 950 °K, respectively.

A particularly interesting transition between ferromagnetic and antiferro-magnetic substances are the so-called ferrites. These are oxides of iron and iron metals which crystallize in the so-called spinel lattice. Their characteristic be-havior is caused by ions of the same or related iron metals with different valen-cies. We may explain this for the case of magnetite if we use, instead of the chemical formula Fe_3O_4, the physically more appropriate formula $Fe^{2+}O^{2-}\text{-}Fe_2^{3+}O_3^{2-}$. In other ferrites, the divalent iron ion is substituted by an ion of manganese, nickel, or some bivalent metal. These ferrites do show ferromagnetic behavior as is well known for magnetite since the beginning of all physics. Their saturation magnetism, however, is rather small, in some cases even zero. According to Néel, this is due to the fact that in ferrites one part of the electrons behaves as in ferromagnetism, another part as in antiferromagnetism. The magnetic behavior of magnetite, for instance, may be understood by assuming that only those electrons that belong to the divalent Fe ions have parallel spin momenta, whereas those of the trivalent Fe ions are mutually antiparallel and therefore do not contribute to the ferromagnetism of magnetite. More specifically, it is the spatial arrangement of the ions and their outer electrons that determines, for a given ferrite, the percentage of electrons behaving ferromagnetically or antiferro-magnetically. Such substances are called *ferri*magnetic. Since the ferrites do not possess metallic conductivity and thus have only very small eddy-current losses, they play an ever larger role in high-frequency equipment.

16. Ferroelectricity

In VII, 6, while discussing the piezoelectric and pyroelectric behavior of some classes of ionic crystals, we pointed out that in a few of these crystals we find a spontaneous parallel orientation of the electric dipole moments of whole crystal domains. This electrical analogy to ferromagnetism was called *ferroelectricity*. Because of this parallel orientation of all dipole moments in an external electric field, such *ferroelectrics are substances with an extremely high value of the dielectric constant ε below a characteristic Curie temperature.*

Ferroelectricity was discovered in 1921 by Valasek for rochelle salt ($KNaC_4H_4O_6 \cdot 4H_2O$) in the temperature range from -20 °C to $+22$ °C. It aroused major interest only after the discovery that barium titanate ($BaTiO_3$) is ferroelectric in the entire temperature range below 118 °C. Besides this broad and favorable temperature range, $BaTiO_3$ has two more advantages: One is its relatively simple lattice structure, namely that of perovskite; the other is the possibility of producing, investigating, and using it as single crystals or as ceramic material.

The dielectric constant ε of $BaTiO_3$ is approximately 500 in the temperature range between -100 °C and -70 °C; it is about 800 between -60 °C and -5 °C.

In the most important range from 20 °C to 80 °C, it is nearly constant at about 1200, and it increases up to 6500 near the Curie point at 118 °C. All these values apply for frequencies below 10^8 cps. With increasing frequency, the dielectric constant decreases steadily so that for frequencies of visible light ε reaches the value 5.76, which corresponds to a refractive index of $n = 2.40$ according to the wellknown formula $\varepsilon = n^2$. This decrease of the dielectric constant at very high frequencies indicates that ferroelectricity cannot be due to a polarization of the electrons (which would be able to follow the field even at very high frequencies) but must be caused by a displacement of the heavy ions.

The different values of the dielectric constant in the different temperature ranges correspond to different lattice structures, all deviating but slightly from the cubic structure, and to different orientations of the dipoles in the crystal. Above the Curie temperature, the lattice is cubic; but by cooling it below 118 °C it is transformed into a slightly distorted tetragonal structure. The spontaneous orientation of the dipole moments in this case is parallel to the so-called c axis, which is the longest of the original cube edges. At about 5 °C, the tetragonal lattice is transformed into a slightly orthorhombic one with the dipoles oriented parallel to a face diagonal. Finally, at -70 °C, a transformation of the ortho-rhombic lattice into a trigonal one occurs in which the electric dipoles are aligned parallel to a space diagonal. In addition to these lattice structures, barium titanate may exist at room temperature also in a non-ferroelectric hexagonal structure. The comparison of the two structures at room temperature, therefore, provides evidence of the structure dependence of ferroelectricity.

We mentioned already that ferroelectric crystals, just as ferromagnetic ones, are built up of domains with different spontaneous orientations of their dipoles. In optically perfect $BaTiO_3$ crystals, this domain structure (and its variation if the orienting electric field is changed) may be observed with impressively vivid colors by observation between crossed polarizers.

If the dielectric polarization is measured as a function of the orienting electric field, hysteresis curves are observed similar to those for ferromagnetic materials, including Barkhausen jumps related to the sudden re-orientation of entire domains (cf. Fig. 254). There seems to be a perfect analogy between ferroelectric and ferro-magnetic phenomena, and this leads to the question how to explain the spontane-ous orientation of the dipole moments of whole domains, and why ferroelectricity is so rare a crystal property.

Both these questions are intimately interrelated. According to the lattice structure of $BaTiO_3$ determined by X-ray analysis, each titanium ion is sur-rounded by an octahedron of doubly negative oxygen ions. From the known size of the ions, we infer that the free space within the octahedron of the O^{--} ions is larger than the Ti ion placed in it. Each titanium ion, therefore, has a certain freedom of motion in its octahedron. The position in the center of the octahedron is, however, energetically less favorable than a position near one of the surround-ing O^{--} ions, as is obvious from bond considerations. Since there is a potential maximum in the center of the octahedron, the titanium ions are situated ex-centrically, and this means that each of them forms an electric dipole with the oxygen ion next to it. The decisive point now is that, because of the electrostatic coupling of all ions in a lattice, *all* titanium ions of a whole crystal domain are attached to those O^{--} ions of their octahedrons which are situated in the same direction away from the symmetry center. Since, on the other hand, all eight positions near an O^{--} ion in each octahedron are equivalent to each other, the application of an external electric field induces all titanium ions to jump to the oxygen ions that are next to them in the direction of the field. At a sufficiently high temperature, however, the vibrational energy of the Ti ion exceeds the height of the central potential hill. The thermal motion then destroys the preference of the Ti ion sites above a certain temperature which is characteristic for each ferroelectric: This is the explanation of the Curie point above which ferro-electricity disappears.

This interpretation makes us understand that the phenomena of ferroelectricity are similar to those of ferromagnetism and also that both phenomena require specific lattice structures. Ferromagnetism, however, is a typical quantum effect, whereas ferroelectricity can be understood on a purely classical basis.

17. Quantum Effects of Many-particle Systems at very Low Temperatures. Superconductivity and Superfluidity

In section VII, 15, we learned how the macroscopic phenomena of ferro-magnetism and antiferromagnetism are caused by an orientation of the magnetic spin momenta, and how this orientation follows from quantum mechanics. In this section we discuss two groups of phenomena which occur at very low tempera-tures only and which again are due to a quantum-mechanical special ordering effect. If the conduction electrons are ordered in a special way, this leads to superconductivity. If large aggregates of atoms are aligned in a special way, we find superfluidity. Both phenomena have yielded decisive information about the quantum-mechanical behavior of many-particle systems.

a) Superconductivity

Back in 1911, KAMMERLINGH ONNES observed that the electric resistivity of pure mercury suddenly decreases to an unmeasurably low value if cooled below a transition temperature at 4.2 °K. Since then, this phenomenon, called super-conductivity, has been found for a large number of metals, alloys, and even metal compounds. Table 23 gives some examples of substances with very high transition

Table 23. *Some Superconducting Metals, Alloys, and Compounds with high Transition Temperatures T_c*

Metal	T_c in °K	Alloy	T_c in °K	Compound	T_c in °K
Tc	11.2	Nb_3Sn	18	Nb_2Sb	18
Nb	9.22	BiPb	8.8	V_3Si	17
Pb	7.26	AsPb	8.4	NbN	16
La	4.71	PPb	7.8	NbH	13 — 14
Ta	4.38	AgPb	7.2	MoN	12.0
V	4.3	LiPb	7.2	NbC	10.1
Hg	4.17	AuPb	7.0	Nb_2N	9.5
Sn	3.69	GaPb	7.0	TaN	9.5
In	3.37	SbPb	6.6	TaC	9.2

temperatures T_c. The residual resistivity of superconductors is still unmeasurably small even for the most refined experimental methods, and this means that it is smaller than the room-temperature resistivity by at least the factor 10^{16}. The ratio of the normal resistivity to that in the superconducting state is about the same as the ratio of resistivities of the best insulators to that of normal metals. The decrease of the resistivity from normal conductivity to superconductivity is virtually discontinuous, but only for purest single crystals and at zero external magnetic field. For very imperfect solids, the resistivity may decrease gradually. According to HILSCH and BUCKEL, the transition temperature of disordered amorphous metal films evaporated at very low temperature is markedly higher than that of the same film if the lattice order has been increased by tempering.

Besides its virtually infinite electric conductivity, the superconducting state is characterized by an abnormal magnetic behavior. From the classical standpoint, we expect the application of an external magnetic field to produce undamped eddy currents in a superconductor. The magnetic field of these currents then screens the interior of the superconductor from the generating field. The magnetic field thus is restricted to a very thin surface layer, *whereas the interior of a superconductor has to be free of currents and fields.* Beyond confirming this expectation, however, MEISSNER and OCHSENFELD found an entirely unexpected effect: *In a superconductor, a magnetic field cannot exist at all.* Even if it existed before cooling below the transition temperature, the field is pushed out of the substance as soon as superconductivity begins. *A superconductor is an ideal diamagnetic material.*

Vanishing electric resistivity and vanishing magnetic permeability thus are the two characteristic properties of the superconducting state. This state is destroyed if a certain critical current density or magnetic field strength is exceeded. Both these critical values are zero at the transition temperature T_c, but they increase nearly quadratically with decreasing temperatures below T_c. Furthermore, while at T_c the transition to the superconducting state occurs *without* any transition heat, there is a transition heat if the superconducting state is produced by turning off a super-critical magnetic field.

Superconductivity is undoubtedly an *electronic* phenomenon. Shortly after its discovery, HABER interpreted superconductivity as the formation of an ordered electron phase caused by a "freezing" of the conduction electrons. Not only fits this interpretation with the fact that all superconductors are metals or electronic semiconductors and that properties not due to electrons are not affected by passing the transition temperature. Also the fact that the electronic heat conductivity in the superconducting state is *smaller* than that in the normal conducting state (which at the same temperature may be reached by applying a magnetic field) and that it decreases to zero for $T \to 0$ agrees with this interpretation.

Observations by different methods show that the energy state of these frozen-in superconducting electrons is separated from that of the ohmic electrons by an energy gap which has a width of about $3.5 \, kT_c$ at $T = 0$ and vanishes at $T = T_c$. That the *superconducting electrons are in an ordered state,* may be inferred from the fact that the entropy of a superconductor is smaller than that of the magnetically induced normal conductor of the same temperature. Finally, it is of interest that the transition temperature of superconducting metals is the higher the larger the spatial density of the conduction electrons of the metal is and that, according to MATHIAS, superconductivity occurs mainly in solids whose atoms have 3, 5, or 7 valence electrons. For alloys or compounds, this number refers to the arithmetic mean of the number of valence electrons of the constituents.

There is no doubt that in addition to the electrons also lattice vibrations contribute to superconductivity. This follows from the *isotope effect,* according to which the transition temperature T_c of isotopes of the same substance depends on the mass M of the lattice constituents as:

$$T_c \sqrt{M} = \text{const}. \tag{7.51}$$

After numerous attempts at an atomistic interpretation of superconductivity, BARDEEN, COOPER, and SHRIEFFER were able to show that at low temperature two effects become significant for the interaction of the conduction electrons, viz., the electrostatic repulsion of neighboring electrons and an *attractive interaction of two electrons of opposite spin which is brought about by the lattice.* These energetically very small effects can be neglected in the normal theory of conductivity. The attractive interaction may be understood by assuming that an electron polarizes the lattice through which it moves in such a way that an attractive force is exerted on a second electron of opposite spin. *If this attractive force is larger than the electrostatic repulsion, superconductivity is possible.* Thus we may understand that metals of high normal conductivity, such as the alkalis, Cu, Ag, and Au do not show superconductivity because the interaction between electrons and lattice is too small. The detailed theory yielded the result that the ground state of a superconductor should be described as consisting of pairs of electrons of opposite spin with 10^{-4} to 10^{-5} cm distance. These pairs all have the same momentum which for the ground state of the superconductor is zero. In this state the system is highly stable against external perturbations. Excited states with only slightly higher energy can be reached from the ground state of this many-electron system by energy absorption, e.g., from an applied electric field. These excited states are formed from electron pairs of equal momenta also, but the momenta are now different from zero. The eigenfunction of this system of metal electrons then corresponds to a supercurrent. If, however, individual electrons are excited in the normal manner to states above the Fermi level, the pair bond must be destroyed and this requires an energy in the order of 10^{-4} ev, which corresponds to the empirically found energy gap between superconducting and normal states of the metal electrons. Such individual excitation then results in a system which contains some free electrons of normal conductivity in addition to the collective of highly correlated electron pairs of equal momenta, in full agreement with observation.

An entirely new development was introduced into superconductivity research by the recent discovery of a group of hard, mechanically often brittle superconductors such as Nb_3Sn of Table 23. In many respects these superconductors behave quite differently than was expected from the Bardeen-Cooper-Shrieffer theory. They have smaller energy gaps, which are strongly structure-sensitive, and also the isotope effect is smaller; occasionally energy gap and isotope effect are even missing completely. Most important is that these *hard superconductors* retain superconductivity up to magnetic field strengths that are by orders of magnitude larger than the limiting field strengths of "normal superconductors". Wires of such hard superconductors, therefore, can carry much higher electric current densities than those of the old soft superconducting materials. Although the technological handling of the hard superconductors still poses considerable problems, it has already been possible to produce magnetic fields of 100,000 gauss with coils of hard superconductors, and values of 300,000 gauss may be foreseen. The way to technical applications of superconductivity, therefore, seems to be opened by this discovery of the hard superconductors. It is not yet quite clear how to explain this phenomenon. But there is much to be said for the hypothesis that hard superconductors have a sponge-like structure of normal-conducting material with very fine filaments of superconducting material.

b) Superfluidity

The second quantum effect of a many-particle system observed at very low temperatures is the *superfluidity of the so-called helium II*, a phenomenon which seems to be entirely different from superconductivity. If helium, which at atmospheric pressure boils at 4.211 °K, is cooled below the so-called λ point of 2.186 °K,

this liquid helium, then called helium II, exhibits a number of very peculiar properties which never have been observed with any other substance. These properties
are collectively called *superfluidity*. The viscosity of helium II decreases with T^6
and apparently vanishes completely at absolute zero. Its thermal conductivity is
abnormally high, 10^8 times larger than that of helium I, and it is not any more
proportional to the temperature gradient. Helium II is able to creep without
viscosity as a film of only some 10^{-6} cm in thickness across the edge of containers
or through extremely thin capillaries or cracks (cf. Fig. 257). The amount of
fluid escaping through such leaks per unit time is by many orders of magnitude
larger than the amount of helium gas that under normal conditions diffuses
through them. Although helium II escapes from such containers, the *total* heat
contents of the liquid remains within the container, an
absolutely unique effect. If two vessels, both filled with
superfluid helium, are connected by an extremely thin
capillary, then a temperature increase of about 10^{-3} degrees produces in the warmer of the two vessels a remarkable pressure increase which can be recognized by conspicuous phenomena such as the so-called fountain effect.
This phenomenon disappears if the thin capillary is substituted by a thicker one. Less sensational but nevertheless of no little theoretical importance is the fact that
periodic temperature changes travel through superfluid
helium nearly undamped as *temperature waves*, and this
even at frequencies up to 10^4 cps, in contrast to what is
possible in all other liquids. Since this kind of waves is
different from that of the normal sound propagation, the
established term "second sound" is not very appropriate.
Its mechanism will be discussed presently.

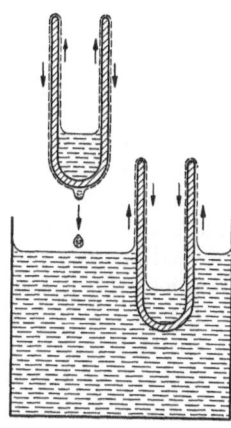

Fig. 257. Schematic representation of an experiment concerning the superfluidity of
helium II

TISZA was the first to point out that the large majority
of these really unique properties of helium II can be
understood if it is assumed that helium II is an interaction-free mixture or *a
solution of two kinds of liquid helium*, one of them being identical with the normal liquid helium I, the other, the ideal superfluid, being distinguished by
vanishing energy content and vanishing viscosity. The real problem for atomic
physics is to explain these last-mentioned properties. TISZA's two-fluid model
is well proven today, and we know from experiments that above the λ point at
2.186 °K the entire liquid consists of helium I, while below the λ point there is
an exponentially increasing percentage of the superfluid helium II, which reaches
99% at 1.0 °K. The two-fluid model, therefore, implies that cooling below the λ
point transforms normal fluid helium progressively into superfluid helium, and
vice versa. Transforming superfluid helium, which does not contain any energy,
into normal fluid helium requires energy, and this explains why the specific heat
of liquid helium, which at 1.0 °K is below 0.1 cal/g degr., increases rapidly to
about 6 cal/g in approaching the λ point above which it decreases as rapidly to
a value of about 0.5 cal/g.

The two-fluid theory explains the majority of the above-described properties
of helium II. Only the real superfluid helium can escape over edges of containers
or through capillaries since it has no viscosity. And since it has zero energy content,
the entire heat content of the initial mixture remains in the vessel. The extremely
high and abnormal thermal conductivity of helium II is not a real diffusion process
but a motion of the superfluid and the normal fluid helium of the mixture in
opposite directions. By this motion, which is similar to osmosis, the normal fluid

helium transports energy to the colder side where it is transformed to superfluid helium and flows back without viscosity. This bi-directional motion, therefore, is due to the fact that in the colder part of the mixture the concentration of the superfluid helium is large, while in the warmer part the normal fluid predominates. This motion and, therefore, the heat flow through helium II is *proportional to the concentration difference of the components and not to the temperature gradient*, in agreement with experience. Finally, the frictionless propagation of heat waves (second sound) is not only in agreement with the two-fluid theory but was even predicted by it before it was actually found by PESHKOV in 1944. Periodic temperature changes of an "emitter" produce in helium II periodic changes of the concentration ratio of superfluid to normal helium. Since these two components move through each other without friction, the concentration changes and the changes of temperature corresponding to them travel as undamped heat waves through helium II. Since the propagation velocity of these heat waves, according to the theory, depends on the concentration ratio of the two components of helium II, its measurement as a function of temperature permits to determine the fraction of the superfluid component in helium II.

What can atomic theory tell us about the nature of superfluid helium and its relation to normal fluid helium? F. LONDON was the first to point out in 1936 that superfluidity may be related to Bose statistics. According to IV, 10, normal He^4 with its two protons, two neutrons, and two electrons is described by a wave function which is symmetrical with respect to particle exchange; it obeys Bose statistics. Since the Pauli principle does not apply to such particles, all He^4 atoms may in principle "condense" in the lowest state of the system composed of them all if absolute zero temperature is approached. In this state they would not carry out any *thermal* motion, but they would nevertheless not form a crystal lattice (in contrast to all other substances) since the zero-point energy of the light helium atoms, according to the uncertainty principle, is larger than their very small van-der-Waals bond energy in the lattice. The vanishing viscosity of superfluid helium, i.e., the vanishing interaction of its helium atoms, would then be due to the impossibility to transfer energy from one atom to another if the thermal energy is smaller than the first energy step of the system, since the energy of the many-particle system ("crystal") of the superfluid helium atoms is quantized. *The atoms of superfluid helium, according to theory, are in the ground state but those of the normal fluid in excited states of the system.* In the superfluid state we thus have to assume an ordered state of all atoms, comparable to that of the electrons of a superconductor.

The relation of the superfluid state to Bose statistics is corroborated by the fact that a liquid consisting of the rare helium atoms of mass 3 does *not* become superfluid. He^3 with its 5 elementary particles (2 electrons, 2 protons, 1 neutron) is described by a wave function which is *anti*symmetric with respect to particle exchange and, therefore, obeys Fermi statistics. Because of this, a "condensation" of all He^3 atoms in the ground state of the system is *impossible*, in contrast to the case of He^4. The difference between Bose and Fermi statistics thus causes a unique difference in the behavior of two isotopes of the same element.

18. Lattice Defects, Diffusion, and Ionic Conductivity in Crystals

In the last sections of this chapter, we shall discuss those properties of solids that depend on structural defects, such as diffusion of atoms or ions and electrolytic conductivity in crystals. In an ideal crystal, each constituent sits in its own potential well and any migration of a particle — for example an exchange of the positions of two particles by selfdiffusion — requires a considerable activation energy because a high potential wall separates the two positions. It is even doubtful whether such an exchange of positions really occurs in solids and whether the spontaneous diffusion is not predominantly due to the progression of whole chains of ions or to the rotation of closed ion rings. In any case, this selfdiffusion depends on temperature exponentially and plays an important role only at relatively high temperatures.

According to three independent groups of observations, however, there is no doubt that ions in solids do migrate far below the temperature required for migration in ideal lattices. First of all, selfdiffusion is confirmed unambiguously for low temperatures by means of radioactive tracers as mentioned in V, 17. The diffusion of radioactive lattice constituents is directly observed by their radiation. It is also well known that two different substances, e.g., silver and gold, when brought into contact, diffuse into one another, and that the diffusion velocity is the larger the less the radii of the two kinds of atoms differ.

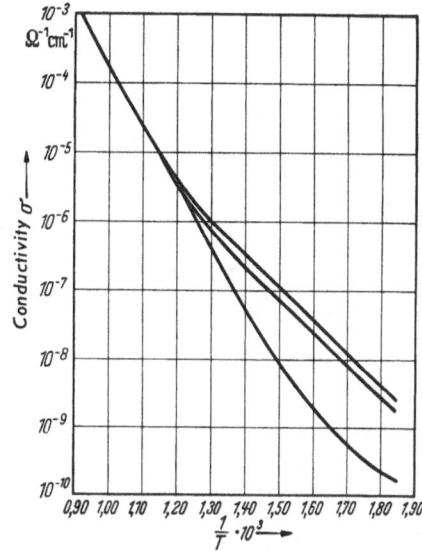

Fig. 258. Temperature dependence of the ionic conductivity σ_i of different rock salt samples. Below 800° K, defect conductivity and, therefore, different behavior of the samples; above 800°K, intrinsic conductivity by the lattice ions. After SMEKAL (simplified)

The second group of observations concerned with the low-temperature diffusion in solids deals with the fundamental processes of corrosion, for instance with the rusting of iron, and therefore is of great technological interest. As soon as metallic iron is covered by a pore-free layer of iron oxide (rust), any further possibility for a *direct* attack of the metal by oxygen is cut off since the metal is separated from the gas by a solid layer. The only way for continued rusting goes via the diffusion of gas ions or metal ions through the crystalline oxide layer. It seems that *metal* ions (and, for reasons of electric compensation, free electrons) migrate from the metal through the solid oxide layer to the surface and there react with gaseous oxygen.

The third group of observations which we regard as evidence for the diffusion of ions in crystals far below the melting point is concerned with the electrolytic conductivity of crystals. Each ionic crystal exhibits electrolytic conductivity. In contrast to the metals, however, in which electrons are responsible for the conductivity, in ionic crystals we observe a migration of ions in the solid with all the properties which are characteristic for electrolytic conduction. Charge transport is associated with a transport of matter to the electrodes, and we find a *negative temperature coefficient of the electric resistance*, which follows directly from the conduction mechanism: The resistivity decreases with increasing temperature because the ions can migrate better if the lattice is loosened up by increased temperature. Fig. 258 shows the logarithm of the ionic conductivity of a number of rock salt crystals versus the reciprocal temperature. The measurements are remarkable in so far as the conductivity of all three NaCl samples is practically the same for temperatures above 800 °K, while each sample has a specific conductivity curve below this temperature.

According to research by SCHOTTKY, WAGNER, FRENKEL, SMEKAL, and others, the diffusion of lattice constituents as well as the low-temperature ionic conductivity are due to *lattice defects* and thus belong to the structure-sensitive crystal properties (cf. VII, 3). Each real crystal has a considerable number of lattice defects, which are not only microscopic cracks, grain boundaries and dislocations but also *atomic* defects, i.e., *empty lattice sites (lattice vacancies)* and ions in *interstitial lattice sites*. Such defects in the lattice arrangement are present in every real crystal independent of the temperature since an ideally regular crystal growth does not occur. The density of the defects increases with increasing temperature, however, because of the loosening-up of the lattice. We shall show below that each crystal in thermal equilibrium has a lattice defect density which depends only on its temperature. That is a consequence of the entropy law according to which the degree of order of *every* thermodynamical system decreases with increasing tem-

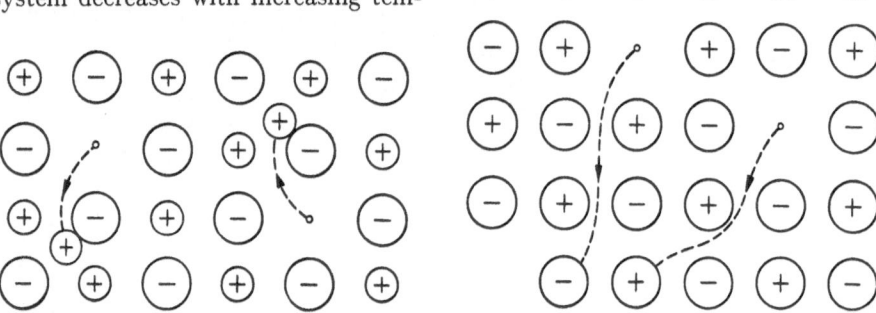

Fig. 259. Positive interstitial ions and lattice vacancies in an ionic crystal: Frenkel defects

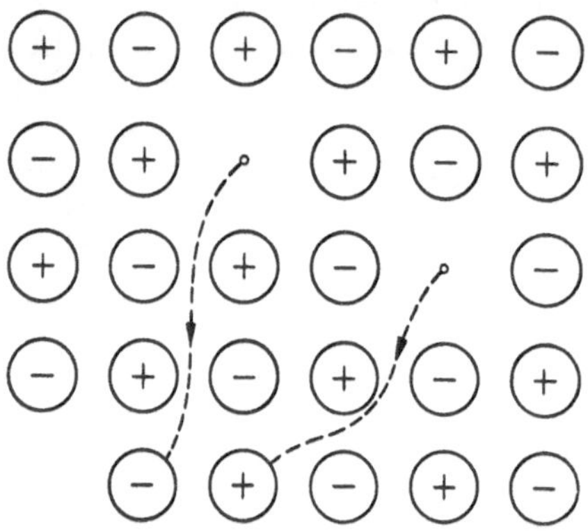

Fig. 260. Positive and negative lattice vacancies: Schottky defects

perature. If a certain defect density in a crystal has been produced by increasing the temperature and if the crystal then is suddenly cooled down, the new equilibrium which corresponds to the lower temperature cannot be reached rapidly enough, and we observe a *"frozen-in" defect density* that corresponds to a higher temperature. The defect density observed at room temperature therefore depends on the thermal history of the crystal.

There are two different types of defect arrangements. According to Fig. 259, normal lattice ions may sit "erroneously" in interstitial places, i.e., they may not have found their correct potential wells but, by deformation of the crystal, have made room for themselves in the neighborhood of their correct positions. By this process, lattice vacancies *and* interstitial ions are produced in the crystal. This type of defect is called *Frenkel defect*. Secondly, ions of *both* signs (since the crystal is electrically neutral in every small crystal volume) may migrate to the crystal surface and thus produce vacancies only (Fig. 260). These defects are called *Schottky defects*. Both types of defects enable lattice constituents to have a much greater mobility than in the ideal crystal since the diffusion can progress stepwise from vacancy to vacancy. We speak of *vacancy diffusion*. For the case of Frenkel defects it is plausible that much less activation energy is required for the motion of an interstitial ion than for an ion in a normal lattice position. Since the defect density depends on the history of the crystal, the *vacancy diffusion* at low temperatures may vary considerably *between different samples of the same crystal type*. This behavior is in contrast to the *intrinsic diffusion of lattice constituents* which is independent of the number of defects and predominates at high temperatures. Thus we may understand Fig. 258, since the ionic conductivity in a crystal is due to a diffusion of the ions drifting in the electric field. According to Fig. 258, we have a migration of lattice ions above 800 °K that is not structure-sensitive, whereas for lower temperatures the concentration of lattice vacancies, which varies from crystal to crystal, causes a correspondingly different conductivity. This may be seen very well if the ionic conductivity for one of the three crystals of Fig. 258 is analytically represented by SMEKAL's formula

$$\sigma(T) = 0.42 \, e^{-10,300/T} + 3.5 \times 10^6 \, e^{-23,600/T}. \tag{7.52}$$

The difference in the order of magnitude of the constants shows that only the 10^7th part of the lattice constituents participates in the conductivity represented by the first term. This term consequently describes the conductivity due to lattice defects which we expect in this order of magnitude. The second term represents the conductivity resulting from the migration of normal lattice ions. In agreement with this explanation, the activation energy for defect conductivity is smaller by the factor 2.3 than that of the second term [cf. the exponents of Eq. (7.52)]. Because of the smaller height of the potential wall (activation energy) in the exponent, the first term of the conductivity formula (vacancy conductivity) predominates at low temperatures in spite of the small numerical factor, whereas at high temperatures the second term (normal ion conductivity) predominates because of its large factor. Since in principle vacancies, interstitial ions, and normal lattice ions of both signs may contribute to the conductivity, the temperature dependence of the conductivity of ionic crystals should be represented by four-term formulae. It turns out, however, that two-term formulae are sufficiently accurate.

For the sake of completeness we mention that special cases of ionic crystals exist in which ionic conductivity is possible due to migration of the smaller ions through the regular lattice of the bigger ions. Silver iodide is an example for this type. The iodine ions form a cubic body-centered lattice; in the interspaces of the I^- ions, the Ag^+ ions find a large number of equivalent sites among which they may migrate by overcoming only low potential barriers.

It is of significance for numerous phenomena of solid-state physics to be discussed presently that in thermal equilibrium the spatial density of the lattice defects is determined by the absolute temperature T and the activation energy E_a required for their production. It may take a very long time, though, to reach the equilibrium, and a temperature increase may also produce irreversible lattice changes, as we shall show presently. If n indicates the number of lattice defects per cm^3, N that of the lattice constituents, and N' the number of possible interstitial sites per cm^3, the equilibrium density of the Schottky defects is given by

$$n_S = N e^{-E_a/kT} \tag{7.53}$$

and that of the Frenkel defects by

$$n_F = \sqrt{N N'} \, e^{-E_a/2kT}. \tag{7.54}$$

The factor 1/2 in the exponent of Eq. (7.54) is due to the fact that the migration of *one* ion into an interstitial site produces *two* lattice defects according to Fig. 259, but only *one* Schottky defect, according to Fig. 260. The two defects behave differently when the crystal is cooled after a previous heating: Frenkel defects may disappear by "recombination" of the interstitial lattice ions with lattice vacancies if the crystal is annealed. It is practically impossible, on the other hand, to heal Schottky defects which result from the migration of ions to the crystal surface. This fact may be used for distinguishing between these two types of defects. A further possibility for distinguishing is based on the fact that the production of Schottky defects is associated with an increase of the crystal volume, which is not true for Frenkel defects. In VII, 1, we pointed out that the formation of vacancies is related to the melting process of solids, and we used this to explain the melting heat.

We just mentioned that, in contrast to Frenkel defects, Schottky defects cannot be healed by cooling. However, lattice vacancies of opposite polarity may associate to form double vacancies or even chains or planes of vacancies. Such double vacancies, similar to Na^+Cl^- molecules, may be electrically neutral. They can then diffuse in the lattice without directly contributing to electrical conductivity. On the other hand, a temperature increase may cause them to dissociate (like the excitons treated in VII, 10c), and then they participate again in the ionic conduction. The dissociation energy of such a double vacancy in an NaCl crystal has been computed to be 0.9 ev. The possible role of chain-like or plane-like associations of lattice vacancies for lattice dislocations, gliding processes, etc. is being much discussed today.

Finally we mention that the influence of artificially produced lattice vacancies of pre-planned density is under investigation. Such vacancies may be produced by substituting the monovalent Na^+ ions of a rock salt lattice by bivalent Ca^{++} ions. Since the whole crystal is electrically neutral, *one* Ca^{++} ion then replaces *two* Na^+ ions, i.e., it "produces" a lattice vacancy next to itself. The second possibility for arbitrarily producing lattice vacancies consists in bombarding the solid by neutrons and, surprisingly, also by X-rays. Very large numbers of such lattice vacancies and other lattice defects are observed in all materials after exposure to fast-neutron and γ-radiation in the interior of nuclear reactors. This so-called radiation damage causes changes of the mechanical strength and of other macroscopic properties of the irradiated materials. It is obvious that the knowledge of these phenomena is of decisive significance for the construction of nuclear power plants.

19. Electrons in Lattice Defects of Ionic Crystals. Color Centers and the Basic Processes of Photography

In the last section we were concerned with the migration of lattice constituents as well as of foreign atoms or ions in the lattice and got acquainted with the decisive influence of lattice defects on this migration. We shall now discuss the effect of light on ionic crystals and the *migration of electrons in such crystals*. The fundamental investigations on these problems were carried out by POHL, HILSCH, and co-workers. In addition to their general interest with respect to the behavior of electrons in insulating crystals, these phenomena are basic for understanding the fundamental processes of photography.

If alkali halide crystals, which normally are transparent in a wide spectral region from the ultraviolet to the infrared, are irradiated by sufficiently short-wave ultraviolet light, by X-rays or cathode rays, or if they are heated in the vapors of their respective metals (e.g., NaCl in Na vapor), the crystals become colored. The characteristic color for each crystal varies from yellow for LiCl to blue for CsCl. The absorption band which corresponds to this color and thus is missing in the untreated crystal was called the *F band* (Farb band or color band) by POHL. The corresponding absorbing centers are called *color centers* or *F centers*.

According to our present knowledge, an F center is an electron which has taken the place of a missing negative halogen ion (Fig. 261) and then, as a non-localized electron (page 417), alternatingly forms an Na atom with one of the six surrounding Na ions. The excitation energy of the first excited state of the F center, computed by wave mechanics to be about 2.6 ev, agrees well with the wavelength of the absorption band. In this excited state, the electron remains bound to its site in the lattice and the crystal remains an insulator. But there are known transitions to higher excited states which are lying in the conduction band so that such a transition will produce freely mobile conduction electrons. In this case, the crystal shows photoconductivity even at lowest temperatures. From the first excited state, the F electron may return either to the ground state under emission of radiation or it may, at a higher temperature, take up about 0.1 ev from thermally excited lattice vibrations and thus reach the conduction band of the crystal. POHL furthermore found that the destruction of F centers by photoelectric release of their electrons not only produced the expected reduction of the absorption intensity of the F band but also the appearance of a new longer-wave absorption band, which he called F' band. The nature of the F' centers which are responsible for this long-wave absorption was explained in the following way: The F' center is an F center that has captured an additional electron (Fig. 261) or, in other words, two electrons instead of one at the position

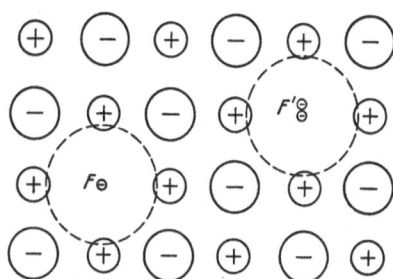

Fig. 261. Schematic representation of F and F' centers in colored alkali-halide crystals: one or two electrons in the lattice place of a missing negative ion

of one missing Cl⁻ ion. The absorption of the F' bands lies in the infrared, i.e., at longer wavelengths than that of the F bands because it requires only a small amount of energy to separate the loosely bound second electron. That absorption here causes separation and not only excitation, is proved by the fact that even at very low temperatures F' absorption leads to photoconductivity. The formation of F' centers according to the present explanation seems clear: An F electron, released from its potential well by the absorption of light with the cooperation of thermal lattice energy, is captured by another F center which is thus transformed into an F' center. In agreement with this explanation, two F centers are destroyed for each quantum absorbed in an F band (quantum yield 2). One is destroyed directly by absorption and one is transformed into an F' center. The fact that the mean free paths of the photoelectrons released from the F centers is inversely proportional to the density of F centers is further direct confirmation of the explanation of the F and F' centers as presented here. In general it is not possible to produce more than 10^{18} F centers per cubic centimeter in a crystal. This is further evidence in favor of the present theory because in general it is not possible to form more F centers than there are Cl⁻ vacancies. If, however, the density of lattice vacancies is increased by mechanical or thermal treatment or by intensive X-ray irradiation, or if experiments are made with a crystal layer evaporated at low temperatures which by itself contains more vacancies, then F center densities of more then 10^{20} cm⁻³ can be obtained. Consequently, *the production of F centers consists in the introduction of free electrons into the halogen vacancies of the lattice.* The free electrons are produced by the absorption of ultra-violet light or X-rays, or by collisions with cathode-ray electrons. *If "additive coloration" is produced by heating the crystal in a metal vapor, the metal ions attach themselves to the crystal surface (together with halogen ions which migrate outward*

and thus produce lattice vacancies), while only their electrons migrate into the crystal.
That actually an electron migration into the crystal produces the F centers and
thus the coloration, was proved by two experiments. In the first place, it does not
matter whether an NaCl crystal is heated in Na vapor or in K vapor; in both
cases identical F centers are produced. Secondly, POHL has shown in a famous
experiment that it is possible to introduce electrons alone into the crystal and thus
to produce color centers. If a potential difference of several hundred volts is put
across a completely transparent KBr crystal and if the lattice is "loosened up"
by raising the temperature to 500 or 600 °C, then electrons migrate from the negative electrode into the crystal, fall into Br⁻ vacancies and thus produce
absorbing F centers. A blue cloud is seen moving into the crystal, which can be
moved back and forth as the polarity of the electric field is reversed (Fig. 262).
The additive coloration produced by heating in a metal vapor or by electron immigration

is a lasting one, in contrast to the coloration by X-ray irradi-
ation where the F electrons may return to their original po-
sitions if the temperature is increased until the crystal is
bleached.

Fig. 262. POHL's experiment
making visible the migration
of electrons in a KBr crystal
at 600 °C. If the electrons migrate in the electric field towards the right, absorbing
("visible") K atoms change
to invisible K⁺ ions: The
"blue cloud" then migrates
to the right

A more detailed spectroscopic investigation of colored
alkali halide crystals revealed that there are further absorbing centers besides the F and F' centers. Their
identification has not yet reached the same degree of
certainty as that of the F and F' centers. There are, for
instance, *associations of F centers*, which sometimes have
a marked (110) or (100) symmetry (A, M, R, N centers).

Furthermore, the existence of a center seems certain which
is the inverse of an F center. It is the so-called V_1 center, which is produced by
a positive electron hole occupying the site of a missing *positive* lattice ion. Just
as the F electron forms an Na atom with one of its surrounding Na⁺ ions, so does
the V_1 electron hole form a Cl atom with a surrounding Cl⁻ ion. Research about
further V centers which partially may be due to associations of V_1 centers is still
in progress. As had been expected, color bands of fundamentally equivalent color
centers were found also in other ionic crystals, in particular after intense X-ray
irradiation, and this also in crystals that have only a weak heteropolar character
such as quartz.

The knowledge gained for the alkali halides has not only been of significance
for solid-state physics in general. *It also opened the path for understanding the
atomic processes on which photography is based.* The fundamental problem here
is to understand how it is possible that an entire photographic "grain" of up to
10^{10} AgBr molecules can be reduced by a chemical developer to metallic silver
after the absorption of only a few photons.

Primarily, the absorption of light produces free electrons in the AgBr microcrystals of the photographic layer. According to a theory by MOTT and GURNEY,
these free electrons are mostly trapped by Ag⁺ ions on surfaces where the ions have
a larger electron affinity than the ions in the interior because they are more
loosely bound. The Ag atoms which are thus formed are negatively charged with
respect to their surroundings and thus attract mobile interstitial Ag⁺ ions. These
in turn attract again photoelectrons until finally a stable silver *crystal seed* is
produced. The growth of silver seeds at the surface of silver bromide crystals may
thus be explained satisfactorily, but not so the growth of seeds *in the interior*
of microcrystals, which nevertheless has been proved to exist for intense irradiation (solarization). Mitchell explains these and related phenomena by assuming
that the defects by which the photoelectrons are trapped are associations of

vacancies. From these an F-center aggregation is supposed to be formed, which alternately attracts electrons and Ag^+ ions and thus, step by step, is transformed into a colloidal silver particle, i.e., into a crystal seed.

Such silver seeds can grow further in the described manner if the irradiation persists until the whole AgBr grain has become metallic silver and all halogen has disappeared. If, however, in the usual way the irradiation is interrupted after the formation of a few seeds, these seeds can be developed *chemically*, a process which transforms the grain into metallic silver while the bromine goes into the developer. This chemical development, according to MOTT, actually is an electrolysis that is possible only after the formation of the metallic silver seed. Because of the electric double layer between silver and developer, the entire silver of the original AgBr grain is attached step by step to the silver seed. The fundamental processes of the production of a photographic image therefore seem to become understandable from the atomistic point of view.

The minimum energy required for the separation of an electron from a halogen ion corresponds to a long wavelength limit beyond which the photographic layer looses its sensitivity for light. Photographic layers thus are sensitive for short wavelengths, but in general not for yellow or red light. It is clear, however, that the presence of easily ionizable particles such as the doubly negative sulfur ions of Ag_2S not only increases the sensitivity of the layer but also shifts the limit of sensitivity towards longer wavelengths. A large shift of this limit, the so-called sensibilization of the layer for long wavelengths of the visible and even for infrared light, is possible by the adsorption of certain dyes, such as the cyanines, at the Ag Br grains. These dyes absorb long-wavelength light, and the electrons excited by this absorption make a radiation-less transition (cf. Fig. 90) into the conduction band of the AgBr. There they initiate the formation of seeds in the same way as "normal" photoelectrons.

20. Semiconductivity

In the early sections of this chapter, we considered rather thoroughly the two extreme cases of solids, the insulating crystals and the conducting metals. As was briefly mentioned on page 448, there are transitional cases of various types between these limiting cases. Among others we know solids with an electronic conductivity which is zero at zero absolute temperature but which, at higher temperatures, bridges the whole gap from the ideal insulator to the metallic conductor. Such solids are called *semiconductors* or, more exactly, *electronic semiconductors*. Their investigation is of ever increasing significance for applied electronics, for wide fields of electrical engineering, and for many branches of instrumentation technique. Also our knowledge of solid-state physics has profited greatly from semiconductor research. All this warrants a rather detailed treatment of electronic semiconductivity and the manifold phenomena related to it.

a) Semiconductor Types and their Charge Carriers

By definition, an electronic semiconductor is a crystal that is an insulator at absolute zero of temperature but which, at higher temperatures, has an electronic conductivity that at first increases very fast with temperature. In order to determine whether a solid is an electronic semiconductor, the Hall effect (see VII, 13) can be used since the highly mobile electrons yield a much larger Hall voltage than the heavier and therefore slower ions. Since, furthermore, the polarity of the Hall voltage depends on the sign of the charge carriers, measurements according to Fig. 248 permit us to distinguish between conduction by negative electrons and by positive holes (cf. VII, 10c). The spatial density and, if the conductivity is known, the mobility of the charge carriers may be determined from Hall effect measurements also.

Electronic conductivity in a non-metallic solid (i.e., a semiconductor) thus is due to electrons which are separated from their respective ions by thermal lattice vibrations (similar to thermal collisions in a gas) and raised to the conduction band which is unoccupied at $T=0$. This is possible in the case of a pure crystal only if the energy difference between the normally empty conduction band and the normally filled energy band of the valence electrons, i.e., the internal ionization energy, is at most of the order of magnitude of one ev as compared with 2 to 20 ev for insulating crystals. Such a crystal with so narrow an energy gap between the two bands that it can be overcome thermally is called an *intrinsic semiconductor*.

A crystal which is not an intrinsic semiconductor can be made into a semiconductor (and this is by far the most frequent case) by introducing into it either

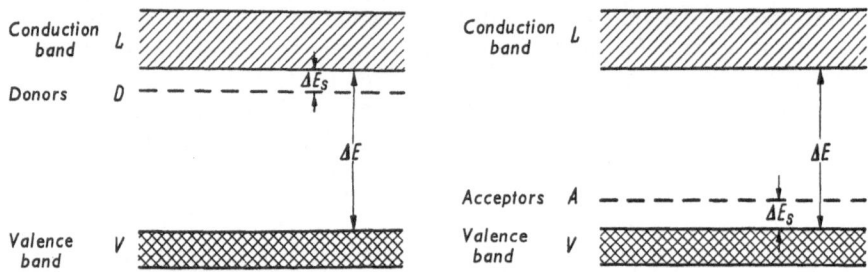

Fig. 263. Energy-band diagram of an excess (*n*-type) semiconductor. V = valence band normally fully occupied by electrons; L = conduction band normally empty; D = localized energy states of impurity or added atoms (donors), from which electrons may easily be transferred to the conduction band L

Fig. 264. Energy-band diagram of a defect (*p*-type) semiconductor. V and L = normally full and empty energy bands, respectively; A = localized energy states of electronegative atoms (acceptors), which can easily be filled from V, thus creating mobile electron holes in V

suitable foreign atoms or an excess of one of the constituent atoms if these have an ionization energy in the crystal lattice (of dielectric constant ε) which is markedly smaller than that of the normal lattice constituents, i.e., if these atoms have a rather loosely bound electron. If, for instance, a small percentage of pentavalent atoms of phosphorus or arsenic are built into the diamond lattice of silicon or germanium, only four of their five valence electrons are used for bonds with the four neighboring lattice atoms, whereas the fifth valence electrons of the P or As atoms does not participate in the lattice bonds and is only loosely bound as an excess electron. Similarly, excess Zn atoms in zinc oxide, ZnO, have very loosely bound outer electrons since these excess electrons cannot form closed shells with valence electrons of corresponding oxygen atoms and therefore cannot form localized bonds. In the energy band model this means that the energy states of such excess atoms or of atoms with excess electrons are localized states D which are situated below the normally unoccupied conduction band L and are separated from it by a small ionization energy ΔE_s (Fig. 263). Since ΔE_s usually is in the order of some hundredths or some tenths ev, electrons from the D states may reach the conduction band L by thermal ionization if the temperature is not too low. The atoms with such D states are therefore called *donors* and this type of semiconductor is called *excess semiconductor* since it requires an excess of electrons not participating in the lattice bonds. It is also designated *n-type semiconductor* because the current in it is carried by *negative* electrons donated by the donors.

The third type of semiconductor, the *p-type* or *deficit semiconductor*, is distinguished by a deficit of binding electrons. It may be made by introducing a small percentage of *trivalent* atoms (boron or gallium) into the silicon or germanium lattice. In this case, one electron is missing which would be needed for the saturation of the bonds with the four lattice neighbors. This can also be expressed by saying that there is an *positive hole* at the site of each trivalent atom. Obviously, a similar effect is obtained if there is a lack of metal atoms or a corresponding excess of electronegative oxygen atoms in a metal-oxide semiconductor. If, for instance, we have an excess of O atoms in the ZnO lattice, electrons needed for saturation of the valency of the O atoms or for the formation of negative ions with closed shells are missing. In other words, there are positive holes near the excess oxygen atoms. These positive holes can be filled by capture of neighboring valence electrons if the temperature is not too low or the kinetic energy of the electrons is not too small. The positive holes then begin to migrate. This electronic process corresponds very closely (but not completely because of the effect discussed on page 442) to that of the migration of lattice vacancies (cf. VII, 18). The energy states of *acceptors* such as the boron atoms in the silicon lattice or the excess O atoms in the ZnO lattice must be situated in the energy band model (as localized A states) closely *above* the valence band V, which at low temperature is fully occupied (Fig. 264). Thermal energy then can raise electrons from the band V, i.e., regular lattice electrons, to these localized acceptors A so that a corresponding number of mobile positive holes is produced in V. Since a positive hole in an electric field obviously moves in a direction opposite to that of a negative electron, *an electron hole behaves like a positive electron*. This is the reason for calling a deficit semiconductor a p-type semiconductor; the current is carried by *positive* carriers, the electron holes. The excess and the deficit semiconductors are also called reduction semiconductors and oxidation semiconductors, respectively, because in the metal-oxide semiconductors the excess of the metal atoms or oxygen atoms is produced by chemical reduction or oxidation (heating in hydrogen or oxygen, respectively). *Consequently, depending on the preliminary treatment, the same crystal may be an excess or a deficit semiconductor, and even both types of conductivity may coexist in the whole crystal or separately in different parts or layers of it*, typical evidence for the "structure dependence" of the properties of a semiconductor. For this reason, the production of identical and reproducible impurity semiconductors (as we may call both excess and defect semiconductors) is quite a technological problem. It is in general difficult to reproduce impurity concentrations exactly, and subsequent thermal treatment may lead to irreversible changes. An example of a semiconductor that can be produced to-day with such high purity that intrinsic semiconduction is obtained is germanium, while germanium or silicon doped with As or Ga atoms are examples for excess and defect semiconductors, respectively. Of particular interest are the compounds of elements of the third and fifth groups of the Periodic Table (discovered by WELKER), all of which are semiconductors and have quite a number of interesting and important properties in magnetic fields and when irradiated by photons or neutrons. Besides, most of them have very high electron mobilities.

We summarize our knowledge with respect to type and origin of the charge carriers in the different types of semiconductors as follows: In an *intrinsic semiconductor*, the electrons released from their bonds by thermal vibrations or collisions come from the valence band V, which contains the bound lattice electrons. Ionization here always produces a mobile electron (in the conduction band L) together with a mobile positive hole (in the valence band V). *Both* carriers contribute to the conductivity. The mobility of the holes and, therefore, also their contribution to conductivity, however, is in most cases by an order of magnitude smaller than that of the electrons. In *n-type semiconductors*, only the electrons in the conduction band L which are released from the donors carry the current, while the corresponding holes are bound to the donor atoms which are localized in the lattice and, therefore, cannot migrate. *Vice versa*, the conduction in *p-type semiconductors* is due to the migration of positive holes in the valence band V, whereas the corresponding electrons are bound to the localized electronegative acceptor atoms in the A states so that they are immobile.

b) The Electric Conductivity of Electronic Semiconductors and its Temperature Dependence

The electric conductivity of all semiconductors, just as that of metals, is given by the relation

$$\sigma = n\,\mu\,e, \qquad (7.55)$$

in which n indicates the number of charge carriers per cm³, μ their mobility, and e the electronic charge. The difficulty of computing the electric conductivity and its temperature dependence, especially for semiconductors, is due to the fact that the number n of charge carriers as well as the mobility μ depend on the temperature in a very complicated manner. The mobility μ in general decreases with increasing temperature, just as in metals, because of the increasing scattering by lattice vibrations. On the other hand, the probability that charge carriers are trapped by lattice defects, as well as the mean duration of capture, decreases with increasing temperature. The influence of these two opposite effects on the effective average mobility is hard to calculate quantitatively.

The charge-carrier density n, i.e., the number of free electrons or mobile positive holes per cm³, can be computed as a function of temperature by means of statistical mechanics. If ΔE is the width of the energy gap ("forbidden zone") between the energy bands V and L (cf. Fig. 263), then we have for an intrinsic semiconductor

$$n = \frac{2(2\pi m k\, T)^{\frac{3}{2}}}{h^3}\, e^{-\frac{\Delta E}{2kT}}, \qquad (7.56)$$

while for a doped semiconductor with n_0 donors or acceptors per cm³ and with a distance ΔE_s of their states from the corresponding band edge the charge carrier density is given by

$$n = \frac{(2\pi m k\, T)^{\frac{3}{4}}}{h^{\frac{3}{2}}}\, n_0^{\frac{1}{2}} e^{-\frac{\Delta E_s}{2kT}}. \qquad (7.57)$$

The relation (7.57), however, is valid only as long as n is small compared with n_0 or as long as only a small percentage of the donors or acceptors is ionized. The application of these formulae and of the conductivities computed from them by means of Eq. (7.55) gets further complicated by the fact that the energy differences ΔE and ΔE_s are not at all independent of temperature and other parameters.

The width ΔE of the energy gap of intrinsic semiconductors decreases with increasing temperature, because of the increasing coupling between electrons and lattice, by about 4×10^{-4} ev per degree for silicon and germanium. For extrinsic semiconductors, the width ΔE_s of the energy gap depends strongly on the defect density n_0. In ZnO, for instance, a decrease of ΔE_s is observed from 0.6 to 0.01 ev for increasing n_0. It is not yet completely clear how to explain these observations.

With respect to the temperature dependence of conductivity, in addition to that of ΔE and ΔE_s, we have to take into account that with increasing temperatures n approaches a saturation state in which all defects (donors or acceptors) are ionized. If in an extrinsic semiconductor the energy gap ΔE between the bands V and L is not too large (e.g., for an extrinsic germanium semiconductor), increasing temperature may furthermore lead to a transition from extrinsic to intrinsic semiconductivity. In this case, electrons are thermally raised from the valence band V to the conduction band L, in addition to the electrons coming from the donors. It is therefore not surprising that the temperature dependence of electric conductivity is rather complicated and may differ very much for different semiconductors. As long as Eqs. (7.56) or (7.57) are valid, we expect a straight line in a diagram $\log \sigma$ versus $1/T$. The slope of this straight line is given by $\Delta E/2k$ or $\Delta E_s/2k$, respectively. Typical curves of this type are shown in Fig. 265. Curve 1 has the expected slope $\Delta E_s/2k$ of an excess semiconductor for low temperatures, and it changes at high temperatures to the steeper slope $\Delta E/2k$ of an intrinsic semiconductor. This transition is reached even before all the donors D are ionized. Curve 2 applies for a case in which ΔE_s is so much smaller than ΔE that saturation $(n=n_0)$ is obtained before intrinsic conduction begins. In Fig. 265 this is to be inferred from the fact that the conductivity in the temperature range in question does not increase any more with the temperature; occasionally, it even decreases with increasing T.

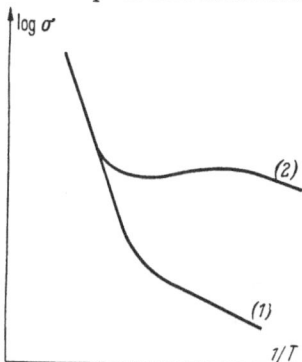

Fig. 265. Temperature dependence of the conductivity of two samples of the same electronic semiconductor, one with donors (or acceptors) of (1) large ionization energy ΔE_s, one (2) with small ionization energy. The branch at the left corresponds to intrinsic semiconductivity

The charge carrier density in semiconductors varies mostly between 10^{13} and 10^{18} cm^{-3}, the electric conductivity of pure germanium being about 10^7 times smaller than that of copper. According to Eq. (4.192) of page 200, the electron gas in the conduction band L is *not* degenerate, and Boltzmann statistics may be used for all computations. For very large defect densities n_0 and for a correspondingly small energy gap ΔE_s between the D levels and the lower edge of L, however, the electron densities of excess semiconductors may increase to more than 10^{19} cm^{-3}. In this case, the electron gas is degenerate at normal temperature and we have quasi-metallic conductivity. *The electronic semiconductors therefore cover the entire region from pure insulators to metallic conductors.*

c) Applications of the Temperature Dependence
of the Conductivity of Electronic Semiconductors

The strong dependence of the electric conductivity on temperature which is characteristic for electronic semiconductors is being applied in many important devices, some of which shall be mentioned here.

Semiconductors are used as very sensitive resistance thermometers or bolometers. Their absolute accuracy in the temperature range from $-25°$ to $+100 °C$ is about $0.01 °C$, while temperature changes as low as $5 \times 10^{-4} °C$ can be measured. The sensitivity of *semiconductor thermometers* thus is by an order of magnitude larger than that of metallic resistance thermometers.

Since the temperature coefficient of the resistance of semiconductors has the opposite sign of that of metals, it is possible to build resistors from a suitable combination of metals and semiconductors with a conductivity which is independent of temperature, at least in a certain temperature range. Such *compensation resistors* are of great significance for measuring devices.

The temperature, and with it the conductivity, of a semiconductor in a circuit depends not only on the ambient temperature but also on the current flowing through it and on the surface area which controls the heat dissipation. By choosing suitable semiconductors with suitable surface shapes, any desired current-voltage characteristic may be obtained. *Current-regulating resistors* carrying a current which is largely independent of the applied voltage are but one example. Another one are the *condensor-protecting resistors* whose conductivity becomes rapidly very large above some critical voltage. They therefore prevent the occurrence of dangerously high voltages across the condensor to be protected if they are connected parallel to this condensor.

The temperature dependence of semiconductors may also serve to suppress high transient current peaks at the moment of applying voltage to electrical devices. After applying a voltage to a semiconductor, some time is needed for heating it up by the current so that its initially high resistance decreases to the lower value which corresponds to the stationary state. A suitably chosen semiconductor resistor in series with an electric motor may prevent the occurrence of undesired current peaks. Such *thermistors* are more and more used in high-frequency instruments for many purposes.

Everything else being equal, the temperature of a semiconductor, and with it its resistance, depends on the thermal conductivity of the ambient medium. Semiconductors, therefore, may be used for measuring the thermal conductivity of liquids and gases, and this leads indirectly to applications for manometers, indicators for the level of a liquid, flow meters, and for similar purposes of physical and chemical instrumentation.

21. Release of Electrons from Semiconductor Surfaces

a) Thermal Electron Emission of Semiconductors and the Mechanism of Electron Emission from Hot Oxide Cathodes

After discussing the electron motion in the interior of semiconductors, we now treat the release of electrons from semiconductor surfaces. We begin by discussing the thermal emission of electrons and then follow up with the photoemission and the secondary electron emission.

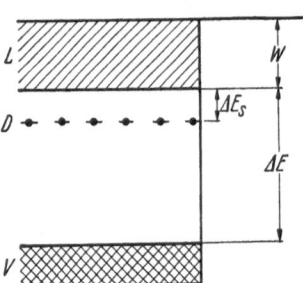

Fig. 266. Schematic diagram illustrating the electron work function of a semiconductor surface

For an n-type semiconductor of low temperature we expect theoretically two different work functions. According to Fig. 266, we need the energy $W + \Delta E_s$ for releasing electrons from donor levels, whereas the energy $W + \Delta E$ is required for releasing electrons from the valence band. These two values have actually been found photoelectrically for the work function at low temperatures. In discussing *thermal* electron emission, however, we have to take into account that at higher temperatures a considerable percentage of the electrons has moved into the conduction band. The Fermi level, from which we have to count the work function, is then situated somewhere between the valence and the conduction band. The saturation-current density of the emitted electrons thus depends on the number n of the electrons occupying the conduction band, which is given by Eqs. (7.56) and (7.57) within the limitations mentioned on page 478. In general, n is so small that the electrons are not degenerate, and we may use Boltzmann statistics rather than Fermi statistics for computing the constant A in RICHARDSON's equation. Surprisingly, this computation leads to the formula

$$j = \frac{4\pi e m k^2}{h^3} T^2 e^{-\frac{W + \Delta E/2}{kT}} \quad [\text{amp/cm}^2] \tag{7.58}$$

which agrees with the Richardson equation for metals. *The emission current density of an intrinsic semiconductor thus is given, just as that of a metal, by a Richardson equation, only with the difference that the effective work function is* $W + \Delta E/2$ as if all electrons had an energy corresponding to the middle of the energy gap between V and L.

For excess semiconductors, for which the conduction electrons come from donor levels, because of the different temperature dependence of the conduction electron density (7.57) the theory leads to the expression

$$j = \frac{3.2 e m^{\frac{1}{4}} k^{\frac{5}{4}}}{h^{\frac{3}{2}}} T^{\frac{5}{4}} n_0^{\frac{1}{2}} e^{-\frac{W + \Delta E_s/2}{kT}}, \tag{7.59}$$

which contains $T^{\frac{5}{4}}$ rather than T^2 and depends on the donor density n_0, as was to be expected. Just as Eq. (7.57), also Eq. (7.59) is valid only as long as $n \ll n_0$, i.e., as long as only a small percentage of the donors has been ionized. It is therefore of interest to note that a formula of general validity is known for the emission current density of all n-type semiconductors, which is very similar to the Richardson equation again:

$$j = \frac{4\pi e m k^2}{h^3} T^2 e^{-\frac{W + U}{kT}}. \tag{7.60}$$

U is here the energetic distance between the lower edge of the conduction band L and the Fermi surface (see IV, 13) of the electrons of the metal electrode in contact with the semiconductor. For n-type semiconductors of low ionization ($n \ll n_0$),

$$U = \frac{\Delta E s}{2} + \frac{kT}{2} \ln \frac{(2\pi m k T)^{\frac{3}{2}}}{h^3 n_0}. \tag{7.61}$$

In the limiting case of very low temperature, U becomes $\Delta E_s/2$, whereas for the limiting case of complete ionization of the donors, i.e., for $n \approx n_0$, we get

$$U = kT \ln \frac{(2\pi m k T)^{\frac{3}{2}}}{h^3 n_0}. \tag{7.62}$$

Since for an intrinsic semiconductor, according to Eq. (7.58), we always have

$$U = \Delta E/2, \tag{7.63}$$

the emission current density for the thermal electron emission by semiconductors can quite generally be expressed by the Richardson equation (Eq. 7.60), wherein U for intrinsic semiconductors is given by Eq. (7.63), for unsaturated excess semiconductors by Eq. (7.61), and for n-type semiconductors close to saturation ionization by Eq. (7.62).

The foregoing considerations form the basis for understanding the mechanism of the oxide cathode which was invented by WEHNELT and which for many decades has been used in electronic instruments of all types. It consists of a metal wire or sheet which is coated with a mixture of metal oxides and is heated to incandescence directly or indirectly. Its thermal electron emission is by many orders of mangitude larger than that of pure metals of the same temperature.

For a particularly well proven and much studied oxide cathode, a sintered mixture of BaO and SrO with small additions of CaO is being used on a nickel base; traces of aluminium or beryllium in the nickel are supposed to be of advantage. A good barium-strontium oxide cathode of this type, operated at approximately 1100 °K, has an emission current density of about 1 amp/cm², which in pulsed operation may be increased to about 70 amp/cm². By using the Richardson equation (7.60), the effective work function $W + U$ is found to amount to only 1 ev; the numerical factor is approximately 100. Freshly produced oxide coatings, however, do not yield this high emission density. They have to be tempered according to a carefully tested recipe, while a slowly increasing current is being drawn. We know today that this tempering transforms the oxide mixture into a semiconductor with just the right donor density. Many empirical results are evidence for the correctness of this interpretation. By tempering, the effective work function decreases from its initial value of 3 ev to 1 ev, while simultaneously an excess of Ba atoms is found which increases up to 10^{18} per cm³, and while the tempered coating exhibits an electronic conductivity which corresponds to an electron density between 10^{12} and 10^{14} per cm³. Below 1100 °K, a hole conductivity becomes apparent which shows that in addition to the donors a smaller concentration of acceptors is present also.

Many empirically known facts concerning the oxide-cathode emission agree well with this interpretation as a semiconductor phenomenon. That oxide cathodes, when operated in certain gases or vapors, get poisoned and that at very high loads the initially high emission density decreases rapidly, seems understandable. At the high operating temperature, a considerable migration of ions (and occluded gases) takes place in the oxide mixture by which type and number of lattice defects and the density of donors and acceptors may easily be changed. Chemical reactions with the discharge gas and the base metal as well as bombardment by the many positive gas ions, which even at the highest possible vacuum are still present, are further reasons for changes of tempered oxide cathodes during operation. For this reason, there is considerable interest in oxide cathodes which are less sensitive against such influences. One of these is the thorium oxide cathode which after suitable tempering also has a donor density n_0 between 10^{17} and 10^{18} per cm³. Since its effective work function is approximately 2.6 ev, it has a higher operating temperature than the Ba-Sr oxide cathode. That the base metal has an influence on the properties of an oxide cathode also, seems understandable since the electrons which are emitted by the semiconductor surface have to be supplied by the current which passes from the base metal into the semiconducting coating. The complicated processes which occur in this boundary layer between metal and semiconductor will be discussed in VII, 22.

In concluding, we want to point out that our discussion was confined to the release of electrons from the interior of the semiconducting mixture. Since atoms and ions in the surface layer of a solid are bound less strongly, we expect the same for electrons in surface levels. Whether these play any essential role in thermal electron emission, is rather doubtful, however, because their number is too small. We shall come back to them in another context in VII, 24.

b) Photoelectric Electron Release from Semiconductor Surfaces

The effective work function of semiconductors as treated in the last section applies for thermal electron emission only. The expressions for the different semiconductor types are the result of statistical computations and take into account the fact that the occupation of donor and acceptor levels just as that of the conduction band in each semiconductor depends strongly on temperature.

For the *photoelectric* electron release from semiconductor *surfaces*, the so-called external photoelectric effect, the situation is quite different because here the temperature is kept constant, mostly at room temperature, whereas we ask for the number and kinetic energy of the electrons emitted as a function of the wavelength of the releasing radiation. For metals, according to VII, 14, the thermal and the photoelectric work function prove to' be the same, because for them the change of the electron energy with temperature (occupation of higher energy states) is small compared with the value of the work function W itself.

We confine our discussion to an n-type semiconductor for which we have to take into account electrons in three different energy regions. The vast majority of all electrons are valence electrons in the valence band V; their work function, according to Fig. 266, is $W + \Delta E$. A much smaller number of electrons, smaller by a factor between 10^4 and 10^7 depending on the impurity level density n_0, is to be found in the donor levels D and the conduction band L. If the energetic distance ΔE_s between D and L is small compared to kT or not much larger, as it seems to be the case for n-type germanium, the majority of these energetic electrons is in the conduction band L. Its photoelectric work function, according to Fig. 266, then is W. If, however, as in many excess semiconductors, ΔE_s is large compared to kT at room temperature, relatively few electrons are in the conduction band L, and the majority of the energetic electrons occupies the donor levels D. Their photoelectric work function obviously is $W + \Delta E_s$.

For the photoelectric emission of semiconductors we thus have to distinguish between three different regions. The long-wave limit of their sensitivity is $h\nu = W$. At this wavelength, we expect a noticeable electron emission only if ΔE_s is of the same order of magnitude or smaller than kT at room temperature and if the crystal or layer exhibits an electric conductivity in the dark. In this case the released electrons come from the conduction band L. With increasing quantum energy, i.e., decreasing wavelength of the incident radiation, the photoelectric efficiency increases suddenly as soon as $h\nu = W + \Delta E_s$. The electrons are then released from the donor levels D. With further decreasing wavelength, a final strong increase of the photoemission of electrons is observed as soon as $h\nu = W + \Delta E$ since then the electrons may be released directly from the valence band V. The last-mentioned strong increase of the photoelectric efficiency due to lattice-electron absorption in general requires violet or ultraviolet radiation, since the energy gap ΔE between V and L for most crystals is well above 1 ev. *Photoelectric studies of semiconductors thus allow us to determine directly their most important energy figures, namely W, ΔE_s, and ΔE, even* though sometimes the emission of photoelectrons from surface states proves to be a perturbing effect which is difficult to eliminate.

For applications in image converters and related electric instruments, there is a considerable interest in *photocathodes*, i.e., semiconductor layers which upon irradiation by visible or infrared light exhibit an electron emission which is proportional to the illumination and may be used for projecting an electron image by electron-optical means. Most photocathode layers used so far are the result of long empirical experience and are understood only partially because of their complicated structure. There is no doubt, however, that they are all typical doped semiconductors. A photocathode for visible radiation is the cesium-antimony layer of the approximate formula Cs_3Sb. It is sensitive in the entire visible region with a maximum of its photoemission at 4000 Å. As a photocathode for the infrared, a very complicated layer is being used which contains Cs, Ag, and oxygen. It shows a maximum of its sensitivity at 8500 Å, whereas its sensitivity reaches up to 1.3 μ.

c) Secondary Electron Emission and Related Phenomena

If we regard the thermal and the photoelectric release of electrons from semiconductors in analogy to the corresponding processes of individual atoms (see III, 6) as ionization of the whole solid, we may raise the question whether a solid may also be ionized by *electron collision*. This is indeed the case, and this phenomenon is called *secondary electron emission*. Though it is by no means an exclusive semiconductor property, since metals and pure insulators may emit secondary electrons also, we treat it here because the most important secondary-electron emitters are semiconductors.

If electrons impinge upon the surface of a solid, a certain fraction is reflected, whereas the rest penetrates into the solid and may cause the emission of secondary electrons. The ratio of the number of electrons which leave the surface, i.e., are reflected *and* emitted, to those which impinge upon the surface is called the secondary-electron emission coefficient δ. We shall see immediately that δ may be smaller or larger than one and that it depends on the solid material, the energy of the primary electrons, and on their angle of incidence. Measured δ values vary between 0.5 and approximately 20. In most cases, maximum δ values are reached with primary electrons of energies from 500 to 1500 ev.

Theoretically, the situation is not quite simple since a considerable number of opposing influences has to be taken into account. It is obvious that, for reasons of momentum conservation, in a collision of a primary electron with a *free* electron, e.g., of a metal, back-scattering is much more improbable than in a collision with a bound valence electron of an insulator or semiconductor. Partly for this reason, metals have a smaller secondary emission coefficient δ (between 0.55 and 1.47) than insulators (mostly between 3 and 6), and δ increases if the primary electrons do not hit perpendicular to but under a small angle with the surface. However, there are quite a number of other reasons for these two results. Since we need an energy between 15 and 30 ev for producing a free electron with an energy sufficient for leaving the solid, every primary electron with an energy E which enters the solid may produce the order of $E/20$ secondary electrons in the interior of the solid. Only those secondary electrons which are produced in the outer layers of the solid, however, have a good chance of leaving it before they have lost their excess energy by interaction with the lattice and the other electrons. Therefore, a small angle of incidence of the primary electrons is advantageous again because the secondary electrons are then produced closer to the surface. For the same reason, however, the secondary emission of metals is smaller than that of insulators and semiconductors, because the secondary electrons in a metal loose their kinetic energy much faster in collisions with its free electrons than in an insulator or semiconductor in collisions with the strongly bound valence electrons. For the same reason, finally, there is an optimum energy of the primary electrons (from 500 to 1500 ev), since a primary of smaller energy releases not enough electrons in the solid, whereas a primary of too high an energy produces its electrons in so deep layers of the solid that the emission probability becomes too small. It is evident that the probability of trapping electrons in surface layers is so large for a large-surface material, such as soot, that more primary electrons may be trapped than secondary electrons are emitted so that δ is very small; it has been found to be about 0.5. Coating of surfaces with soot therefore is an effective method for reducing the secondary electron emission. It is used widely for glass surfaces of certain electron tubes.

The secondary electron emission of intrinsic semiconductors is very similar to that of insulators. For excess semiconductors, however, δ is particularly large because here additional electrons are produced by ionization of donor levels, and because the free mobility of the electrons present in the conduction band prevents the formation of an uncompensated positive space charge which otherwise might reduce further electron emission. In agreement with this reasoning, the secondary electron emission of most insulator crystals increases considerably if they are transformed into semiconductors by doping with donor atoms. In particular, layers such as NaCl or AgS, when doped with cesium, exhibit δ values between 10 and 20. We do not yet understand sufficiently well the phenomenon that the secondary emission coefficient of most semiconductors decreases during continued treatment with fast primary electrons and that simultaneously color centers (see VII, 19) are being produced in alkali halides, which is further evidence for the close relations between all these lattice-defect phenomena. The application of secondary electron emission in *photomultipliers* was mentioned on page 28.

In concluding, we mention two effects which are loosely connected with secondary electron emission, the so-called Malter effect and the mechanism of the crystal counter. If a semiconducting layer which emits secondary electrons is separated from its negatively charged metal basis by a thin insulating layer, the semiconducting layer may become charged positive by 10 or even 100 volts against the metal base due to its secondary electron emission. If the thickness of the insulating layer, which separates the positive semiconductor from the negative metal, is only 10^{-5} cm, we have in front of the metal surface an electric field strength of the order of 10^6 to 10^7 volts per cm. According to VII, 14, this results in a considerable field emission of the metal, which in the semiconductor may produce a real electron avalanche. Such Malter layers behave as if they had a secondary emission coefficient of the order of 10^3 because large quantities of additional electrons are supplied by field emission from the metal.

The crystal counter for fast electrons and nucleons which was mentioned briefly in V, 2 is related to secondary electron emission in so far as energetic electrons, protons, or other charged particles enter the crystal and produce free electrons in the conduction band by collision ionization. Since this production of free electrons occurs mostly in deep layers, the exit probability of these electrons is very small. For that reason, their electric conductivity is used as a measure for its number and indirectly for the primary energy of the ionizing primary particles.

22. Electrical and Optical Phenomena Originating from Boundary Layers in Semiconductors and from Metal Semiconductor Contacts

a) Rectification and Related Phenomena

For most semiconductor phenomena and their technical applications, processes are essential which occur at and near the boundary layers between semiconductors and metal electrodes or between n-conducting and p-conducting regions of a semiconductor. This is obvious since the semiconductor has to be connected to the circuit so that the current has to pass these boundary layers. Because of the different electron arrangement in a metal and a semiconductor or in different regions of a semiconductor, each contact between such regions causes a shifting of charge carriers. Depending on the relative energetic position of the respective energy bands and the work functions on both sides of the contact, the motion of charge carriers across the boundary layer may lead to so-called depletion layers of high ohmic resistance with rectifying properties or to good ohmic contacts.

Let us regard a semiconductor and a metal, both without excess charges and without contact with each other. According to Fig. 267, the relative position of their energy bands is then fixed by the potential outside of both solids, which we choose as zero. The electrons of the metal, just as those of the semiconductor, occupy energy states which lie below this zero potential by their respective work functions W_i. If we now bring the semiconductor and the isolated metal into contact, electrons flow across the boundary layer until an equilibrium state is reached as soon as the Fermi surfaces in the metal and in the semiconductor attain the same level. In any n-type semiconductor, the Fermi surface, according to page 481, is situated below the conduction band by the energy U, i.e., in general lies between the lower edge of the conduction band and the donor levels D. If n-type semiconductor and metal are brought into contact, electrons begin to move from the conduction band of the semiconductor and from its donors D

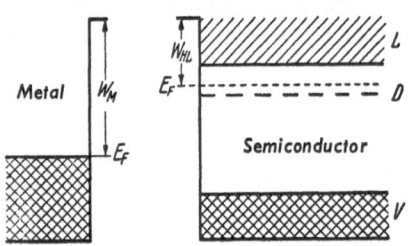

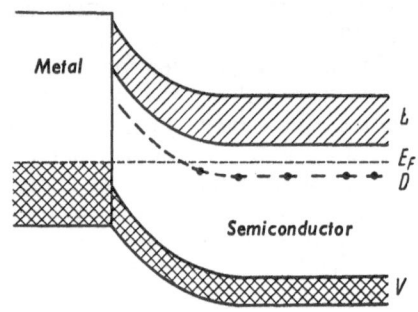

Fig. 267. Potentials of a metal and a semiconductor before contact

Fig. 268. Energy-band diagram for an n-type semiconductor in contact with a metal; formation of a barrier layer with rectifier properties

into the lower conduction band of the metal, if $W_M > W_{HL}$, as in Fig. 267. The metal is thus charged negative against the semiconductor, and we get an electric double layer in the boundary between metal and semiconductor, the electric field of which prevents any further electron flow either way as soon as the Fermi levels of metal and semiconductor are equal. The relative position of the energy bands in this case may be seen in Fig. 268. The bending of the energy bands in the boundary layer corresponds to the electric field in the double layer. Since the donor atoms in the semiconductor region adjacent to the metal have given up their electrons to the metal, and since no electrons can migrate from the right-hand conduction band against the electric field in the boundary layer, this semiconductor layer close to the metal does not show any electronic conductivity and, therefore, is called a depletion layer or *barrier layer*. With respect to the bending of the energy bands, which corresponds closely to the potential-curve concept used so much throughout this book, we want to point out that it always refers to electrons. This means that the bending is plotted in such a way that electrons "on their own" run downwards, whereas positive holes conversely run uphill on their own. This is to be kept in mind in using such diagrams.

If we now apply an alternating electric field to this metal semiconductor contact, the energy bands of the semiconductor are alternatingly shifted upwards and downwards relative to the Fermi level of the metal, because the thickness and with it the resistance and voltage drop of the depletion layer decrease or increase. If the metal is made sufficiently positive, the energy bands of the semiconductor may be lifted so much that the potential barrier which prohibits the electron flow from the semiconductor to the metal, disappears completely. The electrons which are pumped through the circuit into the semiconductor then distribute themselves, depending on the temperature, among the donor levels and the conduction band as in every normal n-type semiconductor and may flow from the conduction band into the metal uninhibitedly. In this case the barrier layer disappears because the donor levels close to the metal contact get occupied due to the statistical exchange of electrons between conduction band and donors. If the polarity is reversed, however, i.e., if the metal is negative relative to the semiconductor, its energy bands are depressed even more than indicated in Fig. 268 so that the potential wall (barrier layer) between metal electron band and conduction band L of the semiconductor does prevent any electron flow from the metal into the semiconductor. A flow of positive holes in the opposite direction would be possible in principle, but according to page 478 *mobile* positive holes do not exist in an n-type semiconductor in any appreciable quantity. *A contact between an n-type semiconductor and a metal thus permits an electron flow from the semiconductor to the metal if the metal is positive relative to the semiconductor, whereas it stops any electron current, if the metal is negative. Such a contact thus acts as a rectifier.* Only if the applied voltage is *very* high, an appreciable current in the reverse direction is observed and many lead to a breakdown of the rectifier. The major part of this reverse current is probably due to collision ionization in the barrier layer. Partly, it may also be due to electrons tunneling through the now rather thin barrier layer from the metal into the semiconductor. Semiconductor rectifiers, consequently, loose their rectifying property beyond a certain critical voltage. For technical applications, a number of rectifying units is being used in series so that each unit remains below the critical voltage.

This rectifying property of semiconductoɪ metal contacts is also the basis for the action of the *crystal detectors* which were used widely as receivers during the first decades of radiotelegraphy and have found new applications in radar technology. They consist of a small semiconductor crystal and a metal tip which touches it lightly. Because of the rectifying property of this metal-semiconductor contact, these detectors rectify the modulated high-frequency current of the receiving antenna and thus transform it into a modulated direct current, which may be heard in a telephone.

Our discussion so far was confined to contacts between a metal and an n-type semiconductor. Contacts between a metal and a p-type semiconductor may be treated analogously, however. *Before* the contact, the acceptor levels of the semiconductor lie below the Fermi surface of the metal electrons. After the contact, some of the electrons flow into the acceptor levels A so that this semiconductor layer close to the metal looses its conductivity and becomes a barrier layer. Simultaneously, an electric double layer is set up by this electron flow which prevents electrons from the metal to reach the semiconductor, as soon as the Fermi surface of the metal electrons becomes equal to that of the p-type semiconductor, which usually lies between V and A. The mobile positive holes, however, which in thermal equilibrium are present in the valence band V, according to page 486 cannot flow downhill and thus cannot enter the metal.

If now an alternating voltage is applied to this contact, it again lifts and depresses alternatingly the energy bands of the semiconductor ɪelative to those of the metal. If the semiconductor is made sufficiently positive, the applied voltage depresses the energy bands to such an extent that the potential wall and with it the barrier layer disappears and a fɪee flow of mobile positive holes from the valence band of the semiconductor to the metal is possible. For reverse polarity (semiconductor negative, metal positive) no flow is possible until, at a sufficiently high supercritical voltage, collision ionization begins or electrons tunnel from the valence band V into the metal. This is identical with stating that then positive holes from the metal flow into the semiconductor due to the high barrier-layer voltage.

If we stick to the concept that charge transport in n-type semiconductors is due to free electrons and in p-type semiconductors due to free positive holes, we thus have the simple rule that *each metal semiconductor contact allows a free flow of charge carriers from the semiconductor to the metal but not in the reverse direction.*

Instead of these metal semiconductor contacts, contacts between n-type and p-type semiconductors are being used increasingly in science and technology. In such a boundary layer, a region of high electron density is in contact with one of high positive-hole density. The high concentration gradient within the boundary layer causes the electrons to diffuse into the p-type region and conversely. The electrons diffusing into the p-region leave behind them an uncompensated positive space charge which forms an electric double layer together with the negative space charge left behind by the positive holes which diffuse into the n-type region. The field strength within this double layer prevents any further charge carrier diffusion as soon as an equilibrium between diffusion current and field current has been established. If a voltage is applied to the pn contact which counteracts the voltage of the double layer, a crossflow of electrons and positive holes becomes possible across this boundary region which before was free of charge carriers. A reversed voltage, on the other hand, which adds to the voltage of the double layer, causes a broadening of the barrier layer which is free of charge carriers and thus increases its resistance. A pn boundary layer thus acts as a very good rectifier.

b) Tunnel Diode, Semiconductor Laser, and Trinistor-controlled Rectifier

The limiting case of a very thin pn boundary layer between highly doped p- and n-regions is of particular interest for semiconductor technology. At a very high density of donors and acceptors, so many electrons occupy the conduction band of the n-type semiconductor even at room temperature and so many positive holes (from the acceptors) the valence band of the p-type region that the Fermi surface of the p-type region lies *with in* the valence band and that of the adjacent n-type region *within* the conduction band. In this case, the density of free electrons and positive holes is so high that electrons and holes are degenerate according to VI, 13. So many electrons then flow across the pn-layer into the p-type region, and so many holes into the n-type region that a double depletion layer of high resistivity is set up. The potential difference which corresponds to this barrier layer, according to Fig. 269, causes such a relative shift of the energy bands of the n-type and the p-type region that the valence band of the p-region overlaps with the conduction band of the n-region. Electrons may then tunnel horizontally across the energy gap from the valence band of the p-region into the conduction band of the n-region, and vice versa. If such a voltage is applied to this *tunnel diode* that the n-type conduction band is lifted relative to the p-type valence band until the overlapping disappears, the electron current goes through a maximum and then towards zero. In this voltage range, the tunnel diode thus has a negative characteristic, which is of high interest for technical applications.

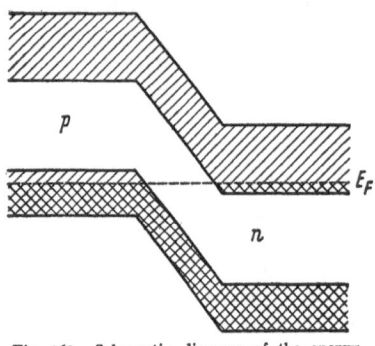

Fig. 269. Schematic diagram of the energy bands of a tunnel diode

pn boundary layers between highly doped p- and n-regions, for instance of GaAs, are used also for the semiconductor LASER or *injection* LASER. If a voltage of a few volts only is applied for short periods in such a way that the n-type region of Fig. 269 is lifted relative to the p-type region at the left until the electrons from the n-conduction band flow into the conduction band of the p-type region at the left, its occupation by electrons may become higher than that of the p-type valence band. This process is called *electric pumping* in contrast to the optical pumping usually applied for the optical LASER according to page 144. Beyond a critical current density of this electron flow from the n-type region to the p-type region, induced transitions from the conduction band to the valence band of the p-type region are possible with emission of recombination radiation, since the electrons from the conduction band recombine with the positive holes of the valence band.

Such an injection LASER may consist of a tiny GaAs cube with a horizontal pn-junction, two horizontal end faces of which carry electrodes for applying the pumping voltage, whereas two vertical end faces are polished so that they partly reflect, partly transmit the emitted radiation. The photons of the recombination radiation then are partly reflected from these end faces back into the crystal and here induce further transitions until a sharply focussed, very

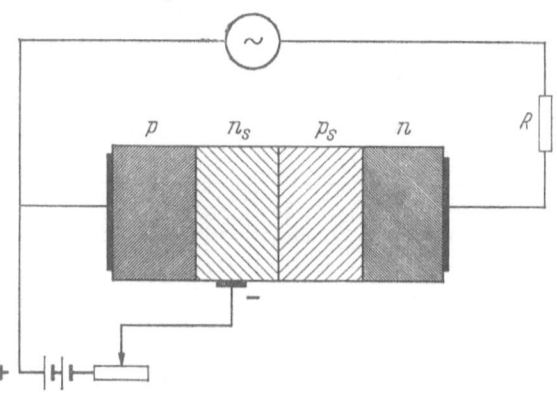

Fig. 270.   Circuit diagram of an electrically controlled four-layer semiconductor switch

intense light beam finally leaves the LASER. The efficiency of this transformation of electric energy into light is extremely high. Theoretically, it might approach 100% if incidental losses are neglected.

An interesting semiconductor device is the semiconductor switch which essentially is an electrically controlled pn junction rectifier. According to Fig. 270 it consists of a $p\,n_s\,p_s\,n$ sequence of semiconducting regions of which the two outer ones contain about 10^{17} acceptors or donors per cm^3, whereas the two inner regions are much less doped and contain only 10^{13} acceptors or donors per cm^3. If we apply negative voltage to the left-hand electrode and positive to the right-hand side, the two inner boundary layers pn_s and p_sn inhibit any electron flow so that the switch is closed. For small reversed voltage, the central boundary layer $n_s p_s$ stops the current flow. If, however, according to Fig. 270, a negative control voltage is applied to the n_s region, electrons flow into the semiconductor in very large numbers and overflow the entire $n_s p_s$ region which originally had a high resistance because of its small donor and acceptor concentration. The switch thus closes as a consequence of the negative control voltage so that the current can pass. Positive control voltage, conversely, stops the current flow in this contactless switch.

So far we spoke mostly of boundary layers of high resistance, i.e., of low carrier concentration. The reverse type of boundary layer is important if we want to connect electrodes to a semiconductor, since here we want no rectification properties. Such a boundary layer of high conductivity occurs always if a metal is contacted with a semiconductor of higher work function.

c) Transistor Physics

Semiconductor physics and its wide applications in engineering got a tremendous boost in 1948 when BARDEEN, BRATTAIN, and SHOCKLEY discovered the *transistor*. Its ever growing importance for the entire field of electrical engineering is based on the fact that it is smaller, less sensitive, and more reliable than the electron tube, but is able to fulfill the same tasks without needing a heating battery and a heat-up time.

The decisive discovery which forms the basis of transistor physics concerns the injection of positive holes through the boundary layer from a metal (or *p*-type

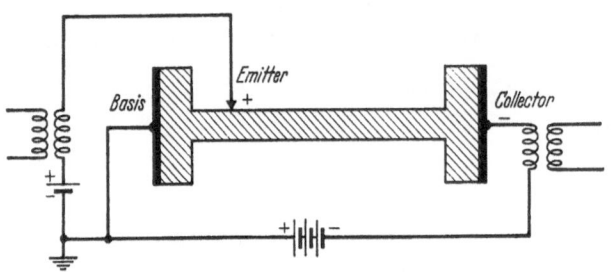

Fig. 271. The band transistor and its circuit. (After SHOCKLEY)

semiconductor) into an *n*-type semiconductor. In the last section we found that for positive metal a current of excess electrosn from the donors may flow through the boundary layer into the metal. It had been overlooked before, however, and was intentionally not mentioned in our discussion, that the high electric field strength under a sharp metal point in contact with a suitable semiconductor surface may pull not only conduction electrons but also valence electrons from the valence band V of the semiconductor into the metal. By this process, mobile positive holes are produced in the *n*-type semiconductor which migrate in the electric field away from the metal into the semiconductor.

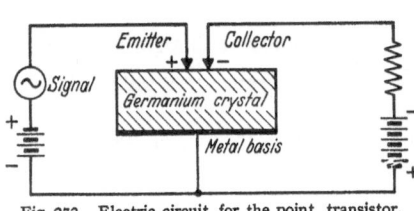

Fig. 272. Electric circuit for the point transistor. (After BARDEEN and BRATTAIN)

For this reason, the process of pulling valence electrons from the semiconductor into the metal is also called *injection of positive holes by the metal point into the semiconductor*. If these positive holes migrate along the field lines, they add to the charge transport until, after an average life time which for germanium is of the order of 10^{-4} sec, they recombine with free electrons. By using wire-like long germanium semiconductors with several metal electrodes, it was possible to show oscillographically the migration of injected positive holes in all detail and thus to measure directly mobility, diffusion constant, and average life time of these positive holes.

By means of Fig. 271 we discuss the effect of injected positive holes on the electron flow in an *n*-type semiconductor. A thin extended germanium crystal is connected to the circuit by means of large non-rectifying metal electrodes. If the metal point, which is called *emitter*, injects positive holes into the semiconductor, these migrate in the strong electric field between emitter and collector towards the right-hand side and thus add to the total current. The decisive point, however, is that these positive holes form a positive space charge which is compensated in the stationary case be an *increased* number of electrons, which flow from the right-hand collector electrode into the semiconductor. In this type of transistor we thus find an *amplification of the current between base electrode and collector, which is controlled by the injection of positive holes from the emitter.* If the large collector electrode is replaced by a sharp metal point, as indicated in Fig. 272, the collector current, for the voltages indicated, passes this contact in the reverse direction, whereas the emitter contact is passed in the flow direction. The resistance of the collector contact therefore is larger by up to two orders of magnitude than that of the emitter contact so that this point transistor acts as *voltage amplifier*. Every change of the hole current injected by the emitter causes a corresponding change of the electron current flowing from the collector to the base electrode just as in the transistor of Fig. 271. Due to the small resistance of the emitter contact, however, a small voltage change is sufficient for producing a given change of the emitter current, whereas the corresponding change of the collector current causes a very large change of the voltage at the collector because of the high resistance of the collector-contact barrier layer.

Futher progress was made by SHOCKLEY's discovery that the sensitive point contacts of the point transistor can be replaced by internal pn junctions within the same semiconductor crystal. Fig. 273 schematically shows the *pn junction transistor*. Its germanium crystal exhibits n-conductivity due to excess donors only in its central region, whereas both outer regions are p-conducting because of excess acceptors. The metal contacts to the three semiconductor regions are large and non-rectifying. From the left-hand p region a current of positive holes is injected into the central region; its intensity depends on the voltage across the left-hand pn junction. Only a small part of these holes is lost by recombination with electrons in the narrow central region. The rest flows across the right-hand junction, which does *not* stop holes flowing in the

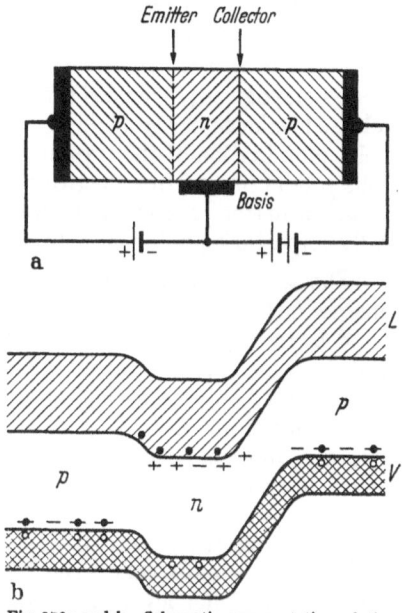

Fig. 273a and b. Schematic representation of the pnp transistor (junction transistor) after SHOCKLEY (a), with energy diagram for this transistor (b)

direction from n to p region, and here produces a voltage change in the collector circuit which is related to the emitter-circuit voltage as the resistances of the right-hand np junction to the left-hand pn junction. If we insert between the right-hand p-region and the collector electrode a further n-region and thus get a $pnpn$ *transistor*, the positive holes which come from the left-hand side stop in front of the last pn junction, which inhibits their flow so that a correspondingly higher electron current from the collector electrode is required for their compensation. This type of transistor, consequently, acts as a *current amplifier*.

Rather than injecting the positive holes through an emitter junction, it is possible to produce them in the semiconductor interior by the internal photoelectric effect to be treated below. Photons of suitable energy are absorbed by valence electrons which thus are lifted into the conduction band L, whereas mobile positive holes are created in the valence band. In this *phototransistor*, the emitter with its variable signal voltage is replaced by an amplitude-modulated light beam which produces a corresponding modulation of the current in the collector circuit of the transistor.

d) Internal Photoelectric Effect, Photoconductivity, and Theory of Semiconductor Photovoltaic Cells

In VII, 20, we designated as electronic semiconductors such crystals which at absolute zero temperature insulate but show an electronic conductivity at higher temperatures. This is due to the fact that electrons are released from their bonds or from donor levels by thermal lattice vibrations, and that analogous thermal effects produce positive holes which carry the electric current also. Free electrons and positive holes, however, may also be produced by photoabsorption. An example for this was discussed in VII, 19 where we learned that absorption of photons from the absorption band of the F' center of colored alkali halide crystals led to free electrons in the crystal which before did not exhibit electronic conductivity. This release of valence electrons by photoabsorption is called *internal photoelectric effect*, in contrast to the external photoelectric effect, discussed on page 483, which is the release of electrons from metal surfaces. Since the released valence electrons become conduction electrons which drift in an electric field, illumination of the crystal by proper wavelengths causes *photoconductivity*. This production of free electrons and positive holes by photoabsorption is possible in *all* insulator or semiconductor crystals. The majority of the good and practically important photoconductors, however, are semiconductors.

The wavelength of the photoelectrically effective radiation depends on the energy distance of the highest occupied electron levels from the conduction band L. In ideal insulators, this is the energy gap ΔE between the energy bands V and L. Since ΔE for most insulator crystals is several ev, we expect photoconductivity only as a consequence of illumination by rather short-wave ultraviolet radiation. According to VII, 2, however, all *real* crystals contain a considerable number of impurity atoms and lattice defects of all types, the energy levels of which are distributed over a wide range of the energy gap. If the levels of these electron donors are situated so closely below the conduction band L that their electrons may be released by interaction with the thermal lattice vibrations, we have called the crystal an electronic semiconductor. Is the distance of the donor levels from the conduction band L large compared to kT but small compared to the energy gap ΔE between V and L, these crystals do not show semiconductivity, but they exhibit a weak absorption on the long-wave side of the lattice absorption, and it is this absorption which causes electrons to reach the conduction band and thus is responsible for photoconductivity. By absorption of the long-wave tail of the lattice absorption, electrons may be lifted, according to VII, 10b, from the valence band V into exciton levels situated below the conduction band, from which they later may reach the conduction band by thermal ionization.

Due to all these processes, *photoconductivity in general is produced by absorption in the long-wave end of the lattice absorption proper*. The lattice absorption itself in most cases is so strong that incident radiation is completely absorbed within a very thin surface layer of the crystal. The high density of electrons and holes here leads to strong recombination, whereas the main volume of the crystal remains non-conducting. This explains why radiation of the short-wave part of the lattice absorption is much less effective in producing photoconductivity than radiation of or near the long-wave limit. Photoconductivity and its wavelength dependence in general are temperature-dependent. This follows not only from the temperature dependence of the energy gap ΔE which was mentioned in VII, 20b, but also from the rather complicated temperature dependence of the electron occupation of the donor levels and the conduction band of *all* semiconductors. And all temperature-dependent photoconductors seem to be semi-conductors.

Related to photoconductivity is also the short-time electronic conductivity which is produced in insulator crystals by ionizing X-ray and γ-radiation as well as by fast electrons (β-rays) and α-particles, and which is the basis of the crystal counters mentioned briefly in V, 2. Whereas the primary process of the electron release (e.g. collision ionization) here is different, the electron drift in the crystal which gives rise to the photocurrent is the same in both cases.

If an electron drifts in the direction of the electric field a distance s and the electrode distance is d, then we measure a charge transport

$$Q = es/d. \tag{7.64}$$

If N electrons are produced per second, the photoelectric current is

$$J = Nes/d. \tag{7.65}$$

Photoconductivity consequently is the larger the larger the average drift length s of each electron in the direction of the field is. This, however, depends not only on the field strength but very essentially on the spatial density of the electron traps in the crystal by which the electrons are trapped at the end of their drift length. For equal illumination and quantum efficiency, photo-conductivity thus is the larger the purer and more ideal the crystal is. This property was mentioned in V, 2 as a requirement for good crystal counters. Actual drift lengths in photoconductors vary between some 10^{-8} cm and about 1 mm.

The primary photoelectrons which drift towards the anode leave behind them in the crystal an uncompensated positive space charge which would make a continuous photocurrent impossible if it could not be compensated. This is easily possible, if the crystal has a certain concentration of free electrons even without illumination, i.e., if the crystal is a semiconductor. In this case the positive space charge is compensated by further electrons emerging from the cathode. These may amplify the primary photocurrent by a large factor, a phenomenon which was called *secondary photoconductivity* by POHL, HILSCH, and SCHOTTKY.

Photoconductivity was first observed in 1873 in selenium, but it was opened to our understanding only when GUDDEN and POHL in 1920 began to study the photoconductivity of single crystals such as diamond and zincblende (ZnS) as well as that of colored alkali halides in a systematic way. Beside these, the crystals of sulfur, tellurium, iodine, phosphorus, silicon, and germanium as well as the oxides, sulfides, selenides, and tellurides of nearly all metals are good photoconductors. It is interesting to note that all these crystals are good semiconductors too. The same applies to most of the crystal phosphors to be discussed in VII, 23. *Electronic semiconductivity, photoconductivity, and phosphorescence thus are three closely related crystal phenomena.* We have to point out, however, that pure intrinsic semiconductors are *not* good photoconductors, whereas a microscopic side-by-side of n-type and p-type regions seems to yield particularly good photoconductivity.

For many applications it is of importance that in semiconductors with small energy gaps photoconductivity may result from absorption of *infrared* radiation. For lead sulfide, PbS, for instance, the maximum of spectral sensitivity is at about 2.3 μ, whereas the sensitivity of lead telluride reaches up to 6 μ, though here the cell has to be cooled by liquid air in order to suppress the conductivity due to thermal ionization. We mentioned on page 49 that these photoconductors are being used widely as radiation detectors in the infrared region, in particular as photoresistive cells in which the change of the resistance by irradiation is used as a measure of the incident radiation.

The photoelectric effect in semiconductors may also be used for producing a photocurrent proportional to the absorbed radiation intensity directly by means of the *photovoltaic cell*. If a layer of a suitable semiconductor, according to Fig. 274, is sandwiched between two metal electrodes one of which is extremely thin and therefore semitransparent, and if these are connected to a circuit with an amperemeter, a current starts flowing in the circuit upon illumination of the metal semiconductor junction. Its intensity is proportional to the intensity of the illumination. This photovoltaic cell is used widely as a solar cell, illumination meter, and for many other purposes. If the circuit is open, illumination causes a

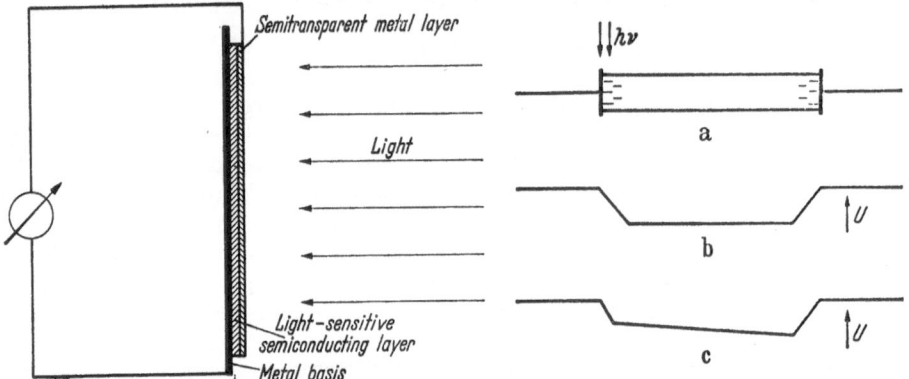

Fig. 274. Photovoltaic cell and its circuit   Fig. 275. Production of photovoltage in a barrier-layer photovoltaic cell

potential difference between the two electrodes, the so-called photo-emf. Its value is 0.6 volt for the much-used selenium photovoltaic cell and 0.9 volt for the GaAs cell which is particularly well suited as a solar cell.

We try to explain the photo-emf by discussing the simple model of Fig. 275. A p-type semiconductor may carry two metal electrodes of the same material. As was discussed on page 487, the acceptor levels of the semiconductor layers close to the electrodes then get filled up with electrons from the metal until the electric field of the resulting electric double layer prevents any further flow of electrons. Using a different language we may say that we then have two mutually compensating contact potentials between the electrodes and the semiconductor so that *no* potential difference exists between the electrodes. This potential is indicated in Fig. 275 b. If now one of the contacts, i.e., metal semiconductor boundary layers, is illuminated by radiation belonging to the wavelengths of the lattice absorption, free electrons and free mobile positive holes are produced in the boundary layer. Because of the electric field of the double layer, the electrons flow towards the left-hand metal electrode and reduce its potential, whereas the positive holes reduce the negative space charge of the semiconductor barrier layer and thus increase its potential, which then looks like Fig. 275 c. The potential distribution near the right-hand electrode remains unchanged because the strong absorption in the crystal prevents the light from reaching the second barrier layer. Continuous illumination results in an equilibrium state because the flow of photoelectrons towards the left-hand metal electrode reduces the electric field of the double layer until the photoelectrically produced positive holes begin to flow against this field and sustain the (lower) positive charge of the electrode. Fig. 275 c shows that illumination of the left-hand barrier layer thus results in a potential difference between the electrodes, the photo-emf. Its value depends on the kinetic energy and the mobility of the photoelectrons and positive holes and, therefore, is a function of the semiconductor material.

For a barrier-layer photovoltaic cell of n-type material we have an analogous situation but with reversed polarity. From our discussion of Fig. 275 follows that in a p-type photovoltaic cell the illuminated electrode is the negative pole; in an n-type cell, the positive pole of the resulting cell. Using our normal concept of conduction by holes in a p-type semiconductor and by electrons in an n-type semiconductor, we may say that *the flow of charge carriers in a photovoltaic cell always occurs from the illuminated electrode towards the adjacent semiconductor.*

The situation is quite analogous if instead of a metal semiconductor junction a pn junction is illuminated, since here too an electric double layer (as treated on page 448) causes a separation of the produced photoelectrons and positive holes and thus leads to a photo-emf.

This theory of the photovoltaic cells, which was first developed by SCHOTTKY and MOTT, evidently requires the photoproduction of free electrons *and* mobile positive holes in the semiconducting layer near the illuminated electrode. According to page 477, such a production of mobile electrons *and* holes is possible only by transitions $V \rightarrow L$, i.e., due to light absorption by *valence electrons*. This conclusion is in best agreement with the empirical result that, e.g., for Cu_2O, the long-wave sensitivity limit of the photovoltaic cells is much shorter than the long-wave limit of the radiation which still produces photoconductivity. *Production of a photo-emf thus requires lattice absorption itself, whereas the longer-wave absorption by impurity atoms and lattice defects which was mentioned on page 492 results in mobile charge carriers of only one polarity and thus produces photoconductivity but no photo-emf.*

23. Crystal Phosphorescence

As our last example of solid-state phenomena which are directly related to crystal imperfections we now treat luminescence. By *luminescence* we mean the phenomena of fluorescence *and* phosphorescence together, i.e., every emission of light by solids as a consequence of previous irradiation by light or particles. Originally, the difference between fluorescence and phosphorescence was seen in the *duration* of the emission. Light emission immediately following the absorption of the exciting radiation was called *fluorescence*, an afterglow of the same or a different wavelength which extends from milliseconds to hours or more was called *phosphorescence*. Nowadays we distinguish the two phenomena by their mechanism. If the electron transition which is coupled with the emission has as its upper energy state that one which was reached directly by the absorption, we speak of fluorescence. If, however, the electron which is excited (or separated from "its" ion) by the absorption of radiation first makes a transition into another energy state (e.g., a metastable state or an electron trap) from which it makes its radiating transition into the ground state later, we speak of phosphorescence. Phosphorescence thus implies a trapping of the absorbed energy, reasons of which will be discussed below.

So far as we know, pure ideal crystals do not show phosphorescence. All phosphorescing crystals rather consist of a basic lattice with built-in activator atoms or radiating centers. The basic material may be a single crystal or it may be a powder. The best-known crystal phosphors are zinc sulfide and cadmium sulfide, ZnS and CdS, which may be used independently or as mixed crystals and are activated by copper, silver, manganese, or other metals. Excess zinc atoms in the ZnS lattice may play the role of activator atoms also. This shows clearly the close relation between phosphors and excess semiconductors (see VII, 20a). Earth alkali sulfides and oxides activated by Cu, Mn, Pb, or rare-earth atoms were phosphors which LENARD studied intensely more than 50 years ago. Alkali halides activated by heavy metals or rare-earth atoms are of particular scientific interest because they may be grown as large single crystals.

The activator atoms may occupy interstitial positions as, for instance, Cu and Ag in ZnS, but they may also substitute for regular lattice ions as in the case of Mn in ZnS. In this case we speak of substitutional phosphors. For these, the best efficiency is reached if 10^{-2} or more regular lattice ions are replaced by activator atoms. The optimum concentration of activators in interstitial phosphors is usually smaller by at least two orders of magnitude. Too high a concentration of activator atoms has a poisoning effect, i.e., it reduces the phosphorescence efficiency. Quite generally, we know atoms which produce phosphorescence as well as those which quench it. In different basic lattices, the same atom may act as activator or as quenching substance.

Just as in producing photoconductivity, radiation of the longwave tail of the lattice absorption is most effective in exciting phosphorescence. Radiation of such wavelengths, however, which can be absorbed by the activator atoms themselves, produces fluorescence and *not* phosphorescence. The excitation spectrum of substitutional phosphors in some cases shows a relation to the absorption spectrum of the activator atoms. In doubly activated phosphors, one type of atom often absorbs the exciting radiation, whereas the second activator emits it. It is most important for our understanding that the phosphorescence radiation is determined mainly by the electron structure of the activator atoms and depends only little on the basic lattice. This can be shown very well by using manganese or rare-earth activators in different basic lattices, because these activators exhibit a very characteristic structure of their emission spectra.

So far as the mechanism of solid-state phosphorescence is concerned, we have to distinguish between two groups of phosphors. In the first group, the electrons are only excited by the absorbed radiation, whereas in the second group they are completely separated from their ions, i.e., lifted into the conduction band. The solid solutions of all unsaturated and aromatic organic compounds belong to this first group. These solids thus remain electric insulators during irradiation and phosphorescence; we call them *molecular phosphors* for reasons to be seen immediately. The phosphors of the second group become photoconducting as a consequence of the absorption; they are the *crystal phosphors* proper.

Since in molecular phosphors no electron migration is connected with the mechanism of phosphorescence, trapping of the excitation energy, which we had recognized as the decisive process for phosphorescence, must occur in or very close to the absorbing center. Here we are not dealing any more with a real solid-state effect in which the lattice as such participates in the process. The lattice of which the molecule is a part only modifies in some way what happens within the molecule. According to Fig. 276, apparently an excited state B of the molecule is reached by absorption of light from the ground state A. From B the molecule may either return into the ground state A directly by emission of fluorescence, or it may make a transition into the metastable state C. After the average life time of these organic phosphors of a second or so, a radiative transition may occur to A or, at a sufficiently high temperature, first a transition back to B and then with emission of light back to the ground state A. The nature of the metastable state C

is not yet quite certain; it might be the lowest triplet state of the molecule which, just as in the helium atom (see Fig. 67), does not combine with the singlet ground state, or only with a 10^8 times smaller probability.

Before we now deal with the activated crystal phosphors proper, in which phosphorescence is always connected with photoconductivity, we have to explain why very pure and ideal, not activated crystals do not show phosphorescence. According to page 445, the reason is the following. The radiative recombination of the free conduction electron originating from the absorption with a freely

mobile positive hole is very improbable because of the difficulty of conserving the momentum, just as it is for the radiative recombination of a free electron and a free ion, according to III, 6. We may see this by using the energy band diagram Fig. 277. If an electron of the valence band V absorbs a photon, it reaches that state in the conduction band L which has the correct quantum number k. If it now does not return immediately into the same energy state in V (a process which would be identical with a mere scattering of the photon), the electron interacts with the lattice and transfers energy to it, thus changing its quantum number k (see VII, 11) by moving towards the lower edge of the conduction band L.

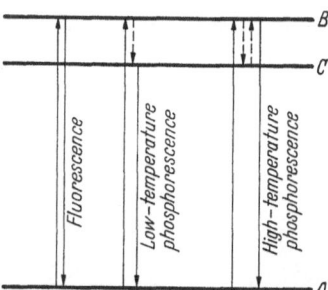

Fig. 276. Energy-level diagram of a phosphorescent organic molecule with a metastable state C and transitions for fluorescence, low-temperature phosphorescence, and high-temperature phosphorescence. (After JABLONSKI)

In general, this transfer of energy to the lattice occurs in a time which is small compared to the average life time of the electron in the conduction band so that the direct return transition is rather improbable. After having arrived at the lower edge of band L, however, the electron cannot return to V any more by

emission of radiation, because the energy state in L with the correct momentum hk is in general occupied. The selection rule for k thus seems to be responsible for the fact that ideal crystal lattices do not exhibit phosphorescence and that recombination radiation of free electrons with free holes does not occur unless there is a *very* high density of electrons and holes, as in the injection LASER (p. 489).

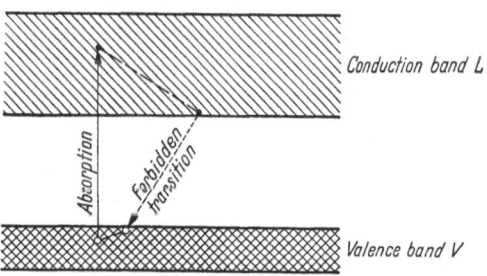

Fig. 277. Energy-band diagram indicating why pure ideal crystals generally do not exhibit luminescence

We now turn to the mechanism of the normal crystal phosphorescence of metal-activated crystals in which absorption primarily leads to photoconductivity. *By absorbing light, an electron, in most cases a valence electron of the basic lattice, reaches the normally empty conduction band L and migrates through the lattice until it comes directly, or after having been trapped in one of the many electron traps for some time, to an activator atom and here gives up its energy or most of it by emission of phosphorescence radiation.*

Most of the empirically well-established properties of the crystal phosphors are easily understood from this model, e.g., the structure-sensitivity of phosphorescence just as its dependence on temperature. At low temperatures, the free electron can move only slowly by diffusion from trap to trap, it may even be "frozen" in some traps with a depth larger than kT. Accordingly, we expect and find at low temperatures a long but not very intense afterglow. At higher temperatures, conversely, the electron mobility in the crystal is larger, the electrons diffuse faster to the emitting centers, and we expect and find an intense but correspondingly shorter phosphorescence. URBACH was able to show that the distribution of the depths of the electron traps in a specific crystal can be measured, if by irradiation at low temperature the electron traps of the phosphor are filled and then its light emission is being measured as a function of the gradually increased temperature. The maxima of these "glow curves" indicate the preferred trap depths.

We now turn to the decisive question ow the excitation energy is being transferred from the electrons to the activator atoms which emit them. No clear decision seems to be possible yet between two different hypotheses.

RIEHL and SCHÖN described the mechanism of phosphorescence by making use of the energy band diagram Fig. 278. In addition to the valence and conduction bands V and L and the electron traps D, this diagram shows the energy states A of the activator atoms which are assumed to be situated immediately *above* the upper edge of the lattice valence band V. The major part of the empirical results on fluorescence and phosphorescence of activated crystal phosphors can be described without any contradictions by this assumption.

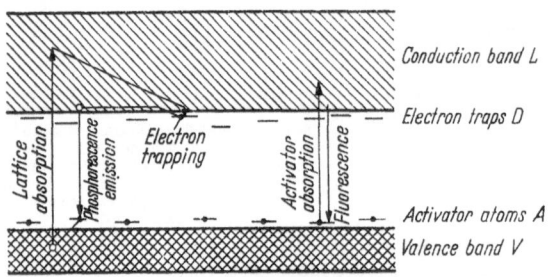

Fig. 278. Energy-band diagram of a crystal with activator atoms and electron traps

If by light absorption a positive hole is created, this is filled by a radiationless transition of an electron from an activator atom into this hole. In this way, a positive activator ion[1] is created with which the free conduction electron can recombine by emission of light as soon as it comes close to the A^+ ion during its migration in the lattice. On the other hand, the conduction electron may be trapped in one of the D levels before it meets an ionized activator atom. It then remains trapped at first, because a transition from D to V is impossible, since the positive hole in V was filled by an electron from A and a transition from D to A is very improbable because trap and acti vator atom are in general not localized at the same spot. *The conduction electron with its energy thus re'mains trapped in D until by trnsfer of thermal energy or absorption of infrared radiation it may go back to L and diffuse to an activator ion.* Here it may recombine with an electron hole in other wor ds, make a radiative transition into the ground state of an activator atom. This would explain that the emission spectrum of this phosphorescence is mainly determined by the activator atom.

The only difficulty with this model is that we do not know a good reason for assuming that the ionization energy of *all* activator atoms in the lattice, according to Fig. 278, should be somewhat smaller than that of the valence electrons. For this reason, FRANCK has suggested that the energy transfer to the activator atoms may be regarded as a process related to sensitized fluorescence (see III, 6a). According to this hypothesis, the positive hole which originates from the primary absorption and migrates through the lattice is being trapped in the distorted, i.e., polarized, neighborhood of an activator atom. At the end of its

---

[1] Actually, the activator in this case becomes a positive ion only if it was a neutral atom before, whereas it becomes a neutral atom if it was a negative ion before.

migration through the lattice, a free electron in the L band would then recombine radiatively with such a positive hole trapped close to an activator atom, and FRANCK points out that this recombination is much more probable from the point of view of momentum conservation than the recombination with a free hole. The energy released by this recombination is easily transferred to the activator atom because of the close interaction in the lattice, i.e., by a process related to a collision of the second kind (see III, 6a). The activator atom is thus excited to emit its own spectral line, broadened to a band by the interaction with the lattice.

The difference between the mechanism of RIEHL-SCHÖN and FRANCK seems to be that the first-named authors regard the A states as energy states of activator atoms themselves, whereas FRANCK regards them as lattice-electron energy states which are disturbed by neighboring activator atoms. It seems reasonable to assume that such disturbed states are states of *higher* potential energy and thus are situated above the unperturbed states of the valence electron band.

We mentioned that the major part of all the experimental results on phosphorescence, though at first sight apparently so confusing, can be explained by means of Fig. 278. It is clear, for instance, that occupied electron traps (D states) are absorbing centers analogous to the color centers which we discussed in VII, 19. Because of the small distance $D-L$, the absorption bands of these traps are often situated in the infrared. Irradiation in these bands evidently releases electrons and thus produces photoconductivity even at low temperatures, for many phosphors also increased phosphorescence, a process discussed by LENARD long ago and called by him "Ausleuchtung" (stimulation). Unexpectedly, on the other hand, a corresponding irradiation of very similar phosphors sometimes leads to a quenching of phosphorescence. This is probably due to some radiationless transfer of the excitation energy to the lattice, the mechanism of which is not yet known in detail. It is understandable, however, that increased temperature leads to increased lattice vibrations and thus to an increasing interaction between lattice and conduction electrons. This in turn explains that the efficiency of phosphorescence of all known crystal phosphors decreases strongly beyond a temperature which is characteristic for each phosphor.

In some phosphors, in particular in strongly copper-activated ZnS, luminescence is excited not only by irradiation of light (*photoluminescence*) but also by applying an electric field (*electroluminescence*). It seems that here in regions of high electric field strength, e.g., in barrier layers or at structural inhomogeneities, electrons can be accelerated to such an extent that they produce secondary electrons and holes by collision ionization. These electrons may then undergo radiative transitions into empty activator levels, and it seems that these transitions often occur after turning off the field or changing its polarity. Many details are not clear yet, among them the problem where the primary ionizing electrons come from, though an injection seems possible as it is known from the transistor. Photoluminescence may also be modified in many ways by electric fields or currents, probably due to a changed occupation of the electron traps in the crystal itself and on its surface. This effect is called *electrophotoluminescence*. Research on these complicated interactions between electron motion and emission in crystal lattices is still going on.

Reviewing the large variety of experimental results and their interpretation, we come to the conclusion that our theoretical concepts often are still too simple. In particular, it might be very important to learn how to combine the concepts of the energy band diagram and of potential curves in order to account for electron migration processes and at the same time take into account the dependence of the electron energy on the motion of the vibrating ions to which they are bound.

24. Atomic Processes at Solid Surfaces

In discussing the release of electrons from metals and semiconductors, we dealt already with surface effects and mentioned the important role which adsorbed gases and other impurities play in all surface phenomena. We now treat a number of further atomic exchange processes at solid-state surfaces and thereby point to their particular structure, which is by no means fully understood.

We begin with the ionization of atoms impinging on incandescent metal surfaces, the so-called Langmuir-Taylor effect. Atoms of the alkali metals potassium, rubidium, and cesium, when impinging on an incandescent tungsten surface, leave it as positive ions, whereas the atoms Li and Na in this case remain neutral atoms. *All* alkali atoms, however, are ionized by an incandescent platinum surface. We may assume that the atoms which impinge on the hot surface are first adsorbed and then, due to its high temperature, evaporate from the surface, either with or without their outermost valence electron. For deciding this alternative, we consider the individual steps of the process, just as we did in Fig. 229 when discussing BORN's cycle. Instead of looking at the evaporation of a neutral atom, we may first consider the release of the positive ion, then furnish the work function energy W for getting the electron out of the metal and finally form the neutral atom by combining the ion with the electron, thus releasing the ionization energy E_i. Actually the process goes in such a way that as little energy as possible is used up. That means that the electron will remain in the metal if

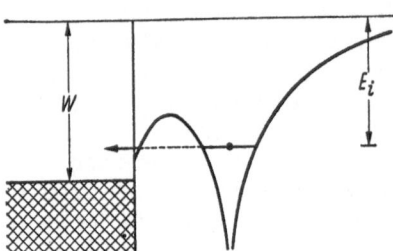

Fig. 279. Schematic representation of potentials of an alkali atom adsorbed at a metal surface, explaining the ionization at a hot metal surface (Langmuir-Taylor effect)

$$W > E_i. \tag{7.66}$$

In this case, the impinging atom will be ionized by the metal. In the normal case, however, in which the ionization energy of the atom is larger than the work function of the metal $(W < E_i)$, energy is gained by combining the ion with the electron after its release from the metal surface so that the atom is not ionized and leaves the incandescent metal surface as a neutral atom. This theoretical prediction agrees well with the experimental result. The heavy alkali atoms K, Rb, and Cs have ionization energies which are smaller than the work function of tungsten (4.5 ev), whereas the ionization energy of Li and Na is larger. The ionization energies of *all* alkali atoms, on the other hand, are smaller than the work function of platinum. Consequently *all* alkali atoms are ionized at an incandescent platinum surface.

The mechanism of the process may be seen from Fig. 279. If Eq. (7.66) applies, the Fermi surface of the metal lies below the energy state of the electron which is bound to its atom. Consequently, the electron may reach the energy band of the metal electrons by tunneling through the potential wall of its atom since the potential is distorted by the electric forces near the solid surface.

The reverse process occurs at the cathode of glow and arc discharges. Here, positive ions from the discharge arrive at the negative cathode and get neutralized by releasing electrons from the cathode which they thus leave as neutral atoms. This neutralization is possible only, as discussed above, if

$$E_i > W. \tag{7.67}$$

If we disregard the alkali-vapor discharges and cathodes made from the few metals of high work function, condition (7.67) is always fulfilled. Even neutral atoms should be able to release electrons from metal surfaces if they are electronegative and if their electron affinity is larger then the work function of the metal. In both cases, the recombination of a positive ion or of an electronegative atom with a metal electron, the excess energy $E_i - W$ is absorbed by the metal lattice. The mechanism of the neutralization of a positive ion may be understood quite easily: Because of the electric field, the negative metal electron is drawn towards the positive ion and combines with it. The situation is more difficult if

$$E_i > 2W, \qquad (7.68)$$

i.e., if the energy which is released by the neutralization of the positive ion is sufficient for the release of *two* electrons from the metal surface. In discharges in the noble gases helium and neon with their very high ionization energies, this *secondary electron emission* by impinging positive ions has been observed. A metastable helium atom (see III, 14) may transfer its excitation energy in a collision of the second kind to the metal surface also and there release one or two secondary electrons. For explaining these processes, presentations similar to Fig. 279 are useful. A positive ion attached to a metal surface combines with one of the conduction electrons, while the binding energy $E_i - W$ which is released by this process is transferred to a second metal electron and enables it to leave the metal. When attached to a metal surface, the excited electron of a metastable atom may undergo a radiationless transition into the ground state of the atom, whereby its excitation energy is transferred to a metal electron which may leave the metal.

It is more difficult to understand how entire atoms are released from the metal lattice by energetic positive ions which impinge upon the cathode of a glow discharge and cause the so-called *cathode sputtering*. Thermal processes such as a local superheating above the critical temperature cannot explain the effect, since investigations with single crystals showed an anisotropy of the angular distribution of the sputtered atoms which depends on the lattice orientation of the sputtered crystals. A direct collision process thus seems to be at least partially responsible for cathode sputtering.

In concluding this section, we want to stress the fact that structure and properties of solid surfaces belong to the most complicated problems of solid-state physics, and our knowledge here is still rather scanty. Work in this field is, however, very important. It is known, for instance, that the catalytic action of metal and semiconductor surfaces is related to electron exchange processes between these surfaces and adsorbed reacting molecules. Details of this process must depend on the specific electron structure of the crystal surface, just as does the work function of tungsten according to page 454. In this context, we may mention also the so-called exo-electron emission, i.e., an exponentially decreasing electron emission from freshly worked metal surfaces, which, at least partly, seems to be due to an emission of electrons from shallow surface traps even at room temperature.

It is obvious that surface atoms of any crystal are bound more or less unilaterally and with a lower strength then inner lattice constituents and that their binding strength depends on their specific orientation with respect to their neighbors. Since the valence electrons are responsible for the binding of the surface atoms, it is equally obvious that the valence electrons themselves form bonds of a different strength with surface atoms than with atoms in the interior of the crystal. This was borne out by detailed calculations of the energy states of surface electrons relative to those of normal valence electrons. The surface-electron states lie in the energy gap between the normal valence band and the conduction band.

With these remarks we conclude our treatment of solid-state physics. We want the reader to see how fruitful the application of the well-established atomistic concepts has proved for the understanding of so many well-known solid state phenomena, though much is still to be learned.

Problems

1. The density of a cubic crystal of NaCl (molecular mass: 58.46) is 2.16 g/cm³. Find from these informations (a) the number of NaCl "molecules" in a one centimeter cube crystal, (b) the number of ions in the one centimeter cube, (c) the distance between adjacent ions in the crystal, and (d) the distance from one ion to the next ion of the same kind.

2. Using the value of the equilibrium distance of Na^+ and Cl^- from problem 1, calculate the energy required per mole to separate a NaCl crystal into its component ions. What fraction of this energy is associated with the repulsive force?

3. Using the relation

$$N(E)\,dE = \frac{8\sqrt{2}\,\pi\,V m^{\frac{3}{2}} E^{\frac{1}{2}}}{[e^{(E-\vartheta)/kT}+1]\,h^3}\,dE$$

for the Fermi-Dirac distribution of electrons in a metal, show that (a) the Fermi energy is given by

$$E_F = \frac{h^2}{8m}\left(\frac{3n}{\pi}\right)^{\frac{2}{3}},$$

where m is the electron mass and n is the number of free electrons per unit volume; (b) evaluate the Fermi energy for silver (atomic mass: 107.88) which has one valence electron per atom and a density of 10.49 g/cm³.

4. The work function (W_0) for electrons in the lowest energy state of the conduction band of aluminum is 14.7 ev, while the effective work function (W) of the metal is 3.0 ev. Calculate the number of free electrons per atom in aluminum having an atomic mass 26.97 and density 2.7 g/cm³.

Literature

General Solid State Physics

Encyclopedia of Physics, Vol. VII/1: Kristallphysik I. Berlin-Göttingen-Heidelberg: Springer 1955.
DEKKER, A. J.: Solid State Physics. New York: Prentice-Hall 1958.
HAUFFE, K.: Reaktionen in und an festen Stoffen. Berlin-Göttingen-Heidelberg: Springer 1955.
HEDVALL, J. A.: Einführung in die Festkörperchemie. Braunschweig: Vieweg 1952.
HUND, F.: Theorie des Aufbaues der Materie. Stuttgart: Teubner 1961.
JAWSON, M. A.: The Theory of Cohesion. London: Pergamon Press 1954.
JONES, H.: The Theory of Brillouin Zones and Electronic States in Crystals. Amsterdam: North Holland 1960.
KITTEL, C.: Introduction to Solid State Physics. New York: Wiley 1953.
LARK-HOROVITZ, K., and V. A. JOHNSON (Editors): Solid State Physics. New York: Academic Press 1959.
PEIERLS, R. E.: Quantum Theory of Solids. Oxford: Clarendon Press 1955.
SACHS, M.: Solid State Theory. New York: McGraw-Hill 1963.
SEITZ, F.: The Modern Theory of Solids. New York: McGraw-Hill 1940.
SLATER, J. C.: Quantum Theory of Matter. New York: McGraw-Hill 1951.
ZWIKKER, C.: Physical Properties of Solid Materials. London: Pergamon Press 1954.

Fluid State Physics

BORN, M., and H. S. GREEN: A General Kinetic Theory of Liquids. Cambridge: University Press 1949.
DARMOIS, E.: L'état Liquide. Paris 1943.
FRENKEL, J.: Kinetic Theory of Liquids. Oxford: Clarendon Press 1946.

GREEN, H. S.: The Molecular Theory of Fluids. Amsterdam: North Holland Publ. Co. 1952.
HIRSCHFELDER, J. O., C. F. CURTISS and R. B. BIRD: Molecular Theory of Gases and Liquids. New York: Wiley 1954.
LEHMANN, O.: Die Lehre von den flüssigen Kristallen. Wiesbaden: J. F. Lehmann 1918.

Crystal Structure

Encyclopedia of Physics, Vol. XXXII: Strukturforschung. Berlin-Göttingen-Heidelberg: Springer 1957.
BIJVOET, J. M., N. H. KOLKMEYER and C. H. McGILLAVRY: X-ray Analysis of Crystals. London: Butterworth 1951.
BRAGG, W. H., and W. L. BRAGG: The Crystalline State. London: Bell 1933.
GLOCKER, R.: Materialprüfung mit Röntgenstrahlen. Berlin-Göttingen-Heidelberg: Springer 1949.
HALLA, F.: Kristallchemie und Kristallphysik metallischer Werkstoffe. Leipzig: Barth 1957.

Piezoelectricity and Ferroelectricity

Encyclopedia of Physics, Vol. XVII: Dielektrika. Berlin-Göttingen-Heidelberg: Springer 1956.
BÖTTCHER, C. J. F.: Electric Polarization. Amsterdam: Elsevier Publ. Co. 1952.
FRÖHLICH, H.: Theory of Dielectrics. Oxford: University Press 1949.
JAYNES, E. T.: Ferroelectricity. Princeton: University Press 1953.
SACHSE, H.: Ferroelektrika. Berlin-Göttingen-Heidelberg: Springer 1956.

Metal Physics

BARRETT, C. S.: Structure of Metals. New York: McGraw-Hill 1952.
BOAS, W.: An Introduction to the Physics of Metals and Alloys. New York: Wiley 1948.
BRANDENBERGER, E.: Grundriß der Allgemeinen Metallkunde. Basel: Reinhardt 1952.
COTTRELL, A. H.: Theoretical Structural Metallurgy. New York: Longman, Green & Co. 1955.
DEHLINGER, U.: Theoretische Metallkunde. Berlin-Göttingen-Heidelberg: Springer 1955.
DEHLINGER, U.: Chemische Physik der Metalle und Legierungen. Leipzig: Akademische Verlagsgesellschaft 1939.
HUME-ROTHERY, W., and G. V. RAYNOR: The Structure of Metals and Alloys. London: The Institute of Metals 1954.
MASING, G.: Lehrbuch der Allgemeinen Metallkunde. 4th Ed. Berlin-Göttingen-Heidelberg: Springer 1955.
MOTT, N. F., and H. JONES: The Theory of the Properties of Metals and Alloys. Oxford: University Press 1936.
ROTHERHAM, L. A.: Creep of Metals. London: Institute of Physics 1951.
WILSON, A. H.: The Theory of Metals. Cambridge: University Press. 2nd Ed. 1953.
ZENER, CL.: Elasticity and Anelasticity of Metals. Chicago: University Press 1948.

Crystal Defects

BUEREN, H. G. VAN: Imperfections in Crystals. 2nd Ed. Amsterdam: North Holland 1962.
COTTRELL, A. H.: Dislocations and Plastic Flow in Crystals. Oxford: University Press 1953.
READ, W. T.: Dislocations in Crystals. New York: McGraw-Hill 1953.
SEEGER, A.: Theorie der Gitterfehlstellen, in Encyclopedia of Physics, Vol. VII/1: Kristallphysik I. Berlin-Göttingen-Heidelberg: Springer 1955.
SHOCKLEY, W. (Editor): Imperfections in Nearly Perfect Crystals. New York: Wiley 1952.

Energy Band Structure and Metallic Bond

Encyclopedia of Physics, Vol. XIX: Elektrische Leitungsphänomene I. Berlin-Göttingen-Heidelberg: Springer 1956.
BRILLOUIN, L.: Wave Propagation in Periodic Structures. New York: McGraw-Hill 1946.

FRÖHLICH, H.: Elektronentheorie der Metalle. Berlin: Springer 1935.
JUSTI, E.: Leitfähigkeit und Leitungsmechanismus fester Stoffe. Göttingen: Vanden-hoeck & Ruprecht 1948.
RAIMES, S.: The Wave Mechanics of Electrons in Metals. Amsterdam: North Holland 1961.
RAYNOR, G. V.: Introduction to the Electron Theory of Metals. London: The Institute of Metals 1947.

Solid-State Magnetism

BECKER, R., u. W. DÖRING: Ferromagnetismus. Berlin: Springer 1939.
BOZORTH, R. M.: Ferromagnetism. New York: Van Nostrand 1951.
KNELLER, E.: Ferromagnetismus. Berlin: Springer 1962.
NÉEL, L.: Magnétisme. Strasbourg 1939.
SNOEK, J. L.: New Developments in Ferromagnetic Materials. Amsterdam: Elsevier Publ. Co. 1947.

Superconductivity and Superfluidity

ATKINS, K. R.: Liquid Helium. New York: Cambridge University Press 1959.
KEESOM, W. H.: Helium. Amsterdam: Elsevier Publ. Co. 1949.
KOPPE, H.: Theorie der Supraleitung. Erg. exakt. Naturw. 23 (1950).
LANE, C. T.: Superfluid Physics. New York: McGraw-Hill 1962.
LONDON, F.: Superfluids. Vol. I. New York: Wiley 1950.
SHOENBERG, D.: Superconductivity. Cambridge: University Press 1952.

Solid-State Diffusion

BARRER, R. M.: Diffusion in and through Solids. Cambridge: University Press 1951.
JOST, W.: Diffusion und chemische Reaktion in festen Stoffen. Dresden u. Leipzig: Steinkopff 1937. Improved English edition New York: Academic Press 1952.
SEITH, W.: Diffusion in Metallen. Berlin-Göttingen-Heidelberg: Springer 1955.
SHEWMON, P. G.: Diffusion in Solids. New York: McGraw-Hill 1963.

Electronic Processes in Ionic Crystals

ANGERER, E. v., u. G. JOOS: Wissenschaftliche Photographie. 6th Ed. Leipzig: Akademische Verlagsgesellschaft 1956.
JAMES, T. H., and G. C. HIGGINS: Fundamentals of Photographic Theory. New York: Wiley 1950.
LIDIARD, A. B.: Ionic Conductivity, in Encyclopedia of Physics, Vol. XX: Elektrische Leitungsphänomene II. Berlin-Göttingen-Heidelberg: Springer 1956.
MOTT, N. F., and R. W. GURNEY: Electronic Processes in Ionic Crystals. 2nd Ed. Cambridge: Oxford Press 1949.
SCHULMAN, J. H., and W. D. COMPTON: Color Centers in Solids. London: Pergamon 1962.
STASIW, O.: Elektronen- und Ionenprozesse in Ionenkristallen. Berlin: Springer 1959.
STUMPF, H.: Quantentheorie der Ionen-Realkristalle. Berlin: Springer 1961.

Semiconductivity and Related Phenomena

BIGUENET, C.: Les Cathodes Chaudes. Paris 1947.
BRUINING, H.: Die Sekundärelektronenemission fester Körper. Berlin: Springer 1942.
DOSSE, J.: Der Transistor. 4th Ed. München: Oldenbourg 1962.
GÄRTNER, W. W.: Einführung in die Theorie des Transistors. Berlin: Springer 1963.
GARLICK, G. F. J.: Photoconductivity, in Encyclopedia of Physics, Vol. XIX: Elektrische Leitungsphänomene I. Berlin-Göttingen-Heidelberg: Springer 1956.
HENISCH, H. K.: Rectifying Semiconductor Contacts. Oxford: Clarendon Press 1957.
HENISCH, H. K. (Editor): Semiconducting Materials. London: Butterworth 1951.
HERMANN, G., u. S. WAGENER: Die Oxydkathode. 2 vols. 2nd Ed. Leipzig: Barth 1948/50.
JOFFÉ, A. F.: Physik der Halbleiter. Berlin: Akademie-Verlag 1958.
MADELUNG, O.: Halbleiter, in Encyclopedia of Physics, Vol. XX: Elektrische Leitungsphänomene II. Berlin-Göttingen-Heidelberg: Springer 1957.
MOSS, T. S.: Photoconductivity in the Elements. London: Butterworth 1952.
MOSS, T. S.: Optical Properties of Semiconductors. New York: Academic Press 1959.
MÜSER, H. A.: Einführung in die Halbleiterphysik. Darmstadt: Steinkopff 1960.
SALOW, H. (Editor): Der Transistor. Berlin: Springer 1963.

SHOCKLEY, W.: Electrons and Holes in Semiconductors. New York: Van Nostrand 1950.

SPENKE, E.: Electronic Semiconductors. New York: McGraw-Hill 1958.

STRUTT, M.: Transistoren. Stuttgart: Hirzel 1953.

WRIGHT, D. A.: Semiconductors. London: Methuen 1950.

ZWORYKIN, V. K., and E. G. RAMBERG: Photoelectricity and its Applications. New York: Wiley 1949.

Luminescence of Solids

BANDOW, F.: Lumineszenz. Stuttgart: Wissenschaftliche Verlagsgesellschaft 1950.

FÖRSTER, TH.: Fluoreszenz organischer Verbindungen. Göttingen: Vandenhoeck & Ruprecht 1951.

GARLICK, G. F. J.: Luminescence, in Encyclopedia of Physics, Vol. XXVI. Berlin: Springer 1957.

LEVERENZ, H. W.: Introduction to the Luminescence of Solids. New York: Wiley 1950.

MATOSSI, F.: Elektrolumineszenz und Elektrophotolumineszenz. Braunschweig: Vieweg 1957.

PRINGSHEIM, P., and M. VOGEL: Luminescence of Liquids and Solids. New York: Interscience 1946.

PRZIBAM, K.: Verfärbung und Lumineszenz. Wien: Springer 1953.

RIEHL, N.: Physikalische und Technische Anwendungen der Lumineszenz. Berlin: Springer 1941.

Appendix

The Most Important Constants and Relations of Atomic Physics

All atomic masses refer to the new C^{12} scale. For the transformation of the old scale into the new one, the relation

$$1 \text{ mu } (C^{12}=12)=1.00031792 \text{ mu } (O^{16}=16)$$

has to be used.

PLANCK's constant	h	$= (6.6256 \pm 0.0005) \cdot 10^{-27}$ erg sec
Velocity of light	c	$= (2.997925 \pm 0.000003) \cdot 10^{10}$ cm/sec
Charge of the electron	e	$= (4.8030 \pm 0.0002) \cdot 10^{10}$ esu
		$= (1.60210 \pm 0.00007) \cdot 10^{-19}$ amp sec
Rest mass of the electron	m_e	$= (9.1091 \pm 0.0004) \cdot 10^{-28}$ g
Atomic weight of the electron	A_e	$= (5.48597 \pm 0.00009) \cdot 10^{-4}$
Rest mass of the proton	m_p	$= (1.67252 \pm 0.00008) \cdot 10^{-24}$ g
Atomic weight of the proton	A_p	$= 1.0072766 \pm 0.0000002$
Rest mass of the neutron	m_n	$= (1.67482 \pm 0.00008) \cdot 10^{-24}$ g
Atomic weight of the neutron	A_n	$= 1.0086654 \pm 0.0000013$
AVOGADRO's number	L	$= (6.0225 \pm 0.0003) \cdot 10^{23}$/mole
FARADAY's constant	F	$= Le = (96487 \pm 2)$ amp sec/mole
BOLTZMANN's constant	k	$= (1.3805 \pm 0.0002) \cdot 10^{-16}$ erg/degree
		$= (8.617 \pm 0.001) \cdot 10^{-5}$ ev/degree
Bohr magneton	μ_B	$= (9.2732 \pm 0.0006) \cdot 10^{-21}$ oersted cm²
Nuclear magneton	μ_k	$= (5.0505 \pm 0.0004) \cdot 10^{-24}$ oersted cm²
Magnetic moment of the proton	μ_p	$= (2.79276 \pm 0.00007) \mu_k$
Magnetic moment of the neutron	μ_n	$= -(1.91315 \pm 0.00007) \mu_k$
Molar volume of an ideal gas (1 atm, 0 °C)	v_{mole}	$= (22414 \pm 3)$ cm³/mole
Number of gas molecules per cm³ (1 atm, 0 °C)	N_0	$= (2.6872 \pm 0.0001) \cdot 10^{19}$ molecules/cm³

Relation between different energy units used in atomic physics:

$$1 \text{ ev} \triangleq 8065.7 \text{ cm}^{-1} \triangleq 1.602 \cdot 10^{-12} \text{ erg} \triangleq 23.04 \text{ kcal/mole}.$$

Relation between atomic mass units and Mev:

$$1 \text{ mu} \triangleq 931.48 \text{ Mev}.$$

Relation between excitation potential V (volt) and wavelength λ of corresponding radiation (Å):

$$\lambda \text{ (Å)} \times V \text{ (volt)} = 12400$$

Units for radioactivity and nuclear radiation:

1 curie (c): Amount of a radioactive substance of which 3.700×10^{10} nuclei disintegrate per second. For radium, this is very nearly 1 gram.

1 roentgen (r): The amount of X-ray or γ-ray emission that produces 2.08×10^9 ion pairs or one electrostatic unit of charge by ionization in one cubic centimeter of dry air of 0 °C and atmospheric pressure. This corresponds to an absorption of 83 ergs per gram of dry air.

Subject Index

510 Subject Index

octet theory 395
odd-odd nuclei 273
oil-drop method 23
Ω^- particle 333
omegatron 41
optical pumping 144
optical vibrations 430
orbital-electron capture 253
orbital magnetism 111
orbital momentum 88
orientation quantum number 115
origin of elements 306
orthohelium 99
orthohydrogen 384
oscillator strength 145
oxide cathode, theory 481

p-type semiconductor 476
Paschen-Back effect 117
pair production 316
parahydrogen 384
paramagnetism of atoms 109
— of metals 456
parhelium 99
parity of elementary particles 327
particle accelerators 219
Pauli principle 121
— — in wave mechanics 185
— — and chemical bond 398
penetrating component of cosmic
  radiation 315
Periodic Table 11
— —, theory 124
phonon 431
phosphorescence 495
photocathodes 483
photoconductivity 492
photoelectric effect 22
— —, internal 492
— — of semiconductor surfaces 482
photofission 288
photography, atomic processes 474
photoionization 72
photoluminescence 499
photomultiplier 28
photon 43
photovoltaic cell 494
π-bonds 403
piezoelectricity 423
π-meson 323
pinch effect 310
pion 323
Planck constant 44
PLANCK's law 13
plasma betatron 226
point transistor 490
polarizability of molecules 348
polarization of nuclei 233
polaron 438
polyatomic molecules, spectra 389
positive rays 29
positron 251, 317
positronium atom 317
potential curve of molecule 361

potential-well model of metal 451
power reactors 298
predissociation 372
preionization 137
primary particles of cosmic radiation 311
principal quantum number, effective 82
principal series 80
proportional counter 217
proton synchroton 228
PROUT's hypothesis 10
pyroelectricity 423

quantization 57
— of wave fields 201
quantum defect 83
quantum electrodynamics 201
quantum mechanics 148
quantum numbers in wave-mechanics 168
quantum statistics 198

Rabi method 235
radiation theory, wave-mechanical 182
radiative recombination 74
radioactivity, natural 239
radiocarbon method 302
radionuclides 251
—, applications 301
radius of nucleus 231
Raman effect 349
Rayleigh scattering 349
reactivity of reactor 296
reactor 294
reactor control 296
recombination, ion-electron 73
recombination of molecules 370
rectification by semiconductors 485
reduction of wave function 170
refractive index of de Broglie waves 194
resonances of elementary particles 328,
  331
Richardson-Einstein-de Haas effect 111
Richardson equation 452
— — for semiconductors 481
Ritz combination principle 55
rotation of molecules 378
rotational quantum number 172
rotation spectra 379
rotational structure of band spectra 380
rotator with free axis 172
— with rigid axis 171
Rutherford scattering 16
Rydberg constant, theory 67
RYDBERG's formula 55

Saha equation 21
Sargent diagram 244
scattering experiments 16
Schmidt curves 278
Schottky defects 470
Schrödinger equation 163
— —, time-dependent 166
scintillation counter 218
second quantization 202
secondary electron emission 484
selection rules in quantum mechanics 184